Principles of Terrestrial Ecosystem Ecology

F. Stuart Chapin, III • Pamela A. Matson
Peter M. Vitousek

Principles of Terrestrial Ecosystem Ecology

Second Edition

Illustrated by Melissa C. Chapin

 Springer

F. Stuart Chapin, III
University of Alaska Fairbanks
Institute of Arctic Biology
Department of Biology & Wildlife
Fairbanks, AK, USA
terry.chapin@alaska.edu

Pamela A. Matson
School of Earth Sciences
Stanford University
Stanford, CA, USA
matson@stanford.edu

Peter M. Vitousek
Department of Biological Sciences
Stanford University
Stanford, CA, USA
vitousek@stanford.edu

ISBN 978-1-4419-9503-2 (hardcover)   e-ISBN 978-1-4419-9504-9
ISBN 978-1-4419-9502-5 (softcover)
DOI 10.1007/978-1-4419-9504-9
Springer New York Dordrecht Heidelberg London

Library of Congress Control Number: 2011935993

Cover illustrations: Temperate forest in the eastern U.S. (North Carolina), showing a complex
multi-layered canopy with sunflecks common in all canopy layers. Cover Photograph courtesy of
Norm Christensen

Printed on acid-free paper

Springer is part of Springer Science+Business Media (www.springer.com)

Preface

Human activities are affecting the global environment in many ways, with numerous direct and indirect effects on ecosystems. The climate and atmospheric composition of Earth are changing rapidly. Humans have directly modified half of the ice-free terrestrial surface and use 40% of terrestrial production. Our actions are causing the sixth major extinction event in the history of life on Earth and radically modify the interactions among forests, fields, streams, and oceans. This book is written to provide a conceptual basis for understanding terrestrial ecosystem processes and their sensitivity to environmental and biotic changes. We believe that an understanding of ecosystem dynamics must underlie our analysis of both the consequences and the mitigation of human-induced changes.

This book is intended to introduce the science of terrestrial ecosystem ecology to advanced undergraduate students, beginning graduate students, and practicing scientists from a wide array of disciplines. We define terrestrial ecosystem ecology to include freshwater ecosystems and their terrestrial matrix. We also include a description of marine ecosystems to provide a broader context for understanding terrestrial ecosystems and as a basis for Earth-System analysis. We provide access to some of the rapidly expanding literature in the many disciplines that contribute to ecosystem understanding. This second edition incorporates new material that accounts for both the substantial scientific advances in ecosystem ecology during the past decade, as well as the evolution of our own understanding.

The first section of this book provides the context for understanding ecosystem ecology. We introduce the science of ecosystem ecology and place it in the context of other components of the Earth System – the atmosphere, ocean, climate and geological systems. We show how these components affect ecosystem processes and contribute to the global variation in terrestrial ecosystem structure and processes. In the second section of the book we consider the mechanisms by which terrestrial ecosystems function and focus on the flow of water and energy and the cycling of carbon and nutrients. We then consider the important role of organisms in ecosystem processes through trophic interactions (feeding relationships), environmental effects, and disturbance. The third section of the book addresses temporal and spatial patterns in ecosystem processes. We finish by considering the integrated effects of these processes at the global scale and their consequences for sustainable use by human societies. Powerpoint lecture notes that include the illustrations in this book are available on the web (http://terrychapin.org/) as supplementary material.

Many people have contributed to the development of this book. We particularly thank our families, whose patience has made the book possible, our students from whom we have learned many of the important ideas that are presented, and Hal Mooney who was a co-author of the first edition. In addition, we thank the following individuals for their constructively critical review of chapters in this book: Richard Bardgett, Dan Binkley, Dave Bowling, Pep Canadell, Mimi Chapin, Doug Cost, Joe Craine, Wolfgang Cramer, Eric Davidson, Sandra Díaz, Jim Elser, Eugenie Euskirchen, Valerie Eviner, Noah Fierer, Jacques Finlay, Doug Frank, Mark Harmon, Sarah Hobbie, Dave Hooper, Bob Howarth, Ivan Janssens, Julia Jones, Bill Lauenroth, Joe McFadden, Dave McGuire, Sam McNaughton, Russ Monson, Deb Peters, Mary Power, Steve Running, Josh Schimel, Ted Schuur, Tim Seastedt, Mark Serreze, Phil Sollins, Bob Sterner, Kevin Trenberth, Dave Turner, Monica Turner, Diana Wall, John Walsh. We also thank Julio Betancourt, Scott Chambers, Norm Christensen, Greg Cortopassi, Steve Davis, Sandra Díaz, Jack Dykinga, Jim Elser, Jim Estes, Peter Franks, Mark Harmon, Al Levno, Mike Kenner, Alan Knapp, Aaryn Olsson, Roger Ruess, Dave Schindler, and David Tongway for the use of their photographs. We particularly thank Joe Craine and Dana Nossov for their constructive comments on the entire book.

Fairbanks, AK, USA   F. Stuart Chapin, III
Stanford, CA, USA   Pamela A. Matson
Stanford, CA, USA   Peter M. Vitousek

Contents

Part I

Context

The Ecosystem Concept

Ecosystem ecology studies the links between organisms and their physical environment within an Earth-System context. This chapter provides background on the conceptual framework and history of ecosystem ecology.

Introduction

Ecosystem ecology addresses the interactions between organisms and their environment as an integrated system. The ecosystem approach is fundamental to managing Earth's resources because it addresses the interactions that link biotic systems, of which people are an integral part, with the physical systems on which they depend. The approach applies at the scale of Earth as a whole, the Amazon River basin, or a farmer's field. An ecosystem approach is critical to the sustainable management and use of resources in an era of increasing human population and consumption and large, rapid changes in the global environment.

The ecosystem approach has grown in importance in many areas. The United Nations Convention on Biodiversity of 1992, for example, promoted an ecosystem approach, including humans, for conserving biodiversity rather than the more species-based approaches that predominated previously. There is growing appreciation for the role that species interactions play in the functioning of ecosystems (Díaz et al. 2006). Important shifts in thinking have occurred about how to manage more sustainably the ecosystems on which we depend for food and fiber. The supply of fish from the sea is now declining because fisheries management depended on species-based stock assessments that did not adequately consider the resources on which commercial fish depend (Walters and Martell 2004). A more holistic view of managed systems can account for the complex interactions that prevail in even the simplest ecosystems. There is also a growing appreciation that a thorough understanding of ecosystems is critical to managing the quality and quantity of our water supplies and in regulating the composition of the atmosphere that determines Earth's climate (Postel and Richter 2003).

A Focal Issue

Human exploitation of Earth's ecosystems has increased more in the last half-century than in the entire previous history of the planet (Steffen et al. 2004), often with unintended detrimental effects. Forest harvest, for example, provides essential wood and paper products (Fig. 1.1). The amount and location of harvest, however, influences other benefits that society receives from forests, including the quantity and quality of water in headwater streams; the recreational and aesthetic benefits of forests; the probability of landslides, insect outbreaks, and forest fires; and the potential of forests to release or sequester carbon dioxide (CO_2), which influences climatic change. How can ecosystems be managed to meet these multiple (and often conflicting) needs? In the Northwestern

F.S. Chapin, III et al., *Principles of Terrestrial Ecosystem Ecology*,
DOI 10.1007/978-1-4419-9504-9_1, © Springer Science+Business Media, LLC 2011

Fig. 1.1 Patch clear-cutting leads to single-species patches in a mosaic of 100 to 500-year native Douglas-fir forests in the Northwestern U.S. The nature and extent of forest clearing influences ecosystem processes at scales ranging from single patches (e.g., productivity and species diversity) to regions (e.g., water supply and fire risk) or even the entire planet (climatic change). Photograph by Al Levno, U.S. Forest Service

U.S., for example, timber was harvested in the second half of the twentieth century more rapidly that it regenerated. Concern about loss of old-growth forest habitat for endangered species such as the spotted owl led to the development of ecosystem management in the 1990s to address the multiple functions and uses of forests (Christensen et al. 1996; Szaro et al. 1999). Ecosystem ecology draws on a breadth of disciplines to provide the principles needed to understand the consequences of society's choices.

Overview of Ecosystem Ecology

The flow of energy and materials through organisms and the physical environment provides a framework for understanding the diversity of form and functioning of Earth's physical and biological processes. Why do tropical forests have large trees but accumulate only a thin layer of dead leaves on the soil surface, whereas tundra supports small plants but an abundance of organic matter at the soil surface?

Why does the concentration of carbon dioxide in the atmosphere decrease in summer and increase in winter? What happens to nitrogen fertilizer that farmers add to their fields but do not harvest with the crop? Why has the introduction of exotic grasses to pastures caused adjacent forests to burn? These are representative of the questions addressed by ecosystem ecology. Answers to these questions require an understanding of the interactions between organisms and their physical environments – both the response of organisms to environment and the effects of organisms on their environment. These questions also require a focus on integrated ecological systems rather than individual organisms or physical components.

Ecosystem analysis seeks to understand the factors that regulate the **pools** (quantities) and **fluxes** (flows) of materials and energy through ecological systems. These materials include carbon, water, nitrogen, rock-derived elements such as phosphorus, and novel chemicals such as pesticides or radionuclides that people have added to the environment. These materials are found in abiotic (nonbiological) pools such as soils, rocks, water, and the atmosphere and in biotic pools such as plants, animals, and soil microorganisms (microbes).

An ecosystem consists of all the organisms and the abiotic pools with which they interact. Ecosystem processes are the transfers of energy and materials from one pool to another. Energy enters an ecosystem when light energy drives the reduction of carbon dioxide (CO_2) to form sugars during photosynthesis. Organic matter and energy are tightly linked as they move through ecosystems. The energy is lost from the ecosystem when organic matter is oxidized back to CO_2 by combustion or by the respiration of plants, animals, and microbes. Materials move among abiotic components of the system through a variety of processes, including the weathering of rocks, the evaporation of water, and the dissolution of materials in water. Fluxes involving biotic components include the absorption of minerals by plants, the fall of autumn leaves, the decomposition of dead organic matter by soil microbes, the consumption of plants by herbivores, and the consumption of herbivores by predators. Most of these fluxes are sensitive to environmental factors such as temperature and moisture, and to biological factors regulating the population dynamics and species interactions in communities. The unique contribution of ecosystem ecology is its focus on biotic and abiotic factors as interacting components of a single integrated system.

Ecosystem processes can be studied at many spatial scales. How big is an ecosystem? Ecosystem processes take place at a wide range of scales, but the appropriate scale of study depends on the question asked (Fig. 1.2). The impact of zooplankton on their algal food might be studied in small bottles in the laboratory. The controls over productivity might be studied in relatively homogeneous patches of a lake, forest, or agricultural field. Questions that involve exchanges occurring over very broad areas might best be addressed at the global scale. The concentration of atmospheric CO_2, for example, depends on global patterns of biotic exchanges of CO_2 and the burning of fossil fuels, which are spatially variable across the planet. The rapid mixing of CO_2 in the atmosphere averages across this variability, facilitating estimates of long-term changes in the total global flux of carbon between Earth and the atmosphere (see Chap. 14).

Some questions require careful measurements of lateral transfers of materials. A watershed is a logical unit to study the impacts of forests on the quantity and quality of the water that supplies a town reservoir. A drainage basin, also known as a catchment or watershed, consists of a stream or river and all the terrestrial surfaces that drain into it. By studying a drainage basin, we can compare the quantities of materials that enter from the air and rocks with the amounts that leave in stream water, just as you balance your checkbook. Studies of input–output budgets of drainage basins have improved our understanding of the interactions between rock weathering, which supplies nutrients, and plant and microbial growth, which retains nutrients in ecosystems (Vitousek and Reiners 1975; Bormann and Likens 1979; Driscoll et al. 2001; Falkenmark and Rockström 2004).

The upper and lower boundaries of an ecosystem also depend on the question asked and the scale that is appropriate to the question. The atmosphere, for example, extends from the gases between soil particles to the edge of outer space. The exchange of CO_2 between a forest and the atmosphere might be measured a few meters above the top of the canopy where variation in CO_2 concentration largely reflects processes occurring within the forest rather than in upwind ecosystems. The regional impact of grasslands on the moisture content of the atmosphere might, however, be measured at a height of several kilometers above the ground, where the moisture released by the ecosystem condenses and returns as precipitation (see Chap. 2). For questions that address plant effects on water and nutrient cycling, the bottom of the ecosystem might be the maximum depth to which roots extend because soil water or nutrients below this depth are inaccessible to plants. Studies of long-term soil development, in contrast, must also consider rocks deep in the soil, which constitute the long-term reservoir of many nutrients that gradually become incorporated into surface soils (see Chap. 3).

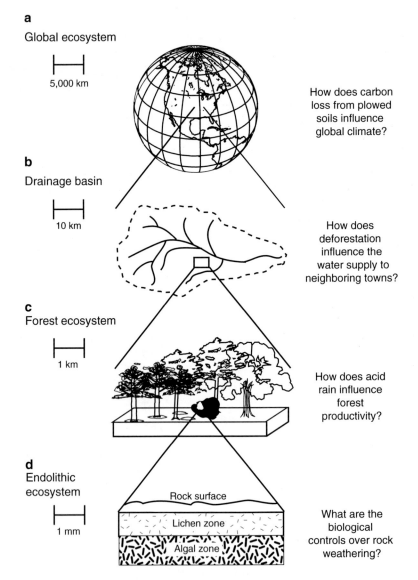

a

Global ecosystem

5,000 km

How does carbon
loss from plowed
soils influence
global climate?

b

Drainage basin

10 km

How does
deforestation
influence the
water supply to
neighboring towns?

c

Forest ecosystem

1 km

How does acid
rain influence
forest
productivity?

d

Endolithic
ecosystem

1 mm

Rock surface

Lichen zone

Algal zone

What are the
biological
controls over rock
weathering?

Fig. 1.2 Examples of ecosystems that range in size by ten orders of magnitude: an endolithic ecosystem in the surface layers of rocks (1×10^{-3} m in height), a forest 1×10^3 m in diameter, a drainage basin (1×10^5 m in length), and Earth (4×10^7 m in circumference). Also shown are examples of questions appropriate to each scale

Ecosystem dynamics are a product of many temporal scales. The rates of ecosystem processes are constantly changing due to fluctuations in environment and activities of organisms on time scales ranging from microseconds to millions of years (see Chap. 12). Light capture during photosynthesis responds almost instantaneously to fluctuations in the light that strikes a leaf. At the opposite extreme, the evolution of photosynthesis two billion years ago added oxygen to the atmosphere over millions of years, causing the prevailing geochemistry of Earth's surface to change from chemical reduction to chemical oxidation (Schlesinger 1997). Microorganisms in the group Archaea evolved in the early reducing atmosphere of Earth. These microbes are still the only organisms that produce methane. They now function in anaerobic environments such as wetland soils or the anaerobic interiors of soil aggregates or animal intestines. Episodes of mountain building and erosion strongly influence the availability of minerals to support plant growth. Vegetation is still migrating in response to the retreat of Pleistocene glaciers 10,000 to 20,000 years ago. After disturbances such as fire or treefall, plant, animal, and microbial communities change gradually over years to centuries. Rates of carbon input to an ecosystem through photosynthesis change over time scales of seconds to decades due to variations in light, temperature, and leaf area.

Many early studies in ecosystem ecology made the simplifying assumption that some ecosystems are in **equilibrium** with their environment. In this perspective, relatively undisturbed ecosystems were thought to have properties that reflected (1) largely closed systems dominated by internal recycling of elements, (2) self-regulation and deterministic dynamics, (3) stable endpoints or cycles, and (4) absence of disturbance and human influence (Pickett et al. 1994; Turner et al. 2001). One of the most important conceptual advances in ecosystem ecology has been the increasing recognition of the importance of past events and external forces in shaping the functioning of ecosystems. In this nonequilibrium perspective, we recognize that most ecosystems exhibit unbalanced inputs and losses; their dynamics are influenced by varying external and internal factors; they exhibit no single stable equilibrium; disturbance is a natural component of their dynamics; and human activities exert a pervasive influence. The complications associated with the current nonequilibrium view require a more dynamic and stochastic perspective on controls over ecosystem processes.

Ecosystems are considered to be at **steady state**, if the balance between inputs and outputs to the system shows no trend with time (Bormann and Likens 1979). Steady state assumptions differ from equilibrium assumptions because they accept temporal and spatial variation as a normal aspect of ecosystem dynamics. Even at steady state, for example, plant growth changes from summer to winter and between wet and dry years (see Chap. 6). At a stand scale, younger individuals replace plants that die from old age or pathogen attack.

At a landscape scale, some patches may be altered by fire or other disturbances, and other patches are in various stages of recovery. These ecosystems or landscapes are in steady state if there is no long-term directional trend in their properties or in the balance between inputs and outputs over the time scale considered.

Ecosystem ecology depends on information and principles developed in physiological, evolutionary, population, and community ecology (Fig. 1.3). The biologically mediated movement of carbon and nitrogen through ecosystems depends on the physiological properties of plants, animals, and soil microbes. The traits of these organisms are the products of their evolutionary histories and the competitive interactions that sort species into communities where they successfully grow, survive, and reproduce (Vrba and Gould 1986). Ecosystem fluxes also depend on the population processes that govern plant, animal, and microbial densities and age structures and on community processes such as competition and predation that determine which species are present and their rates of resource consumption.

The supply of water and minerals from soils to plants depends not only on the activities of soil microbes but also on physical and chemical interactions among rocks, soils, and the atmosphere. The low availability of phosphorus due to the extensive weathering and loss of nutrients in the ancient soils of western Australia, for example, strongly constrains plant growth and the quantity and types of plants and animals that can be supported. Principles of ecosystem ecology must therefore also incorporate the concepts and understanding of disciplines such as geochemistry, hydrology, and climatology that focus on the physical environment (Fig. 1.3).

People interact with ecosystems through both their impacts on ecosystems and their use of **ecosystem services** – the benefits that people derive from ecosystems. The patterns of human engagement with ecosystems reflect a complex suite of social processes operating at many temporal and spatial scales. Ecosystem ecology therefore informs and depends on concepts in the emerging field of **social–ecological stewardship**

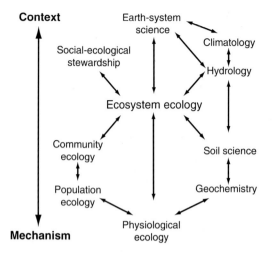

Fig. 1.3 Relationships between ecosystem ecology and other disciplines. Ecosystem ecology integrates the principles of several biological and physical disciplines, determines the resources available to society, and provides the mechanistic basis for Earth-System science

that enables people to shape the trajectory of social–ecological change to enhance ecosystem resilience and human well-being (Fig. 1.3).

Ecosystem ecology provides the mechanistic basis for understanding processes that occur at global scales. Study of Earth as a physical system relies on information about the rates and pathways by which land and water surfaces interact with the atmosphere, rocks, and waters of Earth (Fig. 1.3). Conversely, the global budgets of materials that cycle between the atmosphere, land, and the ocean provide a context for understanding the broader significance of processes studied in a particular ecosystem. Latitudinal and seasonal patterns of atmospheric CO_2 concentration, for example, help define the locations where carbon is absorbed or released from the land and ocean (see Chap. 14).

History of Ecosystem Ecology

Many early discoveries of biology were motivated by questions about the integrated nature of ecological systems. In the seventeenth century, European scientists were still uncertain about the source of materials found in plants.

Plattes, Hooke, and others advanced the novel idea that plants derive nourishment from both air and water (Gorham 1991). Priestley extended this idea in the eighteenth century by showing that plants produce a substance that is essential to support the breathing of animals. At about the same time, MacBride and Priestley showed that breakdown of organic matter caused production of "fixed air" (carbon dioxide) that did not support animal life. In the nineteenth century, De Saussure, Liebig, and others clarified the explicit roles of carbon dioxide, oxygen, and mineral nutrients in these cycles. For example, in 1843, Liebig described the first nitrogen cycle, postulating that nitrogen was fixed by volcanoes, absorbed by plants, and then released to the atmosphere as NH_3 during decomposition, only later to reenter ecosystems with precipitation. Much of the biological research during the nineteenth and twentieth centuries explored the detailed mechanisms of biochemistry, physiology, behavior, and evolution that explain how life functions. Only in recent decades have we returned to the question that originally motivated this research: How are biogeochemical processes integrated in the functioning of natural ecosystems?

Many threads of ecological thought have contributed to the development of ecosystem ecology (Hagen 1992), including ideas relating to **trophic interactions** (the feeding relationships among organisms) and **biogeochemistry** (biological interactions with chemical processes in ecosystems). Early research on trophic interactions emphasized the transfer of energy among organisms. Elton, an English zoologist interested in natural history, described the role that an animal plays in a community (its **niche**) in terms of what it eats and is eaten by (Elton 1927). He viewed each animal species as a link in a **food chain** that describes the movement of matter from one organism to another. Elton's concepts of trophic structure provide a framework for understanding the flow of materials through ecosystems (see Chap. 10).

Hutchinson, an American limnologist, was strongly influenced by the ideas of Elton and the Russian geochemist Vernadsky who described the movement of minerals from soil into vegetation and back to soil. Hutchinson suggested that the resources available in a lake must limit the productivity of algae and that algal productivity, in turn, must limit the abundance of algae-eating animals. Meanwhile, Tansley, a British terrestrial plant ecologist, was also concerned that ecologists focused their studies so strongly on organisms that they failed to recognize the importance of exchange of materials between organisms and their abiotic environment. He coined the term **ecosystem** to emphasize the importance of interchanges of materials between organisms and their environment (Tansley 1935).

Lindeman, another limnologist, was strongly influenced by all these threads of ecological theory. He suggested that energy flow through an ecosystem could be used as a currency to quantify the roles that groups of organisms play in trophic dynamics. Green plants (**primary producers**) capture energy and transfer it to animals (**consumers**) and **decomposers**. At each transfer, some energy is lost from the ecosystem through respiration. Therefore, the productivity of plants constrains the quantity of consumers that an ecosystem can support (see Chap. 10). The energy flow through an ecosystem maps closely to carbon flow in the processes of photosynthesis, trophic transfers, and respiratory release of carbon. Lindeman's dissertation research on "The trophic-dynamic aspect of ecology" was initially rejected for publication because reviewers felt that there were insufficient data to draw such broad conclusions and that it was inappropriate to use mathematical models to infer general relationships based on observations from a single lake. After Lindeman's death, his postdoctoral advisor Hutchinson persuaded the editor to publish this paper, which has been the springboard for many of the basic concepts in ecosystem theory (Lindeman 1942).

H.T. Odum, also trained by Hutchinson, and his brother E.P. Odum further developed the "**systems approach**" to studying ecosystems, emphasizing the general properties of ecosystems without documenting all the underlying mechanisms and interactions. The Odum brothers used radioactive tracers to measure the movement of energy and materials through a coral reef and other systems, enabling them to document the patterns of energy flow and metabolism of whole ecosystems and to suggest generalizations about how ecosystems function (Odum 1969). Ecosystem budgets of energy and materials have since been developed for many freshwater and terrestrial ecosystems (Ovington 1962; Golley 1993), providing information that is essential to generalize about global patterns of processes such as productivity (Saugier et al. 2001; Luyssaert et al. 2007). Some of the questions addressed by systems ecology include information transfer (Margalef 1968), the structure of food webs (Polis 1991), the hierarchical changes in ecosystem controls at different temporal and spatial scales (O'Neill et al. 1986; Peterson et al. 1998; Enquist et al. 2007), and the resilience of ecosystem properties after disturbance (Holling 1973).

We now recognize that element cycles interact in important ways and cannot be understood in isolation. The availability of water and nitrogen are important determinants of the rate at which carbon cycles through the ecosystem. Conversely, the productivity of vegetation strongly influences the cycling rates of nitrogen and water. This **coupling** of biogeochemical cycles is critical to understanding processes ranging from the interactions of plants and fungi on root tips to the responses of terrestrial productivity to human-induced increases in atmospheric CO_2 concentration or nitrogen deposition (see Chap. 9).

Additionally, regional and global changes in the environment have increased ecologists' awareness of the effects of disturbance and other environmental changes on ecosystem processes. **Succession**, the directional change in ecosystem structure and functioning that follows disturbance, is an important framework for understanding these transient dynamics of ecosystems. Early American ecologists such as Cowles and Clements were struck by the relatively predictable patterns of vegetation development after exposure of unvegetated land surfaces. Sand dunes on Lake Michigan, for example, are initially colonized by drought-resistant herbaceous plants that give way to shrubs, then small trees, and eventually forests (Cowles 1899). Clements advanced a theory of community development, suggesting that this vegetation succession is a predictable process that eventually leads, in the absence of disturbance, to a stable community-type characteristic of a particular climate (the **climatic climax**; Clements 1916). He suggested that a community is like an organism made of interacting parts (species) and that successional development toward a climax community is analogous to the development of an organism to adulthood. Clements' ideas were controversial from the outset; other ecologists, such as Gleason (1926), believed that vegetation change was not as predictable as Clements had implied. Instead, chance dispersal events could explain much of the vegetation pattern on the landscape. This debate led to a century of research on the mechanisms responsible for vegetation change (see Chap. 12). Nevertheless, the analogy between an ecological community and an organism laid the groundwork for concepts of ecosystem physiology (e.g., the net exchange of CO_2 and water vapor between the ecosystem and the atmosphere). These measurements of net ecosystem exchange are still an active area of research in ecosystem ecology, although they are now motivated by different questions than those posed by Clements.

Ecosystem ecologists study ecosystems through comparative observations and experiments. The comparative approach originated from studies by plant geographers and soil scientists who described general patterns of variation with respect to climate and geological substrate (Schimper 1898). These studies showed that many of the global patterns of plant production and soil development vary predictably with climate (Jenny 1941; Rodin and Bazilevich 1967; Lieth 1975). The studies also showed that, in a given climatic regime, the properties of vegetation depended strongly on soils and vice versa (Dokuchaev 1879; Jenny 1941; Ellenberg 1978). Process-based studies of organisms and soils provided insight into many of the mechanisms underlying the distributions of organisms and soils along these gradients (Billings and Mooney 1968; Mooney 1972; Paul and Clark 1996; Larcher 2003), providing a basis for extrapolation of processes across complex landscapes to characterize large regions (Woodward 1987; Turner et al. 2001). These studies often relied on field or laboratory experiments that manipulate some ecosystem property (e.g., litter quality or nutrient supply) or process, or on comparative studies across environmental gradients (Vitousek 2004; Turner 2010). Comparative studies have shown, for example, that ecosystems differ substantially in their average productivity and water flux, but that under dry conditions ecosystem are similar in the efficiency with which they use precipitation inputs to support production (Knapp and Smith 2001; Huxman et al. 2004). Paleoecological studies can extend these observations over long time scales and under conditions that do not exist today, using records stored in ice cores, sediments, and tree rings (Webb and Bartlein 1992; Petit et al. 1999).

Manipulations of entire ecosystems provide opportunities to test hypotheses that are suggested by observations (Likens et al. 1977; Schindler 1985; Chapin et al. 1995). These experiments often provide insights that are useful in management. The clear-cutting of an experimental watershed (drainage basin) at Hubbard Brook in the Northeastern U.S., for example, caused a 2–3-fold increase in streamflow and more than 50-fold increase in stream nitrate concentration – to levels exceeding health standards for drinking water (Bormann and Likens 1979). These dramatic results demonstrated the key role of vegetation in regulating the cycling of water and nutrients in forests. The results halted plans for large-scale deforestation that had been planned in order to increase water supplies during a long-term drought. Nutrient addition experiments in the Experimental Lakes Area of southern Canada showed that phosphorus limits the productivity of many lakes (Schindler 1985) and that phosphorus pollution was responsible for algal blooms and fish kills that were common in lakes near densely populated areas in the 1960s. This research provided the basis for regulations that removed phosphorus from detergents and regulated the outflow of sewage effluent.

Changes in the Earth System have led to studies of the interactions among terrestrial ecosystems, the atmosphere, and the ocean.

The dramatic impact of human activities on the Earth System (Steffen et al. 2004; MEA 2005; Ellis and Ramankutty 2008; Rockström et al. 2009) has lent urgency to the need to understand how terrestrial ecosystem processes affect the atmosphere and the ocean. The scale at which these ecosystem impacts are occurring is so large that the traditional tools of ecologists are insufficient. Satellite-based remote sensing of ecosystem properties, global networks of atmospheric sampling sites, and the development of global models are important new tools to address global issues (Goetz et al. 2005; Field et al. 2007; Waring and Running 2007; Bonan 2008). Information on global patterns of CO_2 and pollutants in the atmosphere, for example, provide telltale evidence of the major locations and causes of global problems (Field et al. 2007). This information provides hints about which ecosystems and processes have the greatest impact on the Earth System and therefore where research and management should focus efforts to understand and solve these problems.

The intersection of systems approaches, process understanding, and global analysis is an exciting frontier of ecosystem ecology. How do changes in the global environment alter controls over ecosystem processes? What are the integrated system consequences of these changes? How do these changes in ecosystem properties influence the Earth System? Understanding the rapid changes that are occurring in ecosystems blurs any previous distinction between basic and applied research (Stokes 1997). There is an urgent need to understand how and why the ecosystems of Earth are changing.

Ecosystem Structure and Functioning

Ecosystem Processes

Most ecosystems gain energy from the sun and materials from the air or rocks, transfer these among components within the ecosystem, then release energy and materials to the environment. The essential biological components of ecosystems are plants, animals, and decomposers.

The essential abiotic components of a terrestrial ecosystem are **water**, the **atmosphere**, which supplies carbon and nitrogen, and **soil**, which provides support, storage, and other nutrients required by organisms. **Plants** capture solar energy in the process of bringing carbon into the ecosystem. A few ecosystems, such as deep-sea hydrothermal vents, have no plants but instead have bacteria that derive energy from the oxidation of hydrogen sulfide (H_2S) to produce organic matter. Plants use solar energy to acquire nutrients and assemble organic material.

Decomposer microorganisms (microbes) break down dead organic material, releasing CO_2 to the atmosphere and nutrients in forms that are available to other microbes and plants. If decomposition did not occur, large accumulations of dead organic matter would sequester the nutrients required to support plant growth. **Animals** transfer energy and materials and can regulate the quantity and activities of plants and soil microbes.

An **ecosystem model** describes the major pools and fluxes in an ecosystem and the factors that regulate these fluxes. Carbon, water, and nutrients differ from one another in the relative importance of ecosystem inputs and outputs vs. internal recycling (see Chaps. 4–9). Plants, for example, acquire carbon primarily from the atmosphere, and most carbon released by respiration returns to the atmosphere. Carbon cycling through ecosystems is therefore quite open, with large inputs to, and losses from, the system (see Fig. 6.1). Despite these large carbon inputs and losses, the large quantities of carbon stored in plants and soils of ecosystems buffer the activities of animals and microbes from temporal variations in carbon absorption by plants. The water cycle of ecosystems is also relatively open, with most water entering as precipitation and leaving by evaporation, transpiration, and drainage to groundwater and streams (see Fig. 4.4). In contrast to carbon, most terrestrial ecosystems have a limited capacity to store water in plants and soil, so the activity of organisms is closely linked to water inputs. In contrast to carbon and water, mineral elements, such as nitrogen and phosphorus, are recycled rather tightly within ecosystems, with annual inputs and losses that are small relative to the quantities that annually recycle within the ecosystem (see Fig. 9.17). These differences in the "openness" and "buffering" of cycles fundamentally influence the controls over rates and patterns of cycling of materials through ecosystems.

The pool sizes and rates of cycling of carbon, water, and nutrients differ substantially among ecosystems. Tropical forests have much larger pools of carbon and nutrients in plants than do deserts or tundra. Peat bogs, in contrast, have large pools of soil carbon rather than plant carbon. Ecosystems also differ substantially in annual fluxes of materials among pools, for reasons that we explore in later chapters.

Ecosystem Structure and Constraints

The differences in physical properties between water and air lead to fundamental structural differences between aquatic and terrestrial ecosystems. Due to its greater density, water offers greater physical support for photosynthetic organisms than does the air that bathes terrestrial ecosystems (Table 1.1). The primary producers in **pelagic** (open-water) ecosystems are therefore microscopic photosynthetic organisms (**phytoplankton**) that float near the water surface, where light availability is greatest, whereas terrestrial plants produce elaborate support structures to raise their leaves above neighbors. Plants are often the major habitat-structuring feature on land. Their physical structure governs the patterns of physical environment, organism activity, and ecosystem processes. In the ocean and lakes, however, the environment is physically structured by vertical gradients in light, temperature, oxygen, and salinity. In small lakes and clearwater streams,

benthic (bottom-dwelling) algae account for most primary production (Vander Zanden et al. 2005; Allan and Castillo 2007). Vascular plants are also important primary producers on edges of lakes, streams, rivers, estuaries, and lagoons.

The size of aquatic organisms determines their locomotion strategies. Water is a polar molecule that sticks to the surface of organisms. These viscous forces impede the movement of small organisms and particles. Large organisms, in contrast, can swim, and their speed is largely determined by inertia. The Reynolds' number (Re) is the ratio of inertial to viscous forces and is a measure of the ease with which organisms can move through a viscous fluid like water.

$$Re = \frac{lv}{V_k} \qquad (1.1)$$

The movement of organisms through water is not strongly impeded for organisms with a large length (l) and velocity (v) under conditions of low kinematic viscosity (V_k; Fig. 1.4). Small bacteria and photosynthetic plankton, however, must deal with life at a low Reynolds' number, where viscous forces are much stronger than inertial forces. At these small sizes, diffusion is the main process that moves nutrients to the cell surface, just as with fine roots on land. At slightly larger sizes, zooplankton actively filter feed or swim to acquire their food.

Oxygen and other gases diffuse about 10,000 times more rapidly in air than water, with turbulence and lateral flow enhancing this movement in both air and water. The surface ocean water, for example, has an oxygen concentration 30-fold lower than in air (Table 1.1), and aquatic sediments are much more likely to be anaerobic than are

Table 1.1 Basic properties of water and air at 20°C at sea level that influence ecosystem processes

Property[a]	Water	Air	Ratio (water:air)
Oxygen concentration (ml L^{-1}) at 25°C	7.0	209.0	1:30
Density (kg L^{-1})	1.000	0.0013	800:1
Viscosity (cP)	1.0	0.02	50:1
Heat capacity (cal L^{-1} (°C)$^{-1}$)	1000.0	0.31	3,000:1
Diffusion coefficient (mm s^{-1})			
Oxygen	0.00025	1.98	1:8,000
Carbon dioxide	0.00018	1.55	1:9,000

[a]Data from Moss (1998)

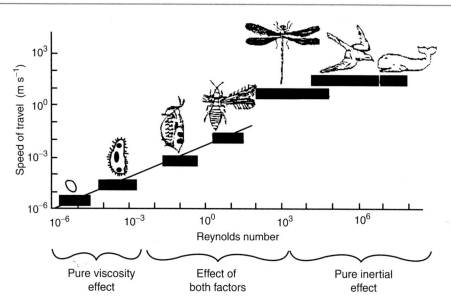

Fig. 1.4 Range of Reynolds numbers for organisms of different lengths and speeds. Small organisms like phytoplankton have small Reynolds numbers and derive their nutrition by diffusion. As size and Reynolds number increase, nutrition based on movement (filter feeding and swimming) becomes progressively more important. Redrawn from Schwoerbel (1987)

terrestrial soils. Aquatic organisms therefore exhibit a variety of adaptations to acquire oxygen and withstand anaerobic conditions. On land, in contrast, the acquisition of water and the avoidance or tolerance of desiccation are more common evolutionary themes.

Streams and rivers are structured by moving water. The physical environment and therefore the biotic structure of stream ecosystems differ dramatically from those of land, lakes, and the ocean. Water constantly moves downstream across the riverbed, bringing in new material from upstream and sweeping away anything that is not attached or able to swim vigorously. Phytoplankton are therefore unimportant in streams, except in slow-moving polluted sites and large rivers. The major primary producers of rapidly moving streams are the algal components of **periphyton**, assemblages of algae, bacteria, and invertebrates that attach to stable surfaces such as rocks and vascular plants. The slippery surfaces of rocks in a riverbed consist of periphyton in a polysaccharide matrix. Submerged or emergent vascular plants and benthic mats become relatively more important in slow-moving sections of a river. Within a given section of river, alternating pools and riffles differ in flow rate and ecosystem structure. Seasonal changes in discharge radically alter the flow regime and therefore structure of rivers and streams. Desert streams, for example, have flash floods after intense rains but may have no surface flow during dry periods (Fisher et al. 1998). Other streams have predictable discharge peaks associated with snowmelt. In general, floods and other high-discharge events are important because they scour sediments and biota from the riverbed and **riparian** (streambank) zones, redistribute logs and other material that structure aquatic habitat, and deposit new soil and create new habitats across floodplains. Some rivers flood annually, so floodplains alternate between being terrestrial and aquatic habitats. Human efforts to prevent flooding by building dams and levees therefore radically alter river and riparian ecosystem structure and dynamics.

Controls Over Ecosystem Processes

Ecosystem structure and functioning are governed by multiple independent control variables. These **state factors**, as Jenny and his

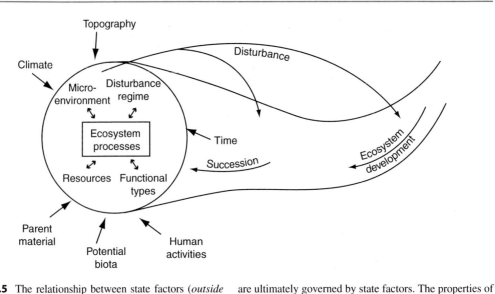

Fig. 1.5 The relationship between state factors (*outside the circle*), interactive controls (*inside the circle*), and ecosystem processes (*inside the box*). The *circle* represents the boundary of the ecosystem, whose structure and functioning respond to and affect interactive controls, which are ultimately governed by state factors. The properties of the ecosystem change through long-term development and shorter-term succession. Modified from Chapin et al. (2006b)

coworkers called them, include **climate**, **parent material** (the rocks that give rise to soils), **topography**, **potential biota** (the organisms present in the region that could potentially occupy a site), and **time** (Fig. 1.5; Jenny 1941; Amundson and Jenny 1997; Vitousek 2004). Together these five factors, among others, set the bounds for the characteristics of an ecosystem.

On broad geographic scales, climate is the state factor that most strongly determines ecosystem processes and structure. Global variations in climate explain the distribution of **biomes** (general categories of ecosystems) such as wet tropical forests, temperate grasslands, and arctic tundra (see Chap. 2). Within each biome, parent material strongly influences the types of soils that develop and explains much of the regional variation in ecosystem processes (see Chap. 3). Topographic relief influences both microclimate and soil development at a local scale. The potential biota governs the types and diversity of organisms that actually occupy a site. Island ecosystems, for example, are often less diverse than climatically similar mainland ecosystems because new species reach islands less often and are more likely to go locally extinct than on the mainland (MacArthur

and Wilson 1967). Time influences the development of soil and the evolution of organisms over long time scales (Vitousek 2004). Time also incorporates the influences on ecosystem processes of past disturbances and environmental changes over a wide range of time scales. State factors are described in more detail in Chap. 3 in the context of soil development.

Late in his life, Jenny (1980) suggested that human activity was becoming so pervasive as to represent a sixth major state factor. Human activities have an increasing impact on virtually all the processes that govern ecosystem properties (MEA 2005). Humans have been a natural component of most ecosystems for thousands of years. Since the beginning of the industrial revolution, however, the magnitude of human impact has been so great and so distinct from that of other organisms that the modern impacts of human activities warrant particular attention (Vitousek et al. 1997b; Steffen et al. 2004). The cumulative impact of human activities extends well beyond an individual ecosystem and affects state factors such as climate (through changes in atmospheric composition) and potential biota (through the introduction and extinction of species; Fig. 1.5). Human activities are causing major changes in the structure and functioning of all ecosystems, resulting in novel conditions that lead to new types of ecosystems (Foley et al. 2005; Ellis and Ramankutty 2008). The major categories of human impact are summarized in the next section.

Jenny's state-factor approach was a major conceptual contribution to ecosystem ecology. First, it emphasized the controls over *processes* rather than simply descriptions of patterns. Second, it suggested a study design to test the importance and mode of action of each control. A logical way to study the role of each state factor is to compare sites that are as similar as possible with respect to all but one factor. A **chronosequence**, for example, is a series of sites of different ages with similar climate, parent material, topography, and potential to be colonized by the same organisms (see Chap. 12). In a **toposequence**, ecosystems differ mainly in their topographic position (Shaver et al. 1991). Sites that differ primarily with respect to climate or parent material allow us to study the impacts of these state factors on ecosystem processes (Vitousek 2004). Finally, a comparison of ecosystems that differ primarily in potential biota, such as the Mediterranean shrublands that have developed on west coasts of California, Chile, Portugal, South Africa, and Australia, illustrates the importance of evolutionary history in shaping ecosystem processes (Mooney and Dunn 1970; Cody and Mooney 1978).

Ecosystem processes both respond to and control the factors that directly govern their activity. Interactive controls are factors that operate at the ecosystem scale and both *control* and *respond to* ecosystem characteristics (Fig. 1.5; Chapin et al. 1996). Important interactive controls include the supply of **resources** to support the growth and maintenance of organisms, **microenvironment** (e.g., temperature, pH) that influences the rates of ecosystem processes, **disturbance regime**, and the **biotic community**.

Resources are the energy and materials in the environment that are used by organisms to support their growth and maintenance (Field et al. 1992). The acquisition of resources by organisms generally depletes their abundance in the environment

and availability to other organisms, although some resources (e.g., atmospheric carbon dioxide) mix so rapidly that they can be considered nondepletable (Rastetter and Shaver 1992). Energy resources can either be chemical energy stored in matter, or incoming solar radiation. Material resources include carbon, oxygen, water, and the other elements that are required for life, which we generically refer to as **nutrients**. In terrestrial ecosystems, these resources are spatially separated, being available primarily either aboveground (light and CO_2) or belowground (water and nutrients). Resource supply is governed by state factors such as climate, parent material, and topography. It is also sensitive to processes occurring within the ecosystem. Light availability, for example, depends on climatic elements such as cloudiness and on topographic aspect but is also sensitive to the degree of shading by vegetation. Similarly, soil fertility depends on parent material and climate, but is also sensitive to ecosystem processes such as erosional loss of soils after overgrazing and inputs of nitrogen from invading nitrogen-fixing species. Soil water availability strongly influences species composition in dry climates. Soil water availability also depends on other interactive controls such as disturbance regime (e.g., compaction by animals) and the types of organisms that are present (e.g., the presence or absence of deep-rooted trees such as mesquite that tap deep groundwater). In aquatic ecosystems, water seldom directly limits the activity of organisms, but light and nutrients are at least as important as on land. Oxygen is a particularly critical resource in aquatic ecosystems because of its low solubility and slow rate of diffusion through water.

The **microenvironment** includes physical and chemical properties like temperature and pH that affect the activity of organisms but, unlike resources, are neither consumed nor depleted by organisms (Field et al. 1992). Microenvironmental factors like temperature vary with climate (a state factor) but are sensitive to ecosystem processes, such as shading and evaporation. Soil pH depends on parent material and time, but also responds to vegetation composition.

Landscape-scale **disturbance** by fire, wind, floods, insect outbreaks, and hurricanes is a critical determinant of the natural structure and process rates in ecosystems (Pickett and White 1985; Peters et al. 2011). Like other interactive controls, disturbance regime depends on both state factors and ecosystem processes. Fire probability and spread, for example, depends on both climate and the quantity and flammability of plants and dead organic matter. Deposition and erosion during floods shape river channels and influence the probability of future floods. Change in either the intensity or frequency of disturbance can cause long-term ecosystem change. Woody plants, for example, often invade grasslands when fire suppression reduces fire frequency.

The nature of the biotic community – i.e., the types of species present, their relative abundances, and the nature of their interactions, can influence ecosystem processes just as strongly as do differences in climate or parent material (see Chap. 11). These species effects can often be generalized at the level of **functional types**, which are groups of species that are similar to one another in their role in a specific community or ecosystem process. Most evergreen tree species, for example, produce leaves that have low rates of photosynthesis and a chemical composition that deters herbivores and slows down decomposition. A shift from one evergreen tree species to another usually has less influence on an ecosystem process than a shift to a deciduous tree species. A gain or loss of key functional types, for example through introduction or removal of species with large ecosystem effects, can permanently change the character of an ecosystem through changes in resource supply or disturbance regime. Introduction of nitrogen-fixing trees onto British mine wastes, for example, substantially increases nitrogen supply, productivity, and rates of vegetation development (Bradshaw 1983). Invasion of grasslands by exotic grasses can alter fire frequency, resource supply, trophic interactions, and rates of most ecosystem processes (D'Antonio and Vitousek 1992; Mack et al. 2001). Elimination of predators can cause an outbreak of deer that overbrowse their food supply (Beschta and Ripple 2009) or move disease-bearing ticks around the landscape (Ostfeld and Keesing 2000). The types of species present in an ecosystem depend strongly on other

interactive controls (see Chap. 11), so functional types respond to and affect most interactive controls and ecosystem processes.

Feedbacks regulate the internal dynamics of ecosystems. A thermostat, for example, causes a furnace to switch on when a house gets cold and to switch off when the house warms to the desired temperature. Natural ecosystems are complex networks of interacting feedbacks (DeAngelis and Post 1991). **Stabilizing feedbacks** (termed negative feedbacks in the systems literature) occur when two components of a system have opposite effects on one another (Fig. 1.6). Consumption of prey by a predator, for example, has a positive effect on the consumer but a negative effect on the prey. The negative effect of predators on prey prevents uncontrolled growth of a prey's population, thereby stabilizing the

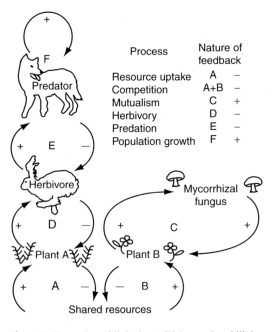

Fig. 1.6 Examples of linked amplifying and stabilizing feedbacks in ecosystems. The effect of each organism (or resource) on other organisms can be positive (+) or negative (–). Feedbacks are amplifying (positive feedbacks) when the reciprocal effects of each organism (or resource) have the same sign (both positive or both negative). Feedbacks are stabilizing (negative feedbacks) when reciprocal effects differ in sign. Stabilizing feedbacks resist tendencies for ecosystems to change, whereas amplifying feedbacks reinforce tendencies to change. Redrawn from Chapin et al. (1996)

population sizes of both predator and prey. There are also **amplifying feedbacks** (termed positive feedbacks in the systems literature) in ecosystems in which both components of a system have a positive effect on one other, or both have a negative effect on one another. Plants, for example, provide their mycorrhizal fungi with carbohydrates in return for nutrients. This exchange of growth-limiting resources between plants and fungi promotes the growth of both components of the symbiosis until they become constrained by other factors.

Stabilizing feedbacks provide resistance to changes in interactive controls and maintain the characteristics of ecosystems in their current state, whereas amplifying feedbacks accentuate changes. The acquisition of water, nutrients, and light to support growth of one plant, for example, reduces availability of these resources to other plants, thereby constraining community productivity (Fig. 1.6). Similarly, animal populations cannot sustain exponential population growth indefinitely because declining food supply and increasing predation reduce the rate of population increase. On the other hand, succession often involves a series of amplifying feedbacks, as plant growth and soil fertility reinforce each other, until another disturbance resets the successional clock. If stabilizing feedbacks are weak or absent (e.g., a low predation rate due to predator control), population cycles can amplify, causing extinction of one or both of the interacting species. Community dynamics, which operate within a single ecosystem patch, primarily involve feedbacks among soil resources and functional types of organisms.

Landscape dynamics, which govern changes in ecosystems through cycles of disturbance and recovery, involve additional feedbacks with microclimate and disturbance regime that link ecosystems across landscapes (see Chap. 13). Post-disturbance vegetation development, for example, is driven by amplifying feedbacks at the ecosystem scale, but also contributes to stabilizing feedbacks in landscapes over longer time periods by maintaining a diversity of successional stages and reducing risks of large-scale spread of disturbances like wildfire or insect outbreaks.

Human-Induced Ecosystem Change

Human Impacts on Ecosystems

Human activities have transformed the land surface, species composition, and biogeochemical cycles at scales that have altered the biogeochemistry and climate of the planet. These **anthropogenic** (human-caused) effects are so profound that the beginning of the industrial revolution (about 1,750) is widely recognized as the start of a new geologic epoch – the **Anthropocene** (see Fig. 2.15; Crutzen 2002).

The most direct and substantial human alteration of ecosystems is through the transformation of land for production of food, fiber, and other goods used by people (Fig. 1.7). People inhabit more than 75% of Earth's ice-free land surface. These inhabited areas include cities and villages (7%), croplands (20%), rangelands (30%), and forests (20%; Fig. 1.8; Foley et al. 2005; Ellis and Ramankutty 2008). The 25% uninhabited lands are primarily barren lands as well as additional forest lands. From inhabited landscapes, people appropriate 25–40% of terrestrial aboveground productivity through human harvest (53% of the human appropriation), land-use change and altered productivity (40%), and human-induced fires (7%; Vitousek et al. 1997b; Haberl et al. 2007).

Human activities have also altered freshwater and marine ecosystems. People currently use about 25% of the runoff from land to the ocean (see Chap. 14; Postel et al. 1996; Vörösmarty et al. 2005). We use about 8% of marine primary production (Pauly and Christensen 1995). Commercial fishing reduces the size and abundance of target species and alters the population characteristics of species that are incidentally caught in the fishery. About 70% of marine fisheries are overexploited, including 25% that have collapsed (defined as greater than 90% reduction in biomass; Mullon et al. 2005). A large proportion of the human population resides within 100 km of a coast, so the coastal margins of the ocean are strongly influenced by human activities. For example, nutrient enrichment of many coastal waters from agricultural runoff and from human and livestock sewage has increased algal production. Decomposition of

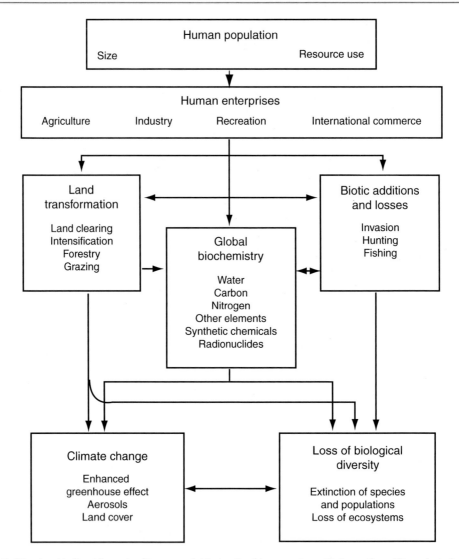

Fig. 1.7   Direct and indirect impacts of human activities on Earth's ecosystems. Redrawn from Vitousek et al. (1997b)

this material depletes oxygen within the water column, creating dead zones where anaerobic conditions kill fish and other animals (see Fig. 9.1; Rabalais et al. 2002).

Land-use change and the resulting loss of habitat are the primary driving forces causing species extinctions and loss of biological diversity (see Chap. 11; Mace et al. 2005). In addition, transport of species around the world increases the frequency of biological invasions, due to the globalization of the economy and increased international transport of people and products. Nonindigenous species now account for 20% or more of the plant

species in many continental areas and 50% or more of the plant species on many islands (Vitousek et al. 1997b). International commerce breaks down biogeographic barriers through both inadvertent introductions and the purposeful selection of species that are intended to grow and reproduce well in their new environment. Many of these introductions, such as agricultural crops and pasture grasses, increase certain ecosystem services, such as food for human consumption. Yet, the addition of new species can also degrade human health (e.g., rinderpest in Africa; Sinclair and Norton-Griffiths 1979) and cause large

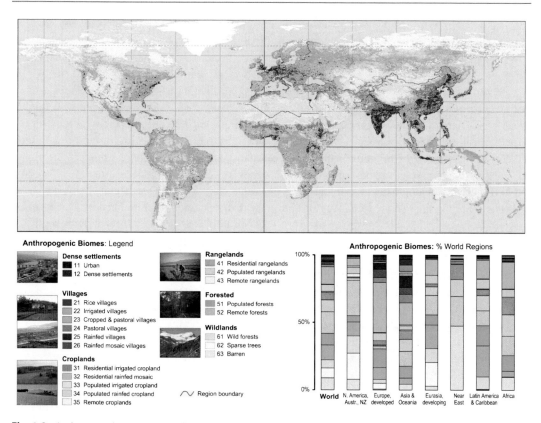

Fig. 1.8 Anthropogenic ecosystems of the world. Human activity has fundamentally altered both the nature of Earth's ecosystems and the way they are conceptualized. Reprinted from Ellis and Ramankutty (2008)

economic losses (e.g., introduction of fire-prone cheatgrass to North American rangelands; Bradley and Mustard 2005). Others alter the structure and functioning of ecosystems, leading to further loss of species diversity. Many biological invasions are irreversible because it is difficult or prohibitively expensive to remove invasive species once they establish.

Human activities have influenced biogeochemical cycles in many ways. Extensive use of fossil fuels and the expansion and intensification of agriculture have increased the concentrations of atmospheric gases, altering global cycles of carbon, nitrogen, phosphorus, sulfur, and water (see Chap. 14). Biogeochemical changes also alter the internal dynamics of ecosystems, as well as downwind ecosystems through atmospheric transport and downstream ecosystems through runoff to lakes, rivers, and the coastal zone of the ocean.

Human activities introduce novel chemicals into the environment. Some apparently harmless anthropogenic gases have had drastic impacts on the atmosphere and ecosystems. Chlorofluorocarbons (CFCs), for example, were first produced in the 1950s as refrigerants, propellants, and solvents. In the upper atmosphere, however, CFCs react with and deplete ozone, which shields Earth's surface from high-energy UV radiation. Ozone depletion was first detected as a dramatic **ozone hole** near the South Pole, but it now occurs at lower latitudes in the southern hemisphere and at high Northern latitudes. Other synthetic organic chemicals include DDT (an insecticide) and PCBs (polychlorinated biphenyls, industrial compounds) that were used extensively in the developed world in the 1960s before their ecological impacts were widely recognized. They are mobile and degrade slowly, causing long-term persistence and transport to ecosystems across the globe. Many of these compounds are fat soluble, so they accumulate in organisms and increase in concentration as they move up food chains (see Chap. 10). When these compounds reach critical concentrations, they can cause reproductive failure (Carson 1962), particularly in higher trophic levels and in animals that feed on fat-rich species. Some processes, such as eggshell formation in birds, are particularly sensitive to pesticide accumulations and have caused population declines in predatory birds like the peregrine falcon, even in regions far removed from the locations of pesticide use.

Atmospheric testing of atomic weapons in the 1950s and 1960s increased atmospheric concentrations of radioactive forms of many elements. Explosions and leaks in nuclear reactors used to generate electricity have also released radioactivity at local to regional scales. The explosion of a power-generating plant in 1986 at Chernobyl in the Ukraine, for example, released radioactivity that directly affected human health in the region and increased the atmospheric deposition of radioactive materials across Eastern Europe and Scandinavia. Some radioactive isotopes of elements such as strontium and cesium, which are chemically similar to calcium and potassium, respectively, are actively accumulated and retained by organisms. Lichens, for example, acquire minerals primarily from the atmosphere and actively accumulate strontium and cesium. Reindeer feeding on lichens further concentrate these minerals, as do people who eat reindeer. For this reason, the input of radioisotopes to the atmosphere or water has had impacts that extend far beyond the regions where they were used.

In other cases, the chemicals that people introduce to ecosystems are much more targeted as in the case of BT-corn, a genetically modified corn variety carrying bacterial genes that cause production of a compound that is toxic to European corn borer. Any introduction of novel chemicals raises questions of toxicity to non-target organisms or the evolution of resistance in target species (Marvier et al. 2007). These questions are amenable to study by ecosystem ecologists.

The growing scale and extent of human activities suggest that *all* ecosystems are being influenced, directly or indirectly, by human actions. No ecosystem functions in isolation, and all are influenced by human activities taking place in adjacent communities and around the world. Human activities are leading to global changes in most major ecosystem controls: climate (global warming), soil and water resources (nitrogen deposition, erosion, diversions), disturbance regime (land-use change, fire suppression), and functional types of organisms (species introductions and extinctions). Many of these global changes interact with one another at regional and local scales (Rockström et al. 2009). All ecosystems are therefore experiencing directional changes in ecosystem controls, creating novel conditions and, in some cases, amplifying feedbacks that lead to novel ecosystems. These changes in interactive controls inevitably alter ecosystem dynamics.

Resilience and Threshold Changes

Despite pervasive human impacts on state factors and interactive controls, ecosystems exhibit a wide range of responses, ranging from substantial resilience to threshold changes. **Resilience** is the capacity of a social–ecological system to maintain similar structure, functioning, and feedbacks despite shocks and perturbations. **Thresholds** are critical levels of one or more ecosystem controls that, when crossed, cause abrupt ecosystem changes. Lakes may, for example, maintain water clarity and support desired fish stocks despite substantial nutrient inputs from agricultural runoff or local septic systems because of stabilizing (negative) feedbacks from lake sediments that bind phosphorus, removing it from the water column, and providing resilience. At some point, however, phosphorus-binding capacity becomes saturated, so sediments become a source of phosphorus to the water column, supporting the growth of nuisance algae that reduce water clarity and trigger a cascade of other events that are not easily reversed (see Chaps. 9 and 12). Biodiversity can also confer resilience because a large number of species is likely to sustain ecosystem processes over a broader range of conditions than would one or a few species (see Chap. 11; Elmqvist et al.

2003; Suding et al. 2008). Social processes that govern the role of people in ecosystems can be a source of resilience (sustainability) or can trigger threshold changes. Ecologists are only beginning to understand the factors that govern ecosystem resilience and threshold change (see Chap. 12). This is emerging as a critical research area in our increasingly human-dominated planet.

Although some pressures on ecosystems are easily observed (e.g., acid rain) or predicted (e.g., rising global temperature that was predicted decades ago and is now being observed), **surprises** that are difficult or impossible to anticipate also occur. Some processes that confer resilience are quite specific to a given driver of change (e.g., sediment sequestration of phosphorus). Others, such as biodiversity or a multiple-use management policy, may confer resilience to a variety of potential changes, some of which may occur unexpectedly.

Degradation in Ecosystem Services

Many ecosystem services have been degraded globally since the mid-twentieth century (Daily 1997; MEA 2005). Society benefits in numerous ways from ecosystems, including (1) **provisioning services** (or **ecosystem goods**), which are products of ecosystems that are directly harvested by people (e.g., food, fiber, and water); (2) **regulating services**, which are the effects of ecosystems on processes that extend beyond their boundaries (e.g., regulation of climate, water quantity and quality, disease, wildfire spread, and pollination); and (3) **cultural services**, which are nonmaterial benefits that are important to society's well-being (e.g., recreational, aesthetic, and spiritual benefits; see Fig. 15.4). Many ecosystem processes (e.g., productivity, nutrient cycling, and maintenance of biodiversity) support these ecosystem services. More than half of these ecosystem services were degraded globally over the last half of the twentieth century – not deliberately, but inadvertently as people sought to meet their material desires and needs (MEA 2005). Change creates both challenges and opportunities. People have amply demonstrated our capacity to alter the life-support system of the planet. With appropriate ecosystem stewardship, this human capacity can be mobilized to not only repair but also enhance the capacity of Earth's life-support system to support societal development. An important challenge for ecosystem ecology is to provide the scientific knowledge to meet this goal.

Summary

Ecosystem ecology addresses the interactions among organisms and their environment as an integrated system through study of the factors that regulate the pools and fluxes of materials and energy through ecological systems. The spatial scale at which we study ecosystems is chosen to facilitate the measurement of important fluxes into, within, and out of the ecosystem. The functioning of ecosystems depends not only on their current structure and environment but also on a legacy of response to past events. The study of ecosystem ecology is highly interdisciplinary, building on many aspects of ecology, hydrology, climatology, geology, and sociology and contributing to current efforts to understand Earth as an integrated system. Many unresolved problems in ecosystem ecology require an integration of systems approaches, process understanding, and global analysis.

Most ecosystems ultimately acquire their energy from the sun and their materials from the atmosphere and rock minerals. Energy and materials are transferred among components within ecosystems and are then released to the environment. The essential biotic components of ecosystems include plants, which bring carbon and energy into the ecosystem; decomposers, which break down dead organic matter and release CO_2 and nutrients; and animals, which transfer energy and materials within ecosystems and modulate the activity of plants and decomposers. The essential abiotic components of ecosystems are the atmosphere, water, and soils. Ecosystem processes are controlled by a set of relatively independent state factors (climate, parent material, topography, potential biota, time, and increasingly human activities) and by a group of interactive controls (including resource supply, microenvironment, disturbance regime, and functional types of organisms) that directly control ecosystem processes. The interactive controls both respond to and affect ecosystem processes, while state factors are considered independent of ecosystems. The stability and resilience of ecosystems depend on the strength and interactions between stabilizing (negative) feedbacks that maintain the characteristics of ecosystems in their current state and amplifying (positive) feedbacks that are sources of renewal and change.

Review Questions

1. What is an ecosystem? How does it differ from a community? What kinds of environmental questions can ecosystem ecologists address that are not easily addressed by community ecologists?

2. What is the difference between a pool and a flux? Which of the following are pools and which are fluxes: plants, plant respiration, rainfall, soil carbon, and consumption of plants by animals?

3. What are the state factors that control the structure and rates of processes in ecosystems? What are the strengths and limitations of the state-factor approach to answering this question?

4. What is the difference between state factors and interactive controls? Why would you treat a state factor and an interactive control differently in developing a management plan for a region?

5. Using a forest or a lake as an example, explain how climatic warming or harvest of trees or fish by people might change the major interactive controls, and how these changes in controls might alter the structure or processes in these ecosystems.

6. Use examples to show how amplifying and stabilizing feedbacks might affect the responses of an ecosystem to climatic change.

Additional Reading

Chapin, F.S., III, G.P. Kofinas, and C. Folke. 2009. *Principles of Ecosystem Stewardship: Resilience-Based Natural Resource Management in a Changing World*. Springer, New York.

Ellis E.C., and N. Ramankutty. 2008. Putting people on the map: Anthropogenic biomes of the world. *Frontiers in Ecology and the Environment* 6:439–447.

Golley, F.B. 1993. *A History of the Ecosystem Concept in Ecology: More than the Sum of the Parts*. Yale University Press, New Haven.

Gorham, E. 1991. Biogeochemistry: Its origins and development. *Biogeochemistry* 13:199–239.

Hagen, J.B. 1992. *An Entangled Bank: The Origins of Ecosystem Ecology*. Rutgers University Press, New Brunswick, New Jersey.

Jenny, H. 1980. *The Soil Resources: Origin and Behavior*. Springer-Verlag, New York.

Lindeman, R.L. 1942. The trophic-dynamic aspects of ecology. *Ecology* 23:399–418.

MEA (Millennium Ecosystem Assessment). 2005. *Ecosystems and Human Well-being: Synthesis*. Island Press, Washington.

Schlesinger, W.H. 1997. *Biogeochemistry: An Analysis of Global Change*. Academic Press, San Diego.

Tansley, A.G. 1935. The use and abuse of vegetational concepts and terms. *Ecology* 16:284–307.

Vitousek, P.M. 2004. *Nutrient Cycling and Limitation: Hawai'i as a Model System*. Princeton University Press, Princeton.

2

Climate is the state factor that most strongly governs the global distribution of terrestrial biomes. This chapter provides a general background on the functioning of the climate system and its interactions with atmospheric chemistry, ocean, and land.

Introduction

Climate exerts a key control over the functioning of Earth's ecosystems. Temperature and water availability govern the rates of many biological and chemical reactions that in turn control critical ecosystem processes. These processes include the production of organic matter by plants, its decomposition by microbes, the weathering of rocks, and the development of soils. Understanding the causes of temporal and spatial variation in climate is therefore critical to understanding the global pattern of ecosystem processes.

The amount of incoming solar radiation, the chemical composition and dynamics of the atmosphere, and the surface properties of Earth determine climate and climate variability. The circulation of the atmosphere and ocean influences the transfer of heat and moisture around the planet and thus strongly influences climate patterns and their variability in space and time. This chapter describes the global energy budget and outlines the roles that the atmosphere, ocean, and land surface play in the redistribution of energy to produce observed patterns of climate and ecosystem distribution.

A Focal Issue

Human activities are modifying Earth's climate, thereby changing fundamental controls over ecosystem processes throughout the planet, often to the detriment of society. Some climatic changes subtly alter the rates of ecosystem process, but other changes, such as the frequency of severe storms have direct devastating effects on society. Climate warming, for example, increases sea-surface temperature, which increases the energy transferred to tropical storms (Fig. 2.1). Although no individual storm can be attributed to climate change, the intensity of tropical storms may increase (IPCC 2007). Other expected effects of climate change include more frequent droughts in drylands such as sub-Saharan Africa, more frequent floods in wet climates and in low-lying coastal zones, warmer weather in cold climates, and more extensive wildfires in fire-prone forests. What determines the distribution of Earth's major climate zones? Why is climate changing, and why do regions differ in the climatic changes they experience? An understanding of the causes of temporal and spatial variation in the climate system facilitates predictions of the changes that are likely to occur in particular places.

Earth's Energy Budget

The sun is the source of the energy available to drive Earth's climate system. The wavelength of energy produced by a body depends on its

F.S. Chapin, III et al., *Principles of Terrestrial Ecosystem Ecology*,
DOI 10.1007/978-1-4419-9504-9_2, © Springer Science+Business Media, LLC 2011

Fig. 2.1 Satellite view of Hurricane Katrina over coastal Louisiana. This tropical storm flooded New Orleans in 2005, killing approximately 1,570 people and causing $40–50 billion of damage. Human-caused ecological changes in coastal Louisiana contributed to the impact of the hurricane. Climate warming is expected to increase the frequency of severe tropical storms like Hurricane Katrina. Image courtesy of NOAA (http://www.katrina.noaa.gov/satellite/satellite.html)

temperature. Because it is hot (6,000°C), the sun emits most energy as high-energy **shortwave radiation** with wavelengths of 0.2–4.0 μm (Fig. 2.2). These include ultraviolet (UV; 8% of the total), visible (39%), and near-infrared (53%) radiation. On average, about 30% of the incoming shortwave radiation is reflected back to space, due to **backscatter** (reflection) from clouds (16%); air molecules, dust, and haze (6%); and Earth's surface (7%; Fig. 2.3). Another 23% of the incoming shortwave radiation is absorbed by the atmosphere, especially by ozone in the upper atmosphere and by clouds and water vapor in the lower atmosphere. The remaining 47% reaches Earth's surface as direct or diffuse radiation and is absorbed there (Trenberth et al. 2009).

Earth also emits radiation, like all bodies, but, due to its lower surface temperature (about 15°C), Earth emits most energy as low-energy **longwave radiation** (Fig. 2.2). Although the atmosphere transmits about half of the incoming shortwave radiation to Earth's surface, **radiatively active gases** (water vapor, CO_2, CH_4, N_2O and industrial products like chlorofluorocarbons [CFCs]) absorb 90% of the outgoing longwave radiation (Fig. 2.3). Of the approximately 10% of longwave radiation that escapes to space, most is in wavelengths where longwave absorption by the atmosphere is small (referred to as atmospheric windows; Fig. 2.2). The energy absorbed by radiatively active gases in the atmosphere is re-radiated in all directions (Fig. 2.3). The portion that is directed back toward the surface contributes to the warming of the planet, a phenomenon known as the **greenhouse effect**. Without these longwave-absorbing gases in the atmosphere, the average temperature at Earth's surface would be about 33°C lower than it is today, and Earth would probably not support life, except perhaps at hydrothermal vents in the deep ocean.

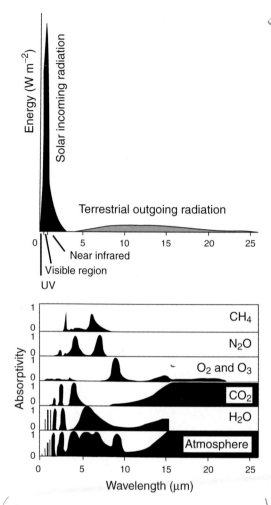

Fig. 2.2 The spectral distribution of solar and terrestrial radiation and the absorption spectra of the major radiatively active gases and of the total atmosphere. These spectra show that the atmosphere absorbs a larger proportion of terrestrial radiation than solar radiation, explaining why the atmosphere is heated from below. Redrawn from Sturman and Tapper (1996) and Barry and Chorley (2003)

As a global long-term average, Earth is normally close to a state of radiative balance, meaning that it emits as much energy back to space (as longwave radiation) as it absorbs. However, human activities are changing the composition of the atmosphere enough to increase the heat retained by the planet, as described later. Assuming balance, the longwave radiation emitted to space must equal the sum of the solar radiation absorbed by both the surface and the atmosphere. The atmosphere is heated by longwave absorption by radiatively active gases and by the absorption of some incoming (short-wave) solar radiation; it is also heated from the surface by non-radiative fluxes of heat that are carried upward by atmospheric **turbulence** (mixing). These include **latent heat flux**, where heat that evaporates water at the surface is subsequently released to the atmosphere as air parcels rise and cool, and the water vapor condenses, forming clouds and precipitation. There is also an upward transfer of heat that is conducted from the warm surface to the air immediately above it and then moved upward by convection of the atmosphere as thermals (**sensible heat flux**). These heat sources collectively sustain the longwave emission to space, as well as a large flux of longwave radiation from the lower atmosphere back to Earth's surface. This back radiation to the surface represents the natural greenhouse effect described earlier.

Long-term records of atmospheric gases, obtained from atmospheric measurements since the 1950s and from air bubbles trapped in glacial ice, show large increases in the major radiatively active gases (CO_2, CH_4, N_2O, and CFCs) since the beginning of the industrial revolution 250 years ago (see Fig. 14.7). Human activities such as fossil fuel burning, industrial activities, animal husbandry, and fertilized and irrigated agriculture contribute to these increases (see Chap. 14). As concentrations of these gases rise, the atmosphere traps more of the longwave radiation emitted by Earth, enhancing the greenhouse effect and increasing Earth's surface temperature. A small imbalance thus exists in the radiative flows shown in Fig. 2.3, estimated to be about 0.26% of the incoming radiation. Most of this excess energy is absorbed in the ocean, causing water to expand and sea level to rise. The warming caused by radiative imbalance also contributes to widespread melting of glaciers and ice sheets (Greenland and Antarctica) and arctic sea ice.

The globally averaged annual energy budget outlined above gives a sense of the critical factors controlling the global climate system. Regional climates, however, reflect spatial variation in energy exchange and in lateral heat transport by the atmosphere and the ocean. Earth is heated more strongly at the equator than at the poles and

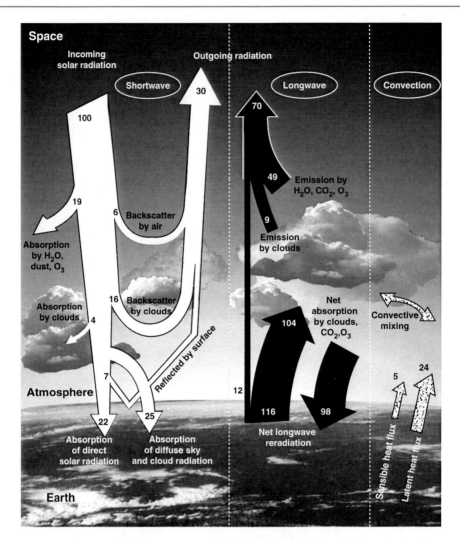

Fig. 2.3 The average annual global energy balance during 2000–2004 for the Earth-atmosphere system. The numbers are percentages of the energy received from incoming solar radiation. At the top of the atmosphere, the incoming solar radiation (100 units or 341 W m^{-2} [global average]) is balanced by reflected shortwave (30 units) and emitted longwave radiation (70 units). Within the atmosphere, the absorbed shortwave radiation (23 units) and absorbed longwave radiation (104 units) and latent + sensible heat flux (29 units) are balanced by longwave emission to space (58 units) and longwave emission to Earth's surface (98 units). At Earth's surface, the incoming shortwave (47 units) and incoming longwave radiation (98 units) are balanced by outgoing longwave radiation (116 units) and latent + sensible heat flux (29 units). Data are from Trenberth et al. (2009)

rotates on an axis that is tilted relative to the plane of its orbit around the sun. Its continents are spread unevenly over the surface, and its atmospheric and oceanic chemistry and physics are dynamic and spatially variable. A more thorough understanding of the atmosphere and ocean is therefore needed to understand the fate and processing of energy and its consequences for Earth's ecosystems.

The Atmospheric System

Atmospheric Composition and Chemistry

The chemical composition of the atmosphere determines its role in Earth's energy budget. The atmosphere is like a giant reaction flask,

Table 2.1 Major chemical constituents of the atmosphere

Compound	Formula	Concentration (%)
Nitrogen	N_2	78.082
Oxygen	O_2	20.945
Argon	Ar	0.934
Carbon dioxide	CO_2	0.039

Data from Schlesinger (1997) and IPCC (2007)

containing thousands of different chemical compounds in gas and particulate forms, undergoing slow and fast reactions, dissolutions, and precipitations. These reactions control the composition of the atmosphere and many of its physical processes, such as cloud formation and energy absorption. The associated heating and cooling, together with the uneven distribution of solar radiation, generate dynamical motions crucial for energy redistribution.

More than 99.9% by volume of Earth's dry atmosphere is composed of nitrogen, oxygen, and argon (Table 2.1). Carbon dioxide (CO_2), the next most abundant gas, accounts for only 0.039% of the atmosphere. These percentages are quite constant around the world and up to 80 km in height above the surface. That homogeneity reflects the fact that these gases have long **mean residence times** (MRT) in the atmosphere. MRT is calculated as the total mass divided by the flux into or out of the atmosphere over a given time period. Nitrogen has an MRT of 13 million years, O_2 10,000 years, and CO_2 5 years (see Chap. 14). Some of the most important radiatively active gases, such as CO_2, nitrous oxide (N_2O), methane (CH_4), and CFCs, react relatively slowly in the atmosphere and have residence times of years to decades. Other gases are much more reactive and have residence times of days to months. Highly reactive gases make up less than 0.001% of the dry volume of the atmosphere and are quite variable in time and space. These reactive gases influence ecological systems through their roles in nutrient delivery, smog, acid rain, and ozone depletion (Graedel and Crutzen 1995). Water vapor is also quite reactive and highly variable both seasonally and spatially.

MRT provides a reasonable estimate of the lifetime of a gas in the atmosphere for those gases like CH_4 and N_2O that undergo irreversible reactions to produce breakdown products. CO_2, however, is not "destroyed" when it is absorbed by the ocean or the biosphere, but continues to exchange with the atmosphere. If all fossil fuel emissions ceased instantly today, the excess fossil-fuel CO_2 in the atmosphere (about 35% higher than the "natural" background) would decline by 50% within 30 years, another 20% within a few centuries, but the remaining 30% excess CO_2 would remain in the atmosphere for thousands of years (IPCC 2007; Archer et al. 2009; see Chap. 14). This will create, from the perspective of a human lifetime, a permanently warmer world (Solomon et al. 2009). The magnitude of this climate warming will depend on the rates at which people reduce their emissions of fossil fuels and other trace gases.

Some atmospheric gases are critical for life. Photosynthetic organisms use CO_2 in the presence of light to produce organic matter that eventually becomes the basic food source for almost all animals and microbes (see Chaps. 5–7). Most organisms also require oxygen for metabolic respiration. Di-nitrogen (N_2) makes up 78% of the atmosphere. It is unavailable to most organisms, but nitrogen-fixing bacteria convert it to biologically available nitrogen that is ultimately used by all organisms to build proteins (see Chap. 9). Other gases, such as carbon monoxide (CO), nitric oxide (NO), nitrous oxide (N_2O), methane (CH_4), and volatile organic carbon compounds like terpenes and isoprene, are the products of plant and microbial activity. Some, like tropospheric ozone (O_3), are produced chemically in the atmosphere as products of chemical reactions involving both **biogenic** (biologically produced) and anthropogenic gases, and can, at high concentrations, damage plants, microbes, and people.

The atmosphere also contains **aerosols**, which are small solid or liquid particles suspended in air. Some aerosol particles arise from volcanic eruptions and from blowing dust and sea salt. Others are produced by reactions with gases from pollution sources and biomass burning. Some aerosols act as **cloud condensation nuclei** around which water vapor condenses to form cloud droplets. Aerosols, together with gases and clouds and characteristics of the surface, determine the reflectivity (**albedo**) of the planet and therefore exert major control over the energy budget and hence climate. The scattering (reflection) of incoming shortwave radiation by some aerosols reduces the radiation reaching Earth's surface and tends to cool the climate. For example, the sulfur dioxide injected into the atmosphere by the volcanic eruption of Mt. Pinatubo in the Philippines in 1991 and the subsequent creation of sulfate aerosols cooled Earth's climate for about a year.

Clouds have complex effects on Earth's radiation budget. All clouds have a high albedo, and hence reflect much more incoming shortwave radiation than does the darker Earth surface. Clouds, however, are composed of water droplets and ice crystals, which are very efficient absorbers of longwave radiation impinging on them from Earth's surface. The first process (reflecting shortwave radiation) has a cooling effect by reflecting incoming energy back to space. The second effect (absorbing longwave radiation) has a warming effect, by preventing energy from escaping to space. The balance of these two effects depends on many factors, including cloud type, temperature, thickness, and height. The reflection of shortwave radiation usually dominates the balance in high clouds, causing cooling, whereas the absorption and re-emission of longwave radiation generally dominates in low clouds, producing a warming effect. While clouds have a net cooling effect globally by reducing solar input, they have a net warming effect in the Arctic and Antarctic, where heat loss predominates.

Atmospheric Structure

Atmospheric pressure and density decline with height above Earth's surface. The average vertical structure of the atmosphere defines four relatively distinct layers characterized by their temperature profiles. The atmosphere is highly compressible, and gravity keeps most of the mass of the atmosphere close to Earth's surface. Pressure, which is related to the mass of the overlying atmosphere, decreases logarithmically

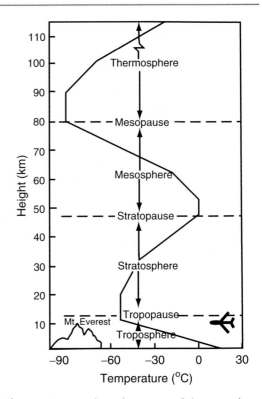

Fig. 2.4 Average thermal structure of the atmosphere, showing the vertical gradients in temperature in Earth's major atmospheric layers. Redrawn from Schlesinger (1997)

with height, as does the density of air. As one moves above the surface toward lower pressure and density, the vertical pressure gradient also decreases. Furthermore, because warm air is less dense than cold air, pressure falls off with height more slowly for warm than for cold air.

The **troposphere** is the lowest atmospheric layer (Fig. 2.4). It contains 75% of the mass of the atmosphere and is heated primarily from the bottom by sensible and latent heat fluxes and by longwave radiation from Earth's surface. Because air heated at the surface cools as it rises and expands, temperature decreases with height in the troposphere.

Above the troposphere is the **stratosphere**, which, unlike the troposphere, is heated from the top, resulting in an increase in temperature with height (Fig. 2.4). Absorption of UV radiation by **ozone** (O_3) in the upper stratosphere warms the air.

Fig. 2.5 Growth in height of the planetary boundary layer (PBL) above the plant canopy between 6 a.m. and noon in the Amazon Basin on a day without thunderstorms. The increase in surface temperature drives evapotranspiration and convective mixing, which causes the boundary layer to increase in height until the rising air becomes cool enough that water vapor condenses to form clouds. Redrawn from Matson and Harriss (1988)

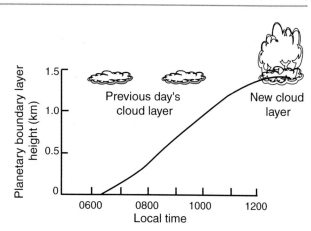

Ozone is most concentrated in the upper stratosphere due to a balance between the availability of shortwave UV necessary to split molecules of O_2 into atomic O and a high enough density of molecules to bring about the required collisions between atomic O and molecular O_2 to form O_3. The ozone layer protects the biota at Earth's surface from UV radiation. Biological systems are very sensitive to UV radiation because it damages DNA, which contains the information needed to drive cellular processes. The concentration of ozone in the stratosphere has been declining due to the production and emission of chlorofluorocarbon chemicals (CFCs) that destroy stratospheric ozone, particularly at the poles. This results in ozone "holes," regions where the transmission of UV radiation to Earth's surface is increased. Because the south polar region is colder and has more stratospheric clouds in which ozone-destroying reactions occur, the ozone hole over Antarctica is much larger than its arctic counterpart. Slow mixing between the troposphere and the stratosphere allows CFCs and other compounds to reach and accumulate in the ozone-rich stratosphere, where they have long residence times.

Above the stratosphere is the **mesosphere**, where temperature again decreases with height. The uppermost layer of the atmosphere, the **thermosphere**, begins at approximately 80 km and extends into space. The thermosphere has a very small fraction of the atmosphere's total mass, composed primarily of O and N atoms that can absorb energy of extremely short wavelengths,

again causing an increase in heating with height (Fig. 2.4). The mesosphere and thermosphere have relatively little impact on the biosphere.

The troposphere is the atmospheric layer where most weather occurs, including thunderstorms, snowstorms, hurricanes, and high and low pressure systems. The troposphere is therefore the portion of the atmosphere that directly responds to and affects ecosystem processes. The **tropopause** is the boundary between the troposphere and the stratosphere. It occurs at a height of about 16 km in the tropics, where tropospheric temperatures are highest and hence where pressure falls off most slowly with height, and at about 9 km in polar regions, where tropospheric temperatures are lowest. The height of the tropopause varies seasonally, being lower in winter than in summer.

The **planetary boundary layer** (PBL) is the lower portion of the troposphere in which air is mixed by surface heating, which creates convective turbulence, and by mechanical turbulence as air moves across Earth's rough surface. The PBL increases in height during the day largely due to convective turbulence. The PBL mixes more rapidly with the free troposphere when the atmosphere is disturbed by storms. The boundary layer over the Amazon Basin, for example, generally grows in height until midday, when it is disrupted by convective activity (Fig. 2.5). The PBL becomes shallower at night when there is no solar energy to drive convective mixing. Air in the PBL is relatively isolated from the free troposphere and therefore functions like a chamber over Earth's surface. The changes in water vapor, CO_2, and other chemical constituents in the PBL therefore serve as an indicator of the biological and physiochemical processes occurring at the surface (Matson and Harriss 1988). The PBL in urban regions, for example, often has higher concentrations of pollutants than the cleaner, more stable air above. At night, gases emitted by the surface, such as CO_2 in natural ecosystems or pollutants in urban environments, often reach high concentrations because they are concentrated in a shallow boundary layer.

Atmospheric Circulation

The fundamental cause of atmospheric circulation is the uneven solar heating of Earth's surface. The equator receives more incoming solar radiation than the poles because Earth is spherical. At the equator, the sun's rays are almost perpendicular to the surface at solar noon. At the lower sun angles characteristic of high latitudes, the sun's rays are spread over a larger surface area (Fig. 2.6), resulting in less radiation received per unit ground area. In addition, the sun's rays have a longer path through the atmosphere at high latitudes, so more of the incoming solar radiation is absorbed, reflected, or scattered before it reaches the surface. This unequal heating of Earth results in higher tropospheric temperatures in the tropics than at the poles, which in turn drives atmospheric circulation and transports atmospheric heat toward the poles. As a consequence of this, the input of shortwave solar radiation exceeds longwave radiation loss to space in the tropics, whereas longwave radiation loss exceeds solar input at temperate and high latitudes (Fig. 2.7).

Atmospheric circulation has both vertical and horizontal components (Fig. 2.8). Surface heating causes the surface air to expand and become less dense than surrounding air, so it rises. As air rises, the decrease in atmospheric pressure with height causes continued expansion, which decreases the average kinetic energy of air molecules, meaning that the rising air becomes cooler.

Cooling causes condensation and precipitation because cool air has a lower capacity to hold water vapor than warm air. Condensation, in turn, releases latent heat, which can cause the rising air to remain warmer than surrounding air, so it continues to rise. The average **lapse rate** (the rate at which air temperature decreases with height) varies regionally depending on the strength of surface heating and the atmospheric moisture content but averages about 6.5°C km^{-1}.

Surface air rises most strongly at the equator because of the intense equatorial heating and the large amount of latent heat released as this moist tropical air rises, expands, cools, and releases heat by condensation of water vapor. This air often rises until it reaches the tropopause. The upward movement and expansion of equatorial air also creates a horizontal pressure gradient that causes the equatorial air aloft to flow horizontally from the equator toward the poles (Fig. 2.8). This poleward-moving air cools because of both emission of longwave radiation to space and mixing with cold air that moves toward the equator from the poles. In addition, the tropical air converges into a smaller volume as it moves poleward because the radius and surface area of Earth decrease from the equator toward the poles. Due to the cooling of the air and its convergence into a smaller volume, the density of air increases, creating a high pressure that causes upper air to subside and warm. Subtropical high-pressure zones typically have clear skies; the resulting high input of solar radiation drives abundant evaporation. This moist subtropical surface air moves back toward the equator to replace the rising equatorial air. Hadley proposed this model of atmospheric circulation in 1735, suggesting that there should be one large circulation cell in the northern hemisphere and another in the southern hemisphere, driven by atmospheric heating and uplift at the equator and subsidence at the poles. Based on observations, Ferrell proposed in 1865 the conceptual model that we still use today, although the actual dynamics are much more complex (Trenberth and Stepaniak 2003). This model describes atmospheric circulation as a series of three circulation cells in each hemisphere. (1) The **Hadley cell** is driven by expansion and uplift of equatorial

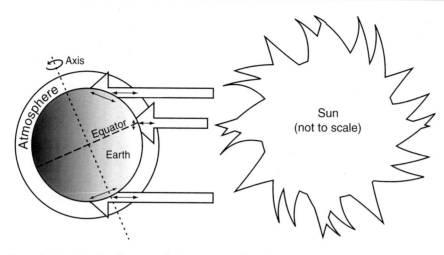

Fig. 2.6 Atmospheric and angle effects on solar inputs to different latitudes. The *arrows* parallel to the sun's rays show the depth of the atmosphere that solar radiation must penetrate. The *arrows* parallel to Earth's surface show the surface area over which a given quantity of solar radiation is distributed. High-latitude ecosystems receive less radiation than those at the equator because radiation at high latitudes has a longer path length through the atmosphere and is spread over a larger ground area

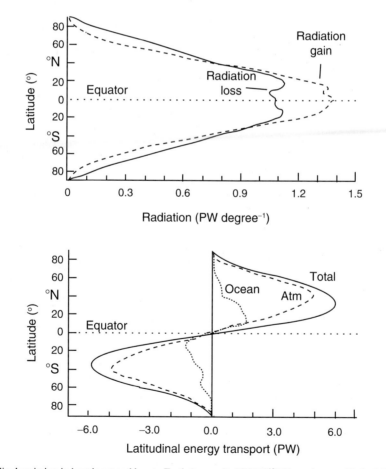

Fig. 2.7 Latitudinal variation in heat input and loss to Earth (*top*; units PW [10^{15} W] per degree of latitude) and in latitudinal heat transport by the ocean and the atmosphere (*bottom*; units PW). Redrawn from Fasullo and Trenberth (2008)

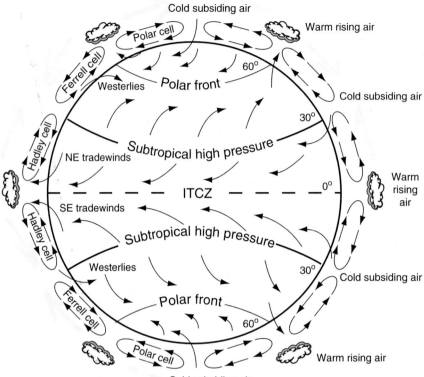

Fig. 2.8 Earth's latitudinal atmospheric circulation is driven by rising air at the equator and subsiding air at the poles. These forces and the Coriolis effect produce three major cells of vertical atmospheric circulation (the Hadley, Ferrell, and polar cells). Air warms and rises at the equator due to intense heating. After reaching the tropopause, the equatorial air moves poleward to about 30°N and S latitudes, where it descends and either returns to the equator, forming the Hadley cell, or moves poleward. Cold dense air at the poles subsides and moves toward the equator until it encounters poleward-moving air at about 60°N and S. Here the air rises and moves either poleward to replace air that has subsided at the poles (the polar cell) or moves toward the equator to form the Ferrell cell. Also shown are the horizontal patterns of atmospheric circulation consisting of the prevailing surface winds (the easterly trade winds in the tropics and the westerlies in the temperate zones). The boundaries between these zones are either low-pressure zones of rising air (the ITCZ and the polar front) or high-pressure zones of subsiding air (the subtropical high-pressure belt and the poles)

air and subsidence of cool dense subtropical air. (2) The **polar cell** is driven by subsidence of cold converging air at the poles. (3) An intermediate **Ferrell cell** is driven indirectly by dynamical processes (Fig. 2.8). The Ferrell cell is actually the long-term average air movement of mid-latitude weather systems rather than a stable permanent atmospheric feature. The circular motion (eddies) of these mid-latitude weather systems produces a net poleward transport of heat. These three cells subdivide the atmosphere into several distinct circulations: tropical air masses between the equator and 30°N and S, temperate air masses between 30° and 60°N and S, and polar air masses between 60°N and S and the poles (Fig. 2.8).

Earth's rotation causes winds to deflect to the right in the northern hemisphere and to the left in the southern hemisphere. Earth and its atmosphere complete one rotation about Earth's axis every day. The direction of rotation is from west to east. Because the atmosphere in equatorial regions is further from Earth's axis of rotation than is the atmosphere at higher latitudes, it has a higher linear velocity than does polar air as it travels around Earth. As parcels of air move north or south, they tend to maintain their **angular**

momentum (M_a), just as a car tends to maintain its momentum, when you try to stop or turn on an icy road. This effect is summarized in the equation:

$$M_a = mvr \qquad (2.1)$$

where m is the mass, v is the velocity, and r is the radius of rotation. If the mass of a parcel of air remains constant, its velocity is inversely related to the radius of rotation (2.1). We know, for example, that a skater can increase her speed of rotation by pulling her arms close to her body, which reduces her effective radius. Air that moves from the equator toward the poles encounters a smaller radius of rotation around Earth's axis. Therefore, to conserve angular momentum, it moves more rapidly (i.e., moves from west to east *relative to Earth's surface*), as it moves poleward (Fig. 2.8). Conversely, air moving toward the equator encounters an increasing radius of rotation around Earth's axis and, to conserve angular momentum, moves more slowly (i.e., moves from east to west *relative to Earth's surface*). There is another effect at work. Air parcels moving eastward relative to the surface are subjected to a larger centrifugal force than parcels at rest with respect to the surface. While this extra centrifugal force acts outward from the axis of Earth's rotation, the fact that Earth's surface is curved means that a component of this centrifugal force is directed toward the equator. The opposite effect occurs if the air is moving east to west relative to the surface. Conservation of angular momentum and the centrifugal force represent the two components of the **Coriolis effect** that work together to deflect moving air parcels to the right in the northern hemisphere and to the left in the southern hemisphere. The Coriolis effect is a "pseudo force" that arises only because we view the motion of the atmosphere relative to Earth's rotating surface. The Coriolis effect explains why midlatitude storms rotate clockwise (counterclockwise) in the northern (southern) hemisphere. The Coriolis effect also explains the rotation of the Hadley cells (Fig. 2.8).

The interaction of vertical and horizontal motions of the atmosphere creates Earth's **prevailing winds**, i.e., the most frequent wind directions. The direction of prevailing winds depends on whether air is moving toward or away from the equator. In the tropics, surface air in the Hadley cell moves from 30°N and S toward the equator, and the Coriolis effect causes these winds to blow from the east, forming easterly **tradewinds** (Fig. 2.8). The region where surface air from northern and southern hemispheres converges is called the **Intertropical Convergence Zone** (ITCZ). Here the rising air creates a zone with light winds and high humidity, known to early sailors as the **doldrums**. Subsiding air at 30°N and S latitudes also produces relatively light winds, known as the **horse latitudes**. The surface air that moves poleward from 30° to 60°N and S is deflected toward the east by the Coriolis effect, forming the prevailing **westerlies**, i.e., surface winds that blow from the west.

At the boundaries between the major cells of atmospheric circulation, relatively sharp gradients of temperature and pressure, together with the Coriolis effect, generate strong winds over a broad height range in the upper troposphere. These are the subtropical and polar **jet streams**. The Coriolis effect explains why these winds blow in a westerly direction, i.e., from west to east.

The locations of the ITCZ and of each circulation cell shifts seasonally because the zone of maximum solar radiation input varies from summer to winter due to Earth's 23.5° tilt with respect to the plane of its orbit around the sun. The seasonal changes in the location of these cells contribute to the seasonality of climate.

The uneven distribution of land and the ocean on Earth's surface creates an uneven pattern of heating that modifies the general latitudinal trends in climate. At 30°N and S, air descends more strongly over the cool ocean than over the relatively warm land because the air is cooler and more dense over the ocean than over the land. The greater subsidence over the ocean creates high-pressure zones over the Atlantic and Pacific (the Bermuda and Pacific highs, respectively) and over the Southern Ocean (Fig. 2.9).

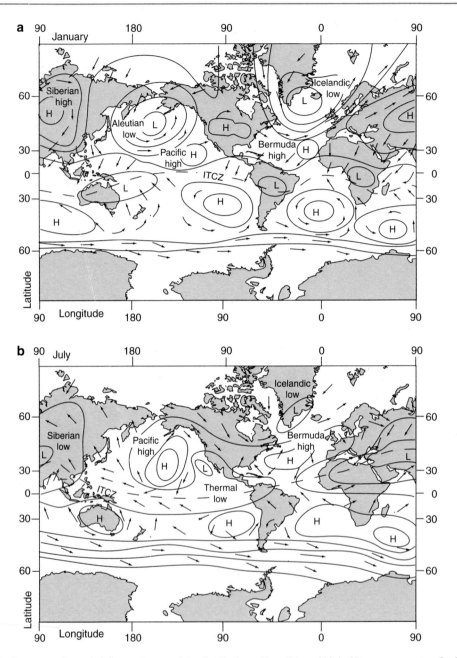

Fig. 2.9 Average surface wind-flow patterns and the distribution of low (L)- and high (H)-pressure centers for January (*top*) and July (*bottom*). Redrawn from Ahrens (1998)

At 60°N, rising air generates semi-permanent low-pressure zones over Iceland and the Aleutian Islands (the Icelandic and Aleutian lows, respectively). These lows are actually time averages of mid-latitude storm tracks, rather than stable features of the circulation. In the southern hemisphere, there is little land at 60°S, leading to a broad trough of low pressure, rather than distinct centers. Air that subsides in high pressure centers spirals outward in a clockwise direction in the northern hemisphere and in a counter-clockwise direction in the southern hemisphere due to an interaction between friction, Coriolis forces, and the pressure gradient force produced by the subsiding air. Winds spiral inward toward low-pressure centers in a counter-clockwise direction in the northern hemisphere and in a clockwise direction in the southern hemisphere. Air in the low-pressure centers rises in balance with the subsiding air in high-pressure centers. The long-term average of these vertical and horizontal motions produces the vertical circulation described by the Ferrell cell (Fig. 2.8) and a horizontal pattern of high- and low-pressure centers commonly observed on weather charts (Fig. 2.9).

These deviations from the expected easterly or westerly direction of prevailing winds are organized on a planetary scale and are known as **planetary waves**. These waves are most pronounced in the northern hemisphere, where there is more land. They are influenced by the Coriolis effect, land–ocean heating contrasts, and the locations of large mountain ranges, such as the Rocky Mountains and Himalayas. These mountain barriers force the northern hemisphere westerlies vertically upward and to the north. Downwind of the mountains, air descends and moves to the south forming a trough, much like the standing waves in the rapids of a fast-moving river that are governed by the location of rocks in the riverbed. Temperatures are comparatively low in the troughs, due to the southward movement of polar air, and comparatively high in the ridges. The trough over eastern North America downwind of the Rocky Mountains (Fig. 2.9), for example, results in relatively cool temperatures and a more southerly location of the arctic tree line in eastern than in western North America. Although planetary waves have preferred locations, they are not static. Changes in their location or in the number of waves alter regional patterns of climate. These step changes in circulation pattern are referred to as shifts in **climate modes**.

Planetary waves and the distribution of major high- and low-pressure centers explain many details of horizontal motion in the atmosphere and therefore the patterns of ecosystem distribution. The locations of major high- and low-pressure centers, for example, explain the movement of mild moist air to the west coasts of continents at 50–60°N and S, where the temperate rainforests of the world occur (the northwestern U.S. and southwestern Chile, for example; Fig. 2.9). The subtropical high pressure centers at 30°N and S cause cool polar air to move toward the equator on the west coasts of continents, creating dry Mediterranean climates near 30°N and S. On the east coasts of continents, subtropical highs cause warm moist equatorial air to move northward at 30°N and S, creating a moist subtropical climate.

The Ocean

Ocean Structure

Like the atmosphere, the ocean maintains rather stable layers with limited vertical mixing between them. The sun heats the ocean from the top, whereas the atmosphere is heated from the bottom. Because warm water is less dense than cold water, the ocean maintains rather stable layers that do not easily mix. The uppermost warm layer of **surface water**, which interacts directly with the atmosphere, extends to depths of 75–200 m, depending on the depth of wind-driven mixing. Most primary production and decomposition occur in the surface waters (see Chaps. 5–7). Another major difference between atmospheric and oceanic circulation is that density of ocean waters is determined by both temperature and salinity, so, unlike warm air, warm water can sink, if it is salty enough.

Relatively sharp gradients in temperature (**thermocline**) and salinity (**halocline**) occur between warm surface waters of the ocean and cooler more saline waters at intermediate depths (200–1,000 m; Fig. 2.10). These two vertical gradients create a gradient in water density (**pycnocline**) that generates a relatively stable vertical stratification of low-density surface water above denser deep water. The deep layer therefore mixes with the surface waters very slowly over hundreds to thousands of years. These deeper layers nonetheless play critical roles in element cycling, productivity, and climate because they are long-term sinks for carbon and the sources of nutrients

Fig. 2.10 Typical vertical
profiles of ocean temperature and salinity. The
thermocline and halocline
(T/H) are the zones where
temperature and salinity,
respectively, decline most
strongly with depth. These
transition zones usually
coincide approximately

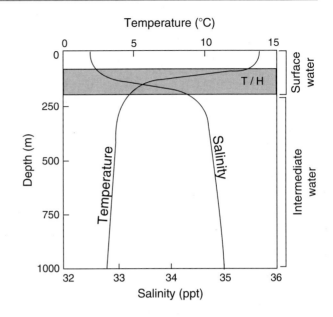

that drive ocean production (see Chaps. 5–9).
Upwelling areas, where nutrient-rich deep waters
move rapidly to the surface, support high levels
of primary and secondary productivity (marine
invertebrates and vertebrates) and are the locations of many of the world's major fisheries.

Ocean Circulation

**Ocean circulation plays a critical role in
Earth's climate system.** The ocean and atmosphere are about equally important in latitudinal
heat transport in the tropics, but the atmosphere
accounts for most latitudinal heat transport at
mid- and high latitudes (Fig. 2.7). The surface
currents of the ocean are driven by surface winds
and therefore show global patterns (Fig. 2.11)
that are generally similar to those of the prevailing surface winds (Fig. 2.9). The ocean currents
are, however, deflected 20–40° relative to the
wind direction by the Coriolis effect. This deflec-
tion and the edges of continents cause ocean currents to be more circular (termed **gyres**) than the
winds that drive them. In equatorial regions, currents flow east to west, driven by the easterly
trade winds, until they reach the continents,
where they split and flow poleward along the
western boundaries of the ocean, carrying warm

tropical water to higher latitudes. On their way
poleward, currents are deflected by the Coriolis
effect. Once the water reaches the high latitudes,
some returns in surface currents toward the tropics along the eastern edges of ocean basins
(Fig. 2.11), and some continues poleward.

Deep-ocean waters show a circulation pattern
quite different from the wind-driven surface circulation. In the polar regions, especially in the
winter off southern Greenland and off Antarctica,
cold air cools the surface waters, increasing their
density. Formation of sea ice, which excludes salt
from ice crystals (**brine rejection**), increases the
salinity of surface waters, also increasing their
density. The high density of these cold saline
waters causes them to sink. This **downwelling** to
form the North Atlantic Deep Water off of
Greenland, and the Antarctic Bottom Water off of
Antarctica drives the global **thermohaline circulation** in the mid and deep ocean that ultimately transfers water among the major ocean
basins (Fig. 2.12). The descent of cold dense
water at high latitudes is balanced by the upwelling of deep water on the eastern margins of ocean
basins at lower latitudes, where along-shore
surface currents are deflected offshore by the
Coriolis effect and easterly trade winds. There is
a net transfer of North Atlantic Deep Water to
other ocean basins, particularly the eastern Pacific

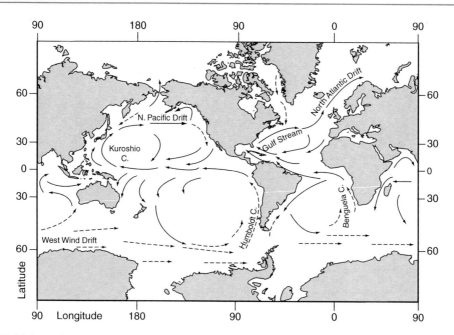

Fig. 2.11 Major surface ocean currents. Warm currents (C) are shown by *solid arrows* and cold currents by *dashed arrows*. Redrawn from Ahrens (1998)

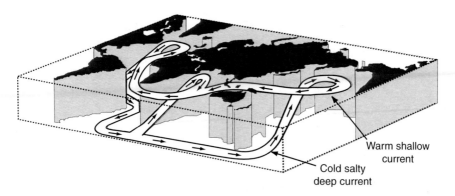

Fig. 2.12 Circulation patterns of deep and surface waters among the major ocean basins

and Indian Oceans, where old phosphorus-rich waters emerge at the surface. Net poleward movement of warm surface waters balances the movement of cold deep water toward the equator. Changes in the strength of the thermohaline circulation can have significant effects on climate because of its control over latitudinal heat transport. In addition, the thermohaline circulation transfers carbon to depth, where it remains for centuries (see Chap. 14).

The ocean, with its high heat capacity, heats up and cools down much more slowly than does the land and therefore has a moderating influence on the climate of adjacent land. Wintertime temperatures in Great Britain and Western Europe, for example, are much milder than at similar latitudes on the east coast of North America due to the warm **North Atlantic drift** (the poleward extension of the Gulf Stream; Fig. 2.11). Conversely, cold upwelling currents or currents moving toward the equator from the poles cool adjacent landmasses in summer. The cold California current, for example, which runs north to south along the west coast of the U.S., keeps summer temperatures in Northern California lower than at similar latitudes along the east coast of the U.S. These temperature differences play critical roles in determining the distribution of different kinds of ecosystems across the globe.

Landform Effects on Climate

The spatial distribution of land, water, and mountains modify the general latitudinal trends in climate. The greater heat capacity of the ocean has short-term regional as well as long-term global consequences. The ocean warms more slowly than land during the day and in summer and cools more slowly than land at night and in winter, influencing atmospheric circulation at local to continental scales. The seasonal reversal of winds (**monsoons**) in eastern Asia, for example, is driven largely by the temperature difference between the land and the adjacent seas. During the northern-hemisphere winter, the land is colder than the ocean, giving rise to cold dense continental air that flows southward from Siberian high-pressure centers across India to the ocean (Figs. 2.9 and 2.13). In summer, however, the land heats relative to the ocean, forcing the air to rise, in turn drawing in moist surface air from the ocean. Condensation of water vapor in the rising moist air produces large amounts of precipitation. Northward migration of the trade winds in summer enhances onshore flow of air, and the mountainous topography of northern India enhances vertical motion, increasing the proportion of water vapor that is converted to precipitation. Together, these seasonal changes in winds give rise to predictable seasonal patterns of temperature and precipitation that strongly influence the structure and functioning of ecosystems.

At scales of a few kilometers, the differential heating between land and ocean produces **land and sea breezes**. During the day, strong heating over land causes air to rise, drawing in cool air from the ocean (Fig. 2.13). The rising of air over the land increases the height at which a given pressure occurs, causing this upper air to move from land toward the ocean, if the large-scale prevailing winds are weak. The resulting increase in the mass of atmosphere over the ocean raises the surface pressure, which causes surface air to flow from the ocean toward the land. The resulting circulation cell is similar in principle to that which occurs in the Hadley cell (Fig. 2.8) or Asian monsoons (Fig. 2.13). At night, when the ocean is warmer than the land, air rises over the ocean, and the surface breeze blows from the land to the ocean, reversing the circulation cell. The net effect of sea breezes is to reduce temperature extremes and increase precipitation on land near the ocean or large lakes.

Mountain ranges affect local atmospheric circulation and climate through several types of **orographic effects**, which are effects due to the presence of mountains. As winds carry air up the windward sides of mountains, the air cools, and water vapor condenses and precipitates. Therefore, the windward side tends to be cold and wet. When the air moves down the leeward side of the mountain, it expands and warms, increasing its capacity to absorb and retain water. This creates a **rain shadow**, i.e., a zone of low precipitation downwind of the mountains. The rain shadow of the Rocky Mountains extends 1,500 km to the east, resulting in a strong west-to-east gradient in annual precipitation from eastern Colorado (300 mm) to Illinois (1,000 mm; see Fig. 13.3; Burke et al. 1989). Deserts or desert grasslands (steppes) are often found immediately downwind of the major mountain ranges of the world. Mountain systems can also influence climate by channeling winds through valleys. The Santa Ana winds of Southern California occur when high pressure over the interior deserts funnels warm dry winds through valleys toward the Pacific coast, creating dry windy conditions that promote intense wildfires.

Sloping terrain creates unique patterns of microclimate at scales ranging from anthills to mountain ranges. Slopes facing toward the equator (south-facing slopes in the northern hemisphere and north-facing slopes in the southern hemisphere) receive more radiation than opposing

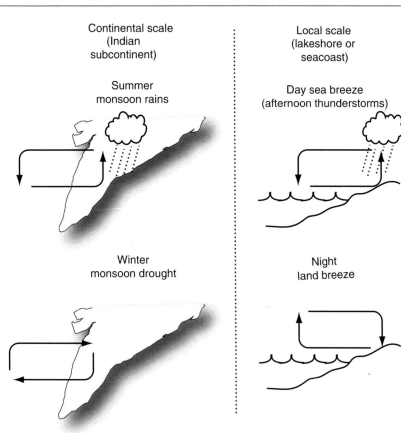

Fig. 2.13 Effects of land–sea heating contrasts on winds and precipitation at continental and local scales. At the continental scale, the greater heating of land than sea during summer causes air to rise, drawing in cool moist ocean air over India that fuels precipitation. In winter, the ocean is warmer than the land, reversing these wind patterns. At the local scale, similar heating contrasts in coastal or lakeshore areas cause sea breezes and afternoon thunderstorms during the day and land breezes at night

slopes, creating warmer drier conditions. In cold or moist climates, the warmer microenvironment on equator-facing slopes provides conditions that enhance productivity, decomposition, and other ecosystem processes, whereas in dry climates, the greater drought on these slopes limits production. Microclimatic variation associated with slope and **aspect** (the compass direction that a slope faces) allows representatives of an ecosystem type to exist hundreds of kilometers beyond its major zone of distribution. These outlier populations are important sources of colonizing individuals during times of rapid climatic change and are therefore important in understanding species migration and the long-term dynamics of ecosystem changes (see Chap. 12).

Topography also influences climate through drainage of cold dense air. When air cools at night, it becomes denser and tends to flow downhill (**katabatic winds**) into valleys, where it accumulates. This can produce temperature **inversions** (cool air beneath warm air, a vertical temperature profile reversed from the typical pattern in the troposphere of decreasing temperature with increasing elevation; Fig. 2.4). Inversions occur primarily at night and in winter, when heating from the sun is insufficient to promote convective mixing. Clouds also tend to inhibit the formation of winter and nighttime inversions because they increase longwave emission to the surface. Increases in solar heating or windy conditions, such as might accompany the passage of frontal systems, break up inversions. Inversions are climatically important because they increase the seasonal and diurnal temperature extremes experienced by ecosystems in low-lying areas. In cool climates, inversions greatly reduce the length of the frost-free growing season.

Vegetation Influences on Climate

Vegetation influences climate through its effects on the surface energy budget. Climate is quite sensitive to regional variations in vegetation and water content at Earth's surface. The albedo (the fraction of the incident shortwave radiation reflected from a surface) determines the quantity of solar energy absorbed by the surface, which is subsequently available for transfer to the atmosphere as longwave radiation and turbulent fluxes of sensible and latent heat. Water generally has a low albedo, so lakes and the ocean absorb considerable solar energy. At the opposite extreme, snow and ice have a high albedo and hence absorb little solar radiation, contributing to the cold conditions required for their persistence. Vegetation is intermediate in albedo, with values generally decreasing from grasslands, with their highly reflective standing dead leaves, to deciduous forests to dark conifer forests (see Chap. 4). Recent land-use changes have substantially altered regional albedo by increasing the area of exposed bare soil. The albedo of soil depends on soil type and wetness but is often higher than that of vegetation in dry climates. Consequently, overgrazing often increases albedo, reducing energy absorption and the transfer of energy to the atmosphere. This leads to cooling and subsidence, so moist ocean air is not drawn inland by sea breezes. This can reduce precipitation and the capacity of vegetation to recover from overgrazing (Foley et al. 2003a). The large magnitude of many land-surface feedbacks to climate suggests that land-surface change can be an important contributor to regional climatic change (Foley et al. 2003b).

Ecosystem structure influences the efficiency with which turbulent fluxes of sensible and latent heat are transferred to the atmosphere. Wind passing over tall uneven canopies creates mechanical turbulence that increases the efficiency of heat transfer from the surface to the atmosphere (see Chap. 4). Smooth surfaces, in contrast, tend to heat up because they transfer their heat only by convection and not by mechanical turbulence.

The effects of vegetation structure on the efficiency of water and energy exchange influence regional climate. About 25–40% of the precipitation in the Amazon basin comes from water that is recycled from land by evapotranspiration. (Costa and Foley 1999). Simulations by climate models suggest that, if the Amazon basin were completely converted from forest to pasture, this would lead to a permanently warmer drier climate over the Amazon basin (Foley et al. 2003b). The shallower roots of grasses would absorb less water than trees, leading to lower transpiration rates (Fig. 2.14). Pastures would therefore release more of the absorbed solar radiation as sensible heat, which directly warms the atmosphere. There are many uncertainties, however. Changes in cloudiness, for example, can have either a positive or a negative effect on radiative forcing, depending on cloud properties and height.

Changes in albedo caused by vegetation change can create amplifying feedbacks. At high latitudes, for example, tree-covered landscapes absorb more solar radiation prior to snowmelt than does snow-covered tundra. Model simulations suggest that the northward movement of the tree line 6,000 years ago could have reduced the regional albedo and increased energy absorption enough to explain half of the climate warming that occurred at that time (Foley et al. 1994). The warmer regional climate would, in turn, favor tree reproduction and establishment at the tree line (Payette and Filion 1985), providing an amplifying (positive) feedback to regional warming (see Chap. 12). Predictions about the impact of future climate on vegetation should therefore also consider ecosystem feedbacks to climate (Field et al. 2007; Chapin et al. 2008).

Albedo, energy partitioning between latent and sensible heat fluxes, and surface structure also influence the amount of longwave radiation emitted to the atmosphere (Fig. 2.3). Longwave radiation depends on surface temperature, which tends to be high when the surface absorbs large

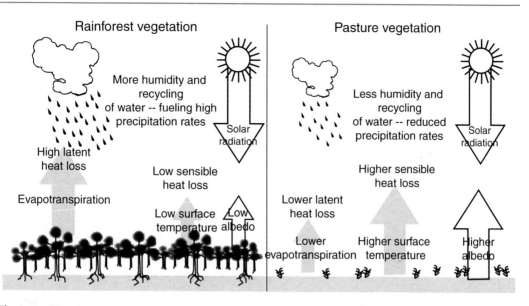

Fig. 2.14 Climatic consequences of tropical deforestation and conversion to pasture. In forested conditions, the low albedo provides ample energy absorption to drive high transpiration rates that cool the surface and supply abundant moisture to the atmosphere to fuel high precipitation rates. In pasture conditions that develop after deforestation, low vegetation cover and shallow roots restrict transpiration and therefore the moisture available to support precipitation. This, together with high sensible heat flux leads to a warm, dry climate. Based on Foley et al. (2003b)

amounts of incoming radiation (low albedo), has little water to evaporate, or has a smooth surface that is inefficient in transferring turbulent fluxes of sensible and latent heat to the atmosphere (see Chap. 4). Deserts, for example, experience large net longwave energy losses because their dry smooth surfaces lead to high surface temperatures, and little moisture is available to support evaporation that would otherwise cool the soil.

Temporal Variability in Climate

Long-Term Changes

Millennial-scale climatic change is driven primarily by changes in the distribution of solar input and changes in atmospheric composition. Earth's climate is a dynamic system that has changed repeatedly, producing frequent, and sometimes abrupt, changes in climate, including dramatic glacial periods (Fig. 2.15) and sea-level changes. Volcanic eruptions and asteroid impacts alter climate on short time scales through changes

in absorption or reflection of solar energy. Continental drift and mountain building and erosion have modified the patterns of atmospheric and ocean circulation on longer time scales. The primary force responsible for the evolution of Earth's climate, however, has been changes in the input of solar radiation, which has increased by about 30% over the past four billion years, as the sun matured (Schlesinger 1997). On millennial time scales, the distribution of solar input has varied primarily due to predictable variations in Earth's orbit.

Three types of variations in Earth's orbit influence the amount of solar radiation received at the surface at different times of the year and at different latitudes: **eccentricity** (the degree of ellipticity of Earth's orbit around the sun), **tilt** (the angle between Earth's axis of rotation relative to the plane of its orbit around the sun), and **precession** (a "wobbling" in Earth's axis of rotation with respect to the stars, determining the time of year when different locations on Earth are closest to the sun). The periodicities of these orbital parameters (eccentricity, tilt, and precession) are

Fig. 2.15 Geological time periods in Earth's history, showing major glacial events (*dark bars*) and ecological events that strongly influenced ecosystem processes. Note the changes in time scale in units of millions of years (Ma). The most recent geologic epoch (the Anthropocene) began about 1750 with the beginning of the industrial revolution and is characterized by human domination of the biosphere (Crutzen 2002). Modified from Sturman and Tapper (1996)

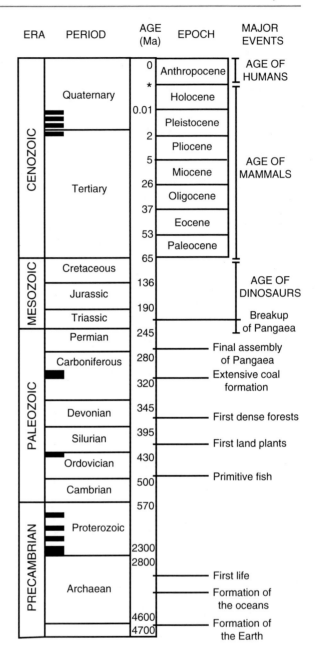

approximately 100,000, 41,000, and 23,000 years, respectively. Interactions among these cycles produce **Milankovitch cycles** of solar input that correlate with the glacial and interglacial cycles. Analysis of these cycles indicates that Earth would not naturally enter another ice age for at least 30,000 years, so natural cycles in solar input will not substantially offset human-driven warming of climate (IPCC 2007). Ice ages are triggered by minima in northern high-latitude summer radiation that enable winter snowfall to persist through the year and build northern-hemisphere ice sheets that reflect incoming radiation. These changes become globally amplified by feedbacks in Earth's climate system (such as changes in atmospheric CO_2 concentration) to cause large climatic changes throughout the planet.

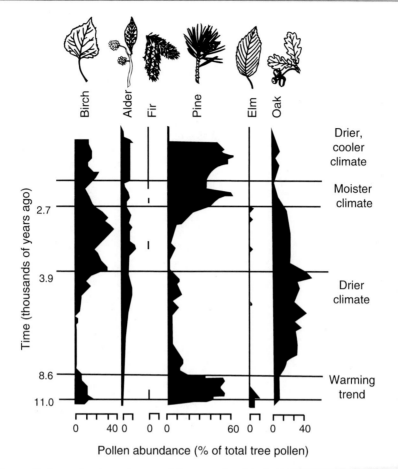

Fig. 2.16 Pollen profile from a bog in northwestern Minnesota showing changes in the dominant tree species over the past 11,000 years. Redrawn from McAndrews (1966)

The chemistry of ice and trapped air bubbles provide a paleorecord of the climate when the ice formed. Ice cores drilled in Antarctica and Greenland indicate considerable climate variability over the past 650,000 years, in large part related to the Milankovitch cycles (see Fig. 14.6). Analysis of bubbles in these cores indicates that past warming events have been associated with increases in CO_2 and CH_4 concentrations, providing circumstantial evidence for a past role of radiatively active gases in climate change. The unique feature of the recent anthropogenic increases in these gases is that they are occurring during an *interglacial* period, when Earth's climate is already relatively warm. These cores indicate that the CO_2 concentration of the atmosphere is higher now than at any time in at least the last 650,000 years (IPCC 2007). Fine-scale

analysis of ice cores from Greenland suggests that large changes from glacial to interglacial climate can occur in decades or less. Such rapid transitions in the climate system to a new state may be related to sudden changes in the strength of the thermohaline circulation that drives oceanic heat transport from the equator to the poles.

Past climates can also be reconstructed from other paleorecords. Tree-ring records, obtained from living and dead trees, provide information about climate during the last several thousand years. Variation in the width of tree rings records temperature and moisture, and chemical composition of wood reflects the characteristics of the atmosphere at the time the wood was formed. Pollen preserved in low-oxygen sediments of lakes provides a history of plant taxa and climate over the past tens of thousand years (Fig. 2.16).

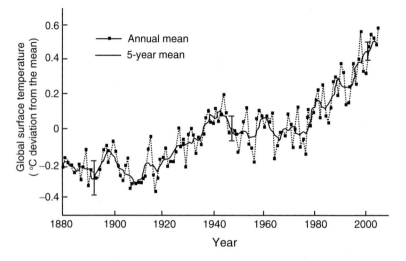

Fig. 2.17 Time course of the average surface temperature of Earth from 1850 to 2005 (relative to the average temperature for this time period). Redrawn from IPCC (2007)

Pollen records from networks of sites can be used to construct maps of species distributions at various times in the past and provide a history of species migrations across continents after climatic changes (COHMAP 1988). Other proxy records provide measures of temperature (species composition of Chironomids), precipitation (lake level), pH, and geochemistry.

The combination of paleoclimate proxies indicates that climate is inherently variable over all time scales. Atmospheric, oceanic, and other environmental changes that are occurring now due to human activities must be viewed as overlays on the natural climate variability that stems from long-term changes in Earth's surface characteristics and orbital geometry.

Anthropogenic Climate Change

Earth's climate during the last half of the twentieth century was warmer than during any 50-year interval in the last 500 years and probably the last 1,300 years or longer (Fig. 2.17; IPCC 2007; Serreze 2010). This warming is most pronounced near Earth's surface, where its ecological effects are greatest. A small amount of the recent warming reflects an increase in solar input, but most of the warming results from human activities that increase the concentrations of radiatively active gases in the atmosphere (Fig. 2.18). These gases trap more of the longwave radiation emitted by Earth's surface and warm the atmosphere, which retains more water vapor (another potent greenhouse gas) and further increases the trapping of longwave radiation. As a result, Earth is no longer in radiative equilibrium but is losing less energy to space than it is absorbing from the sun. Consequently, Earth's surface warmed about 0.7°C from 1880 to 2008 (Fig. 2.17) and is projected to warm an additional three to four times that amount by the end of the twenty-first century (Serreze 2010).

Climate models and recent observations indicate that warming will be most pronounced in the interiors of continents, far from the moderating effects of the ocean, and at high latitudes. The high-latitude warming reflects an amplifying feedback. As climate warms, the snow and sea ice melt earlier in the year, which replaces the reflective snow or ice cover with a low-albedo land or water surface. These darker surfaces absorb more radiation and transfer this energy to the atmosphere, which amplifies the rate of climate warming. Clouds, increases in water vapor, and increases in poleward energy transport also contribute to polar warming. Those changes in the climate system that occur over years to

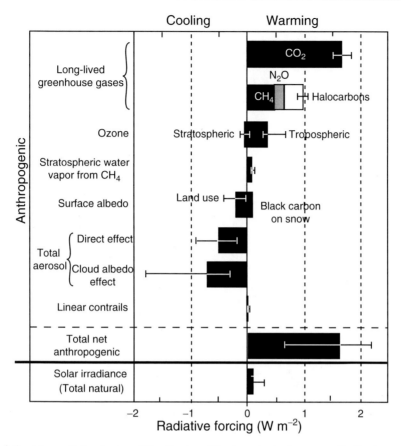

Fig. 2.18 Global average radiative forcing of the climate system (i.e., external forces that modify the climate system) estimated for 2005. Some changes in the climate system lead to net warming; others lead to net cooling. The largest single cause of climate warming is the increased concentration of atmospheric CO_2, primarily as a result of burning fossil fuels. Redrawn from IPCC (2007)

decades are dominated by amplifying feedbacks, such as the ice–albedo feedback, causing anthropogenic warming to accelerate (Serreze 2010).

As climate warms, the air has a higher capacity to hold water vapor, so there is greater evaporation from the ocean and other moist surfaces. In areas where rising air leads to condensation, this leads to greater precipitation. Continental interiors are less likely to experience large precipitation increases but will be dried by increasing evaporation. Consequently, soil moisture and runoff to streams and rivers are likely to increase in coastal regions and mountains and to decrease in continental interiors. In other words, wet regions will likely become wetter and dry regions drier. Winter warming is likely to reduce the snowpack in mountains and therefore the spring runoff that fills reservoirs on which many cities depend for water supply. The complex controls and nonlinear feedbacks in the climate system make detailed climate projections problematic and are active areas of research (IPCC 2007).

Interannual Climate Variability

Much of the interannual variation in regional climate is associated with large-scale changes in the atmosphere–ocean system. Superimposed on long-term climate variability are interannual variations that have been noted by farmers, fishermen, and naturalists for centuries. Some of this variability exhibits repeating geographic and temporal patterns. For example, **El Niño/ Southern Oscillation** or ENSO (Webster and Palmer 1997; Federov and Philander 2000) events are part of a large-scale, air–sea interaction that couples atmospheric pressure changes (the Southern Oscillation) with changes in ocean temperature (El Niño) over the equatorial Pacific Ocean. ENSO events have occurred, on average, every 3–7 years over the past century, with considerable irregularity (Trenberth and Haar 1996). No events occurred between 1943 and 1951, for example, and three major events occurred between 1988 and 1999.

In most years, the easterly trade winds push the warm surface waters of the Pacific westward, so the layer of warm surface waters is deeper in the western Pacific than in the east (Figs. 2.8 and 2.19). The resulting warm waters in the western Pacific are associated with a low-pressure center and promote convection and high rainfall in Indonesia. The offshore movement of surface waters in the eastern Pacific promotes upwelling of colder, deeper water off the coasts of Ecuador and Peru. These cold, nutrient-rich waters support a productive fishery (see Chap. 9) and promote subsidence of upper air, leading to the development of a high-pressure center and low precipitation. At times, however, the eastern-Pacific high-pressure center, Indonesian low-pressure center, and the easterly trades all weaken. The warm surface waters then move eastward, forming a deep layer of warm surface water in the eastern Pacific. This reduces or shuts down the upwelling of cold water, promoting atmospheric convection and rainfall in coastal Ecuador and Peru. The colder waters in the western Pacific, in contrast, inhibit convection, leading to droughts in Indonesia, Australia, and India. This pattern is commonly termed **El Niño**. Periods in which the "normal" pattern is particularly strong, with relatively cool surface waters in the eastern Pacific, are termed **La Niña**. The trigger for changes in this ocean–atmosphere system are uncertain, but may involve large-scale ocean waves, known as **Kelvin waves**, that travel back and forth across the tropical Pacific.

ENSO events have widespread climatic, ecosystem, and societal consequences. Strong El Niño phases cause dramatic reductions in anchovy fisheries in Peru with corresponding reproductive failure and mortality in sea birds and marine mammals. For the past four centuries, Peruvian potato farmers detected incipient El Niño conditions by looking at the brightness of stars in the summer, which corresponds to the high cirrus clouds that accompany El Niño (Orlove et al. 2000). This enabled them to adjust planting dates for their most critical crop. Similarly, annually variable harvest of shearwater chicks by New Zealand Maori provided early detection of El Niño events (Lyver et al. 1999). Extremes in precipitation linked to ENSO cycles are also evident in areas distant from the tropical Pacific. El Niño events bring hot, dry weather to the Amazon Basin, potentially affecting tree growth, soil carbon storage, and fire probability. Northward extension of warm tropical waters to the Northern Pacific brings rains to coastal California and high winter temperatures to Alaska. An important lesson from ENSO studies is that strong climatic events in one region have climatic consequences throughout the globe due to the dynamic interactions (termed **teleconnections**) associated with atmospheric circulation and ocean currents.

The Pacific North America (PNA) pattern is another large-scale pattern of climate variability. The positive mode of the PNA is characterized by above-average atmospheric pressure with warm, dry weather in western North America and below-average pressure and low temperatures in the east. Another large-scale climate pattern is the Pacific Decadal Oscillation (PDO), a multi-decadal pattern of climate variability that appears to modulate ENSO events. More El Niño events tend to occur when the PDO is in its positive phase, as during the last 25 years of the twentieth century. The North Atlantic Oscillation (NAO) is still another large-scale circulation pattern. Positive phases of the NAO are associated with a strengthening of the pressure gradient between the Icelandic low- and the Bermuda high-pressure systems (Fig. 2.9). This increases heat transport

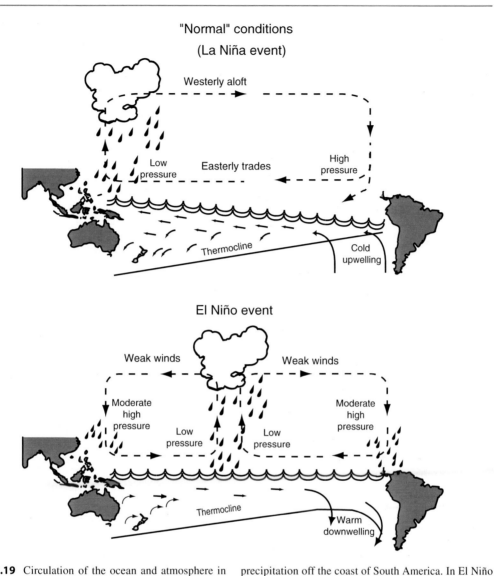

Fig. 2.19 Circulation of the ocean and atmosphere in the tropical Pacific between South America and Indonesia during "normal (La Niña) years" and during El Niño years. In normal years, strong easterly trade winds push surface ocean waters to the west, producing deep, warm waters and high precipitation off the coast of Southeast Asia and cold, upwelling waters and low precipitation off the coast of South America. In El Niño years, however, weak easterly winds allow the surface waters to move from west to east across the Pacific Ocean, leading to cooler surface waters and less precipitation in Southeast Asia and warmer surface waters and more precipitation off South America. Redrawn from McElroy (2002)

to high latitudes by wind and ocean currents, leading to a warming of Scandinavia and western North America and a cooling of eastern Canada. Although the factors that initiate these large-scale climate features are poorly understood, the patterns themselves and their ecosystem consequences are becoming more predictable. Future climatic changes will likely be associated with changes in the strength and frequencies of certain phases of these large-scale climate patterns rather than simple linear trends in climate. Climate warming, for example, might increase the frequency of El Niño events and positive phases of the PDO and NAO.

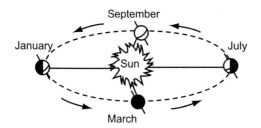

Fig. 2.20 Earth's orbit around the sun, showing that the zone of greatest heating (the ITCZ) is south of the equator in January, north of the equator in July, and at the equator in March and September

Seasonal and Daily Variation

Seasonal and daily variations in solar input have profound but predictable effects on climate and ecosystems. Perhaps the most obvious variations in the climate system are the patterns of seasonal and diurnal change. Earth rotates on its axis at 23.5° relative to its orbital plane about the sun. This tilt in Earth's axis results in strong seasonal variations in day length and solar **irradiance**, i.e., the quantity of solar energy received at Earth's surface per unit time. During the spring and autumn **equinoxes**, the sun is directly overhead at the equator, and the entire earth surface receives approximately 12 h of daylight (Fig. 2.20). At the northern-hemisphere summer **solstice**, the sun's rays strike Earth most directly in the northern hemisphere, and day length is maximized. At the northern-hemisphere winter solstice, the sun's rays strike Earth most obliquely in the northern hemisphere, and day length is minimized. The summer and winter solstices in the southern hemisphere are 6 months out of phase with those in the north. Variations in incident radiation become increasingly pronounced as latitude increases. Thus, tropical environments experience relatively small seasonal differences in solar irradiance and day length, whereas such differences are maximized in the Arctic and Antarctic. Above the Arctic and Antarctic circles, there are 24 h of daylight at the summer solstice, and the sun never rises at the winter solstice. The relative homogeneity of temperature and light throughout the year in the tropics contributes to their high productivity and diversity.

At higher latitudes, the length of the warm season strongly influences the life forms and productivity of ecosystems.

Variations in light and temperature play an important role in determining the types of plants that grow in a given climate and the rates at which biological processes occur. Almost all biological processes are temperature dependent, with slower rates occurring at lower temperatures. Seasonal variations in day length (**photoperiod**) provide important cues that allow organisms to prepare for seasonal variations in climate.

In aquatic ecosystems, seasonal changes in irradiance influence not only the temperature and light environment but also the fundamental structure of the ecosystem. Both lakes and the ocean are heated from the top, with most solar radiation absorbed and converted to heat in the upper centimeters to meters of the water column. This surface heating tends to **stratify** lakes and the ocean, with warmer, less dense water at the surface (Fig. 2.21). This tendency for stratification is counter-balanced by turbulent mixing from wind, river inflow, and the cooling of surface waters that occurs at night and during periods of cold weather. Stratification is least pronounced in wind-exposed lakes or lakes with large river inputs (e.g., many reservoirs) where turbulence mixes water to substantial depth. In the open ocean, the turbulent mixed layer is often 100–200 m in depth. In shallow lakes, turbulence often mixes the entire water column.

Lake stratification is most stable in the temperate zone between about 25–40° N and S latitude (Kalff 2002). In colder climates, cold surface waters reduce the temperature (and therefore density) gradients from the surface to depth. In the tropics, deep waters are warm throughout the year, so there is only a weak temperature gradient (often about 1°C) from the surface to depth. Seasonal fluctuations in wind-driven evaporation and cloudiness account for much of the seasonal variation in surface-water temperatures of tropical lakes.

Stratification of nontropical lakes develops during summer, when the heating of surface waters is most intense. Weakly stratified lakes often mix water throughout the water column even

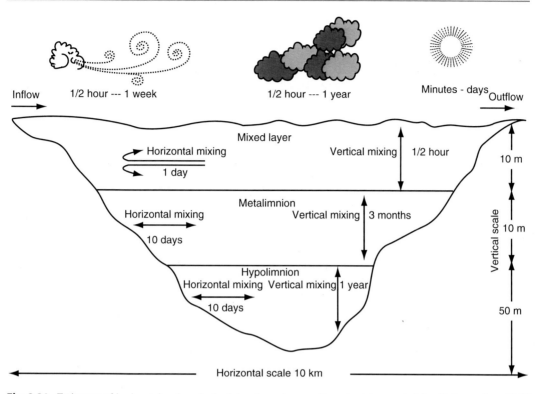

Fig. 2.21 Estimates of horizontal and vertical mixing times in a medium-sized temperate lake. Redrawn from Kalff (2002)

during the summer. In these lakes, mixing may occur at night, if air temperatures are cooler than the surface waters, or during storms, when wind-driven mixing is more intense. In lakes that are more stably stratified (e.g., temperate lakes that are deep or protected from wind), two relatively discrete layers develop: an **epilimnion** at the surface that is heated by absorbed radiation and mixed by wind and a **hypolimnion** at depth that is colder, more dense, and unaffected by surface turbulence (Fig. 2.21). **Turnover** of these stably stratified temperate lakes occurs in the autumn, when air temperature declines below the temperature of the epilimnion, causing the epilimnion to cool. This surface cooling reduces the density gradient from the surface to depth so that wind-driven turbulence mixes waters more deeply in the lake. Even in wind-protected lakes, nighttime cooling makes surface waters cooler and denser, causing the water column to mix to depth.

Stratification is important because it separates a well-lighted surface layer where photosynthesis exceeds respiration from a deeper, poorly illuminated hypolimnion where respiration exceeds photosynthesis. This spatial separation of these key ecosystem processes results in surface oxygenation and nutrient depletion and nutrient enrichment and oxygen depletion at depth. Seasonal and wind-driven mixing events are critical for resupplying nutrients to the epilimnion and oxygen to the hypolimnion. Lakes often experience a spring algal bloom when increases in light and temperature enable algae to take advantage of the nutrients that are resupplied to the epilimnion during autumn and winter. Eutrophication of lakes by nutrient inputs from fertilizers or sewage reduces water clarity, which concentrates the heating of water near the surface and reduces the depth of the epilimnion. Increased surface production also increases the rain of dead organic matter to depth, which depletes oxygen from the water column, making eutrophic lakes less suitable for fish despite their high algal productivity.

Storms and Weather

Storms, droughts, and other unpredictable weather events strongly influence ecosystems. Because extreme events, by definition, occur infrequently, it is generally impossible to explain unambiguously the climatic cause of a particular event. The intensity of hurricanes and other tropical storms, for example, depends on sea-surface temperature, so it is not surprising that ocean warming is associated with an increase in hurricane intensity (IPCC 2007). Nonetheless, we cannot say that climate warming causes any particular event, such as Hurricane Katrina, which flooded New Orleans in 2005 (Fig. 2.1). Rather, intense hurricanes of that sort will probably occur more often, if climate continues to warm. Increased latitudinal heat transport associated with climate warming has also caused a strengthening and poleward shift in westerly winds, increasing the frequency of intense storms at high latitudes. These tropical and high-latitude storms are important agents of disturbance, so changes

in their intensity are likely to alter the structure and long-term dynamics of ecosystems (see Chap. 12).

Relationship of Climate to Ecosystem Distribution and Structure

Climate is the major determinant of the global distribution of biomes. The major types of ecosystems show predictable relationships with climatic variables such as temperature and moisture (Fig. 2.22; Holdridge 1947; Whittaker 1975; Bailey 1998). An understanding of the causes of geographic patterns of climate (Fig. 2.23), as presented in this chapter, therefore allows us to predict the distribution of Earth's major biomes (Fig. 2.24).

Tropical wet forests (rainforests) occur from 12°N to 3°S and correspond to the ITCZ. Day length and solar angle show little seasonal change within this zone, leading to consistently high temperatures (Figs. 2.22–2.25). High solar

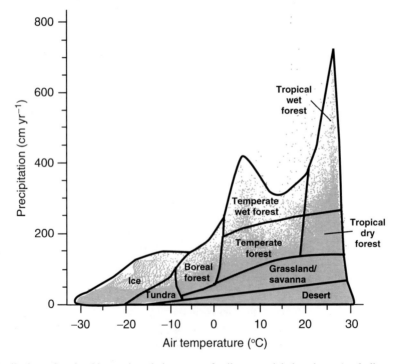

Fig. 2.22 Distribution of major biomes in relation to average annual air temperature and total annual precipitation. *Gray dots* show the temperature–precipitation regime

of all terrestrial locations (excluding Antarctica) at 18.5-km resolution (data from New et al. (2002)). Diagram kindly provided by Joseph Craine and Andrew Elmore

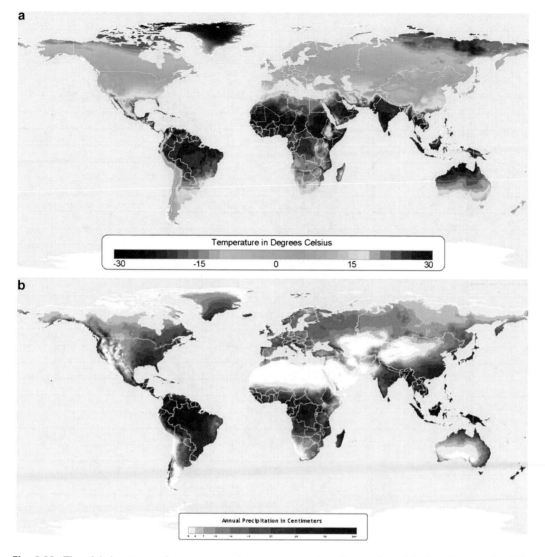

Fig. 2.23 The global patterns of average annual temperature and total annual precipitation (New et al. 1999). Reproduced from the Atlas of the Biosphere (http://www.sage.wisc.edu/atlas/)

radiation and convergence of the easterly trade winds at the ITCZ promote strong convective uplift leading to high precipitation (175–400 cm annually). Periods of relatively low precipitation seldom last more than 1–2 months. **Tropical dry forests** (Fig. 2.26) occur north and south of tropical wet forests. Tropical dry forests have pronounced wet and dry seasons because of seasonal movement of ITCZ over (wet season) and away from these forests (dry season). **Tropical savannas** (Fig. 2.27) occur between the tropical dry forests and deserts. These savannas are warm

and have low precipitation that is highly seasonal. **Subtropical deserts** (Fig. 2.28) at 25–30°N and S have a warm, dry climate because of the subsidence of air in the descending limb of the Hadley cell.

Mid-latitude deserts, grasslands, and shrublands (Fig. 2.29) occur in the interiors of continents, particularly in the rain shadow of mountain ranges. They have low unpredictable precipitation, low winter temperatures, and greater temperature extremes than tropical deserts. As precipitation increases, there is a gradual transition from desert

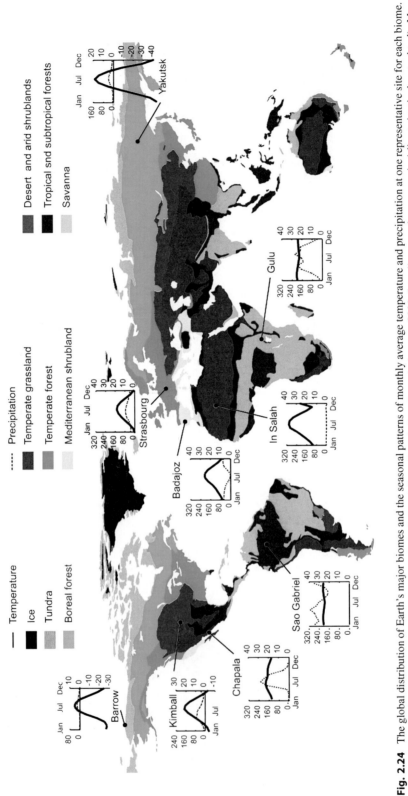

Fig. 2.24 The global distribution of Earth's major biomes and the seasonal patterns of monthly average temperature and precipitation at one representative site for each biome. Climate data are monthly averages of the entire period of record for selected sites through the year 2000 (http://www.ncdc.noaa.gov/oa/climate/stationlocator.html). Map redrawn from Bailey (1998)

Fig. 2.25 Tropical wet forest in Brazil. It is characterized by a diversity of life forms and species, including vines, epiphytes, and broadleafed evergreen trees. Photograph by Peter Vitousek

Fig. 2.26 Subtropical dry forest in Chamela, western Mexico in the wet and dry seasons. The forest is dominated by drought-deciduous trees. Photograph by Peter Vitousek

Fig. 2.27 Subtropical savanna in Kruger National Park, South Africa, showing a diversity of plants (grasses, shrubs, and trees) and grazing mammals. These fine-leaf savannas burn frequently, permitting both trees and grasses to coexist. Photograph courtesy of Alan K. Knapp

Fig. 2.28 Sonoran desert landscape in the Superstition Mountains of Arizona, showing a diversity of drought-adapted life forms, with substantial bare ground between plants. Photograph courtesy of Jim Elser

Fig. 2.29 Mid-latitude Kansas grassland (tallgrass prairie) in early summer with bison grazing. This landscape was burned early in the spring. Here, trees are restricted to the wetter portions of the landscape where they are also protected from fire. Photograph courtesy of Alan K. Knapp

Fig. 2.30 Mediterranean shrubland in the Santa Monica Mountains of coastal California. It occurs on steep slopes with shallow soils and supporting drought-adapted deciduous and evergreen shrubs. Photograph courtesy of Stephen Davis

to grassland to shrubland. **Mediterranean shrublands** (Fig. 2.30) are situated on the west coasts of continents. In summer, subtropical oceanic high-pressure centers and cold upwelling coastal currents produce a warm dry climate. In winter, as wind and pressure systems move toward the equator, storms produced by polar fronts provide unpredictable precipitation. **Temperate forests** (Fig. 2.31) occur in mid-latitudes, where there is enough precipitation to support trees. The polar front, the boundary between the polar and subtropical air masses, migrates north and

Fig. 2.31 Temperate forest in the eastern U.S. (North Carolina), showing a complex multi-layered canopy with sunflecks common in all canopy layers. Photograph courtesy of Norm Christensen

Fig. 2.32 Temperate wet forest in the Valley of the Giants in the Oregon Coast Range of the western U.S. The stand contains a range of tree ages up to five centuries. The understory has coarse woody debris and a flora of shrubs, ferns, herbs, mosses, and tree seedlings. Photograph courtesy of Mark E. Harmon

south of these forests from summer to winter, producing a strongly seasonal climate. **Temperate wet forests** (rainforests; Fig. 2.32) occur on the west coasts of continents at 40–65°N and S, where westerlies blowing across a relatively warm ocean provide an abundant moisture source, and migrating low-pressure centers associated with the polar front promote high precipitation. Winters are mild, and summers are cool.

The boreal forest (taiga; Fig. 2.33) occurs in continental interiors at 50–70°N. The winter climate is dominated by polar air masses and the

Fig. 2.33 Boreal forest on the Tanana River of Interior Alaska. The landscape contains a spectrum of stand ages, ranging from early successional shrub stands on the point bar in the lower left and in the clearcut in the upper left to mature white spruce stands in the center of the photograph to muskegs on terraces in the distance that are thousands of years old. Photograph courtesy of Roger Ruess

summer climate by temperate air masses, producing cold winters and mild summers. The distance from oceanic moisture sources results in low precipitation. The subzero average annual temperature leads to **permafrost** (permanently frozen ground) that restricts drainage and creates poorly drained soils and peatlands in low-lying areas. **Arctic tundra** (Fig. 2.34) is a zone north of the polar front in both summer and winter, resulting in a climate that is too cold to support growth of trees. Short cool summers restrict biological activity and limit the range of life forms that can survive.

Vegetation structure varies with climate both among and within biomes. Predictable growth forms of plants dominate each biome type. Broadleaved evergreen trees, for example, dominate tropical wet forests, whereas areas that are periodically too cold or dry for growth of these trees are dominated by deciduous forests or, under more extreme conditions, by tundra or desert, respectively. Biomes are not discrete units with sharp boundaries but vary continuously in structure along climatic gradients. Along a moisture gradient in the tropics, for example, vegetation changes from tall evergreen trees in the wettest sites to a mix of evergreen and deciduous trees in areas with seasonal drought (Fig. 2.35; Ellenberg 1979). As the climate becomes still drier, the stature of the trees and shrubs declines because of less light competition and more competition for water (Fig. 2.36). Ultimately, this leads to a shrubless desert with herbaceous perennial herbs in dry habitats. With extreme drought, the dominant life form becomes annuals and bulbs (herbaceous perennials in which aboveground parts die during the dry season). A similar gradient of growth forms, leaf types, and life forms occurs along moisture gradients at other latitudes.

The diversity of growth forms within some ecosystems can be nearly as great as the diversity of dominant growth forms across biomes. In tropical wet forests, for example, continuous seasonal growth in a warm, moist climate produces large trees with dense canopies that intercept, and compete for, a large fraction of

Fig. 2.34 Arctic tundra near Toolik Lake in northern foothills of the Brooks Range of Alaska. The landforms were shaped by Pleistocene glaciations, and the soils are kept wet and cold by a continuous layer of permafrost 30–50 cm beneath the surface. Photograph by Stuart Chapin

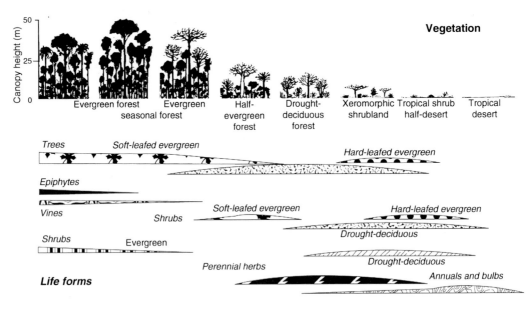

Fig. 2.35 The change in life-form dominance along a tropical gradient where precipitation changes but temperature is relatively constant. Redrawn from Ellenberg (1979)

the incoming radiation. Light then becomes the main driver of diversity within the ecosystem. Plants that reach the canopy and have access to light compete well with tall trees. These growth forms include vines, which parasitize trees for support without investing carbon in strong stems. Epiphytes are also common in the canopies of tropical wet forests where they receive abundant light, but, because their roots are restricted to the canopy, their growth is often water-limited. Epiphytes have therefore evolved various specializations to trap water and nutrients. There is a wide range of sub-canopy trees, shrubs, and herbs that are adapted to grow slowly under the low-light conditions beneath the canopy (Fig. 2.35). Light is the most important general driver of structural diversity in the dense forests of wet tropical regions.

Fig. 2.36 Patagonian steppe in cold, arid mountains of Argentina. Steppe is an example of a cold, dry ecosystem type intermediate between widespread "biomes." Photograph courtesy of Sandra Díaz

What determines structural diversity where moisture, rather than light, is limiting? Deserts, particularly warm deserts, have a great diversity of plant forms, including evergreen and deciduous small trees and shrubs, succulents, herbaceous perennials, and annuals. These growth forms do not show a well-defined vertical partitioning but show consistent horizontal patterns related to moisture availability. Trees and tall shrubs, for example, predominate adjacent to seasonal streams, evergreen shrubs in clay-rich soils that retain water, and succulents in the driest habitats. Competition for water results in diverse strategies for gaining, storing, and using the limited water supply. This leads to a wide range of rooting strategies and capacities to avoid or withstand drought.

Species diversity declines from the tropics to high latitudes and in many cases from low to high elevation. Species-rich tropical areas support more than 5,000 species of plants in a 10,000-km^2 area, whereas the high arctic has fewer than 200 species in the same area. Many animal groups show similar latitudinal patterns of diversity, in part because of their dependence on the underlying plant diversity. Climate, the evolutionary time available for species radia-

tion, productivity, disturbance frequency, competitive interactions, land area available, and other factors have all been hypothesized to contribute to global patterns of diversity (Heywood and Watson 1995). Models that include only climate, acting as a filter on the plant functional types that can occur in a region, can reproduce the general global patterns of structural and species diversity (Fig. 2.37; Kleiden and Mooney 2000). The actual causes for geographic patterns of species diversity are undoubtedly more complex, but these models and other analyses suggest that human-induced changes in climate, land use, and invasions of exotic species may alter future patterns of diversity.

Summary

The balance between incoming and outgoing radiation determines Earth's energy budget. The atmosphere transmits about half of the incoming shortwave solar radiation to Earth's surface but absorbs 90% of the outgoing longwave radiation emitted by Earth. This causes the atmosphere to be heated primarily from the bottom

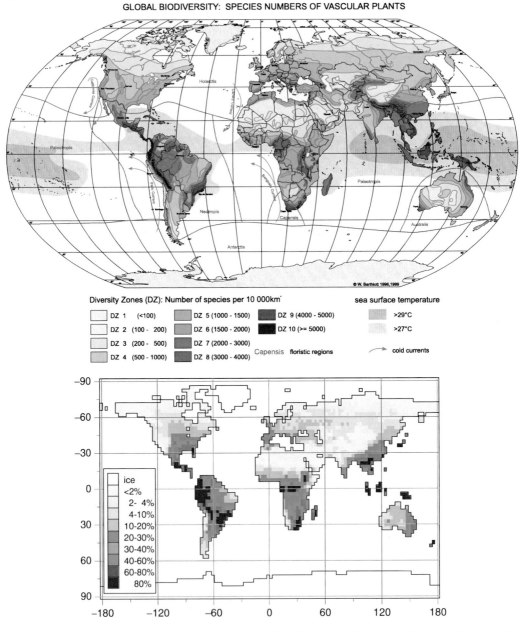

GLOBAL BIODIVERSITY: SPECIES NUMBERS OF VASCULAR PLANTS

Diversity Zones (DZ): Number of species per 10 000km²

		sea surface temperature
DZ 1 (<100)	DZ 5 (1000 - 1500)	DZ 9 (4000 - 5000)	>29°C
DZ 2 (100 - 200)	DZ 6 (1500 - 2000)	DZ 10 (>= 5000)	>27°C
DZ 3 (200 - 500)	DZ 7 (2000 - 3000)		
DZ 4 (500 - 1000)	DZ 8 (3000 - 4000)	Capensis floristic regions	cold currents

Fig. 2.37 Global distribution of species richness based on observations (*top*; units number of species per 10,000 km²) and on model simulations (*bottom*; units % of maximum diversity simulated) that use climate as a filter to reduce the number of allocation strategies. Reprinted from Kleiden and Mooney (2000)

and generates convective motion in the atmosphere. Large-scale patterns of atmospheric circulation occur because the tropics receive more energy from the sun than they emit to space, whereas the poles lose more energy to space than they receive from the sun. The resulting circulation cells transport heat from the equator to the poles to balance these inequalities. In the process, they create three relatively distinct air masses in each hemisphere, a tropical air mass (0–30°N and S), a temperate air mass (30–60°N and S), and a polar air mass (60–90°N and S). There are four major areas of high pressure (the two poles and 30°N and S), where air descends, and precipitation is low. The subtropical high-pressure belts are the zones of the world's major deserts. There are three major zones of low pressure (the equator and 60°N and S), where air rises, and precipitation is high. These areas support the tropical rainforests at the equator and the temperate rainforests of northwestern North America and southwestern South America. Ocean currents account for about 40% of the latitudinal heat transport from the equator to the poles. These currents are driven by surface winds and by the downwelling of cold saline waters at high latitudes, balanced by upwelling at lower latitudes.

Regional and local patterns of climate reflect heterogeneity in Earth's surface. Uneven heating between the land and the ocean modifies the general latitudinal patterns of climate by generating zones of prevailing high and low pressure. These pressure centers are associated with storm tracks that are guided by major mountain ranges in ways that strongly influence regional patterns of climate. The ocean and large lakes also moderate climate on adjacent lands because their high heat capacity causes them to heat or cool more slowly than land. These heating contrasts produce predictable seasonal winds (monsoons) and daily winds (land/sea breezes) that influence the adjacent land. Mountains also create heterogeneity in precipitation and in the quantity of solar radiation intercepted.

Vegetation influences climate through its effects on surface albedo, which determines the quantity of incoming radiation absorbed by the surface, and energy released to the atmosphere via longwave radiation and turbulent fluxes of latent and sensible heat. Sensible heat fluxes and longwave radiation directly heat the atmosphere, and latent heat transfers water vapor to the atmosphere, influencing local temperature and moisture sources for precipitation.

Climate is variable over all time scales. Long-term variations in climate are driven largely by changes in solar input and atmospheric composition. Superimposed on these long-term trends are predictable daily and seasonal patterns of climate, as well as repeating patterns such as those associated with El Niño/Southern Oscillation. These oscillations cause widespread changes in the geographic pattern of climate on time scales

of years to decades. Future changes in climate may reflect changes in the frequencies of these large-scale climate modes.

Review Questions

1. Describe the energy budget of Earth's surface and the atmosphere. What are the major pathways by which energy is absorbed by Earth's surface? By the atmosphere? What are the roles of clouds and radiatively active gases in determining the relative importance of these pathways?

2. Why is the troposphere warmest at the bottom but the stratosphere is warmest at the top? How does each of these atmospheric layers influence the environment of ecosystems?

3. Explain how unequal heating of Earth by the sun and the resulting atmospheric circulation produces the major latitudinal climate zones, such as those characterized by tropical forests, subtropical deserts, temperate forests, and arctic tundra.

4. How do the rotation of Earth (and the resulting Coriolis effect) and the separation of Earth's surface into the ocean and continents influence the global patterns of climate?

5. How does the chemical composition of Earth's atmosphere influence the climate of Earth?

6. What causes the global pattern in surface ocean currents? Why are the deep-water ocean currents different from those at the surface? What is the nature of the connection between deep- and surface-ocean currents?

7. How does ocean circulation influence climate at global, continental, and local scales?

8. How does topography affect climate at continental and local scales?

9. What are the major causes of long-term changes in climate? How would you expect future climate to differ from that of today in 100 years? 100,000 years? 1 billion years? Explain your answers.

10. Explain how the interannual variations in climate of Indonesia, Peru, and California are interconnected.

11. Explain the climatic basis for the global distribution of each major biome type. Use maps of global winds and ocean currents to explain these distributions.

12. Describe the climate of your birthplace. Using your understanding of the global climate system, explain why this location has its characteristic climate.

Additional Reading

Ahrens, C.D. 2003. *Meteorology Today: An Introduction to Weather, Climate, and the Environment*. 7th edition. Thomson Learning, Pacific Grove, CA.

Foley, J.A., M.H. Costa, C. Delire, N. Ramankutty, and P. Snyder. 2003. Green surprise? How terrestrial ecosystems could affect earth's climate. *Frontiers of Ecology and the Environment* 1:38–44.

Graedel, T.E. and P.J. Crutzen. 1995. *Atmosphere, Climate, and Change*. Scientific American Library. New York.

Oke, T.R. 1987. *Boundary Layer Climates*. 2nd Edition. Methuen, London.

Skinner, B.J., S.C. Porter, and D.B. Botkin. 1999. *The Blue Planet: An Introduction to Earth System Science*. 2nd Edition. Wiley, New York.

Serreze, M.C. 2010. Understanding recent climate change. *Conservation Biology* 24:10–17.

Sturman, A.P., and N.J. Tapper. 1996. *The Weather and Climate of Australia and New Zealand*. Oxford University Press, Oxford.

Trenberth, K.E. and D.P. Stepaniak. 2004. The flow of energy through the earth's climate system. *Quarterly Journal of the Meteorological Society* 30: 2677–2701.

Within a given climatic regime, soil properties
are the major factor governing ecosystem processes. This chapter provides background on
the factors regulating those soil and sediment
properties that most strongly influence ecosystems as well as the transport of materials from
land to rivers, lakes, and the ocean.

Introduction

Soils form a thin film over Earth's surface in
which geological and biological processes
intersect. The soil consists of solids, liquids,
and gases, with solids typically occupying about
half the soil volume, and liquids and gases each
occupying 15–35% of the volume (Ugolini and
Spaltenstein 1992). The physical soil matrix
provides a source of water and nutrients to plants
and microbes and is the physical support system
in which terrestrial vegetation is rooted. It is the
medium in which most decomposer organisms
and many animals live. For these reasons, the
physical and chemical properties of soils
strongly influence all aspects of ecosystem functioning, which, in turn, feed back to influence
the physical, structural, and chemical properties
of soils (see Fig. 1.5; Amundson et al. 2007).
Soils play such an integral role in ecosystem
processes that it is difficult to separate the study
of soils from that of ecosystem processes. In
open-water (**pelagic**) ecosystems, phytoplankton cannot directly tap resources from sediments, so sediment processes provide nutrient
resources to primary producers only indirectly
through mixing of the water column.

Soils are also a critical component of the total
Earth System. They mediate many of the key
reactions in the giant global reduction–oxidation
cycles of carbon, nitrogen, and sulfur and provide
essential resources to biological processes that
drive these cycles. Soils represent the intersection
of the "bio," "geo," and chemistry in biogeochemistry. Many of the later chapters in this book
address the short-term dynamics of soil processes, particularly those processes that occur on
timescales of hours to centuries. This chapter
emphasizes soil processes that occur over longer
timescales or that are strongly influenced by
physical and chemical interactions with the environment. This is essential background for understanding the dynamics of ecosystems.

A Focal Issue

Human activities have massively increased
nutrient and sediment inputs from terrestrial
to aquatic ecosystems. Soils that developed over
thousands of years can be eroded away in years to
decades, causing loss of productive capacity in
upland ecosystems and accumulation in reservoirs, lowland floodplains, estuaries, and coastal
waters. On human timescales, this is an essentially
permanent restructuring of regional landscapes.
The extensive cultivation of drought-sensitive
crops on marginal lands in the U.S. in the 1920s,
for example, created a landscape vulnerable to

F.S. Chapin, III et al., *Principles of Terrestrial Ecosystem Ecology*,
DOI 10.1007/978-1-4419-9504-9_3, © Springer Science+Business Media, LLC 2011

Fig. 3.1 Extensive cultivation replaced drought-resistant native vegetation with drought-sensitive crops in the midwestern U.S. in the 1920s. In the "Dustbowl" era, a drought in the 1930s killed these crops and generated massive dust storms, such as this one approaching Stratford Texas in 1935. Photograph courtesy of NOAA, http://www.photolib.noaa.gov/htmls/theb1365.htm

drought. Hot, dry weather combined with strong winds in the 1930s caused extensive wind erosion that reduced the productive potential of soils, modified regional climate, and triggered land abandonment and human migration (Fig. 3.1; see Chap. 12; Peters et al. 2004; Schubert et al. 2004). Erosion of the loess plateau in China and drylands in sub-Saharan Africa are current issues that threaten livelihoods of millions of people over extensive regions. What properties of vegetation and soils cause some soils to be more susceptible to erosion than others? Why are topsoils, which are the first layers to be eroded, so much more fertile than deeper soils? What are the consequences of wind and water erosion for those ecosystems where soil particles are deposited? What management practices sustain the productivity soils and reduce erosion rates? This chapter addresses these questions and other issues that are important for sustainability of ecosystems and managed landscapes.

Controls Over Soil Formation

The soil properties of an ecosystem result from the dynamic balance of two opposing forces: soil formation and soil loss. State factors differ in their effects on these opposing processes and therefore on soil and ecosystem properties (Jenny 1941; Amundson and Jenny 1997).

Parent Material

The physical and chemical properties of rocks and the rates at which they are uplifted and weathered strongly influence soil properties. The dynamics of the rock cycle, operating over billions of years, govern the variation and distribution of geological materials on Earth's surface. The rock cycle describes the cyclic process by which rocks are formed and **weathered**, i.e., chemically and physically altered near Earth's

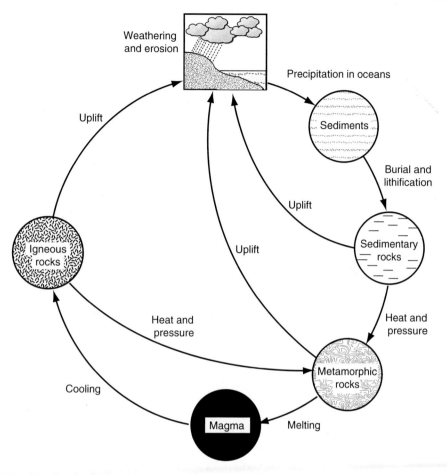

Fig. 3.2 The rock cycles as proposed by Hutton in 1785. Rocks are weathered to form sediment, which is then buried. After deep burial, the rocks undergo metamorphosis or melting, or both. Later, they are deformed and uplifted into mountain chains, only to be weathered again and recycled. Redrawn from Press and Siever (1986)

surface (Fig. 3.2). The rock cycle produces minerals that buffer the biological acidity that accounts for much of rock weathering but also provides many of the nutrients that allow biology to produce this acidity. The compounds produced by weathering move via rivers to lakes, reservoirs, and the ocean where they are deposited to form sediments, which are then buried to form **sedimentary rocks. Igneous rocks** form when **magma** from deep within Earth moves upward toward the surface in cracks or volcanoes. Either sedimentary or igneous rocks can be modified under heat or pressure to form **metamorphic rocks**. With additional heat and pressure, metamorphic rocks melt and become magma. Any of these rock types can be raised to the surface via uplift, after which the material is again subjected to weathering and erosion (Fig. 3.2). Earth's crust cycles through the rock cycle every 100–200 million years, i.e., two to four times since plants first colonized the land (see Fig. 2.15). The timing and locations of uplift and the type of rock uplifted ultimately determine the distribution of different types of bedrock across Earth's surface.

Plate tectonics are the driving force behind the rock cycle. The **lithosphere** or crust, the strong outermost shell of Earth that rides on partially molten material beneath, is broken into large rigid plates, each of which moves independently. Where the plates converge and collide,

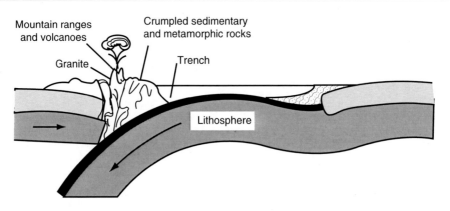

Fig. 3.3 Cross section of a zone of plate collision, in which the oceanic plate is subducted beneath a continental plate, forming an ocean trench in the zone of subduction and mountains and volcanoes in the zone of uplift. Redrawn from Press and Siever (1986)

portions of the lithosphere buckle downward and are **subducted**, leading to the formation of ocean trenches, while the overriding plate is **uplifted**, causing the formation of mountain ranges (Fig. 3.3). Regions of plate collision and active mountain building coincide with Earth's major earthquake belts. The Himalayan Mountains, for example, are still rising due to the collision of the Indian subcontinent with Asia 40 million years ago. If plates converge in one place, they must diverge or separate elsewhere. Throughout Earth history, massive super-continents have formed and broken apart, with continents rafting to new locations and forming new super-continents. This occurred most recently when the super-continent of Pangaea broke up 50–200 million years ago to form Eurasia, Africa, Antarctica, and the Americas. Australia, for example, is moving from its point of origin in Antarctica toward Southeast Asia at 5–6 cm year^{-1}. The mid-Atlantic and mid-Pacific ridges are zones of active divergence of today's ocean plates. Continental drift has rafted the world's biota and soils through multiple climate zones during their evolutionary history.

Climate

Temperature, moisture, carbon dioxide, and oxygen influence rates of chemical reactions that govern the rate and products of weathering, as well as biological activity, and therefore the development of soils from rocks. Temperature, moisture, and oxygen also influence biological processes such as the production of organic matter by plants and its decomposition by microbes and therefore the amount and quality of organic matter in the soil (see Chaps. 5–7). Soil carbon, for example, increases with decreasing temperature and with increasing precipitation along global and regional climate gradients (Post et al. 1982; Burke et al. 1989; Jobbágy and Jackson 2000). Precipitation is one pathway by which materials enter ecosystems. **Oligotrophic** (nutrient-poor) bogs are isolated from mineral soils and depend entirely on precipitation to supply new minerals. The movement of water is also crucial in determining whether the products of weathering accumulate or are lost from a soil and transported to other places. In summary, climate affects virtually all soil properties at scales ranging from local to global.

Topography

Topography influences soils through its effect on climate, moisture availability, and differential transport of fine soil particles. Topographic gradients form a hillslope complex or **catena** from ridge top to valley bottom. These gradients and the **aspect** (compass direction) of the slope strongly influence soil properties (Amundson and

Fig. 3.4 Relationship between hillslope position, likelihood of erosion or deposition, and soil organic carbon concentration. Redrawn from Birkeland (1999)

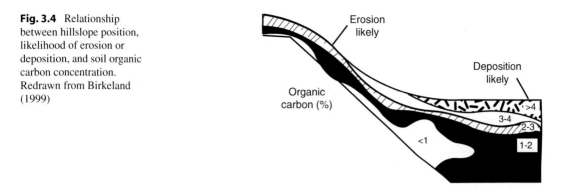

Jenny 1997). Erosion, for example, preferentially moves fine-grained materials downslope and deposits them at lower slope positions, where they tend to form deep fine-textured soils with a high soil organic content (Fig. 3.4) and high water-holding capacity. These valley-bottom soils supply more resources to plants and microbes and provide greater physical stability, typically leading to higher rates of most ecosystem processes than on ridges or shoulders of slopes. Soils in lower slope positions in sagebrush ecosystems, for example, have greater soil moisture, higher soil organic matter content, and higher rates of nitrogen mineralization and gaseous losses than do upslope soils (Burke et al. 1990; Matson et al. 1991).

The aspect of a slope influences solar input (see Chap. 2) and therefore soil temperature, rates of evapotranspiration, and soil moisture. At high latitudes and in wet climates, the cool wet environment of poleward-facing slopes reduces rates of decomposition and mineralization (Van Cleve et al. 1991). At low latitudes and in dry climates, the greater retention of soil moisture on these slopes allows a longer growing season and supports forests, whereas slopes facing the equator are more likely to support desert or shrub vegetation (Whittaker and Niering 1965).

Finally, slope position determines patterns of snow redistribution in cold climates, with deepest accumulations beneath ridges and in the protected lower slopes. These differential accumulations alter effective precipitation and length of growing season enough to influence plant and microbial processes well into the summer.

Time

Many soil-forming processes occur slowly, so the time over which soils develop influences their properties. Rocks and minerals are weathered over time, and important nutrient elements are transferred among soil layers or transported out of the ecosystem. Hillslopes erode, and valley bottoms accumulate materials, and biological processes add organic matter and critical nutrient elements like carbon and nitrogen. Phosphorus availability is high early in soil development and declines in availability over time due to losses from the system and phosphorus fixation in mineral forms that are unavailable to plants (Fig. 3.5; Walker and Syers 1976). This process plays out over millions of years of soil development in Hawai'i, despite a warm moist climate, changing the system from nitrogen limitation on young soils to phosphorus limitation on older soils (Hobbie and Vitousek 2000; Vitousek 2004).

Some changes in soil properties happen relatively quickly. Retreating glaciers and river floodplains often deposit phosphorus-rich till. If seed sources are available, these soils are colonized by plants with symbiotic nitrogen-fixing microbes, allowing these ecosystems to accumulate their maximum pool sizes of carbon and nitrogen within 50–100 years (Crocker and Major 1955; Van Cleve et al. 1991). Other soil-forming processes occur slowly. Young marine terraces in coastal California have relatively high phosphorus availability but low carbon and nitrogen content. Over at least tens of thousands of years, these terraces accumulate organic matter and

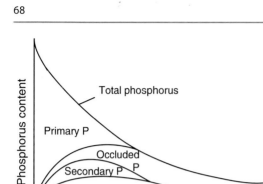

Fig. 3.5 Effects of long-term weathering and soil development on the distribution and availability of phosphorus (P). Newly exposed geologic substrate is relatively rich in weatherable minerals, which release phosphorus. This release leads to accumulation of both organic and readily soluble forms (secondary phosphorus such as calcium phosphate). As primary minerals disappear and secondary minerals capable of sorbing phosphorus accumulate, an increasing proportion of the phosphorus remaining in the system is held in unavailable (occluded) forms. Availability of phosphorus to plants peaks relatively early in this sequences and declines thereafter. Redrawn from Walker and Syers (1976)

nitrogen, causing a change from coastal grassland to productive redwood forest (Jenny et al. 1969). Over several 100,000 years, silicates are leached out, leaving behind a hardpan of iron and aluminum oxides with very low fertility and seasonally anaerobic soils. The pygmy cypress forests that develop on these old terraces have very low productivity. The phenolic compounds produced by these trees as defenses against herbivores also retard decomposition, further reducing soil fertility (see Chap. 7; Northup et al. 1995).

Potential Biota

The past and present organisms at a site strongly influence soil chemical and physical properties. Most soil development occurs in the presence of living organisms. Plants are the sources of organic carbon that enter soils, and functionally different types of plants (e.g., grasses, deciduous trees, evergreen shrubs) strongly influence the amount and especially the depth distribution of soil carbon (Jobbágy and

Jackson 2000). Carbon-containing soil organic matter, in turn, influences most functional properties of soils, as described later.

Plants also strongly influence mineral properties of soils. They are geochemical pumps that remove bio-essential elements from soils, store them in tissues, and return them to the soil through litterfall and decomposition (Amundson et al. 2007). In the process, soluble forms of rock-derived minerals such as phosphorus, calcium, potassium, and silicon can be moved upward in the soil profile and are most available in the upper portion of the soil. This is partially offset by downward leaching. Upward movement generally predominates unless minerals precipitate out in less available forms at depth (e.g., calcium in desert soils or iron and aluminum in wet soils), as described later. CO_2 from plant and microbial respiration and the organic acids produced by many plants generate soil acidity that contributes to rock weathering. Vegetation differences in either absorption of minerals or release of organics strongly influences soil properties (see Chap. 7). It is often difficult, however, to separate the chicken from the egg. Did the vegetation determine soil properties or vice versa (Berner et al. 2004; Dietrich and Perron 2006; Amundson et al. 2007)?

One approach to determining vegetation effects on soils has been to plant monocultures or species mixes into initially homogeneous sites. Rapidly growing grasses in a nitrogen-poor perennial grassland enhanced the nitrogen mineralization (or reduced microbial immobilization) of nitrogen by soils within 3 years (see Fig. 11.5; Wedin and Tilman 1990), as did deep-rooted forbs in an annual grassland (Hooper and Vitousek 1998). Another approach is to examine the consequences of species invasions or extinctions on soil processes. The invasion of a non-native nitrogen fixer into Hawaiian rainforests, for example, increased nitrogen inputs to the system more than fivefold, altering the characteristics of soils and the colonization and competitive balance among native plant species (see Fig. 11.3; Vitousek et al. 1987). Yet another approach is to examine weathering and erosion rates in places without biota (Mars or early Precambrian soils) or with minimal biotic effects (e.g., Antarctic dry valleys; Amundson et al. 2007).

Animals also influence soil properties. Earthworms, termites, and invertebrate shredders, for example, stimulate decomposition (see Chap. 7), thereby modifying soil properties that are influenced by soil organic content. Grazers such as North American bison concentrate sodium in their wallows, which disperses clays and creates water-holding pans. Other grazers like African rhinos generate large dung middens that concentrate nutrients, whereas termites form large termitaria that concentrate soil resources and vertically redistribute nutrients. Microorganisms also influence the structure and properties of soils through the types of organic compounds they release into the soil environment.

Human Activities

Over the past 40 years, the doubling of human population and associated agricultural and industrial activities have strongly influenced soil development worldwide. Human activities directly influence soils through changes in nutrient inputs, irrigation, alteration of soil microenvironment, and increased erosional loss of soils. Human activities also indirectly affect soils through changes in other drivers, including changes in atmospheric composition and the additions and deletions of species.

Controls Over Soil Loss

Soil formation depends on the balance between deposition, erosion, and soil development (i.e., the changes that soils undergo in place). Soil thickness varies with hillslope position, with erosion dominating on steep slopes, deposition in valley bottoms, and soil development on gentle slopes and terraces where the lateral transport of materials is minimal (Fig. 3.4). Much of Earth's surface is in hilly or mountainous terrain where erosion and deposition are important processes. Erosion removes the products of weathering and biological activity. In young soils, erosional

losses reduce soil fertility by removing clays and organic matter that store water and nutrients. On highly weathered landscapes, however, erosion renews soil fertility by removing the highly weathered remnants (sands and iron oxides) that contribute little to soil fertility and exposing less weathered materials that provide a new source of essential nutrients (Porder et al. 2005).

The dominant erosional processes depend on topography, the properties of surface materials, and the pathways by which water leaves the landscape. Mass wasting is a major erosional process in most regions. This is the downslope movement of soil or rock material under the influence of gravity without the direct aid of other media such as water, air, or ice. Mass wasting includes both fine-scale processes such as the movement of individual soil particles (**soil creep**) and massive events such as landslides or debris flows that can rapidly transport cubic meters to cubic kilometers of material. Mass wasting occurs most rapidly on steep slopes, regardless of the underlying mechanism. Any process that moves soil particles (e.g., freeze–thaw events or animal burrowing) contributes to their net downhill movement. Erosion caused by soil creep is the aggregate result of millions of tiny events. Gophers, for example, as a result of their preference for deep soils, burrow more actively and increase erosion from deep soils, reducing the variability in soil thickness across landscapes (Yoo et al. 2005). Landslides, on the other hand, are rare but massive events. The probability of a landslide depends on the **shear stress** that the soil experiences, i.e., the force parallel to the slope that drives mass wasting events such as landslides. It is the balance between the gravitational driving force for downslope movement (F_t) and the friction that resists this movement (F_n; Fig. 3.6).

Many factors influence the **shear strength** of a soil mass (i.e., the shear stress that a soil can sustain without slope failure; Selby 1993). Sometimes the sliding friction between the material and some well-defined plane (such as a frozen soil layer) determines whether a landslide occurs. More commonly, however, it is the internal friction among individual components within the soil

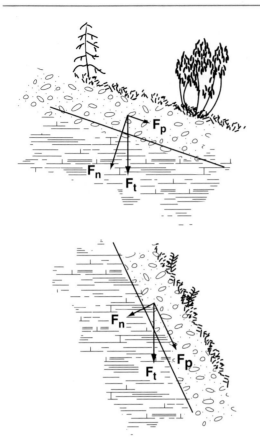

Fig. 3.6 The effect of slope angle on the partitioning of the total gravitational force (F_t) into a component (F_n) that is normal (perpendicular) to the slope (and therefore contributes to friction that resists erosion) and a component (F_p, shear stress) that is parallel to the slope (and therefore promotes erosion). Steep slopes have a larger value of F_p and lower values of F_n and therefore a greater tendency for mass wasting

matrix that largely determines its resistance to mass wasting. Cohesion among soil particles and water molecules enhances the internal friction that resists mass wasting. A small amount of water enhances cohesion among particles, explaining why sand castles are easier to make with moist than with dry sand. High water content, however, increases the weight of the soil, makes soil grains more buoyant, and reduces the frictional strength. Wet soils become unstable, leading to liquefaction of the soil mass, which can flow down slope. Fine-particle soils have lower slope thresholds of instability and are more likely to lead to slope failure than are coarse-textured

soils. Roots also increase the resistance of soils to downslope movement, so deforestation and other land-use changes that reduce root biomass increase the probability of landslides.

The pathways by which water leaves the landscape strongly influence erosion. Water can leave a landscape via several pathways: evaporation and transpiration to the atmosphere, groundwater flow, shallow subsurface flow, and overland flow (when precipitation exceeds infiltration rate; see Fig. 4.4). The relative importance of these pathways depends on topography, vegetation, and material properties such as the hydraulic conductivity of soils. Groundwater and shallow subsurface flow dissolve and remove ions and small particles. At the opposite extreme, overland flow causes erosion primarily by surface sheet wash, rills, and rain splash. This often occurs in sparsely vegetated arid and semi-arid soil-mantled landscapes and on disturbed ground. Overland flow rates of 0.15–3 cm s^{-1} are enough to suspend clay and silt particles and move them downhill (Selby 1993). As water collects into gullies, its velocity, and therefore erosion potential, increases. A doubling of velocity causes a 60-fold increase in the size of particles that can be eroded. Vegetation and a litter layers greatly increase infiltration into the soil by reducing the velocity with which raindrops hit the soil, thereby preventing surface compaction by raindrops. Vegetated soils are also less compact because roots and soil animals create channels in the soil. In these ways, vegetation and a litter layer substantially increase infiltration and therefore groundwater and subsurface flow.

High wind speeds at the soil surface are another important agent of erosion. This often occurs after vegetation removal. Some agricultural areas in China have lost meters of soil to wind erosion and have become a major source of iron to phytoplankton in the Pacific Ocean (see Chap. 9).

Streams and rivers play an important role in soil redistribution across landscapes. At the scale of large river basins, three broad geomorphic zones can be identified (Naiman et al. 2005): an **erosional zone**, where erosion dominates over deposition, a **transfer zone**, where erosion and

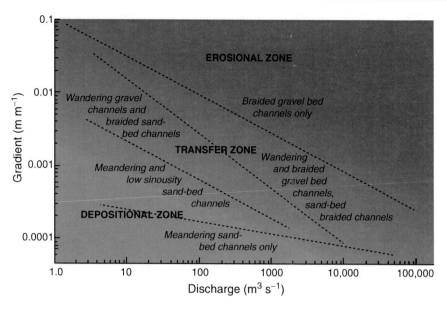

Fig. 3.7 Effects of river discharge and elevational gradient on patterns of river channel morphology. There is a gradual transition from a predominance of erosion in streams with steep gradients or high discharge (Erosional Zone shown in *red*) to a predominance of deposition in streams with shallow gradients or low discharge (Depositional Zone, shown in *green*). Modified from Church (2002)

deposition are in dynamic balance over long timescales, and a **depositional zone**, where deposition rate exceeds erosion rate and the capacity of the river to transport suspended materials (Fig. 3.7). Most sediments delivered to the ocean originate in the erosional zone (Milliman and Syvitski 1992). Here, slopes become steeper as headwater streams downcut into their beds, increasing the shear stress on adjacent soils and the rate of mass wasting. As materials are delivered to the stream by mass wasting and erosion of the streambed, they are transported downstream at a rate that depends on flow velocity and grain size of the sediments, with fine particles moving downstream faster than gravel and boulders. Glaciers, mining, or vegetation removal substantially augments sediment delivery in the erosional zone.

In the transfer zone, there is less delivery of primary sediments to the stream or river and the dominant processes are the sorting of sediments according to grain size and the downstream transport of materials as a result of a balance of erosion and deposition. When stream energy

increases, for example during a flood, progressively larger particles are mobilized, and, as river energy declines, the larger particles are deposited first. This produces a heterogeneous patchwork of gravel bars, sand bars, and silt-filled side channels (Naiman et al. 2005). Stream energy, and therefore the size of particles transported, is greater during flood events, in steep gradients (e.g., riffles), and in deep narrow channels. The transfer zone that links zones of erosion and deposition may shift through time as a result of (1) mountain uplift or sea-level change, which together determine the vertical gradient in the river basin; (2) discharge, which depends on prevailing climate and water inputs or removals from streams; and (3) sediment inputs, which may be influenced by human activities and other factors. Floodplains form during periods when deposition predominates, and channel incision occurs when erosion predominates.

In the depositional zone, rivers tend to meander and develop broad alluvial floodplains and deltas. Rivers in the depositional zone tend to show larger peak discharges (floods) than upstream

Table 3.1 Climatic and topographic effects on long-term erosion rates

Climate zone	Relief	Erosion rate[a] (mm century^{-1})
Glacial	Gentle (ice sheets)	5–20
	Steep (valleys)	100–500
Polar montane	Steep	1–100
Temperate maritime	Mostly gentle	0.5–10
Temperate continental	Gentle	1–10
	Steep	10–20+
Mediterranean	–	1–?
Semi-arid	Gentle	10–100
Arid	–	1–?
Wet subtropics	–	1–100?
Wet tropics	Gentle	1–10
	Steep	10–100

Data from Selby (1993)

[a]Erosion rates are estimated from average sediment yields of rivers in different climatic and topographic regimes. Extreme uncertainty in maximum values is indicated (?)

because of the accumulation of water from a large drainage basin into a single channel. During these floods, the river overflows its banks and fills low-lying areas. Flooding accounts for most of the deposition in this river zone. In the Amazon, for example, more sediment is transported laterally to the floodplain than to the ocean (Dunne et al. 1998). Other finer-scale dynamics that occur within the floodplain involve the erosion of sediments on the outer bends of meanders, where river velocity is greatest, and deposition as new sand or silt bars on the inner sides of river bends. These dynamics cause the river to redistribute materials within the floodplain, creating habitat mosaics of different-aged stands.

Sediments that enter the ocean are deposited near the river mouth, forming a delta or tidal mudflats or are redistributed by coastal currents. Soft (non-rocky) coastlines, including sandy beaches and barrier islands are maintained by the dynamic balance between the delivery of sediments to the coastal zone, their horizontal redistribution by coastal currents and storms, and export (particularly of fine particles offshore). Dredging of harbors to maintain shipping channels and "armoring" of coastlines to prevent erosion in one location reduces sediment inputs elsewhere, often with disastrous unintended consequences. Redistributing sediment delivery from the Mississippi River by routing river flow offshore, for example, contributed to subsidence of wetlands and loss of barrier islands that would otherwise have helped to protect New Orleans during the 2005 Hurricane Katrina.

Erosion of landscapes results from the combined action of wind, water, ice, and mass wasting. On average, erosion of terrestrial material to the ocean is about 1–10 mm century^{-1} (Selby 1993). However, erosion rates vary regionally by two to three orders of magnitude, depending on topography, climate, human activities, and the sensitivity of rocks and soils to erosion (Table 3.1). Erosion rates tend to approach rates of tectonic uplift, so regions with active tectonic uplift and steep slopes generally have higher erosion rates than flat, weathered terrain. Climate influences erosion primarily through its effects on vegetation cover. In arid, semi-arid, and polar regions with minimal vegetation, for example, surface wash from raindrop impacts and overland flow during intense rains cause most erosion. In contrast, ecosystems with greater vegetative cover lose material primarily through the dissolution of rocks (weathering) to produce soluble compounds that leach out of the system. Low vegetation cover also makes lands more prone to soil loss from wind erosion. The contribution of large rare events like landslides to long-term erosion rates is poorly known. They may be more important in redistributing materials within a drainage basin than in causing loss from the land to the ocean. For example, 90% of the materials eroded from the upland Piedmont region in the southeastern U.S. since 1700 is still stored on hillslopes, valley bottoms, and reservoirs (Selby 1993). At a global scale, human activities have increased erosion and sediment flux in rivers by 2.3 billion metric tons per year, but have reduced sediment flux to the ocean by 1.4 billion metric tons per year because of sediment trapping in reservoirs (Syvitski et al. 2005). These patterns are regionally variable, however. Indonesia, for example, has considerable land-use change and sediment transport but very few reservoirs to prevent these sediments from reaching the ocean. Much of the erosion on natural landscapes probably occurs during high-rainfall events or after disturbances have reduced vegetation cover rather than during average conditions.

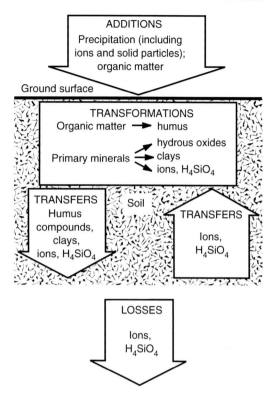

Fig. 3.8 Processes leading to additions, transformations, transfers, and losses of materials from soils. Redrawn from Birkeland (1999)

Development of Soil Profiles

Soils develop through the additions of materials to the system, transformations of those materials within the system, transfers down and up in the soil profile, and losses of materials from the system (Fig. 3.8; Richter and Markewitz 2001).

Additions to Soils

Direct inputs to the soil system come from both outside and inside the ecosystem. Inputs from *outside* the ecosystem come from precipitation and wind, which deposit ions and dust particles, and floods and tidal exchange, which deposit sediments and solutes (see Chap. 9). The source of these materials determines their size distribution and chemistry, leading to the development of soils with specific textural and chemical characteristics. Sometimes these inputs are huge, for example, hundreds to thousands of g m^{-2} of dust inputs to loess-accumulating regions of North America and Asia during the Pleistocene (Sun et al. 2000; Bettis et al. 2003). Organisms *within* the ecosystem add organic matter and nitrogen to

the soil as dead organic matter, including the above- and belowground portions of plants, animals, and soil microbes.

Soil Transformations

Within the soil, materials are transformed through an interaction of physical, chemical, and biological processes. Freshly deposited dead organic matter is transformed in the soil by **decomposition** to soil organic matter, releasing carbon dioxide and nutrients such as nitrogen and phosphorus (see Chap. 7). Recalcitrant plant and microbial organic compounds undergo physicochemical interactions with soil minerals that contribute to the long-term storage of soil organic matter. The quantity of soil carbon in deep soils, for example, correlates more closely with clay content than with climate (Jobbágy and Jackson 2000).

Weathering is the change of parent rocks and minerals to produce more stable forms. This occurs when rocks and minerals become exposed to physical and chemical conditions different from those under which they formed (Ugolini and Spaltenstein 1992). Weathering involves both physical and chemical processes and is influenced by characteristics of the parent material, environmental conditions (temperature and moisture), and the activities of organisms. **Physical weathering** is the fragmentation of parent material without chemical change. This can occur when rocks are fractured by expansion and contraction during cycles of freeze–thaw, heating–cooling, or wetting–drying or when roots grow into rock fissures. Fire, for example, is a potent force for physical weathering because it rapidly heats exposed rock surfaces while leaving the deeper layers cool. In addition, soil particles and rock fragments are abraded by wind, or ground against one another by glaciers, landslides, or floods. Physical weathering is especially important in extreme and highly seasonal climates. Wherever it occurs, it opens channels in rocks for penetration by water and air and increases the surface area for chemical weathering reactions.

Chemical weathering occurs when parent rock materials react with acidic or oxidizing substances, usually in the presence of water. During chemical weathering, **primary minerals** (unmodified minerals present in the rock or unconsolidated parent material) dissolve, releasing ions and forming **secondary minerals** (insoluble reaction products of weathering). Chemical weathering most commonly involves the reaction of water and acid on a mineral. Carbonic acid is the most important of these acids. It forms through the reaction of CO_2 with water and then ionizes to produce a hydrogen ion and a bicarbonate ion. The CO_2 concentration in soil, which drives the formation of carbonic acid, is 10- to 500-fold higher than in air, due to the **respiration** (CO_2 production) by plants, soil animals, and microbes and the low diffusivity of gases in soil. Weathering rates are particularly high adjacent to roots because high rates of biological activity produce abundant CO_2 and organic acids in the **rhizosphere**, the zone of soil that is directly influenced by roots. Carbonic acid, for example, attacks potassium feldspar, which is converted to a secondary mineral, kaolinite by the removal of soluble silica and potassium (3.1).

$$2KAlSi_3O_8 + 2(H^+ + HCO_3^-) + H_2O \rightarrow Al_2Si_2O_5(OH)_4 + 4SiO_2 + 2K^+ + 2HCO_3^- \qquad (3.1)$$

Other sources of acidity that promote chemical weathering include organic acids, nitric acid, sulfuric acid, and the hydrogen ions excreted by plant roots when cations are absorbed (Richter and Markewitz 2001). Plant roots and microbes secrete many organic acids into the soil, which influence chemical weathering through their contribution to soil acidity and their capacity to **chelate** ions. In the chelation process, organic acids combine with metallic ions, such as Fe^{3+} and Al^{3+}, making them soluble and mobile. Chelation lowers the concentration of unchelated inorganic ions at the mineral surface, so dissolved and primary mineral forms are no longer in equilibrium with one another. This accelerates the rate of weathering.

Warm climates promote chemical weathering because temperature speeds chemical reactions and enhances the activities of plants and microbes. Wet conditions promote weathering through their

direct effects on weathering reactions and their effects on biological processes. Not surprisingly, the hot, wet conditions of the humid tropics yield the highest rates of chemical weathering.

The physical and chemical properties of rock minerals determine their susceptibility to weathering and the chemical products that result. Sedimentary rocks like shale that form by chemical precipitation, for example, have more basic cations like calcium (Ca^{2+}), sodium (Na^+), and potassium (K^+) than do igneous rocks. Sedimentary rocks tend to produce soils with a relatively high pH and a high capacity to supply mineral cations to plants. Igneous rocks form more acidic soils.

Minerals weather in the same order in which they crystallized during formation (Schlesinger 1997; Birkeland 1999). Olivine, for example, is one of the first minerals to crystallize as magma

Table 3.2 Stability of common minerals under weathering conditions at Earth's surface

Most stable	Fe^{3+} oxides	Secondary mineral
	Al^{3+} oxides	Secondary mineral
	Quartz	Primary mineral
	Clay minerals	Secondary mineral
	K$^+$ feldspar	Primary mineral
	Na$^+$ feldspar	Primary mineral
	Ca^{2+} feldspar	Primary mineral
Least stable	Olivine	Primary mineral

Data from Press and Siever (1986)

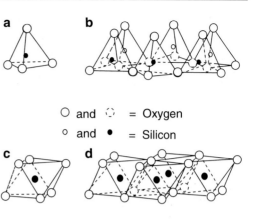

O and ◌ = Oxygen
o and ● = Silicon

O and ◌ = Hydroxyl
● = Aluminum, magnesium, etc.

Fig. 3.9 Diagram showing the molecular structure of a simple clay layer: (**a**) a tetrahedral unit, (**b**) a tetrahedral sheet, (**c**) an octahedral unit, and (**d**) an octahedral sheet. Redrawn from Grim (1968)

cools, forms relatively few bonds, and weathers easily. Feldspar forms and weathers more slowly than olivine, and quartz is one of the last minerals to crystallize, has strong bonds that create a crystalline structure, and is highly resistant to weathering (Table 3.2). Secondary minerals such as the silicate clay minerals and iron and aluminum oxides are among the most resistant minerals to weathering. Differences among elements in their susceptibility to weathering and solubility in water leads to the following sequence in which elements are weathered from rocks and leached into rivers:

$$Cl > SO_4 > Na > Ca > Mg > K > Si > Fe > Al$$

$$(3.2)$$

Moderately weathered soils therefore have relatively high concentrations of Ca$^+$, Mg$^+$, and K$^+$ (elements essential for plant growth) and low concentrations of soluble Al^{3+} (a slowly weathered element often toxic to plants). In contrast, in the ancient soils of the wet tropics, the relatively mobile ions of Si and Mg^{2+} as well as Ca^{2+}, K$^+$, and Na$^+$ have leached away, leaving behind the less mobile ions of Al^{3+} and Fe^{3+}.

The secondary minerals formed in weathering reactions play critical roles in soils and ecosystem processes. Insoluble products of chemical weathering are fine clay particles consisting of hydrated silicates of aluminum, iron, and magnesium arranged in layers (sheets). Two types of sheets make up these minerals: A tetrahedral sheet consists of units with one silicon atom surrounded by four O$^-$ groups (Fig. 3.9a). An octahedral sheet consists of units with six O$^-$ or OH$^-$ groups surrounding an Al^{3+}, Mg^{2+}, or Fe^{3+} ion (Fig. 3.9c). Various combinations of these

sheets give rise to a wide variety of clay minerals with different exchange properties. Montmorillonite or illite, for example, which have 2:1 ratios of silica- to aluminum-dominated layers, have a higher **cation exchange capacity** (CEC) than does kaolinite, which has a 1:1 ratio of silica- to aluminum-dominated layers (Fig. 3.10). Some exchange sites on soil minerals, particularly silicate clays with surface oxygen layers, have a permanently fixed charge. Other exchange sites, particularly iron and aluminum clays with surface hydroxyl layers, vary between positive and negative depending on pH.

In tropical climates, silica is preferentially leached from secondary clay minerals, producing red iron and aluminum oxide clays like gibbsite, which has only aluminum-dominated octahedral sheets. Highly weathered minerals dominated by octahedral sheets strongly bind anions like phosphate. In cold, wet climates, however, iron and aluminum are preferentially leached, leaving behind silica-dominated quartz sand. CEC tends to decline with weathering, whereas anion exchange capacity increases (Fig. 3.10), as discussed later. Most soils contain mixtures of several secondary minerals. The structure and concentration of clay minerals strongly influence the CEC, water-holding capacity, and other characteristics of soils.

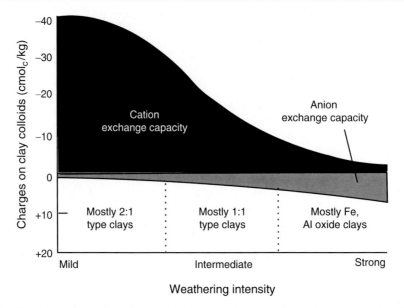

Fig. 3.10 The effect of weathering intensity on the charges on clay minerals and in turn on their cation and anion exchange capacities. Redrawn from Brady and Weil (2008)

Secondary minerals that form in soils can be either **crystalline**, with highly regular arrangements of atoms, as in the silicate clay minerals described earlier, or **amorphous**, with no regular arrangement of atoms. Allophane ($Al_2O_3 \cdot 2SiO_2 \cdot nH_2O$), for example, is an amorphous secondary mineral characteristic of volcanic ash deposits. With time, allophane transforms through loss of silica to crystalline aluminum oxide minerals like gibbsite ($Al(OH)_3$). Allophane has a high anion exchange capacity due to a surplus of positive charges. It also strongly binds phosphorus and can lead to phosphorus limitation in relatively young volcanic soils.

Soil Transfers

Vertical transfers of materials through soils generate distinctive soil profiles, i.e., the vertical layering of soils. These transfers typically occur by **leaching** (the downward movement of dissolved materials) and particulate transport in water. Soluble ions that are added in precipitation or released by weathering in upper layers of the soil profile can move downward in solution until a change in chemical environment causes them to become reactants in chemical processes, leading

to insoluble products, or until dehydration causes them to precipitate out of solution. The quantity of base cations in secondary minerals therefore often increases with depth within the upper meter of soil. These cations are leached from upper layers (termed **horizons**) and form new minerals under the new conditions of pH and ionic content encountered at depth. Chelated complexes of organic compounds and iron or aluminum ions are also water soluble and can move in water to deeper layers of the soil profile. Slight changes in ionic content or the microbial breakdown of the organic matter are among the processes that can cause the metal ions to precipitate as oxides. Clay particles like silicates and iron and aluminum oxides can also be transported downward in solution, sometimes forming deep horizons with high clay content in wet climates. Soil texture affects the rate and depth of leaching and thus the translocation and accumulation of materials in soil profiles. Constituents released during weathering of coarse-textured glacial till, for example, may be leached from the soil before they have a chance to chemically react to form secondary minerals.

Soils of arid and semi-arid environments also accumulate materials in specific horizons. These systems often have a hard calcium carbonate-rich

calcic horizon or caliche. Downward-moving soil water carries dissolved Ca^{2+} and bicarbonate (HCO_3^-). Precipitation as calcium carbonate occurs under conditions of increasing pH, which drives reaction (3.3) to the left. Precipitation can also occur under saturating concentrations of carbonate, or with evaporation of soil water.

$$CaCO_3 + H_2CO_3 \leftrightarrow Ca^{2+} + 2HCO_3^- \quad (3.3)$$

Although most of the transfers in soils occur through the downward movement of water, materials can also move upward in water. The capillary rise of water from a shallow water table, for example, transfers water and ions from lower to upper soil layers (see Chap. 4). Because capillary water movement depends on adhesive properties of soil particles, the potential distance for capillary rise is greater in clay soils with small pore sizes than in sandy soils (Birkeland 1999), as explained later. Soluble ions or compounds may accumulate in layers at the top of the capillary fringe. **Salt pans**, for example, form at the soil surface in low-lying areas of deserts, forming extensive **salt flats**, where the water evaporates rather than running off. Minerals that are added to soils in irrigation water in dry regions can also accumulate at the soil surface, as the water evaporates. This **salinization** has led to widespread abandonment of farmland in dry regions of the world, as in many parts of Australia.

Some minerals accumulate at the interface between waterlogged and aerobic soils. Poor drainage often leads to low oxygen availability because oxygen diffuses 10,000 times more slowly in water than in air and is easily depleted in waterlogged soils by root and microbial respiration. Low oxygen concentration creates reducing conditions that convert ions with multiple oxidation states to their reduced forms. Iron and manganese, for example, are more soluble in their reduced states (Fe^{2+} and Mn^{2+}, respectively) than in their oxidized states (Fe^{3+} and Mn^{4+}, respectively). Fe^{2+} and Mn^{2+} diffuse through waterlogged soils to the surface of the water table, where there is enough oxygen to convert them to their oxidized forms. Here they precipitate out of solution to form a distinct iron- and manganese-rich layer. This layering of iron and

manganese is particularly pronounced in lake sediments where there is a strong gradient in oxygen concentration from the sediment surface. The conversion from ferric (Fe^{3+}) to ferrous (Fe^{2+}) iron gives rise to the characteristic gray and bluish colors of waterlogged **gley soils**.

Soils that are subjected to repeated wetting and drying and saturation during some seasons can also develop characteristic accumulations of minerals. **Plinthite**, for example, is an iron- and aluminum-rich material in tropical soils that can harden irreversibly with repeated cycles of wetting and drying. Depending on their location within the profile, these layers can impede water drainage and root growth.

The actions of plant roots and soil animals transfer materials up and down the soil profile (Paton et al. 1995). Organic matter inputs to soil occur primarily at the surface and in upper soil horizons. When leaves or roots are shed or plants die, the minerals acquired by deep roots are also deposited on or near the soil surface. This contributes, for example, to the base-rich soils and unique ground flora beneath deep-rooted oak trees in southern Sweden (Andersson 1991) or dogwood trees in the eastern U.S. (Thomas 1969). Tree windthrow, which occurs when large trees are toppled by strong winds, also redistributes roots and associated soil upward. Finally, animals such as gophers transfer materials up and down in the soil profile as they tunnel and feed on plant roots. Earthworms in temperate soils and termites in tropical soils are particularly important in transferring surface organic matter deep into the soil profile and, at the same time, bringing mineral soil from depth to the surface. These processes play critical roles in the redistribution of nutrients and in the control of net primary productivity.

Losses from Soils

Materials are lost from soil profiles primarily as solutions and gases. The quantity of minerals leached from an ecosystem depends on both the amount of water flowing through the soil profile and its solute concentration. Many factors influence these concentrations, including plant demand, microbial mineralization rate, cation or anion exchange capacity, and previous losses via leaching or gas fluxes. As water moves through the soil, exchange reactions with mineral and organic surfaces replace loosely bound ions on the exchange complex with ions that bind more tightly, as explained later. In this way, **monovalent** (ions with a single charge) cations such as Na^+, NH_4^+, and K^+ and anions such as Cl^- and NO_3^- are easily released from the exchange complex into the soil solution and are particularly prone to leaching loss. The maintenance of charge balance of soil solutions requires that the leaching of negatively charged ions (**anions**) be accompanied by an equal charge of positive ions (**cations**). Inputs of H_2SO_4 in acid rain therefore increase leaching losses of readily exchangeable base cations like Na^+, NH_4^+, and K^+, which leach downward with SO_4^{2-}.

Materials can also be lost from soils as gases. Gas emissions depend on the rate of gas production by microbes, the diffusional paths through soils, and the exchange at the soil–air interface (Livingston and Hutchinson 1995). The controls over these losses are discussed in Chap. 9.

Soil Horizons and Soil Classification

Ecosystem differences in additions, transformations, transfers, accumulations, and loss give rise to distinct soils and soil profiles. Soils include organic, mineral, gaseous, and aqueous constituents arranged in a relatively predictable vertical structure. The number and depth of **horizons** (layers) and the characteristics of each layer in a soil profile vary widely among soils. Nonetheless, a series of horizons can be described that is typical of many soils (Fig. 3.11). The organic or **O horizon** of soil consists of organic material that accumulates above the mineral soil. This organic layer is derived from the **litter** of dead plants and animals and can be subdivided based on the degree of decomposition that most material has undergone, with the lower portion of the organic horizon being more decomposed. The **A horizon** is the uppermost mineral soil horizon. Being adjacent to the O horizon, it typi-

cally contains substantial organic matter and is therefore dark in color. The O and A horizons are the zones of most active plant and microbial processes and therefore have highest nutrient supply rates (see Chap. 9). Many soils in wet climates have an **E horizon** beneath the A horizon that is strongly leached. Most clay minerals and iron and aluminum oxides have been leached from the horizon, leaving behind resistant minerals like quartz, among other sand and silt-size particles. The **B horizon** beneath the A and E is the zone of maximum accumulation of iron and aluminum oxides and clays. Salts and precipitates sometimes also accumulate here, especially in arid and semi-arid environments. The **C horizon** lies beneath the A and B horizons. Although it may accumulate some of the leached material from above, it is relatively unaffected by soil-forming processes and typically includes a significant portion of unweathered parent material. Finally, at some depth, there is an unweathered layer of bedrock (R). Leaching and cation loss predominate in wet environments, producing acid soils. Salt inputs and accumulation predominate in dry environments, producing basic soils.

Despite the large variation among the world's soils, they can be classified into major groups that have formed in response to similar soil-forming factors and processes and therefore share many of the same properties. Soil classification systems rely on the diagnostic characteristics of specific horizons and on organic matter content, base saturation, and properties that indicate wetness or dryness. The soil taxonomy used in the U.S. recognizes 12 major soil groups, called **soil orders** (Table 3.3). Most agronomic and ecosystem studies classify soils to the level of a **soil series**, a group of soil profiles with similar profile characteristics such as type, thickness, and properties of the soil horizons. Soil series can be further subdivided into **types** based on the texture of the A horizon, and into **phases** based on information such as landscape position, stoniness, and salinity. A comparison of soil profiles from the major soil orders illustrates the impact of different climatic regimes on soil development (Figs. 3.12 and 3.13). More detailed descriptions of soil orders are presented by Brady and Weil (2008).

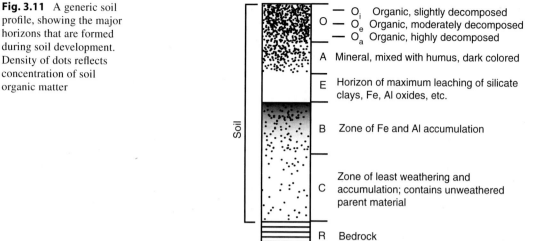

Fig. 3.11 A generic soil profile, showing the major horizons that are formed during soil development. Density of dots reflects concentration of soil organic matter

Table 3.3 Names of the soil orders in the U.S. soil taxonomy and their characteristics and typical locations

Soil order	Area (% of ice-free land)	Major characteristics	Typical occurrence
Rock and sand	14.1		
Entisols	16.3	No well-developed horizons	Sand deposits, plowed fields
Inceptisols	9.9	Weakly developed soils	Young or eroded soils
Histosols	1.2	Highly organic; low oxygen	Peatland, bog
Gelisol	8.6	Presence of permafrost	Tundra, boreal forest
Andisols	0.7	From volcanic ejecta; moderately developed horizons	Recent volcanic areas
Aridisols	12.1	Dry soils with little leaching	Arid areas
Mollisols	6.9	Deep, dark-colored A horizon with >50% base saturation	Grasslands, some deciduous forests
Vertisols	2.4	High content (>30%) of swelling clays; crack deeply when dry	Grassland with distinct wet and dry seasons
Alfisols	9.7	Enough precipitation to leach clays into a B horizon; >50% base saturation	Moist forests; shrublands
Spodosols	2.6	Sandy leached (E) horizon; acidic B horizon; surface organic accumulation	Cold, wet climates, usually beneath conifer forests
Ultisols	8.5	Clay-rich B horizon, low base saturation	Wet tropical/subtropical climate; forest or savanna
Oxisols	7.6	Highly leached horizon on old landforms	Hot, humid tropics beneath forests

Data from Miller and Donahue (1990) and Brady and Weil (2008)

Entisols are soils with minimal soil development. They occur either because the soils are recent, or processes that disrupt soil structure dominate over soil-forming processes. This is the most widespread soil type in the world, occupying 16% of the ice-free surface. **Inceptisols**, in which the soil profile has only begun to develop, occupy an additional 10% of the ice-free surface. Thus, including rock and shifting sand, about 40% of the ice-free surface of Earth shows minimal soil development (Table 3.3; Fig. 3.12).

Histosols are highly organic soils that develop in any climate zone under waterlogged conditions that restrict oxygen diffusion into the soil, leading to slow rates of decomposition and accumulation

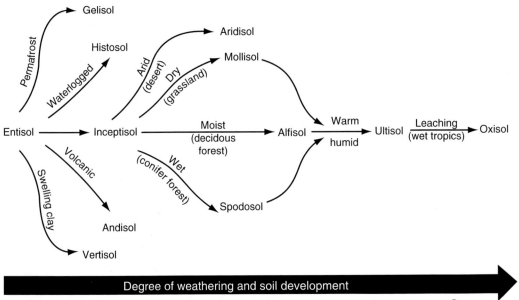

Fig. 3.12 Relationships among the major soil orders, showing the conditions under which they form, relative time required for formation, and the types of ecosystems with which they are most commonly associated. Based on Birkeland (1999) and Brady and Weil (2008)

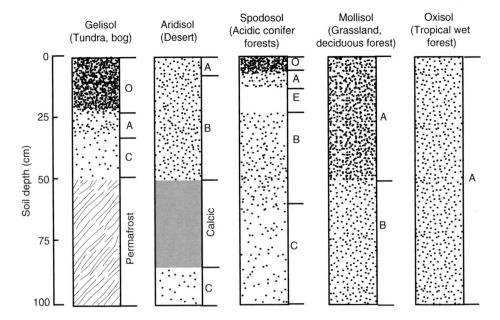

Fig. 3.13 Typical profiles of five contrasting soil orders, showing differences in the types and depths of horizons. Symbols as in Fig. 3.11

Fig. 3.14 Diagram showing the general soil moisture and temperature regimes that characterize the most extensive soils of seven soil orders. Soils of other soil orders (Andisols, Entisols, Inceptisols, and Histosols) can occur across this entire spectrum of environmental conditions. Vertisols (not shown) occur only where clay materials are abundant, under intermediate temperature, and moisture conditions. Data from Brady and Weil (2008)

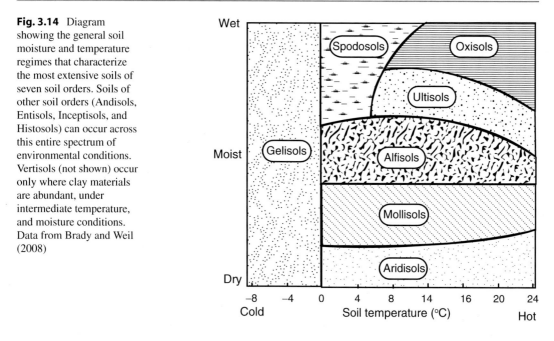

of organic matter. Histosols have a well-developed O horizon of undecomposed organic material where most plants are rooted. The high water table prevents the vertical leaching required for soil development, so these soils have weak development of mineral soil horizons. **Gelisols** are soils that develop in climates with an average annual temperature below 0°C that are underlain by a layer of permanently frozen soil (permafrost). They typically have a surface organic horizon or are frost churned (Figs. 3.13 and 3.14).

Andisols are young soils on volcanic substrates that tend to produce amorphous clays.

Aridisols, as the name implies, develop in arid climates. The low rainfall minimizes weathering and deep leaching, causing accumulation of soluble salts. There is no surface O horizon. The shallow A horizon has little organic matter due to low productivity and rapid decomposition. Low precipitation results in a poorly developed B horizon. Many of these soils form a calcic layer of calcium and magnesium carbonates that precipitate at depth because there is insufficient water to leach them out of the system. Desert calcic layers can greatly reduce root penetration, restricting the roots of many desert plants to surface soils. Aridisols are a widespread soil type, accounting

for 12% of the terrestrial surface (Miller and Donahue 1990).

Vertisols are characterized by swelling and shrinking clays. These clay-rich soils tend to occur in warm regions with a moist to dry climate, often on limestone or other base-rich parent materials. Vertisols often have no B horizon because the swelling and shrinking leads to a vertically well-mixed soil. Trees are often excluded from vertisols due to the frequent soil disturbance.

Mollisols are fertile soils that develop beneath grasslands and some deciduous forests. They have a deep, organic-rich A horizon with a high nutrient content that grades into a B horizon. Due to their high fertility, mollisols have been extensively cultivated and support the major grain-growing regions of the world. They account for 22% of U.S. soils and 7% of soils worldwide (Miller and Donahue 1990).

Spodosols (or podzols by European terminology) are highly leached soils that develop most commonly in cool, wet climates, usually beneath conifer stands. Beneath the A horizon is usually a highly leached, almost white, E horizon and a dark brown or black B horizon, where leaching products accumulate. These soils are often coarse textured and acidic. **Alfisols** usually develop beneath temperate and subtropical forests, especially deciduous forests that receive less precipitation. They are less strongly leached than spodosols and have a base-rich zone of clay accumulation in the B horizon.

Ultisols develop in warm, wet climates, where there is substantial leaching. The B horizon of these soils often has a high clay content and a low base saturation. **Oxisols** are the most highly weathered and leached group of soils. They occur on old landforms in the wet tropics. The A horizon is so highly weathered that it contains iron and aluminum oxides, largely as clay particles with very little silica and extremely low fertility. This horizon often extends several meters in depth.

Four generalizations emerge from this broad comparison of soil orders:

1. Nearly half (40%) of Earth's soils show minimal soil development and therefore largely reflect the properties of their parent material and current climate.
2. Wet environments tend to produce acidic leached soils, whereas dry environments produce basic ones in which cations accumulate.
3. Weathering and soil formation occur most rapidly in warm, wet climates, where plant productivity is greatest. Weathering is accentuated with time.
4. The quantity, quality, and turnover rate of soil organic matter are sensitive to climate and strongly influence soil fertility and other soil properties

Soil Properties and Ecosystem Functioning

Soil Physical Properties

Spatial and temporal variations in soil development generate large variations in soil properties. In the following paragraphs, we discuss how the properties of soil particles and the configuration of intervening spaces govern the availability of water and nutrient resources for plant growth and therefore their cycling through ecosystems.

Particle size distribution (**soil texture**) is important because it determines the surface area in a given soil volume. Soil texture is defined by the relative proportion of three sizes of particles: **clay** (<0.002 mm), **silt** (0.002–0.05 mm), and **sand** (0.05–2.0 mm; Fig. 3.15). **Loam** soils, which constitute the majority of soils, are mixtures of these three size classes and exhibit some properties of each size class. Rocks and gravel are larger (>2 mm) particles that also occupy a substantial proportion of the volume of many soils. Most gravel and sand particles are unweathered primary minerals, whereas clay particles are mostly secondary minerals. Silt particles are intermediate in composition (Fig. 3.16).

Soil texture depends on the balance between soil development that occurs in place, deposition by wind or water, and erosional loss of materials. As soils weather in place, the conversion of primary to secondary minerals (mostly small particles) increases the proportion of small soil particles. For this reason, high-latitude soils, with their slow rates of chemical weathering, often have low clay content, often about 10%, compared to temperate or tropical soils. Weathering rate and texture also depend on parent material, as discussed earlier. Small particles are particularly susceptible to erosion by wind or water. Water erosion transports clay from hilltops to valley bottoms, producing fine-textured soils in river valleys and leaving coarser-textured soils on the slopes. If river valleys are poorly vegetated, as in braided rivers that drain glaciated landscapes, wind can then move fine particles back to hillslopes to form **loess** soils with a high silt content. Over millions of years, minerals dissolve and are lost from the soil.

Clay particles have about 10,000 times greater surface area than the same weight of medium-sized sand particles (Brady and Weil 2008). Organic matter also has a high surface-volume ratio. Surface area, in turn, determines the amount of water that adsorbs to particle surfaces and therefore the capacity of soils to retain water. Surface charge and CEC also depend on particle surface area, as described later. Soil texture influences these and so many other important soil characteristics that it is a good general

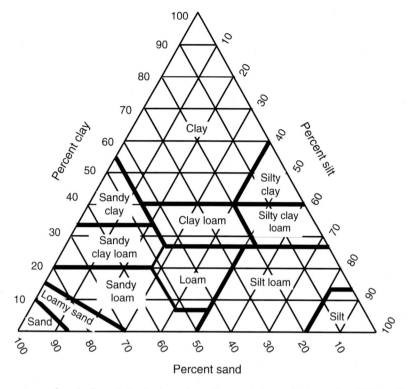

Fig. 3.15 Percentages of sand, silt, and clay in the major soil textural classes. Redrawn from Birkeland (1999)

Fig. 3.16 General relationship between particle size and kinds of minerals present. Redrawn from Brady and Weil (2008)

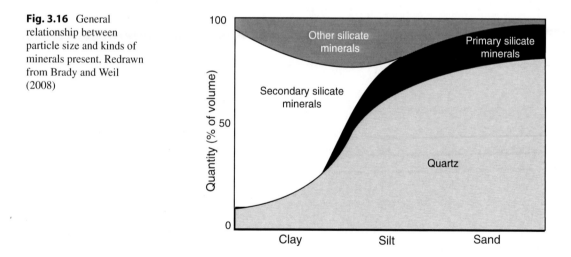

predictor of many ecosystem properties (Parton et al. 1987).

The physical properties of soils depend on the properties of both particles and the spaces between them. **Bulk density**, the mass of dry soil per unit volume, is an easily measured index of the relative proportion of particles and **voids** (spaces) in the soil. Bulk densities of mineral soil horizons (1.0–2.0 g cm^{-3}) are typically higher than those of organic horizons (0.05–0.4 g cm^{-3}).

Fig. 3.17 Volume distribution of organic matter, sand, silt, clay, micropores, and macropores in a representative silt loam. Redrawn from Brady and Weil (2008)

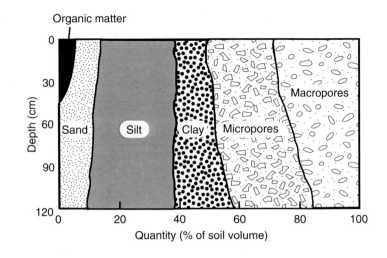

Fine-textured soils usually have aggregated groupings of particles, as described later, with intervening spaces and therefore generally have greater pore volume and lower bulk density than coarse-textured soils. If compacted, however, clay soils can have higher bulk density than coarse-textured soils.

Soil structure reflects the aggregation of soil particles into larger units. **Aggregates** form when soil particles become cemented together and then crack into larger units as soils dry or freeze. Soil aggregates are hierarchically structured with large aggregates (>3 mm diameter) consisting of progressively smaller aggregates and sub-aggregates down to clusters of a few clay and humus particles less than 0.001 mm in diameter. Aggregates form more easily in loam and clay soils than in sandy soils. Materials that glue soil particles together to form aggregates include organic matter, iron oxides, polyvalent cations, and silica. Iron and aluminum oxides are particularly important to aggregate formation in highly weathered tropical soils. For example, iron oxides can cement clay particles to produce very stable **pseudosand** aggregates that improve soil drainage in some clay-rich tropical oxisols and ultisols. In contrast, organic matter, such as polysaccharide secreted by roots and bacteria, is more important in aggregate formation in temperate soils. Fungal hyphae also contribute strongly to aggregation in many soils. For these reasons,

disturbances that reduce soil organic content and its associated microbes can lead to a loss of soil structure, which contributes to further soil degradation. Earthworms and other soil invertebrates contribute to aggregate formation by ingesting soil and producing feces that retain a coherent structure. Plant species and their microbial associates differ in the capacity of their exudates to form aggregates. Some mycorrhizal fungi, for example, produce a glycoprotein **glomalin** that is particularly effective in cementing microaggregates to form macroaggregates (Wilson et al. 2009). In summary, soil texture, mineral chemistry, organic matter content, and species composition all influence soil structure.

The pore structure of soils is critical to their functioning. Pores account for about half the soil volume (Fig. 3.17) and range in size from micropores that are too small for bacteria or root hairs to penetrate to macropores (>0.08 mm diameter). Micropores include both the original spaces between soil particles and the spaces that form as clays swell and shrink due to absorption and loss of water between clay platelets. This swelling and shrinking creates a wide range of pore sizes, from the smallest micropores to large macropores between soil aggregate. Macropores also form when roots and soil animals, especially earthworms, move through the soil, often along previously formed cracks or fractures. The resulting cracks and channels are important pathways for water infiltration, gas diffusion, and root growth, thus affecting water availability, soil aeration, oxidation–reduction processes, and plant growth. The fine-scale heterogeneity in pore structure is critical to the functioning of soils. Slow gas diffusion through the partially cemented pores within aggregates creates anaerobic conditions immediately adjacent to aerobic surfaces of soil pores. This allows anaerobic processes (e.g., denitrification) that require the products of aerobic processes (nitrification, in this case) to occur even in well-aerated soils (see Chap. 9). The surfaces of macropores are hot spots of biological activity, including the secretion of exudates by roots, rapid growth and turnover of bacterial biofilms, and predation by soil animals (see Chap. 7).

Human activities substantially alter the soil structure of many ecosystems. Compaction by animals and machinery compresses many of the larger cracks and pores between aggregates, reducing oxygen diffusion into the soil. Compaction also reduces infiltration of rainwater, increasing the likelihood of overland flow and erosion. Conversely, plowing mechanically disrupts aggregates, creates new macropores, and disrupts the macropores that were previously present. Depending on the initial condition of the soil, plowing can either improve or degrade soil structure. In native prairie, for example, plowing reduces total pore volume by disrupting aggregates and initial macropores, whereas plowing of compacted soils may increase macropore volume (Brady and Weil 2008). Regardless of the effect of plowing on macropore volume, the breaking up of previously cemented aggregates increases oxygen diffusion and decomposition of soil organic matter that was previously "protected" from decomposition by anaerobic conditions, leading to loss of soil carbon (Fisher and Binkley 2000; Baker et al. 2007).

Water is a critical resource for most ecosystem processes, and its availability depends critically on soil structure. In soils, water is held in the pore spaces as films of water adsorbed to soil particles. The soil is **water saturated** when all pore spaces are filled with water. When the larger pores fill with water, water begins to drain under the influence of gravity (**saturated flow**), even when some of the smaller pores within aggregates have not yet filled. Water drainage continues until, often after several days, the adhesive forces that hold water in films on soil particles equals the gravitational pressure. At this point, called **field capacity**, water no longer freely drains.

At water contents below field capacity, water moves through the soil by **unsaturated flow** in response to gradients of **water potential**, i.e., the potential energy of water relative to pure water (see Chap. 4). When plant roots absorb water from the soil to replace water that is lost in transpiration, this reduces the thickness of water films adjacent to roots, causing the remaining water to adhere more tightly to soil particles. The net effect is to reduce the soil water potential at the root surface. Water moves along water films through the soil pores toward the root in response to this gradient in water potential. As plants continue to transpire, water continues moving toward the root until some minimal water potential is reached, when roots can no longer extract water from the particle surfaces or the continuity of the water film is broken (see Chap. 4). This point is called the **permanent wilting point. Water-holding capacity** is the difference in water content between field capacity and permanent wilting point (see Fig. 4.8). Water-holding capacity is substantially enhanced by presence of clay and soil organic matter because of their large surface area. The water-holding capacity of an organic soil might, for example, be 300% (3 kg H_2O per kg dry soil), while that of a clay soil may be 30% and that of a sandy soil could be less than 20%. On a volumetric basis, water-holding capacity is normally highest in loam soils. One consequence of this difference is that, for a given amount of rainfall, coarse-textured soils will be wetted more deeply than soils without large pores (e.g., many clay soils) but will retain less water in surface soil horizons that are accessible to most plants. The water-holding characteristics of soils help determine the amount of water available for plant absorption and growth and for microbial processes, including decomposition and nutrient cycling and loss.

Soil Chemical Properties

In addition to its effects as a resource that supports plant growth, water strongly influences chemical and biological processes in soils through its effects on oxygen availability. Oxidation–reduction reactions involve the transfer of electrons from one reactant to another, yielding chemical energy that can be used by organisms (Schlesinger 1997). In these reactions, the energy source (often organic matter) gives up one or more electrons (**oxidation**). These electrons are transferred to electron acceptors (**reduction**). A handy mnemonic is: "LEO (loss of electrons = oxidation) the lion says GER (gain of electrons = reduction)." **Redox potential** is the tendency of an environment to receive or supply electrons (Schlesinger 1997; Fisher and Binkley 2000). More precisely, it is the net oxidation state of a pair of chemicals such as sulfate and sulfide or water and oxygen. Any soil or sediment has a mixture of chemicals that support complex patterns of electron transfer among chemicals. In addition, soils and sediments differ widely in redox potential (net oxidation state) due to their chemical composition and oxygen availability. Under the most aerobic conditions, which occur inside the mitochondria of live, eukaryotic cells, redox reactions transfer electrons from carbohydrates through a series of reactions to oxygen. This series of reactions releases the energy that supports cellular growth and maintenance. Other redox reactions occur in the cells of soil or benthic organisms, when electrons are transferred from electron donors to acceptors other than oxygen (Table 3.4). Organisms harvest the most energy by transferring electrons to oxygen, so this reaction predominates when oxygen is present. However, under anaerobic conditions, which commonly occur in flooded soils with high organic matter contents, in the interior of soil aggregates, or in lake or coastal ocean sediments, electrons must be transferred to other electron acceptors, with progressively less energy being released with transfer to each of the following electron acceptors (Table 3.4):

$$O_2 > NO_3^- > Mn^{4+} > Fe^{3+} > SO_4^{2-} > CO_2 > H^+$$

$$(3.4)$$

Table 3.4 Sequence of H^+-consuming redox reactions that occur with progressive declines in redox potential

Reaction[a]	Redox potential[b] (mV)	Energy release[b] (Kcal mol^{-1} per e$^-$)
Reduction of O_2 $O_2 + 4H^+ + 4e^- \rightarrow 2H_2O$	812	29.9
Reduction of NO_3^- $NO_3^- + 2H^+ + 2e^- \rightarrow NO_2^- + H_2O$	747	28.4
Reduction of Mn^{4+} to Mn^{2+} $MnO_2 + 4H^+ + 2e^- \rightarrow Mn^{2+} + 2H_2O$	526	23.3
Reduction of Fe^{3+} to Fe^{2+} $Fe(OH)_3 + 3H^+ + e^- \rightarrow Fe^{2+} + 3H_2O$	−47	10.1
Reduction of SO_4^{2-} to H_2S $SO_4^{2-} + 10H^+ + 8e^- \rightarrow H_2S + 4H_2O$	−221	5.9
Reduction of CO_2 to CH_4 $CO_2 + 8H^+ + 8e^- \rightarrow CH_4 + 2H_2O$	−244	5.6

Data from Schlesinger (1997)

[a]The reactions at the top of the table occur in soils with high redox potential and release more energy (and are therefore favored) when the electron acceptors are available. The reactions at the bottom of the table release less energy and therefore occur only if other electron acceptors are absent or have already been consumed by redox reactions. Abbreviations include electrons (e$^-$), nitrite ion (NO_2^-), manganese dioxide (MnO_2), ferric hydroxide ($Fe(OH)_3$), organic matter (CH_2O), universal gas constant (R), temperature (T), and equilibrium constant (K)

[b]Assumes that all reactants and products are at molar concentrations, which is seldom true, and complete coupling to the oxidation reaction:

$CH_2O + H_2O \rightarrow CO_2 + 4H^+ + 4e^-$ and that the energy released $= RT \ln(K)$

Organic matter is abundant enough in most soils and sediments to serve as the major electron donor, although reduced iron, sulfide, etc. are also important electron donors. In the absence of oxygen, reduced compounds such as Fe^{2+} become increasingly important electron donors. As soil oxygen becomes depleted and redox potential declines, the preferred electron acceptors are gradually consumed (Table 3.4). A similar gradient in redox potential and preferred electron acceptors occurs with depth in flooded soils or in lake sediments. As oxygen becomes depleted with depth or time, for example, the redox reaction that generates the most energy is initially denitrification (transfer of electrons to nitrate) followed by reduction of Mn^{4+} to Mn^{2+}. These reactions are typically carried out by facultatively anaerobic bacteria, i.e., bacteria that can metabolize and grow under either aerobic or anaerobic conditions. Facultative anaerobes use oxygen when present because of the greater energy return (Table 3.4), but switch to nitrate or Mn^{4+} as electron acceptors when oxygen is depleted by decomposer respiration. Nitrate is produced in soils or water through nitrification by obligate aerobic nitrifying bacteria (see Chap. 9). Denitrification is therefore most important in redox reactions in situations with substantial temporal or spatial variation in oxygen availability or external nitrate inputs. Denitrification is particularly important, for example, in the interior of soil aggregates, in seasonally flooded soils, and in the hypolimnion (bottom water) and sediments of seasonally stratified lakes.

Below the zone of Mn^{4+} reduction (or after Mn^{4+} has been depleted), most redox reactions are performed by obligately anaerobic bacteria or occur abiotically (Howarth 1984; Schlesinger 1997). Under these conditions, Fe^{3+} is reduced to Fe^{2+}, with organic matter as the most common electron donor in bacterial reduction. Alternatively, in salt marshes sulfate is reduced to sulfide by sulfate-reducing bacteria, and sulfide serves as the electron donor in abiotic reduction of Fe^{3+} (Howarth 1984). In either case, there is a visible transition from red (Fe^{3+}) to black or gray (Fe^{2+}) color of the soil or sediment. These reactions provide less than half as much energy per unit of organic matter decomposed as does denitrification or manganese reduction (Table 3.4), so iron- and sulfate-reducing bacteria can compete effectively only when nitrate and Mn^{4+} have been depleted from soil. Finally, as other electron acceptors are depleted, methanogenic bacteria reduce CO_2 to methane (CH_4), often in combination with continued sulfate reduction. In general, **hypoxic** (weakly oxygenated) environments often support high rates of denitrification because of the juxtaposition in space or time of aerobic and anaerobic microenvironments, whereas environments that are permanently anaerobic are more important in sulfate reduction (marine environments or salt marshes) or methanogenesis (low-sulfur environments). In coastal marine sediments, methanogenesis and sulfate reduction are linked, with methanogenesis producing methane that is consumed by sulfate-reducing bacteria; this results in net reduction of sulfate and very little net emission of methane (Howarth 1984). Many soils, deep lake waters, and some sediments experience substantial seasonal fluctuations in oxygen availability and therefore in the relative importance of each redox reaction. The biological bases of redox reactions are described in Chap. 7, and their role in element cycles is described in Chap. 9.

Soil organic matter content is a critical component of soils and sediments. It provides the energy and carbon base for heterotrophic soil organisms (see Chap. 7) and is an important reservoir of essential nutrients required for plant growth (see Chap. 8). In addition, it strongly affects rates of weathering and soil development, soil water-holding capacity, soil structure, and nutrient retention. Soil organic matter originates from dead plant, animal, and microbial tissues, but includes materials ranging from new, undecomposed plant tissues to charcoal to resynthesized humic substances that are thousands of years old, whose origins are chemically and physically unrecognizable (see Chap. 7). Because soil organic matter is critical to so many soil properties, loss of soil organic matter through inappropriate land management is a major cause of land degradation and loss of biological productivity.

The capacity of ecosystems to provide cations to support biological activity depends, over the long term, on parent material and rates of weathering and loss, as discussed earlier. Over days to decades, however, the cations that are loosely bound to the soil exchange complex (primarily clay particles and soil organic matter) are the primary source of supply. **Cation exchange capacity (CEC)** reflects the capacity of a soil to form loose electrostatic bonds between positively charged cations and the negatively charged sites on the surfaces of soil minerals and organic matter. Cation exchange occurs when a cation in solution displaces a cation on the exchange complex. Values for CEC vary more than 100-fold among clay minerals and tend to decline with weathering (Fig. 3.10). The negative charge on clay minerals originates from an excess of negative charges on their surfaces and exposed edges. Soil organic matter also has a very high CEC due to the presence of –OH and –COOH groups and contributes substantially to the total CEC of some soils. Organic matter, for example, accounts for most CEC in those tropical soils that consist primarily of iron and aluminum oxides and 1:1 silicate clay minerals, which have a relatively low CEC. High-latitude soils also derive a large proportion of their CEC from organic matter due to their high organic content and low clay content. The pool of exchangeable cations in the soil is many times larger than the pool of soluble cations and represents the major short-term reservoir of cations for plant and microbial absorption.

Base saturation is the percentage of the total exchangeable cation pool that is accounted for by **base cations** (the non-hydrogen, non-aluminum cations). The identity of the cations on the exchange sites depends on the concentrations of cations in the soil solution and on the strength with which different cations are held to the exchange complex. In general, cations occupy exchange sites and displace other ions in the sequence

$$H(Al^{3+}) > H^+ > Ca^{2+} > Mg^{2+} > K^+ \approx NH_4^+ > Na^+$$

$$(3.5)$$

so leached soils tend to lose Na^+ and NH_4^+ but retain Al^{3+} and H^+. This displacement series is a consequence of differences among ions in charge and hydrated radius. Ions with more positive

charges bind more tightly to the exchange complex than do ions with a single charge. Ions with a smaller hydrated radius have their charge concentrated in a smaller volume and also tend to bind tightly to the exchange complex.

Minerals and organic matter have both positively and negatively charged groups and therefore electrostatically bind both anions and cations, although CEC is generally much greater than **anion exchange capacity** (Fig. 3.10). In some soils, especially those in the tropics, iron and aluminum oxide minerals have a positive surface charge at their typical pHs. In these soils, there is enough anion exchange capacity to attract anions more strongly than cations (Uehara and Gillman 1981). As with cations, anion adsorption depends on the concentration of anions and their relative capacities to be held or to displace other anions. Anions generally occupy exchange sites and displace other ions in the sequence

$$PO_4^{3-} > SO_4^{3-} > Cl^- > NO_3^- \qquad (3.6)$$

so leached soils tend to lose NO_3^- and Cl^- but retain PO_4^{3-} and SO_4^{3-}.

In addition to weak electrostatic bonds associated with cation and anion exchange, minerals can strongly bind both cations (e.g., K^+) and anions (e.g., PO_4^{3-}). The ecologically most important of these strong chemical bonds causes **phosphorus fixation**, which is particularly pronounced in highly weathered tropical soils and in some volcanic soils, explaining why ecosystems with these soils often show strong phosphorus limitation of plant growth and decomposition (Uehara and Gillman 1981). Phosphorus fixation is also sensitive to pH, causing phosphorus availability in soils to decline substantially at both high pH (e.g., limestone soils) and low pH (e.g., highly weathered soils).

Ecosystems often maintain a relatively stable pH despite continuous inputs of H^+ from precipitation, decomposition, and more recently from anthropogenic acid rain. This **buffering capacity** results from a multitude of soil chemical reactions that produce or consume H^+. These include reactions of H^+ with aluminum compounds like gibbsite at low pH and with carbonates at high pH. Many of these reactions are a normal component of chemical weathering. Reactions of H^+ with organic matter occur over a wide pH range. Exchange of H^+ with cations on organic and mineral exchange complexes also contributes to buffering, particularly at intermediate pH. Soils differ in the relative importance and the capacity of these reactions to buffer pH, but soils with a high CEC and base saturation often have the greatest buffering capacity. Buffering capacity is important because it maintains soil pH within a relatively narrow range for long periods even with chronic exposure to acid rain. When the buffering capacity is exceeded, the soil pH begins to drop, which can solubilize $Al(OH)_x$, Al^{3+}, and other cations, with potentially toxic effects in both terrestrial and downstream aquatic ecosystems (Schulze 1989; Driscoll et al. 2001). Acidic temperate and tropical soils, for example, have a relatively low CEC and buffering capacity, and some of the reactions that consume H^+ release aluminum in solution, making these soils toxic to those plants and microbes that are not adapted to acidic conditions.

Summary

Five state factors control the formation and characteristics of soils. (1) Parent material is generated by the rock cycle, in which rocks are formed, uplifted, and weathered to produce the materials from which soil is derived. (2) Climate is the factor that most strongly determines the rates of soil-forming processes and therefore rates of soil development. (3) Topography modifies these rates at a local scale through its effects on microclimate and the balance between soil development and erosion. (4) Organisms also strongly influence soil development through their effects on the physical and chemical environment. (5) Time integrates the impact of all state factors in determining the long-term trajectory of soil development. In recent decades, human activities have modified the relative importance of these state factors and substantially altered Earth's soils.

The development of soil profiles represents the balance between profile development, soil mixing, erosion, and deposition. Profile development occurs through the input, transformation, vertical transfer, and loss of materials from soils. Inputs to soils come from both outside the ecosystem (e.g., dust or precipitation inputs) and inside the ecosystem (e.g., litter inputs). The organic matter inputs are decomposed to produce CO_2 and nutrients or are transformed into recalcitrant organic compounds. The carbonic acid derived from CO_2 and the organic acids produced during decomposition convert primary minerals into secondary clay minerals with greater surface area and CEC. Water moves these secondary minerals and the soluble weathering products down through the soil profile until new chemical conditions cause them to become reactants or precipitate out of solution. Leaching of materials into groundwater or erosion and gaseous losses to the atmosphere are the major avenues of loss of materials from soils. The net effect of these processes is to form soil horizons that vary with climate, parent material, biota, and soil age and have distinctive physical, chemical, and biological properties.

Review Questions

1. What processes are responsible for the cycling of rock material in Earth's crust?
2. At large geographic scales, which state factors control soil formation? How might interactive controls modify the effects of these state factors?
3. What processes determine erosion rate? Which of these processes are most strongly influenced by human activities?
4. What processes cause soil profiles to develop? Explain how differences in climate, drainage, and biota might affect profile development.
5. What are the processes involved in physical and chemical weathering? Give examples of each. How do plants and plant products contribute to each?
6. How does soil texture affect other soil properties? Why does it influence ecosystem processes so strongly?
7. What is cation exchange capacity (CEC), and what determines its magnitude in temperate soils? How would you expect the determinants of CEC to differ between histosols, alfisols, and oxisols?

8. In a warm climate, how do soil processes and properties differ between sites with extremely high and extremely low precipitation? In a moist climate, how do soil processes and properties differ between sites with extremely high and extremely low soil temperature?

9. If global warming caused only an increase in temperature, how would you expect this to affect soil properties after 100 years? After a million years?

Additional Reading

Amundson, R., and H. Jenny. 1997. On a state factor model of ecosystems. *BioScience* 47:536–543.

Birkeland, P.W. 1999. *Soils and Geomorphology.* 3rd Edition. Oxford University Press, New York.

Brady, N.C. and R.R. Weil. 2008. *The Nature and Properties of Soils.* 14th edition. Pearson Education, Inc., Upper Saddle River, NJ.

Chadwick, O.A., L.A. Derry, P.M. Vitousek, B.J. Huebert, and L.O. Hedin. 1999. Changing sources of nutrients during 4 million years of soil and ecosystem development. *Nature* 397:491–497.

Jenny, H. 1941. *Factors of Soil Formation.* McGraw-Hill, New York.

Selby, M.J. 1993. *Hillslope Materials and Processes.* 2nd edition. Oxford University Press, Oxford.

Ugolini, F.C. and H. Spaltenstein. 1992. Pedosphere. Pages 123–153 in S.S. Butcher, R.J. Charlson, G.H. Orians, and G.V. Wolfe, editors. *Global Biogeochemical Cycles.* Academic Press, London.

Vitousek, P.M. 2004. *Nutrient Cycling and Limitation: Hawai'i as a Model System.* Princeton University Press, Princeton.

Part II

Mechanisms

The hydrologic cycle, driven by solar energy, is the master cycle that drives all other biogeochemical cycles. This chapter describes ecosystem energy budgets and other controls over the hydrologic cycle.

Introduction

Water and solar energy are essential for life. Their uneven distribution across Earth's surface largely account for the large-scale patterns of ecosystem structure and functioning and are therefore central to an understanding of ecosystem dynamics. Water and energy cycles are so tightly intertwined that they cannot be treated separately. Solar energy drives the hydrologic cycle through the vertical transfer of water from Earth to the atmosphere via **evapotranspiration**, the sum of surface evaporation and the loss of water from plant leaves (**transpiration**). Conversely, evapotranspiration accounts for 80% of the turbulent energy transfer (i.e., latent plus sensible heat flux) from Earth to the atmosphere and is therefore a key process in Earth's energy budget (see Fig. 2.3). The hydrologic cycle also controls Earth's biogeochemical cycles by dissolving nutrients and transferring them within and among ecosystems. Water and nutrients, in turn, provide the soil resources that support the growth of organisms. The movement of materials that are dissolved and suspended in water links ecosystems within a landscape.

A Focal Issue

Human activities have substantially altered Earth's hydrologic cycle at regional to global scales. People now use about 50% of Earth's available renewable fresh water, but this proportion exceeds 100% in some dry regions (Oki and Kanae 2006; Carpenter and Biggs 2009). This human use of fresh water affects land and water management, the movement of pollutants among ecosystems, and, indirectly, ecosystem processes in unmanaged ecosystems. Land-use changes have altered terrestrial water and energy budgets enough to change regional and global climate (Fig. 4.1; Chase et al. 2000; Foley et al. 2005). In Australia, for example, a decade-long drought at the end of the twentieth century reduced water availability below levels required for agriculture, and in dry portions of the midwestern U.S., irrigated farming is drawing on "fossil groundwater" that is depleted much more rapidly than it can be replenished by rainfall in the current climate. How much precipitation is needed to meet the water needs of different crops or other ecosystem types, and how is this influenced by plant and soil properties? What determines the proportion of incoming precipitation that enters water supplies and is potentially available to support societal needs? Evaporation of water is also one of the primary fates of energy from incoming solar radiation, which affects both air and water circulation (Fig. 4.1). What happens to the energy

F.S. Chapin, III et al., *Principles of Terrestrial Ecosystem Ecology*,
DOI 10.1007/978-1-4419-9504-9_4, © Springer Science+Business Media, LLC 2011

Fig. 4.1 Land-use change in southwestern Australia from a dark native heath vegetation to a wheatland that reflects more incoming radiation causes greater surface heating over the heath. This causes air to rise, drawing moisture-laden air from the wheatland and forming clouds that increase precipitation over the heath. The 30% reduction in precipitation over the wheatland reduces the viability of agriculture in this dry region (see Chap. 13)

that is absorbed by an ecosystem, if there is insufficient water to cool the canopy by evaporation?

Finally, human activities alter the capacity of the atmosphere to hold water vapor. Water vapor is *the* major greenhouse gas. It is transparent to shortwave radiation from the sun but absorbs longwave radiation from Earth (see Fig. 2.2) and thus provides an insulative thermal blanket. Climate warming caused by emissions CO_2 and other greenhouse gases increases the quantity of water vapor in the atmosphere and therefore the efficiency with which the atmosphere traps longwave radiation. This **water vapor feedback** explains why climate responds so sensitively to emissions of other greenhouse gases (see Chap. 2). Warming accelerates the hydrologic cycle, increasing evaporation and rainfall at the global scale (see Chap. 14). Warming also causes sea level to rise, mainly (so far) due to the thermal expansion of the ocean and secondarily to melting of glaciers and ice caps. Rising sea level endangers the coastal zone, where most of the world's major cities are located. How much of the coastal zone or city near you is likely to be flooded with projected levels of sea level rise (e.g., http://flood.firetree.net/)? Given the key role of water and energy in ecosystem and global processes, it is critical that we understand the controls over water and energy exchange and the extent to which they have been modified by human actions.

Surface Energy Balance

Radiation Budget

The radiative energy absorbed by a surface is the balance between incoming and outgoing radiation. Here we focus on ecosystem-scale radiation budgets, although the same general principles apply at any scale, ranging from the surface of a leaf to the surface of the globe (see Fig. 2.3). The two major components of the radiation budget are shortwave radiation (K), the high-energy radiation emitted by the sun, and longwave

Fig. 4.2 Radiation budget of a Douglas fir forest during the summer. Redrawn from Oke (1987)

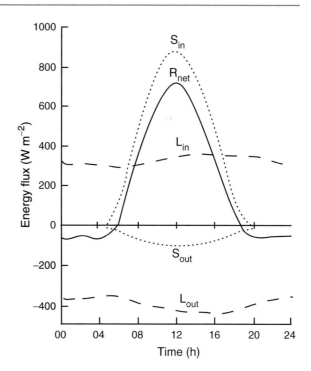

radiation (*L*), the thermal energy emitted by all bodies (see Chap. 2). **Net radiation** (R_{net}) is the balance between the inputs and outputs of short-wave and longwave radiation, measured in W m^{-2} (Fig. 4.2).

$$R_{net} = (K_{in} - K_{out}) + (L_{in} - L_{out}) \quad (4.1)$$

At noon on a clear day, **direct radiation** from the sun accounts for 90% of the shortwave input to an ecosystem (see Fig. 2.3). Additional input of shortwave radiation comes as **diffuse radiation** that is scattered by particles and gases in the atmosphere and **reflected radiation** from clouds and surrounding landscape units such as lakes, dunes, or snowfields. Diffuse radiation becomes a larger proportion of incoming shortwave radiation on cloudy or polluted days or near dawn or dusk when sun angles are lower.

The proportion of the incoming shortwave radiation that is absorbed depends on the **albedo** (*a*) or shortwave reflectance of the ecosystem surface. Albedo varies at least 10-fold among ecosystems, ranging from highly reflective

surfaces such as fresh snow to surfaces with a low reflectance such as wet soils or the water in lakes and the ocean (Table 4.1). Conifer canopies, for example, have a lower albedo (i.e., absorb a larger proportion of incoming radiation) than deciduous forests, and grasslands with standing dead leaves have relatively high albedo. Albedo depends not only on the reflectance of individual leaves, stems, and soil but also on ecosystem structure. A complex canopy has a lower albedo (less reflectance) than do individual leaves because much of the light reflected or transmitted by one leaf is absorbed by other leaves and stems, resulting in efficient light capture by the canopy as a whole. For this reason, deep, uneven canopies of conifer forests have a low albedo. In contrast, a relatively smooth canopy, such as a crop or grassland, reflects more of the incoming shortwave radiation from upper leaves directly back to space (Baldocchi et al. 2004).

Changes in ecosystem albedo explain in part why high-latitude regions are warming more rapidly than low latitudes. As climate warms, snow, lake ice, and sea ice melt earlier in the spring,

Table 4.1 Typical values of albedo for the major surface types on Earth

Surface type	Albedo
Ocean and lakes	0.03–0.10[a]
Bare soil	
Wet, dark	0.05
Dry, dark	0.13
Dry, light	0.40
Evergreen conifer	0.08–0.11
Deciduous conifer	0.13–0.15
Evergreen broadleaf	0.11–0.13
Deciduous broadleaf	0.14–0.15
Arctic tundra	0.15–0.20
Grassland	0.18–0.21
Savanna	0.18–0.21
Agricultural crops	0.18–0.19
Desert	0.20–0.45
Sea ice	0.30–0.45
Snow	
Old	0.40–0.70
Fresh	0.75–0.95

Data from Oke (1987), Sturman and Tapper (1996), Eugster et al. (2000), Hollinger et al. (2010)
[a]Albedo of water increases greatly (0.1–1.0) at solar angles less than 30°

replacing a reflective snow-covered surface with a dark absorptive surface (Euskirchen et al. 2007). This process, together with the resulting change in surface temperature, is referred to as the **snow (or ice) albedo feedback**. Over longer time scales, the northward movement of trees into tundra causes an additional reduction in regional albedo because the dark forest canopy masks the underlying snow-covered surface. As tree line moves north, the land surface absorbs more energy, which is then transferred to the atmosphere, causing an amplifying (positive)

feedback to regional warming (Foley et al. 1994; Chapin et al. 2005). Albedo also changes in response to short-term changes in solar input. Canopies absorb a larger proportion of incoming radiation (lower albedo) at midday than at dawn or dusk and during cloudy (more diffuse radiation) than during clear conditions (Hollinger et al. 2010).

Across all vegetation types, albedo increases with increasing leaf nitrogen up to about 2.5% nitrogen and is relatively insensitive to further increases in leaf nitrogen (Hollinger et al. 2010). The increase in albedo with increasing nitrogen may result from the large surface area for gas exchange between photosynthetic cells and internal air spaces in high-nitrogen leaves (see Chap. 5; Hollinger et al. 2010). Each time radiation passes between water and air, a large proportion of the near infrared portion of the spectrum is reflected, so high-nitrogen leaves reflect more shortwave radiation than do low-nitrogen leaves. Regardless of the mechanism, this relationship suggests that, despite their low photosynthetic rates (see Chap. 5), low-nitrogen canopies absorb a larger proportion of total incoming radiation than do canopies on high-fertility sites.

The amount of longwave (thermal) radiation emitted by an object depends on its temperature and **emissivity**, a coefficient that describes the capacity of a body to emit radiation. Most absorbed radiation is emitted (emissivity of about 0.98 in vegetated ecosystems), so longwave radiation balance depends primarily on the temperature of the sky, which determines L_{in}, and the temperature of the ecosystem surface, which determines L_{out} (4.2).

$$R_{net} = (K_{in} - K_{out}) + (L_{in} - L_{out}) = (1 - \alpha) K_{in} + \sigma (\varepsilon_{sky} T_{sky}^4 - \varepsilon_{surf} T_{surf}^4) \qquad (4.2)$$

where α is the surface albedo, σ is the Stefan–Boltzman constant (5.67×10^{-8} W m^{-2} K^{-4}), T is absolute temperature (°K), and ε is emissivity. Clouds and water vapor are warmer than the upper atmosphere and trap longwave emissions from the surface, so ecosystems receive more longwave radiation under cloudy than clear skies and under humid conditions. This explains why

cloudy nights are warmer than clear ones and why cloudless dry conditions make deserts cold at night, despite the high inputs of solar energy during the day.

Longwave radiation emitted by the ecosystem (L_{out}) depends on surface temperature, which, in turn, depends on the quantity of radiation received by the surface and the efficiency

Fig. 4.3 Average daily sensible and latent heat fluxes during summer for different ecosystems measured by eddy covariance. *Dashed lines* show different Bowen ratios (β, ratio of sensible to latent heat flux). Ecosystems include deciduous forests, conifer forests, agricultural crops, tundra and grasslands. Also shown are a variety of ecosystems from a Mediterranean climate, which have high fluxes due to their high input of net radiation. Redrawn from Wilson et al. (2002)

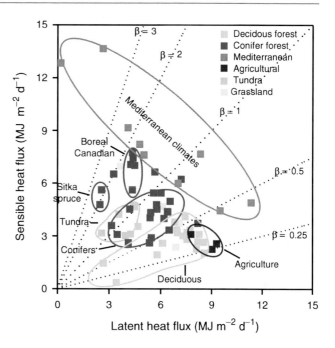

with which this energy is transferred to the air and soil by turbulent processes, as described in the next section. Surfaces that absorb a large amount of radiation, due to high solar inputs or low albedo, tend to be warmer and therefore emit more longwave radiation. Dry surfaces and leaves with low transpiration rates tend to be particularly warm because they are not cooled by the evaporation of water. Desert sands, recent burn scars, and city pavement, for example, are generally hot (Bonan 2008). Conversely, a well-watered lawn is much cooler than an ecosystem that is dry or dominated by plants with low transpiration rates. In general, shortwave radiation input, albedo, surface roughness (see next section), and surface temperature are the parameters that most strongly influence radiation balance and therefore net radiation (4.2).

Partitioning of Absorbed Radiation

Net radiation, the radiative energy absorbed by an ecosystem, is approximately balanced by energy that is transferred out of the ecosystem by non-radiative processes. These non-radiative processes include **ground heat flux** (*G*), which conducts energy into the soil, and turbulent transfer of energy from the surface to the air as evapotranspiration of water (**latent heat flux**, *LE*) or heat (**sensible heat flux**, *H*). A small amount of energy (generally less than 10% of daily net radiation) may be stored (Δ*S*) in the ecosystem as chemical energy through photosynthesis and as a temperature increase of the plant biomass. At other times, stored energy is released by respiration and declines in biomass temperature. Although the energy trapped by photosynthesis is *the* major energetic engine that drives the carbon cycle of ecosystems, it is only a tiny part (<5%) of the total energy budget of ecosystems. Because ecosystem energy storage is usually small, energy absorbed by the surface as net radiation approximately equals energy loss by non-radiative processes over a day (4.3).

$$R_{net} = H + LE + G + \Delta S \qquad (4.3)$$

where *L* is the **latent heat of vaporization** (2.45 MJ kg^{-1} at 20°C), and *E* is the rate of evapotranspiration. As presented in this equation, R_{net} is positive when directed toward the surface; H, LE, G, and ΔS are positive when directed away from the surface. Ecosystems with high net radiation, such as those in Mediterranean climates, have higher sensible and/or latent fluxes than do ecosystems with low net radiation, such as arctic tundra or temperate rainforests (Fig. 4.3; Wilson et al. 2002).

Ground heat flux (G) is negligible over a day in most ecosystems because the heat conducted down into the soil during the day is balanced by heat conducted back up to the surface at night. The magnitude of ground heat flux depends on the thermal gradient between the soil surface and deep soils and the thermal conductivity of soils, which is greatest in soils that are wet and have a high bulk density (fewer insulating air pockets). The steepest thermal gradients and greatest ground heat flux occur in regions with permafrost. In the arctic, for example, approximately 10–20% of the energy absorbed during summer is consumed by thawing of frozen soil. This energy is released back to the atmosphere the next winter, when the soil refreezes (Chapin et al. 2000a).

Lakes and the ocean also have substantial summer "ground heat flux" because solar inputs penetrate beneath the surface, and the high heat capacity and turbulent mixing of water efficiently move heat away from the surface. In clear lakes, about half of the incoming shortwave radiation is absorbed and converted to heat in the top 10 cm, with the remaining heat conversion occurring at greater depths (Kalff 2002). Less transparent lakes convert shortwave radiation to heat closer to the surface, causing these lakes to stratify earlier in the spring and to have a colder hypolimnion than in more transparent lakes. The longer period of stratification increases the likelihood of anoxia in the hypolimnion, thereby altering all aspects of the biotic environment (Kalff 2002).

In contrast to soil heat flux, heat transfer to the atmosphere occurs primarily by **turbulence**, the irregular velocities of air movement between the surface and the **bulk air** (i.e., air above the canopy that is not strongly influenced by the canopy).

Two processes generate this turbulence. **Convective turbulence** results from conduction (diffusion) of sensible heat over 1–2 mm from the surface to the near-surface air. The warm air expands, and the resulting increase in buoyancy causes this low-density warm air to rise, creating convective turbulence. A second more efficient process of energy transfer involves **mechanical turbulence**, when horizontally moving air slows down unevenly as it moves across an irregular surface. Tall uneven canopies such as conifer forests are aerodynamically **rough** compared to short smooth crop canopies. The mechanical turbulence generated by airflow across uneven topographic and vegetation surfaces creates eddies of air that sweep down into the canopy, transporting bulk air inward and canopy air out. These eddies transfer energy away from the surface and mix it with the atmosphere (Jarvis and McNaughton 1986; Bonan 2008). Conversely, air flowing across short, smooth canopies such as grasslands or crops tends to be less turbulent, so these canopies are less efficient in shedding the energy that they absorb, i.e., they are less tightly **coupled** to the bulk atmosphere. Because smooth canopies are less efficient in shedding heat, they tend to have higher surface temperatures during the day and greater longwave emissions than do forest canopies.

Turbulence transfers not only sensible heat but also the latent heat contained in water vapor that is transpired by plants or evaporates from leaf or soil surfaces. This energy is released when water vapor condenses to form cloud droplets. Dewfall represents a small latent heat flux from the atmosphere to the ecosystem at night under conditions of high relative humidity and cold leaf or soil surfaces.

Latent and sensible heat fluxes from ecosystems interact in ways that depend on surface moisture. The consumption of heat by evaporation cools the surface, thereby reducing the temperature differential between the surface and the air that drives sensible heat flux. Conversely, the warming of surface air by sensible heat flux increases the quantity of water vapor that the air can hold and causes convective movement of

Table 4.2 Representative Bowen ratios (ratio of sensible to latent heat flux) of different ecosystem types

Surface type	Bowen ratio
Desert	>10
Semi-arid landscape	2–6
Arctic tundra	0.3–2.0
Temperate forest and grassland	0.4–0.8
Boreal forest	0.5–1.5
Forest, wet canopy	−0.7–0.4
Water-stressed crops	1.0–1.6
Irrigated crops	−0.5–0.5
Tropical wet forest	0.1–0.3
Tropical ocean	<0.1

Data from Jarvis (1976), Oke (1987), Eugster et al. (2000)

moist air away from the evaporating surfaces. Both of these processes increase the vapor pressure gradient that drives evaporation. Because of these interdependencies, surface moisture has a strong impact on the **Bowen ratio**, i.e., the ratio of sensible to latent heat flux.

Bowen ratios vary by more than two orders of magnitude among ecosystems, indicating that either latent heat flux or sensible heat flux can dominate the turbulent energy transfer from ecosystems to the atmosphere (Table 4.2, Fig. 4.3; Wilson et al. 2002). In general, energy flux from wet ecosystems (e.g., open water and ecosystems whose canopy is often wet) is dominated by evapotranspiration (Bowen ratio < 0.5), whereas energy flux from other ecosystems, especially dry ones, is dominated by sensible heat flux (Bowen ratio > 0.5). Species characteristics also influence Bowen ratio, with greater evapotranspiration (lower Bowen ratio) from ecosystems dominated by rapidly growing plants with high rates of photosynthesis and transpiration (Table 4.2; see Chap. 5). Deciduous forests, for example, have higher transpiration rates and lower Bowen ratios than do conifer forests (Fig. 4.3). Strong winds or rough canopies, which generate atmospheric turbulence, reduce surface temperature, thereby reducing sensible heat flux and Bowen ratio. For these reasons, energy partitioning varies substantially both seasonally and among ecosystems. The Bowen ratio determines the strength of the linkage between the energy and water budgets of ecosystems, with wet ecosystems (low Bowen ratios) having a larger proportion of turbulent energy exchange occurring as evapotranspiration and therefore a tighter linkage between water and energy budgets (Box 4.1).

The spatial configuration of ecosystems on a landscape influences energy partitioning because heating contrasts between adjacent ecosystems create convective turbulence. This turbulence, and therefore sensible and latent heat fluxes, is greater at boundaries than in the centers of ecosystems (see Chap. 13). Most evaporation from large lakes, for example, occurs near their edges, rather than in the center, where the overlying air is so stable that it saturates rapidly and supports a relatively low evaporation rate. For the same reason, a mosaic of crops and fallow fields would support greater evapotranspiration than large homogeneous areas that contained the same proportions of crop and fallow. When ecosystem patches that differ strongly in albedo or energy partitioning are larger in diameter than the depth of the planetary boundary layer (> ≈10 km), they can modify mesoscale atmospheric circulations and cloud and precipitation patterns (Fig. 4.1; see Chap. 13; Pielke and Avisar 1990; Weaver and Avissar 2001).

Snow-covered surfaces experience threshold changes in energy exchange at the time of snowmelt. The high albedo of snow-covered surfaces minimizes energy absorption until snowmelt occurs, at which time there is a dramatic increase in the energy absorbed by the surface and transferred to the atmosphere. This often results in abrupt increases in regional air temperature after snowmelt. Leaf-out also alters energy exchange by both changing albedo and increasing evapotranspiration at the expense of sensible heat flux. Because of the dramatic difference in energy budget between snow-covered and snow-free seasons, recent advances in the date that snow melts on land or ice melts on lakes or the ocean create a strong amplifying (positive) feedback to high-latitude warming (Euskirchen et al. 2007).

Box 4.1 The Energetics of Water Movement

Water and energy participate in two of the most dynamic cycles on the planet in terms of both quantities moved and their rapidity of turnover. The energetics of water movement are critical to understanding both the linkage between these cycles and their underlying controls. Evapotranspiration is one of the largest terms in both the water and energy budgets of ecosystems, so factors governing the magnitude of evapotranspiration determine the tightness of the linkage between the water and energy cycles.

Due to its high **specific heat** – the energy required to warm 1 g of a substance by 1°C – water changes temperature relatively slowly for a given energy input. It takes four times more energy to raise the temperature of water by 1°C than an equivalent mass of air. Consequently, the summer temperature near large water bodies fluctuates less and is generally cooler than in inland areas. A wet surface also heats more slowly but evaporates more water than a dry surface.

Massive amounts of energy are absorbed or released when water changes state. It takes 580 times more energy (2.45 MJ kg^{-1}) to vaporize 1 g of water at 20°C than to increase its temperature by 1°C. Evapotranspiration therefore has a powerful cooling effect on transpiring leaves or other evaporating surfaces. Conversely, condensation of water vapor to form clouds has a powerful warming effect on the atmosphere, providing the added buoyancy that forms tall thunderheads (see Chap. 2).

Vapor pressure is the partial pressure exerted by water molecules in the air. The air immediately adjacent to an evaporating surface is approximately saturated at the temperature of the surface, for example the cell walls of a photosynthetic cell inside a leaf. The **vapor pressure deficit** (VPD) is the difference between the actual vapor pressure of air and the vapor pressure of saturated air at the same temperature. VPD is the driving force for evapotranspiration and indeed for the movement of water from soil through plants to the atmosphere. It is loosely used to describe the difference in vapor pressure between the air immediately adjacent to an evaporating surface and the bulk atmosphere, although, strictly speaking, the air masses are at different temperatures. **Conductance** of water vapor (the inverse of resistance) is the flux of water vapor per unit driving force (VPD).

Overview of Ecosystem Water Budgets

The water available to support the productivity of ecosystems depends on the balance between inputs and outputs. Water is the resource that most strongly constrains the productivity of the biosphere and therefore plays a central role in the dynamics of ecosystems. In addition, water increasingly constrains the opportunities for sustainable development of human societies in many parts of the world (Rockström et al. 1999; Vörösmarty et al. 2005; Carpenter and Biggs 2009). It is therefore important to understand water budgets to wisely manage the movement of water into, through, and out of ecosystems to meet the needs of both nature and society.

An ecosystem behaves like a bucket that is filled by precipitation and emptied by evapotranspiration and runoff. Lakes, for example, are filled by precipitation and by inflow from streams and adjacent ecosystems; water leaves by surface evaporation. When water inputs from precipitation and inflow exceed evaporation, the excess "overflows the bucket" and leaves as outflow. In many temperate lakes, evaporation is similar to precipitation, so the outflow from lakes is similar to the inflow. In warm, dry climates,

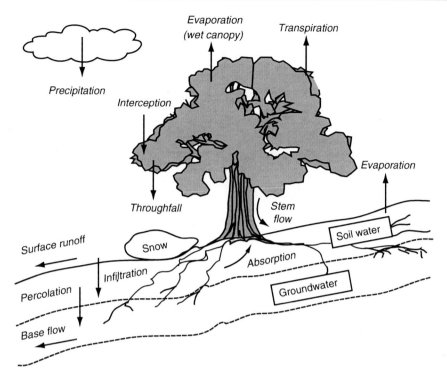

Fig. 4.4 Major water fluxes in an ecosystem

where evaporation exceeds inputs, outflow is less than inflow to lakes. Extremely dry climates, such as deserts, have such high evaporation rates that outflow seldom occurs, creating **closed-basin lakes**.

Terrestrial ecosystems also behave like a bucket in which water accumulates in the ecosystem until the water-holding capacity of soils is exceeded (Fig. 4.4). At this point, the excess water drains to groundwater or runs over the ground surface. The water losses from the ecosystem move laterally to other ecosystems such as streams and lakes. **Blue water** is the liquid water in rivers, lakes, reservoirs, and groundwater aquifers that is potentially available to society. Evaporation from the soil surface and transpiration by plants (**green water**) are the other major avenues of water loss from the soil reservoir. These processes continue only as long as the soil contains water that plants can tap, just as evaporation from a bucket continues only as long as the bucket contains water. Green water fluxes of the terrestrial biosphere exceed the blue water fluxes (Fig. 4.5; see Fig. 14.3).

Water Inputs to Ecosystems

Precipitation is the major water input to most terrestrial ecosystems. Global and regional controls over precipitation therefore determine the quantity and seasonality of water inputs to most ecosystems (see Chap. 2). In ecosystems that receive some precipitation as snow, however, the water contained in the snowpack does not enter the soil until snowmelt, often months after the precipitation occurs. This causes the seasonality of water input to soils to differ from that of precipitation.

Vegetation in some ecosystems, particularly in riparian zones, accesses additional groundwater that flows laterally through the ecosystem. Desert communities of **phreatophytes** (deep-rooted plants that tap groundwater), for example, may absorb enough groundwater that the ecosystem loses more water in transpiration than it receives in precipitation. Lakes and streams also receive most of their water inputs from groundwater or

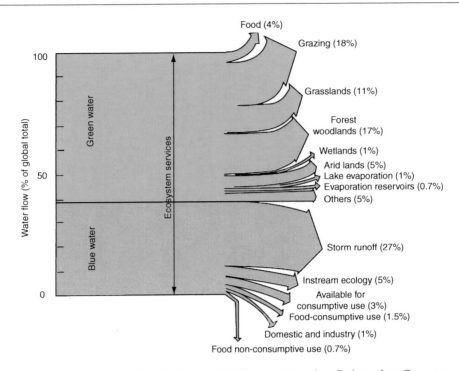

Fig. 4.5 Blue water and green water flows that support global ecosystem services. Redrawn from Carpenter and Biggs (2009) based on data from Rockström et al. (1999)

runoff that drains from adjacent terrestrial ecosystems. Water inputs to freshwater ecosystems are therefore linked only indirectly to precipitation.

In ecosystems with frequent fog, **canopy interception** of fog increases the water inputs to ecosystems, when cloud droplets that might not otherwise precipitate are deposited on leaf surfaces and drip from the canopy to the soil. The coastal redwood trees of California, for example, depend on fog-derived water inputs during summer, when precipitation is low, but fog occurs frequently (Ewing et al. 2009). Similarly, in areas that are climatically marginal for Australian rainforests, the capture of fog and mist by trees can augment rainfall by 40% (Hutley et al. 1997), just as in New Zealand high-elevation tussock grasslands (Mark and Dickinson 2008). In the absence of fog, however, canopy interception generally *reduces* the proportion of precipitation that enters the ecosystem because of canopy evaporation, as described in the next section.

Water Movements Within Ecosystems

Water Movement from the Canopy to the Soil

In closed-canopy forests, a substantial proportion of incoming precipitation lands in the canopy (Fig. 4.4). This precipitation can be evaporated directly back to the atmosphere, absorbed by the leaves, drip to the ground (**throughfall**), or run down stems to the ground (**stemflow**). **Canopy interception** is the fraction of precipitation that does not reach the ground. It is commonly about 10–20% for closed-canopy ecosystems (Bonan 2008). After light rain or snowfall, a substantial proportion of the precipitation may evaporate and return directly to the atmosphere without entering the soil. Throughfall is the process that delivers most of the water from the canopy to the soil.

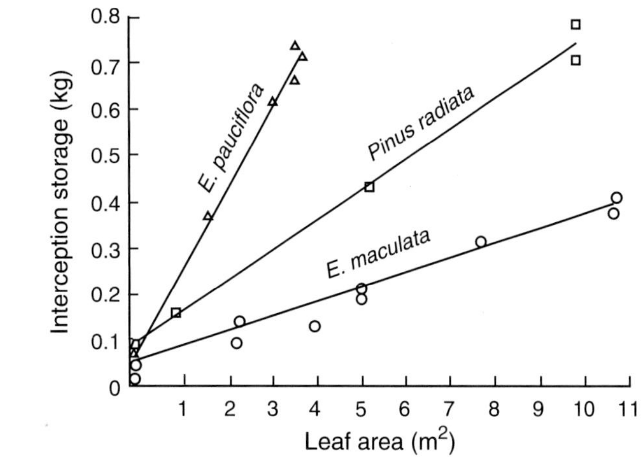

Fig. 4.6 Interception storage capacity of *Eucalyptus* species with different leaf areas. Redrawn from Aston (1979)

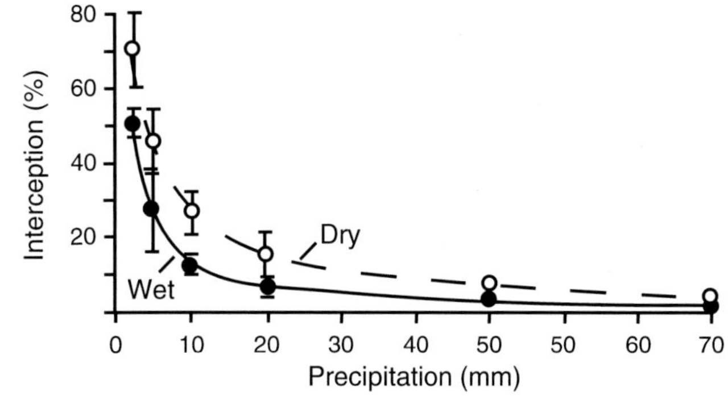

Fig. 4.7 Interception of water by dry and wet grass canopies in western Amazonia in relationship to precipitation inputs. Redrawn from Ataroff and Naranjo (2009)

The capacity of the canopy to intercept and store water differs among ecosystems. It depends primarily on canopy surface area, particularly the surface area of leaves (Fig. 4.6). Forests, for example, often store 0.8, 0.3, and 0.25 mm of precipitation on leaves, branches, and stems, respectively. Conifer forests typically store somewhat more water than deciduous forests (Waring and Running 2007; Bonan 2008). Epiphytes, which are rooted in the canopy, depend entirely on canopy interception for their water supply and increase canopy interception. Factors such as stand age and epiphyte load influence canopy interception through their effects on canopy surface area.

The bark texture and architecture of stems and trunks influences the amount and direction of stemflow. Trees and shrubs with smooth bark have greater stem flow (about 12% of precipitation)

than do rough-barked plants such as conifers (about 2% of precipitation; Waring and Running 2007). In the *Eucalyptus* mallee in southwestern Australia, as much as 25% of the incoming precipitation runs down stems, due to the parachute-shaped architecture of these shrubs. The stemflow then penetrates to depth in the soil profile through channels at the soil–root interface (Nulsen et al. 1986).

In grasslands, where precipitation is generally less than in forests, interception is often 30–40% of precipitation, a larger proportion than in forests (Seastedt 1985; Ataroff and Naranjo 2009). For small precipitation events, 70% of the precipitation can be intercepted by a dry grassland canopy, with the fraction of intercepted precipitation declining with increasing event size (Fig. 4.7; Ataroff and Naranjo 2009). Factors such as grazing or burning that alter canopy structure influence the amount of water intercepted by a grassland. Burned prairies, for example, intercept about half as much of the growing season precipitation as do unburned prairies, where standing dead leaves intercept a large proportion of the precipitation (Seastedt 1985; Gilliam et al. 1987).

In general, canopy interception reduces water input to soils, especially from light rains. Only in the presence of fog does canopy interception augment water inputs to soils. However, beyond these simple generalizations, relatively little is known about variations among ecosystems in canopy effects on water inputs to soils.

Water Storage and Movement in the Soil

Soil water is stored primarily in thin water films on the surfaces of soil particles. The water-holding capacity of a soil depends on its total pore volume and the surface area of the surrounding particles (see Chap. 3). Pore volume, in turn, depends on soil depth and the proportion of the soil volume occupied by pores. Shallow soils on ridge tops, for example, hold less water than deep valley-bottom soils. Rocky or sandy soils, in which soil solids occupy much of the soil volume and particles have a low surface-to-volume ratio, hold less water than fine-textured soils.

Water moves along a gradient from high to low potential energy. The energy status of water depends on its concentration and various pressures. The pressures in natural systems can be described in terms of either hydrostatic pressures or matric forces (Passioura 1988). The major hydrostatic pressures in natural systems are: (1) gravitational pressure, which depends on height, and (2) pressures that are generated by evaporation and by physiological processes in organisms. Matric forces result from the adsorption of water to the surfaces of cells or soil particles. The thinner the water film, the more tightly the water molecules are held to surfaces by matric forces.

We can consider these forces simultaneously by expressing them in units of **water potential**, i.e., the potential energy of water relative to pure water at the soil surface. The total water potential (ψ_t) is the sum of the individual potentials.

$$\psi_t = \psi_p + \psi_o + \psi_m \qquad (4.4)$$

The **pressure potential** (ψ_p) is generated by gravitational forces and physiological processes of organisms; the **osmotic potential** (ψ_o) reflects presence of substances dissolved in water; the **matric potential** (ψ_m) is caused by adsorption of water to surfaces. In some treatments, matric potential is considered a component of pressure potential (Passioura 1988; Lambers et al. 2008). By convention, the water potential of pure water under no pressure at the soil surface is given a value of zero. Water potentials are positive if they have a higher potential energy than this reference or negative if they have a lower potential energy. Water potentials are negative in most parts of an ecosystem because water is held under tension in soils and stems and because it contains dissolved solutes.

Pressure gradients associated with gravity and matric forces control most water movement through soils. The rate of water flow through the soil (J_s) depends on the driving force (the gradient in water potential) and the resistance to water movement. This resistance, in turn, depends on the **hydraulic conductivity** (L_s) of the soil, and the path length (l) of the column through which the water travels.

$$J_s = L_s \frac{\Delta \psi_t}{l} \qquad (4.5)$$

This simple relationship describes most of the patterns of water movement through soils, including the **infiltration** of rainwater or snowmelt into the soil and the movement of water from the soil to plant roots. Soils differ strikingly in hydraulic conductivity due to differences in soil texture and aggregate structure (see Chap. 3). For this reason, water moves much more readily through coarse-textured sandy soils than through clay soils or compacted soils. The rate of water flow in saturated soils, for example, differs by three orders of magnitude between fine- and coarse-textured soils (<0.25 to >250 mm h^{-1}).

Fig. 4.8 Plant-available water at field capacity as a function of soil texture. Redrawn from Kramer and Boyer (1995)

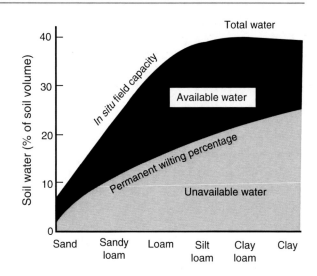

Infiltration of rainwater into the soil depends not only on hydraulic conductivity but also on preferential flow through **macropores** created by cracks in the soil or channels produced by plant roots and soil animals (Dingman 2001). Variation in flow paths in the surface millimeters of soil can have large effects on infiltration. Impaction by raindrops on an unprotected mineral soil, for example, can reduce hydraulic conductivity dramatically. Once water enters the soil, it moves downward under the force of gravity until the matric forces, which account for the adsorption of water to soil particles, exceed the gravitational potential. Water that is not retained by matric forces drains through the soil to groundwater and then to streams and lakes. The **field capacity** of a soil is the quantity of water retained by a soil after gravitational water has drained.

At field capacity, the water potential of a soil is about −0.03 MPa, i.e., close to the water potential of pure water (0.00 MPa). As a soil dries, the films of soil water become thinner, and the remaining water is held more tightly to particle surfaces. The **permanent wilting point** is the soil water potential (about −1.5 MPa) at which most mesic plants wilt because they cannot obtain water from soils. Many drought-adapted plants, however, can obtain water from soils at water potentials as low as −3.0 to −8.0 MPa (Larcher 2003). A second consequence of thin water films

in dry soils is that water cannot move directly across air-filled soil pores but must move through water films around the edges of pores along a much longer, more tortuous path. For this reason, the hydraulic conductivity of soil declines dramatically as the soil dries. The difference in the water content between field capacity and permanent wilting point (water-holding capacity) provides an estimate of the plant-available water (Fig. 4.8), although some of this water is held in such small pores that it moves very slowly to roots. Vegetation often extracts 65–75% of the plant-available water before there are signs of water stress (Waring and Running 2007). The total quantity of water available to vegetation is the available water content per unit soil volume times the volume exploited by roots.

Water Movement from Soil to Roots

Water moves from soil to the roots of transpiring plants by flowing from high to low water potential. Water moves from the soil into the root whenever the root has a lower water potential than the surrounding soil. Movement of water into the root along a water-potential gradient causes the water film on adjacent soil particles to become thinner. This remaining water is adsorbed more tightly to soil particles and therefore has a lower water potential. The localized reduction in water potential near the root causes water to move along soil films toward the root. In this way, a root can access most available water within a radius of about 6 mm. As the soil dries, hydraulic conductivity declines, and the root accesses water less rapidly. In saline soils, the osmotic potential of the soil solution reduces total soil water potential, so roots with a given water potential can absorb less water from saline than from nonsaline soils.

A continuous pathway for water movement from the soil to the root is provided by root hairs and mycorrhizal hyphae that extend into the soil and by carbohydrates secreted by the root that maximize contact between the root and the soil. The root cannot absorb water if this root–soil contact is interrupted by the shrinking of drying soil or by the consumption of root hairs and root cortical cells by soil animals.

Rooting depth reflects a compromise between water and nutrient availability. Most plant roots are in the upper soil horizons where nutrient inputs are greatest and where nutrients are generally most available (see Chap. 9). In a given ecosystem, short-lived herbs are generally more shallow rooted than long-lived shrubs and trees and depend more on surface moisture (Fig. 4.9; Schenk and Jackson 2002). In arid ecosystems, surface evaporation and transpiration dry out the surface soils. For this reason, deserts, arid shrublands, and tropical savannas have many species with deep roots (Fig. 4.10). Phreatophytes are an extreme example of deep-rooted plants. Roots of these desert plants extend to the water table, often a depth of tens of meters. These plants have no physiological adaptations to drought and have high transpiration rates. Even wet ecosystems such as tropical rainforests have dry seasons that explain the occurrence of deep-rooted tropical trees that tap water from depths of more than 8 m (Nepstad et al. 1994). Relatively deep water (2–8 m depth) accounts for more than 75% of the water transpired by these forests. Deep-rooted plants may be more common and play a larger role in ecosystem water budgets than is generally appreciated.

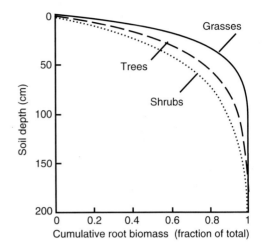

Fig. 4.9 The cumulative fraction of roots found at different soil depths for three plant growth forms averaged over all biomes. Redrawn from Jackson et al. (1996)

Rooting depth has important ecosystem consequences because it determines the soil volume that is exploited by vegetation (see Chap. 11). California grassland soils below a meter depth, for example, remain moist even at the end of the summer drought, whereas an adjacent chaparral shrub community uses water to a depth of 2 m. This greater rooting depth contributes to the longer growing season and greater productivity of the chaparral. Even in the chaparral, species differences in rooting depth lead to differences in water supply and drought stress.

Water Movement Through Plants

The vapor-pressure gradient from the leaf surface to the atmosphere is the driving force for water movement through plants. Water transport from the soil through the plant to the atmosphere takes place in a soil–plant–atmosphere continuum that is interconnected by a continuous film of liquid water. Water moves from the soil through the plant to the atmosphere along a gradient in water potential. The low water potential of *unsaturated* air outside leaves, relative to the water potential of *saturated* air inside leaves, is the major driving force for water loss from leaves

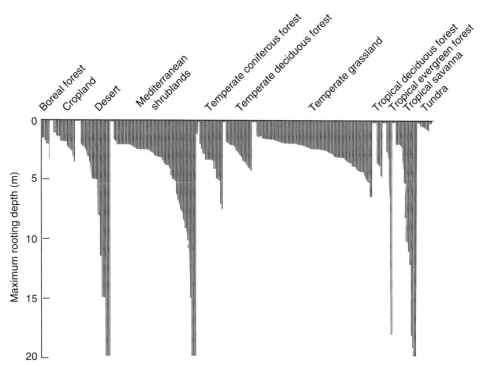

Fig. 4.10 Maximum rooting depths of selected species in the major biome types of the world. Species in each biome differ widely in rooting depth. Woody species in dry environments are often deeply rooted. Redrawn from Canadell et al. (1996)

(Box 4.1), which in turn drives water transport along a water-potential gradient from the roots to the leaves, which in turn drives water movement from the soil into the plant. Water moves through the plant under tension (negative pressure) as it is "sucked up" through xylem vessels to replace water that is lost by transpiration at the leaf surface. The rate of water movement through the plant (J_p) is determined by the water-potential gradient (the driving force; $\Delta\psi_t$) and the resistance to water movement, just as described for water movement through soils (Eq. 4.5). As in soils, the resistance to water movement through the plant depends on hydraulic conductivity (or conductance; L_p) and path length (l). The movement of water into and through the plant is driven entirely by the physical process of evaporation from the leaf surface and involves no direct expenditure of metabolic energy by the plant except to produce the roots. This contrasts with the acquisition of carbon and nutrients for which the plant directly expends considerable metabolic energy.

Roots

Water moves through roots along a water-potential gradient from moist soils to the atmosphere during the day and sometimes to dry surface soils at night. In moist soils, the cell membranes, which are composed of hydrophobic lipids, constitute the greatest resistance to water movement through roots (see Fig. 8.5). This membrane resistance to water flow is greatest under conditions of low root temperature or low oxygen, so plants that are not adapted to these conditions experience substantial water stress in cold or saturated soils. In dry soils, gaps between the root and the soil or breakage of water columns within the root, as described later, account for the greatest resistance to water flow through the plant. Plants overcome these disruptions in the water pathway from soil to leaves primarily

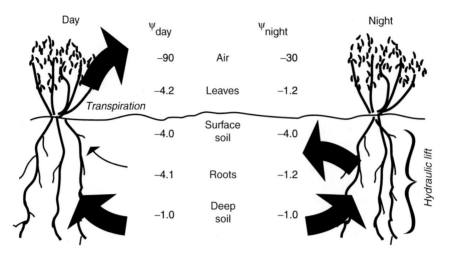

Fig. 4.11 Representative patterns of soil-water potential and water movement in arid environments during the day and at night. During the day, water moves from soils (especially deep soils) to the atmosphere in response to the strong water-potential gradient from the plant to the atmosphere. At night, when stomata are closed, water moves from wet soils at depth to dry surface soils through the root system, the process of hydraulic lift

by producing new roots, whose water transport pathway are intact and whose carbohydrates, root hairs, and mycorrhizae improve the contact with soil-water films.

In dry environments, there is a strong vertical gradient in soil-water potential due to the low water potential of dry surface soils. However, water moves slowly through the soil because of the low hydraulic conductivity of dry soils. During the day, when plants lose water through transpiration, plant-water potential is lower than soil-water potential, so water moves from the soil into the plant, particularly from deep soils where water is most available (highest soil-water potential; Fig. 4.11). At night, when stomata close and transpiration ceases, plant-water potential equilibrates with the water potential of deep soils. When surface soils are drier than those at depth, the water-potential gradient is from deep to shallow soils. Because roots have much higher hydraulic conductivity than dry soils, this gradient in water potential drives **hydraulic lift**, the vertical movement of water from deep soils *through roots* to shallow soils along a water-potential gradient (Caldwell and Richards 1989). This water movement can be documented by measuring changes in the isotopic composition of water (Box 4.2). Hydraulic lift occurs in most arid ecosystems and in many moist forests. Sugar maple trees, for example, acquire all their moisture from deep roots during dry periods, but 3–60% of the water used by shallow-rooted herbs in these forests comes from water that has been hydraulically lifted by the maple trees (Dawson 1993). In the Great Basin deserts of western North America, 20–50% of the water used by shallow-rooted grasses comes from water that is hydraulically lifted by deep-rooted sagebrush shrubs. The water provided by hydraulic lift stimulates decomposition and mineralization in dry, shallow soils, augmenting the supplies of both water and nutrients to shallow-rooted species. Because deep-rooted plants both provide water to, and remove water and nutrients from, shallow soils, hydraulic lift complicates the interpretation of species interactions in many ecosystems. When surface soils are wetter than deep soils after rain, roots provide an avenue to recharge deep soils (Burgess et al. 1998). Thus roots provide an avenue for rapid water transport from soil of high to low water potential, regardless of the vertical direction of the water-potential gradient.

Box 4.2 Tracing Water Flow Through Ecosystems

Stable isotopes are useful tools for tracing the movement of elements or compounds through ecosystems (Dawson and Siegwolf 2007). The source of water used by plants, for example, can be determined from its isotopic composition. Two isotopes of an element differ in the number of neutrons in the nucleus and therefore differ in their physical properties more strongly than their chemical properties. The ratio of the concentration of deuterium (D) to hydrogen (H) provides a useful signature of different water sources. These ratios are often expressed relative to the ratio in some standard substance (such as the ocean in the case of water). Therefore, water that has more D than ocean water has a positive value, and water with less D than ocean water has a negative value. Evaporation discriminates against the heavier isotope (deuterium), causing the isotopic ratio of D/H in water vapor to decline (become more negative), relative to the water source that gave rise to evaporation (Fig. 4.12). Condensation, on the other hand, raises the D/H ratio, causing rainfall to have a less negative hydrogen isotopic ratio than its parent air mass. The D/H ratio of water vapor remaining in the atmosphere therefore declines (becomes more negative) with sequential rainfall events. There is also a positive linear relationship between air temperature at the time of precipitation and the D/H ratio, so summer precipitation has a higher D/H ratio than winter precipitation. These changes in D/H ratio with evaporation and condensation generate characteristic signatures of different pools of water in ecosystems. During the growing season, for example, deep water is more likely to be derived from winter precipitation (low D/H ratio) and shallow water from summer precipitation (high D/H ratio). These isotopic signatures can be used to identify the sources of water used by plants (Fig. 4.13). The isotopic ratio of xylem water, for example, can show the relative proportions of deep vs. shallow water used by plants and therefore their dependence on winter vs. summer precipitation. Similarly, D/H ratios show that some plants such as redwood trees derive most of their water from fog, whereas others use soil water or ground water (Dawson 1993; Limm et al. 2009). D/H ratios of stream water identify the relative contributions of soil water from recent precipitation

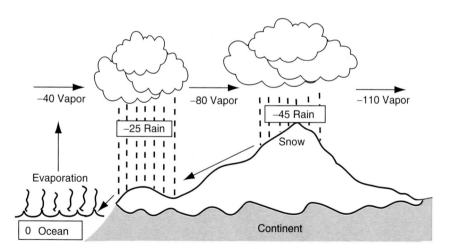

Fig. 4.12 The effect of evaporation and subsequent condensation during rainfall on the ratio of hydrogen isotopes. Redrawn from Dawson (1993)

(continued)

Box 4.2 (continued)

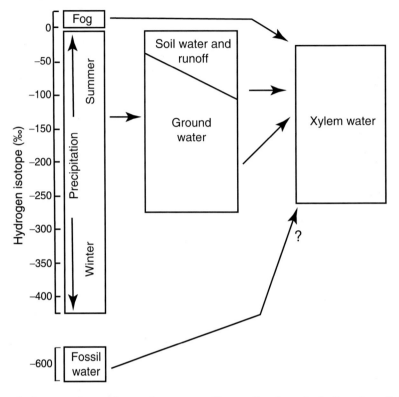

Fig. 4.13 Isotopic signature of water from various sources. By sampling the water in the xylem of plants, one can determine the main water supply used by a plant. Redrawn from Dawson (1993)

events vs. ground water. Oxygen isotope ratios in water show patterns of variation similar to those of hydrogen and have been particularly useful in estimating the atmospheric temperatures associated with the snowfall that produced glacial ice in the distant past (see Fig. 14.2).

Stems

Water moves through stems to replace water lost by transpiring leaves. The water-conducting tissues in the xylem are narrow capillaries of dead cells that extend from the roots to the leaves. Water is "sucked up" through these capillary tubes in response to the water-potential gradient created by transpirational water loss. The cohesion of water molecules to one another and their adhesion to the walls of the narrow capillary tubes allow these water columns to be raised under tension (a negative water potential) as much as 100 m in tall trees.

There is a tradeoff between hydraulic conductivity of xylem vessels and their risk of **cavitation,** i.e., the breakage of water columns under tension (Jackson et al. 2000; Sperry et al. 2008). Hydraulic conductivity of stems varies with the fourth power of capillary diameter, so a small increase in vessel diameter greatly increases hydraulic conductivity. For example, vines, which have relatively narrow stems and rely on other plants for physical support, have large-diameter xylem vessels. This allows rapid water transport through narrow stems but increases the risk of cavitation and may explain why vines are most

Fig. 4.14 The relationship between the water potential at which a plant loses all xylem conductivity due to cavitation and the minimum water potential observed in nature. Each data point represents a different species. The 1:1 (*dashed*) line is the line expected if there were no safety factor, i.e., if each species lost all conductivity at the lowest water potential observed in nature. Species that naturally experience low water potentials exhibit a greater margin of safety (i.e., a greater departure from the 1:1 line). Redrawn from Sperry (1995)

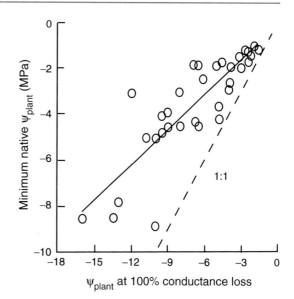

common in moist environments such as tropical wet forests. The stems of tropical vines, for example, have hydraulic conductivities and velocities of sap flow that are 50- to 100-fold higher than those of conifers (Larcher 2003). Broad-leaved deciduous trees are intermediate. Many plants in moist environments, particularly herbaceous plants, function close to the water potential where cavitation occurs, suggesting that they invest just enough in water transport tissues to allow water transport for the growing season (Sperry 1995; Sperry et al. 2008). Plants from dry environments produce stems with a larger safety factor, i.e., stems that resist cavitation at much lower water potentials than the plants commonly experience (Fig. 4.14).

Fine roots appear to be even more prone to cavitation than are stems due to their relatively large vessel diameters (Jackson et al. 2000). Due to the greater root surface area than stem cross-sectional xylem area, however, these two organs probably limit water transport to a similar degree at the whole-plant level (Craine 2009).

Plants in cold environments suffer cavitation from freezing. Trees adapted to these cold environments typically produce abundant small-diameter vessels that can, in some species, refill after cavitation (diffuse-porous species).

In contrast, many trees in warm environments produce small-diameter vessels as well as large-diameter vessels that cannot be refilled after cavitation and therefore function for only a single growing season (ring-porous species).

The water transported by a stem depends on both the hydraulic conductivity of individual conducting elements and the total quantity of conducting tissue (the **sapwood**). There is a strong linear relationship between the cross-sectional area of sapwood and the leaf area supported by a tree (Fig. 4.15). However, the slope of this relationship varies strikingly among species and environments. Drought-resistant species generally have less leaf area per unit of sapwood than do drought-sensitive species because of the small vessel diameter (lower conductance) of drought-resistant species. The ratio of leaf area to sapwood area, for example, is generally more than twice as great in trees from mesic environments as in trees from dry environments (Margolis et al. 1995). Any factor that enhances the productivity of a tree increases its ratio of leaf area to sapwood area. This ratio increases, for example, with improvements in nutrient or moisture status and is greater in dominant than subdominant individuals of a stand.

Water storage in stems buffers the plant from imbalances in water supply and demand.

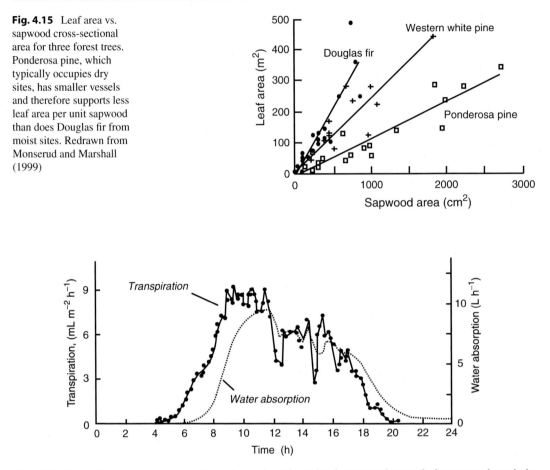

Fig. 4.15 Leaf area vs. sapwood cross-sectional area for three forest trees. Ponderosa pine, which typically occupies dry sites, has smaller vessels and therefore supports less leaf area per unit sapwood than does Douglas fir from moist sites. Redrawn from Monserud and Marshall (1999)

Fig. 4.16 Diurnal time course of water absorption and water loss by Siberian larch. During morning, transpiration is supported by water loss from stems, creating a lower water potential in stems and roots, which generates the water-potential gradient to absorb water from the soil. Absorption is measured as vertical transport through the stem, and water loss as transpiration per unit leaf area. The water stored in stems is replenished at night. Redrawn from Schulze et al. (1987)

The water content of tree trunks generally decreases during the day, causing water absorption by roots to lag behind transpirational water loss by about 2 h (Fig. 4.16). The quantity of water stored in sapwood is substantial, equivalent to as much as 5–10 days of transpiration. This sapwood water, however, exchanges relatively slowly, so stores of water in sapwood seldom account for more than 10% of transpiration. In tropical dry forests, where trees lose their leaves during the dry season, this stored water is critical to support flowering during the dry season. Trees with low-density wood and large stem water storage can flower during the dry season, whereas trees with high-density wood and low stem water storage can flower only during the wet season (Borchert 1994). Water stored by desert succulents may allow transpiration to continue for several weeks after water absorption from the soil has ceased.

Leaves

Water loss from leaves is controlled by the evaporative potential of the air, the water supply from the soil, and the stomatal conductance of leaves. Soil water supply and the evaporative potential of the air are the major environmental controls over water loss from

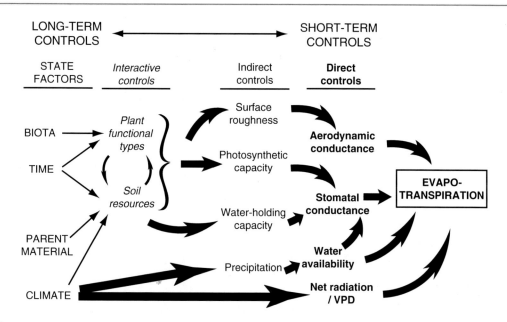

Fig. 4.17 The major factors governing temporal and spatial variation in evapotranspiration from a plant canopy. These controls range from the proximate controls, which determine the diurnal and seasonal variations in evapotranspiration, to the interactive controls and state factors, which are the ultimate causes of ecosystem differences in evapotranspiration. The thick *arrows* indicate that all these controls are important, with stomatal conductance most important in dry conditions and aerodynamic conductance in moist conditions. Net radiation is the strongest driver of evapotranspiration for smooth canopies and VPD for rough canopies

leaves (Fig. 4.17). **Stomata** (or stomates) are pores in the leaf surface that can be opened or closed by the plant to regulate the rates at which CO_2 enters the leaf and water is lost (the process of transpiration). Stomata determine the **conductance** of water vapor between the plant and the air, i.e., the flux of water vapor per unit driving force (vapor pressure gradient). When the stomata are open, leaves have a high conductance, and water vapor is rapidly lost. When stomata close, conductance declines to very low levels, and little water loss occurs. In dry soils, the low hydraulic conductivity of the soil minimizes the amount of water that can move directly from soil to the air by surface evaporation. The extensive root systems of plants and the high hydraulic conductivity of plant xylem make plants an effective conduit for moving water from the soil to the atmosphere. Plants adjust the size of stomatal openings to regulate the loss of water from leaves. Because stomatal conductance also determines the rate of CO_2 entry into leaves, there is an inevitable tradeoff between carbon gain and water loss by leaves (see Chap. 5).

Diurnal and climatic differences in air temperature and humidity determine the driving force for transpiration. Air inside the leaf is always saturated with water vapor because it is adjacent to moist cell surfaces. On a sunny day, air temperature rises to a maximum shortly after midday, allowing the air to hold more water. This rise in air temperature and the radiation absorbed by the leaf increases the temperature of the leaf and therefore the water vapor concentration of air spaces inside the leaf. The water vapor concentration of the external air increases less than that inside the leaf. The resulting increase in the gradient in water vapor concentration (i.e., VPD, Box. 4.1) between the inside and the outside of the leaf increases the transpirational water loss from the leaf. In the evening, the temperature decreases, causing a decline in the water vapor concentration inside the leaf and a decline in transpiration. Variations in weather or climate that cause an increase in air temperature or a decrease in atmospheric moisture content also enhance the driving force for transpirational water loss. The evaporative potential of desert air is therefore extremely high because it is both hot and dry. Cloud forests generally have low evaporative potential because the air is saturated and clouds reduce radiation input. Cold climates have low evaporative potential because cold air holds relatively little water vapor.

Stomatal conductance is *the* major control that plants exert over water loss from a leaf. Some plants reduce stomatal conductance when leaves are exposed to warm, dry air that would otherwise cause high transpirational water loss. Species differ considerably in their sensitivity of stomatal conductance to the evaporative potential of the air. Both the mechanism and ecological patterns in the sensitivity of stomatal conductance to atmospheric humidity are poorly understood.

Stomatal conductance declines in response to drought because plants sense the soil moisture content of their root systems. Roots exposed to low soil moisture produce **abscisic acid** (ABA), a hormone that is transported from roots to leaves and causes a reduction in stomatal conductance. The degree of coupling between soil moisture and stomatal conductance depends on specific plant adaptations. **Isohydric** plants, which tend to grow in moist environments, close their stomata at relatively high soil moistures before they experience large changes in plant-water potential. This stops photosynthesis, so they must rely on stored reserves to meet their energy demands during dry periods, but it prevents hydraulic failure. In contrast, **anisohydric** plants, which tend to grow in dry sites, show less response of stomatal conductance to soil drying and therefore continue to photosynthesize and to absorb and lose water as the soil dries (McDowell et al. 2008). These plants therefore maintain greater physiological activity in dry soils than do plants adapted to moist habitats, and, in the process, they transfer more water to the atmosphere under dry conditions. Although anisohydric plants are relatively drought tolerant, they are predisposed to hydraulic failure under extreme

drought because they operate closer to their safety limits under these conditions. This in turn reduces their resistance to insect outbreaks and other indirect effects of drought (McDowell et al. 2008).

Species differ in stomatal conductance under favorable conditions. Stomatal conductance is highest in rapidly growing plants adapted to moist fertile soils (see Chap. 5; Körner et al. 1979; Schulze et al. 1994).

Water Losses from Ecosystems

Water input is the major determinant of water outputs from ecosystems. The water loss from ecosystems equals the input in precipitation (P) adjusted for any changes in water storage (ΔS). The major avenues of loss are evapotranspiration (E) and runoff (R).

$$P \pm \Delta S = E + R \qquad (4.6)$$

Just as in the case of carbon and energy, the changes in water storage are generally small relative to inputs and outputs, when averaged over long time periods (multiple years). In ecosystem comparisons therefore the quantity of water entering the ecosystem largely determines water output, just as GPP (carbon input) is the major determinant of ecosystem respiration (carbon output; see Chap. 7).

The route by which water leaves an ecosystem depends on the partitioning between evapotranspiration and runoff. This partitioning has a critical impact on regional hydrologic cycles because green water that returns to the atmosphere is available to support precipitation in the same or other ecosystems. In contrast, runoff supplies the blue water input to aquatic ecosystems and provides most of the water used by people (Fig. 4.5). In a sense, runoff is the "left-overs" of water that entered in precipitation and was not transferred to the atmosphere by evapotranspiration. In summary, controls over evapotranspiration largely determine the partitioning between evapotranspiration and runoff.

Evaporation from Wet Canopies

Evaporation of water intercepted by the canopy is greatest in ecosystems with a high surface roughness. Forests have high rates of evaporation from wet canopies, primarily because the efficient mixing that occurs in rough forest canopies promotes rapid evaporation from each leaf (Kelliher and Jackson 2001). The large water storage capacity of forest canopies is less important than its **surface roughness** (vertical irregularities in the height of the canopy surface) in explaining the quantity of water evaporated from wet canopies. The evaporation rate from a wet canopy depends primarily on the climatic conditions that drive evaporation (primarily VPD) and the degree to which environmental conditions in the canopy are coupled by turbulence to conditions in the atmosphere. Turbulence, in turn, is greatest in ecosystems with a tall, aerodynamically rough canopy. In forests, which are tightly coupled to atmospheric conditions, wet canopy evaporation is largely independent of net radiation and is similar during the day and night. In grasslands, which are less tightly coupled to the atmosphere, wet canopy evaporation depends on net radiation as well as VPD and is greater during the day than at night. Due to differences in canopy roughness, forests have greater wet-canopy evaporation than do shrublands or grasslands, and conifer forests evaporate more water from wet canopies than do deciduous forests.

Climate is the other factor that governs evaporation from wet canopies. Climate determines the frequency with which the canopy intercepts precipitation or dew and the conditions that drive evaporation. Ecosystems in wet climates generally have greater canopy evaporation because of the more frequent capture of rainfall by the canopy, even though the low VPD of wet climates causes this evaporation to occur slowly. The *frequency* of rainfall and dew formation is generally more important than total precipitation in governing the annual flux of wet-canopy evaporation (Rutter et al. 1971). The canopy acts like a bucket that stores water from a given rain or dew event until its storage capacity is exceeded, at which point water moves to the ground as throughfall or stem

flow. Canopy evaporation increases exponentially with air temperature because of the temperature effects on VPD (Box 4.1; McNaughton 1976), so ecosystems generally lose more intercepted water through canopy evaporation in warm than in cold climates. Despite these generalizations, the interactions among multiple controls over wet-canopy evaporation are so complex that they are best addressed through physically based models that consider all these factors simultaneously (Waring and Running 2007; Monteith and Unsworth 2008).

Canopies that intercept precipitation as snow or ice often store twice as much water equivalent as when precipitation is received in liquid form. Snow interception and subsequent **sublimation** (vaporization of a solid) from the canopy is greatest in ecosystems with a high **leaf area index** (LAI, the leaf area per unit ground area). Most snow usually falls to the ground, however, where low net radiation and low wind speeds minimize sublimation. In tundra, where there is no canopy in winter to shade the snow, or in continental boreal forests with low precipitation and low wind speeds, sublimation can account for 30% and 50%, respectively, of winter precipitation (Liston and Sturm 1998; Pomeroy et al. 1999; Sturm et al. 2001).

Evapotranspiration from Dry Canopies

Water moves from a dry canopy to the atmosphere above the canopy in two consecutive steps: diffusion and turbulent mixing. These two steps in the hydrologic pathway are controlled by quite different processes. **Surface conductance** determines the flux of water vapor from inside the leaf or soil to the near-surface air and is controlled primarily by leaf stomata and soil surface properties, respectively. **Aerodynamic conductance**, also termed boundary-layer conductance, determines the flux of water vapor from the air near the leaf or soil surface to the bulk air above the canopy and is controlled primarily by turbulent mixing within the canopy. Ecosystem structure and soil moisture determine the relative importance of these two controls.

Vegetation structure and climate govern evapotranspiration rate when soil moisture is adequate. Under moist-soil conditions, turbulent mixing between bulk and canopy air largely determines the rate of water loss because open stomata and soil evaporation allow rapid diffusion of water vapor to the air immediately above these surfaces. The aerodynamic conductance, which defines the potential for turbulent mixing, depends on wind speed and the size and number of roughness elements, such as trees. Aerodynamic conductance is greatest when surface turbulence mixes large quantities of air from the bulk atmosphere with air inside the canopy and couples the evaporation at the leaf or soil surfaces with the atmospheric moisture content above the canopy. Ecosystems such as forests with tall, aerodynamically rough canopies therefore have a higher aerodynamic conductance and reduce soil moisture more rapidly than do grasslands or crops (Mark and Dickinson 2008).

Vegetation structure also determines which climatic variables regulate evapotranspiration. In aerodynamically rough, well-coupled canopies, the moisture content of canopy air is similar to that above the canopy, so the moisture content of the bulk air is the main determinant of evapotranspiration (Waring and Running 2007). In canopies that are short, smooth, and weakly coupled, by contrast, the air adjacent to leaves mixes less readily with the bulk air, so evapotranspiration moistens the canopy air and reduces the driving force for diffusion through stomata. In these smooth canopies, evapotranspiration is determined more by net radiation than by the moisture content of the bulk air because net radiation determines surface temperature and therefore the driving force for water vapor diffusion through stomata to the near-surface air. The **decoupling coefficient**, which indicates the degree of canopy decoupling from the bulk air (Table 4.3; Jarvis and McNaughton 1986), is determined primarily by canopy height. In summary, the moisture content of the bulk air (as measured by VPD) is the dominant control in tall, well-coupled canopies, whereas net radiation is the dominant driver of evapotranspiration in short, weakly coupled canopies (Waring and Running 2007). These patterns

Table 4.3 Decoupling coefficient of vegetation canopies in the field under conditions of adequate moisture supply

Vegetation	Decoupling coefficient[a]
Alfalfa	0.9
Strawberry patch	0.85
Permanent pasture	0.8
Grassland	0.8
Tomato field	0.7
Wheat field	0.6
Prairie	0.5
Cotton	0.4
Heathland	0.3
Citrus orchard	0.3
Forest	0.2
Pine woods	0.1

Data from Jarvis and McNaughton (1986) and Jones (1992)

[a] A completely smooth surface has a decoupling coefficient of 1.0, and a canopy in which the air is identical to that in the atmosphere has a decoupling coefficient of zero

of environmental control over evapotranspiration from ecosystems with moist soils and dry canopies are identical to those that we described earlier for wet-canopy evapotranspiration.

Under moist conditions where turbulent mixing within the canopy is the rate-limiting step, ecosystem differences in surface conductance are surprisingly small (Kelliher et al. 1995). In sparse vegetation, evaporation from the soil surface is the major avenue of water loss. As leaf area increases, transpiration increases (more leaf area to transpire), which is counteracted by a decrease in soil evaporation (more shading and less turbulent exchange at the soil surface). Consequently, surface conductance is relatively insensitive to the quantity of leaf area present. Vegetation affects maximum surface conductance primarily through its effects on stomatal conductance (Kelliher et al. 1995). However, even this effect is often relatively small. Maximum stomatal conductance of individual leaves is relatively similar among natural ecosystems (Körner 1994; Kelliher et al. 1995). Woody and herbaceous ecosystems, for example, have similar stomatal conductance of individual leaves (Körner 1994) and similar surface conductance of entire ecosystems (Kelliher et al. 1995). Crops, however, which have about 50% higher stomatal conductance

Fig. 4.18 Response of plant-water potential and transpiration to soil moisture (Sucoff 1972; Gardner 1983; Waring and Running 2007). Soil moisture has little effect on plant-water potential or transpiration until about 75% of the available water has been removed from the rooting zone

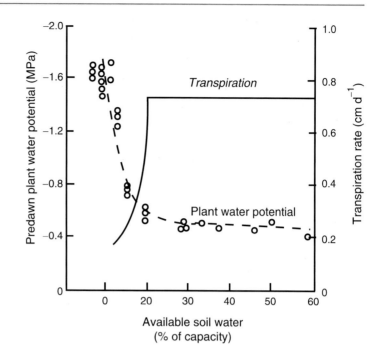

than does natural vegetation, also have about 50% higher surface conductance (Schulze et al. 1994; Kelliher et al. 1995). In summary, under moist-soil conditions, evapotranspiration is controlled much more strongly by surface roughness effects on aerodynamic conductance than by leaf area or maximum stomatal conductance.

As soil moisture declines, the control over evapotranspiration shifts from canopy structure to soil moisture. Plant-water potential and transpiration rate are surprisingly insensitive to water availability until plants have depleted about 75% of the plant-available soil water (Fig. 4.18). Evapotranspiration from dry canopies is therefore relatively insensitive to precipitation over a broad range of soil moisture (Fig. 4.19). Over this range of soil moisture, aerodynamic conductance remains the primary control over evapotranspiration. As soils continue to dry, however, their hydraulic conductivity declines. This creates a relatively abrupt threshold of soil moisture, below which the rate of water supply to roots declines, and plants experience water stress (low water potential; Fig. 4.18). Under these circumstances,

plant stomata close, reducing surface conductance and evapotranspiration below their physiological maxima, just as described earlier for individual leaves. Under these dry-soil conditions, surface conductance limits water movement from the ecosystem to the atmosphere and is controlled primarily by the effects of soil moisture on stomatal conductance, as described earlier.

In summary, aerodynamic conductance, which depends on plant height and the number of roughness elements, is the main control over evapotranspiration from dry canopies under conditions of adequate water supply. Stomatal conductance exerts an increasingly important control over evapotranspiration as soil moisture declines below the point where soil hydraulic conductance is substantially reduced. In other words, stomatal conductance (and therefore surface conductance) accounts for *temporal* variation in evapotranspiration in response to soil drying, but surface roughness (and therefore aerodynamic conductance) is the major factor explaining *ecosystem differences* in evapotranspiration under *moist* conditions.

Fig. 4.19 Relationship between annual water input (precipitation) and output (evapotranspiration and streamflow) from a temperate forest watershed (Hubbard Brook in the U.S.) over a 19-year period. In this moist forest, evapotranspiration varies little among years, whereas streamflow is quite sensitive to the quantity of precipitation. Redrawn from Bormann and Likens (1979)

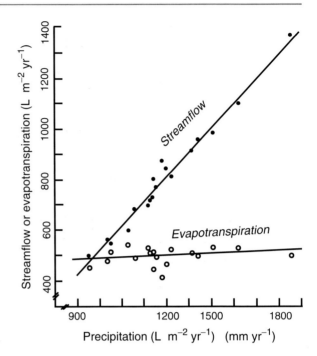

Changes in Storage

Water inputs that exceed outputs replenish water that is stored in soil and groundwater. Water that enters the soil is retained until the soil reaches field capacity. Additional water moves downward to groundwater. In cold climates in winter, most of the precipitation input is stored above ground in the snowpack. The snowpack substantially increases the quantity of water that an ecosystem can store and the residence time of water in the ecosystem. Stored water supports evapotranspiration at times when evapotranspiration exceeds precipitation; the declines in soil moisture during these times draw down water storage. The seasonal recharge and depletion of stored water are important controls over evapotranspiration and NPP in many ecosystems.

Groundwater, i.e., the water beneath the rooting zone, is a large pool that is inaccessible to plants in many ecosystems. The size of this pool depends on the depth to impermeable layers and the porosity of materials in this layer. Porosity governs the pore volume available to hold water and the resistance to lateral drainage of water.

The groundwater pool has a relatively constant size, so, when new water enters groundwater from the top, it displaces older water that drains laterally to streams, lakes, and the ocean. The time lag between inputs to groundwater and outputs can be substantial (months to millennia) because of the large size of this pool.

People modify groundwater pools by changing the vegetation and associated rooting depth and by tapping groundwater to support human activities. Introduction of deep-rooted exotic species in arid regions as shade trees often allows the ecosystem to tap groundwater that was previously inaccessible. This can cause the water table to drop. The introduction of deep-rooted *Tamarix* in North American deserts, for example, caused the water table to drop so much that desert ponds have dried, endangering endemic fish species (Berry 1970).

Removal of vegetation causes the water table to rise because surface water is no longer tapped to support evapotranspiration. The clearing of heathlands for agriculture in Western Australia, for example, reduced the depth of the rooting zone, causing naturally saline groundwater to rise close to the surface. This reduced the productive potential of the crops, further reducing evapotranspiration and the depth to groundwater. Finally, evaporation from the soil surface increased soil salinity to the point that soils no longer supported crop growth in many areas nor could they be recolonized by native heath vegetation (Nulsen et al. 1986). Planting of salt-tolerant eucalypt forests in these saline soils increased evapotranspiration and reduced the height of the water table, thereby reducing soil salinity within and adjacent to forests (Jackson et al. 2005). In these ways, human modification of vegetation substantially alters the hydrologic cycle and all aspects of ecosystem structure and functioning.

Expansion of human populations into arid regions is often subsidized by tapping groundwater that would otherwise be unavailable to surface organisms. Irrigated agriculture often uses 80–90% of the water and is highly productive because of warm temperatures and high solar radiation, when the natural constraints of water limitation are removed (see Fig. 14.1). These irrigated lands are important sources of fruits, vegetables, cotton, rice, and other high-value crops. Conversion of arid regions to irrigated agriculture, however, reduces the amount of water available for runoff. Human use of water in the arid southwestern U.S., for example, converted the Rio Grande River from a major river to a small stream with intermittent flow during some times of year. Irrigation also increases soil evaporation, which increases soil salinity in a fashion similar to that described for Western Australia.

In cases where evapotranspiration of irrigated agriculture exceeds precipitation, there is not only a decrease in runoff but also a depletion of the groundwater pool. The Ogallala aquifer in the north-central U.S., for example, accumulated water when the climate was much wetter than today. Tapping of this "fossil water" has increased depth to water table substantially. Continued drawdown of this aquifer cannot be sustained indefinitely because current water sources cannot replenish it as rapidly as it is being depleted to support irrigation.

Runoff

Runoff from terrestrial ecosystems is the difference between precipitation inputs, changes in storage, and losses to evapotranspiration (4.6). Average runoff (or **discharge**) from a drainage basin depends primarily on precipitation and evapotranspiration because long-term changes in storage are usually negligible. Runoff responds to variation in precipitation much more strongly than does evapotranspiration (Fig. 4.19) because runoff constitutes the leftovers after the water demands for evapotranspiration and groundwater recharge have been met. Runoff is therefore greater in wet than in dry climates or seasons. Over hours to weeks, runoff generally increases after rainfall events and decreases during dry periods. Changes in water storage buffer this linkage between precipitation and runoff. The recharge of soil moisture in grasslands, shrublands, and dry forests, for example, may prevent large increases in streamflow after a rain when soils are dry, whereas streamflow may increase rapidly after a storm when soils are wet or shallow (Jones 2000). In ecosystems with a small capacity to store water such as deserts with coarse-textured soils and a calcic layer or ecosystems underlain by permafrost, runoff responds almost immediately to precipitation, and rainstorms can cause flash floods. Conversely, slowly draining groundwater provides a continued source of water to streams (**base flow**) even at times without precipitation. In this way, water balance determines the distribution and abundance of freshwater ecosystems and their temporal variability (Kalff 2002).

In ecosystems that develop a snowpack in winter, precipitation inputs are stored in the ecosystem during winter, causing winter stream flows to decline, regardless of the seasonality of precipitation. During spring snowmelt, this stored water recharges aquifers or moves directly to streams, causing large spring runoff events. Glacial rivers, for example, have greatest runoff in midsummer, when warm temperatures cause greatest melting, whereas non-glacial rivers in the same climate zone have peak flow in early spring after snowmelt. When climate warming changes snowfall to winter rains, this increases winter runoff and reduces the spring snowmelt pulse and summer runoff, which are important water sources for many cities.

River flow integrates the precipitation, evapotranspiration, and changes in storage throughout the drainage basin. In large rivers, the seasonal variations in flow often reflect patterns of precipitation and evapotranspiration that occur upstream, hours to weeks previously. These integrative effects of runoff from large drainage basins make runoff a good indicator of long-term changes in the hydrologic cycle.

Seasonal variations in streamflow are a major determinant of the structure and seasonality of ecosystem processes in streams and rivers. Periods of high flow in streams and rivers, for example, scour stream channels, removing or redistributing sediments, algae, and detritus (Power 1992a). In undammed rivers, high flow events may lead to predictable patterns of bank erosion and deposition. Life histories of river biota are adapted to natural flow regimes (Poff et al. 1997; Lytle and Poff 2004). Dams that reduce the intensity or seasonality of high-flow events therefore dramatically alter the natural disturbance regime and functioning of freshwater ecosystems.

Vegetation strongly influences the quantity of runoff. Because evapotranspiration is such a large component of the hydrologic budget of an ecosystem, any vegetation change that alters evapotranspiration inevitably affects runoff. Deforested drainage basins, for example, exhibit increased annual runoff, although this often lasts only a few years (Fig. 4.20; see Chap. 12; Trimble et al. 1987; Moore and Wondzell 2005). In contrast, planting of new forests reduces runoff (Jackson et al. 2005; Mark and Dickinson 2008; NRC 2008). Planting forests to sequester carbon can therefore have unintended side effects of reducing water yields and availability of freshwater (Jackson et al. 2005; Mark and Dickinson 2008). On average, plantation forests have 38% less runoff than the non-forest vegetation they replace, and in 13% of the cases, streams dried up completely in at least 1 year (Jackson et al. 2005). More subtle vegetation changes also alter runoff.

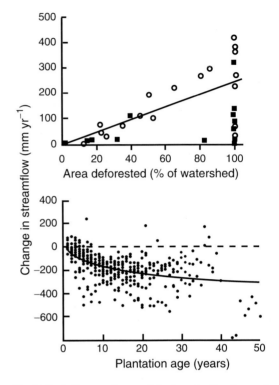

Fig. 4.20 Influence of removal (*top*) or planting (*bottom*) of trees on changes in streamflow. Streamflow in the southeastern U.S. (*open symbols*) and the more arid southwest (*closed symbols*) increases linearly with the proportion of the drainage basin that is deforested. Increased streamflow after forest harvest is least pronounced in arid ecosystems. Data from NRC (2008). Conversely, planting trees in previously unforested watersheds reduces stream flow almost immediately in watersheds sampled throughout the world. Redrawn from Jackson et al. (2005)

Conifer forests produce less runoff than deciduous forests because of their greater leaf area for interception and their longer season for evapotranspiration (Swank and Douglass 1974; Jones and Post 2004). Changes in climate, fire regime, insect outbreaks that alter vegetation structure and composition generally have predictable effects on evapotranspiration and runoff (NRC 2008).

Vegetation also influences the seasonality of runoff. Deforestation, for example, typically increases overland flow and reduces infiltration, causing larger peak flows of streams during storms and reduced flow between precipitation events. This increases the risk of flooding and reduces water flows during dry periods.

Summary

The energy and water budgets of ecosystems are inextricably linked because net radiation is the major driving force for evapotranspiration, and evapotranspiration is a large component of both water and energy flux from ecosystems. Net radiation is the balance between incoming and outgoing short- and longwave radiation. Ecosystems affect net radiation primarily through albedo (shortwave reflectance), which depends on the reflectance of individual leaves and other surfaces and on canopy roughness, which depends on canopy height and complexity. Most absorbed energy is released to the atmosphere as latent heat flux (evapotranspiration) and sensible heat flux. Latent heat flux cools the surface and transfers water vapor to the atmosphere, whereas sensible heat flux warms the surface air. The Bowen ratio, i.e., the ratio of sensible to latent heat flux, determines the strength of the coupling of the water cycle to the energy budget. This coupling is strongest in moist ecosystems.

Water enters terrestrial ecosystems primarily as precipitation and leaves as evapotranspiration and runoff. Water moves through ecosystems in response to gradients in water potential. Water enters the ecosystem and moves down through the soil in response to gravity. Available water in the soil moves along a film of liquid water through the soil–plant–atmosphere continuum in response to a gradient in water potential that is driven by transpiration (evaporation from the cell surfaces inside leaves). Evapotranspiration from canopies depends on the driving forces for evaporation (net radiation and VPD of the air) and two conductance terms, the aerodynamic and the surface conductance. Aerodynamic conductance depends on the degree to which the canopy is coupled to the atmosphere, which varies with canopy height and aerodynamic roughness. Surface conductance depends on the stomatal conductance of leaves in the canopy and on soil evaporation in sparsely vegetated ecosystems. Stomatal and surface conductances are relatively similar among natural ecosystems, but are somewhat higher in crop systems. Climate influences evapotranspiration both directly and through its effect on soil water availability, which determines stomatal conductance. Vegetation influences evapotranspiration through its effect on plant height and canopy roughness (which govern aerodynamic conductance) and on stomatal conductance (which influences surface conductance and the plant response to soil moisture).

The partitioning of water loss between evapotranspiration and runoff depends primarily on water storage in the rooting zone and the rate of evapotranspiration. Runoff is the leftover water that drains from the ecosystem at times when precipitation exceeds evapotranspiration plus any increase in water storage. Human activities alter the hydrologic cycle primarily through changes in land cover and use, which affect evapotranspiration and soil-water storage.

Review Questions

1. What climatic and ecosystem properties govern energy absorbed by an ecosystem?

2. What are the major avenues by which energy absorbed by an ecosystem is exchanged with the atmosphere? What determines the total energy exchange? What determines the relative importance of the pathways by which energy is exchanged?

3. What are the consequences of transpiration for ecosystem energy exchange and for the linkage between energy and water budgets of an ecosystem?

4. How might global changes in climate and land use alter the components of energy exchange in an ecosystem?

5. What determines the balance among the major pathways of water movement in an ecosystem, for example between evaporation, transpiration, and runoff? How do climate, soils, and vegetation influence the pools and fluxes of water in an ecosystem?

6. What are the mechanisms driving water absorption and loss from plants? How do plant properties influence water absorption and loss?

7. How do the controls over water loss from plant canopies differ from the controls at the level of individual leaves?

8. Describe how grassland and forests differ in properties that influence wet-canopy evaporation, transpiration, soil evaporation, infiltration, and runoff. What will be the consequences for runoff and for regional climate of a policy that encourages the replacement of grasslands with forests so as to increase terrestrial carbon storage?

Additional Reading

Bonan, G.B. 2008. *Ecological Climatology: Principles and Applications.* 2nd edition. Cambridge University Press, Cambridge.

Dawson, T.E. 1993. Water sources of plants as determined from xylem-water isotopic composition: Perspectives on plant competition, distribution, and water relations. Pages 465–496 *in* J.R. Ehleringer, A.E. Hall, and G.D. Farquhar, editors. *Stable Isotopes and Plant Carbon-Water Relations.* Academic Press, San Diego.

Jackson, R.B., J.S. Sperry, and T.E. Dawson. 2000. Root water uptake and transport: Using physiological processes in global predictions. *Trends in Plant Science* 5:482–488.

Jarvis, P.G., and K.G. McNaughton. 1986. Stomatal control of transpiration: Scaling up from leaf to region. *Advances in Ecological Research* 15:1–49.

Kelliher, F.M., R. Leuning, M.R. Raupach, and E.-D. Schulze. 1995. Maximum conductances for evaporation from global vegetation types. *Agricultural and Forest Meteorology* 73:1–16.

NRC. 2008. *Hydrologic Effects of a Changing Forest Landscape.* National Academies Press, Washington.

Oke, T.R. 1987. *Boundary Layer Climates.* 2nd Edition. Methuen, London.

Schulze, E.-D., F.M. Kelliher, C. Körner, J. Lloyd, and R. Leuning. 1994. Relationship among maximum stomatal conductance, ecosystem surface conductance, carbon assimilation rate, and plant nitrogen nutrition: A global ecology scaling exercise. *Annual Review of Ecology and Systematics* 25:629–660.

Sperry, J.S. 1995. Limitations on stem water transport and their consequences. Pages 105–124 *in* B.L. Gartner, editor. *Plant Stems: Physiology and Functional Morphology.* Academic Press, San Diego.

Waring, R.H. and S.W. Running. 2007. *Forest Ecosystems: Analysis at Multiple Scales.* 3rd edition. Academic Press, San Diego.

Carbon Inputs to Ecosystems

5

Photosynthesis by plants provides the carbon and energy that drive most biological processes in ecosystems. This chapter describes the controls over carbon input to ecosystems.

Introduction

The energy fixed by photosynthesis directly supports plant growth and produces organic matter that is consumed by animals and soil microbes. The carbon derived from photosynthesis makes up about half of the organic matter on Earth; hydrogen and oxygen account for most of the rest. Human activities have radically modified the rate at which carbon enters the terrestrial biosphere by changing most of the controls over this process. We have increased the quantity of atmospheric CO_2 by 35% to which terrestrial plants are exposed. At regional and global scales, we have altered the availability of water and nutrients, the major soil resources that determine the capacity of plants to use atmospheric CO_2. Finally, through changes in land cover and the introduction and extinction of species, we have changed the regional distribution of the carbon-fixing potential of the terrestrial biosphere. Because of the central role that carbon plays in the climate system (see Chap. 2), the biosphere, and society, it is critical that we understand the factors that regulate its cycling through plants and ecosystems. We

address carbon inputs to ecosystems through photosynthesis in this chapter and the carbon losses from plants and ecosystems in Chaps. 6 and 7, respectively. The balance of these processes governs the patterns of carbon accumulation and loss in ecosystems and the carbon distribution between the land, atmosphere, and ocean.

A Focal Issue

Carbon and water exchange through pores (stomata) in the leaf surface governs the efficiency with which increasingly scarce water resources support food production for a growing human population. Open stomata (Fig. 5.1) maximize carbon gain and productivity when water is abundant, but at the cost of substantial water loss. Partial closure of stomata under dry conditions reduces carbon gain but increases the efficiency with which water supports plant growth. What constrains the capacity of the biosphere to gain carbon? Where and in what seasons does most photosynthesis occur? How do plants regulate the balance between carbon gain and water loss? Application of current understanding of the controls over tradeoffs between carbon gain and water loss could reduce the likelihood of a "train wreck" resulting from current trends in increasing food demands and declining availability of freshwater to support agricultural production.

F.S. Chapin, III et al., *Principles of Terrestrial Ecosystem Ecology*,
DOI 10.1007/978-1-4419-9504-9_5, © Springer Science+Business Media, LLC 2011

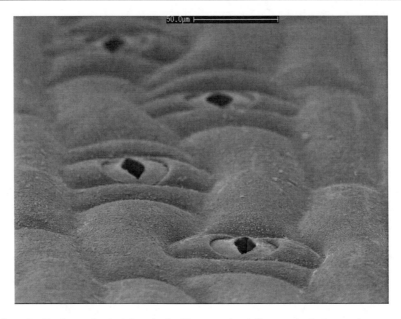

Fig. 5.1 Surface of a *Tradescantia virginiana* leaf with open stomatal pores. Selection for plants that differ in stomatal density and physiological regulation of stomatal opening influences both the maximum rate and the efficiency with which plants use water to gain carbon. Photograph courtesy of Peter Franks

Overview of Carbon Inputs to Ecosystems

Photosynthesis is the process by which most carbon and chemical energy enter ecosystems. The proximate controls over photosynthesis at the cellular or leaf level are the availability of photosynthetic reactants such as light energy and CO_2; temperature, which governs reaction rates; and the availability of nitrogen, which is required to produce photosynthetic enzymes. Photosynthesis at the scale of ecosystems is termed **gross primary production** (GPP). Like photosynthesis by individual cells or leaves, GPP varies diurnally and seasonally in response to variations in light, temperature, and nitrogen supply. *Differences among ecosystems* in annual GPP, however, are determined primarily by the quantity of photosynthetic tissue and the duration of its activity (Fig. 5.2). These, in turn, depend on the availability of soil resources (water and nutrients), climate, and time since disturbance. In this chapter, we explore the mechanisms behind these causal relationships.

Carbon is the main element that plants reduce with energy derived from the sun. Carbon and energy are therefore tightly linked as they enter, move through, and leave ecosystems. Photosynthesis uses light energy (i.e., radiation in the visible portion of the spectrum) to reduce CO_2 and produce carbon-containing organic compounds. This organic carbon and its associated energy are then transferred among components within the ecosystem and are eventually released to the atmosphere by respiration or combustion.

The energy content of organic matter varies among carbon compounds, but for whole tissues, it is relatively constant at about 20 kJ g^{-1} of ash-free dry mass (Golley 1961; Larcher 2003; Fig. 5.3). The carbon concentration of organic matter is also variable but averages about 45% of dry weight in herbaceous tissues and 50% in wood (Gower et al. 1999; Sterner and Elser 2002). Both the carbon and energy contents of organic matter are greatest in materials such as seeds and animal fat that have high lipid content and are lowest in tissues with high concentrations of minerals or organic acids. Because of the relative constancy

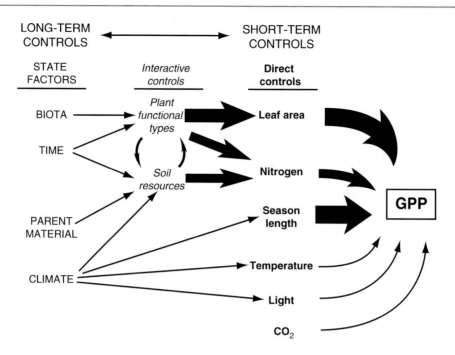

Fig. 5.2 The major factors governing temporal and spatial variation in gross primary production (GPP) in ecosystems. These controls range from proximate controls, which determine the diurnal and seasonal variations in GPP, to the interactive controls and state factors, which are the ultimate causes of ecosystem differences in GPP. Thickness of the *arrows* indicates the strength of the direct and indirect effects. The factors that account for most of the variation among ecosystems in GPP are leaf area and length of the photosynthetic season, which are ultimately determined by the interacting effects of soil resources, climate, vegetation, and disturbance regime

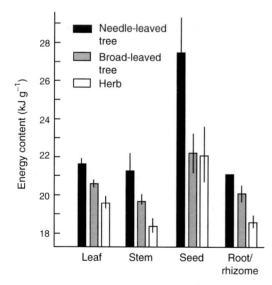

Fig. 5.3 Energy content of major tissues in conifer trees, broad-leaved trees, and broad-leaved herbs. Compounds that contribute to a high energy content include lipids (seeds), terpenes and resins (conifers), proteins (leaves), and lignin (woody tissues). Values are expressed per gram of ash-free dry mass. Data from Larcher (2003)

of the carbon and energy contents of organic matter, carbon, energy, and biomass have been used interchangeably as currencies of the carbon and energy dynamics of ecosystems. The preferred units differ among subfields of ecology, depending on the processes that are of greatest interest or are measured most directly. Production studies, for example, typically focus on biomass, trophic studies on energy, and gas exchange studies on carbon.

Biochemistry of Photosynthesis

The biochemistry of photosynthesis governs the environmental controls over carbon inputs to ecosystems. Photosynthesis involves two major groups of reactions: The **light-harvesting reactions** (or light-*dependent* reactions) transform light energy into temporary forms of chemical energy (ATP and NADPH; Lambers et al. 2008). The

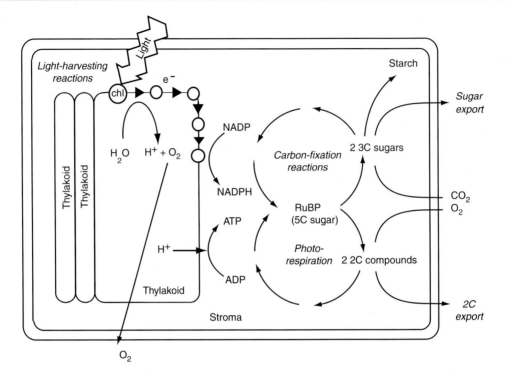

Fig. 5.4 A chloroplast, showing the location of the major photosynthetic reactions. The light-harvesting reactions occur in the **thylakoid** membranes; chlorophyll (chl) absorbs visible light and funnels the energy to reaction centers, where water inside the thylakoid is split to H⁺ and O₂, and resulting electrons are passed down an electron-transport chain in the thylakoid membrane, ultimately to NADP, producing NADPH. During this process, protons move across the thylakoid membrane to the **stroma**, and the proton (H⁺) gradient drives the synthesis of ATP. ATP and NADPH provide the energy to synthesize ribulose-bisphosphate (RuBP), which reacts either with CO_2 to produce sugars and starch (carbon-fixation reactions of photosynthesis) or with O_2 to produce two-carbon intermediates (photorespiration) and ultimately CO_2. Through either carbon fixation or photorespiration, ADP and NADP are regenerated to become reactants in the production of additional ATP and NADPH. The net effect of photosynthesis is to convert light energy into chemical energy (sugars and starches) that is available to support plant growth and maintenance

carbon-fixation reactions (or light-*independent* reactions, sometimes called the dark reaction) use the products of the light-harvesting reactions to convert CO_2 into sugars, a more permanent form of chemical energy that can be stored, transported, or metabolized. Both groups of reactions occur simultaneously in the light in **chloroplasts**, which are organelles inside photosynthetic cells (Fig. 5.4). In the light-harvesting reactions, **chlorophyll** (a light-absorbing pigment) captures energy from visible light. Absorbed radiation is converted to chemical energy (NADPH and ATP), and oxygen is produced as a waste product. Visible radiation accounts for 40% of incoming solar radiation (see Chap. 2), which places an upper limit on the

potential efficiency of photosynthesis in converting solar radiation into chemical energy.

The carbon-fixation reactions of photosynthesis use the chemical energy (ATP and NADPH) from the light-harvesting reactions to reduce CO_2 to sugars. The rate-limiting step in the carbon-fixation reactions is the reaction of a five-carbon sugar (ribulose-bisphosphate [RuBP]) with CO_2 to form two three-carbon organic acids (phosphoglycerate), which are then reduced using ATP and NADPH from the light reactions to form three-carbon sugars (glyceraldehyde 3-phosphate). The initial attachment of CO_2 to a carbon skeleton is catalyzed by the enzyme ribulose-bisphosphate carboxylase-oxygenase (**Rubisco**). The rate of this reaction is generally limited by the products of the light-harvesting reaction and by the concentration of CO_2 in the chloroplast. A surprisingly high concentration of Rubisco is required for carbon fixation. Rubisco accounts for about 25% of the nitrogen in photosynthetic cells, and other photosynthetic enzymes make up an additional 25%. The remaining enzymatic steps in the carbon-fixation reactions use ATP and NADPH from the light-harvesting reactions to convert some molecules of the three-carbon sugar (glyceraldehyde 3-phosphate) to RuBP, thus closing the photosynthetic carbon reduction cycle, and convert the rest to the six-carbon sugar, glucose, that is transported out of the chloroplast (Fig. 5.4). The most notable features of the carbon-fixation reactions are: (1) their large nitrogen requirement for Rubisco and other photosynthetic enzymes; (2) their dependence on the products of the light-harvesting reactions (ATP and NADPH), which in turn depend on **irradiance**, i.e., the light received by the photosynthetic cell; and (3) their frequent limitation by CO_2 delivery to the chloroplast. The basic biochemistry of photosynthesis therefore dictates that this process must be sensitive to light and CO_2 availability over timescales of milliseconds to minutes and sensitive to nitrogen supply over timescales of days to weeks (Fig. 5.2; Evans 1989).

Rubisco is both a **carboxylase**, which initiates the carbon-fixation reactions of photosynthesis, and an **oxygenase**, which catalyzes the reaction between RuBP and oxygen (Fig. 5.4). Early in the evolution of photosynthesis on Earth, oxygen concentrations were very low, and CO_2 concentrations were high, so the oxygenase activity of this enzyme occurred at negligible rates (Sage 2004). The oxygenase initiates a series of steps that break down sugars to CO_2. This process of **photorespiration** immediately respires away 20–40% of the carbon fixed by photosynthesis and regenerates ADP and NADP in the process. Why do plants have such an inefficient system of carbon acquisition, by which they immediately lose a third of the carbon that they acquire through photosynthesis? Photorespiration is best viewed as a carbon recovery process. Photorespiration recycles about 75% of the carbon processed by the oxygenase activity of Rubisco at a cost of two ATPs and one NADPH to produce one CO_2 and

one three-carbon acid (phosphoglycerate), which can be recycled back to RuBP. If the plant were to acquire this phosphoglycerate solely through assimilation of three new CO_2 molecules, the cost would be 9 ATP and 6 NADPH. Photorespiration may also act as a safety valve by providing a supply of reactants (ADP and NADP) to the light reaction under conditions in which an inadequate supply of CO_2 limits the rate at which these reactants can be regenerated by carbon-fixation reactions. In the absence of photorespiration, continued light harvesting produces oxygen radicals that destroy photosynthetic pigments.

Plants have additional lines of defense against excessive energy capture that are at least as important as photorespiration. Terrestrial plants and algae in shallow coral reefs, for example, have a **photoprotection** mechanism involving changes in pigments of the **xanthophyll cycle**. When excitation energy in the light-harvesting reactions exceeds the capacity of these reactions to synthesize ATP and NADPH, the xanthophyll pigment is converted to a form that receives this excess absorbed energy from the excited chlorophyll and dissipates it harmlessly as heat (Demming-Adams and Adams 1996). This processing of excess energy under high light prevents **photodestruction** of photosynthetic pigments under these conditions.

The photosynthetic reactions described above are known collectively as C_3 **photosynthesis** because two molecules of the three-carbon acid, phosphoglycerate are the initial products of carbon fixation. *C_3 photosynthesis is the fundamental photosynthetic pathway of all photosynthetic organisms on Earth*, although there are important variations on this theme that we discuss later. Plant chloroplasts, for example, have many similarities to, and probably evolved from, symbiotic bluegreen photosynthetic bacteria. Other carbon-fixation reactions contribute to the photosynthesis of some terrestrial plants (**C_4 photosynthesis** and **Crassulacian Acid Metabolism** or **CAM**). These reactions initially produce a four-carbon acid that is subsequently broken down to release CO_2 that enters the normal C_3 photosynthetic pathway to produce three-carbon sugars. However, the bottom line is that

C_3 photosynthesis is the fundamental mechanism by which carbon enters all ecosystems, so an understanding of its environmental controls provides considerable insight into the carbon dynamics of ecosystems.

Net photosynthesis is the net rate of carbon gain measured at the level of individual cells or leaves. It is the balance between simultaneous CO_2 fixation and respiration of photosynthetic cells in the light (including both photorespiration and mitochondrial respiration). Respiration rate is proportional to protein content, so photosynthetic cells and leaves with a high capacity for photosynthesis (lots of photosynthetic protein), also lose a lot of carbon due to their high respiration rate. The **light compensation point** (irradiance at which photosynthesis just balances respiration) is therefore higher in cells or leaves that have a high photosynthetic capacity. There is therefore a tradeoff between the capacity of plants to photosynthesize at high light (lots of protein and high photosynthetic capacity) and their performance at low light (less protein, lower respiration rate, and positive net photosynthesis at low light availability, i.e., a low light compensation point).

Plants adjust the components of photosynthesis, so the energy trapped by light-harvesting reactions closely matches the energy needed for the CO_2-fixation reactions. As plants produce new cells over days to weeks, protein synthesis is distributed between light-harvesting vs. carbon-fixing enzymes so that capacities for light harvesting and carbon fixation are approximately balanced under the typical light and CO_2 environment of the cell or leaf. Plants increase their investment in *light-harvesting capacity* in low-light environments and their *carbon-fixing capacity* at high light. Total **photosynthetic capacity** reflects the quantity of photosynthetic enzymes, which depends on nitrogen acquisition from their environment. Once a photosynthetic cell is produced, there is limited capacity to adjust the proportions of light-harvesting and carbon-fixing enzymes.

At low light, where the supply of ATP and NADPH from the light-harvesting reactions limits the rate of carbon fixation, net photosynthesis

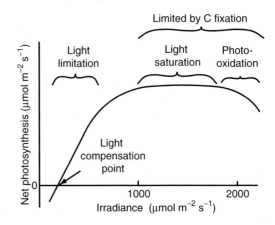

Fig. 5.5 Relationship of net photosynthetic rate to photosynthetically active radiation and the processes that limit photosynthesis at different irradiances. The linear increase in photosynthesis in response to increased light (in the range of light limitation) indicates relatively constant light-use efficiency. The light compensation point is the minimum irradiance at which the leaf shows a net gain of carbon

increases linearly with increasing light (Fig. 5.5). The slope of this line (the **quantum yield** of photosynthesis) is a measure of the efficiency with which photosynthetic cells use absorbed light to produce sugars. Quantum yield is similar (about 1–4% of the incoming light energy) among all C_3 plants (both aquatic and terrestrial) at low light in the absence of environmental stress (Kalff 2002; Lambers et al. 2008). At high irradiance, photosynthesis becomes **light saturated**, i.e., it no longer responds to changes in light supply, due to the finite capacity of light-harvesting reactions to capture light. As a result, light energy is converted less efficiently into sugar energy at high light. **Photosynthetic capacity** (maximum photosynthetic rate measured at light saturation) depends on the quantity of photosynthetic enzymes in the cell and is generally higher in large-celled algal species and rapidly growing terrestrial species that characterize nutrient-rich waters and lands, respectively. Photosynthesis declines at extremely high light, when the xanthophylls cycle photoprotective process in the chloroplast are overwhelmed, due to **photo-oxidation** of photosynthetic enzymes and pigments (Kalff 2002; Mann and Lazier 2006; Lambers et al. 2008).

In the next sections, we describe how environmental controls over photosynthesis operate in aquatic and terrestrial ecosystems. We begin with aquatic systems, where most primary producers are single-celled organisms (phytoplankton), and water seldom limits photosynthesis, thus simplifying the nature of environmental controls over carbon entry to the ecosystem. We then add the additional complexities found in terrestrial ecosystems.

Pelagic Photosynthesis

Light Limitation

Photosynthesis in pelagic (open-water) ecosystems of lakes and the ocean depends on light availability and phytoplankton biomass. Light enters water at the surface of lakes and the ocean and decreases exponentially with depth:

$$I_z = I_o e^{-kz} \qquad (5.1)$$

where I is the **irradiance** (the quantity of radiant energy received at a surface per unit time) at depth z (m), I_o is the irradiance at the water surface; and k is the extinction coefficient. Light reduction through the water column results from absorption by water, chlorophyll, dissolved organic substances, and organic or sediment particles. In the

clear water of the open-ocean and **oligotrophic** (low-nutrient) lakes, water accounts for most of the energy absorption, and high-energy blue light penetrates to the greatest depth, up to 50–100 m in clear lakes (Kalff 2002) and 200 m in the open ocean (Fig. 5.6; Valiela 1995). In **eutrophic** (high-nutrient) lakes and rivers, chlorophyll absorbs most of the light, which may penetrate only a few meters or less. Tannins absorb most light in tea-colored oligotrophic lakes in acidic low-nutrient landscapes. The depth of light penetration has two important consequences for pelagic ecosystems. First, it determines the depth of the **euphotic zone**, where there is enough light to support phytoplankton growth, i.e., where their photosynthesis exceeds respiration (see Chap. 6). This is often defined arbitrarily as the depth at which light is 1% of that available at the surface, although some phytoplankton photosynthesis occurs at even lower light intensities (Kalff 2002). In small, shallow lakes, which are by far the most numerous, the euphotic zone extends to the lake bottom, and much of the production occurs on the lake bottom, particularly in nutrient-poor settings (Vadeboncoeur et al. 2002; Vander Zanden et al. 2006; Vadeboncoeur et al. 2008). Second, the depth of light penetration in lakes influences stratification because most of the absorbed solar radiation is converted to heat, which reduces water density and promotes stratification (warmer less dense water at the surface). Eutrophic lakes

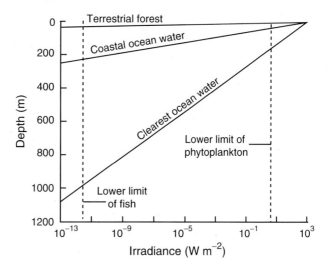

Fig. 5.6 Light availability at different distances beneath the surface of a forest canopy (Chazdon and Fetcher 1984) and the coastal and open ocean (Valiela 1995). Modified from Valiela (1995)

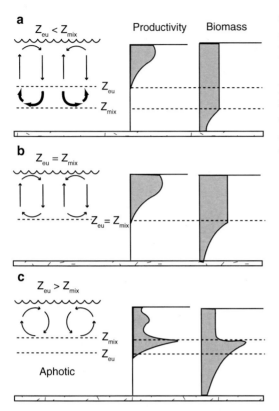

Fig. 5.7 Influence of the relative depths of the euphotic zone (z_{eu}) and mixed layer (z_{mix}) on the vertical distribution of phytoplankton and biomass. Redrawn from Thornton et al. (1990)

with shallow light penetration therefore tend to show greatest stratification and are most resistant to wind-driven mixing.

The distribution of photosynthesis through the water column depends on the depth distribution of phytoplankton and their photosynthetic response to light intensity (Valiela 1995; Kalff 2002). Mixing of the surface water typically occurs more rapidly (e.g., an hour or less) than phytoplankton can produce new cells (about a day; see Fig. 2.21), so turbulent mixing rather than cellular production or death determines the vertical distribution of phytoplankton and therefore the depth distribution of photosynthetic potential in the water column (Fig. 5.7; Thornton et al. 1990). When winds are calm and in sheltered lakes, other factors that influence the vertical distribution of phytoplankton include rates of cell production and mortality and the rates at which algae sink or swim. Large-bodied algae and dia-

toms with silica skeletons sink more rapidly than other phytoplankton (Kalff 2002; Mann and Lazier 2006).

Phytoplankton are like the terrestrial shade plants that will be described later. Due to their relatively low concentration of photosynthetic enzymes, they have both a low photosynthetic capacity and a low respiration rate. They therefore maintain positive net photosynthesis at the low light levels that characterize most of the water column and the depths at which cells spend most of their lives. Maximum photosynthesis in marine phytoplankton typically occurs at 5–25% of full sun, a few meters below the water surface (Valiela 1995; Mann and Lazier 2006). High light intensities that occur near the water surface on clear days reduce photosynthetic rate, but, due to turbulent mixing, phytoplankton spend relatively little time near the surface. Below the depth of maximum photosynthesis, carbon uptake declines with depth in parallel with the exponential decline in light intensity.

The depth of the euphotic zone is often similar to or less than the mixing depth of surface waters. In this case, there is a relatively uniform depth distribution of phytoplankton biomass, and the depth distribution of photosynthesis can be readily predicted from the light response curve of photosynthesis and the depth profile of light availability (Fig. 5.7b). In strongly stratified or extremely clear lakes, light sometimes penetrates more deeply than the mixed layer. In this case, there is an additional peak in phytoplankton biomass and photosynthesis at the base of the euphotic zone driven by the greater nutrient availability below the mixed layer (Fig. 5.7c). The actual depth distribution of photosynthesis is more complex than these simple rules imply because variability in mixing creates vertical and horizontal patchiness in the distribution of nutrients and phytoplankton.

In the ocean and clear lakes at high latitudes, UV-B may also contribute to low photosynthetic rates in surface waters, raising questions about whether aquatic production may have been reduced by high-latitude increases in UV-B (the "ozone holes" caused by anthropogenic CFCs; see Chap. 1). Colored dissolved organic compounds absorb UV-B radiation, so changes in these dissolved organics will likely mediate any potential UV-B impacts on aquatic ecosystems (Williamson et al. 1996; Kalff 2002). Photosynthesis at the ocean or lake surface appears to be light-limited mainly at high latitudes during winter due to low solar angles, short days, and snow-covered ice. At depth, light limits photosynthesis in all pelagic habitats.

CO$_2$ Supply

Photosynthesis is less often carbon-limited in aquatic than in terrestrial ecosystems. In marine pelagic ecosystems, for example, only 1% of the carbon in a given water volume is involved in primary production, whereas the nitrogen in this water may cycle through primary production 10–100 times a year (Thurman 1991). One reason for the apparently low responsiveness of pelagic photosynthesis to carbon supply is that inorganic carbon is available in substantial concentrations in several forms, including CO_2, bicarbonate, carbonate, and carbonic acid. When CO_2 dissolves in water, a small part is transformed to carbonic acid, which in turn dissociates to bicarbonate, carbonate, and H^+ ions with a concomitant drop in pH.

$$H_2O + CO_2 \leftrightarrow H_2CO_3 \leftrightarrow H^+ + HCO_3^- \leftrightarrow 2H^+ + CO_3^{2-} \tag{5.2}$$

As expected from these equilibrium reactions, the predominant forms of inorganic carbon are free CO_2 and carbonic acid at low pH (the equation driven to the left), soluble bicarbonate at about pH 8 (typical of ocean waters), and carbonates at high pH (equation driven to the right). Fossil-fuel emissions to the atmosphere have increased the CO_2 inputs to the ocean, driving (5.2) to the right. The resulting 30% increase in ocean acidity (H^+) tends to dissolve the carbonate shells of marine invertebrates and calcareous phytoplankton (coccolithophores) with potentially profound impacts on the functioning of marine ecosystems (see Chap. 14). Bicarbonate accounts for 90% of the inorganic carbon in most marine waters. Despite the predominance of bicarbonate in the ocean, phytoplankton in pelagic ecosystems use CO_2 as their primary carbon source. As CO_2 is consumed, it is replenished from bicarbonate (5.2). Some marine algae in the littoral zone, such as the macroalga, *Ulva*, also use bicarbonate.

It is still actively debated the extent to which marine productivity will respond directly to increasing atmospheric CO_2. Phytoplankton with low affinity for bicarbonate and most phytoplankton under eutrophic conditions increase photosynthesis and growth in response to added CO_2 (Schippers et al. 2004). **Daily photosynthesis in unpolluted freshwater ecosystems is seldom carbon-limited, just as in the ocean.** Groundwater entering freshwater ecosystems is super-saturated with CO_2 derived from root and microbial respiration in terrestrial soils (Kling et al. 1991; Cole et al. 1994). Most streams, rivers, and lakes are net sources of CO_2 to the atmosphere because the CO_2 input from groundwater generally exceeds the capacity of aquatic primary producers to use the CO_2. In addition, aquatic decomposition of both aquatic and terrestrially derived organic carbon generates a large CO_2 source within lakes and rivers (see Chap. 7; Kortelainen et al. 2006; Cole et al. 2007). Eutrophic lakes with their high plankton biomass have a greater demand for CO_2 to support photosynthesis than do oligotrophic systems, but their organic accumulation and high decomposition rate in sediments also contribute a large CO_2 input to the water column from depth. This creates a strong vertical gradient in CO_2 in stratified eutrophic lakes, with CO_2 being absorbed from the atmosphere during the day and returned at night (Carpenter et al. 2001), just as in terrestrial ecosystems. Some freshwater vascular plants such as *Isoetes* use CAM photosynthesis to acquire CO_2 at night and refix it by photosynthesis during the day (Keeley 1990). Other freshwater vascular plants transport CO_2 from the roots to the canopy to supplement CO_2 supplied from the water column.

Nutrient Limitation

Nutrients limit phytoplankton photosynthesis primarily through their effects on the production of new cells. Productivity and photosynthesis are closely linked in all ecosystems through a system of amplifying (positive) feedbacks (see Chap. 6): Photosynthesis provides the carbon and energy to produce new photosynthetic cells, which increases the quantity of photosynthesis that can occur. This feedback is particularly strong in pelagic systems, where most primary production is by phytoplankton through the production of new photosynthetic cells. Nutrients strongly limit productivity in most unpolluted aquatic ecosystems, both freshwater and marine. As nutrient availability increases, the rate of production of new cells increases but each cell maintains a relatively modest concentration of photosynthetic enzymes, which accounts for their low photosynthetic capacity and low light compensation point. In other words, phytoplankton respond to nutrient supply primarily by increasing photosynthetic biomass, not by increasing the photosynthetic capacity of individual cells. This increases the amount of phytoplankton biomass distributed through the water column but enables each cell to function in the low-light environment in which it spends most of its life (due to its low light compensation point, which is a consequence of its low photosynthetic capacity).

Phytoplankton species differ somewhat in photosynthetic capacity. Large-celled species with a high photosynthetic capacity dominate eutrophic waters, whereas small-celled **nanoplankton** (2–20 μm in diameter) and **picoplankton** (<2 μm in diameter) dominate oligotrophic waters. As described in Chaps. 6 and 9, large-celled species have an advantage in producing biomass rapidly when nutrients are readily available. In contrast, small-celled species, with their higher surface-to-volume ratio, are less limited by nutrient diffusion to the cell surface and are competitively favored in nutrient-poor waters.

Pelagic GPP

Total photosynthesis of pelagic ecosystems integrates the effects of nutrients on phytoplankton biomass and the effects of light and other environmental factors on the photosynthetic activity of individual cells. GPP is the rate of photosynthesis integrated through the water column, typically over time steps of days to a year (e.g., g C m^{-2} of ecosystem yr^{-1}). Ecosystem modeling and remote sensing have played a major role in estimating GPP in aquatic ecosystems. Turbulent mixing maintains a relatively homogeneous distribution of photosynthetic capacity throughout the surface mixed layer (constant photosynthetic capacity and light compensation point), although the efficiency with which chlorophyll traps light adjusts relatively rapidly and is greater at depth than at the surface (Flynn 2003; Mann and Lazier 2006). Because of the relatively homogeneous photosynthetic capacity through the mixed layer, chlorophyll content is a useful indicator of phytoplankton biomass. In the ocean, the vertical distribution of light absorption by chlorophyll can be estimated from satellite-derived color images of the ocean surface using **SeaWiFS** (Sea-viewing Wide Field-of-view Sensor). SeaWiFS estimates the depth profile of radiation absorbed by chlorophyll because different wavelengths of light penetrate to different depths.

As discussed earlier, the shape of photosynthesis-depth curve depends on the intensity and depth of turbulent mixing and the depth of light penetration (Fig. 5.7; Thornton et al. 1990; Kalff 2002; Mann and Lazier 2006). Lakes accumulate carbon when the total photosynthesis integrated through the water column (GPP) exceeds the total respiration. The **compensation depth** is the depth at which GPP equals phytoplankton respiration integrated through the water column. If the mixing depth is below the compensation depth, phytoplankton respiration beneath this depth exceeds photosynthesis, and they lose carbon. In the most productive pelagic ecosystems, such as eutrophic lakes and upwelling systems, the mixing depth is considerably shallower than the compensation depth.

Living on the Edge: Streams and Shorelines

Streams and littoral (shoreline) habitats have properties that depend on both terrestrial and aquatic components. On the terrestrial side, riparian vegetation benefits from a stable water supply and what is often a relatively favorable nutrient environment (see Chap. 13; Naiman et al. 2005). For this reason, salt marshes, freshwater marshes, and emergent vegetation along stable lakeshores often support high rates of photosynthesis and productivity (Valiela 1995). On the aquatic side, shading by emergent vascular plants and terrestrial vegetation largely defines the light environment of headwater streams and stable lake and stream banks, as described later.

Lotic (flowing-water) ecosystems such as streams and rivers have unique properties that distinguish them from both lakes and terrestrial systems. Primary producers of streams include **macrophytes** (large plants) such as vascular plants and mosses, **benthic** (bottom-dwelling) algae, **epiphytic** algae that attach to the surface of vascular plants, moss and macroalgae, and planktonic algae that float in slow-moving waters. The relative contribution of different primary producers to photosynthesis differs among geomorphic zones (erosional, transfer, and depositional) within the river basin and depends on patterns of flow rate, flood frequency, and substrate stability (see Chap. 3). Small headwater streams in the erosional zone of a drainage basin are often shaded by riparian vegetation, have relatively high flows (at least in some seasons), and variable nutrient inputs, depending on the dynamics of adjacent terrestrial ecosystems. Attached algae (**periphyton**), mosses, and liverworts on rocks and stable sediments generally account for most of the photosynthesis in headwater streams (Allan and Castillo 2007). As headwater streams join to form larger rivers, the greater solar input supports more photosynthesis by macrophytes along shallow stable riverbanks and by periphyton on stable riverbeds (Fig. 5.8). During periods of low flow,

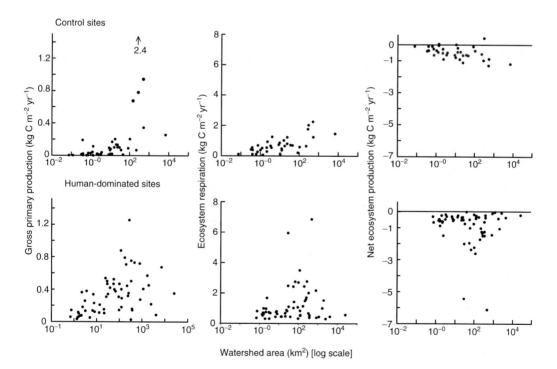

Fig. 5.8 Gross primary production, ecosystem respiration, and net ecosystem production of rivers and streams that differ in watershed area. Redrawn from Finlay 2011

benthic algae such as *Cladophora* can form extensive mats (Power 1992b). Benthic mosses are important in many cold water streams and rivers. In slow-moving rivers polluted by wastewater or agricultural runoff, pelagic algae can dominate if the doubling rate of algae is more rapid than their rate of downstream export (Allan and Castillo 2007). In general, GPP increases with increasing stream size, although it is quite variable, especially in large human-dominated drainage basins (Finlay 2011).

The controls over photosynthesis in streams and rivers vary depending on primary producer type and environment. Benthic algae in forested headwater streams, for example, have relatively low rates of photosynthesis because of low light availability, just as on the adjacent terrestrial forest floor. Removal of riparian trees and shrubs often increases photosynthesis and production in deforested headwater streams (Allan and Castillo 2007). In other cases, nutrients so strongly limit algal growth that algae show relatively little response to added light. In general, nutrients influence benthic photosynthesis primarily through their effect on the rate of production of new photosynthetic cells rather than on the photosynthetic properties of those cells, just as in lakes and the ocean. As discussed later, the high turbulence of flowing waters reduces limitation by nutrient diffusion to algal cells, so nutrient limitation tends to be less pronounced in flowing water than in pelagic ecosystems. Because of the super-saturation of groundwater with CO_2, photosynthesis in the streams that receive this groundwater is seldom CO_2-limited.

Stream macrophytes generally contribute a relatively small proportion of the photosynthetic carbon inputs to flowing-water ecosystems because of the small proportion of the stream surface area that they usually occupy. Mosses tend to dominate in shaded headwater streams, especially when waters are cold, and floating or emergent vascular plants dominate in lowland floodplain rivers and estuaries with slower currents, greater sediment accumulation, and higher light availability.

The phytoplankton present in the water column of slow-moving eutrophic rivers often originate from permanent populations in slow-moving side channels, lakes, reservoirs, or pools and get swept into the river channel. Since the maximum doubling time of most phytoplankton is once or twice per day, there is a strong inverse relationship between discharge and phytoplankton biomass in rivers. River phytoplankton populations can be self-sustaining if the currents are slow enough and nutrients are abundant enough to support rapid production throughout the year. In other cases, the rivers are seasonally seeded with phytoplankton from river-associated lakes and side channels. The roles of light and nutrients in controlling photosynthesis of river phytoplankton are similar to those in lakes. The total photosynthesis (GPP) in a section of river depends not only on the light environment and photosynthetic properties of the plants in that ecosystem but also on algal transport from upstream river segments, as discussed in Chaps. 7 and 9.

Terrestrial Photosynthesis

Photosynthetic Structure of Terrestrial Ecosystems

The physical differences between air and water account for the major photosynthetic differences between terrestrial and aquatic ecosystems. Aquatic algae are bathed in water that physically supports them and brings CO_2 and nutrients directly to photosynthetic cells. Water turbulence continuously mixes planktonic algae to different positions in the vertical light gradient. In contrast, the leaves of terrestrial plants are suspended from elaborate support structures and remain at fixed locations in the canopy. These leaves and their support structures create and respond to the vertical light gradient in terrestrial canopies. Thus, in contrast to phytoplankton, terrestrial leaves have opportunities to adjust photosynthesis to a particular light environment. Photosynthetic cells in the leaves of terrestrial plants are encased in waxy cuticles to minimize water loss, but this impermeable coating also slows CO_2 diffusion to the sites of carbon fixation in chloroplasts. Terrestrial leaves thus face tradeoffs between water loss and

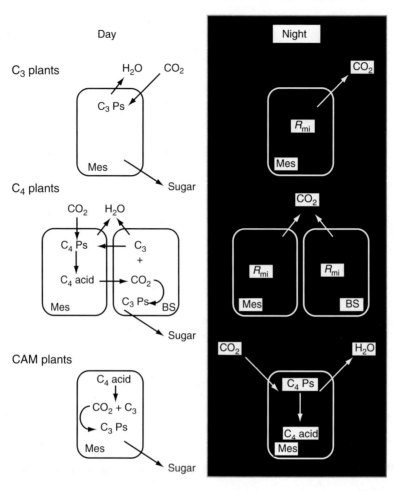

Fig. 5.9 Cellular location and diurnal timing of CO_2 fixation and water exchange in leaves with C_3, C_4, and Crassulacean Acid Metabolism (*CAM*) photosynthetic pathways. In C_3 and CAM plants, all photosynthesis occurs in mesophyll (*Mes*) cells. In C_4 plants, C_4 carbon fixation (C_4 *Ps*) occurs in mesophyll cells and C_3 fixation (C_3 *Ps*) occurs in bundle sheath (*BS*) cells. Mitochondrial respiration (R_{mi}) occurs at night. Exchanges with the atmosphere of CO_2 and water vapor occur during the day in C_3 and C_4 plants and at night in CAM plants

CO_2 absorption that are not an issue in aquatic ecosystems.

In terrestrial plants, the CO_2 used in photosynthesis diffuses along a concentration gradient from the atmosphere outside the leaf to the chloroplast. CO_2 first diffuses across a layer of relatively still air close to the leaf surface (the leaf **boundary layer**) and then through the **stomata** (small pores in the leaf surface), the diameter of which is regulated by the plant (Figs. 5.1, 5.9; Lambers et al. 2008). Once inside the leaf, CO_2 diffuses through air spaces between cells, dissolves in water on the cell surfaces, and diffuses the short distance from the cell surface to the chloroplast. C_3 leaf chloroplasts contain an enzyme, carbonic anhydrase that catalyzes the conversion of bicarbonate to dissolved CO_2, maximizing the concentration of the form of carbon (CO_2) that is fixed by Rubisco. The boundary layer, stomata, and cellular water all influence the overall diffusion of CO_2 from the free air to Rubisco, but stomata are the largest (and most variable) component of this resistance. The thin, flat shape of most leaves and the abundance of air spaces inside leaves maximize the rate of CO_2 diffusion from the bulk air to the chloroplast.

Cell walls inside the leaf are coated with a thin film of water that facilitates the efficient transfer of CO_2 from the air to the interior of cells. This water readily evaporates, and water vapor diffuses out through the stomata across the boundary layer to the atmosphere. The open stomata that are necessary for plants to gain carbon are therefore also an avenue for water loss (see Chap. 4). In other words, terrestrial plants face an inevitable tradeoff between CO_2 absorption (which is necessary to drive photosynthesis) and water loss (which must be replaced by absorption of water from the soil). This tradeoff can be as high as 400 molecules of water lost for each molecule of CO_2 absorbed. Plants regulate CO_2 absorption and water loss by changing the size of stomatal openings, which regulates **stomatal conductance**, the flux of water vapor, or CO_2 per unit driving force (i.e., for a given concentration gradient). When plants reduce stomatal conductance to conserve water, photosynthesis declines, reducing the efficiency with which plants convert light energy to carbohydrates. Plant regulation of CO_2 delivery to the chloroplast is therefore a compromise between maximizing photosynthesis and minimizing water loss and depends on the relative supplies of CO_2, light, and mineral nutrients, as described later. We now describe two photosynthetic pathways that enhance plant performance in warm, high-light environments (C_4 photosynthesis) and dry environments (CAM photosynthesis).

C_4 Photosynthesis

C_4 photosynthesis adds an additional set of carbon-fixation reactions that enable some plants to increase net photosynthesis in warm, high-light environments by reducing photorespiration. About 85% of vascular-plant species fix carbon by the C_3 photosynthetic pathway, in which Rubisco is the primary carboxylating enzyme. The first biochemically stable products of C_3 photosynthesis are three-carbon organic acids. About 3% of the global flora photosynthesizes by the **C_4 photosynthetic pathway** (Sage 2004), contributing about 23% of terrestrial GPP

(Still et al. 2003). C_4 species dominate many warm, high-light environments, particularly tropical grasslands and savannas. C_4-dominated ecosystems account for nearly a third of the ice-free terrestrial surface (see Table 6.6) and are therefore quantitatively important in the global carbon cycle. In C_4 photosynthesis, phosphoenolpyruvate (PEP) is first carboxylated by **PEP carboxylase** in mesophyll cells to produce four-carbon organic acids (Fig. 5.9). These organic acids are transported to specialized **bundle sheath cells**, where they are decarboxylated. The CO_2 released from the organic acids then enters the normal C_3 pathway of photosynthesis to produce sugars that are exported from the leaf. There are three ecologically important features of the C_4 photosynthetic pathway:

First, C_4 acids move to the bundle sheath cells, where they are decarboxylated, concentrating CO_2 at the site where Rubisco fixes carbon. This increases the efficiency of carboxylation by Rubisco because it increases the concentration of CO_2 relative to O_2, which would otherwise compete for the active site of the enzyme. Apparent photorespiration measured at the leaf level is low in C_4 plants because most of the RuBP in the bundle sheath chloroplasts reacts with CO_2 rather than with O_2 and because the PEP carboxylase in the mesophyll cells scavenges any photorespired CO_2 that diffuses away from the bundle sheath cells.

Second, PEP carboxylase draws down the concentration of CO_2 inside the leaf to a greater extent than does Rubisco. This increases the CO_2 concentration gradient between the external air and the internal air spaces of the leaf. A C_4 plant can therefore absorb CO_2 with more tightly closed stomata than can a C_3 plant, thus reducing water loss.

Third, the net cost of regenerating the carbon acceptor molecule (PEP) of the C_4 pathway is two ATPs for each CO_2 fixed, a 30% increase in the energy requirement of photosynthesis compared to C_3 plants.

The major advantage of the C_4 photosynthetic pathway is increased carboxylation under conditions that would otherwise favor photorespiration (Sage 2004). Due to their lack of photorespiration, which increases exponentially with rising temperature, C_4 plants maintain higher rates of net photosynthesis at high temperatures than do C_3 plants; this explains the success of C_4 plants in warm environments. C_4 photosynthesis initially evolved with similar frequency in mesic, arid, and saline environments, and today's C_4 plants appear to be no more drought tolerant than C_3 plants (Sage 2004). Nonetheless, the low stomatal conductance of C_4 plants appears to pre-adapt them to dry conditions, so C_4 genera now occur in a wider range of dry habitats than their C_3 counterparts (Osborne and Freckleton 2009). The main disadvantage of the C_4 pathway is the additional energy cost for each carbon fixed by photosynthesis, which is best met under high-light conditions (Edwards and Smith 2010). The C_4 pathway is therefore most advantageous in warm, high-light conditions, such as tropical grasslands and marshes. The C_4 pathway occurs in 18 plant families and has evolved independently at least 45 times (Sage 2004). C_4 species first became abundant in the late Miocene 6–8 million years ago, probably triggered by a global decline in atmospheric CO_2 concentration (Cerling 1999). C_4 grasslands expanded during glacial periods, when CO_2 concentrations declined, and retracted at the end of glacial periods, when atmospheric CO_2 concentration increased, suggesting that the evolution of C_4 photosynthesis was tightly tied to variations in atmospheric CO_2 concentration. However, there is little geographic variation in atmospheric CO_2 concentration, so the current geographic distribution of C_4 plants appears to be controlled primarily by temperature and light availability, rather than by CO_2 concentration.

C_4 plants have an isotopic signature that allows tracking of their past and present role in ecosystems. C_4 plants incorporate a larger fraction of ^{13}C than do C_3 plants during photosynthesis (Box 5.1) and therefore have a distinct isotopic signature that characterizes any organic matter that originated by this photosynthetic pathway, including animals and soil organic matter. Isotopic measurements are a valuable tool in studying ecological processes in ecosystems where the relative abundance of C_3 and C_4 plants has changed over time (Ehleringer et al. 1993).

Crassulacean Acid Metabolism

Crassulacean acid metabolism (CAM) is a photosynthetic pathway that enables plants to gain carbon under extremely dry conditions. Succulent plant species (e.g., cactuses) in dry environments, including many epiphytes in the canopies of tropical forests, gain carbon through CAM photosynthesis. CAM accounts for a small proportion of terrestrial carbon gain because it is active only under extremely dry conditions. Even in these environments, some CAM plants switch to C_3 photosynthesis when enough water is available.

In CAM photosynthesis, plants close their stomata during the day, when high tissue temperatures and low relative humidity of the external air would otherwise cause large transpirational water loss (Fig. 5.9). At night, they open their stomata, and CO_2 enters the leaf and is fixed by PEP carboxylase. The resulting C_4 acids are stored in vacuoles until the next day when they are decarboxylated, releasing CO_2 to be fixed by normal C_3 photosynthesis. Thus, in CAM plants there is a *temporal* (day-night) separation of C_3 and C_4 CO_2 fixation, whereas in C_4 plants there is a *spatial* separation of C_3 and C_4 CO_2 fixation between bundle sheath and mesophyll cells. CAM photosynthesis is energetically expensive, like C_4 photosynthesis; it therefore occurs primarily in dry, high-light environments such as deserts, shallow rocky soils, and canopies of tropical forests. CAM photosynthesis allows some plants to gain carbon under extremely dry conditions that would otherwise preclude carbon fixation in ecosystems.

CO_2 Limitation

Plants adjust the components of photosynthesis, so physical and biochemical processes co-limit carbon fixation. Photosynthesis operates most efficiently when the rate of CO_2 diffusion into the leaf matches the biochemical capacity of the leaf to fix CO_2. Terrestrial plants regulate the components of photosynthesis to approach this balance, as seen from the response of photosynthesis

Box 5.1 Carbon Isotopes

The three isotopic forms of carbon (^{12}C, ^{13}C, and ^{14}C) differ in their number of neutrons but have the same number of protons and electrons. The additional atomic mass causes the heavier isotopes to react more slowly in some reactions, particularly in the carboxylation of CO_2 by Rubisco. Carboxylating enzymes preferentially fix the lightest of these isotopes of carbon (^{12}C). C_3 plants generally have a relatively high CO_2 concentration inside the leaf, due to their high stomatal conductance. Under these circumstances, Rubisco **discriminates** against the heavier isotope ^{13}C, causing $^{13}CO_2$ to accumulate within the airspaces of the leaf. $^{13}CO_2$ therefore diffuses out of the leaf through the stomata along a concentration gradient of $^{13}CO_2$ at the same time that $^{12}CO_2$ is diffusing into the leaf. In C_4 and CAM plants, in contrast, PEP carboxylase has such a high affinity for CO_2 that it reacts with most of the CO_2 that enters the leaf, resulting in relatively little discrimination against $^{13}CO_2$. Consequently, the ^{13}C concentrations of CAM and C_4 plants are much higher (less negative isotopic ratios) than those of C_3 plants (Table 5.1).

This difference in isotopic composition among C_3, C_4, and CAM plants remains in any organic compounds derived from these plants. This makes it possible to calculate the relative proportions of C_3 and C_4 plants in the diet of animals by measuring the ^{13}C content of the animal tissue; this can be done even in fossil bones such as those of early humans. Changes in the isotopic composition of fossil bones are a clear indicator of changes in diet. In situations where vegetation has changed from C_3 to

Table 5.1 Representative ^{13}C concentrations (‰) of atmospheric CO_2 and selected plant and soil materials

Material	∂^{13}C (‰)[a]
PeeDee limestone standard	0.0
Atmospheric CO_2	−8
Plant material	
Unstressed C_3 plant	−27
Water-stressed C_3 plant	−25
Unstressed C_4 plant	−13
Water-stressed C_4 plant	−13
CAM plant[b]	−27 to −11
Soil organic matter	
Derived from unstressed C_3 plants	−27
Derived from C_4 or CAM plants	−13

Data from O'Leary (1988) and Ehleringer and Osmond (1989)

[a] The concentrations are expressed relative to an internationally agreed-on standard (PeeDee belemnite):

$$\partial^{13}C_{std} = 1000\left(\frac{R_{sam}}{R_{std}} - 1\right)$$

where ∂^{13}C is the isotope ratio in delta units relative to a standard, and R_{sam} and R_{std} are the isotope abundance ratios of the sample and standard, respectively (Ehleringer and Osmond 1989)

[b] Values of −11 under conditions of CAM photosynthesis; many CAM plants switch to C_3 photosynthesis under favorable moisture regimes, giving an isotopic ratio similar to that of unstressed C_3 plants

C_4 dominance (or vice versa), the organic matter in plants differs in its isotopic composition from that of the soil (and its previous vegetation). Changes in the carbon isotope composition of soil organic matter over time then provides a tool to estimate the current rates of turnover of soil organic matter that formed beneath the previous vegetation.

to the CO_2 concentration inside the leaf (Fig. 5.10). When the internal CO_2 concentration is low, photosynthesis increases approximately linearly with increasing CO_2 concentration. Under these circumstances, the leaf has more carbon-fixation capacity than it can use, and photosynthesis is limited by the rate of diffusion of CO_2 into the

leaf. The plant can increase photosynthesis only by opening stomatal pores. Alternatively, if CO_2 concentration inside the leaf is high, photosynthesis shows little response to variation in CO_2 concentration (the asymptote approached in Fig. 5.10). In this case, photosynthesis is limited by the rate of regeneration of RuBP (the compound

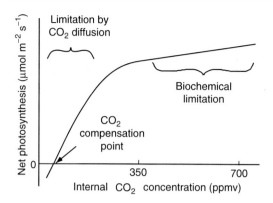

Fig. 5.10 Relationship of the net photosynthetic rate to the CO_2 concentration inside the leaf. Photosynthetic rate is limited by the rate of CO_2 diffusion into the chloroplast in the initial (*left-hand side*) linear portion of the CO_2 response curve and by biochemical processes at higher CO_2 concentrations. The CO_2 compensation point is the minimum CO_2 concentration at which the leaf shows a net gain of carbon

that reacts with CO_2), and changes in stomatal opening have little influence on photosynthesis. At high internal CO_2 concentrations, carboxylation may be limited by: (1) insufficient light (or light-harvesting pigments) to provide energy, (2) insufficient nitrogen invested in photosynthetic enzymes to process the ATP, NADPH, and CO_2 present in the chloroplast, or (3) insufficient phosphate or sugar phosphates to synthesize RuBP.

Under a wide variety of circumstances, terrestrial plants adjust the components of photosynthesis, so CO_2 diffusion and biochemistry are about equally limiting to photosynthesis (Farquhar and Sharkey 1982), causing plants to respond to both CO_2 availability and biochemical limitations (light, nitrogen, or phosphorus). Plants make this adjustment by altering stomatal conductance, which occurs within minutes, or by changing the concentrations of light-harvesting pigments or photosynthetic enzymes, which occurs over days to weeks. The general principle of co-limitation of photosynthesis by biochemistry and diffusion provides the basis for understanding most of the adjustments by individual leaves to minimize the environmental limitations of photosynthesis. Stomatal conductance is regulated, so photosynthesis usually occurs near

the break point of the CO_2-response curve (Fig. 5.10; Körner et al. 1979), where CO_2 supply and carbon-fixation capacity are about equally limiting to photosynthesis.

Changes in stomatal conductance by leaves minimize the effects of CO_2 supply on photosynthesis. The free atmosphere is so well mixed that its CO_2 concentration varies globally by only 4% – not enough to cause significant regional variation in photosynthesis. In dense canopies, photosynthesis reduces CO_2 concentration somewhat within the canopy, and soil respiration is a source of CO_2 at the base of the canopy. However, the shade leaves in the lower canopy tend to be light-limited and therefore relatively unresponsive to CO_2 concentration. Consequently, vertical variation in CO_2 concentration within the canopy has relatively little effect on whole-ecosystem photosynthesis (Field 1991).

Although spatial variation in CO_2 concentration does not explain much of the global variation in photosynthetic rate, the 35% increase in atmospheric CO_2 concentration since the beginning of the industrial revolution has caused a general increase in carbon gain by ecosystems (see Chap. 7; Canadell et al. 2007). In both growth-chamber and field studies, a doubling of CO_2 concentration increases photosynthetic rate by 30–50% (Curtis and Wang 1998; Ainsworth and Long 2005). This enhancement of photosynthesis by elevated CO_2 is most pronounced in C_3 plants, especially woody species (Ainsworth and Long 2005). Over time, most plants acclimate to elevated CO_2 by reducing photosynthetic capacity and stomatal conductance, as expected from our hypothesis of co-limitation of photosynthesis by biochemistry and diffusion. This **down-regulation** of CO_2 absorption in response to elevated CO_2 enables plants to sustain carbon uptake, while reducing transpiration rate and their water demand from soils. In this way, elevated CO_2 often stimulates plant growth more strongly by reducing moisture limitation than by its direct effects on photosynthesis. C_4 plants are often just as sensitive to the *indirect* effects of CO_2 as are C_3 plants, so the long-term effects of elevated CO_2 on the competitive balance between C_3 and C_4 plants are difficult to predict (Mooney et al. 1999).

Light Limitation

Physical environment determines light inputs to ecosystems, and leaf area governs the distribution of light within the canopy. Leaves experience large fluctuations (10- to 1,000-fold) in incident light due to changes in sun angle, cloudiness, and the location of **sunflecks** (patches of direct sunlight that penetrate a plant canopy; Fig. 5.11). The vertical distribution of leaf area, however, is the major factor governing the light environment of individual leaves. Light distribution within terrestrial canopies is approximated by an empirical relationship identical to that observed in aquatic ecosystems:

$$I_z = I_o e^{-kLz} \qquad (5.3)$$

where I is irradiance at height z (m) beneath the canopy surface, I_o is the irradiance at the top of the canopy, k is the extinction coefficient per unit leaf area, and L is the **leaf area index** (LAI; the projected leaf area per unit of ground area) above the point of measurement. The actual distribution of light through the canopy is more complex and depends on the balance of direct and diffuse radiation. LAI is a key parameter governing ecosystem processes because it determines both the area that is potentially available to absorb light and the degree to which light is attenuated through the canopy. LAI is equivalent to the total upper surface area of all leaves per area of ground (or the projected leaf area in the case of cylindrical needle-like leaves).

LAI varies widely among ecosystems but typically has values of 1–8 m² leaf m⁻² ground for ecosystems with a closed canopy. The **extinction coefficient** is a constant that describes the exponential decrease in irradiance through a canopy. It is low for vertically inclined or small leaves (e.g., 0.3–0.5 for grasses), allowing substantial penetration of direct radiation into the canopy, but high for near-horizontal leaves (0.7–0.8). Clumping of leaves around stems, as in conifers, and variable leaf angles is associated with intermediate values for k. Equation (5.3) indicates that light is distributed unevenly in an ecosystem and that the leaves near the top of the canopy capture

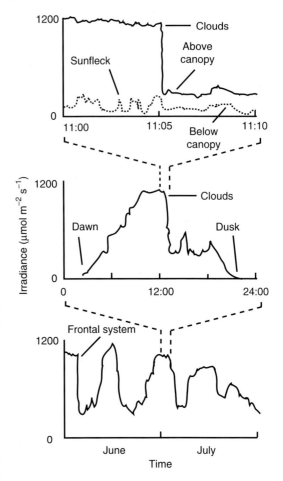

Fig. 5.11 Hypothetical time course of photosynthetically active radiation above and below the canopy of a temperate forest over minutes, hours, and months. Over the course of a few minutes, light at the top of the canopy varies with cloudiness. Beneath the canopy, light also varies due to the presence or absence of sunflecks of direct irradiance, which can last tenths of seconds to minutes. During a day, changes in solar angle and passing clouds cause large changes in light. Convective activity often increases cloudiness in the afternoon. During the growing season, seasonal changes in the solar angle and the passage of frontal systems are the major causes of variation in light. Some times of year have greater frequency of cloudiness than others due to changes in directions of the prevailing winds and the passage of frontal systems

most of the available light. Irradiance at the ground surface of a forest, for example, is often only 1–2% of that at the top of the canopy, similar to the light available at the bottom of aquatic euphotic zones (Fig. 5.6).

The shape of the light-response curve of photosynthesis in terrestrial plants is identical to that of aquatic algae (Fig. 5.5). Under light-limiting conditions, photosynthesis increases linearly with increasing light availability (constant quantum yield or light-use efficiency). As the light-harvesting capacity of chlorophyll becomes light saturated, photosynthesis reaches its maximum rate (photosynthetic capacity). At extremely high light, photosynthesis may decline due to photo-oxidation of pigments and enzymes, just as in phytoplankton (Fig. 5.5).

In response to fluctuations in light availability over minutes to hours (Fig. 5.11), plants alter stomatal conductance to adjust CO_2 supply to meet the needs of carbon-fixation reactions (Pearcy 1990; Chazdon and Pearcy 1991). Stomatal conductance increases in high light, when CO_2 demand is high, and decreases in low light, when photosynthetic demand for CO_2 is low. These stomatal adjustments result in a relatively constant CO_2 concentration inside the leaf, as expected from our hypothesis of co-limitation of photosynthesis by biochemistry and diffusion. It allows plants to conserve water under low light and to maximize CO_2 absorption at high light.

Over longer time scales (days to months), plants respond to variations in light availability by producing leaves with different photosynthetic properties. This *physiological adjustment by an organism* in response to a change in some environmental parameter is known as **acclimation**. Leaves at the top of the canopy (**sun leaves**) have more cell layers, are thicker, and therefore have a higher photosynthetic capacity per unit leaf area than do **shade leaves** produced under low light (Terashima and Hikosaka 1995; Walters and Reich 1999). The respiration rate of a tissue depends on its protein content (see Chap. 6), so the low photosynthetic capacity and protein content of shade leaves are associated with a lower respiration rate per unit area than in sun leaves. For this reason, shade leaves maintain a positive carbon balance (photosynthesis minus respiration) under lower light levels than do sun leaves (Fig. 5.12).

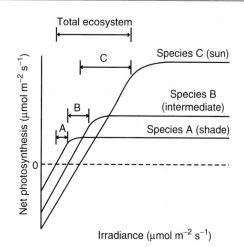

Fig. 5.12 Light response curves of net photosynthesis in plants adapted (or acclimated) to low, intermediate, and high light. *Horizontal arrows* show the range of irradiance over which net photosynthesis is positive and responds linearly to irradiance for each species and for the ecosystem as a whole. Acclimation increases the range of light availability over which net photosynthesis responds linearly to light, i.e., has a constant light-use efficiency

Plants can also produce shade leaves as a result of **adaptation**, the *genetic adjustment by a population* to maximize performance in a particular environment. Species that are *adapted* to high light and intolerant of shade typically have a higher photosynthetic capacity per unit mass or area than do shade-tolerant species, even when growing in the shade (Walters and Reich 1999). The main disadvantage of the high protein and photosynthetic rate typical of shade-intolerant species is that they also have a higher respiration rate, due to their higher protein content. Species that are adapted to low light and are tolerant of shade have a low photosynthetic capacity, but can photosynthesize at lower light levels than shade-intolerant species. In other words, they have a low light compensation point. At the light compensation point, leaf respiration completely offsets photosynthetic carbon gain, resulting in zero net photosynthesis (Fig. 5.5). A mature shaded leaf typically does not import carbon from the rest of the plant, so the leaf senesces and dies if it falls below the light compensation point for a long time. This puts an upper limit on the leaf area that an ecosystem can support, regardless of how favorable the climate and supply of soil resources may be. On average, the leaf-level light compensation point of shade-tolerant species is about half of that of shade-intolerant species (Craine and Reich 2005).

Variations in leaf angle also influence the efficiency with which a plant canopy uses light. At high light, plants produce leaves that are steeply angled, so they absorb less light (see Chap. 4). This is advantageous because it reduces the probability of overheating or photo-oxidation of photosynthetic pigments at the top of the canopy. At the same time, it allows more light to penetrate to lower leaves. Leaves at the bottom of the canopy, on the other hand, are more horizontal in orientation to maximize light capture and are produced in an arrangement that minimizes overlap with other leaves of the plant (Craine 2009).

Do differences in light availability explain the differences among ecosystems in carbon gain? In midsummer, when plants of most ecosystems are photosynthetically active, the daily input of visible light is nearly as great in the Arctic as in the tropics but is spread over more hours and is more diffuse at high latitudes (Billings and Mooney 1968). The greater daily carbon gain in the tropics than at high latitudes is therefore unlikely to be a simple function of the light available to drive photosynthesis. Neither can variation in light availability due to cloudiness explain differences among ecosystems in energy capture. The most productive ecosystems on Earth, the tropical and temperate rainforests, have a high frequency of cloudiness, whereas arid grasslands and deserts, which are less cloudy and receive nearly 10-fold more light annually, are less productive. Seasonal and interannual variations in irradiance can, however, contribute to temporal variation in carbon gain by ecosystems. Aerosols emitted by volcanic eruptions and fires, for example, can reduce solar irradiance and photosynthesis over large areas in particular years. Similarly, photosynthesis (GPP) of the Amazon rainforest is greater in the dry season than under the cloudy conditions of the wet season (Saleska et al. 2007). In summary, light availability strongly influences daily and seasonal patterns of carbon input and the distribution of

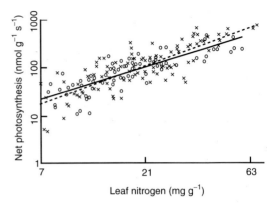

Fig. 5.13 Relationship between leaf-nitrogen concentration and maximum photosynthetic capacity (photosynthetic rate measured under favorable conditions) for plants from Earth's major biomes. *Circles* and the *solid regression line* are for 11 species from 6 biomes using a common methodology. *Crosses* and the *dashed regression line* are data from the literature. Redrawn from Reich et al. (1997)

photosynthesis within the canopy, but it is only a minor factor explaining regional variations in annual carbon inputs to ecosystems (Fig. 5.2).

Nitrogen Limitation and Photosynthetic Capacity

Vascular plant species differ 10 to 50-fold in their photosynthetic capacity. Photosynthetic capacity is the photosynthetic rate per unit leaf mass measured under favorable conditions of light, moisture, and temperature. It is a measure of the carbon-gaining potential *per unit of biomass invested in leaves*. Photosynthetic capacity correlates strongly with leaf nitrogen concentration (Fig. 5.13; Field and Mooney 1986; Reich et al. 1997, 1999; Wright et al. 2004) because photosynthetic enzymes account for a large proportion of the nitrogen in leaves (Fig. 5.2). Many ecological factors can lead to a high leaf-nitrogen concentration and therefore a high photosynthetic capacity. Plants growing in high-nitrogen soils, for example, have higher tissue nitrogen concentrations and photosynthetic rates than do the same species growing on less fertile soils. This acclimation of plants to a high nitrogen supply contributes to the high photosynthetic rates in agricultural fields and other ecosystems with a rapid nitrogen

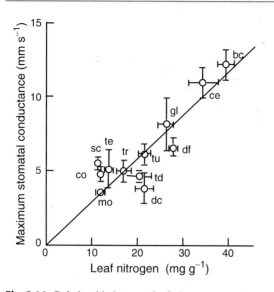

Fig. 5.14 Relationship between leaf-nitrogen concentration and maximum stomatal conductance of plants from Earth's major biomes. Each point and its standard error represent a different biome: bc, broad-leafed crops; ce, cereal crops; co, evergreen conifer forest; dc, deciduous conifer forest; df, tropical dry forest; gl, grassland; mo, monsoonal forest; sc, sclerophyllous shrub; sd, dry savanna; sw, wet savanna; tc, tropical tree crop; td, temperate deciduous broadleaved forest; te, temperate evergreen broadleaved forest; tr, tropical wet forest; tu, herbaceous tundra. Redrawn from Schulze et al. (1994)

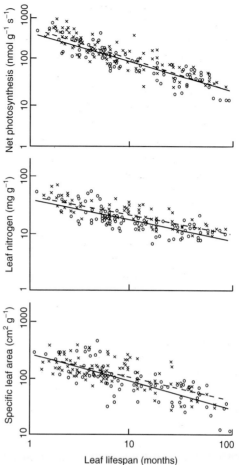

Fig. 5.15 The effect of leaf life span on photosynthetic capacity, leaf-nitrogen concentration, and specific leaf area. Symbols as in Fig. 5.13. Redrawn from Reich et al. (1997)

turnover. Many species differ in their leaf-nitrogen concentration, even when growing in the same soils. Species adapted to productive habitats usually produce leaves that are short-lived and have high tissue-nitrogen concentrations and high photosynthetic rates. Nitrogen-fixing plants also typically have high leaf-nitrogen concentrations and correspondingly high photosynthetic rates. In summary, regardless of the cause of variation in leaf-nitrogen concentration, there is always a strong positive correlation between leaf-nitrogen concentration and photosynthetic capacity (Fig. 5.13; Reich et al. 1997; Wright et al. 2004).

Plants with a high photosynthetic capacity have a high stomatal conductance, in the absence of environmental stress (Fig. 5.14), as expected from our hypothesis of co-limitation of photosynthesis by biochemistry and diffusion. This enables plants with a high photosynthetic capacity to absorb CO_2 rapidly, despite high rates of water loss.

Conversely, species with a low photosynthetic capacity conserve water as a result of their lower stomatal conductance.

There appears to be an unavoidable tradeoff between traits that maximize photosynthetic rate and traits that maximize leaf longevity (Fig. 5.15; Reich et al. 1997, 1999; Wright et al. 2004). Many plant species that grow in low-nutrient environments produce long-lived leaves because nutrients are insufficient to support rapid leaf turnover (Chapin 1980; Craine 2009). Shade-tolerant species also produce longer-lived leaves than do shade-intolerant species (Reich et al. 1999; Wright et al. 2004). Long-lived leaves typically have a low leaf-nitrogen concentration and a low photosynthetic capacity; they must therefore photosynthesize for a relatively long time to break even in their lifetime carbon budget (Gulmon and Mooney 1986; Reich et al. 1997). To survive, long-lived leaves must have enough structural rigidity to withstand drought and winter desiccation. These structural requirements cause leaves to be dense, i.e., to have a small surface area per unit of biomass, termed **specific leaf area** (SLA). Long-lived leaves must also be well defended against herbivores and pathogens, if they are to persist. This requires substantial allocation to lignin, tannins, and other non-nitrogenous compounds that deter herbivores, but also contribute to tissue mass and a low SLA.

The broad relationship among species with respect to photosynthetic rate and leaf life span is similar in all biomes; a twofold decrease in leaf life span gives rise to about a fivefold increase in photosynthetic capacity (Reich et al. 1999; Wright et al. 2004).

Plants in productive environments produce short-lived leaves with a high tissue-nitrogen concentration and a high photosynthetic capacity; this allows a large carbon return per unit of biomass invested in leaves, if enough light is available. These leaves have a high SLA, which maximizes the quantity of leaf area displayed and the light captured per unit of leaf mass. The resulting high rates of carbon gain support a high maximum relative growth rate in the absence of environmental stress or competition from other plants (Fig. 5.16; Schulze and Chapin 1987). Many early successional habitats, such as recently abandoned agricultural fields, canopy gaps, or post-fire sites, have enough light, water, and nutrients to support high growth rates and are characterized by species with short-lived leaves, high tissue-nitrogen concentration, high SLA, and high photosynthetic rate (see Chap. 12). Even in late succession, environments with high water and nutrient availability are characterized by canopy species with relatively high nitrogen concentration and photosynthetic rate. Plants in the canopy of these habitats can grow quickly to replace leaves removed by herbivores or to fill canopy gaps produced by death of branches or individuals.

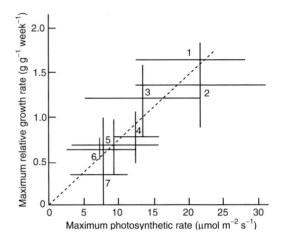

Fig. 5.16 Relationship between maximum photosynthetic rate and maximum relative growth rate for major plant growth forms: (1) agricultural crop species, (2) herbaceous sun species, (3) grasses and sedges, (4) summer deciduous trees, (5) evergreen and deciduous dwarf shrubs, (6) herbaceous shade species and bulbs; (7) evergreen conifers. Redrawn from Schulze and Chapin (1987)

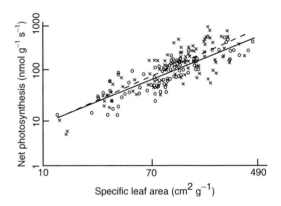

Fig. 5.17 The relationship between specific leaf area (SLA) and photosynthetic capacity. The consistency of this relationship makes it possible to use SLA as an easily measured index of photosynthetic capacity. Symbols as in Fig. 5.13. Redrawn from Reich et al. (1997)

In summary, plants produce leaves with a continuum of photosynthetic characteristics, ranging from short-lived, low-density leaves with a high nitrogen concentration and high photosynthetic rate to long-lived, dense leaves with a low nitrogen concentration and low photosynthetic rate. These correlations among traits are so consistent that SLA is often used in ecosystem comparisons as an easily measured index of photosynthetic capacity (Fig. 5.17).

There is only modest variation in photosynthetic capacity per unit leaf area because leaves with a high photosynthetic capacity per unit leaf biomass also have a high SLA. Photosynthetic capacity or assimilation rate *per unit leaf area* (A_{area}) is a measure of the capacity of leaves to capture a unit of incoming radiation. It is calculated by dividing photosynthetic (assimilation) rate per unit leaf mass (A_{mass}) by SLA.

$$A_{area} = \frac{A_{mass}}{SLA} \qquad (5.4)$$

$$(g cm^{-2} s^{-1}) = (g g^{-1} s^{-1}) / (cm^2 g^{-1})$$

There is relatively little variation in A_{area} among plants from different ecosystems (Lambers and Poorter 1992). In productive habitats, both mass-based photosynthesis and SLA are high (Fig. 5.15). In unproductive habitats, both of these parameters are low, resulting in modest variation in area-based photosynthetic rate (Lambers and Poorter 1992). To the extent that A_{area} varies among plants, it tends to be higher in species with short-lived leaves (Reich et al. 1997). Mass-based photosynthetic capacity is a good measure of the physiological potential for photosynthesis (the photosynthetic rate per unit of biomass invested in leaves). Area-based photosynthetic capacity is a good measure of the efficiency of these leaves at the ecosystem scale (photosynthetic rate per unit of available light). Variation in soil resources has a much greater effect on the quantity of leaf area produced than on the photosynthetic capacity per unit leaf area.

Water Limitation

Water limitation reduces the capacity of individual leaves to match CO_2 supply with light availability. Water stress is often associated with high light because sunny conditions correlate with low precipitation (low water supply) and with low humidity (high rate of water loss). High light also leads to an increase in leaf temperature and water vapor concentration inside the leaf and therefore greater vapor pressure deficit and water loss by transpiration (see Chap. 4). The high-light conditions in which a plant would be expected to increase stomatal conductance to minimize CO_2 limitations to photosynthesis are therefore often the same conditions in which the resulting transpirational water loss is greatest and most detrimental to the plant. This tradeoff between a response that maximizes carbon gain (stomata open) and one that minimizes water loss (stomata closed) is typical of the physiological compromises faced by plants whose physiology and growth may be limited by more than one environmental resource (Mooney 1972). When water supply is abundant, leaves typically open their stomata in response to high light, despite the associated high rate of water loss. As leaf water stress develops, stomatal conductance declines to reduce water loss (see Fig. 4.17). This decline in stomatal conductance reduces photosynthetic rate and the efficiency of using light to fix carbon (i.e., **light-use efficiency** [LUE]) below levels found in unstressed plants.

Plant acclimation and adaptation to low water is qualitatively different than adaptation to low nutrients (Killingbeck and Whitford 1996; Cunningham et al. 1999; Wright et al. 2001; Craine 2009). Plants in dry habitats typically have thicker leaves, similar leaf-nitrogen concentration, and therefore more nitrogen per unit leaf area than do plants in moist habitats. Dry-site plants also have a low stomatal conductance. This combination of traits enables dry-site plants to maintain higher rates of photosynthesis at a given rate of water loss compared to plants in moist sites (Cunningham et al. 1999; Wright et al. 2001). Dry-site leaves basically service more photosynthetic cells and photosynthetic capacity for a given stomatal conductance.

Plants in dry areas minimize water stress by reducing leaf area (by shedding leaves or producing fewer new leaves). Some drought-adapted plants produce leaves that minimize radiation absorption; their leaves reflect most incoming radiation or are steeply inclined toward the sun (see Chap. 4; Ehleringer and Mooney 1978). High radiation absorption is a *disadvantage* in dry environments because it increases leaf temperature, which increases respiratory carbon loss

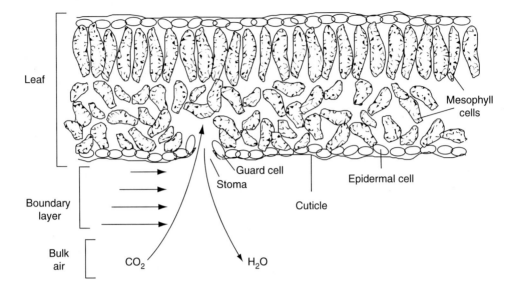

Fig. 5.18 Cross section of a leaf, showing the diffusion pathways of CO_2 and H_2O into and out of the leaf, respectively. Length of the *horizontal arrows* outside the leaf is proportional to wind speeds in the boundary layer

(see Chap. 6) and transpirational water loss (see Chap. 4). Thus plants in dry environments have several mechanisms by which they reduce radiation absorption to conserve water and carbon. The low leaf area, the reflective nature of leaves, and the steep angle of leaves are the main factors accounting for the low absorption of radiation and low carbon inputs in dry environments. In other words, plants adjust to dry environments primarily by altering leaf area and radiation absorption rather than by reducing photosynthetic capacity per unit leaf area.

Water-use efficiency (WUE) of photosynthesis is defined as the carbon gain per unit of water lost. WUE is quite sensitive to the size of stomatal openings because stomatal conductance has slightly different effects on the rates of CO_2 entry and water loss. Water leaving the leaf encounters two resistances to flow: the stomata and the boundary layer of still air on the leaf surface (Fig. 5.18). Resistance to CO_2 diffusion from the bulk air to the site of photosynthesis includes the same stomatal and boundary layer resistances *plus* an additional internal resistance associated with diffusion of CO_2 from the cell surface into the chloroplast and any biochemical limitations associated with carboxylation. Because of this additional resistance to CO_2 movement into the leaf, any change in stomatal conductance has a *proportionately* greater effect on water loss than on carbon gain. In addition, water diffuses more rapidly than does CO_2 because of its smaller molecular mass and the steeper concentration gradient that drives diffusion across the stomata. For all these reasons, as stomata close, water loss declines to a greater extent than does CO_2 absorption. The low stomatal conductance of plants in dry environments results in less photosynthesis per unit of time but greater carbon gain per unit of water loss, i.e., greater WUE. Plants in dry environments also enhance WUE by maintaining a somewhat higher photosynthetic capacity than would be expected for their stomatal conductance, thereby drawing down the internal CO_2 concentration and maximizing the diffusion gradient for CO_2 entering the leaf (Wright et al. 2001). Carbon isotope ratios in plants provide an integrated index of WUE during plant growth because the ^{13}C concentration of newly fixed carbon increases under conditions of low internal CO_2 concentration (Box 5.1; Ehleringer 1993). C_4 and CAM photosynthesis are additional adaptations that augment the WUE of plants, and ultimately ecosystems.

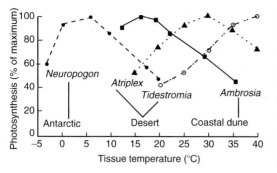

Fig. 5.19 Temperature response of photosynthesis in plants from contrasting temperature regimes. Species include antarctic lichen (*Neuropogon acromelanus*), a cool coastal dune plant (*Ambrosia chamissonis*), an evergreen desert shrub (*Atriplex hymenelytra*), and a summer-active desert perennial (*Tidestromia oblongifolia*). Redrawn from Mooney (1986)

Temperature Effects

Extreme temperatures limit carbon absorption. Photosynthetic rate is typically highest near leaf temperatures commonly experienced on sunny days (Fig. 5.19). Leaf temperature may differ substantially from air temperature due to the cooling effects of transpiration, the effects of leaf surface properties on energy absorption, and the influence of adjacent surfaces on the thermal and radiation environment of the leaf (see Chap. 4). At low temperatures, photosynthesis is limited directly by temperature, as are all chemical reactions. At high temperatures, photosynthesis also declines, due to increased photorespiration and, under extreme conditions, enzyme inactivation and destruction of photosynthetic pigments. Temperature *extremes* often have a greater effect on photosynthesis than does *average* temperature because of damage to photosynthetic machinery (Berry and Björkman 1980; Waring and Running 2007).

Several factors minimize the sensitivity of photosynthesis to temperature. The enzymatically controlled carbon-fixation reactions are typically more sensitive to low temperature than are the biophysically controlled light-harvesting reactions. Carbon-fixation reactions therefore tend to limit photosynthesis at low temperature. Plants adapted to cold climates compensate for this by producing leaves with high concentrations of leaf

nitrogen and photosynthetic enzymes, which enable carboxylation to keep pace with the energy supply from the light-harvesting reactions (Berry and Björkman 1980). This explains why arctic and alpine plants typically have high leaf-nitrogen concentrations despite low soil-nitrogen availability (Körner and Larcher 1988). Plants in cold environments also have hairs and other morphological traits that raise leaf temperature above air temperature (Körner 1999). In hot environments with an adequate water supply, plants produce leaves with high photosynthetic rates. The associated high transpiration rate cools the leaf, often reducing leaf temperature below air temperature.

In hot, dry environments, plants close stomata to conserve water, and the cooling effect of transpiration is reduced. Plants in these environments often produce small leaves, which shed heat effectively and maintain temperatures close to air temperature (see Chap. 4). In summary, despite the sensitivity of photosynthesis to short-term variation in temperature, leaf properties minimize the differences in leaf temperature among ecosystems, and plants acclimate and adapt so there is no clear relationship between temperature and average photosynthetic rate of leaves in the field, when ecosystems are compared.

Pollutants

Pollutants reduce carbon gain, primarily by reducing leaf area or photosynthetic capacity. Many pollutants, such as SO_2 and ozone, reduce photosynthesis through their effects on growth and the production of leaf area. Pollutants also directly reduce photosynthesis by entering the stomata and damaging the photosynthetic machinery, thereby reducing photosynthetic capacity (Winner et al. 1985). Plants then reduce stomatal conductance to balance CO_2 absorption with the reduced capacity for carbon fixation. This reduces the entry of pollutants into the leaf, reducing the vulnerability of the leaf to further injury. Plants growing in low-fertility or dry conditions are pre-adapted to pollutant stress because their low stomatal conductance minimizes the quantity of pollutants entering leaves. Pollutants therefore affect these plants less than they affect rapidly growing crops and other plants with high stomatal conductance.

Terrestrial GPP

GPP of terrestrial ecosystems integrates the effects of environmental factors and leaf photosynthetic properties through the canopy. GPP is the sum of the net photosynthesis by all photosynthetic tissue measured at the ecosystem scale. The controls over GPP in terrestrial ecosystems are more complex than in aquatic systems for at least three reasons: (1) Unlike aquatic systems, both the quantity and photosynthetic properties of terrestrial photosynthetic tissues change from the top to the bottom of the canopy. (2) In addition to light and nutrients, which influence photosynthesis in all ecosystems, terrestrial photosynthesis is sensitive to the availability of water and the delivery of CO_2 to photosynthetic cells. (3) The structure of the plant canopy influences the delivery of light and CO_2 to, and the loss of water from, photosynthetic cells. Despite these complexities, recent technological developments allow measurement of fluxes of CO_2 and other compounds at scales of tens to thousands of square meters, making it possible to measure whole-ecosystem carbon fluxes even in large-statured ecosystems like forests (Baldocchi 2003). These measurements, when combined with simulation modeling, permit estimation of GPP and other ecosystem carbon fluxes (see Box 7.2). In this chapter, we focus on ecological controls over GPP and consider its role in the ecosystem carbon balance in Chap. 7.

Canopy Processes

The vertical profile of leaf photosynthetic properties in a canopy maximizes GPP in terrestrial ecosystems. In contrast to pelagic ecosystems, leaves in terrestrial canopies remain fixed in the same vertical location throughout their lives. Their photosynthetic properties are therefore adapted and acclimated to the environment where they are situated. In most closed-canopy ecosystems, for example, photosynthetic capacity of individual leaves decreases exponentially through the canopy in parallel with the exponential decline in irradiance (Eq. (5.3); Hirose and Werger 1987). This is radically different from aquatic ecosystems, where turbulence causes regular mixing of the algal cells in surface waters, and algae at all depths have a low photosynthetic capacity typical of shade plants. The matching of photosynthetic capacity to light availability in terrestrial ecosystems is the response we expect from individual leaves within the canopy because it maintains the co-limitation of photosynthesis by diffusion and biochemical processes in each leaf. The matching of photosynthetic capacity to light availability occurs through the preferential transfer of nitrogen to leaves at the top of the canopy. At least three processes cause this to occur. (1) New leaves are produced primarily at the top of the canopy where light availability is highest, causing nitrogen to be transported to the top of the canopy (Field 1983; Hirose and Werger 1987). (2) Leaves at the bottom of the canopy senesce when they become shaded below their light compensation point. Much of the nitrogen resorbed from these senescing leaves (see Chap. 8) is transported to the top of the canopy to support the production of young leaves with high photosynthetic capacity. (3) Sun leaves at the top of the canopy develop more cell layers than shade leaves and therefore contain more nitrogen per unit leaf area. The accumulation of nitrogen at the top of the canopy is most pronounced in dense canopies, which develop under circumstances of high water and nitrogen availability (Field 1991). In environments where leaf area is limited by water, nitrogen, or time since disturbance, there is less advantage to concentrating nitrogen at the top of the canopy because light availability is high throughout the canopy. In these sparse canopies, light availability, nitrogen concentrations, and photosynthetic rates show a more uniform vertical distribution.

Canopy-scale relationships between light and nitrogen occur even in multi-species communities. In a single individual, there is an obvious selective advantage to optimizing nitrogen distribution within the canopy because this provides the greatest carbon return per unit of nitrogen invested in leaves. We know less about the factors governing carbon gain in multi-species

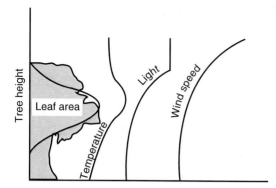

Fig. 5.20 Typical vertical gradients in leaf area, temperature, light, and wind speed in a forest. Temperature is highest in the mid-canopy where most energy is absorbed. Based on Landsberg and Gower (1997)

stands. In such stands, the individuals at the top of the canopy account for most of the photosynthesis and may be able to support greater root biomass to acquire more nitrogen, compared to smaller subcanopy or understory individuals. This specialization and competition among individuals probably contributes to the vertical scaling of nitrogen and photosynthesis observed in multi-species stands (Craine 2009).

Vertical gradients in other environmental variables often reinforce the maximization of carbon gain near the top of the canopy. The canopy modifies not only light availability but also other variables that influence photosynthetic rate, including wind speed, temperature, relative humidity, and CO_2 concentration (Fig. 5.20). The most important of these effects is the decrease in wind speed from the free atmosphere to the ground surface. The friction of air moving across Earth's surface causes wind speed to decrease exponentially from the free atmosphere to the top of the canopy. In other words, Earth's surface creates a boundary layer similar to that which develops around individual leaves (Fig. 5.18). Wind speed continues to decrease from the top of the canopy to the ground surface in ways that depend on canopy structure. Smooth canopies, characteristic of crops or grasslands, show a gradual decrease in wind speed from the top of the canopy to the ground surface, whereas rough canopies, characteristic of many forests, create more friction and turbulence that increases the vertical mixing of air within the canopy (see Chap. 4; McNaughton and

Jarvis 1991). For this reason, gas exchange in rough canopies is more tightly **coupled** to conditions in the free atmosphere than in smooth canopies.

Wind speed is important because it reduces the thickness of the boundary layer of still air around each leaf, producing steeper gradients in temperature and in concentrations of CO_2 and water vapor from the leaf surface to the atmosphere. This speeds the diffusion of CO_2 into the leaf and the loss of water from the leaf, enhancing both photosynthesis and transpiration. A reduction in thickness of the leaf boundary layer also brings leaf temperature closer to air temperature. The net effect of wind on photosynthesis is generally positive at moderate wind speeds and adequate moisture supply, enhancing photosynthesis at the top of the canopy, where wind speed is highest. When low soil moisture or a long pathway for water transport from the soil to the top of the canopy reduces water supply to the uppermost leaves, as in tall forests, the uppermost leaves reduce their stomatal conductance, causing the zone of maximum photosynthesis to shift farther down in the canopy. Although multiple environmental gradients within the canopy have complex effects on photosynthesis, they probably enhance photosynthesis near the top of canopies in those ecosystems with enough water and nutrients to develop dense canopies. Variations in light and water availability and leaf-nitrogen concentrations then cause diurnal and seasonal shifts the height of maximum photosynthesis within the canopy.

Canopy properties extend the range of light availability over which the light-use efficiency (LUE) of the canopy remains constant. The light-response curve of canopy photosynthesis, measured in closed canopies (LAI > ≈3), saturates at higher irradiance than does photosynthesis by a single leaf (Fig. 5.21) for several reasons (Jarvis and Leverenz 1983). The more vertical angle of leaves in the upper canopy reduces the probability of their becoming light saturated and increases light penetration into the canopy. The clumped distribution of leaves in shoots, branches, and crowns also increases light penetration into the canopy. Conifer canopies are particularly effective in distributing light through the canopy due to the clumping of needles around stems. This could explain why conifer forests often

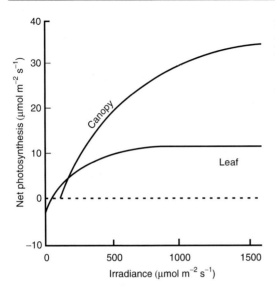

Fig. 5.21 Light response curve of photosynthesis in a single leaf and a forest canopy. Canopies maintain a relatively constant LUE (linear response of photosynthesis to light) over a broader range of light availability than do individual leaves. Redrawn from Ruimy et al. (1995)

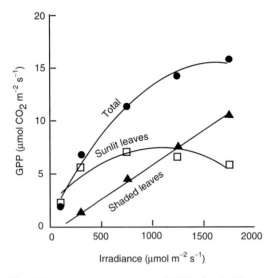

Fig. 5.22 Light response curve of GPP of a deciduous forest, showing the contributions of shaded and sunlit leaves, as calculated by the model CANVEG. Redrawn from Law et al. (2002)

fact, at high light (and correspondingly high temperature and vapor pressure deficit), photosynthesis may decline in the upper canopy, causing shaded leaves to account for most of the total canopy photosynthesis under some circumstances (Fig. 5.22; Law et al. 2002).

In most ecosystems, including all forests that have been measured, GPP approaches a plateau at high light, indicating a decline in LUE at high light (Fig. 5.23; Ruimy et al. 1995; Law et al. 2002; Turner et al. 2003b). This decline in LUE at high light is most pronounced in low-resource environments with sparse canopies, where canopy photosynthetic capacity is low, and all leaves experience a similar light regime (Gower et al. 1999; Baldocchi and Amthor 2001; Turner et al. 2003b). In other words, canopy photosynthetic response to light mirrors a photosynthetic response that is similar to that of all individual leaves. In dense canopies, more leaves are shaded and operate in the linear portion of the light-response curve, increasing LUE of the canopy as a whole (Fig. 5.23; Teskey et al. 1995; Turner et al. 2003b).

Leaf Area

Variation in soil resource supply accounts for much of the spatial variation in leaf area and GPP among ecosystem types. Analysis of satellite imagery shows that about 70% of the ice-free terrestrial surface has relatively open canopies (LAI < 1; Fig. 5.24; Graetz 1991). GPP correlates closely with leaf area below an LAI of about 4 (Schulze et al. 1994), suggesting that leaf area is a critical determinant of GPP on most of Earth's terrestrial surface, just as algal biomass or chlorophyll is a key determinant of pelagic GPP (Fig. 5.1). GPP is less sensitive to LAI in dense canopies because the leaves in the middle and bottom of the canopy contribute relatively little to GPP over the course of a day or year. The availability of soil resources, especially water and nutrient supply, is a critical determinant of LAI for two reasons: (1) Plants in high-resource environments produce a large amount of leaf biomass, and (2) leaves produced in these environments have a high SLA, i.e.,

support a higher LAI than deciduous forests. The light compensation point also decreases from the top to the bottom of the canopy (Fig. 5.12), so lower leaves maintain a positive carbon balance, despite their relatively low light availability. In

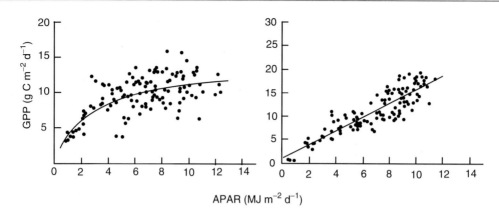

Fig. 5.23 Response of GPP to absorbed photosynthetically active radiation (APAR) in a Massachusetts deciduous forest (*left*) and a Kansas grassland (*right*). The forest maintains a relatively constant light-use efficiency up to 30–50% of full sun, although there is considerable variability. The grassland maintains a constant light-use efficiency over the entire range of naturally occurring irradiance. Redrawn from Turner et al. (2003b)

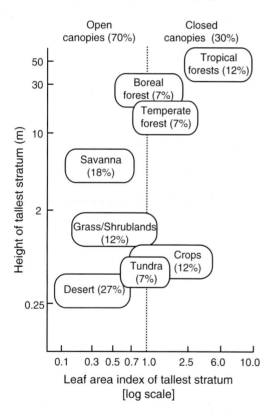

Fig. 5.24 LAI and canopy height of the major biomes. Typical values for that biome and the percentage of the terrestrial surface that it occupies are shown. The vertical line shows 100% canopy cover (LAI = 1). Redrawn from Graetz (1991)

a large leaf area per unit of leaf biomass. As discussed earlier, a high SLA maximizes light capture and therefore carbon gain per unit of leaf biomass

(Fig. 5.17; Lambers and Poorter 1992; Reich et al. 1997; Wright et al. 2004).

Disturbances, herbivory, and pathogens reduce leaf area below levels that resources can support. Soil resources and light extinction through the canopy determine the upper limit to the leaf area that an ecosystem can support. However, many factors regularly reduce leaf area below this potential LAI. Drought and freezing are climatic factors that cause plants to shed leaves. Other causes of leaf loss include physical disturbances (e.g., fire and wind) and biotic agents (e.g., herbivores and pathogens). After major disturbances, the remaining plants may be too small, have too few meristems, or lack the productive potential to quickly produce the leaf area that could potentially be supported by the climate and soil resources of a site. For this reason, LAI tends to increase with time after disturbance to an asymptote, then (at least in forests) often declines in late succession (see Chap. 12).

Human activities increasingly affect the leaf area of ecosystems in ways that cannot be predicted from climate. Overgrazing by cattle, sheep, and goats, for example, directly removes leaf area and causes shifts to vegetation types that are less productive and have less leaf area than would otherwise occur in that climate zone (Reynolds and Stafford Smith 2002). Acid rain and other pollutants can also cause leaf loss. Nitrogen deposition can stimulate leaf production above levels that would be predicted from climate and soil type, just as nutrient and water additions to agricultural fields augment LAI and therefore GPP. Because of human activities, LAI cannot be estimated simply from correlations with climate. Fortunately, satellites provide the opportunity to estimate LAI directly, although the technology is still improving. Satellites tend to underestimate the LAI of dense canopies because they cannot "see" all the leaves. LIDAR (Light Detection and Ranging) uses reflection of light pulses (lasers) to detect three-dimensional canopy structure, much like radar, and shows promise in improving remote-sensing estimates of LAI. Fortunately, most of the world's canopies are relatively open, so their LAI can be estimated relatively accurately from satellites. Information about global distribution of LAI is an important input to models that calculate regional patterns of carbon input to terrestrial ecosystems (Running et al. 2004).

Length of the Photosynthetic Season

The length of the photosynthetic season accounts for much of the ecosystem differences in GPP. Most ecosystems experience times that are too cold or too dry for significant photosynthesis to occur. During winter in cold climates and times with negligible soil water in dry climates, plants either die (annuals), lose their leaves (deciduous plants), or become physiologically dormant (some evergreen plants). During these times, there is negligible carbon absorption by the ecosystem, regardless of light availability and CO_2 concentration. In a sense, the non-photosynthetic season is simply a case of extreme environmental stress. At high latitudes and altitudes and in dry ecosystems, this is probably *the* major constraint on carbon inputs to ecosystems (Fig. 5.2; see Chap. 6; Körner 1999). For annuals and deciduous plants, the lack of leaf area is sufficient to explain the absence of photosynthetic carbon gain in the nongrowing season. Lack of water or extremely low temperatures can, however, prevent even evergreen plants from gaining carbon. Some evergreen species partially disassemble their photosynthetic machinery during the nongrowing season. These plants require some time after the return of favorable environmental conditions to reassemble their photosynthetic machinery (Bergh and Linder 1999), so not all early-season irradiance is used efficiently

to gain carbon (Xiao et al. 2010). In tropical ecosystems, however, where conditions are more continuously favorable for photosynthesis, leaves maintain their photosynthetic machinery from the time they are fully expanded until they are shed. Models that simulate GPP often define the length of the photosynthetic season in terms of thresholds of minimum temperature or moisture below which plants do not produce leaves or do not photosynthesize (Running et al. 2004).

Environmental controls over GPP during the growing season are similar to those described for net photosynthesis of individual leaves. Soil resources (nutrients and moisture) influence GPP primarily through their effects on photosynthetic potential and leaf area rather than through variations in the efficiency of converting light to carbohydrates (Turner et al. 2003b). Consequently, *ecosystem differences* in GPP depend more strongly on differences in the quantity of light absorbed and length of photosynthetic season than on the efficiency of converting light to carbohydrates (i.e., LUE).

The *seasonal changes* in GPP depend on both the seasonal patterns of leaf area development and loss and the photosynthetic response of individual leaves to variations in light and temperature, which influence LUE. These environmental factors have a particularly strong effect on leaves at the top of the canopy, which account for most GPP. The thinner boundary layer and greater distance for water transport from roots, for example, makes the uppermost leaves particularly sensitive to variation in temperature, soil moisture, and relative humidity.

LUE varies diurnally, being lowest at times of high light. Seasonal patterns of LUE are more complex because they depend not only on light availability but also on seasonal variations in leaf area, canopy nitrogen, and various environmental stresses such as drought and freezing. LUE is highest in high-resource ecosystems such as crops with a high LAI and photosynthetic capacity. LUE is lowest in low-resource ecosystems such as the boreal forest and arid grasslands (Turner et al. 2003b). LUE also declines with increasing temperature (reflecting increases in photorespiration; Lafont et al. 2002; Turner et al. 2003b) and is strongly reduced at extremely low temperatures (Teskey et al. 1995). The detailed patterns and causes of temporal and spatial patterns of LUE and GPP are active research areas that promise to provide important advances in understanding and predicting patterns of carbon inputs to ecosystems (Running et al. 2004; Luyssaert et al. 2007; Waring and Running 2007).

Satellite-Based Estimates of GPP

Satellite-based estimates of absorbed radiation and LUE allow daily mapping of GPP at global scales. An important conclusion of leaf- and canopy-level studies of photosynthesis is that many factors cause convergence of ecosystems toward a relatively similar efficiency of converting absorbed light energy into carbohydrates. (1) All C_3 plants have a similar quantum yield (LUE) at low to moderate irradiance. (2) Penetration of light and vertical variations in photosynthetic properties through a canopy extend the range of irradiance over which LUE remains relatively constant. (3) LUE of a given ecosystem varies primarily in response to light intensity and short-term environmental stresses that reduce stomatal conductance. Over the long term, however, plants respond to environmental stresses by reducing leaf area and the concentrations of photosynthetic pigments and enzymes so photosynthetic capacity matches stomatal conductance. In other words, plants in low-resource environments reduce the amount of light absorbed more strongly than they reduce the efficiency with which absorbed light is converted to carbohydrates. Modeling studies and field measurements suggest that ecosystems differ much more strongly in leaf area and photosynthetic capacity than in LUE (Field 1991; Turner et al. 2003b).

If LUE is indeed similar and shows predictable patterns among ecosystems, GPP can be estimated from satellite measurements of light absorption by ecosystems, and correcting this for known causes of variation in LUE. Leaves at the top of the canopy have a disproportionately large effect on the light that is both absorbed and reflected by the ecosystem. Satellites can measure the incoming and reflected radiation. This similarity in bias between the vertical distribution of absorbed and reflected radiation makes satellites an ideal tool for estimating canopy photosynthesis. The challenge, however, is to estimate the fraction of absorbed radiation that has been absorbed by leaves rather than by soil or other non-photosynthetic surfaces. Vegetation has a different spectrum of absorbed and reflected radiation than does the atmosphere, water, clouds, or bare soil. This occurs because chlorophyll and associated light-harvesting pigments or accessory pigments, which are concentrated at the canopy surface, absorb visible light (VIS) efficiently. The optical properties that result from the cellular structure of leaves, however, make them highly reflective in the near infrared (NIR) range. Ecologists have used these unique properties of vegetation to generate an index of vegetation "greenness": the **normalized difference vegetation index (NDVI)**.

$$NDVI = \frac{(NIR - VIS)}{(NIR + VIS)} \qquad (5.5)$$

NDVI is approximately equal to the fraction of incoming photosynthetically active radiation (PAR) that is absorbed by vegetation (FPAR):

$$FPAR \approx NDVI \approx APAR / PAR \qquad (5.6)$$

where APAR is the absorbed photosynthetically active radiation (Running et al. 2004). FPAR can also be measured directly in ecosystems, knowing the irradiance at the top (I_o) and bottom (I_z) of the canopy or the relationship between I_o and leaf area index (LAI, L):

$$FPAR = 1 - (I_z / I_o) \qquad (5.7)$$

where $I_z = I_o\, e^{-kLz}$, and k is the extinction coefficient (5.3). Sites with a high rate of carbon gain generally have a high NDVI because of their high chlorophyll content (low reflectance of VIS) and high leaf area (high reflectance of NIR). Species differences in leaf structure also influence infrared reflectance (and therefore NDVI). Conifer forests, for example, generally have a lower NDVI than deciduous forests despite their greater leaf area. Consequently, NDVI must be used cautiously when comparing ecosystems dominated by structurally different types of plants (Verbyla 1995). The maximum NDVI measured by satellites is very similar to that measured on the ground (Fig. 5.25). If LUE is known, GPP can be calculated from irradiance (PAR) and FPAR or NDVI:

$$GPP = LUE \times FPAR \times PAR \approx LUE \times NDVI \times PAR \qquad (5.8)$$

MODIS (Moderate Resolution Imaging Spectroradiometer) sensors carried aboard satellites directly measure reflectance from space, allowing calculation of NDVI. Ecosystem models

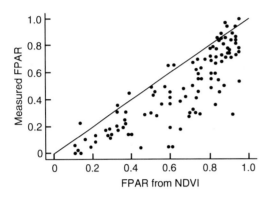

Fig. 5.25 Relationship between FPAR (the fraction of photosynthetically active radiation absorbed by vegetation) estimated from satellite measurements of NDVI (X-axis) and FPAR measured in the field (Y-axis). Data were collected from a wide range of ecosystems, including temperate and tropical grasslands and temperate and boreal conifer forests. Satellites provide an approximate measure of the photosynthetically active radiation absorbed by vegetation and therefore the carbon inputs to ecosystems. Redrawn from Los et al. (2000)

have estimated LUE for different biomes, under varying conditions of vapor pressure deficit and temperature (Running et al. 2000; White et al. 2000). Using these modeled LUE values (g carbon MJ^{-1}) and observed climate, NDVI and PAR (MJ m^{-2}), daily GPP (g carbon m^{-2}) can now be calculated globally at a 1-km scale (Running et al. 2004). These calculations are based on daily observations of weather, weekly estimates of NDVI, and annual estimates of biome distributions. The methodology for estimating global patterns of GPP is continually being tested and improved. Currently, differences in the scale at which weather observations are made account for much of the discrepancy between GPP estimates from satellites and those measured at specific field sites. Other sources of variation include the controls over GPP that were described in the previous section (Turner et al. 2005; Heinsch et al. 2006). In the conterminous U.S. summer, GPP is highest in fertile moist ecosystems like croplands and deciduous forests and lowest in dry ecosystems like grasslands and forests (Fig. 5.26). Evergreen forests have modest mid-summer GPP but continue photosynthesizing during the winter.

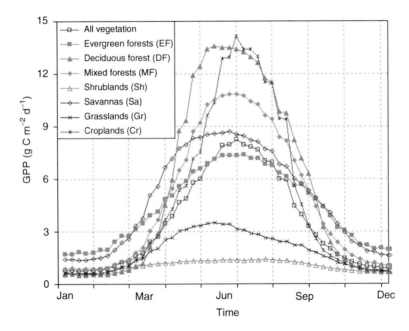

Fig. 5.26 Predicted seasonal pattern of GPP in different biomes of the U.S. averaged from 2001 to 2006, based on a regression model that uses AmeriFlux (a network of ecosystem flux studies) GPP measurements and MODIS satellite imagery. Redrawn from Xiao et al. (2010)

These seasonal and ecosystem differences in GPP are the major factors explaining ecosystem differences in NPP (see Chap. 6) and carbon accumulation (see Chap. 7).

Summary

Most carbon enters terrestrial ecosystems through photosynthesis mediated by primary producers (plants on land and phytoplankton in aquatic ecosystems). The light-harvesting reactions of photosynthesis transform light energy into chemical energy, which is used by the carbon-fixation reactions to convert CO_2 to sugars. The enzymes that carry out these reactions account for about half of the nitrogen in photosynthetic cells.

In pelagic ecosystems, phytoplankton are relatively well mixed throughout the euphotic zone and have photosynthetic properties similar to shade plants. Aquatic GPP depends on the quantity of phytoplankton and the vertical profile of light and other physical factors. Nutrient availability, as affected by stratification and vertical mixing, strongly influences phytoplankton abundance and therefore GPP.

Plants on land adjust the components of photosynthesis so physical and biochemical processes co-limit carbon fixation. At low light, for example, plants reduce the quantity of photosynthetic machinery per unit leaf area by producing thinner leaves. As atmospheric CO_2 concentration increases, plants reduce stomatal conductance. The major environmental factors that explain differences *among* ecosystems in carbon gain are the length of time during which conditions are suitable for photosynthesis and the soil resources (water and nutrients) available to support the production and maintenance of leaf area. Environmental stresses, such as inadequate water supply, extreme temperatures, and pollutants, reduce the efficiency with which plants use light to gain carbon. Plants also respond to these stresses by reducing leaf area and nitrogen content so as to maintain a relatively constant efficiency in the use of light to fix carbon. Consequently, ecosystem differences in photosynthesis at the ecosystem scale (GPP) are determined primarily by leaf area and secondarily by environmental stresses that reduce the efficiency with which these leaves convert light to chemical energy.

Review Questions

1. How do light, CO_2, and nitrogen interact to influence the biochemistry of photosynthesis in C_3 plants? What biochemical adjustments occur when each of these resources declines in availability?
2. Describe the environmental controls over photosynthesis in pelagic ecosystems in terms of the photosynthetic response of individual cells (e.g., light response curve) and ecosystem-scale photosynthesis (GPP).
3. How does each major environmental variable (CO_2, light, nitrogen, water, temperature, pollutants) affect photosynthetic rate in terrestrial plants in the short term? How do plants adjust to changes in each factor over the long term?
4. How does the response of photosynthesis to one environmental variable (e.g., water or nitrogen) affect the response to other environmental variables (e.g., light, CO_2, or pollutants)? Considering these interactions among environmental variables, how might anthropogenic increases in nitrogen inputs affect the response of Earth's ecosystems to rising atmospheric CO_2?
5. How do environmental stresses affect light-use efficiency in the short term? How does vegetation adjust to maximize LUE in stressful environments over the long term?
6. What factors are most important in explaining differences among ecosystems in GPP? Over what timescale does each of these factors have its greatest impact on GPP? Explain your answers.
7. What factors most strongly affect leaf area and photosynthetic capacity of vegetation?
8. How do the factors regulating photosynthesis in a forest canopy differ from those in individual leaves? How do availability of soil resources (water and nutrients) and the structure of the canopy influence the importance of these canopy effects?

Additional Reading

Craine, J.M. 2009. *Resource Strategies of Wild Plants.* Princeton University Press, Princeton.

Lambers, H., F.S. Chapin, III, and T.L. Pons. 2008. *Plant Physiological Ecology.* 2nd edition. Springer, New York.

Law, B.E., E. Falge, L. Gu, D.D. Baldocchi, P. Bakwin et al. 2002. Environmental controls over carbon dioxide and water vapor exchange of terrestrial vegetation. *Agricultural and Forest Meteorology* 113:97–120.

Running, S.W., R.R. Nemani, F.A. Heinsch, M. Zhao, M. Reeves, and H. Hashimoto. 2004. A continuous satellite-derived measure of global terrestrial primary production. *BioScience* 54:547–560.

Sage, R.F. 2004. The evolution of C_4 photosynthesis. *New Phytologist* 161:341–370.

Schulze, E.-D., F. M. Kelliher, C. Körner, J. Lloyd, and R. Leuning. 1994. Relationship among maximum stomatal conductance, ecosystem surface conductance, carbon assimilation rate, and plant nitrogen nutrition: A global ecology scaling exercise. *Annual Review of Ecology and Systematics* 25:629–660.

Turner, D.P., S. Urbanski, D. Bremer, S.C. Wofsy, T. Meyers et al. 2003. A cross-biome comparison of daily light use efficiency for gross primary production. *Global Change Biology* 9:383–395.

Waring, R.H. and S.W. Running. 2007. *Forest Ecosystems: Analysis at Multiple Scales.* 3rd edition. Academic Press, San Diego.

Wright, I.J., P.B. Reich, M. Westoby, D.D. Ackerly, Z. Barusch et al. 2004. The world-wide leaf economics spectrum. *Nature* 428:821–827.

Plant Carbon Budgets

6

The balance between carbon inputs through gross primary production (GPP) and carbon losses through plant respiration and tissue turnover govern the carbon balance of plants. This chapter describes the factors that regulate this balance.

Introduction

Plant production determines the amount of energy available to sustain all organisms, including people. We depend on plant production directly for food and fiber and indirectly because of the critical role of plants in all ecosystem processes. About half of gross primary production (GPP) is respired by plants to provide the energy that supports their growth and maintenance (Schlesinger 1997; Waring and Running 2007). Net primary production (NPP) is the net carbon gain by plants and equals the difference between GPP and plant respiration. Plants lose carbon through several pathways besides respiration (Fig. 6.1). These include the death of plants or plant parts (e.g., leaves); the consumption of plants by herbivores; the secretion of water-soluble or volatile organic compounds into the environment; and the targeted transfer of carbon to symbiotically associated microbes (e.g., mycorrhizal fungi and nitrogen-fixing bacteria). Finally, carbon can be removed from plants by fire, human harvest, and other disturbances.

A Focal Issue

The productivity of the biosphere is concentrated in areas undergoing rapid land-use change. Tropical wet forests, for example, occupy 12% of terrestrial land area but account for a third of terrestrial primary production (Fig. 6.2). They are being rapidly cleared, much of it by illegal logging (Sampson et al. 2005). Similar high rates of deforestation occurred in the temperate zone centuries earlier and are now returning to forest or being converted to cities (see Chap. 12). Land-use change is equally important at the unproductive end of the spectrum, where lands that are cold and dry (tundra, desert, grasslands, and shrublands) occupy half the terrestrial land area and together contribute about as much productivity as tropical forests. What environmental factors govern the productivity of these changing landscapes? If they are replaced by different vegetation, will they be as productive? The coastal zones of the ocean, which are the marine equivalent of tropical wet forests, are also undergoing rapid changes due to overfishing and nutrient runoff from the land. A clear understanding of factors governing Earth's primary productivity is

F.S. Chapin, III et al., *Principles of Terrestrial Ecosystem Ecology*,
DOI 10.1007/978-1-4419-9504-9_6, © Springer Science+Business Media, LLC 2011

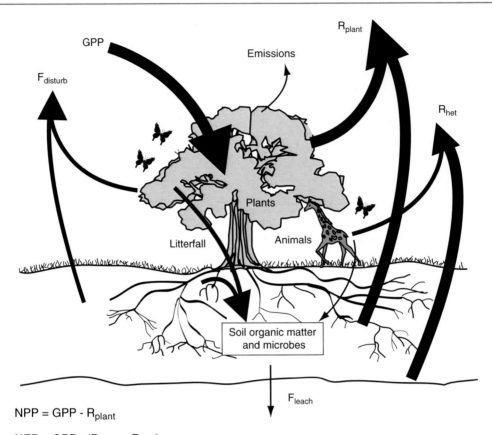

NPP = GPP - R$_{plant}$

NEP = GPP - (R$_{plant}$ + R$_{het}$)

Fig. 6.1 Overview of the major carbon fluxes of an eco-system. Carbon enters the ecosystem as gross primary pro-duction (*GPP*), through photosynthesis by plants. Roots and aboveground portions of plants return about half of this carbon to the atmosphere as plant respiration (R_{plant}). Net primary production (*NPP*) is the difference between carbon gain by GPP and carbon loss through R_{plant}. Most *NPP* is transferred to soil organic matter as litterfall, root death, root exudation, and root transfers to symbionts; some *NPP* is eaten by animals and sometimes is lost from the ecosystem through disturbance (wildfire or harvest).

Animals also transfer some carbon to soils through excretion and mortality. Most carbon entering the soil is lost through microbial respiration (which, together with animal respi-ration, is termed heterotrophic respiration: R_{het}). Net eco-system production (*NEP*) is the balance between GPP and plant-plus-heterotrophic respiration. Additional carbon is lost from soils through leaching and disturbance. Net eco-system carbon balance (*NECB*) is the net carbon accumu-lation by an ecosystem; it equals the carbon inputs from *GPP* minus the various avenues of carbon loss (respiration, leaching, disturbance, etc.; see Fig. 7.23)

essential to meet the needs for nature and for human livelihoods in a rapidly changing world.

Plant Respiration

Respiration provides the energy for a plant to acquire nutrients and to produce and main-tain biomass. Plant respiration is the carbon released by mitochondrial respiration. It is not "wasted" carbon. It serves the essential function

of providing energy for growth and maintenance, just as it does in animals and microbes. We can separate total plant respiration (R_{plant}) into three functional components: growth respiration (R_{growth}), maintenance respiration (R_{maint}), and the respiratory cost of ion absorption (R_{ion}).

$$R_{plant} = R_{growth} + R_{maint} + R_{ion} \qquad (6.1)$$

Each of these respiratory components involves mitochondrial oxidation of carbohydrates to

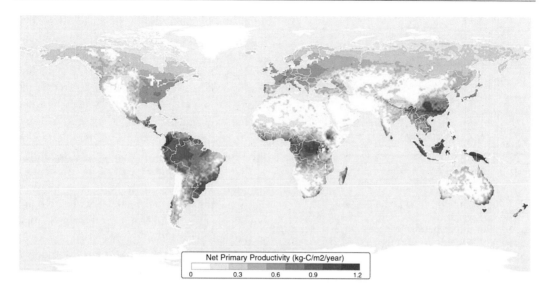

Fig. 6.2 The global pattern of net primary productivity (Foley et al. 1996; Kucharik et al. 2000). The patterns of productivity correlate more closely with precipitation than with temperature (see Fig. 2.23), indicating a strong role of moisture in regulating the productivity of the biosphere. Reproduced from the Atlas of the Biosphere (http://atlas.sage.wisc.edu)

Table 6.1 Concentration and carbon cost of major chemical constituents in a sedge leaf[a]

Component	Concentration (%)	Cost (mg C g^{-1} product)	Total cost[b] (mg C g^{-1} tissue)
Sugar	11.9	438	52
Nucleic acid	1.2	409	5
Polysaccharide	9.0	467	42
Cellulose	21.6	467	101
Hemicellulose	31.0	467	145
Amino acid	0.9	468	4
Protein	9.7	649	63
Tannin	4.8	767	37
Lignin	4.2	928	39
Lipid	5.7	1,212	69
Total cost			557

[a] Data from Chapin (1989)
[b] The four most expensive constituents account for 37% of the cost of synthesis but only 24% of the mass of the tissue. The total cost of production (557 mg C g^{-1} tissue) is equivalent to 1.23 g carbohydrate per gram of tissue, with 20% of this being respired and 80% incorporated into biomass

produce ATP. They differ only in the *functions* for which ATP is used by the plant. Separation of respiration into these functional components allows us to understand the ecological controls over plant respiration.

All plants are similar in their efficiency of converting sugars into new biomass. Growth of new tissue requires biosynthesis of many classes of chemical compounds, including cellulose, proteins, nucleic acids, and lipids (Table 6.1). The carbon cost of synthesizing each compound includes the carbon that is incorporated into that compound plus the carbon oxidized to CO_2 to provide the ATPs that drive biosynthesis. These carbon costs can be calculated for each class of compound from knowledge of its biosynthetic pathway (Penning de Vries et al. 1974; Amthor 2000). The cost of producing a gram of tissue can then be calculated from the

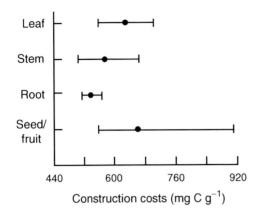

Fig. 6.3 Range of construction costs for a survey of leaves ($n = 123$), stems ($n = 38$), roots ($n = 35$), and fruits or seeds ($n = 31$). Values are averages with 10th and 90th percentiles in units of mg C g^{-1} dry mass. The carbon cost of producing new biomass differs little among plant parts, except for those fruits and seeds that store lipid and have a higher cost of synthesis than do other plant parts. Redrawn from Poorter (1994)

concentration of each class of chemical compound in a tissue and its carbon cost of synthesis.

There is a threefold range in the carbon cost of synthesis of the major classes of chemical compounds found in plants (Table 6.1). The most energetically expensive compounds in plants are proteins, tannins, lignin (vascular land plants only), and lipids. In general, metabolically active tissues, such as leaves, have high concentrations of proteins, tannins, and lipids. The tannins and lipophilic substances such as terpenes serve primarily to defend protein-rich tissues from herbivores and pathogens (see Chap. 10). Structural tissues have high lignin and low protein, tannin, and lipid concentrations. Leaves of rapidly growing species with high protein concentration have higher tannin and lower lignin concentrations than leaves with low protein concentrations. Consequently, most plant tissues contain some expensive constituents, although the nature of these constituents differs among plant parts and species. In fact, the carbon cost of producing plant tissue is surprisingly similar across species, tissue types, and ecosystems (Fig. 6.3; Chapin 1989; Poorter 1994; Villar et al. 2006). These general patterns are observed in both phytoplankton (Hay and Fenical 1988) and terrestrial plants (Chapin 1989).

On average, about 20% of the energy expended in growth is expended as growth respiration, and the remaining 80% is incorporated into new biomass (Table 6.1). The rates of growth and therefore of growth respiration measured at the ecosystem scale (g C m^{-2} day^{-1}) increase when temperature and moisture favor growth, but growth respiration is a relatively constant fraction of NPP, regardless of environmental conditions.

The total respiratory cost of ion absorption probably correlates with NPP. Ion transport across membranes is energetically expensive and may account for 25–50% of the respiration in roots or phytoplankton cells (Lambers et al. 2008). Several factors cause this cost of ion absorption to differ among ecosystems. The quantity of nutrients absorbed is greatest in productive environments, although the respiratory cost per unit of absorbed nutrients may be greater in unproductive environments (Lambers et al. 2008). The respiratory cost of nitrogen absorption and use depends on the form of nitrogen absorbed because nitrate must be reduced to ammonium (an exceptionally expensive process) before it can be incorporated into proteins or other organic compounds. The cost of nitrate reduction is also variable among terrestrial plant species and ecosystems, depending on whether the nitrate is reduced in roots or leaves (see Chap. 8). In general, we expect R_{ion} to correlate with the total quantity of ions absorbed and therefore to show a positive relationship with NPP.

Maintenance respiration: How variable is the cost of maintaining plant biomass? All live cells, even those that are not actively growing, require energy to maintain ion gradients across cell membranes and to replace degraded proteins, membranes, and other constituents. Maintenance respiration provides the ATP for these maintenance and repair functions. Laboratory experiments suggest that about 85% of maintenance respiration is associated with the turnover of proteins (about 2–5% turnover per day), explaining why there is a strong correlation between protein concentration and whole-tissue respiration rate in nongrowing tissues (Penning de Vries 1975). We therefore expect maintenance respiration to be greatest in ecosystems with high tissue-nitrogen concentrations or a large plant biomass and thus

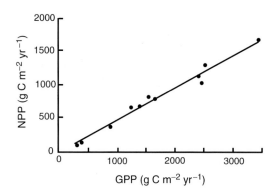

Fig. 6.4 Relationship between GPP and NPP in 11 forests from the U.S, Australia, and New Zealand. These forests were selected from a wide range of moisture and temperature conditions. GPP and NPP were estimated using a model of ecosystem carbon balance. The simulations suggest that all these forests show a similar partitioning of GPP between plant respiration (53%) and NPP (47%), despite large variations in climate. Redrawn from Waring et al. (1998)

to be greatest in productive ecosystems. Simulation models suggest that maintenance respiration may account for about half of total plant respiration; the other half is associated with growth and ion absorption (Lambers et al. 2008).

Maintenance respiration depends on environment as well as tissue chemistry. It increases with temperature because proteins and membrane lipids degrade and must be replaced more rapidly at high temperatures. Drought also imposes short-term metabolic costs associated with synthesis of osmotically active organic solutes (see Chap. 4). These effects of environmental stress on maintenance respiration are the major factors that alter the partitioning between growth and respiration and therefore are the major sources of variability in the efficiency of converting GPP into NPP. Maintenance respiration increases during times of environmental change but, after acclimation, maintenance respiration returns to values close to those predicted from biochemical composition (Semikhatova 2000). Over the long term therefore maintenance respiration may not be strongly affected by environmental stress except in strongly fluctuating environments.

Plant respiration is a relatively constant proportion of GPP, when ecosystems are compared. Although the respiration rate of any given plant increases exponentially with ambient temperature, acclimation and adaptation counterbalance this direct temperature effect on respiration. Plants from hot environments have lower respiration rates at a given temperature than do plants from cold places (Billings and Mooney 1968). The net result of these counteracting temperature effects is that plants from different thermal environments have similar respiration rates, when measured at their average habitat temperature (Semikhatova 2000).

In summary, studies of the basic components of respiration associated with growth, ion absorption, and maintenance suggest that total plant respiration should be a relatively constant fraction of GPP. In phytoplankton, for example, the heat produced by respiration is proportional to biomass (carbon content) across five orders of magnitude in cell mass (Johnson et al. 2009). The predictions are also consistent with more mechanistic modeling of plant carbon balance, which shows that total plant respiration is about half (48–60%) of GPP, when a wide range of ecosystems is compared (Fig. 6.4; Ryan et al. 1994; Landsberg and Gower 1997). In other words, plants have a growth efficiency of about 40–50% – the proportion of GPP that is converted to NPP. Variation in maintenance respiration is the most likely cause for variation in this efficiency. Microbes have a similar growth efficiency (about 40%; see Chap. 9) of producing biomass from their substrates, despite very different mechanisms of acquiring carbon and nitrogen from the environment. This apparent similarity may reflect a common underlying biochemistry of costs of synthesis and maintenance. However, there are too few studies to know how variable this efficiency is among seasons, years, organisms, and ecosystems.

What Is NPP?

Net primary production is the net carbon gain by plants. It is the balance between the carbon gained by GPP and carbon released by plant mitochondrial respiration.

$$NPP = GPP - R_{plant} \qquad (6.2)$$

Like GPP, NPP is generally measured at the ecosystem scale, usually over relatively long time intervals, such as a year (g biomass or g C m^{-2} year^{-1}). NPP includes the new biomass produced by plants, the soluble organic compounds that diffuse or are secreted into the environment (**root or phytoplankton exudation**), the carbon transfers to microbes that are symbiotically associated with roots (e.g., mycorrhizae and nitrogen-fixing bacteria), and the volatile emissions that are lost from leaves to the atmosphere (Clark et al. 2001). Most field measurements of NPP document only the new plant biomass produced and therefore probably underestimate the true NPP by at least 30% (Table 6.2). Root exudates are rapidly taken up and respired by microbes adjacent to roots and are generally measured in field studies as a portion of root respiration. Similarly, pelagic phytoplankton and bacteria often attach to surfaces of organic particles, where bacteria absorb and respire phytoplankton exudates (Mann and Lazier 2006). Volatile emissions are also rarely measured, but are generally a small fraction (<1–5%) of NPP and thus are probably not a major source of error (Guenther et al. 1995). Some biomass dies or is removed by herbivores before it can be measured, so even the new biomass measured in field studies is an underestimate of biomass production. For some purposes, these errors may not be too important. A frequent objective of measuring terrestrial NPP, for example, is to estimate the

rate of biomass increment. Root exudates, transfers to symbionts, losses to herbivores, and volatile emissions are lost from plants and therefore do not directly contribute to biomass increment. Consequently, failure to measure these components of NPP does not bias estimates of biomass accumulation. However, these losses of NPP from plants fuel other ecosystem processes such as herbivory, decomposition, and nutrient turnover and are therefore important components of the overall carbon dynamics of ecosystems and a critical carbon source for microbes (Schlesinger 1997; Mann and Lazier 2006).

Some components of NPP, such as root production, are particularly difficult to measure and have sometimes been assumed to be some constant ratio (e.g., 1:1) of aboveground production (Fahey et al. 1998). Fewer than 10% of the studies that report terrestrial NPP actually measure belowground production (Clark et al. 2001). Estimates of aboveground NPP sometimes include only large plants (e.g., trees in forests) and exclude understory shrubs or mosses, which can account for a substantial proportion of NPP in some ecosystems. Most published summaries of NPP do not state explicitly which components of NPP have been included (or sometimes even whether the units are grams of carbon or grams of biomass). For these reasons, considerable caution must be used when comparing data on NPP or biomass among studies. In general, we know less about the true magnitude of terrestrial NPP than the extensive literature on the topic would suggest.

Table 6.2 Major components of NPP and representative values of their relative magnitudes

Components of NPP[a]	% of NPP
New plant biomass	40–70
Leaves and reproductive parts (fine litterfall)	10–30
Apical stem growth	0–10
Secondary stem growth	0–30
New roots	30–40
Root secretions	20–40
Root exudates	10–30
Root transfers to mycorrhizae	15–30
Losses to herbivores and mortality	1–40
Volatile emissions	0–5

[a] Seldom, if ever, have all of these components been measured in a single study

Marine NPP

The large area of the ocean is offset by their low average productivity per unit area, so the ocean and the land each contribute about half of global NPP. Although the ocean covers 70% of Earth's surface, the average NPP per unit area is only 20% of that on land (Table 6.3). Aquatic productivity is, however, highly variable, just as on land. The most productive aquatic ecosystems, such as coral reefs, kelp forests, and eutrophic lakes, can be at least as productive as the most productive terrestrial ecosystems (Fig. 6.5). NPP in the open ocean, which accounts for 90% of the ocean area, however, is similar to that of terrestrial deserts and tundra. Because of its large area, the open ocean accounts for 60% of marine production, with picoplankton accounting for about 90% of this production (Valiela 1995).

Table 6.3 Characteristics of the ocean and continents[a]

Unit	The ocean	Continents
Surface area (% of Earth)	71	29
Volume of life zone (%)	99.5	0.5
Living biomass (10^{12} kg C)	2	560
Living biomass (10^3 kg km^{-2})	5.6	3,700
Dead organic matter (10^6 kg km^{-2})	5.5	10
Net primary production (10^3 kg C km^{-2} year^{-1})	69	330
Residence time of C in living biomass (year)	0.08	11.2

[a]Data from Cohen (1994)

The small size and lack of non-photosynthetic support structures in marine phytoplankton mean that marine primary producers require relatively little biomass to support a given photosynthetic capacity. The average primary producer biomass per unit area on land, for example, is 660-fold greater than in the ocean, although the average NPP per unit area on land is only fivefold greater than in the ocean (Table 6.3; Cohen 1994). Phytoplankton biomass of the ocean and lakes turns over 20–40 times per year, or even daily under conditions that are favorable for growth, whereas turnover for terrestrial plant biomass generally occurs over years to decades (Valiela 1995).

Ocean productivity is ultimately limited by the rate of nutrient supply from the land or deep ocean waters. For this reason, productivity is greater in coastal waters than in the open ocean. Tidal mixing of sediment nutrients into the water

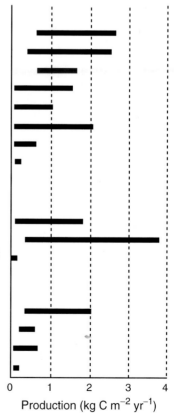

Fig. 6.5 Comparison of NPP among selected marine, freshwater, and terrestrial ecosystems. Marine and freshwater ecosystems exhibit the same range of NPP that occurs on land, but unproductive marine ecosystems (the open ocean) are much more extensive. Redrawn from Valiela (1995)

column and oxygenation of the water column contribute to the high productivity of estuaries and intertidal and near-shore marine ecosystems that constitute the **Coastal Boundary Zone Biome** (Nixon 1988; Longhurst 1998). Coral reefs are among the most productive ecosystems on Earth (Fig. 6.5). Frequent tidal flushing supplies nutrients to algae that grow on the surfaces of dead corals. These algae have high turnover rates because fish constantly graze them. The biomass of algae in this ecosystem is therefore small, just like the biomass of phytoplankton in pelagic ecosystems. Human activities have massively increased nutrient inputs to the coastal zone, particularly in estuaries, where rivers deliver nutrients derived from agricultural runoff, sewage, and erosion. This eutrophication disrupts the normal balance between algae, grazers, and decomposers (see Chap. 7; see Fig. 9.1).

In pelagic ecosystems, upwelling near the west coasts of continents provides the greatest rate of nutrient supply. Upwelling supports some of Earth's major fisheries off Peru, northwest Africa, eastern India, southwest Africa, and the western U.S. (Fig. 6.6; Valiela 1995). In these areas, Coriolis forces cause winds and surface waters to move offshore (see Fig. 2.11). These surface waters are replaced by nutrient-rich waters from depth. Upwelling also occurs in the open ocean where major ocean currents diverge (Mann and Lazier 2006). This occurs, for example, in the Equatorial Pacific, where ocean currents diverge to the north and south and in the Southern Ocean, the North Atlantic, and the North Pacific (Valiela 1995). These regions have relatively high nutrient availability and productivity.

Vertical gradients in water density also influence nutrient transport from subsurface to surface waters. In the **Trades Biome** of the central subtropical ocean basins, high solar input creates a strong vertical temperature gradient with an extremely stable **thermocline**, in which low-density warm water is underlain by high-density cold water (see Chap. 2;

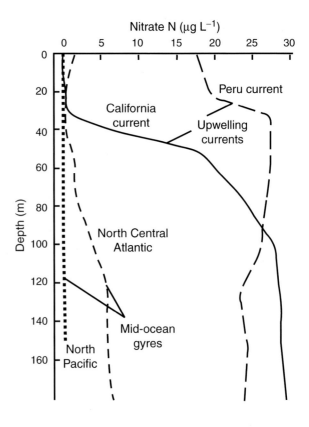

Fig. 6.6 Depth profiles of nitrate in mid-ocean gyres and upwelling zones of the ocean. Redrawn from Dugdale (1976)

Longhurst 1998). The vertical stability is reinforced by a stable halocline in which high-density saline waters lie beneath less saline surface waters. This stable stratification of water minimizes vertical mixing by waves and ocean currents, so nutrient availability and productivity of the subtropical ocean are extremely low.

As latitude increases, surface ocean temperature declines. This weakens the vertical density gradient, so storm waves and currents are more effective in mixing deep nutrient-rich waters to the surface. The strong westerly winds and storm tracks associated with the polar jet also contribute to effective mixing in the temperate/high-latitude **Westerlies Biome** (Longhurst 1998). Temperate and polar ocean waters are therefore more nutrient rich and productive than are tropical open ocean waters. The upward mixing of nutrients is greatest during winter, when surface waters are coldest, and the vertical stratification is least stable. Winter is also the time of year when strong equator-to-pole heating gradients generate the strongest winds (see Chap. 2). During winter, turbulent mixing disperses the phytoplankton deep within the water column where there is not enough light to support growth. In spring, however, an increase in solar radiation heats the surface waters and reduces the depth of the mixed layer. This concentrates phytoplankton within the euphotic zone, leading to a spring **bloom** of phytoplankton (Mann and Lazier 2006). The bloom ends when nutrients are depleted by production, and most phytoplankton have been consumed by zooplankton grazers. A second bloom sometimes occurs in autumn when a decline in surface stratification increases nutrient mixing into surface waters.

In the **Polar Biome**, surface waters have low salinity because of the large freshwater input from rivers and melting sea ice, leading to a strong stratification of the water column. As the snow-covered sea ice melts, light availability increases, and wind-driven mixing augments upwelling, leading to a summer bloom of productivity (Carmack and Chapman 2003; Mann and Lazier 2006).

The high productivity of high-latitude ocean basins supports rich fisheries, although many of these have been depleted by overfishing (Pauly

et al. 2005). The latitudinal variation in pelagic productivity also explains several other interesting ecological patterns, such as the annual migration of many whales and sea birds between the Antarctic and the Arctic Oceans to capitalize on summer blooms of polar productivity and spring blooms of productivity in the Westerlies Biome. In addition, a high proportion of fish species at high latitudes have an **anadromous** life history, in which they exploit the productive marine environment to support growth during the adult phase and use the relatively predator-free freshwater environment to reproduce. This anadromous life history strategy is increasingly favored as latitude increases because marine productivity increases with increasing latitude, whereas terrestrial productivity declines with increasing latitude (Gross et al. 1988).

In summary, NPP is greatest and least nutrient-limited in the coastal zone. In the open ocean, nutrient limitation is most extreme in zones of greatest surface heating (in the tropics and during summer) because heating reduces the density of surface water, which inhibits the upward mixing of dense, nutrient-rich waters from depth. Conditions that are conducive to deep mixing (strong winds, cold–dense surface waters, tidal mixing, etc.) reduce the magnitude of nutrient limitation to the point that other environmental factors such as light or temperature limit NPP. We discuss the influence of interactions among different nutrients on NPP in Chap. 9.

Lake NPP

The productivity of unpolluted lakes, like that in the open ocean, is generally nutrient-limited. The controls over pelagic productivity of lakes are quite similar to those in the ocean, with nutrient inputs from land and mixing strongly influencing productivity, just as described for NPP of the ocean and GPP of lakes (see Chap. 5). In winter, solar radiation is low at higher latitudes, leading to a shallow euphotic zone. In addition, weak stratification and deep mixing carry phytoplankton below the base of the euphotic zone, leading to low productivity. Light input is further reduced in those lakes that have snow-covered ice. In spring, the increase in solar radiation deepens the euphotic zone and warms the surface water, leading to a shallower mixing depth and a concentration of phytoplankton within the euphotic zone (Kalff 2002). Favorable light and temperature conditions enable phytoplankton to exploit the nutrients that mix into surface waters over winter, leading to a spring phytoplankton bloom. Just as in the ocean, the bloom ends when phytoplankton have depleted the surface nutrients and grazers reduce phytoplankton biomass. Also, as in the ocean, small phytoplankton (pico- and nano-plankton) dominate pelagic production of lakes under low-nutrient conditions (oligotrophic lakes and mid-summer conditions), and large algal cells dominate under high-nutrient conditions. Small phytoplankton tend to be more readily consumed by zooplankton grazers, so "bottom-up" (nutrient) effects interact with "top-down" (grazing) effects on lake NPP. In general, nutrients appear to explain much of the variation in phytoplankton productivity and biomass among lakes, and temperature influences the rate at which this biomass is attained (Kalff 2002). About 13% of GPP is exuded by phytoplankton into their environment (Kalff 2002). This does not directly contribute to phytoplankton biomass accumulation but may be critical in stimulating decomposition and nutrient mineralization by nearby bacteria (see Chap. 7).

Most lakes differ from the open ocean in supporting substantial benthic primary production. This is true for all small (<1 km^2) lakes, which account for 43% of total lake area (Downing et al. 2006), and even for many large lakes, which often have a large proportion of their benthic area within the euphotic zone. Benthic production is particularly important in unpolluted clearwater lakes, where it often accounts for half of NPP and an even larger proportion of the energetic base (phytoplankton plus bacteria) that feeds fish production (Vander Zanden et al. 2005, 2006). Many studies of aquatic production overlook benthic production and therefore underestimate the energy available at the base of the food chain (see Chap. 10)

Lakes are generally small aquatic patches in a terrestrial matrix, so they are strongly influenced by nutrient inputs from groundwater and streams (Schindler 1978). The granitic bedrock of the Canadian Shield, from which soils were scraped away by Pleistocene glaciers, for example, have low rates of nutrient input from watersheds to lakes. The strong nutrient limitation of many of these lakes makes them vulnerable to changes in nutrient inputs from agriculture or acid rain (Driscoll et al. 2001). Trout and other top predators in oligotrophic lakes may require decades to reach a large size, whereas this may occur in a few months or years in eutrophic lakes.

The physical properties of lakes also influence the degree of nutrient limitation of NPP. In general, weakly stratified lakes mix nutrients more readily from depth and are therefore less likely to be nutrient-limited. Deep mixing and weak nutrient limitation characterize wind-exposed lakes, large lakes, and tropical lakes with weak vertical temperature gradients and larger nutrient inputs from sediments. Some of the most productive lakes are shallow lowland lakes with naturally high rates of nutrient input (Kalff 2002).

Anthropogenic addition of nutrients to lakes often causes **eutrophication**, a nutrient-induced increase in lake productivity. Eutrophication radically alters ecosystem structure and functioning. Increased phytoplankton biomass reduces water clarity, thereby reducing the depth of the euphotic zone (see Fig. 8.2; Kalff 2002). This in turn reduces the oxygen available at depth. The increased productivity also increases the demand for oxygen to support the decomposition of the large detrital inputs. If mixing is insufficient to provide oxygen at depth, the deeper waters no longer support fish and other oxygen-requiring heterotrophs. This situation is particularly severe in winter, when low temperature limits oxygen production from photosynthesis. In ice-covered lakes, ice and snow reduce light inputs that drive photosynthesis (providing oxygen) and prevent the surface mixing of oxygen into the lake. Lakes in which the entire water column becomes anaerobic during winter do not support fish. Even during summer, the accumulation of algal detritus at times of low surface mixing can deplete oxygen from the water column, leading to high fish mortality.

In summary, nutrient limitation of NPP is widespread in lakes and changes seasonally as a result of wind-driven mixing, just as in the ocean.

Stream and River NPP

The controls over NPP in streams and rivers vary depending on stream size and environment. In general, the factors that govern NPP are similar to the controls over stream GPP (see Chap. 5) because of the tight amplifying (positive) feedbacks between photosynthesis and production of new photosynthetic cells in stream and river ecosystems. Nutrients, light, and warmth enhance GPP and NPP, whereas substrate instability, current velocity, suspended sediments, and grazing reduce plant biomass and therefore GPP and NPP (Fig. 6.7; Biggs 1996). Just as for GPP, NPP in forested headwater streams is about half that in larger open streams (Webster et al. 1995; Mulholland et al. 2001). In many river systems, NPP increases from small headwater streams to larger, more open streams and rivers, just like GPP (see Fig. 5.8). Large rivers are quite variable in NPP (Webster et al. 1995; McTammany et al. 2003; Allan and Castillo 2007), just as described for GPP (see Chap. 5).

The controls over NPP in streams and rivers differ substantially from those in pelagic ecosystems

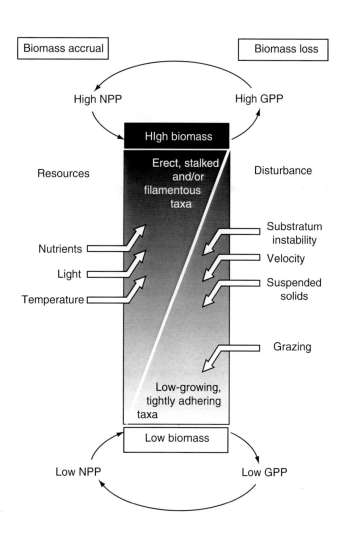

Fig. 6.7 Factors controlling the biomass and physical structure of periphyton in streams. Adapted from Biggs (1996)

of the ocean and lakes. In rapidly flowing streams and rivers, light is more often limiting to NPP than in pelagic ecosystems because of shading by streamside vegetation, suspended sediments, or (in slow-moving eutrophic waters) phytoplankton. In addition, flowing water replenishes nutrients at the surface of algal cells more rapidly than in the wind-driven mixing of the ocean and lakes, thereby reducing the degree of nutrient limitation (see Chap. 8). Finally, in slow-moving rivers, where phytoplankton become an important contributor to NPP, downstream export of phytoplankton cells limits the rate of accumulation of photosynthetic biomass and therefore NPP. This contrasts with lakes and the ocean where grazing exerts a more important control over phytoplankton accumulation (Allan and Castillo 2007). In general, the NPP by phytoplankton of slow-moving rivers is much less than in lakes with a similar nutrient and temperature regime.

Temporal and spatial heterogeneity generate tremendous variation in NPP and its controls in stream ecosystems. The biophysical differences between adjacent pools and riffles can be just as extreme as the average differences between lakes and streams. Similarly, most streams and rivers experience periodic floods followed by periods of low flow (or no flow at all). This radically alters both the conditions that influence NPP and the dislodging of primary producer biomass that supports GPP and NPP. These pulse-release properties of streams are much more extreme than the patterns of temporal variation in lakes or the ocean (Kalff 2002; Allan and Castillo 2007).

Terrestrial NPP

The nature of environmental regulation of terrestrial NPP differs substantially from that in aquatic ecosystems. Whereas phytoplankton cells are directly bathed in water and nutrients, terrestrial plants must acquire these resources from a soil medium, where there is no light to power photosynthesis. This complicates the amplifying (positive) feedback between photosynthesis and NPP because much of the new biomass produced by terrestrial plants is roots and support structures

that do not directly enhance the photosynthetic capacity of the plant. In addition, the NPP of terrestrial plants often responds to availability of CO_2 and water, which seldom limit the NPP of aquatic plants. This adds to the number and potential interactions of environmental controls over NPP. Finally, aquatic NPP is the simple balance between photosynthetic carbon gain by all cells during the day and their respiration at night, whereas on land, non-photosynthetic tissues respire both day and night. This complicates the diurnal patterns of carbon use in terrestrial plants.

Physiological Controls Over NPP

Photosynthesis, NPP, and respiration: Who is in charge? NPP is the balance of carbon gained by GPP and the carbon lost by respiration of all plant parts (Fig. 6.1). However, this simple equation (6.2) does not tell us whether the conditions governing photosynthesis dictate the amount of carbon that is available to support growth or whether conditions influencing growth rate determine the magnitude of photosynthesis – in other words whether photosynthesis "pushes" growth or whether growth "pulls" photosynthesis. On short timescales (seconds to days), environmental controls over photosynthesis (e.g., light and water availability) strongly influence photosynthetic carbon gain (photosynthesis "pushes" growth). However, on monthly to annual timescales, plants adjust leaf area and photosynthetic capacity so carbon gain matches the soil resources that are available to support growth (growth "pulls" photosynthesis; see Fig. 5.2). Plant carbohydrate concentrations are usually lowest when environmental conditions favor rapid growth (i.e., carbohydrates are drawn down by growth) and tend to accumulate during periods of drought or nutrient stress or when low temperature constrains NPP (Chapin 1991a). If the products of photosynthesis directly controlled NPP, we would expect high carbohydrate concentrations to coincide with rapid growth or to show no consistent relationship with growth rate.

Results of growth experiments also indicate that growth is not simply a consequence of the controls over photosynthetic carbon gain. Terrestrial plants

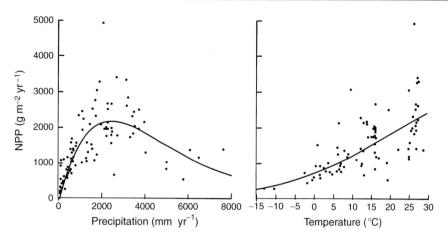

Fig. 6.8 Relationship of aboveground NPP (in units of biomass) with average annual temperature and total annual precipitation. NPP is greatest in warm, moist environments such as tropical wet forests and lowest in cold or dry ecosystems such as tundra and deserts. In tropical forests, NPP declines at extremely high precipitation (>3 m year⁻¹), due to indirect effects of excess moisture, such as low soil oxygen and loss of nutrients through leaching. Redrawn from Schuur (2003)

respond to low availability of water, nutrients, or oxygen in their rooting zone by producing hormones that reduce growth rate. The decline in growth subsequently leads to a decline in photosynthesis (Gollan et al. 1985; Chapin 1991a; Davies and Zhang 1991). The general conclusion from these experiments is that plants actively sense the resource supply in their environment and adjust their growth rate accordingly. These changes in growth rate then change the **sink strength** (demand) for carbohydrates and nutrients, leading to changes in photosynthesis and nutrient absorption (Chapin 1991a; Lambers et al. 2008). The resulting changes in growth and nutrition determine the leaf area index (LAI) and photosynthetic capacity, which, as we have seen, largely account for ecosystem differences in carbon input (see Fig. 5.2; Gower et al. 1999).

The feedbacks from sink strength to photosynthesis are not 100% effective. Leaf carbohydrate concentrations increase during the day and decline at night, allowing plants to maintain a relatively constant supply of carbohydrates to non-photosynthetic organs. Similarly, carbohydrate concentrations increase during periods (hours to weeks) of sunny weather and decline under cloudy conditions. Over these short timescales, the conditions affecting photosynthesis are the primary determinants of the carbohydrates available to support growth. The short-term

controls over photosynthesis by environment probably determine the hourly to weekly patterns of NPP, whereas soil resources govern annual carbon gain and NPP and the patterns of variation in NPP across landscapes and biomes.

Environmental and Species Controls Over NPP

The climatic controls over NPP are mediated primarily through the availability of belowground resources. At a global scale, the largest ecosystem differences in NPP are associated with variation in climate. NPP is greatest in warm, moist environments, where tropical rainforests occur, and is least in climates that are dry (e.g., deserts) or cold (e.g., tundra; Fig. 6.2; see Fig. 2.23). NPP correlates most strongly with precipitation; NPP is highest at about 2–3 m year⁻¹ of precipitation (typical of rainforests) and declines at extremely low or high precipitation (Fig. 6.8; Gower 2002; Schuur 2003; Huxman et al. 2004; Luyssaert et al. 2007). When dry ecosystems (i.e., deserts) are excluded, NPP also increases exponentially with increasing temperature. The largest differences in NPP reflect biome differences in both climate and vegetation structure. When ecosystems are grouped into biomes, there is a 14-fold range in average NPP (Table 6.4).

Table 6.4 Net primary production (NPP) of the major biome types based on biomass harvests[a]

Biome	Aboveground NPP (g m^{-2} year^{-1})	Belowground NPP (g m^{-2} year^{-1})	Belowground NPP (% of total)	Total NPP[b] (g m^{-2} year^{-1})
Tropical forests	1,400	1,100	44	2,500
Temperate forests	950	600	39	1,550
Boreal forests	230	150	39	380 (670)[b]
Mediterranean shrublands	500	500	50	1,000
Tropical savannas/grasslands	540	540	50	1,080
Temperate grasslands	250	500	67	750
Deserts	150	100	40	250
Arctic tundra	80	100	57	180
Crops	530	80	13	610

[a] NPP is expressed in units of dry mass. NPP estimated from harvests excludes NPP that is not available to harvest, due to consumption by herbivores, root exudation, transfer to mycorrhizae, and volatile emissions
[b] Data from Saugier et al. (2001). These estimates are generally intermediate among estimates from other NPP compilations (Scurlock and Olson 2002; Zheng et al. 2003), except for boreal forests, where NPP estimates are 75% greater than those of Saugier et al. (2001). Therefore, boreal NPP may be underestimated relative to other biomes

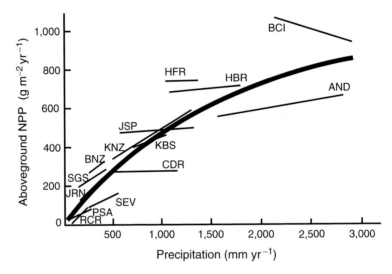

Fig. 6.9 Relationship of aboveground NPP with total annual precipitation across 14 sites. The thick curved line shows the relationship between average aboveground NPP and average precipitation across all sites. The thin straight lines show the interannual variation in aboveground NPP and annual precipitation at a given site. Sites include deserts (RCR), grasslands and steppe (PSA, SEV, JRN, SGS, CDR, KNZ, KBS, and JSP), and forests (BNZ, HBR, HFR, AND, and BCI), mostly from Long-Term Ecological Research sites. Redrawn from Huxman et al. (2004)

Do these correlations of NPP with climate reflect simple direct effects of temperature and moisture on plant growth, or are other factors involved?

Comparisons *among ecosystems* show that NPP increases most strongly with increasing precipitation in dry sites (the left-hand end of the curve in Figs. 6.8 and 6.9), suggesting that NPP is most moisture-limited in dry sites. These dry sites also show greatest sensitivity of NPP to *interannual variation* in rainfall (the slope of the thin straight lines in Fig. 6.9) and to experimental additions of water (Huxman et al. 2004). Within any given site, NPP responds most strongly to experimental addition of water in dry years and to nutrient additions in wet years. Even deserts respond to nutrient addition in wet seasons and years (Gutierrez and Whitford 1987). In dry sites, NPP responds more strongly to water than to nutrient addition in most years, and in **mesic** (moist) sites, NPP responds more strongly to nutrient addition than to water addition in most years (Huxman et al. 2004). In summary, (1) over the long term, NPP of most ecosystems is limited by multiple belowground resources (both water and nutrients, and sometimes oxygen in very wet sites). (2) The nature of environmental limitation varies among years, being most moisture-limited in dry years, and most nutrient-limited in wet years. (3) Moisture limitation of NPP occurs most often in dry sites, and nutrient limitation occurs most often in mesic sites. Thus a simple characterization of dry sites as water-limited and mesic sites as nutrient-limited is a reasonable first approximation, but it ignores the broader range of environmental limitations that all sites experience from time to time.

The types of plants that occupy an ecosystem also influence its productivity. Any given ecosystem shows a much narrower range of NPP in response to interannual variation in environment (the straight lines in Fig. 6.9) than does the *average* NPP of the full range of sites (the curved line in Fig. 6.9). Thus, a desert or grassland can never be as productive as a mesic forest, no matter how much water and nutrients it receives, because the plants lack the productive potential (capacity to produce leaf area) of large trees. Even among grasslands, the range of variation in NPP from wet to dry years is less for a given site than across all grassland sites (e.g., SGS, CDR, KNZ, JSP in Fig. 6.9) because species that dominate dry grasslands have lower productive potential than do those in mesic grasslands and cannot take full advantage of moist years (Lauenroth and Sala 1992). On the other hand, plants in dry grasslands are better adapted to dry conditions and are less likely to die in response to severe drought (see Chap. 4). Thus, long-term environmental change affects NPP in at least two ways: (1) through direct effects on the balance between water and nutrient limitation, and (2) particularly through changes in species composition and therefore the environmental tolerances and productive potential of the species present in the ecosystem.

What about cold sites, where the climate correlations suggest that NPP should be temperature-limited? In the tundra, NPP increases more in response to added nitrogen than to experimental increases in temperature (Chapin et al. 1995; McKane et al. 1997). Thus, in tundra, the climate–NPP correlation probably reflects temperature effects on nitrogen supply (see Chap. 9) or length of growing season more than a direct temperature effect on NPP. Similarly, NPP in the boreal forest correlates closely with soil temperature, but soil-warming experiments show that this effect is mediated primarily by enhanced decomposition and nitrogen supply (Van Cleve et al. 1990).

In summary, in ecosystems where climate–NPP correlations suggest a strong climatic limitation of NPP, experiments and observations show that this is mediated primarily by climatic effects on belowground resources.

What constrains NPP in warm, moist climates where temperature and moisture appear optimal for growth? Tropical forests typically have higher NPP than other terrestrial biomes (Table 6.4). Among tropical forests, litter production tends to correlate with the supply of nutrients, especially phosphorus (Vitousek 1984), suggesting that NPP in tropical forests may also be limited by the supply of belowground resources. NPP in tropical dry forests is moisture-limited, but in extremely wet climates (>3 m year^{-1} of precipitation, Fig. 6.8), NPP declines in response to increasing precipitation, probably due to oxygen limitation to roots and soil microbes and to leaching loss of essential nutrients (Schuur 2003). NPP in tropical forests is therefore probably also limited by the supply of belowground resources, including nutrients and sometimes water (relatively dry forests) or oxygen (relatively wet forests).

In temperate salt marshes, where water and nutrients are abundant, NPP responds directly to increases in CO_2 (Drake et al. 1996), as do crops that are supplied with a high nutrient supply. However, NPP is enhanced by nutrient additions even in the most fertile agricultural systems (Evans 1980), indicating the widespread occurrence of nutrient limitation to NPP (see Fig. 8.1).

In summary, experiments and observations in a wide range of ecosystems provide a relatively consistent picture. Over the range of conditions that an ecosystem encounters through time, its NPP might be limited by multiple factors. However, the supply of belowground resources is generally among the most important constraints on NPP. The factors determining the supply and acquisition of belowground resources and the productive potential of vegetation are generally the major *direct* controls over NPP and therefore the carbon input to ecosystems.

The importance of belowground resources and species traits in controlling NPP is consistent with our earlier conclusion that GPP is governed more by leaf area and length of the photosynthetic season than by the direct effects of temperature and CO_2 on photosynthesis (see Chap. 5). In fact, modeling studies suggest that NPP is a surprisingly constant fraction (40–52%) of GPP across broad environmental gradients (Fig. 6.4; Landsberg and Gower 1997; Waring and Running 2007). This is consistent with our conclusion that GPP and NPP are controlled by the same factors.

Allocation

Allocation of NPP

Patterns of biomass allocation minimize resource limitation and maximize resource capture and NPP. Our discussion of the controls over NPP suggests an interesting paradox: A high leaf area is necessary to maximize NPP, yet the major factors that constrain NPP are belowground resources. The plant is faced with a dilemma of how to distribute biomass between leaves (to maximize carbon gain) and roots (to maximize acquisition of belowground resources). Plants exhibit a consistent pattern of **allocation** – the distribution of growth among plant parts – that maximizes growth in response to the balance between aboveground and belowground resource supply rates (Garnier 1991).

In general, plants allocate production to minimize limitation by any single resource. Plants allocate new biomass preferentially to roots when water or nutrients limit growth. They allocate

new biomass preferentially to shoots when light is limiting (Reynolds and Thornley 1982). Plants can increase acquisition of a resource by producing more biomass of the appropriate tissue, by increasing the activity of each unit of biomass, or by retaining the biomass for a longer time. A plant can, for example, increase carbon gain by increasing leaf area or photosynthetic rate per unit leaf area or by retaining the leaves for a longer time before they are shed. Similarly, a plant can increase nitrogen absorption by altering root morphology or by increasing root biomass, root longevity, nitrogen absorption rate per unit root, or extent of mycorrhizal colonization. Changes in allocation and root morphology have a particularly strong effect on nutrient absorption. It is the integrated activity (mass multiplied by acquisition rate per unit biomass multiplied by time) that must be balanced between shoots and roots to maximize growth and NPP (Garnier 1991). These allocation rules are key features of all simulation models of NPP (Reynolds et al. 1993) and in the differing allocation responses to low water, low nutrients, and low light (Craine 2009).

Observations in ecosystems are generally consistent with allocation theory. Tundra, grasslands, and shrublands, for example, allocate a larger proportion of NPP below ground than do forests (Table 6.4; Gower et al. 1999; Saugier et al. 2001). Crops, with their relatively favorable water and nutrient supplies, show least allocation below ground. More subtle apparent differences in belowground NPP allocation (Table 6.4) should be interpreted cautiously because belowground NPP is difficult to measure and is sensitive to the methods used and to assumptions made about turnover of fine roots.

Allocation Response to Multiple Resources

NPP in most ecosystems is limited most strongly by a single resource but also responds to other resources. If plants were perfectly successful in allocating biomass to acquire the most limiting resource, they would be equally limited by all resources (Bloom et al. 1985; Rastetter and

Shaver 1992). As we have seen, this is seldom the case. NPP in most ecosystems responds most strongly to a particular resource, for example to water in deserts and in arid grasslands and shrublands; to nitrogen in tundra and many boreal and temperate forests; and to phosphorus in many tropical wet and dry forests. Thus, as a first approximation, deserts are water-limited ecosystems, and temperate forests are nitrogen-limited ecosystems. In many ecosystems, however, NPP responds to increased availability of more than one resource. Why does this occur?

The simplest view of environmental limitation is that growth is limited by a single resource at any moment in time. Another resource becomes limiting only when the supply of the first resource increases above the point of limitation (Liebig's **law of the minimum**). At least five processes contribute to the multiple resource limitation observed in many ecosystems: (1) Plants adjust allocation to maximize capture of (and minimize limitation by) the most limiting resource. (2) Changes in the environment (e.g., rainstorms or wet years, pulses of nutrient supply) alter the relative abundance of resources so different factors limit NPP at different times. (3) Plants exhibit mechanisms that increase the supply of the most limiting resource. (4) Organisms retain a larger proportion of some resources (e.g., nutrients) when they are in short supply. (5) Different resources limit different species in an ecosystem, so ecosystem-scale NPP responds to the addition of more than one resource. Each of these processes contributes to the response of ecosystems to multiple resources.

Plants adjust resource acquisition to maximize capture of (and minimize limitation by) the most limiting resource. As discussed earlier, plants adjust allocation of new production to roots vs. shoots to minimize limitation by belowground vs. aboveground resources, respectively. Plants also alter allocation within the root system to maximize capture of the most limiting belowground resource (Rastetter and Shaver 1992). For example, in deserts nutrient availability is greatest close to the soil surface, whereas water is generally more consistently available at depth. The amount of nutrient or water that a new root

acquires therefore depends on the depth at which roots are produced. To acquire water, some desert plants produce coarse, deep water-roots that efficiently conduct water but have low rates of nutrient absorption. Other plants produce only shallow roots and remain active only when surface water is available.

The biochemical investment by roots is specific for each nutrient. Nitrogen absorption, for example, requires synthesis of specific enzymes to absorb nitrogen, reduce nitrate, and assimilate reduced nitrogen into amino acids, whereas different enzymes are required to absorb phosphorus (see Chap. 8). This biochemical allocation to absorption of specific nutrients fine-tunes the capacity of plants to absorb those specific nutrients that most strongly limit growth.

Changes in the environment (e.g., rain storms, pulses of nutrient supply) change the relative abundance of resources so different resources limit NPP at different times. Most ecosystems experience temporal changes in the factor that most limits NPP because essential resources do not become equally available at the same time. Light, for example, decreases but water increases during rainy periods. Many ecosystems experience a pulse of nutrient availability at the beginning of the growing season, when temperatures may be suboptimal for growth. Because all the major factors that determine NPP change dramatically over several timescales, it would be surprising if there were not corresponding changes in the relative importance of these factors in limiting NPP (Huxman et al. 2004).

Temporal changes in the limitation of NPP are buffered by storage. Plants accumulate carbohydrates or nutrients during times when their availability is high and use their stores to support growth when the supply declines (Chapin et al. 1990). Over seasonal timescales, plants use stored carbohydrates and nutrients to support their burst of spring growth and replenish these stores at other times when photosynthesis and nutrient absorption exceed the demands for growth (see Chap. 8). Other than trees, most plants have very little capacity to store water, relative to their daily water demand and are therefore less buffered against variation in water than in light or nutrients

(Craine 2009). Some desert succulents do, however, have substantial water storage capacity (see Chap. 4). In summary, storage enables plants to acquire resources when they are readily available and use them at times of low supply, thus reducing temporal variation in the identity of the limiting resource.

In the case of nutrients, plants can increase the supply of the most limiting resource. Plants that have symbiotic associations with nitrogen-fixing microbes directly promote nitrogen inputs to ecosystems (see Chap. 8). Some ericoid and ectomycorrhizal associates of other plant species break down proteins and transport the resulting amino acids to plants (Read 1991). Some plants enhance the supply of phosphorus through the production of organic chelates that solubilize mineral phosphorus or through the production of phosphatases that cleave organic phosphates in the soil. Plants also exude carbohydrates that enhance mineralization near the root (see Chap. 9). Analogously, plants with fine leaves intercept fog, which increases water inputs to foggy ecosystems (see Chap. 4; Mark and Dickinson 2008).

Organisms retain a larger proportion of some resources (e.g., nutrients) when these resources are in short supply. Preferential retention and recycling of growth-limiting nutrients by plants, animals, and microbes retains these nutrients in ecosystems. Those nutrients that are present in excess of the biological requirements of organisms, as when nitrogen deposition saturates the nitrogen demands of vegetation, are more likely to be leached or lost as trace gases to the atmosphere (see Chap. 9; Vitousek and Reiners 1975).

Species differ in the resources that limit their growth, so ecosystem-scale NPP responds to the addition of more than one resource. Many species in an ecosystem have slightly different environmental requirements and therefore are limited by different resource combinations. Tundra species in the same ecosystem, for example, differ in their response to temperature, light, and nutrients (Chapin and Shaver 1985), and in some cases to the addition of nitrogen vs. phosphorus. Some desert species respond to summer

rain and others to winter rain. These differences among plant species in the factors that limit or stimulate growth contribute to the coexistence of species in a variable environment (Tilman 1988). This may be particularly important in explaining why species differ in their productivity response to interannual variation in weather and why the productivity of ecosystems varies less among years than does the productivity of any of the component species (Chapin and Shaver 1985). Spatial heterogeneity in the supply of potentially limiting resources also contributes to spatial variation in resource response.

Diurnal and Seasonal Cycles of Allocation

Photosynthesis and growth are highly resilient to daily and seasonal variations in the environment. Daily and seasonal variations in the environment are two of the most predictable perturbations experienced by ecosystems. Many organisms adjust their physiology and behavior based on innate **circadian** (about 24 h) **rhythms** that lead to 24-h cycles. Stomatal conductance and carbon gain, for example, show a circadian rhythm even under constant conditions because stomata have an innate ~24-h cycle of stomatal opening and closing. Plants store starch in the leaves during the day and break it down at night, so the rate of carbohydrate transport to roots is nearly constant over the course of a day (Lambers et al. 2008). Thus belowground processes, such as root exudation and carbon transport to mycorrhizae, are buffered from diurnal variations in photosynthetic carbon gain.

Organisms adjust seasonally in response to changing **photoperiod** (day length). Many temperate plants, for example, exhibit a relatively predictable pattern of **phenology**, the seasonal timing of production and loss of leaves, flowers, fruits, etc. Plant leaves begin to senesce and reduce their rates of photosynthesis when day length or other environmental cues signal the characteristic onset of winter. During physiologically programmed senescence, plants break down many of the compounds in the senescing tissue and transport about half of the nitrogen and phosphorus and some of the carbon from the senescing tissue to storage organs. This **resorption** minimizes nutrient loss during senescence (see Chap. 8; Chapin and Moilanen 1991). These stores provide resources to support plant growth the next spring, so NPP does not depend entirely on acquisition of new resources at times when no leaves are present. Other ecosystem processes change as either direct consequences of changes in environment (e.g., the decline in decomposition during winter due to lower temperatures) or indirect consequences of changes in other processes (e.g., the pulse of litter input to soil after leaf senescence). Ecosystem processes largely recover after each period of the cycle due to the predictable nature of diurnal and seasonal perturbations and the resilience of most processes to these changes. It is therefore unnecessary to consider explicitly the physiological basis of circadian and photoperiodic controls in order to predict ecosystem processes over longer timescales (see Chap. 12). In contrast to temperate ecosystems, tropical wet forests exhibit a less well-defined seasonality. Individual species often shed their leaves synchronously, but species differ in their timing of senescence, so the ecosystem as a whole shows less pronounced seasonality of production and senescence.

The seasonality of plant growth depends on the seasonality of leaf area and factors regulating photosynthesis. Spring growth of plants is initially supported by stored reserves of carbon and nutrients that were acquired in previous years. Leaves quickly become a net source of carbon for the rest of the plant, and growth during the remainder of the growing season is largely supported by the current year's photosynthate. There is often competition among plant parts for allocation of a limited carbohydrate supply early in the growing season, resulting in a seasonal progression of production of different plant parts, for example, with leaves produced first, followed by roots, and then by wood (Kozlowski et al. 1991). Plants species differ, however, in their seasonal allocation calendars. Plants with evergreen leaves may allocate NPP to root growth earlier than would deciduous plants because they already

have a leaf canopy that can provide carbon (Kummerow et al. 1983). Ring-porous temperate trees must first allocate carbon to xylem production in spring to develop a functional water transport system. The water columns in their large-diameter vessels **cavitate** (break) during winter freezing, so xylem vessels remain functional for only a single growing season. This large carbon requirement to rebuild xylem vessels each spring may explain the northern boundary of ring-porous species such as oaks (Zimmermann 1983). Seedlings in dry environments often depend entirely on their cotyledons for photosynthesis during the first weeks of growth and allocate all NPP to root growth to explore for a dependable water supply. The allocation calendar of a plant provides a general seasonal framework for allocation. Fluctuations in environment cause plants to modify this allocation calendar to achieve the appropriate balance of carbon and nutrients.

Tissue Turnover

The balance between NPP and biomass loss determines the annual increment in plant biomass. Plants retain only part of the biomass they produce. Plants regulate some of this biomass loss, for example the senescence of leaves in autumn. Senescence occurs throughout the growing season in grasslands but occurs as pulses during autumn or at the beginning of the dry season in many ecosystems. Other losses (e.g., to herbivores and pathogens, windthrow, and fire) are more strongly determined by environment, although even these tissue losses are influenced by plant properties such as anti-fungal compounds or fire-resistant bark. Still other biomass transfers to the soil result from mortality of entire plants. Given the substantial, although incomplete, physiological control over tissue loss, why do plants dispose of the biomass in which they invested so much carbon, water, and nutrients to produce?

Tissue loss is an important mechanism by which plants balance resource requirements with resource supply from the environment. Plants depend on regular large inputs of carbon, water, and, to a lesser extent, nutrients to maintain vital processes. For example, once biomass is produced, it requires continued carbon inputs to support maintenance respiration. If the plant (or organ) cannot meet these carbon demands, the plant (or organ) dies. Similarly, if the plant cannot absorb enough water to replace the water that is inevitably lost during photosynthesis, it must shed transpiring organs (leaves) or die. The plant must therefore shed biomass whenever resources decline below some threshold needed for maintenance. Senescence is just as important as production in adjusting to changes in resource supply and is the *only* mechanism by which plants can reduce biomass and maintenance costs when resources decline in availability.

Senescence is the programmed breakdown of tissues. The location of senescence is physiologically controlled to eliminate tissues that are least useful to the plant. Grazing of aboveground tissues, for example, causes a decline in root production so that normal rates of root senescence reduce root biomass (Ruess et al. 1998). Similarly, grazing of belowground tissues reduces leaf longevity, which reduces leaf biomass (Detling et al. 1980). Although the controls over senescence and mortality of belowground tissues are poorly understood, these patterns of variation in production and senescence appear to maintain the functional balance between leaves and roots in response to environmental variation (Garnier 1991).

Growth and senescence together enable individual plants to explore new territory. Leaf and shoot growth generally occurs at the top of the canopy or in canopy gaps, where light availability is highest. This is balanced by senescence of leaves and stems in less favorable light environments (Bazzaz 1996). This balance between biomass production and loss allows trees and shrubs to grow toward the light. Similarly, roots often proliferate in areas of nutrient enrichment or where there is minimal competition from other roots, and root death is greatest in zones of local water or nutrient depletion (see Chap. 8). This exploration of unoccupied habitat by shoots and roots requires senescence and tissue loss in less favorable microsites to reduce maintenance costs of less productive tissues and to provide the nutrient capital to produce new tissues. The exploration of new territory through synchronized growth and senescence reduces spatial variability in ecosystems by filling canopy gaps and exploiting nutrient-rich patches of soil.

Senescence causes tissue loss at times when maintenance costs greatly exceed resource gain. In seasonally variable environments, there are extended periods of time when temperature or moisture is predictably unfavorable. In these ecosystems, the cost of producing tissues that can withstand the rigors of this unfavorable period and of maintaining tissues when they provide negligible benefit to the plant may exceed the cost of producing new tissues when conditions again become favorable (Chabot and Hicks 1982). Arctic, boreal, and temperate ecosystems, for example, predictably experience seasons that are too cold for plants to acquire resources and grow. There is a pulse of autumn senescence of leaves and roots, often triggered by some combination of photoperiod and low temperature (Ruess et al. 1996). Dry ecosystems experience similar pulses of leaf and root senescence with the onset of drought. Senescence and tissue loss are therefore highly pulsed in most ecosystems and occur just before the period when conditions are least favorable for resource acquisition and growth. These seasonal pulses of senescence account for most tissue loss in highly seasonal environments.

Leaf longevity varies among plant species from a few weeks to several years or decades. In general, plants in high-resource environments produce short-lived leaves with a high specific leaf area (SLA) and a high photosynthetic rate per leaf area, but they have little resistance to environmental stresses. These "disposable leaves" are typically shed when conditions become unfavorable (winter or dry season) and are replaced the next spring. The greater longevity of leaves from low-resource environments reduces the nutrient requirement by plants to maintain leaf area (see Chap. 8). We know much less about the controls over senescence and turnover of roots than of leaves. Roots appear to die when they are attacked by herbivores or pathogens or encounter unfavorable environmental conditions without a programmed pattern of senescence and redistribution of materials to other parts of the plant.

Senescence enables plants to shed parasites, pathogens, and herbivores. Because leaves and fine roots represent relatively large packets of nutrients and organic matter, they are constantly under attack by pathogens, parasites, and herbivores. **Phyllosphere fungi**, for example, begin colonizing and growing on leaves shortly after budbreak, initially as parasites and later as part of the decomposer community when the leaf is shed (see Chap. 7). These fungi account for the mottled appearance of many older leaves. Pathogenic root fungi are a major cause of reduced yields in agro-ecosystems and are common in natural ecosystems. Plants have a variety of mechanisms for detecting natural enemies and respond initially through the production of induced chemical defenses (see Chap. 10) and, in the case of severe attack, by shedding tissues.

Large unpredictable biomass losses occur in most ecosystems. Windstorms, fires, herbivore outbreaks, and epidemics of pathogens often cause large tissue losses that are unpredictable and occur before any programmed senescence of tissues and associated nutrient resorption. Due to nutrient resorption during senescence, these unpredictable biomass losses incur approximately twice the nutrient loss per gram to the plant as that occurring after senescence (see Chap. 8). They often increase spatial heterogeneity of light and nutrient resources in the ecosystem through patchy pulses of litter input and creation of gaps that range in scale from individual leaves to entire stands. All ecosystems are at some stage in the regrowth after biomass losses occurring at multiple timescales (see Chap. 12).

Global Distribution of Biomass and NPP

Biome Differences in Biomass

The plant biomass of an ecosystem is the balance between NPP and tissue turnover. NPP and tissue loss are seldom in perfect balance. NPP tends to exceed tissue loss shortly after disturbance; at other times, tissue loss may exceed NPP (see Chap. 12). Ecosystems that are close to steady state, however, often show a consistent relationship between plant biomass and climate. Total plant biomass varies 60-fold among Earth's major terrestrial biomes (Table 6.5). Forests have the most biomass. Among forests, average biomass declines 4.5-fold from the tropics to the low-statured boreal forest, where NPP is low and stand-replacing fires often remove biomass. Deserts and tundra have only 1% as much aboveground biomass as do tropical forests. In any biome, disturbance often reduces plant biomass below levels that the climate and soil resources could support. Crops, for example, from which biomass is regularly removed, have a biomass similar to that of tundra or desert, despite more favorable growing conditions. When disturbance frequency declines, for example, through fire prevention in grasslands and savannas, biomass often increases as a result of changes in both production and longevity of leaves and roots. Biomass can also change through invasion of shrubs and trees (see Chap. 12).

Patterns of biomass allocation reflect the factors that most strongly limit plant growth in ecosystems (Table 6.5). About 70–80% of the biomass in forests is above ground because forests characterize sites with relatively abundant supplies of water and nutrients, so light often limits the growth of individual plants. In shrublands, grasslands, and tundra, however, water or nutrients more severely limit production, and the majority of biomass occurs below ground. Crops maintain the smallest proportion of biomass as roots because of their favorable water and nutrient regimes.

Tropical forests account for about half of Earth's total plant biomass, although they occupy only 13% of the ice-free land area; other forests contribute an additional 30% of global biomass (Table 6.6). Non-forested biomes therefore account for less than 20% of total plant biomass, although they occupy 70% of the ice-free land surface. Crops for example, account for only 1% of terrestrial biomass although they occupy more than 10% of the ice-free land area. Thus, most of the terrestrial surface has relatively low biomass (see Fig. 5.24). This observation alone raises concerns about deforestation in the tropics where

Table 6.5 Biomass distribution of the major terrestrial biomes[a]

Biome	Shoot (g m^{-2})	Root (g m^{-2})	Root (% of total)	Total (g m^{-2})
Tropical forests	30,400	8,400	22	38,800
Temperate forests	21,000	5,700	21	26,700
Boreal forests	6,100	2,200	27	8,300
Mediterranean shrublands	6,000	6,000	50	12,000
Tropical savannas/grasslands	4,000	1,700	30	5,700
Temperate grasslands	250	500	67	750
Deserts	350	350	50	700
Arctic tundra	250	400	62	650
Crops	530	80	13	610

Data from Saugier et al. (2001)
[a] Biomass is expressed in units of dry mass

Table 6.6 Global extent of terrestrial biomes and their total carbon in plant biomass and NPP[a]

Biome	Area (10^6 km^2)	Total plant C pool (Pg C)	Total NPP (Pg C year^{-1})
Tropical forests	17.5	320	20.6
Temperate forests	10.4	130	7.6
Boreal forests	13.7	54	2.4
Mediterranean shrublands	2.8	16	1.3
Tropical savannas/grasslands	27.6	74	14.0
Temperate grasslands	15.0	6	5.3
Deserts	27.7	9	3.3
Arctic tundra	5.6	2	0.5
Crops	13.5	4	3.9
Ice	15.5		
Total	149.3	615	58.9

Calculated from Saugier et al. (2001)
[a] Biomass and NPP are expressed in units of carbon, assuming that plant biomass is 47% carbon (Gower et al. 1999; Sterner and Elser 2002; Zheng et al. 2003)

ecosystem biomass is greatest, independent of the associated species losses.

Biome Differences in NPP

The length of the growing season is the major factor explaining biome differences in NPP. Most ecosystems experience times that are too cold or too dry for significant photosynthesis or plant growth to occur. When NPP of each biome is adjusted for the length of the growing season, all forested ecosystems have similar NPP (about 5 g m^{-2} day^{-1}), and there is only about a threefold difference in NPP between deserts and tropical forests (Table 6.7). These calculations suggest that the length of the growing season accounts for much of the biome differences in NPP (Bonan

1993; Gower et al. 1999; Körner 1999; Chapin 2003; Kerkhoff et al. 2005). When adjusted for length of growing season, aboveground NPP of the world's biomes shows no relationship to temperature, although deserts and tundra are less productive than forests (Fig. 6.10; Kerkhoff et al. 2005).

Leaf area accounts for much of the biome differences in carbon gain during the growing season. Average total LAI varies about sixfold among biomes; the most productive ecosystems generally have the highest LAI (Table 6.7; see Chap. 5). When NPP is adjusted for differences in both length of growing season and leaf area, unproductive ecosystems such as tundra or desert do not differ consistently in NPP from more productive ecosystems (Table 6.7). If anything, the less productive ecosystems may have higher NPP per unit of leaf area and growing-season length

Table 6.7 Productivity per day and per unit leaf area[a]

Biome	Season length[b] (days)	Daily NPP per ground area ($g\ m^{-2}\ day^{-1}$)	Total LAI[c] ($m^2\ m^{-2}$)	Daily NPP per leaf area ($g\ m^{-2}\ day^{-1}$)
Tropical forests	365	6.8	6.0	1.14
Temperate forests	250	6.2	6.0	1.03
Boreal forests	150	2.5	3.5	0.72
Mediterranean shrublands	200	5.0	2.0	2.50
Tropical savannas/grasslands	200	5.4	5.0	1.08
Temperate grasslands	150	5.0	3.5	1.43
Deserts	100	2.5	1.0	2.50
Arctic tundra	100	1.8	1.0	1.80
Crops	200	3.1	4.0	0.76

[a] Calculated from Table 6.4. NPP is expressed in units of dry mass
[b] Estimated
[c] Data from Gower (2002)

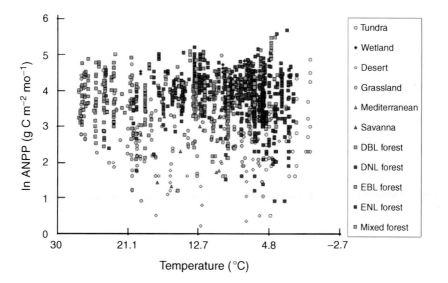

Fig. 6.10 Relationship of aboveground NPP per month of growing season (log scale) to the average growing-season temperature (graphed from high to low) for the world's ecosystems. When adjusted for length of growing season, aboveground NPP (ANPP) shows no relationship to growing-season temperature. Redrawn from Kerkhoff et al. (2005)

than do crops and forests. On average, plants in most biomes produce 1–3 g total biomass m^{-2} leaf day^{-1} during the growing season. This is equivalent to a GPP of about 1–3 g carbon m^{-2} leaf day^{-1} because NPP is about half of GPP, and biomass is about 50% carbon. Apparent differences among biomes in these values reflect substantial uncertainty in the underlying data. At this point, there is little evidence for strong ecological patterns in NPP per unit leaf area and length of growing season.

LAI is both a cause and a consequence of differences in NPP, just as in aquatic ecosystems. LAI is determined largely by the availability of soil resources (mainly water and nutrients). Tropical wet forests, for example, occur in a warm, moist climate that provides adequate water and nutrient release to support a large leaf area. These leaves remain photosynthetically active throughout the year because there are no long periods of unfavorable weather causing massive leaf loss, and plants can tap stores of deep groundwater during dry months (Woodward 1987). Deserts, in contrast, produce little leaf area because of inadequate precipitation and water storage, and arctic tundra supplies nitrogen too slowly to produce a large leaf area. In both deserts and tundra, the short growing season gives little time for leaf production, and unfavorable conditions between growing seasons limit leaf survival. The resulting low leaf area that generally characterizes these ecosystems is a major factor accounting for their low productivity (Table 6.7).

Disturbances modify the relationship between climate and NPP. There is substantial variability in NPP among sites within a biome. Some of this variability reflects variation in state factors such as climate and parent material. However, disturbance also affects NPP substantially, in part through changes in resource supply and LAI. Forest NPP, for example, often declines immediately after disturbance due to loss of LAI and then increases until the canopy closes and the available light is more fully utilized (see Fig. 12.13; Ryan et al. 1997). In later successional forests, NPP declines for a variety of reasons.

About half (50–60%) of the NPP of the biosphere occurs on land; the rest occurs in aquatic ecosystems (see Chap. 14). When summed at the global level, tropical forests account for about a third of Earth's terrestrial NPP; all forests account for about half of terrestrial NPP (Table 6.5). Grasslands and savannas account for an additional third of terrestrial NPP; these ecosystems are much more important in their contribution to terrestrial production than to biomass. Crops contribute to terrestrial NPP in proportion to their areal extent; they account for about 10% of terrestrial production and occupy 10% of the ice-free land surface.

Summary

Plant respiration provides the energy to acquire nutrients and to produce and maintain biomass. All plants are similar in their efficiency of converting sugars into biomass. Therefore, ecosystem

differences in plant respiration largely reflect differences in the amount and nitrogen content of biomass produced and, secondarily, in the effects of environmental stress, particularly temperature and moisture, on maintenance respiration. Most ecosystems appear to exhibit a similar efficiency of converting photosynthate (GPP) into NPP; half of the carbon gained through GPP becomes NPP, and the other half returns to the atmosphere as plant respiration.

NPP is the net carbon gained by plants. It includes new plant biomass produced, exudation, carbon transfers to symbionts, and the emission of volatile organic compounds by plants. Differences in NPP among marine and lake ecosystems depend primarily on physical forces that govern nutrient resupply from depth. NPP varies seasonally in these ecosystems in response to changes in light, temperature, and mixing. Light, nutrients, current, and disturbance interact to determine NPP of flowing waters. Differences among terrestrial biomes in NPP correlate with climate at the global scale largely because temperature and precipitation determine the availability of soil resources required to support plant growth. Plants actively sense the availability of these resources and adjust leaf longevity, leaf area, and photosynthesis to match this resource supply. For this reason, NPP is greatest in environments with high availability of belowground resources. After disturbance, leaf area and NPP are often reduced below levels that the environment could potentially support. Plants maximize production by allocating new growth to tissues that acquire the most limiting resources. Constantly shifting patterns of allocation reduce the degree of limitation of NPP by any single resource and make NPP in most ecosystems responsive to more than one resource. Tissue loss is just as important as NPP in explaining changes in plant biomass. Programmed loss of tissues provides a supply of plant nutrients that supports new production. Biomass and NPP are greatest in warm, moist environments and least in environments that are cold or dry. The length of the photosynthetic season and leaf area are the two strongest determinants of the global patterns in NPP. Most ecosystems have a similar (1–3 g biomass m^{-2} of leaf day^{-1}) daily NPP per unit leaf area.

Review Questions

1. What controls the partitioning of carbon between growth and respiration? Explain why the efficiency of converting sugars into new biomass is relatively constant.
2. What factors influence the variability in maintenance respiration?
3. Describe how climate influences seasonal variation in NPP of the ocean and lakes through its effects on surface heating and vertical mixing.
4. How do light and nutrients interact to influence NPP in the ocean, lakes, and flowing waters?
5. Describe the multiple ways in which climate affects the NPP of grasslands or tundra.
6. There is generally a close correlation between GPP and NPP. Describe the mechanisms that account for short-term variations in GPP and NPP (e.g., diurnal and seasonal variations).
7. Describe the mechanisms that account for the relationship between GPP and NPP when terrestrial ecosystems from different climatic regimes are compared.
8. How does allocation to roots vs. shoots respond to shade, nutrients, water, CO_2, or grazing?
9. How does variation in allocation influence resource limitation, resource capture, and NPP?
10. Why do plants senesce tissues in which they have invested carbon and nutrients rather than retaining tissues until they are removed by disturbance or herbivory?
11. Describe the carbon budget of a terrestrial plant in terms of GPP, respiration, and production. How would you expect each of these parameters to respond to changes in temperature, water, light, and nitrogen?

Additional Reading

Chapin, F.S., III, E.-D. Schulze, and H.A. Mooney. 1990. The ecology and economics of storage in plants. *Annual Review of Ecology and Systematics* 21: 423–448.

Clark, D.A., S. Brown, D.W. Kicklighter, J.Q. Chambers, J.R. Thomlinson, et al. 2001. Measuring net primary production in forests: A synthesis of current concepts and field methods. *Ecological Applications* 11: 356–370.

Kalff, J. 2002. *Limnology*. Prentice-Hall, Upper Saddle River, NJ.

Kerkhoff, A.J., B.J. Enquist, J.J. Elser, and W.F. Fagan. 2005. Plant allometry, stoichiometry and the temperature-dependence of primary productivity. *Global Ecology and Biogeography* 14:585–598.

Mann, K.H. and J.R.N. Lazier. 2006. *Dynamics of Marine Ecosystems: Biological-Physical Interactions in the Oceans*. 3rd edition. Blackwell Publishing, Victoria, Australia.

Poorter, H. 1994. Construction costs and payback time of biomass: a whole plant perspective. Pages 111–127 *in* J. Roy and E. Garnier, editors. *A Whole-Plant Perspective on Carbon-Nitrogen Interactions*. SPB Academic Publishing, The Hague.

Rastetter, E.B., and G.R. Shaver. 1992. A model of multiple element limitation for acclimating vegetation. *Ecology* 73:1157–1174.

Saugier, B., J. Roy, and H.A. Mooney. 2001. Estimations of global terrestrial productivity: Converging toward a single number? Pages 543–557 *in* J. Roy, B. Saugier, and H.A. Mooney, editors. *Terrestrial Global Productivity*. Academic Press, San Diego.

Schuur, E.A.G. 2003. Productivity and global climate revisited: The sensitivity of tropical forest growth to precipitation. *Ecology* 84:1165–1170.

Waring, R.H., and S.W. Running. 2007. *Forest Ecosystems: Analysis at Multiple Scales*. 3rd Edition. Academic Press, New York.

Decomposition and Ecosystem Carbon Budgets

7

Decomposition breaks down dead organic matter, ultimately releasing carbon to the atmosphere and nutrients in forms that can be used for plant and microbial production. This chapter describes the key controls over decomposition and the carbon balance of ecosystems.

Introduction

Decomposition is the physical and chemical breakdown of detritus (i.e., dead plant, animal, and microbial material). Decomposition causes a decrease in detrital mass, as materials are fragmented, converted to other organic compounds, and ultimately to inorganic nutrients and CO_2. If there were no decomposition, ecosystems would quickly accumulate large quantities of detritus, leading to a sequestration of nutrients in forms that are unavailable to plants and a depletion of atmospheric CO_2. Eventually, many biological processes would grind to a halt. Although this has never happened, there have been times such as the Carboniferous period (see Fig. 2.15) when decomposition did not keep pace with primary production, leading to vast accumulations of carbon- and nitrogen-containing coal and oil. The *balance* between primary production and decomposition therefore strongly influences carbon and nutrient cycling at ecosystem and global scales.

If the climate warming associated with anthropogenic CO_2 emissions were to cause even small changes in the balance between NPP and

decomposition, this could greatly alter the CO_2 concentration of the atmosphere and therefore the rate of climate warming. Because of the many critical roles of carbon balance in the biosphere and the Earth System, substantive changes in carbon cycling of plants and ecosystems are an issue of fundamental societal importance.

A Focal Issue

Mismanagement of carbon sequestration of the biosphere amplifies human impacts on climate change. The capacity of natural ecosystems to sequester carbon is generally degraded when lands are cleared for agriculture. Clearing tropical rainforests for oil palm plantations (Fig. 7.1) or plowing prairies to plant corn reduces the capacity of these systems to sequester carbon, and the energy-intensive management of these crops often consumes as much fossil fuels as are offset by the biofuels produced. What happens to soil carbon, when lands are cleared? Is the activity of decomposer organisms determined primarily by environment, by vegetation, or by their own community composition? If the activity of both plants and decomposer microbes is favored by warm, moist conditions, how does climatic change influence the net carbon balance of ecosystems? Given that ecosystems differ in their capacity to store and release carbon, in what locations are human-induced changes in land use and climate likely to have greatest impact on the carbon balance of the biosphere? A clear

F.S. Chapin, III et al., *Principles of Terrestrial Ecosystem Ecology*, DOI 10.1007/978-1-4419-9504-9_7, © Springer Science+Business Media, LLC 2011

Fig. 7.1 Land-use change greatly alters ecosystem carbon balance. Tropical rainforests, which sequester substantial carbon, have been cleared to grow oil palm, which is used extensively as a food product and increasingly as a source of bioethanol (a substitute for fossil fuels). The loss of potential to sequester carbon due to clearing of rain forest is greater than the climatic benefits due to the substitution of bioethanol for fossil fuels. Extensive development of oil palm plantations is driven more strongly by policy-driven economics than by the logical management of ecosystems for climate regulation. Photograph courtesy of World Land Trust

understanding of the role of the biosphere in the carbon cycle of ecosystems and the planet is essential to any strategy that effectively addresses the accelerating rates of climate change.

Overview of Decomposition and Ecosystem Carbon Balance

The leaching, fragmentation, and chemical alteration of dead organic matter by decomposition ultimately convert detritus to CO_2 and mineral nutrients and a remnant pool of complex organic compounds that resist further microbial breakdown. Most decomposition occurs in the litter layer and in the organic and mineral horizons of the soil (see Chap. 3). Decomposition is a consequence of interacting physical and chemical processes occurring inside and outside of living soil microbes and animals. Decomposition results from three types of processes, each with unique controls and consequences. (1) **Leaching** by water transfers soluble materials away from decomposing organic matter into the environment. These soluble materials are either absorbed by organisms, react with the mineral phase of soil or sediments, or are lost from the system in solution. (2) **Fragmentation** by soil animals breaks large pieces of organic matter into smaller ones, which they eat, and, in the process, create fresh surfaces for microbial colonization. Soil animals also mix the decomposing organic matter into the soil and return organic matter to the soil or sediments as fecal pellets, which have a higher surface-to-volume ratio and provide a more favorable environment for soil microbes (i.e., bacteria and fungi) than does the original material consumed. (3) **Chemical alteration** of dead organic matter results primarily from the activity of soil microbes, although some chemical reactions also occur spontaneously in the soil without microbial mediation.

Dead plant material (leaf, stem, and root **litter**) and animal residues are gradually decomposed until their original identity is no longer recognizable, at which point they are considered **soil organic matter** (SOM). Most compounds in litter are too large and insoluble to pass through microbial membranes. Microbes therefore secrete **exoenzymes** (extracellular enzymes) into their environment to initiate breakdown of litter. These exoenzymes convert macromolecules into soluble products that can be absorbed and metabolized by microbes. Microbes also secrete waste products of metabolism, such as CO_2 and inorganic nitrogen, and produce polysaccharides that enable them to attach to soil particles. When microbes die, their bodies become part of the organic substrate available for decomposition.

Decomposition is largely a consequence of the feeding activity of soil animals (fragmentation) and heterotrophic microbes (chemical alteration). The evolutionary forces that shape decomposition are those that maximize the growth, survival, and reproduction of soil organisms. In other words, decomposition occurs to meet the energetic and nutritional demands of decomposer organisms, not as a community service for the carbon cycle. The ecosystem consequences of decomposition are the **mineralization** of organic matter to inorganic components (CO_2, mineral nutrients, and water) and the **transformation** of some organic matter into complex organic compounds that are often **recalcitrant**, that is, resistant to further microbial breakdown.

The controls over organic matter breakdown change radically once SOM becomes incorporated into mineral soil. The soil moisture, oxygen, and thermal regimes of mineral soil are quite different than in the litter layer. In the mineral soil, SOM can complex with clay minerals or undergo nonenzymatic chemical reactions to form more complex compounds. **Humus**, for example, is a complex mixture of soil organic compounds with highly irregular structure. The long-term persistence of organic matter in soils depends upon chemical recalcitrance, sorption of organic compounds to clay surfaces, and other controls over microbial activity, although the relative importance of these processes is uncertain (Schmidt et al. in press).

Microbes and animals feed on live and dead organic matter to support their energetic and nutritional demands. The associated **heterotrophic respiration** accounts for about half of the CO_2 released from ecosystems to the atmosphere. Carbon is also transferred to the atmosphere through the production of carbon-containing trace gases such as methane and by combustion in wildfires. Finally, carbon leaches from ecosystems in dissolved and particulate forms and moves laterally through erosion and deposition of soil, movement of animals, etc. These lateral fluxes of carbon from terrestrial ecosystems are critical energy subsidies to aquatic ecosystems and constitute a significant component of the carbon budgets of many ecosystems.

In this chapter, we first describe decomposition in terrestrial ecosystems. We then describe important differences in decomposition between terrestrial and aquatic systems and finally integrate carbon loss pathways with carbon inputs to ecosystems (see Chaps. 5 and 6) to assess **net ecosystem carbon balance**.

Leaching of Litter

Leaching is the rate-determining step for mass loss of plant litter when it first senesces. **Leaching** is the physical process by which mineral ions and small water-soluble organic compounds dissolve in water and flow out of the detritus. Leaching begins when tissues are still alive and is most important during and shortly after tissue senescence (see Chap. 8). Soluble compounds are a larger proportion of the mass (and therefore account for more leaching loss) in leaf and fine root litter than in woody stems and roots. Leaching losses from litter are proportionally more important for nutrients than for carbon. Leaching loss from fresh litter occurs most rapidly (minutes to hours) in environments with high rainfall and is negligible under dry conditions. Compounds leached from litter include sugars, amino acids, and other organic compounds that are **labile** (easily broken down) or are absorbed intact by soil microbes.

Litter Fragmentation

Fragmentation creates fresh surfaces for microbial colonization and increases the proportion of the litter mass that is accessible to microbial attack. Fresh detritus is initially covered by a protective layer of cuticle or bark on plants or of skin or exoskeleton on animals. These outer coatings are designed, in part, to protect tissues from microbial attack. Within plant tissues, the labile cell contents are further protected from microbial attack by lignin-impregnated cell walls. Fragmentation of litter greatly enhances microbial decomposition by piercing these protective barriers, by increasing the ratio of litter surface area to mass, and by inoculating the residual mass with soil microbes.

Animals are the main agents of litter fragmentation, although freeze–thaw and wetting–drying cycles can also disrupt the cellular structure of litter. Animals fragment litter as a by-product of their feeding activities. Bears, voles, and other mammals tear apart wood or mix the soil as they search for insects, plant roots, and other food. Soil invertebrates fragment the litter to produce particles that are small enough to ingest. Enzymes in animal guts digest the microbial "jam" that coats the surface of litter particles, providing energy and nutrients to support animal growth and reproduction. The presence of soil invertebrates has a major effect on decomposition rate in moist temperate and tropical ecosystems, but is less important where temperature or moisture strongly constrains decomposition (Wall et al. 2008). The species composition of the invertebrate community, however, causes only a modest (7%) variation in decomposition rate (Wall et al. 2008). Apparently, different soil animals have roughly equivalent effects on fragmentation rates.

Chemical Alteration

Fungi

Fungi and bacteria are the main initial decomposers of terrestrial dead plant material, accounting for about 95% of the total decomposer biomass and respiration. Fungi consist of networks of **hyphae** (i.e., filaments that enable them to grow into new substrates and transport materials through the soil over distances of centimeters to meters). These hyphal networks enable fungi to acquire their carbon in one place and their nitrogen in another, much as plants gain CO_2 from the air but water and nutrients from the soil. Fungi that decompose fresh leaf or woody litter, for example, may acquire carbon from the surface litter and nitrogen from deeper, more decomposed soil horizons. Fungi secrete enzymes that enable them to penetrate the cuticle of dead leaves or the suberized exterior of roots to gain access to the interior of a dead plant organ. Here they proliferate within and between dead plant cells. At a smaller scale, some fungi gain access to the nitrogen, lignin-encrusted cellulose, and other labile constituents of dead cells by breaking down the lignin in cell walls. The large energy investment in lignin-degrading enzymes serves primarily to gain access to these relatively labile compounds.

Fungi produce dense networks of hyphae when resources are plentiful, allowing efficient access to these resources, but sparse hyphal networks when resources are scarce, reallocating resources from one part of the network to exploration of new litter and soil. This flexible growth strategy enables fungi to grow into new areas to explore for substrate, even when current substrates are exhausted. A substantial proportion (perhaps 25%) of the carbon and nitrogen used to support fungal growth are transported from elsewhere in the hyphal network, rather than being absorbed from the immediate environment where the fungal growth occurs (Mary et al. 1996).

Fungi have enzyme systems capable of breaking down all classes of plant compounds. They have a competitive advantage over many bacteria in decomposing tissues with low nutrient concentrations because of their ability to import nitrogen and phosphorus. In addition, fungi typically require less nitrogen per unit biomass than bacteria (i.e., the C:N ratio of fungi is often higher than the C:N ratio of bacteria). This may explain why fungal:bacterial ratios are typically higher in soils with high C:N ratios (Fierer et al. 2009a). White-rot fungi specialize on lignin degradation in wood, whereas brown-rot fungi cleave some of the side chains of lignin but leave the phenol units behind (giving the wood a brown color). White-rot fungi are generally outcompeted by more rapidly growing microbes when nitrogen is abundant, so nitrogen additions have little effect (or sometimes a negative effect) on white-rot fungal decomposition of wood (Waldrop and Zak 2006; Janssens et al. 2010). Most fungi lack a capacity for anaerobic metabolism and are therefore absent from or dormant in anaerobic soils and aquatic sediments.

Mycorrhizae are a symbiotic association between plant roots and fungi in which the plant gains nutrients from the fungus in return for carbohydrates (see Chap. 8). Although mycorrhizal fungi get most of their carbon from plant roots, they also play a role in decomposition by breaking down proteins into amino acids. These amino acids support fungal growth but are also transferred to their host plants (Read 1991; Finlay 2008). Mycorrhizal fungi also produce cellulases to gain entry into plant roots and participate in the breakdown of SOM, but the extent to which mycorrhizal cellulases participate in decomposition of dead organic matter is uncertain.

In the few ecosystems where fungal diversity has been examined using modern molecular techniques, there are 10- to 100-fold more fungal than plant taxa (Fierer et al. 2007; Taylor et al. 2010). Fungal taxa differ in trophic role (mycorrhizal or **saprotrophic** – eating dead organic matter), soil horizon, season of activity, and many other, as yet unknown, dimensions of their ecological niches, with these local sources of diversity often greater than variation among ecosystems (Fierer et al. 2007).

Bacteria and Archaea

The small size and large surface:volume ratio of bacteria and archaea enable them to rapidly absorb soluble substrates and to grow and divide quickly in substrate-rich zones. Archaea are structurally similar but evolutionarily distinct from bacteria. Like bacteria, they are metabolically diverse. The opportunist strategy of bacterial and archaea (which, for convenience, we will label as "bacteria") explain their dominance in the **rhizosphere** (the zone of soil directly influenced by plant roots) and in dead animal carcasses, where labile substrates are abundant. Bacteria are also important in breaking down live and dead bacterial and fungal cells. The major functional limitation resulting from their small size is that each bacterium depends mostly on the substrates that move toward it. Some of these substrates are products of bacterial exoenzymes. These products diffuse to the bacterium along a concentration gradient created by (1) the activity of the exoenzymes, which produce soluble substrates, and (2) the absorption of substrates by the bacterium, which reduces substrate concentrations at the bacterial surface. Other soluble substrates flow past the bacterium in water moving through the soil. This water movement is driven by gradients in water potential associated with plant transpiration, evaporation at the soil surface, and gravitational water movement (see Chap. 4). Water movement (and therefore substrate supply) is most rapid in **macropores** (relatively large air or water spaces between soil aggregates). Bacteria therefore often line macropore surfaces and absorb substrates from the flowing water, just as fishermen net salmon migrating up a stream. Macropores are also preferentially exploited by roots because of the reduced physical resistance to root elongation, providing an additional source of labile substrates to bacteria. Bacteria attached to the exposed surfaces of macropores are vulnerable to predation by protozoa and nematodes, which use the water films in macropores as highways to move through the soil. This leads to rapid bacterial turnover on exposed particle surfaces.

A wide range of bacterial types is present in soils; indeed, we are just beginning to characterize their abundance and diversity through molecular methods, and we can expect much more information to become available as these techniques are refined and applied more widely. Rapidly growing gram-negative bacteria specialize on labile substrates secreted by roots. Actinobacteria are slow-growing, gram-positive bacteria that have a filamentous structure similar to that of fungal hyphae. Like fungi, actinobacteria produce lignin-degrading enzymes and can break down relatively recalcitrant substrates. They often produce antibiotics to reduce competition from other microbes. The best predictor of bacterial community composition and exoenzyme activities appears to be soil pH (Sinsabaugh et al. 2008; Fierer et al. 2009a).

The bacterial communities that coat soil aggregates have a surprisingly complex structure. They often occur as **biofilms**, microbial communities embedded in a matrix of polysaccharides secreted by bacteria. This microbial "slime" protects bacteria from grazing by protozoa and reduces bacterial water stress by holding water like a sponge. The matrix also increases the efficiency of bacterial exoenzymes by preventing them from being swept away by moving water. The bacteria in biofilms often act as a **consortium**, that is, a group of genetically unrelated bacteria, each of which produces only some of the enzymes required to break down complex macromolecules. The breakdown of these molecules to soluble products requires the coordinated production of exoenzymes by several types of bacteria. This is analogous to an assembly line, in which the final product depends on the coordinated action of several consecutive steps, and no bacterium benefits unless all the steps are in place to produce the final product. The evolutionary forces and population interactions that shape the composition and functioning of microbial consortia are poorly understood. Consortia are particularly important in the breakdown of pesticides and other organic residues that people have added to the environment.

Because most bacteria are immobile, a bacterial colony eventually exhausts the substrates in its immediate environment, especially within soil aggregates that have restricted water movement. When this occurs, they become inactive and reduce their respiration to negligible rates. Bacteria may remain inactive for years. Live bacteria have been recovered from permafrost that is three million years old (Gilichinsky et al. 2008). About 50–80% of the bacteria in soils are metabolically inactive (Norton and Firestone 1991). Inactive bacteria reactivate in the presence of labile substrates, for example, when a root grows through the soil and exudes carbohydrates.

The inactive bacteria in soils represent a reservoir of decomposition *potential,* analogous to the buried seed pool that provides a source of plant colonizers after a disturbance. Like the buried seed pool, the enzymatic potential of these inactive bacteria may differ from the enzymes produced by the active bacterial community. Consequently, DNA probes or microbiological culturing techniques are better indices of what the soil *could* do (its metabolic diversity and enzymatic potential) than of its current metabolic activity.

Bacterial, archaeal, viral, and fungal communities living in soil are highly diverse (Fierer et al. 2007). However, bacteria and archaea can thrive in a broader range of microenvironments than fungi, including habitats that are anaerobic, have little available carbon or nitrogen, are contaminated with toxic heavy metals, or experience extremes in temperature or UV radiation.

Soil Animals

Soil animals influence decomposition by fragmenting, transforming, and transporting litter, grazing populations of bacteria and fungi, and altering soil structure. Microfauna are the smallest animals (<0.1 mm diameter). They include nematodes, protozoans, such as ciliates and amoebae, and rotifers (Fig. 7.2; Wallwork 1976; Lousier and Bamforth 1990). Protozoans are single-celled animals that ingest their prey primarily by **phagocytosis**, that is, by enclosing them in a membrane-bound structure within the cell. Protozoans are usually mobile and are voracious predators of bacteria and other microfauna (Lavelle et al. 1997). Protozoans are particularly important predators in the rhizosphere and other soil microsites with rapid bacterial growth rates (Coleman 1994). The preferential grazing by protozoa on bacteria (even on particular species of bacteria) tends to reduce bacterial:fungal ratios compared to soils from which protozoa are excluded. Nematodes are an abundant and trophically diverse group in which each species specializes on bacteria, fungi, roots, or other soil animals. Bacterial-feeding nematodes in forest litter, for example, can consume about 80 g m^{-2} year^{-1} of

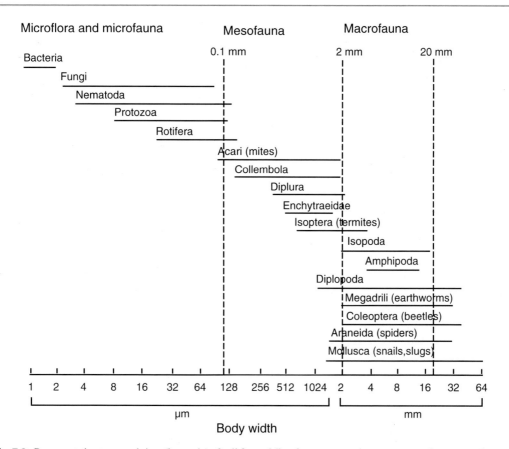

Fig. 7.2 Representative types and sizes (log scale) of soil fauna. Microfauna are most important as predators; mesofauna as organisms that fragment litter; and macrofauna as ecosystem engineers. Redrawn from Swift et al. (1979)

bacteria, resulting in the mineralization of $2–13 \, \mathrm{g \, m^{-2} \, year^{-1}}$ of nitrogen, a substantial proportion of the nitrogen that annually cycles through the soil (Anderson et al. 1981). Protozoans and nematodes are aquatic animals that move through water-filled pores between soil particles and are therefore more sensitive to water stress than are fungi and the meso- and macrofauna that fragment soil particles. Their populations fluctuate dramatically, both spatially and temporally, due to drying–wetting events and predation (Beare et al. 1992). When protozoans die, their bodies are rapidly broken down by soil microbes, especially by bacteria.

The **mesofauna** are a numerically abundant and taxonomically diverse group of soil animals 0.1–2 mm in diameter (Fig. 7.2). These are the animals with the greatest effect on decomposition.

They fragment and ingest litter coated with microbial biomass, producing large amounts of fecal material with a greater surface area and moisture-holding capacity than the original litter (Lavelle et al. 1997). This altered litter environment is more favorable for decomposition. Mesofauna selectively feed on litter that has been conditioned by microbial activity and also selectively feed on soil fungi, causing changes in fungal community structure. Collembola are small insects that feed primarily on fungi, whereas mites (Acari) are a more trophically diverse group of spider-like animals that consume decomposing litter or feed on bacteria, fungi, or soil animals. Feeding by micro- and mesofauna can significantly alter the biomass and activity of microbial communities and therefore rates of decomposition and nutrient turnover (Bardgett 2005).

Large soil animals (**macrofauna**), such as earthworms and termites, are **ecosystem engineers** that alter resource availability by modifying the physical properties of soils and litter (Jones et al. 1994). Some of them, like the mesofauna, fragment litter (Lavelle et al. 1997). Others burrow or ingest soil, reducing soil bulk density, breaking up soil aggregates, and increasing soil aeration and water infiltration (Beare et al. 1992). The passages created by earthworms create channels in the soil that water and roots easily penetrate. They create patterns of soil structure that promote or constrain the activities of soil microbes and other soil animals. In temperate pastures, earthworms may process 4 kg m^{-2} year^{-1} of soil, moving 3–4 mm of new soil to the ground surface each year (Paul and Clark 1996). This is a geomorphic force that, integrated over time, is orders of magnitude larger than landslides or surface soil erosion (see Table 3.1). In temperate forests, exotic earthworms have substantially reduced soil carbon storage (Bohlen et al. 2004). Soil mixing by earthworms tends to disrupt the formation of distinct soil horizons. Once the soil enters the digestive tract of an earthworm, mixing and secretions by the earthworms stimulate microbial activity, so soil microbes act as gut mutualists. Many of the soil organisms are digested during passage through the gut, which absorbs the resulting products. Earthworms are most abundant in the temperate zone, whereas termites and ants are the dominant ecosystem engineers in tropical soils. Termites eat plant litter directly, digest the cellulose with the aid of mutualistic protozoans and bacteria in their guts, and mix the organic matter into the soil. Dung beetles in tropical grasslands perform a similar function with mammalian dung. This burial of surface organic matter places it in a humid environment where decomposition occurs more rapidly.

The soil fauna is critical to the carbon and nutrient dynamics of soils. Microbes constitute 70–80% of the labile carbon and nitrogen in soils, so exclusion of soil animals from soils or natural variation in their predation on microbes significantly alters carbon and nitrogen turnover in soils, although their net effect is relatively modest (up to 30%; Swift et al. 1979; Verhoef and Brussaard 1990). Sometimes, soil animals *inhibit* decomposition through their direct consumption of microbial biomass, and sometimes they *stimulate* decomposition by reducing the density of microbial predators (Bardgett et al. 2005b).

Because of their high respiration rate, soil animals metabolize much of the microbial carbon they consume to CO_2 and excrete the microbial nitrogen and phosphorus that exceeds their requirements for growth and reproduction. These nutrients become available for absorption by plants or microbes (see Chap. 8). Soil animals account for only about 5% of soil respiration, so their major effect on decomposition is the enhancement of microbial activity through fragmentation (Wall et al. 2001), rather than their own processing of energy derived from detritus.

Temporal and Spatial Heterogeneity of Decomposition

Temporal Pattern

The predominant controls over decomposition change with time. Decomposition is the consequence of the interactions of leaching, fragmentation, and chemical alteration. As soon as a leaf unfolds, it is colonized by aerially borne bacteria and fungal spores that begin breaking down the cuticle and leaf surfaces that have been exposed by herbivores, pathogens, or physical breakage (Haynes 1986). This **phyllosphere decomposition** of live leaves is generally ignored because it is not easily separated from plant-controlled changes in leaf mass and chemistry. Other bacteria and fungi live inside live leaves, producing toxins that reduce herbivory, thereby altering the properties and functioning of leaves (Clay 1990). Both groups provide a microbial inoculum that rapidly initiates decomposition of labile substrates when the leaf falls to the ground. Similarly, the root cortex begins to break down while the conducting tissues of roots still function in water and nutrient transport, blurring the distinction between live and dead roots.

Litter mass initially decreases rapidly as it decomposes, and decomposition rate declines

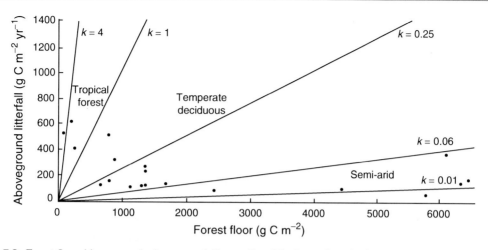

Fig. 7.3 Forest-floor biomass and aboveground litter inputs for selected evergreen forests. Lines show the relationship between forest floor mass and aboveground litterfall for selected decomposition constants (k). Redrawn from Olsen (1963)

as litter ages (Haynes 1986; Harmon et al. 2009). This is often described as an exponential relationship, implying that a constant *proportion* of the litter is decomposed each year.

$$L_t = L_0 e^{-kt} \qquad (7.1)$$

$$\ln \frac{L_t}{L_0} = -kt \qquad (7.2)$$

where L_0 is the litter mass at time zero and L_t is the mass at time t. The **decomposition rate constant**, k, is an exponent that characterizes the decomposition rate of a particular material. The mean **residence time**, that is, the time required for the litter to decompose under steady-state conditions, equals $1/k$. The residence time of litter can also be estimated as the average pool size of litter divided by the average annual input. Residence time differs substantially among biomes (Fig. 7.3).

$$\frac{l}{k} = \frac{\text{litter pool}}{\text{litterfall}} \quad \text{or} \quad k = \frac{\text{litterfall}}{\text{litter pool}} \qquad (7.3)$$

The calculation of residence time from pools and fluxes assumes that the measurements made at a particular time are representative of the steady state, which is seldom the case (see Chap. 12).

Year-to-year variation in weather or directional changes in climate cause more rapid changes in litterfall than in the litter pool, creating challenges in estimating residence time. The decomposition constant varies widely with substrate composition. Sugars, for example, have a residence time of hours to days, whereas lignin has a residence time of months to decades, depending on the ecosystem. Plant and animal tissues differ substantially in their chemical composition and therefore in their decay constants. Taken as a whole, leaf and fine-root litter generally has a residence time of months to years, logs a residence time of years to centuries, and organic material mixed with mineral soil a residence time of years to millennia.

The exponential model of decomposition (7.1), which implies a constant proportion of litter decomposed each year, is therefore only a rough approximation of the actual pattern of decline in litter mass with time. The process is more accurately described by multiple curves that describe at least four phases (Fig. 7.4; Adair et al. 2008; Harmon et al. 2009). Leaching of cell solubles dominates the first phase. Fresh leaf or fine-root litter, for example, can lose 5% of its mass in 24 h due to leaching alone. The second phase of decomposition occurs more slowly and involves a combination of fragmentation by soil animals, chemical alteration by soil microbes, and leaching of decay products from the litter. During the

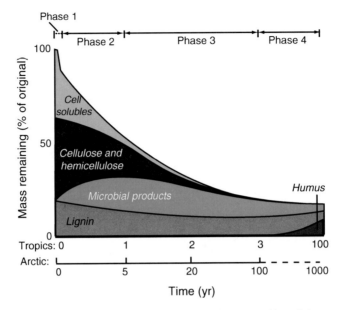

Fig. 7.4 Representative time course of leaf-litter decomposition, showing the major chemical constituents (cell solubles, cellulose and hemicellulose, lignin, microbial products, and humus), the four major phases of litter decomposition, and the timescales commonly found in warm (tropical) and cold (arctic) environments. Leaching dominates the first phase of decomposition. Substrate composition of litter changes during litter decomposition of phases 2 and 3 because labile substrates, such as cell solubles, are broken down more rapidly than recalcitrant compounds such as lignin and microbial cell walls. In phase 4, litter particles contact mineral surfaces, forming soil organic matter

second phase of decomposition, relatively labile substrates are decomposed, leaving behind more recalcitrant ones. The third phase of decomposition involves the same *processes* as the second phase but occurs more slowly because the remaining compounds are recalcitrant and decompose slowly. Decomposition during the second and third phases is often measured as mass loss from dead leaves (Aerts 1997), roots (Berg et al. 1998), or twigs that are tethered on threads or placed in mesh **litterbags** and weighed periodically (Vogt et al. 1986; Robertson and Paul 2000). Logs can be placed on the ground surface. Exponential models of decomposition have been applied primarily to the second and third phases (Harmon et al. 2009). The fourth and final phase of decomposition occurs quite slowly and involves the chemical alteration of organic matter that is mixed with mineral soil and the leaching of breakdown products to other soil layers. Decomposition during this final phase is often estimated from measurements of soil respiration or isotopic tracers, given that mass loss is very slow (Box 7.1; Schlesinger 1977; Trumbore and Harden 1997).

The decomposition rate and decomposition rate constant (k in 7.1) gradually decline through these four phases of decomposition.

In seasonal environments, microbial respiration often occurs over a longer time period and peaks later in the season than does plant growth. Like plant growth, microbial respiration is favored by warm, moist conditions and is therefore greatest during the season of maximum plant growth. Heterotrophic respiration, however, typically begins earlier in the season and ends later than does plant growth for at least three reasons: (1) Microbial respiration typically occurs over a broader range of temperatures (e.g., −10–40°C) and soil moistures than does plant growth. (2) The soil is buffered from temperature extremes that aboveground parts of plants must cope with. (3) Soil temperature lags behind air temperature, so microbial respiration remains high in late summer and autumn at times when plant activity has begun to decline (Davidson and Janssens 2006). Microbial activity is also influenced by the seasonality of plant activity. Root turnover and exudation are often greatest in mid-season when

Box 7.1 Isotopes and Soil Carbon Turnover

The quantity of soil carbon differs dramatically among ecosystems (Post et al. 1982). The total quantity of carbon in an ecosystem, however, gives relatively little insight into its dynamics. Tropical forests and tundra, for example, have similar quantities of soil carbon, despite their radically different climates and productivities. The simplest measure of soil carbon turnover is its residence time estimated from the pool size and carbon inputs (7.3). These measurements show that, even though tropical forests and arctic tundra have similar-sized soil carbon pools, the turnover may be 500 times more rapid in the tropical forest. More sophisticated approaches to estimating soil carbon turnover using carbon isotopes (Ehleringer et al. 2000) lead to a similar conclusion. In the tropics, 85% of the ^{14}C that entered ecosystems during the era of nuclear testing in the 1960s has been converted to humus, whereas this proportion is only 50% in temperate soils and close to zero in boreal soils (Trumbore 1993; Trumbore and Harden 1997). This comparison clearly indicates more rapid turnover of SOM in the tropics than at high latitudes.

Carbon isotopes can also be used to estimate the impacts of land-use change on carbon turnover in situations where the vegetation change is associated with a change in carbon isotopes. In Hawai'i, for example, replacement of C_3 forests by pastures dominated by C_4 grasses causes a gradual change in the carbon isotope ratio of SOM from values similar to C_3 plants toward values similar to C_4 plants (Townsand et al. 1995). This information can be used to estimate the quantity of the original forest carbon that remains in the ecosystem:

$$\%C_{SI} = \frac{C_{S2} - C_{V2}}{C_{VI} - C_{V2}} \cdot 100 \qquad \text{(B7.1)}$$

where $\%C_{SI}$ is the percentage of soil derived from the initial ecosystem type, C_{S2} is the ^{13}C content of soil from the second soil type, C_{V2} is the ^{13}C content of soil from the second vegetation type, and C_{VI} is the ^{13}C content of vegetation from the initial ecosystem type.

photosynthesis is high, contributing to the mid-season peak in soil respiration. Autumn or dry-season senescence provides an additional input of substrates that supports late-season soil respiration.

Vertical Distribution

Most decomposition occurs near the soil surface, where litter inputs are concentrated. Most aboveground litter (leaves and wood) decomposes and releases nutrients on or near the soil surface. Roots therefore tend to grow in surface soils in order to access these nutrients. Thus most root litter is also produced in surface soils, reinforcing the surface localization of decomposition. There are some deep roots, however, and soil mixing by animals, especially termites and earthworms, as well as leaching of dissolved organic matter to depth. About half of the soil organic carbon therefore is typically below 20 cm depth, even though only a third of the roots are below that depth (Fig. 7.5; Jobbágy and Jackson 2000). On average, the deep-soil carbon is older, more recalcitrant, and more tightly bound to soil minerals than is surface carbon (Trumbore and Harden 1997), but a small fraction of the deep soil C is modern, coming mostly from turnover of deep roots.

Decomposition rates are spatially heterogeneous at several scales. The surface litter layer exhibits large daily changes in temperature and moisture. Decomposition in this layer is dominated by fungi that import nitrogen from below. This is a radically different environment than the mineral soil, where temperature and moisture are more stable, some of the organic matter is humified and recalcitrant, and mineral soil surfaces

Fig. 7.5 Globally averaged depth profiles of soil organic matter and roots in the top meter of soil. Redrawn from Jobbágy and Jackson (2000)

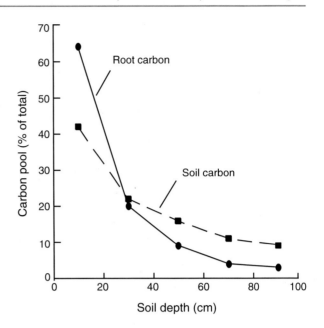

Factors Controlling Decomposition

Decomposition in ecosystems is controlled by three major factors: substrate quality, characteristics of the microbial community, and physical environment (Swift et al. 1979; Allison 2006). About 75% of terrestrial organic carbon is dead organic matter in soils (see Chap. 14) and represents potential food for decomposer organisms. Given the potent capacity of soil microbes to grow and to break down SOM, why don't they consume it all? In other words, why is the world brown (Allison 2006)? There are multiple contributing factors, but the most important of these appear to be substrate quality, physical environment, and microbial community composition (Allison 2006).

Litter Quality

The availability of belowground resources is the factor governing ecological patterns in litter quality. Plants that grow rapidly, both because of the environment in which they grow and their species properties, typically produce litter that

bind dead organic matter and microbial enzymes. At a finer scale, the rhizosphere around roots is a carbon-rich microenvironment that supports much higher microbial activity than the bulk soil. Finally, the interior of soil aggregates is more likely to be anaerobic than are the surfaces of soil pores. Movement within the soil by roots, water, and soil animals is constantly changing the spatial configuration of these different decomposition environments.

In some ecosystems, such as tropical wet forests, significant quantities of aboveground litter are caught on epiphytes and branches of the canopy. In these wet ecosystems, substantial decomposition, nutrient release, and nutrient absorption by rooted epiphytes occur in the canopy, thereby short-circuiting the soil phase (Nadkarni 1981). Some terrestrial litter and dissolved organic carbon (DOC) also enters streams and lakes, where they become important energy sources for aquatic food webs, as described later. In low-nutrient ecosystems, much of the DOC that enters streams is so recalcitrant that it remains largely unprocessed, leading to the "black-water" rivers that characterize many tropical and boreal forests and temperate swamps.

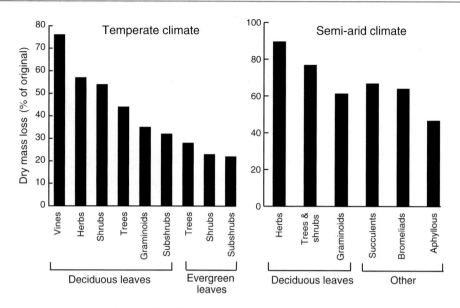

Fig. 7.6 Decomposition rate of leaves of British deciduous and evergreen plant species (*left*) and of deciduous plants and arid-zone plants in Argentina (*right*). Data from Cornelissen (1996) and Perez-Harguindeguy et al. (2000)

decomposes quickly because the same morphological and chemical traits that promote NPP also regulate decomposition (Hobbie 1992; De Deyn et al. 2008). Both NPP and decomposition are enhanced by a high allocation to leaves and by the production of nutrient-rich leaves with a short life span. These tissues decompose rapidly because they have high concentrations of labile compounds such as proteins and low concentrations of recalcitrant cell-wall components such as lignin (Reich et al. 1997). Consequently, species from productive sites produce fine litter that decomposes rapidly (Fig. 7.6; Cornelissen 1996; De Deyn et al. 2008). Species differences in litter quality are an important mechanism by which plant species affect ecosystem processes (see Chap. 11; Hobbie 1992) and are excellent predictors of landscape patterns of initial litter decomposition (Flanagan and Van Cleve 1983). Ecosystems such as forests that produce large quantities of woody stems and roots produce distinct litter types with quite different litter qualities and decomposition rates, with wood decomposing much more slowly than fine litter.

Carbon quality of substrates is the primary chemical determinant of the decomposition of fresh litter. In controlled experiments, differences in substrate **quality**, that is, susceptibility of a substrate to decomposition, give rise to a five- to tenfold range in litter decomposition rate. Animal carcasses decompose more rapidly than plants; leaves decompose more rapidly than wood; deciduous leaves decompose more rapidly than evergreen leaves; and leaves from high-nutrient environments decompose more rapidly than leaves from infertile sites (Figs. 7.6 and 7.7). These differences in decomposition rate are a logical consequence of litter chemistry. Litter compounds can be categorized roughly as: (1) labile metabolic compounds, such as sugars and amino acids, (2) moderately labile structural compounds, such as cellulose and hemicellulose, and (3) recalcitrant structural material, such as lignin, suberin, and cutin. Rapidly decomposing litter generally has higher concentrations of labile substrates and lower concentrations of recalcitrant compounds than does slowly decomposing litter.

Five interrelated chemical properties of organic matter determine substrate quality (J. Schimel, personal communication): size of molecules, types of chemical bonds, regularity of

Fig. 7.7 Time course of decomposition of a deciduous leaf, a conifer needle, and wood in a Canadian temperate forest. Data from MacLean and Wein (1978)

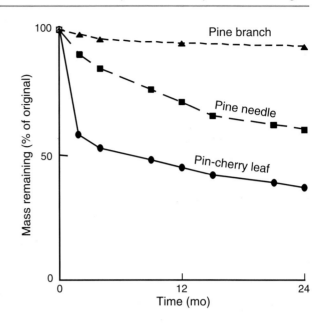

structures, toxicity, and nutrient concentrations. (1) Large molecules cannot pass through microbial membranes so they must be processed outside microbial cells by exoenzymes. This limits the degree of control that a given microbe can exert over the detection of substrate availability, delivery of enzymes in response to substrate supply and efficient utilization of breakdown products. Due to differences in molecular size, sugars and amino acids are metabolized more easily than cellulose and proteins, respectively. (2) Some chemical bonds are easier to break than others. Ester linkages that bind phosphate to organic skeletons or peptide bonds that link amino acids to form proteins, for example, are easier to break than the double bonds of aromatic rings. For these reasons, the nitrogen in proteins is much more available to microbes than nitrogen contained in aromatic rings. (3) Compounds like lignin that have a highly irregular structure do not fit the active sites of most enzymes, so they are broken down slowly by nonspecific enzymes (e.g., peroxidases) compared to compounds like cellulose that consist of chains of regularly repeating glucose units. (4) Some soluble compounds such as phenolics and alkaloids are toxic and kill or reduce the activity of microbes that absorb them. (5) Organic compounds containing

nitrogen and phosphorus are the major nutrient source supporting microbial growth, so organic matter such as straw that contains low concentrations of these elements may not provide enough nutrients to allow microbes to use fully the carbon present in the litter.

The effects of nutrients on decomposition are largely indirect, mediated by carbon quality of substrates. Although decomposition rates are slow in low-nutrient environments, direct effects of nutrient concentrations in litter or in the soil are seldom seen (Fog 1988; Hobbie 2008). For example, placing the same litter in soils of different nitrogen availability does not consistently alter decomposition, and litters of similar carbon chemistry but different nitrogen concentrations do not differ consistently in decomposition rate (Haynes 1986; Prescott 1995; Prescott et al. 1999; Hobbie and Vitousek 2000; Knorr et al. 2005; Hobbie 2008). Nonetheless, litter with a low ratio of carbon concentration to nitrogen concentration (low **C:N ratio**; high nitrogen concentration) generally decomposes quickly, especially in the early stages of decomposition (Enríquez et al. 1993; Gholz et al. 2000), indicating that C:N ratio is a good *predictor* of initial rates of decomposition. Initial **lignin:nitrogen ratio** of litter is also a good *predictor* of initial

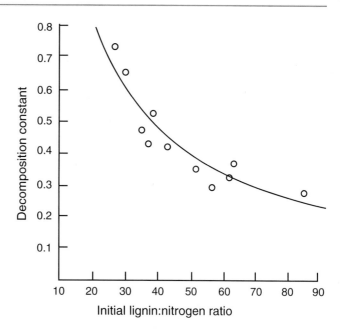

Fig. 7.8 Relationship between the lignin:nitrogen ratio of litter and its decomposition constant. Redrawn from Melillo et al. (1982)

rates of decomposition rate (Fig. 7.8; Berg and Staaf 1980; Melillo et al. 1982; Taylor et al. 1989). To the extent that nutrients influence decomposition, nitrogen is likely to stimulate decomposition of low-lignin litter and inhibit decomposition of high-lignin litter, with no significant overall effect of nitrogen on decomposition (Fog 1988; Allison 2006; Janssens et al. 2010).

Litter nutrient concentrations may influence the fate of carbon metabolized by microbes. Microbes that decompose nitrogen-rich litter, for example, release a larger proportion of the carbon in respiration rather than retaining it in microbial biomass (Manzoni et al. 2008). This may cause high-nitrogen litter to lose its labile carbon so quickly that the remaining litter decomposes quite slowly in the later stages of decomposition (Berg and Meentemeyer 2002).

Both the age and the initial quality of SOM influence its decomposition rate. As litter decomposes, its decomposition rate declines because microbes first consume the more labile substrates, leaving behind more recalcitrant compounds (Fig. 7.4). As microbes die, chitin and other recalcitrant components in their cell walls comprise an increasing proportion of the litter mass. Species effects on litter decomposition rate gradually decline through time as labile

substrates are depleted. In addition, older litter fragments that mix downward into mineral soil undergo abiotic chemical reactions and interactions with mineral surfaces that further reduce decomposition rate (Allison 2006). Rates of these later phases of decomposition are difficult to predict (Currie et al. 2010).

The SOM in mineral soils is a mixture of organic compounds of different ages and chemical compositions. It includes fragments of recently shed root, stem, and leaf litter, together with SOM that is thousands of years old (Oades 1989). These different aged components of SOM can be partially separated by density centrifugation because recently produced particles are less dense than older ones and are less likely to be bound to mineral particles. Soils in which a large proportion of the SOM is in the light fraction generally have higher decomposition rates (Robertson and Paul 2000). Alternatively, soil can be chemically separated into distinct fractions, such as water-soluble compounds, **humic acids**, and **fulvic acids** that differ in average age and ease of breakdown. SOM as a whole typically has a mean residence time of 20–50 years, although this can range from 1 to 2 years in cultivated fields to thousands of years in environments where decomposition occurs slowly.

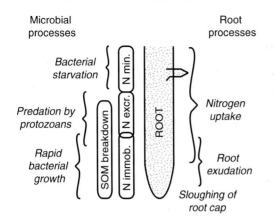

Fig. 7.9 Root and microbial processes in the rhizosphere and the resulting impacts on soil organic matter breakdown and nitrogen dynamics in the rhizosphere

Rhizosphere Stimulation of Decomposition

Plants stimulate decomposition near their roots. The **rhizosphere** is the soil within a few millimeters of plant roots. The rhizosphere comprises virtually all the soil in fine-rooted grasslands, where the average distance between roots is about 1 mm, whereas forests are less densely rooted (often 10 mm between roots; Newman 1985). Roots alter the chemistry of the rhizosphere by secreting carbohydrates and absorbing nutrients. These processes are most active in the zone behind the tips of actively growing roots (Fig. 7.9; Jaeger et al. 1999). The growth of bacteria in the zone of exudation (Norton and Firestone 1991) is supported by high carbon availability (20–40% of NPP; see Table 6.2) and is therefore limited most strongly by nutrients (Cheng et al. 1996). Bacteria use the labile carbon to produce enzymes that "mine" SOM for nutrients.

Microbial immobilization of nutrients in the rhizosphere benefits the plant only if these nutrients are subsequently released and become available to the root. Two processes contribute to the release of nutrients from rhizosphere microbes. First, protozoa and nematodes may graze rhizosphere bacteria, using bacterial carbon to support their high energetic demands and excreting the excess nutrients (Clarholm 1985). Second, as the

root matures and exudation rate declines, those bacteria that survive predation may become energy-limited and break down nitrogen-containing compounds in the soil to meet their energy demands, excreting the nitrogen into the rhizosphere as ammonium (Kuzyakov et al. 2000). Food-web interactions among multiple trophic levels complicate these rhizosphere dynamics (Moore et al. 2003).

The relative contribution of grazing and starvation in rhizosphere nutrient release is unknown, but net nitrogen mineralization in the rhizosphere has been estimated to be 30% higher than in bulk soil. In general, the presence of plant roots can stimulate the decomposition of SOM up to threefold, depending on the type of plant species and soil conditions (Cheng et al. 2003). Rhizosphere decomposition may be more sensitive to factors influencing plant carbohydrate status (e.g., light and grazing) than to soil physical environment (Craine et al. 1999; Bardgett et al. 2005a), so the nature of controls over decomposition (soil environment vs. plant carbohydrate status) could differ substantially among ecosystems, depending on the extent of rhizosphere decomposition and the nature of its ecological controls.

Mycorrhizal fungi are functionally an extension of the root system, allowing the root-fungal symbiosis to absorb nutrients at a distance from the root. The **mycorrhizosphere** around mycorrhizal fungal hyphae rapidly moves plant carbon into the bulk soil through a combination of hyphal turnover and exudation (Norton et al. 1990; Finlay 2008). This might also stimulate decomposition, just as in the rhizosphere of roots.

Microbial Community Composition and Enzymatic Capacity

The activity of soil microbes is more important than their biomass in determining decomposition rate. Microbial biomass is a relatively constant proportion (about 2%) of total soil carbon and therefore has the largest pool size (g m^{-2}) in those stands with the largest quantities of soil carbon (Fierer et al. 2009a); these tend to be the stands with lowest productivity and slowest decomposition (Vance and Chapin 2001). In agricultural soils, microbial biomass also tends to be higher in extremely wet or dry soils, where decomposition is slow, than in moderately moist soils with higher decomposition rate (Insam 1990). Since most microbial biomass is inactive, it is probably more important as a reservoir of nutrients (see Chap. 9) than as a predictor of decomposition rate. This differs from the controls over carbon inputs to ecosystems, where the quantities of plant biomass and leaf area are extremely important determinants of GPP. Those microbial processes like nitrification that are conducted by a restricted number of microbial groups, on the other hand, appear to be sensitive to the population sizes of these groups.

Soil enzyme activity sometimes depends on microbial community composition. The composition of the microbial community is potentially important for decomposition because it influences the types and rates of enzyme production and therefore the rates at which substrates are broken down. Enzymes that break down common substrates like proteins and cellulose are universally present in soils because of their production by most types of microbes (Schimel 2001). Microbial communities that are quite different in composition therefore often have relatively similar decomposition rate and exoenzyme composition (Kemmitt et al. 2008; Fierer et al. 2009b). On the other hand, enzymes involved in processes that occur only in specific environments, such as the anaerobic process of methane production, appear more sensitive to microbial community composition (Gulledge et al. 1997; Schimel 2001). Litter that is decomposed in soils associated with the plant that produces the litter decomposes about 10% faster than in soils from other places (Ayres et al. 2009). This "home field advantage" of decomposition results from the development of distinct microbial communities that are adapted to the litter that they most frequently encounter. These effects of microbial community composition on decomposition are small, however, compared to environmental and substrate-quality effects (Parton et al. 2007; Fierer et al. 2009a).

Soil enzyme activity is also influenced by the binding of enzymes to surfaces or their breakdown by soil proteases. Binding of an enzyme to the external surface of roots or microbes often prolongs its activity in soil, whereas binding to mineral particles can alter the enzyme configuration or block its active site, thereby reducing activity. A brief description of a few soil enzyme systems illustrates some of the microbial and soil controls over exoenzyme activity.

Most soil microbes, including ericoid and ectomycorrhizal fungi, produce enzymes (proteases and peptidases) that break down proteins to amino acids, which are easily absorbed by microbes and used either to produce microbial protein or to provide respiratory energy. Because proteases are subject to attack by other proteases, their lifetime in the soil is short, and soil protease activity tends to mirror microbial activity. Phosphatases, which cleave phosphate from organic phosphate compounds, are, however, more long lived, so their activity in soil is correlated more strongly with the availability of organic phosphate in soil than with microbial activity (Kroehler and Linkins 1991).

Cellulose is the most abundant chemical constituent of plant litter. It consists of chains of glucose units, often thousands of units in length, but none of this glucose is available to support microbial metabolism until acted upon by exoenzymes. Cellulose breakdown requires three separate enzyme systems (Paul and Clark 1996): **Endocellulases** break down the internal bonds to disrupt the crystalline structure of cellulose. **Exocellulases** then cleave off disaccharide units from the ends of chains, forming cellobiose, which is absorbed by microbes and broken down intracellularly to glucose by **cellobiase**. Some soil microbes, including most fungi, produce the entire suite of cellulase enzymes. Other organisms, such as some bacteria, produce only some cellulase enzymes and must function as part of microbial consortia to gain energy from cellulose breakdown.

Lignin is degraded slowly because only some organisms (primarily fungi) produce the necessary enzymes, and these microbes produce enzymes only when nitrogen is unavailable.

Sometimes this is mediated by competition between rapidly growing bacteria that break down labile organic matter and release nitrogen that inhibits more slow-growing lignolytic fungi. Lignin forms non-enzymatically by condensation reactions with phenols and free radicals, creating an irregular structure that does not fit the specificity required by the active site of most enzymes. For this reason, lignin-degrading enzymes use hydrogen peroxide to generate free radicals, which have a low specificity for substrates but are very powerful oxidizers. Oxygen is required to generate the hydrogen peroxide and the subsequent free radicals, so lignin breakdown does not occur in anaerobic soils. Decomposers generally invest more energy in producing lignin-degrading enzymes than they gain by metabolizing its breakdown products (Coûteaux et al. 1995). Lignin appears to be degraded to gain access to the nitrogen in the interior of lignified dead cells or to provide access to lignin-encrusted cellulose (Coûteaux et al. 1995; Adair et al. 2008). Because of the generation of free radicals, some of the enzymes involved in lignin breakdown also modify existing organic matter and generate more complex soil humus.

As discussed earlier, predation by soil animals generally has only a modest effect on decomposition. In the ocean, viruses and other diseases exert an important control over decomposition, but little is known about the role of disease as a "top-down" control over terrestrial decomposition (Allison 2006).

The Environment

Moisture
Decomposition increases with increasing moisture, until soils become so waterlogged that anaerobic conditions inhibit decomposition. Decomposers, like plants, are most productive under warm, moist conditions, if enough oxygen is available. This accounts for the high decomposition rates in tropical forests (Gholz et al. 2000). Decomposition rate of mineral soil generally declines at soil moistures less than 30–50% of dry mass (Haynes 1986), due to the reduced

thickness of moisture films on soil surfaces and therefore the rate of diffusion of substrates to microbes (Stark and Firestone 1995). Osmotic effects further restrict the activity of soil microbes in extremely dry or **saline** (salty) soils. Bacteria function at lower water availability than do plant roots, so decomposition continues even in soils that are too dry to support plant activity, perhaps contributing to the low soil organic content of arid ecosystems. Rewetting of very dry soils by dew or rain can influence decomposition by creating an osmotic shock that stresses microbial cells, causing a flush of available carbon. The net effect of drying–wetting cycles is a stimulation of decomposition, if the cycles are infrequent (as generally occurs in soils), but frequent cycles, as in the litter layer, can reduce microbial populations enough to reduce decomposition rates (Clein and Schimel 1994). Drying–wetting cycles tend to stimulate the decomposition of labile substrates (e.g., hemicellulose), which are broken down largely by rapidly growing bacteria, and to retard the decomposition of recalcitrant ones (e.g., lignin; Haynes 1986), which are broken down by slow-growing fungi.

Decomposition is also reduced at high soil moisture (e.g., >100–150% of soil dry mass in mineral soils; Haynes 1986). Oxygen diffuses 10,000 times more slowly through water than through air, so water acts as a barrier to oxygen supply in wet soils or logs, or in wet microsites within aggregates of well-drained soils. Oxygen limitation to decomposition can occur under many circumstances, including topographic controls over drainage, presence of hardpans or permafrost, high clay content, or compaction by animals and agricultural equipment. Irrigation or rain events can lead to short-term oxygen depletion. In warm environments, the solubility of oxygen in water is low, and oxygen is rapidly depleted by root and microbial respiration, making decomposition particularly sensitive to high soil moisture. NPP is generally less limited by high soil moisture than is decomposition because many plants that are adapted to these conditions transport oxygen from leaves to roots. The large accumulations of SOM in histosol soils of swamps and bogs at all latitudes (see Chap. 3)

a

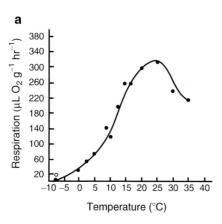

b

Fig. 7.10 Relationship between temperature and soil respiration in (*left*) laboratory incubations of tundra soils and (*right*) field measurements of soil respiration in 15 globally distributed studies, where data have been fitted to have the same respiration rate at 10°C. Redrawn from Flanagan and Veum (1974) and Lloyd and Taylor (1994)

clearly indicate the importance of oxygen limitation to decomposition.

Decaying logs create their own unique microenvironment and generally have a higher moisture content than adjacent surface litter. In moist environments, log decomposition rate may therefore be limited by oxygen supply at times when microbes in neighboring surface litter are moisture-limited. Oxygen often diffuses along cracks and insect galleries and therefore penetrates to the interior of logs more rapidly than might be predicted from log moisture content (Hicks and Harmon 2002).

Temperature
Although microbial respiration and decomposition increase with temperature in the short term, indirect effects constrain their temperature sensitivity over annual to decadal timescales. Microbial enzyme activity and respiration increase exponentially with short-term increases in temperature over a broad temperature range (Fig. 7.10), speeding up the mineralization of organic carbon to CO_2. The decomposition of recalcitrant substrates is particularly temperature sensitive. A temperature increase from 10 to 20°C, for example, increases the decomposition of the biochemically labile citric acid twofold, the more biochemically recalcitrant tannic acid threefold, and recalcitrant SOM fivefold (Fierer et al. 2005).

Several processes, operating at different timescales, constrain this apparently high temperature sensitivity (Davidson and Janssens 2006). Over days to weeks, microbes may acclimate to higher temperatures by down-regulating respiration (Bradford et al. 2008). Substrate pools decline faster at warmer temperatures, reducing carbon availability and limiting available energy to microbes. Seasonal shifts in microbial community composition to guilds that remain active in each season further reduce the seasonal variation in decomposition rate.

Temperature has many indirect effects on decomposition that act through its effects on other environmental variables (Fig. 7.11). In wet soils or microsites (e.g., aggregates), temperature stimulation of respiration consumes enough oxygen to reduce its availability and therefore microbial respiration. Over longer timescales, however, high temperature reduces soil moisture by augmenting evapotranspiration, which enhances oxygen diffusion. Similarly, at high latitudes, warming thaws the permafrost, improving drainage and the environment for decomposition. Over still longer timescales, vegetation changes alter the quantity and quality of organic matter inputs to soils (see Chap. 12). In summary, the temperature response of decomposition is far from simple. The stimulation of decomposition by warming that is consistently observed on

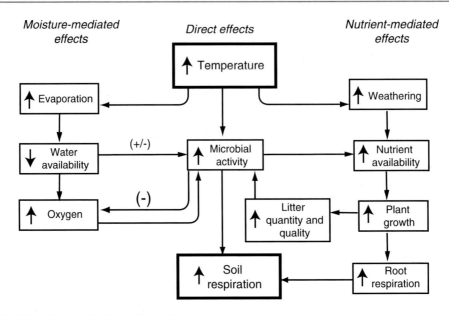

Fig. 7.11 Major direct and indirect effects of temperature on soil respiration

hourly to weekly timescales is increasingly modified by indirect effects over longer timescales, suggesting that the long-term effects of climate warming on decomposition is a fertile topic for future research (Davidson and Janssens 2006; Currie et al. 2010).

Soil Organic Matter

Decomposition of soil organic matter (SOM) is strongly influenced by its reactions with soil minerals. Up to this point, we have focused primarily on the factors controlling the breakdown and loss of litter. Equally important are the processes that reduce rates of decomposition and foster organic accumulation in mineral soils.

Soil Properties
Clay minerals reduce the decomposition rate of soil organic matter, thereby increasing soil organic content. Clays alter the physical environment of soils by increasing water-holding capacity (see Chap. 3). The resulting restriction in oxygen supply can reduce decomposition in wet clay soils. Even at moderate soil moisture, clays enhance organic accumulation by binding

SOM (making it less accessible to microbial enzymes), binding microbial enzymes (reducing their capacity to attach to substrates), and binding the soluble products of exoenzyme activity (making these products less available for absorption by soil microbes). This binding of organic matter to clays occurs because the high density of negatively charged sites on clay minerals attract the positive charges on the organic matter (amine groups) or form bridges with polyvalent cations (Ca^{2+}, Fe^{2+}, Al^{3+}, Mn^{4+}) that bind to negative groups (e.g., carboxyl groups) on organic matter (Fig. 7.12). The net effect of this binding by clay minerals is to "protect" SOM and reduce its decomposition rate. SOM protection by clay minerals is most important in ecosystems such as grasslands or tropical forests, where decomposition is relatively rapid and where soil animals rapidly mix fresh litter with mineral soil. Mineral protection of SOM is less important in conifer forests or tundra where much of the decomposition occurs above the mineral soil in a well-developed organic soil horizon (O horizon).

Both the type and quantity of clay influence decomposition. Many tropical clay minerals have a high aluminum concentration that binds tightly

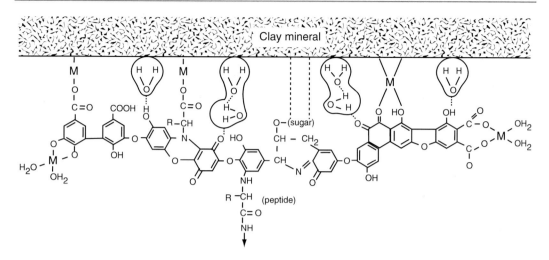

Fig. 7.12 Schematic diagram of the interactions between soil organic matter and clay particles, as mediated by water (H–O–H) and metal ions (M). Redrawn from Stevenson (1994)

to organic matter through covalent bonds. Clays with a multi-layered lattice structure bind organic compounds between the silicate layers, making them particularly effective in SOM protection (see Chap. 3).

All else being equal, soil organic matter decomposes more slowly in acidic than in neutral soils, largely due to indirect effects. Many processes can acidify soils, including cation leaching, acid deposition, and the accumulation of organic acids in highly organic soils. These conditions tend to be associated with low nutrient availability (and therefore low litter quality) and with levels of aluminum that may be toxic to many microbes, especially bacteria.

Soil Disturbance

Soil disturbance increases decomposition by promoting aeration and exposing new surfaces to microbial attack. The mechanism by which disturbance stimulates decomposition is basically the same at all scales, ranging from the movement of earthworms through soils to tillage of agricultural fields. Disturbance disrupts soil aggregates so the organic matter contained within them becomes more exposed to oxygen and microbial colonization. This disturbance effect is most pronounced in warm, wet soils, where the increased aeration has greatest effect on decom-

position. In a soil converted to irrigated cotton, for example, tillage caused loss in 3–5 years of half its organic content that had required centuries to millennia to accumulate (Haynes 1986). Similarly, carbon sequestered in soils of restored prairies over 10–20 years (West and Post 2002) can be lost rapidly if these soils are returned to agricultural tillage. The loss of organic matter and disruption of aggregates by plowing eventually impedes the drainage of water, the growth of roots, and the mineralization of soil nutrients.

Humus Formation

In climates that are favorable for decomposition, substantial quantities of carbon persist in mineral soils for thousands of years. It has long been thought that this is primarily soil humus that accumulates due to its recalcitrance (Oades 1989). Recent research suggests, however, that sorption to soil minerals may be a more important protective mechanism and that simple compounds may be just as persistent in soils as the complex ones (Schmidt et al. in press). To the extent that humus formation occurs, the following steps (Fig. 7.13) have been implicated (Zech and Kogel-Knabner 1994):

1. **Selective preservation**. Decomposition selectively degrades labile compounds in detritus, leaving behind recalcitrant materials like

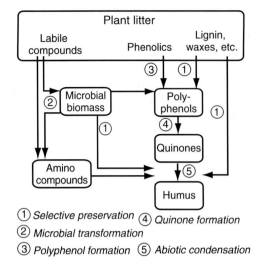

① Selective preservation ④ Quinone formation
② Microbial transformation
③ Polyphenol formation ⑤ Abiotic condensation

Fig. 7.13 Principle pathways of humus formation. The major steps in humus formation are described in the text

waxes, cutins, suberin, lignin, chitin, and microbial cell walls. Partial microbial breakdown of these recalcitrant "leftovers" may produce compounds with reactive groups and side chains that are reactants in nonspecific soil reactions. There is, however, little direct evidence that lignin is any more persistent than more simple compounds in mineral soils (Schmidt et al. in press).

2. **Microbial transformation**. Enzymatic breakdown of SOM produces low-molecular-weight water-soluble products, some of which may react in humus formation. Amino compounds such as amino acids from protein breakdown and sugar amines from degradation of microbial cell walls may be particularly important (see step 5; Fig. 7.13).

3. **Polyphenol formation**. Soluble phenolic compounds may also be important reactants in humus formation. They come from at least three sources (Haynes 1986): (1) microbial degradation of plant lignin, (2) the synthesis of phenolic polymers by soil microbes from simple non-lignin plant precursors, and (3) polyphenols produced by plants as defenses against herbivores and pathogens.

4. **Quinone formation**. The polyphenol oxidase and peroxidase enzymes produced by fungi to break down lignin and other phenolic compounds also convert polyphenols into highly reactive compounds called quinones.

5. **Abiotic condensation**. Quinones undergo spontaneous condensation reactions with many soil compounds, especially compounds with which they react readily (e.g., compounds with amino groups) or that are abundant (e.g., recalcitrant compounds that accumulate in soils).

The chemical nature of persistent SOM differs among ecosystems (Haynes 1986; Paul and Clark 1996). Forest organic matter includes insoluble compounds with extensive networks of aromatic rings and few side chains. This reflects an abundance of phenolic compounds in leaves and wood that defend plants against herbivores and pathogens. In grasslands, a larger proportion of SOM is water soluble due to extensive side chains and many charged groups.

Peat Accumulation and Trace Gas Emissions

Wet soils contain about a third of Earth's store of soil organic matter (Schlesinger 1997). In environments where low oxygen availability inhibits decomposition, organic matter accumulates in a relatively undecomposed state. This organic matter accumulates, not because it is chemically recalcitrant, but because environmental conditions constrain the activity of decomposers more strongly than they constrain carbon inputs by plants. In these wet ecosystems, SOM is often quite labile and decomposes quickly whenever soils dry enough for oxygen diffusion to overcome the "environmental protection" of this organic matter (Neff and Hooper 2002). It is important to understand the controls over decomposition in wetland soils because of the large soil carbon reservoir they contain and the sensitivity of this reservoir to environmental change. In addition, anaerobic decomposition in wetlands often releases trace gases (methane and nitrous oxide) that have about 23- and 300-fold, respectively, greater warming effect on the atmosphere per molecule than does CO_2 (see Chap. 2; IPCC 2007).

Within a poorly aerated soil, there is a gradient in decomposition rate from well-oxygenated to oxygen-depleted zones that occur at depth or within the interior of soil aggregates. This gradient in redox potential (see Chap. 3) determines the availability of electron acceptors that organisms can use to support their growth and respiration. Those microbes that transfer electrons from their food (soluble organic compounds they have absorbed) to oxygen, for example, capture the most energy to support their metabolism and growth. As oxygen is depleted, however, only those microbes that are able to transfer electrons from their food (organic substrates) to *other* electron acceptors can metabolize (decompose) organic matter and grow. The amount of energy released to support microbial growth declines progressively with transfer from organic matter to each of the following electron acceptors:

$$O_2 > NO_3^- > Mn^{4+} > Fe^{3+} > SO_4^{2-} > CO_2 > H^+$$

$$(7.4)$$

Many soil organisms carry out only one or a few redox reactions. Temporal and spatial variations in the availability of these electron acceptors therefore determine the competitive balance among these organisms and their contribution to decomposition. Organisms that derive more energy from their redox reactions (e.g., aerobic decomposers relative to denitrifiers) have a competitive advantage, when there is an adequate supply of their preferred electron acceptor because they are able to support more growth per unit of organic substrate consumed.

In flooded soils and sediments, there is a dynamic equilibrium determined by the supply of oxygen as an oxidant at the surface and buried organic carbon, which serves as a source of reducing power. As this organic matter is decomposed, microbes consume oxygen and other electron acceptors to support their metabolism and growth. Therefore, aerobic decomposition predominates at the surface and near oxygen-transporting roots, whereas other energy-producing processes become important only when oxygen has been depleted. This results in a vertical zonation of decomposition processes, with aerobic decomposition at the surface, then a zone of denitrification, then zones of manganese and iron reduction, then sulfate reduction, then methane production. Depending on the availability of each of these electron acceptors, the zone can occupy either a significant portion of the vertical profile (and therefore account for substantial decomposition) or can be of negligible importance. Denitrification, for example, is the second most energetically favorable redox reaction, after oxygen has been depleted (7.4). During denitrification, denitrifiers transfer electrons from organic matter to nitrate, producing the gases nitric oxide, nitrous oxide, and di-nitrogen, as waste products. Nitrate availability is often low in anaerobic environments, however, because nitrification, which produces nitrate, is an *aerobic* process (see Chap. 9). Denitrification is therefore relatively unimportant in most wetlands but is important where soil aeration is patchy, for example in anaerobic interiors of soil aggregates of an otherwise aerobic soil, or where water table fluctuates, as in irrigated fields or rice paddies.

As nitrate is depleted, other bacteria, using other electron acceptors, ferment labile organic compounds to produce acetate, other simple organic compounds, and hydrogen. If sulfate is available, as in estuaries, salt marshes, and ocean sediments, sulfate reducers transfer electrons from simple organic compounds to sulfate (7.4), producing hydrogen sulfide and decomposing the organic matter to support their metabolism and growth.

The concentrations of both nitrate and sulfate are low in most non-coastal wetlands and lake sediments (Schlesinger 1997), so methane production is often the predominant mode of anaerobic decomposition in these ecosystems. Conversely, in marine sediments, where sulfate is abundant, methane production is less important. **Methanogens** produce methane (CH_4) when other electron acceptors have been depleted (7.4). Methane production can occur through several pathways. Some methanogens split acetate into CO_2 and CH_4 (7.5). Others use hydrogen (H_2), which is a by-product of fermentation, as an energy source and bicarbonate (derived from CO_2) as an electron acceptor (7.6), much the way NO_3^- or SO_4^{2-} serve as electron acceptors in denitrification and sulfate reduction, respectively.

$$CH_3COOH \rightarrow CO_2 + CH_4 \qquad (7.5)$$

$$CO_2 + 4H_2 \rightarrow CH_4 + 2H_2O \qquad (7.6)$$

Methane is even more highly reduced than are carbohydrates, so it is a good energy source for organisms that have access to oxygen. Another group of bacteria (**methanotrophs**) that occur in the surface soils of wetlands use methane as an energy source and consume much of the methane as it diffuses from depth toward the atmosphere. Therefore, not all methane produced within an ecosystem actually leaves the system. Methane flux *from the ecosystem* is usually highest when methane escapes through plant gas transport tissues or as bubbles that bypass the zone of methane consumption by methanotrophs (Walter et al. 2006).

Enzymes that convert ammonium to nitrate as part of the nitrogen cycle (see Chap. 9) also react with methane, causing well-aerated soils to be a net sink for methane. Even in wetlands that produce substantial methane, more carbon is generally released as CO_2 by decomposers near the soil surface than as methane by methanogens at depth, so aerobic respiration is still the dominant pathway of carbon return to the atmosphere. Methane is quantitatively more important in its role as a greenhouse gas (see Chap. 2) rather than as a component of the carbon cycle (see Chap. 14).

In summary, conditions that reduce the rate of decomposition (either humification of organic matter under environmental conditions that are favorable for decomposition or peat accumulation in waterlogged soils) contribute to long-term carbon storage in ecosystems. In the next sections, we put these controls over decomposition into the context of whole-ecosystem carbon budgets.

Heterotrophic Respiration

Heterotrophic respiration by soil microbes and animals is one of the largest avenues of carbon loss from ecosystems. Decomposer microbes and their predators account for most of this respiration. Annual heterotrophic respiration correlates closely with NPP across carbon cycling rates that vary at least tenfold globally, suggesting

that, on average, respiration by decomposers and other heterotrophs breaks down about the same amount of organic matter that enters the ecosystem each year (Fig. 7.14). This relationship occurs by definition in ecosystems close to steady state (i.e., where there are no large gains or losses of SOM). Both concurrent carbon inputs (e.g., daily GPP) and long-term site productivity (as reflected in LAI) are important predictors of heterotrophic respiration (Migliavacca et al. 2010). Measurements of soil respiration, which includes both heterotrophic and root respiration, are consistent with this generalization. Both soil respiration and heterotrophic respiration (Figs. 7.14 and 7.15) correlate closely with NPP (Raich and Schlesinger 1992; Janssens et al. 2010). About half (25–65%) of soil respiration derives from roots, and the rest comes from decomposition (Raich and Schlesinger 1992; Högberg et al. 2001; Bhupinderpal-Singh et al. 2003).

Heterotrophic respiration shows little relationship with the total quantity of organic matter in soils because most soil carbon is sorbed to mineral surfaces, chemically recalcitrant or in an unfavorable soil environment (e.g., low temperature or low oxygen availability). This means that total soil organic content is not a good predictor

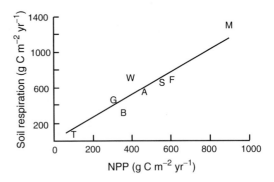

Fig. 7.14 Relationship between average annual NPP and average annual soil respiration rate for Earth's major biomes. Ecosystem types are agricultural lands (A), boreal forest and woodland (B), desert scrub (D), temperate forest (F), temperate grassland (G), tropical wet forest (M), tropical savanna and dry forest (S), tundra (T), and Mediterranean woodland and heath (W). Root respiration probably accounts for the 25% greater soil respiration than NPP at any point along this regression line. Redrawn from Raich and Schlesinger (1992)

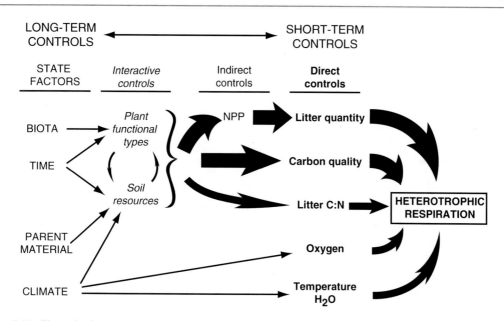

Fig. 7.15 The major factors governing temporal and spatial variation in heterotrophic respiration in ecosystems. These controls range from proximate controls over seasonal variation in heterotrophic respiration to the state factors and interactive controls that are the ultimate causes of ecosystem differences in heterotrophic respiration. Thickness of the *arrows* indicates the strength of the direct and indirect effects. The factors that account for most of the variation in heterotrophic respiration among ecosystems are the quantity and carbon quality of litter inputs, which are ultimately determined by the interacting effects of soil resources, climate, vegetation, and disturbance regime

of stand-level carbon loss (Clein et al. 2000). In fact, the largest soil carbon accumulations often occur in ecosystems such as peat bogs with low NPP but even slower decomposition.

Although nitrogen concentration of litter has a small and inconsistent influence on decomposition, addition of nitrogen to temperate forest soils reduces heterotrophic respiration at the ecosystem scale (Janssens et al. 2010). This is most pronounced in productive sites, where nitrogen limitation of plant production is least likely to occur (Fig. 7.16) and explains why organic matter tends to accumulate in response to nitrogen deposition (Magnani et al. 2007; Sutton et al. 2008; Liu and Greaver 2009). Nitrogen inhibition of heterotrophic respiration is probably the result of multiple effects, including a decline in microbial biomass, particularly of decomposer and mycorrhizal fungi, a reduction in exudation by roots and mycorrhizae, and a decline in the production of lignin-degrading enzymes (Fog 1988; Treseder 2008; Janssens et al. 2010).

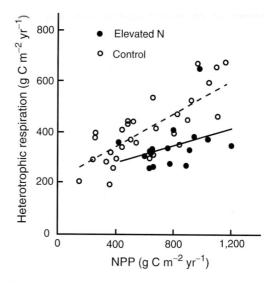

Fig. 7.16 Relationship of heterotrophic respiration to NPP in forests exposed to background or elevated (>0.55 g N m^{-2} year^{-1}) nitrogen deposition. The strong dependence of heterotrophic respiration on NPP is reduced by nitrogen deposition, particularly in productive forests. Redrawn from Janssens et al. (2010)

The linkage of carbon and nutrient cycles constrains potential imbalances between NPP and decomposition. Both plants and microbes require carbon (energy) and nutrients for growth. For example, microbes decomposing fresh litter acquire nitrogen from the substrate or the soil to meet their nitrogen needs for growth (nitrogen **immobilization**). This nitrogen is subsequently released (**mineralized**) when the microbes break down nitrogen-containing compounds to meet their energy needs (see Chap. 9). The magnitude and timing of nitrogen immobilization and release depend on substrate chemistry. Litter produced by nitrogen-limited plants, for example, has a relatively low nitrogen concentration and high concentrations of recalcitrant compounds. Microbes that decompose this litter mineralize nitrogen slowly, constraining the nitrogen supply to plants and therefore NPP. A second important linkage between carbon and nitrogen cycles is mediated by mycorrhizae, whose growth is supported directly by GPP and which mineralize nutrients to support their growth and that of their host plants (Högberg et al. 2001; Finlay 2008). This physiological requirement for both carbon and nutrients for life's processes imposes an inevitable linkage between carbon and nutrient cycles and therefore a rough long-term balance between NPP and decomposition in ecosystems. In Chap. 9, we discuss the processes that modify the balance between carbon and nutrient cycles and therefore the strength of this linkage.

Net Ecosystem Production (NEP)

On short timescales, GPP and respiration typically dominate the carbon balance of terrestrial ecosystems. Their balance is termed **net ecosystem production** (NEP).

$$NEP = GPP - R_{ecosyst} = NPP - R_{het} \quad (7.7)$$

where

$$R_{ecosyst} = R_{plant} + R_{het} \quad (7.8)$$

Ecosystem respiration ($R_{ecosyst}$) is the sum of the respiration from plants (R_{plant}) and heterotrophs (R_{het}) – that is, microbes and animals. NEP is a valuable concept because it addresses the major processes by which organisms gain carbon and energy (GPP) and use this energy through respiration to support their growth and maintenance ($R_{ecosyst}$). NEP thus explicitly links the physiology of organisms to the carbon balance of ecosystems (Woodwell and Whittaker 1968; Chapin et al. 2006a; Luyssaert et al. 2007). It is analogous to NPP (GPP$- R_{plant}$) of plants and can be readily incorporated into process-based models that address the physiology of all organisms in ecosystems.

As discussed later, it is virtually impossible to measure NEP directly. However, in terrestrial ecosystems, gaseous exchange with the bulk atmosphere supplies most of the CO_2 that supports GPP and removes most of the respiratory CO_2. This net CO_2 exchange of the entire ecosystem, termed **net ecosystem exchange**, NEE, is therefore usually a reasonable approximation of NEP, when measured over short time periods. NEE is now being measured in a wide range of ecosystems (Box 7.2). NEE may systematically overestimate NEP in terrestrial ecosystems and underestimate it in freshwater ecosystems, as discussed later, but it probably provides a reasonable proxy for *geographic patterns* of NEP and their environmental controls in those ecosystems that are close to steady state (Baldocchi et al. 2001; Luyssaert et al. 2007; Xiao et al. 2008).

NEE is defined, by convention, as CO_2 flux *from the ecosystem to the atmosphere*. It corresponds to a *negative* carbon input to ecosystems. NEE is defined in this way because atmospheric scientists, who originated the term, seek to document net sources of CO_2 to the atmosphere (i.e., NEE) that account for rising atmospheric CO_2 concentration. Therefore, CO_2 input to the ecosystem is a *negative* NEE.

NEP is determined by factors that cause an imbalance **between GPP and** R$_{ecosyst}$. In ecosystems that have not been recently disturbed, NEP is a small difference between two very large fluxes (Fig. 7.17): (1) photosynthetic carbon gain and (2) carbon loss through respiration (primarily

Box 7.2 Measuring Carbon Fluxes of Ecosystems and Regions

Photosynthesis (GPP) and respiration are usually the largest carbon fluxes between terrestrial ecosystems and the atmosphere. As turbulent eddies of air move across the surface of an ecosystem, like balls rolling across a lawn, the downward-moving limb of the eddy carries atmospheric air into the ecosystem, and the upward-moving limb transports ecosystem air to the free atmosphere. The **eddy covariance technique** takes rapid measurements (about ten times per second) of vertical wind speed and the CO_2 content of upward and downward moving parcels of air. The CO_2 flux can be calculated directly from these measurements (the minute, instantaneous changes in CO_2 concentration times the instantaneous changes in vertical wind velocity that occur as turbulent eddies pass the sensors). When these fluxes are summed over an hour, a day, or a year, they represent the net CO_2 flux between the ecosystem and the atmosphere (i.e., NEE) over that time period (see Fig. 7.22). The technology for measuring these fluxes and correcting for potential artifacts is rapidly improving (Baldocchi 2003). Comparisons of long-term NEE measurements across networks of sites provide a basis for understanding and generalizing about the controls over temporal and spatial variations in NEE among terrestrial ecosystems. This understanding has

been incorporated into models that estimate various carbon fluxes (e.g., GPP, ecosystem respiration, and NEP) based on ecosystem properties (e.g., ecosystem type and leaf area) and environmental conditions (e.g., temperature) that can be remotely sensed from space (Running et al. 2004), leading to estimates of carbon fluxes across broad regions (see Fig. 7.19).

NEE measurements can be complemented by measurement of other fluxes, such as CO_2 from wildfire or carbon transfers to groundwater and aquatic ecosystems (Cole et al. 2007). The integration of all these fluxes provides an estimate of net ecosystem carbon balance (NECB) – the rate of carbon accumulation or loss by an ecosystem (Randerson et al. 2002; Chapin et al. 2006a).

An independent check of these flux estimates comes from large-scale atmospheric measurements (see Fig. 7.27). Atmospheric circulation models can calculate, based on measurements, the change in quantity of CO_2 contained in an air mass, as it moves across a continent or ocean. From this information, the net regional flux (regional NEE) between the surface and the atmosphere can be calculated and compared with estimates made from surface measurements and models (Fan et al. 1998; Gurney et al. 2002; Schuh et al. 2010).

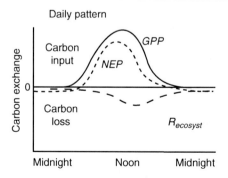

Fig. 7.17 Idealized daily (season of active plant growth) and seasonal pattern of gross primary production (*GPP*), ecosystem respiration (*R*$_{ecosyst}$), and net ecosystem production (*NEP*) of an ecosystem. *NEP* is the difference between two large fluxes (carbon input as GPP and carbon loss through respiration). In these diagrams, *NEP* is shown as positive over the diurnal cycle (*GPP* > ecosys-

tem respiration during the season of active plant growth) and close to zero over the annual cycles, assuming that the ecosystem is at steady state. The actual pattern of these fluxes varies with environmental conditions, successional status, and other factors (see text). Carbon losses due to leaching and disturbance are assumed to be zero in these diagrams

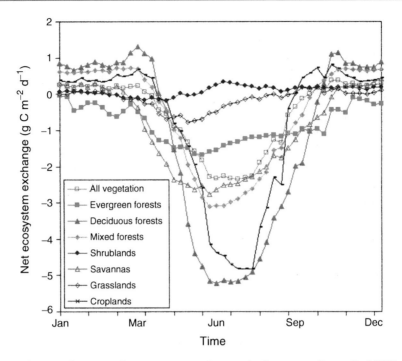

Fig. 7.18 Predicted seasonal pattern of net ecosystem exchange (NEE) of major U.S. vegetation types in 2005, based on a regression model that uses AmeriFlux (a network of ecosystem flux studies) NEE measurements and MODIS satellite imagery. Redrawn from Xiao et al. (2008)

by plants and microbes). In general, GPP is closely correlated with ecosystem respiration on timescales of days to weeks (Migliavacca et al. 2010) because both plant respiration and heterotrophic respiration are strongly affected by the quantity of carbon that enters ecosystems through GPP, as discussed earlier. When GPP exactly equals ecosystem respiration, NEP is, by definition, zero. There is therefore no reason to expect NEP to correlate in any simple way with GPP, NPP, or ecosystem respiration. However, GPP and respiration are seldom perfectly balanced. During the day, photosynthesis exceeds respiration, with the reverse occurring at night. Similarly, during the growing season, NEP is positive because photosynthesis exceeds respiration as plants accumulate biomass. In nongrowing seasons, when photosynthesis is low, heterotrophic respiration dominates, and NEP is negative. This gives rise to very simple and predictable daily and seasonal patterns of NEP (Fig. 7.17).

Consistent with this expected seasonal pattern (Fig. 7.17), NEP is generally positive (or NEE

negative) during seasons favorable for plant growth (GPP > ecosystem respiration) and negative (NEE positive) during seasons unfavorable for plant growth (GPP < ecosystem respiration; Fig. 7.18). The magnitude of the seasonal changes in NEP differs among ecosystems. Within the U.S., deciduous forests have the largest positive growing-season NEP and most negative nongrowing-season NEP, and shrublands show least seasonal variation in NEP (Xiao et al. 2008). Coastal evergreen forests show a modest positive NEP throughout the year. Not surprisingly, positive NEP (negative NEE) is most pronounced during summer in the eastern U.S., where deciduous forests dominate, is more evenly distributed throughout the year in coastal evergreen forests of the Pacific Northwest, and is negative (NEE positive) during summer in arid regions of the southwestern U.S. (Fig. 7.19). Midwestern croplands also have a strong positive NEP (negative NEE) during summer. In general, these seasonal variations in NEP are driven more strongly by GPP than by ecosystem respiration because both

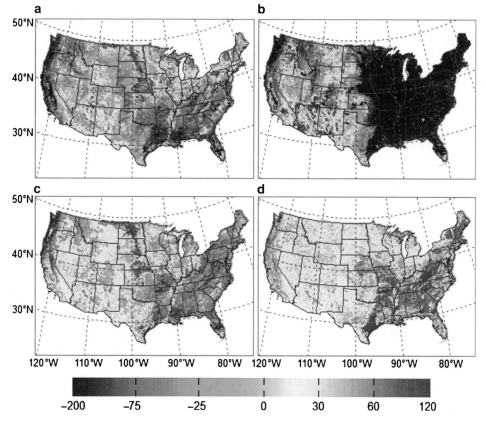

Fig. 7.19 Maps of predicted net ecosystem exchange (NEE) based on a regression model that uses AmeriFlux NEE measurements and MODIS satellite imagery during (a) spring (March-May), (b) summer (June-August), (c) autumn (September-November), and (d) winter (December-February). Redrawn from Xiao et al. (2008)

GPP and ecosystem respiration are generally highest during the growing season and GPP declines more strongly than respiration during the nongrowing season.

NEP also varies with time since disturbance. NEP is expected to decline with disturbances such as logging, hurricanes, or wildfire that reduce plant biomass and GPP (see Chap. 12). In addition, heterotrophic respiration often increases after disturbance because of transfer of aboveground biomass to the ground surface (e.g., hurricanes) or environmental changes that favor decomposition. NEP should recover as biomass and GPP increase, then approach zero as GPP comes into equilibrium with ecosystem respiration. What is surprising, however, is that NEP often remains positive, even in forests more than a century old (Fig. 7.20; Luyssaert et al. 2007). Across the U.S., all ecosystem types except

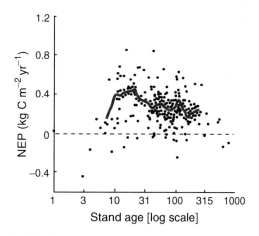

Fig. 7.20 Observed relationship of forest net ecosystem production (NEP) to stand age. Positive values indicate that the forest is a sink for carbon and negative values that it is a source. The line shows the average value of NEP (*n* = 500 forest plots). Maximum NEP occurs at about 20–30 years but usually remains positive for hundreds of years. Redrawn from Luyssaert et al. (2007)

shrublands show, on average, a positive NEP in the absence of recent disturbance (Fig. 7.19; Xiao et al. 2008), as does the terrestrial biosphere as a whole (Le Quéré et al. 2009).

There are at least four potential explanations for the generally positive NEP documented in a wide range of terrestrial ecosystems: (1) This may reflect the expected successional pattern of NEP, with ecosystems being carbon sinks for a very long time after disturbance. In other words, ecosystems may seldom reach steady state before a new disturbance occurs (Luyssaert et al. 2007; Xiao et al. 2008). (2) Recent environmental changes, such as increased atmospheric CO_2 and nitrogen deposition, may have stimulated photosynthesis and reduced respiration, leading to greater carbon sequestration (Magnani et al. 2007; de Vries et al. 2009; Liu and Greaver 2009; Janssens et al. 2010). (3) Carbon loss through leaching and other transfers may be important (but unmeasured) components of the net carbon balance from terrestrial ecosystems; these non-gaseous carbon losses would not be detected in measurements of NEE, leading to potential over-estimates of NEP (Kling et al. 1991; Randerson et al. 2002; Cole et al. 2007). (4) There may be unintended biases in site selection, measurements, or models (Baldocchi 2003; Sutton et al. 2008). Ecologists are vigorously debating the magnitude and relative importance of these potential explanations for a generally positive NEP measured in terrestrial ecosystems. We now explore these issues in more detail.

NEP generally follows the expected successional pattern. It declines with disturbance. Insect outbreaks, for example, reduce NEP as a result of declines in leaf area (and therefore GPP) and increases in heterotrophic respiration (Kurz et al. 2008). As vegetation recovers, GPP increases more strongly than respiration, leading to increased NEP (Fig. 7.20; see Fig. 12.13). After about 80 years, however, NEP begins to decline as the forest ages (Magnani et al. 2007). As pointed out earlier, NEP seldom declines to zero, even in old forests (Luyssaert et al. 2007; Xiao et al. 2008). About half of the carbon accumulation in forests occurs belowground in roots and soils, and, of the aboveground portion, about

two-thirds accumulates in coarse woody debris and the rest in live stems (see Chap. 6). However, even ecosystems such as arctic tundra that seldom experience large-scale disturbances and post-disturbance succession appear to have a positive NEP (McGuire et al. 2009), suggesting that successional dynamics are not the only explanation for the generally positive NEP observed on land.

Global increases in atmospheric CO_2 and nitrogen inputs to ecosystems augment NEP because they stimulate GPP more strongly than ecosystem respiration. Nitrogen deposition associated with acid rain, for example, stimulates carbon storage by about 6% in forests and 2% in agricultural fields, with no detectable change in other natural ecosystems (Sutton et al. 2008; Liu and Greaver 2009). It is more difficult to assess the effects of rising CO_2 on NEP because the CO_2 increase has been relatively uniform across the planet. Experimental studies, however, show that elevated CO_2 concentrations often stimulate NEP, especially in more fertile ecosystems or ecosystems to which nutrients have been added to simulate the effects of nitrogen deposition (McGuire et al. 1995a; Ciais et al. 2005a). Anthropogenic changes in the environment have therefore often enhanced NEP in undisturbed terrestrial ecosystems.

The effects of temporal and spatial variation in climate on NEP are not easy to predict because warm temperatures, improved soil aeration, and improved moisture availability stimulate all components of NEP: GPP, plant respiration, and microbial respiration. In southern Europe, for example, GPP and ecosystem respiration are both strongly moisture-limited. Both of these fluxes increase to a similar extent in moist years or sites, so there is no significant relationship between moisture supply and NEP (Fig. 7.21; Reichstein et al. 2007). Similarly, in northern sites, where GPP and ecosystem respiration are primarily temperature-limited, both of these fluxes increase to a similar extent in warm years and sites, so there is no significant relationship between temperature and NEP (Fig. 7.21; Luyssaert et al. 2007; Magnani et al. 2007; Reichstein et al. 2007; Piao et al. 2009). When considered together, moisture has a stronger effect on NEP than does temperature, primarily because of the strong

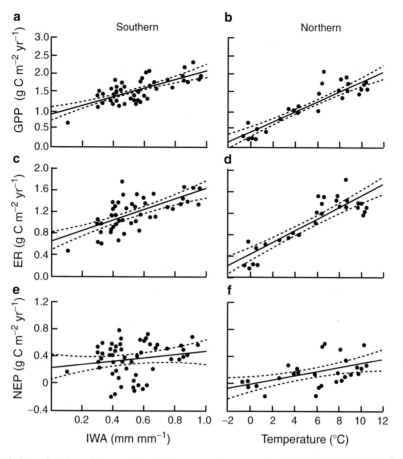

Fig. 7.21 Correlation of carbon fluxes with an index of water availability (IWA; ratio of actual to potential evapotranspiration) in southern European forests and average annual temperature in northern European forests. Fluxes shown are gross primary production (GPP; plots **a** and **b**); ecosystem respiration (ER; plots **c** and **d**); and net ecosystem production (NEP; plots **e** and **f**). The solid and dashed lines are the average and 95% confidence intervals, respectively. GPP and ER are much more strongly correlated with environmental controls than is NEP. Redrawn from Reichstein et al. (2007)

effects of drought in reducing GPP (Reichstein et al. 2007). Thus, interannual variations in climate affect NEP primarily through their effects on GPP rather than ecosystem respiration (Ciais et al. 2005a; Groendahl et al. 2007; Luyssaert et al. 2007; Reichstein et al. 2007), just as observed among seasons within a year.

Despite the modest sensitivity of NEP to variations in temperature and moisture, *changes* in climate drivers may alter NEP over the long term. The magnitude of warming over the last two decades of the twentieth century, for example, explains much of the current variation in NEP among forested sites, even though these sites show little sensitivity of NEP to current year's temperature (Piao et al. 2009). Short-term climate shocks can also have a long-term impact. A severe drought in Europe in July 2003, for example, reduced NEP enough to offset the previous 4 years of carbon sequestration (Ciais et al. 2005b). The seasonality of warming is also important. Spring warming, for example, increases GPP and NEP by advancing the date of snowmelt and the onset of plant growth and photosynthesis (Euskirchen et al. 2006; Lafleur and Humphreys 2007; Piao et al. 2008). In the autumn, however, when sun angle is lower and soils are warmer, warming increases ecosystem respiration more strongly than GPP and therefore reduces autumn NEP. For these reasons, pronounced spring warming in Eurasia leads to increased annual NEP, whereas pronounced autumn warming in North America has reduced annual NEP (Piao et al. 2008).

Changes in water table and soil aeration also cause complex changes in NEP. Drainage of waterlogged peatlands initially reduces NEP because GPP declines (Chivers et al. 2009) and in some cases, heterotrophic respiration increases (Silvola et al. 1996). Over the longer term, invasion of more productive non-peatland species enhance leaf area and GPP, often leading to positive NEP (carbon sequestration; Minkkinen et al. 2002; Laiho et al. 2003). Thawing of permafrost in response to recent climate warming causes ecosystem respiration to increase more strongly than GPP, causing a loss of carbon that accumulated thousands of years ago (Schuur et al. 2009). This negative NEP could become a strong amplifying (positive) feedback to climate warming, given that there is twice as much carbon in the permafrost as in the atmosphere (Zimov et al. 2006; Schuur et al. 2008).

In summary, natural post-disturbance successional processes, climate variations, and human impacts on the atmosphere all influence NEP, primarily through their effects on GPP. Current evidence suggests that human activities substantially influence these controls over the NEP of the biosphere. These effects are exerted through disturbance and land cover change, which can either increase or reduce NEP; nitrogen deposition and increased atmospheric CO_2, which generally increase NEP; and anthropogenic

climate warming, which has variable effects on NEP. The net effect of changes in NEP on the climate system and the biosphere depends on the overall changes in ecosystem carbon stocks, as explained in the next section.

Net Ecosystem Carbon Balance

Net ecosystem carbon balance (NECB) is the net rate of carbon accumulation by an ecosystem. It is the balance between carbon entering and leaving the ecosystem, that is, the change in ecosystem carbon stock through time:

$$NECB = \frac{dC}{dt} \qquad (7.9)$$

To understand NECB, it is useful to visualize the ecosystem as a defined volume with explicit top, bottom, and sides (Fig. 7.22; Chapin et al. 2006a). The top of this ecosystem "box" in terrestrial ecosystems is above the canopy, the bottom is below the rooting zone, and the sides define the area to be analyzed. Most carbon enters the ecosystem as gross primary production (GPP) and leaves through several processes, including plant and heterotrophic respiration, leaching of DOC and DIC, emissions of volatile organic compounds, methane flux, and disturbance. Lateral transfers such as erosion/deposition, animal movements, or harvest can bring additional carbon into or out of the ecosystem (Fig. 7.22). NECB is the increase (positive value) or loss (negative value) in the quantity of carbon in this ecosystem box.

$$NECB = (gaseous\ inputs - losses) + (dissolved\ inputs - losses) + (particulate\ inputs - losses) \quad (7.10)$$

$$NECB = (-NEE + F_{CO} + F_{CH4} + F_{VOC}) + (F_{DIC} + F_{DOC}) + F_{POC} \qquad (7.11)$$

Gaseous Carbon Fluxes

GPP and ecosystem respiration are the dominant gaseous carbon fluxes most of the time. However, wildfire is an additional large

episodic cause of CO_2 loss from some ecosystems, and CH_4 and CO fluxes are additional climatically important gaseous emissions. Combustion of organic matter by wildfire is a non-respiratory loss of CO_2 from ecosystems to

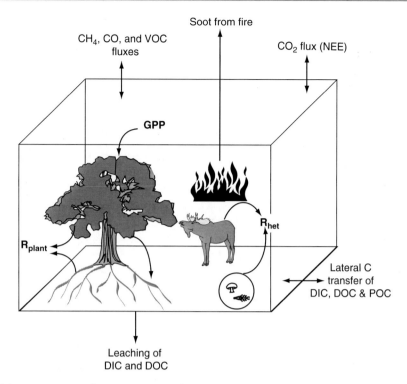

Soot from fire

CH$_4$, CO, and VOC fluxes

CO$_2$ flux (NEE)

GPP

R$_{het}$

R$_{plant}$

Lateral C transfer of DIC, DOC & POC

Leaching of DIC and DOC

Fig. 7.22 Major components of net ecosystem carbon balance (NECB). The fluxes that determine net ecosystem production (NEP) are shown in bold. The box repre-sents the ecosystem. Fluxes contributing to NECB and NEP are defined in the text. Redrawn from Chapin et al. (2006a)

the atmosphere. Wildfire is therefore an impor-tant component of NEE and NECB whenever it occurs, particularly when NEE and NECB are integrated over timescales long enough to incor-porate disturbance. Wildfire is *not* a component of NEP (i.e., the balance of GPP and respiration). In many cases, the carbon losses with wildfire are significant components of long-term carbon bud-gets (Figs. 7.22 and 7.23). Carbon losses during fires in the Canadian boreal forest, for example, are equivalent to about 6–30% of average NPP (Harden et al. 2000; McGuire et al. 2010). Because of their sensitivity to successional status, NECB and NEE estimated at the regional scale depend on the relative abundance of stands of dif-ferent ages. At times of increasing disturbance frequency, NECB is likely to be negative, as with recent increases in wildfire in western North America. Conversely, areas that have experienced widespread abandonment of agricultural lands in the last century, as in Europe or the northeastern

U.S., may experience a positive NECB. Inadequate information on the regional variation in distur-bance frequency and NECB is one of the greatest sources of uncertainty in explaining recent changes in the global carbon cycle (see Chap. 14).

Non-CO$_2$ gaseous fluxes can be large com-ponents of NECB in ecosystems where net CO$_2$ flux is small. In permafrost- and ice-dominated portions of the northern hemisphere (arctic and boreal lands and the Arctic Ocean), for example, the land and ocean are modest carbon sinks. Large methane emissions from wetlands cause the region to exert a positive greenhouse-gas warming effect on climate (McGuire et al. 2009; McGuire et al. 2010). In addition, the emissions of carbon monoxide from wildfires (47 Tg C year^{-1}) and methane from wetlands and wildfires (31 Tg C year^{-1}) are similar in magnitude to the net sequestration of CO$_2$ (51 Tg C year^{-1}), indicating the importance of multiple gases in regional carbon balance (McGuire et al. 2010).

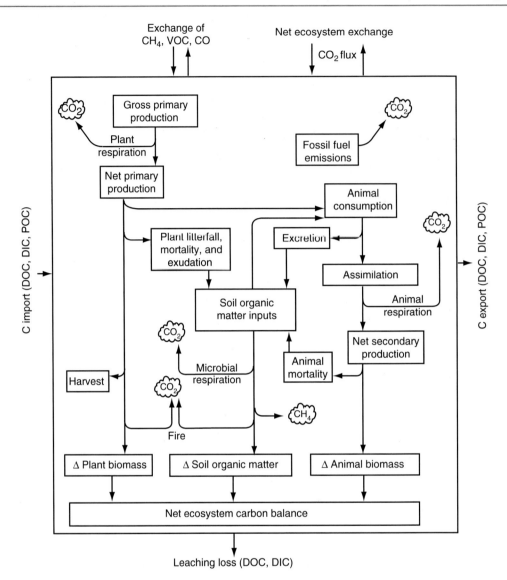

Fig. 7.23 Overview of the carbon fluxes of an ecosystem. The large box represents the ecosystem, which exchanges carbon with the atmosphere, other ecosystems, and groundwater

Disturbances that redistribute carbon within the ecosystem affect NECB only indirectly **because the carbon remains within the ecosystem**. Hurricanes or insect outbreaks, for example, transfer carbon from live plants to soil or to standing dead plants without the carbon being lost from the ecosystem. These disturbances can *indirectly* affect NECB, however, by reducing photosynthetic capacity and increasing the food available to decomposers. In other words, these changes in carbon balance affect NECB through their effects on NEP (= GPP – ecosystem respiration).

Fossil fuel combustion by people is an increasing source of CO_2 to the atmosphere. It is a large carbon flux in industrial agriculture and in many ecosystems such as towns and cities that are occupied by people (see Chap. 14). Fossil-fuel combustion represents a transfer from previously inert geological pools of organic carbon (coal, oil, and natural gas) to the atmosphere.

Particulate Carbon Fluxes

Lateral transfer of particulate carbon into or out of ecosystems can be important to the long-term carbon budgets of ecosystems. Carbon can move laterally into or out of ecosystems through erosion and deposition by wind or water or by movement of animals, including people (Figs. 7.22 and 7.23). In many ecosystems, these lateral transfers are so small that they are undetectable in most years. Over long time periods or during extreme events, such as floods, landslides, or forest harvest, lateral transfers can, however, be quantitatively important. Observations of NEE of Europe, for example, could not be explained based on measured ecosystem fluxes without accounting for food imports from other countries (Ciais et al. 2008). Similarly, within the crop-producing states of the Midwestern U.S., the eastern-most states export most of their crops and are a net CO_2 sink, whereas the western-most states feed these crops to animals, which respire the carbon to the atmosphere, causing little net carbon sequestration (Schuh et al. 2010). Lateral transfers of carbon-containing biomass are significant components of NECB in managed forests, agricultural and grazing ecosystems, and other human-modified ecosystems, which now occupy much of the terrestrial surface (Ellis and Ramankutty 2008).

Dissolved Carbon Fluxes

Leaching of dissolved organic and inorganic carbon (DOC and DIC, respectively) to groundwater and streams is a quantitatively important avenue of carbon loss from some ecosystems (Figs. 7.22 and 7.23). We discuss these in the next section in the context of the carbon balance of streams and rivers.

In summary, fluxes in addition to GPP and ecosystem respiration are important fluxes in most ecosystems, especially over long time periods. Therefore, changes in NEP and NEE tell only part of the story about changes in the carbon balance of terrestrial ecosystems.

Stream Carbon Fluxes

Stream Decomposition

The horizontal flow of carbon in streams is similar to its vertical movement through the soil on land but occurs over much larger distances. The basic steps in decomposition are identical on land and in aquatic ecosystems (Valiela 1995; Wagener et al. 1998; Gessner et al. 2010). These steps include leaching of soluble materials from detritus (up to 25% of initial dry mass in 24 h), fragmentation of litter into small particles by invertebrates and physical processes, and microbial decomposition of labile and recalcitrant substrates (Allan and Castillo 2007). On land, these processes begin at the soil surface, and organic matter moves downward in the soil profile due to mixing by soil invertebrates, burial by new litter, downward leaching, and other processes (Wagener et al. 1998). In stream ecosystems, the same processes occur, but cycling materials are also carried downstream tens of kilometers in the process. Energy and nutrients therefore **spiral** down streams, rather than cycling vertically as they tend to do in most terrestrial ecosystems (Fisher et al. 1998).

In forest headwater streams, the dominant energy input is terrestrial detritus that enters as **coarse particulate organic matter** (CPOM; particles >1 mm) such as leaves and wood (Fig. 7.24). Low light availability limits algal production in these streams (see Chap. 6). The controls over the processing of CPOM are remarkably similar to those that occur on land. Fine litter that enters the stream becomes lodged behind rocks or coarse woody debris, is leached by flowing water and colonized by invertebrates that fragment and ingest small particles, increasing the surface area for microbial colonization (Fig. 7.25). The leaching of **dissolved organic carbon** (DOC; particles <0.5 μm) and export of **fine particulate organic matter** (FPOM; particles >0.5 μm and <1 mm) from leaf packs in the stream leads to an exponential pattern of mass loss with time (Eq. 7.1), just as on land (Allan and Castillo 2007). Decomposition in

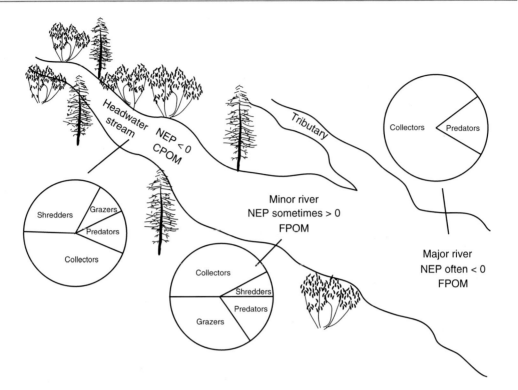

Fig. 7.24 The river continuum concept of a representative river system. Headwater streams have little instream GPP and high heterotrophic respiration, so NEP is negative. Coarse particulate organic matter (CPOM) dominates the detrital pool. Shredders and filter feeders (collectors) are the dominant invertebrates. In middle sections of rivers, more light is available, and GPP sometimes exceeds ecosystem respiration (NEP sometimes positive).

Fine particulate organic matter (FPOM) is the dominant form of organic matter, and filter feeders and grazers are the dominant organisms. Large rivers accumulate considerable organic-rich sediments dominated by filter feeders feeding on FPOM from upstream. Ecosystem respiration often exceeds GPP (NEP often negative). Ecosystem respiration generally exceeds GPP (NEP negative) for entire river systems. Based on Vannote et al. (1980)

streams is more rapid than on land because water never limits microbial activity, and fine particles are swept downstream. Most fine (nonwoody) litter in temperate streams loses half its mass in less than a year (Webster and Benfield 1986), whereas it takes more than a year in corresponding terrestrial environments (Fig. 7.6). Stream decomposers are aquatic specialists rather than organisms that enter the streams on the leaves.

Fungi are the dominant decomposers in flowing water, and bacteria dominate in poorly aerated stream sediments (Allan and Castillo 2007). Fungi can either enhance bacterial activity (Gulis and Suberkropp 2003) or compete with bacteria (Wright and Covich 2005; Allan and Castillo 2007). The stimulatory effects of invertebrates on microbial decomposition of litter and the chemical effects of litter quality on decomposition in streams (Allan and Castillo 2007) are virtually identical to the patterns described earlier for terrestrial ecosystems (Gessner et al. 2010).

FPOM comes primarily from within the stream through the processing of CPOM into small particles and feces by invertebrates, release of asexual spores by aquatic fungi, the abrasion of periphyton from rocks, the absorption of DOC by microbes, and other processes. About a third of the leaf material consumed by shredders, for example, is released into the stream as FPOM (Giller and Malmqvist 1998). Stream invertebrates assimilate a relatively small proportion (10–20%) of the organic matter that they ingest (see Chap. 10), resulting in a substantial production of feces. FPOM is generally more recalcitrant than CPOM

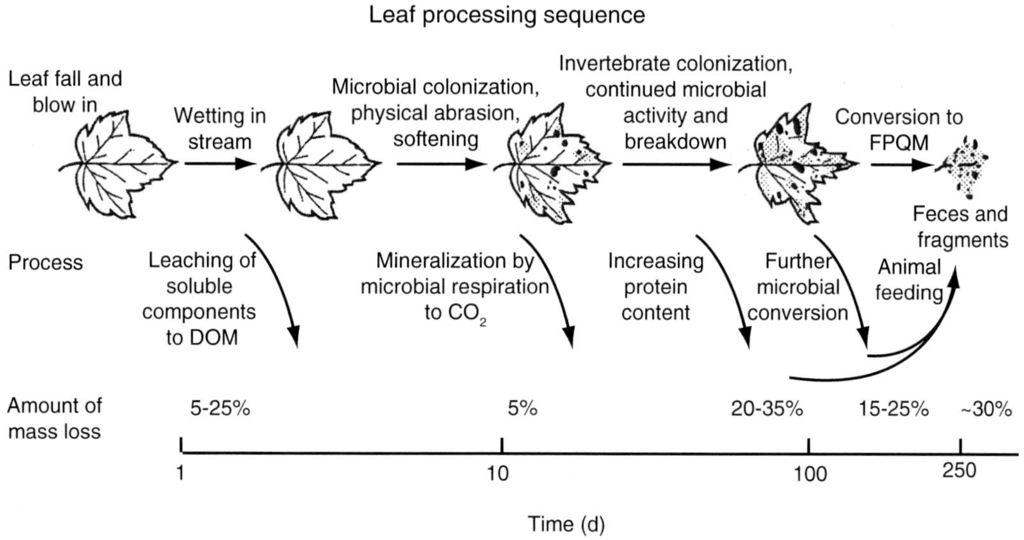

Leaf processing sequence

Fig. 7.25 Processes that "condition" a leaf of moderate litter quality in a temperate stream. Redrawn from Allan and Castillo (2007)

because organisms have already removed the labile substrates (Allan and Castillo 2007). In fact, most FPOM appears to be produced by stream organisms rather than being fine particles of terrestrial origin (e.g., leaf fragments). Bacteria are the dominant decomposers of FPOM (Findlay et al. 2002; Allan and Castillo 2007).

DOC is the largest pool of organic carbon in most streams (Karlsson et al. 2005; Allan and Castillo 2007). It derives from both instream processes, such as leaching of fresh litter and excretion by algae, higher plants, and microbes, and as inputs from terrestrial wetlands and riparian areas. Stream DOC contains a diversity of compounds that vary widely in decomposability. Labile DOC is an important energy source for decomposers and higher trophic levels (Allan and Castillo 2007). During spring algal blooms, for example, stream DOC increases by as much as 37% during the day as a result of algal exudation (Kaplan and Bott 1989; Allan and Castillo 2007). Annual pulses of DOC often come from leaching of fish carcasses and autumn leaves. DOC is removed from the stream primarily by microbial absorption but also by abiotic processes such as photo-oxidation and binding to clay particles (Allan and Castillo 2007). Periphyton, the biofilm

mixtures of algae and bacteria that create the slimy films on rocks, wood, and macrophytes, appear particularly important in the efficient transfer of algal DOC to bacterial decomposers. In tropical blackwater rivers and boreal peatlands, much of the DOC is tannins and recalcitrant humic and fulvic acids leached from soils. These compounds are processed slowly in streams.

Rivers and streams have a belowground component analogous to terrestrial soils. In the **hyporheic zone**, groundwater moves downstream within the streambed. Substantial decomposition occurs in the hyporheic zone, releasing nutrients that support instream algal production. In intermittent streams, the hyporheic zone is all that remains of the stream during dry periods. Water moves more slowly and therefore has a shorter processing length in the hyporheic zone than in the stream channel, so the spiraling length is much shorter (Fisher et al. 1998).

Stream Carbon Budgets

There is a continuum in stream metabolism from headwaters to the ocean. Stream ecosystems differ dramatically from their terrestrial counterparts in the importance of *lateral linkages* of biogeochemical processes as materials spiral downstream. The **river continuum concept** integrates stream size, energy sources, food webs, and nutrient processing into a longitudinal model of river metabolism from headwaters to the ocean (Fig. 7.24; Vannote et al. 1980). Detrital food webs and heterotrophic processing of energy dominate many headwater streams, particularly in forests, because large terrestrial inputs of litter provide lots of food for microbes, and low light availability limits algal production. These headwater streams therefore have a negative NEP (GPP < ecosystem respiration) and export considerable organic material downstream (Webster and Meyer 1997; Mulholland et al. 2001; Allan and Castillo 2007). Even unshaded headwater streams of tundra, boreal forest, and wetlands are generally heterotrophic because of large inputs of terrestrial organic matter and nutrient limitation of algal production (Peterson et al. 1986). Most headwater streams are dominated by invertebrate **shredders** that break leaves and other detritus into pieces and digest the microbial jam on the surface of these particles, just as occurs in the soil (Wagener et al. 1998). This creates fresh surfaces for microbial attack and produces feces and other fine material that are carried downstream. Desert streams are a major exception to the heterotrophic dominance of headwater streams. Streams in arid environments receive very little litter input or shade and are therefore dominated by algal production and have a positive NEP (GPP > ecosystem respiration; Fisher et al. 1982; Jones et al. 1997).

Downstream, where rivers are wide enough to receive substantial light input, GPP is greater than in headwaters (see Chap. 6), but heterotrophic respiration still generally dominates (negative NEP; Webster and Meyer 1997), depending on light availability, water clarity, and water depth, which influence GPP, and on detrital inputs from upstream (FPOM) or adjacent riverbanks, which influence ecosystem respiration (Howarth et al. 1996b). Some of the fine particles are consumed in suspension by filter feeders like black fly larvae or from benthic sediments by **collectors** like oligochaete worms. The abundance of algae and

their **grazers** depends on light availability. Finally, large rivers in their deposition zone have sediments that support substantial hyporheic decomposition. These rivers also support both algal production and bacterial decomposition in the water column if these organisms can reproduce fast enough to offset their downstream export. The shallow gradient and low current velocity in some channels in the deposition zone often allow this to occur (Allan and Castillo 2007). In those rivers where suspended sediments and low water clarity limit algal production, detrital processing tends to dominate (negative NEP; GPP < ecosystem respiration).

Rivers and streams are highly pulsed systems, leading to large temporal fluctuations in carbon metabolism. Seasonal pulses of litterfall cause large seasonal variation in organic matter inputs to streams, just as on land. Snowmelt or heavy rains increase runoff through surface litter and increase the suspension and transport of terrestrial organic and mineral particles, substantially increasing the transfer of organic matter and sediments to streams. In many headwater streams, storm events that account for 1% of the annual discharge transport 70–80% of the annual FPOM throughput of streams (Bilby and Likens 1980; Webster et al. 1990; Allan and Castillo 2007). Finally, flood events dislodge primary producers and transport sediments, woody debris, and other organic matter downstream. Since algal biomass is a strong determinant of GPP and NPP (see Chap. 6), floods constrain the potential of streams to support GPP and a positive NEP. In large unregulated rivers such as the Amazon, flooding converts much of the floodplain from a terrestrial to an aquatic habitat, and 70–90% of the annual carbon inputs to the system come from floodplain inputs during flooding (Bayley 1989; Meyer and Edwards 1990; Lewis et al. 2001; Allan and Castillo 2007). Within a given climatic regime, disturbances that radically reduce primary producer biomass tend to occur much more often in streams than on land. This contributes to the dominance of heterotrophic processes in stream and river ecosystems.

The carbon metabolism of a stream segment (**reach**) is strongly influenced by the site itself (e.g., shade, temperature, and inputs of terrestrial litter) and by upstream processes. Organic matter in rivers typically travels 10–100 km (its **turnover length**) before it is broken down and lost by respiration (Allan and Castillo 2007; Webster 2007). Because heterotrophic headwater streams account for about 85% of the total length of most river systems (Peterson et al. 2001), export of dissolved and particulate carbon from headwater streams has a huge effect on the metabolism of the entire river system.

Taken as a whole, river systems are generally heterotrophic, that is, have a negative NEP (GPP < ecosystem respiration; Cole et al. 2007). This differs strikingly from the generally positive NEP (GPP > ecosystem respiration) of most terrestrial ecosystems. This fundamental difference in carbon metabolism reflects the important role in landscape metabolism of carbon transfer from terrestrial to aquatic systems (see Chap. 13). Clearly, some of terrestrial NEP does not represent carbon sequestered on land but is transferred to aquatic systems where it returns to the atmosphere as CO_2, is stored in sediments of lakes and reservoirs, or is transported to the ocean (Cole et al. 2007).

The terrestrial-to-aquatic carbon transfer has two important components. The first is the transfer of particulate and DOC that supports aquatic heterotrophic respiration, as described earlier. In addition, groundwater that enters streams has extremely high CO_2 concentrations, about 75% of which comes from root and microbial respiration in soils, and 25% from weathering of rocks (Schlesinger 1997; Cole et al. 2007). About 20% of the carbon that appears to be sequestered on land (i.e., positive NEP) moves as DIC (dissolved inorganic carbon [DIC]) to aquatic systems, where it is degassed and returns to the atmosphere (Kling et al. 1991; Algesten et al. 2003; Kortelainen et al. 2006; Cole et al. 2007). In other words, much of the CO_2 release from aquatic ecosystems actually derives from terrestrial respiration, and much of the positive NEP (negative NEE) on land does *not* contribute to terrestrial carbon accumulation (positive NECB).

Of the carbon that enters aquatic systems from the land (as dissolved CO_2 and dead organic matter), about 40% returns to the atmosphere as CO_2, 12% is stored in sediments of lakes and reservoirs, and the remainder (about half) is transported to the ocean, roughly equally as organic and inorganic carbon (Cole et al. 2007).

Lake Carbon Fluxes

Decomposition in lakes is faster than in streams or on land because of the high litter quality of algae. Lignin, which is important for structural support of land plants and which contributes to slow decomposition of terrestrial litter, is not needed in lakes, where primary producers (algae) float in the water or are attached to the bottom. In addition, as in streams, moisture never limits decomposition in lakes. Decomposition in lakes is therefore more rapid than on land, and 70–85% of the decomposition in lakes occurs in the water column before dead organic matter sinks to the sediments (Kalff 2002). An intermediate-sized dead algal cell (nanoplankton of 10 μm diameter) would sink at a rate of about 0.25 m day^{-1} and would require 40 days to sink 10 m (Baines and Pace 1994; Kalff 2002). Since the mixing time of water is on the order of half an hour in the mixed layer, a year in bottom waters, and 3 months in the intermediate **metalimnion** (see Fig. 2.21), most detrital particles are repeatedly mixed back into the water column where decomposition continues before they can sink to depth. The only reason that particles can sink in such a turbulent environment is that there is a gradual transition from turbulent flow in the mixed layer to laminar flow at the base of the mixed layer to very little flow at depth. The loss rate of particles from the mixed layer to the metalimnion depends on particle abundance (a function of productivity) and sinking rate of particles just above this boundary layer (Kalff 2002). In lakes with a thin mixed layer (i.e., lakes that are small, protected from wind, or highly stratified), a larger proportion of detrital particles enter the boundary between mixed layer and the nonturbulent waters below and are therefore likely to sink to the bottom. Decomposition continues as particles sink through deeper waters, so the quantity of "lake snow" that reaches the sediments and sediment organic content are lower in deep than in shallow lakes. In summary, lake decomposition is strongly influenced by lake physical properties such as turbulence, stratification, and lake depth. Dead organic matter flux to lake sediments is greatest in lakes that are eutrophic, small, shallow, and protected from winds.

Large particles that are likely to sink out of the mixed layer of lakes derive from large algae, fecal pellets of zooplankton, and the aggregation and flocculation of detrital materials. Large algae dominate in eutrophic lakes and in lakes with abundant zooplankton that consume small algal cells. Large algae tend to be less edible than small algae and are therefore more likely to die before being eaten. Fecal pellets of zooplankton are relatively dense and can sink >100 m day^{-1}, 400 times faster than an intermediate-sized dead algal cell. Aggregation of dissolved organic matter to particles also influences both its decomposition rate and its probability of sinking out of the mixed layer. Aggregation occurs because organic compounds with their charged groups (e.g., carboxyl and amine groups) interact directly or through cation bridges and are often stabilized by bacterial secretions, just as in soils (see Chap. 3). Aggregation speeds decomposition because it increases the encounter rate between bacteria and their substrate. In small lakes, as much as half the organic matter that enters sediments comes from terrestrial inputs from streams or from the littoral zone rather than from algal production (Kalff 2002).

Grazing influences lake decomposition in complex ways. First grazing "competes" with decomposition by consuming algal cells before they die. Lakes differ dramatically from terrestrial ecosystems in that more energy goes through grazing than through detrital pathways (see Chap. 10). Second, by producing dense fecal pellets and by eating small edible algae, grazers increase the size and sinking rate of dead organic matter and therefore the probability of dead organic matter reaching the sediments. Finally, detrital and plant-based trophic systems are tightly intertwined in pelagic food webs because most grazers select food based more strongly on size than on quality and therefore do not strongly differentiate among live algal cells, dead algal cells, and organic aggregates of appropriate size (see Chap. 10). Grazers therefore contribute directly to the decomposition of dead organic matter in lakes.

About 15–30% of lake decomposition occurs in the sediments. Sediment decomposition is particularly important in lakes that are eutrophic, shallow, or small. Here the controls over decomposition are similar to those in wetland soils and are strongly influenced by oxygen availability. In poorly oxygenated sediments, redox reactions determine the pathway of energy release and whether the product of decomposition is CO_2 or CH_4. In oxygenated sediments, most decomposition occurs aerobically, and mollusks and worms exert important controls over sediment aeration and therefore decomposition, as in coastal ocean sediments.

Transfer of organic matter from the water column to sediments is not a one-way path. Sediment resuspension can return a substantial proportion of surface sediments, particularly recently deposited, loosely consolidated organic matter, to the water column. Turbulence usually drives sediment resuspension and is greatest in shallow waters (e.g., <15 m depth; Kalff 2002). Resuspension is greatest during storms, when water turbulence is high, and during periods of weak stratification, when the mixing depth is greatest. Thus many temperate lakes often experience spring and autumn peaks in resuspension. Development of algal mats and littoral macrophyte beds reduce the magnitude of resuspension.

Sediment resuspension influences not only the interaction between the water column and sediments but also the lateral movement of sediments within the lake basin. Shallow sediments are often resuspended, removing fine particles and leaving behind coarse sediments that facilitate oxygen diffusion. Over the long term, sediments move from shallow depths either to deeper portions of the lake or to littoral macrophyte beds where vascular plants and algal mats stabilize the sediments. The boundary between zones of net resuspension of sediments and net accumulation depends on the turbulence dynamics of the lake (and therefore on size, depth, stratification, and protection from wind). This transition can occur at <3 m in shallow wind-protected lakes with a gradual underwater slope to >40 m in large deep lakes with steep slopes. Sediment accumulation zones are the major locations of carbon storage in lakes.

Lakes are the main sites of carbon sequestration in freshwater ecosystems. On average, about 12% of the terrestrial carbon that enters freshwater systems is deposited in lake sediments (Cole et al. 2007). Reservoirs are particularly important sites of carbon sequestration because former terrestrial soils are suddenly placed in a low-oxygen environment that reduces decomposition rate and favors carbon release as CH_4, a powerful greenhouse gas, rather than as CO_2. In addition, reservoirs are more effective than natural lakes in trapping organic particles that enter from rivers, due to low resuspension rates and long water residence times. Consequently, reservoirs currently bury more carbon than all natural lakes combined and 1.5-fold more carbon than is exported to ocean sediments (Dean and Gorham 1998, Cole et al. 2007). Similarly, sediment delivery from land to the ocean has declined, despite increased sediment delivery to rivers because of sediment capture by reservoirs (see Chap. 3; Syvitski et al. 2005). This illustrates ways in which human activities can inadvertently alter the carbon dynamics and geomorphic processes of landscapes to a degree that, in their aggregate, are important at global scales.

Ocean Carbon Fluxes

Patterns of ocean decomposition are qualitatively similar to those in lakes. This decomposition occurs relatively quickly because the carbon substrates are mostly labile organic compounds of low molecular weight (Fenchel 1994) in contrast to the structurally complex, carbon-rich compounds (cellulose, lignin, phenols, tannins) that dominate terrestrial detritus. Marine decomposition is characterized by rapid leaching of dead cells followed by chemical transformation. This is identical to the decomposition of terrestrial litter, except that the initial "litter" (dead cells) is so

small that no invertebrate fragmentation occurs. The chemical controls over decomposition are also very similar to those observed on land (Valiela 1995). Viruses play an important role in planktonic food webs, lysing both phytoplankton and bacteria. Viral lysis may account for 5–25% of bacterial mortality in pelagic ecosystems (Valiela 1995). Dissolved organic matter that is excreted by phytoplankton (about 10% of NPP) or released by lysis of phytoplankton and bacteria or during grazing tends to aggregate into particles that are colonized by bacteria (Valiela 1995), just as in lakes. Pelagic phytoplankton, bacteria, viruses, and particulate dead organic matter are grazed by small (nanoplankton) flagellate protozoans, which in turn are fed upon by larger zooplankton. The detritus-based food web (see Chap. 10) is therefore tightly interwoven with the phytoplankton-based trophic system in pelagic food webs and contributes substantially to the energy and nutrients that support marine fisheries. This **microbial loop** in pelagic ecosystems recycles most of the carbon (80–95%) and nutrients within the euphotic zone before being lost to depth (Fig. 7.26).

Pelagic carbon cycling pumps carbon and nutrients from the ocean surface to depth (Fig. 7.26). Although most of the planktonic carbon acquired through photosynthesis returns to the environment in respiration, just as in terrestrial and freshwater ecosystems, marine pelagic ecosystems also transport 5–20% of the carbon fixed in the euphotic zone into the deeper ocean (Valiela 1995), a somewhat smaller proportion than occurs in most lakes. This process is called the **biological pump**. The carbon flux to depth correlates closely with primary production, so the environmental controls over NPP largely determine the rate of carbon export to the deep ocean. This carbon export consists of particulate dead organic matter (feces and dead cells) and the carbonate exoskeletons that provide structural rigidity to many marine organisms. Carbonate accounts for about 25% of the biotically fixed carbon that rains out of the euphotic zone (Howarth et al. 1996b). The carbonates redissolve under pressure as they sink to depth. Only relatively large particles sink fast enough to reach the sediments before being mostly decomposed.

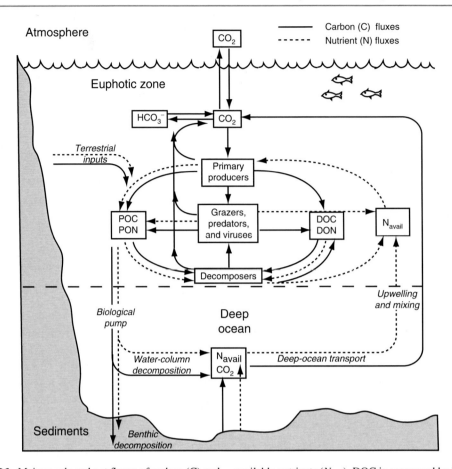

Fig. 7.26 Major pools and net fluxes of carbon (C) and nutrients (N) in the ocean. CO_2 in the euphotic zone equilibrates with bicarbonate (HCO_3^-) in ocean water and with CO_2 in the atmosphere. CO_2 is depleted by photosynthesis by primary producers and is replenished by respiration of organisms and by upwelling and mixing from depth. Grazers consume primary producers and bacteria and are eaten by other animals and lysed by viruses. Each of these organisms releases dissolved and particulate forms of carbon and nutrients (DOC, DON; POC, PON). Animals and decomposers also release available nutrients (N_{avail}). DOC is consumed by bacteria, and available nutrients are absorbed by primary producers. Particulate carbon and nutrients produced by feces and dead organisms sink from the euphotic zone toward the sediments; as they sink, they decompose, releasing CO_2 and available nutrients. Benthic decomposition also releases CO_2 and available nutrients. Bottom waters, which are relatively rich in CO_2 and available nutrients, eventually return to the surface through mixing and upwelling; this augments the supply of available nutrients in the euphotic zone

Over decades to centuries, some of this carbon in deep waters recirculates to the surface through upwelling and mixing. This long-term circulation pattern will cause the impacts of the current increase in atmospheric CO_2 to affect marine biogeochemistry for centuries after its impacts are felt in terrestrial ecosystems. The net effect of the biological pump is to move carbon from the atmosphere to the deep waters and to ocean sediments. Carbon accumulation in mid-ocean sediments is slow (about 0.01% of NPP) because most decomposition occurs in the water column before organic matter reaches the sediments and because these well-oxygenated sediments support decomposition of much of the remaining carbon (Valiela 1995).

The biological pump that transports carbon to depth carries with it the nutrients contained in dead organic matter. The rapid (about weekly) turnover of carbon and nutrients in phytoplankton in the euphotic zone (Falkowski et al. 1998) makes these nutrients vulnerable to loss from the ecosystem and contributes to the relatively open nutrient cycles of pelagic ecosystems. The longer-lived and larger primary producers on land can store and internally recycle nutrients for years. This reduces the proportion of nutrients that are annually cycled and contributes to the tightness of terrestrial nutrient cycles.

Benthic decomposition is more important in estuaries and continental shelves than in the deep ocean because the coastal pelagic system is more productive, generating more detritus, and receives terrestrial organic matter inputs from rivers. In addition, the dead organic matter has less time to decompose before it reaches the sediments. Here oxygen consumption by decomposers depletes the oxygen enough that decomposition becomes oxygen-limited, and organic matter accumulates or becomes a carbon source for anaerobic decomposers such as sulfate reducers, methanogens, and denitrifiers, just as described for terrestrial wetlands. Filter-feeding benthic invertebrates that feed by irrigating their burrows facilitate aerobic decomposition by creating a large surface area for oxygen exchange between the water and

the anaerobic sediments. Eutrophication of rivers greatly stimulates the productivity of many estuaries and increases the rain of dead organic matter to the sediments. This augments the oxygen depletion by benthic decomposers, creating dead zones that no longer support fish and macroinvertebrates (see Chap. 9; Howarth et al. 2011). Two-thirds of the estuaries in the U.S. have been degraded in this fashion, and dead zones are becoming more common in estuaries and coastal zones throughout the world (Howarth et al. 2011).

Carbon Exchange at the Global Scale

Seasonal and latitudinal variations in the CO_2 concentration of the atmosphere provide a clear indication of global-scale variation of NEE (Fung et al. 1987; Keeling et al. 1996a; Piao et al. 2008). At high northern latitudes, conditions are warm during summer, and photosynthesis exceeds total respiration (positive NEP, negative NEE), causing a decline in the concentration of atmospheric CO_2 (Fig. 7.27). Conversely, in winter, when photosynthesis is reduced by low

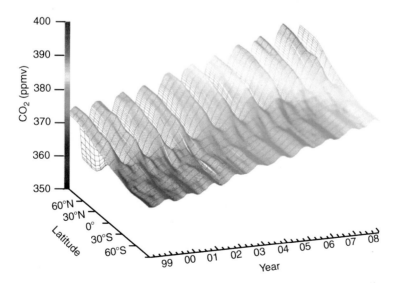

Fig. 7.27 Seasonal and latitudinal variations in the concentration of atmospheric CO_2. Seasonal and latitudinal variations in CO_2 concentration reflect primarily the balance of terrestrial photosynthesis and respiration. The upward trend in concentration across years results from anthropogenic CO_2 inputs to the atmosphere. Figure courtesy of Pieter Tans, http://www.esrl.noaa.gov/gmd/ccgg/

temperature and shedding of leaves, respiration becomes the dominant carbon exchange (positive NEE), causing an increase in atmospheric CO_2. These seasonal changes in the balance between photosynthesis and respiration occur synchronously over broad latitudinal bands, giving rise to regular annual fluctuations in atmospheric CO_2, literally the breathing of the **biosphere** (i.e., all live organisms on Earth; Fung et al. 1987).

Latitudinal variations in climate modify these patterns of annual carbon exchange. In contrast to the striking seasonality of NEE at north temperate and high latitudes, the concentration of atmospheric CO_2 remains nearly constant in the tropics because carbon gain by photosynthesis is balanced by approximately equal carbon loss by respiration throughout the year. In other words, NEP and NEE are close to zero in all seasons. Seasonal changes in atmospheric CO_2 concentration are also relatively small at high southern latitudes where the ocean occupies most of Earth's surface. Carbon exchange with the ocean is largely determined by physical factors, such as wind, temperature, and CO_2 concentration in the surface waters (see Chap. 14), which show

less seasonal variation. In summary, the global patterns of variation in atmospheric CO_2 concentration provide convincing evidence that carbon exchange by terrestrial ecosystems is large in scale and sensitive to climate.

The final general pattern evident in the atmospheric CO_2 record is a gradual increase in CO_2 concentration from one year to the next (Fig. 7.27), primarily a result of fossil fuel inputs to the atmosphere that began with the industrial revolution in the nineteenth century (see Chap. 14). The rising concentration of atmospheric CO_2 is an issue of international concern because CO_2 is a greenhouse gas that contributes to climate warming (see Chap. 2). Note that the within-year variation in CO_2 concentration caused by biospheric exchange is about ten times larger than the annual CO_2 increase. If the net carbon gain by ecosystems could be increased over the long term, this might reduce the rate of climate warming. Unfortunately, the capacity of terrestrial and marine ecosystems to remove CO_2 from the atmosphere appears to be declining (Fig. 7.28), as terrestrial vegetation becomes less carbon-limited (see Chap. 5) and as CO_2 saturates the capacity of

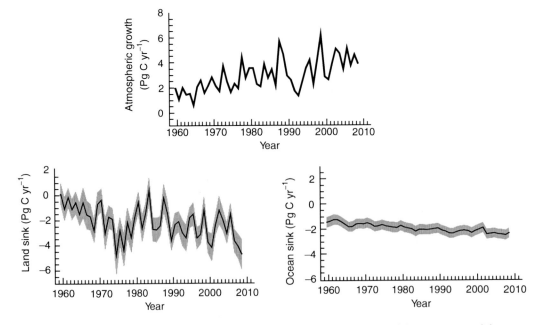

Fig. 7.28 Fraction of fossil fuel CO_2 emissions that remains in the atmosphere, terrestrial ecosystems, and the ocean. Data from Canadell et al. (2007)

the ocean to dissolve more CO_2 (see Chap. 14; Canadell et al. 2007). Ecosystem ecologists are playing a key role in global policy as they seek to link changes in the climate system to carbon fluxes from ecosystems (Fig. 7.22). These advances come through the integration, using computer simulations, of measurements made at leaf-to-global scales.

Summary

Decomposition ultimately converts dead organic matter into CO_2 and inorganic nutrients through leaching, fragmentation, and chemical alteration. Leaching removes soluble materials from decomposing organic matter. Fragmentation by animals breaks large pieces of organic matter into smaller ones that they eat, creating fresh surfaces for microbial colonization. Fragmentation in terrestrial systems also mixes the decomposing organic matter into the soil. Bacteria and fungi carry out most of the chemical alteration of dead organic matter, although some chemical reactions occur spontaneously without microbial mediation.

Decomposition rate is controlled by substrate quality, the physical environment, and composition of the microbial community. Carbon chemistry is a strong determinant of litter quality. Labile substrates, such as sugars and proteins, decompose more rapidly than recalcitrant ones, such as lignin and microbial cell walls. Plants in high-resource environments produce high-quality litter and therefore support rapid decomposition rates. Decomposition rate declines with time, as labile substrates are depleted. Soil animals influence decomposition by fragmenting litter, consuming soil microbes, and mixing the litter into mineral soil. The environmental factors that favor NPP (warm, moist, fertile soils) also promote decomposition, so there is no clear relationship between the amount of carbon that accumulates in soils with either NPP or decomposition rate.

NECB is the rate at which carbon accumulates in ecosystems. This accumulation occurs through gaseous, dissolved, and particulate exchanges with the atmosphere and with other ecosystems. In the absence of large disturbances, net ecosystem production (NEP) – the balance between GPP and ecosystem respiration – is the largest determinant of NECB. This is closely approximated in terrestrial ecosystems by measurement of net ecosystem exchange (NEE) and in aquatic ecosystems by fluxes of DIC. NEP is influenced more strongly by time since disturbance than by the environment. Surprisingly, most terrestrial ecosystems appear to be active sinks for carbon for reasons that are vigorously debated. Some disturbances such as wildfire cause large non-respiratory carbon losses that are not a component of NEP. Inclusion of these disturbances in estimates of NECB provides a more complete accounting of the interactions of ecosystems with the atmosphere. Human activities are altering most of the major controls over NECB at a global scale in ways that are altering global climate.

In contrast to terrestrial ecosystems, most streams and rivers have a negative NEP because of large terrestrial-to-aquatic transfers of organic matter and dissolved CO_2. Decomposition in streams is similar to that on land, except that the products spiral downriver linking stream metabolism horizontally throughout entire river systems. Of the carbon that enters streams from the land (as dissolved CO_2 and dead organic matter), about 40% returns to the atmosphere as CO_2, 12% is stored in sediments of lakes and reservoirs, and the remainder (about half) is transported to the ocean. In lakes and the ocean, most decomposition occurs in the water column, leading to rapid recycling of the nutrients from dead organic matter. About 25% of the carbon sinks to depth (the biological pump).

Review Questions

1. What is decomposition, and why is it important to the functioning of ecosystems?
2. What are the three major processes that contribute to decomposition? What are the major controls over each of these processes? Which of these processes is directly responsible for most of the mass loss from decomposing litter?

3. How do bacteria and fungi differ in their environmental responses and their roles in decomposition?

4. What roles do soil animals play in decomposition? How does this role differ between protozoans and earthworms?

5. Why do decomposer microbes secrete enzymes into the soil rather than breaking down dead organic matter inside their bodies?

6. What chemical traits determine the quality of soil organic matter? How do carbon quality and C:N ratio differ between litter of plants growing on fertile vs. infertile soils?

7. Describe the mechanisms by which temperature and moisture affect decomposition rate.

8. How do roots influence decomposition rate? How does decomposition in the rhizosphere differ from that in the bulk soil? Why?

9. What controls the carbon input to headwater streams? Why is this important to the carbon balance of linked terrestrial–aquatic landscapes?

10. How do the controls over NEP and NECB differ from the controls over GPP and decomposition. Why are these controls different?

Additional Reading

Adair, E.C., W.J. Parton, S.J. Del Grosso, W.L. Silver, M.E. Harmon, et al. 2008. A simple three-pool model accurately describes patterns of long-term, global litter decomposition in the Long-term Intersite Decomposition Experiment Team (LIDET) data set. *Global Change Biology* 14:2636–2660.

Bardgett, R. D. 2005. *The Biology of Soil: A Community and Ecosystem Approach*. Oxford University Press, Oxford.

Canadell, J.G., C. Le Quéré, M.R. Raupach, C.B. Field, E.T. Buitehuls, et al. 2007. Contributions to accelerating atmospheric CO_2 growth from economic activity, carbon intensity, and efficiency of natural sinks. *Proceedings of the National Academy of Sciences, USA* 104:10288–10293.

Chapin, F.S., III, G.M. Woodwell, J.T. Randerson, G.M. Lovett, E.B. Rastetter et al. 2006. Reconciling carbon-cycle concepts, terminology, and methods. *Ecosystems* 9:1041–1050.

Chapin, F.S., III, J. McFarland, A.D. McGuire, E.S. Euskirchen, R.W. Ruess, et al. 2009. The changing global carbon cycle: Linking plant-soil carbon dynamics to global consequences. *Journal of Ecology* 97:840–850.

Davidson, E.A. and I.A. Janssens. 2006. Temperature sensitivity of soil carbon decomposition and feedbacks to climate change. *Nature* 440:165–173.

De Deyn, G.B., J.H.C. Cornelissen, and R.D. Bardgett. 2008. Plant functional traits and soil carbon sequestration in contrasting biomes. *Ecology Letters* 11:516–531.

Janssens, I.A., W. Dieleman, S. Luyssaert, J.-A. Subke, M. Reichstein, et al. 2010. Reduction of forest soil respiration in response to nitrogen deposition. *Nature Geoscience* 3:315–322.

Mary, B., S. Recous, D. Darwis, and D. Robin. 1996. Interactions between decomposition of plant residues and nitrogen cycling in soil. *Plant and Soil* 181:71–82.

Paul, E.A., and F.E. Clark. 1996. *Soil Microbiology and Biochemistry*. 2nd Edition. Academic Press, San Diego.

Schlesinger, W.H. 1977. Carbon balance in terrestrial detritus. *Annual Review of Ecology and Systematics* 8:51–81.

Plant Nutrient Use

8

Nutrient absorption, use, and loss by plants are key steps in the mineral cycling of ecosystems. This chapter describes the factors that regulate nutrient cycling through vegetation.

Introduction

Nutrient supply constrains the productivity of the biosphere. Experimental addition of nutrients increases productivity of most ecosystems, both aquatic and terrestrial, indicating the widespread nutrient limitation of primary production (Vitousek and Howarth 1991; Elser et al. 2007). Although water availability may be the primary constraint on terrestrial production (see Chap. 4), within any climatic zone, there is usually a strong positive correlation between nutrient availability and plant production. In lakes and the ocean, the productivity of fisheries closely corresponds to nutrient supply and primary production. Intensive agriculture also depends on nutrient additions for continued production (Fig. 8.1). Given the widespread occurrence of nutrient limitation, an understanding of the controls over acquisition, use, and loss of nutrients by plants is essential to characterizing the controls over plant production and other ecosystem processes.

A Focal Issue

Agricultural and urban runoff of nutrients has increased algal production of many lakes and streams, reducing their water quality and recreational value. Nutrient pollution that enhances production substantially above naturally occurring levels causes a cascade of effects that propagate through all ecosystem components. In clearwater lakes, for example, nutrient enrichment often causes nuisance algal blooms (Fig. 8.2) and greatly alters or eliminates fish populations. High productivity is therefore not always a good thing. Why are some lakes more sensitive to nutrient inputs than others? Which nutrients have the greatest long-term impact on lake productivity, and how can these inputs from the land be reduced in managed landscapes? Understanding and managing the nutrient controls over primary production are critical, not only to specific ecosystems of concern but also to the carbon dynamics and climate of the planet.

Overview

Plants require more of some nutrients than others. Primary **macronutrients** are the nutrients needed in the largest amounts. Macronutrients that commonly limit plant growth include nitrogen, phosphorus, and potassium (Table 8.1). Plants also require calcium, magnesium, and sulfur in large quantities, but these nutrients less often limit plant growth. **Micronutrients** are also essential for plants but are only needed in small quantities. These include boron, chloride, copper, iron, manganese, molybdenum, and zinc. **Beneficial nutrients** enhance growth under specific conditions or for specific groups of plants

F.S. Chapin, III et al., *Principles of Terrestrial Ecosystem Ecology*,
DOI 10.1007/978-1-4419-9504-9_8, © Springer Science+Business Media, LLC 2011

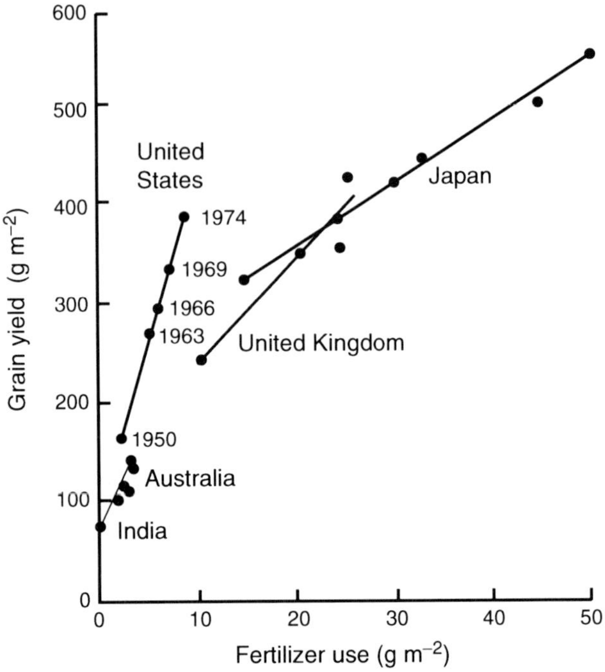

Fig. 8.1 Response of grain
yield of cereal crops to
fertilizer addition. These
studies were conducted
during the green revolution.
Yield is most responsive to
nutrient addition at low
nutrient addition rates; it
often saturates with further
nitrogen additions.
Redrawn from Evans
(1980)

Fig. 8.2 Experimental
stimulation of lake
productivity by phosphorus
addition. An experimental
curtain separates the two
halves of this lake in the
experimental lakes area of
Canada, with phosphorus
having been added to the
section on the lower right
(Schindler 1974).
Phosphorus addition
stimulated the production
of algae and a nitrogen-
fixing
cyanobacterium,
transforming the lake from
clear water to a thick algal
soup. Photograph courtesy
of David Schindler

(Marschner 1995). Ferns, for example, require aluminum, nitrogen-fixing symbionts need cobalt, and diatoms need silicon (Larcher 2003). Other nutrients are not required or are required in such small amounts that even modest levels are harmful (toxic – e.g., selenium). Roots typically

Table 8.1 Nutrients required by plants and their major functions

Nutrient	Role in plants
Macronutrients	Required by all plants in large quantities
Primary	Usually most limiting because used in largest amounts
Nitrogen (N)	Component of proteins, enzymes, phospholipids, and nucleic acids
Phosphorus (P)	Component of proteins, coenzymes, nucleic acids, oils, phospholipids, sugars, starches
	Critical in energy transfer (ATP)
Potassium (K)	Component of proteins
	Role in disease protection, photosynthesis, ion transport, osmotic regulation, enzyme catalyst
Secondary	Major nutrients but less often limiting
Calcium (Ca)	Component of cell walls
	Regulates structure and permeability of membranes, root growth
	Enzyme catalyst
Magnesium (Mg)	Component of chlorophyll
	Activates enzymes
Sulfur (S)	Component of proteins and most enzymes
	Role in enzyme activation, cold resistance
Micronutrients	Required by all plants in small quantities
Boron (B)	Role in sugar translocation and carbohydrate metabolism
Chloride (Cl)	Role in photosynthetic reactions, osmotic regulation
Copper (Cu)	Component of some enzymes, role as a catalyst
Iron (Fe)	Role in chlorophyll synthesis, enzymes, oxygen transfer
Manganese (Mn)	Activates enzymes, role as a catalyst
Molybdenum (Mo)	Role in N fixation, NO_3 enzymes, Fe absorption, and translocation
Zinc (Zn)	Activates enzymes, regulates sugar consumption
Beneficial nutrients	Required by certain plants or by plants under specific environmental conditions
Aluminum (Al)	
Cobalt (Co)	
Iodine (I)	
Nickel (Ni)	
Selenium (Se)	
Silicon (Si)	
Sodium (Na)	
Vanadium (V)	

Reprinted from Chapin and Eviner (2004)

exclude these nutrients, although some plants have evolved tolerance and may even accumulate them to high levels as a defense against pathogens and herbivores (Boyd 2004).

The quantity of nutrients that cycle through vegetation depends on the dynamic balance between nutrient supply from the environment and nutrient requirements to support plant growth. The ratio of nutrients required to support maximal growth is similar in most plants (Ingestad and Ågren 1988; Sterner and Elser 2002). Any nutrient present in less than the optimal balance is likely to limit growth and is therefore likely to be absorbed preferentially by plants. Nutrients present in excess of plant requirements are absorbed more slowly. Nutrients in plants therefore converge toward a common ratio, though with persistent variation that reflects differences in how different plants use nutrients or differences in nutrient supply. One consequence of this convergence is that plant growth often responds to addition of more than one nutrient (multiple nutrient limitation; Rastetter and Shaver 1992; Elser et al. 2007; Vitousek et al. 2010).

Second, where one element is present in much lower relative abundance than other essential nutrients, the supply of that limiting nutrient determines cycling rates of most nutrients. This element stoichiometry influences cycling rates of most essential nutrients in ecosystems (see Chap. 9; Sterner and Elser 2002). A key to understanding nutrient cycling is therefore to determine which nutrient(s) limit plant growth and the factors controlling the cycling of those nutrients. In some cases, plant growth is primarily limited by factors other than nutrient supply, in which case the flux of nutrients through vegetation depends on plant nutrient requirement rather than directly on supply rate. Plant-available nutrients that are not absorbed by vegetation are often susceptible to loss from the ecosystem (see Chap. 9). We begin this chapter with a discussion of the nature of nutrient limitation to plant growth, then discuss marine and freshwater ecosystems where the absorption of nutrients links directly to their use in growth. We then move to the more complex nutrient relations of terrestrial ecosystems and explore the controls over nutrient absorption by vegetation, the relationship of nutrient content to production, and finally the controls over nutrient loss from plants.

At its most basic level, nutrient limitation to plant growth is defined operationally, as occurring where additions of a nutrient (or nutrients) enhance the growth of plants. There are numerous predictors and indicators of nutrient limitation, including element ratios in plant tissues, measures of nutrient supply rates in soils, and the root allocation that plants make to acquire nutrients. These indicators can provide good evidence for the existence of nutrient limitation within a given geographical, botanical, or environmental space, where they are calibrated with experimental nutrient additions.

Despite the straightforward empirical nature of this definition, several factors complicate its application to understanding nutrient limitation in plant communities. First, some plants are inherently less responsive to added nutrients. For example, species adapted to and occupying nutrient-poor sites may respond to a pulse of added nutrients by storing most of them, responding with only a small

(though sustained) increase in growth. In contrast, plants adapted to and occupying fertile sites may take a similar pulse of nutrients and allocate most of them to increased growth (Chapin et al. 1986b). Plants in both situations are nutrient-limited, but the plants in the nutrient-rich site appear more limited than those in the nutrient-poor site.

Second, not all of the nutrient limitation identified by short-term nutrient addition experiments is equivalent. Addition of some nutrients may boost plant growth temporarily without fundamentally changing plant communities, whereas addition of other nutrients can transform communities and ecosystems, and it may not be possible to tell which is which in the short term. An illustration of this distinction that had strong practical implications is a controversy in the 1970s over which nutrients were capable of driving the **eutrophication** (excessive enrichment and transformation) of lake ecosystems. Different segments of society had interests in different elements. Detergents were a major source of phosphate, agriculture of nitrate, and sewage treatment plants of dissolved organic carbon that could decompose and supply CO_2. Experimental studies of short-term nutrient additions to **oligotrophic** (low-nutrient), low-alkalinity lake water demonstrated that plankton growth responded to additions of nitrogen, phosphorus, or CO_2, proving that plankton in the lake water was nutrient-limited, but giving little insight into which of these nutrients might drive eutrophication. However, a series of whole-lake experiments by David Schindler (1971) demonstrated that additions of phosphorus (and only phosphorus) were necessary and sufficient to drive lake eutrophication.

What accounts for the disconnect between short-term bioassays of nutrient limitation and the response of whole lakes to nutrient additions? In this case, phosphorus additions favored the growth of nitrogen-fixing cyanobacteria, whose activity brought nitrogen into lakes. The nitrogen- and phosphorus-stimulated growth of plankton depleted CO_2, causing greater CO_2 limitation in the short term, but steepening the diffusion gradient and CO_2 flux between atmosphere and lake water. With all three elements enriched, the lake was transformed from oligotrophic to eutrophic – from clear to pond scum (Fig. 8.2). However, while phosphorus additions could bring more nitrogen and carbon into a lake, adding nitrogen or carbon could not bring more phosphorus into lake. All three were limiting – but only phosphorus additions could transform a lake from oligotrophic to eutrophic. We consider an element with the ability to transform a community or ecosystem to be an **ultimate limiting nutrient**, while any nutrient whose addition enhances growth in the short term is a **proximate limiting nutrient**.

Both nitrogen and phosphorus represent proximate limiting nutrients in many ecosystems (Elser et al. 2007). Either may function as ultimate limiting nutrient in terrestrial or aquatic ecosystems under some circumstances, as is illustrated above for phosphorus in lake ecosystems. The dynamics of terrestrial ecosystems differ, in that terrestrial ecosystems are open systems that can accumulate nutrients from inputs (uplift of unweathered rock, rain, dust) over many decades, and in which plants and microorganisms that are limited by a particular nutrient are good at taking up and retaining that nutrient within ecosystems. Proximate limitation by phosphorus can occur for many reasons, including an increase in the supply of nitrogen or other resources. However, ultimate limitation by phosphorus is likely to occur where long-term weathering and leaching deplete the supply of available phosphorus in ancient soils or where weathering of parent material cannot supply enough phosphorus to match the supply of other limiting resources (either because the parent material contains little phosphorus, or because the phosphorus it contains is recalcitrant to weathering (Vitousek et al. 2010)).

Identifying and explaining ultimate nitrogen limitation is more challenging due to the potential of biological nitrogen fixation to use the vast and accessible pool of N_2 in the atmosphere to bring biologically available nitrogen into circulation in ecosystems. For ultimate limitation by nitrogen to occur, two conditions must be met. First, there must be a pathway of loss of nitrogen from ecosystems that cannot be prevented by those organisms that are limited by nitrogen supply. The loss via leaching of some forms of dissolved organic nitrogen represents one such pathway (Hedin et al. 1995). Second, some factor or factors must constrain biological nitrogen fixation even where nitrogen is limiting. Possible factors include the energetic cost of nitrogen fixation (which could keep plants with nitrogen-fixing symbioses from reaching through a closed plant canopy), disproportionate limitation of nitrogen fixers by phosphorus or another element, and preferential grazing on the typically nitrogen-rich tissues of symbiotic nitrogen fixers (see Chap. 9; Vitousek and Field 1999). Similar factors can drive ultimate nitrogen limitation in estuaries, where the combination of a relatively short residence time for water and constraints to nitrogen-fixing cyanobacteria by a combination of iron limitation and grazing can sustain nitrogen limitation (Howarth and Marino 2006).

Both proximate and ultimate nutrient limitations are important to the functioning of ecosystems, and we will consider both of them (together with the mechanisms that drive them) in the remainder of this chapter.

Ocean Ecosystems

The euphotic zone of the open ocean is generally nutrient poor. The open ocean is a nutritional desert, remote from the benthic supply of nutrients and distant from terrestrial inputs. This differs strikingly from estuaries and zones of coastal upwelling, where nutrient return from sediments or deep water enriches surface waters. It also differs from terrestrial ecosystems in which roots are situated in the most active zone of nutrient supply, and transport tissues carry nutrients directly to photosynthetic cells in the canopy. Nutrient availability in the open ocean is therefore generally low.

Because of their small size and therefore the strong viscous forces that bind them to water molecules (see Chap. 1), phytoplankton cannot swim (flagellates or ciliates), sink (through changes in buoyancy), or float fast enough to significantly increase their encounter rate with nutrient ions or molecules in the water. Diffusion of nutrients to the cell surface is therefore the rate-limiting process in nutrient absorption by phytoplankton

(Mann and Lazier 2006). Phytoplankton (algae and cyanobacteria) create a diffusion gradient by actively absorbing nutrients and thereby reducing the nutrient concentration at the cell surface. The small size of pelagic phytoplankton (high surface-to-volume ratio) reduces their degree of diffusion limitation. Pico-and nanoplankton (< 2 μm and 2–20 μm in diameter, respectively) dominate oligotrophic marine biomes. In contrast, larger and less edible phytoplankton are most abundant in nutrient-rich waters, where grazing is a stronger influence on community composition. These large phytoplankton have vacuoles that store nutrients when available, giving them a competitive advantage in nutrient-rich waters (Falkowski et al. 1998).

Small phytoplankton absorb nutrients from oligotrophic ocean waters at concentrations that are chemically undetectable. How do they do it? The answer is still unclear, but many phytoplankton attach to aggregates of organic particles, where they are in close proximity to bacteria that are mineralizing dead organic matter (Mann and Lazier 2006). These fine-scale processes could be important in what looks like a homogenous open ocean.

The *magnitude* of nutrient limitation in the open ocean reflects the balance between stratification that results from surface heating and turbulent mixing by winds and ocean currents. Large areas of the open ocean, particularly in the Trades Biome of the tropics (see Chap. 6), are permanently stratified with a warm, nutrient-impoverished surface layer sharply separated from cold, salty, nutrient-rich deeper waters. A sharp transition in water density (the **pycnocline**) between these layers prevents upward mixing of nutrients or downward mixing of phytoplankton (Mann and Lazier 2006). Much of the production is supported by ammonium that is recycled within the water column by grazing and detrital food webs (see Chap. 10). Large-scale currents driven by tradewinds or periodic storms mix some nutrients upward, compensating for the nutrients that sink to depth in fecal pellets and dead cells (see Chap. 7). The ratio of nitrate to ammonium is usually much greater in deep waters (see Chap. 9). Some picoplankton specifically require ammonium regenerated by surface-layer phytoplankton

turnover. Other plankton use nitrate mixed upward from depth. The ratio of these phytoplankton types is a good indicator of the relative importance of nutrient regeneration within the euphotic zone (**regenerated production**) vs. supply by vertical mixing from below (**new production**; Dugdale and Goering 1967; Mann and Lazier 2006). For example, *Prochlorococcus*, a cyanobacterium and the smallest and probably most abundant photosynthetic organism on Earth, often occurs at depth in oligotrophic surface waters. Here, where light intensity is low, it meets its energy requirements by both photosynthesis and absorption of dissolved organic carbon (DOC). It does not have enough energy to reduce nitrate, so it is an obligate user of recycled nitrogen in the form of ammonium. Although it was not discovered until the 1980s because of its small size, *Prochlorococcus* is now thought to account for 60% of the biomass of the North Pacific gyre and perhaps half of the production of the world's oligotrophic ocean waters (Mann and Lazier 2006). This example illustrates how new discoveries are still revolutionizing our understanding of the controls over ecosystem processes.

In the Westerlies and Polar Biomes of temperate and high-latitude regions of the open ocean, deep mixing during winter, when waters are least stratified, brings nutrients upward from depth (see Chap. 6). Deep mixing also disperses phytoplankton throughout a very large volume, so they spend much of their time beneath the euphotic zone, where they lack the energy to absorb nutrients and grow (i.e., are light- rather than nutrient-limited). As in the tropics, the pycnocline (i.e., the density gradient that results from the thermocline and halocline) between the surface and deep waters prevents phytoplankton from sinking to deeper waters (Mann and Lazier 2006). In spring, the higher sun angle heats the surface waters, causing the thermocline to rise and the phytoplankton to become concentrated in a thinner well-lighted surface layer. This leads to a spring bloom of nutrient absorption and production (Mann and Lazier 2006). The relatively cool temperatures of these springtime waters constrain the growth of grazers and give phytoplankton a head start in growth. Eventually, grazers eat most of

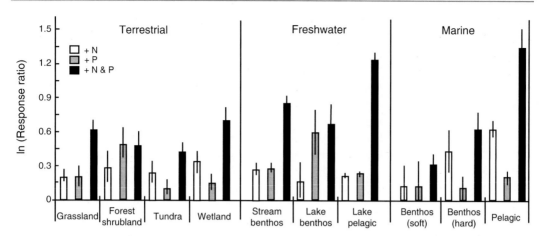

Fig. 8.3 Relative response of plant production to addition of nitrogen or phosphorus or to both nutrients in major habitat types of terrestrial, freshwater, and marine ecosystems. Relative response is calculated as the biomass or production in the enriched treatment divided by its value in the control treatment and then ln-transformed. Redrawn from Elser et al. (2007)

the phytoplankton, and most of the nutrients sink to depth, ending the spring bloom. An autumn bloom often occurs in temperate but not in polar waters.

In the Coastal Boundary Zone Biome, waters are generally well mixed throughout the water column as a result of several processes, including tidal flushing, river inputs, upwelling, and mixing by coastal currents. These nutrient-rich waters support rapid nutrient absorption and growth through much of the year (see Chap. 6), explaining why these zones support some of Earth's most productive fisheries (see Chap. 10).

The *balance* **of nutrients is often just as important as total quantities in explaining patterns of nutrient absorption and production in the open ocean.** Most marine phytoplankton have an N:P ratio (ratio of nitrogen atoms to phosphorus atoms) of about 16:1 (or 7.2:1 by mass, the **Redfield ratio**; Redfield 1958; Sterner and Elser 2002). The marine phytoplankton N:P ratio varies in time and space (range 5–30), mainly due to variation in phosphorus concentration, but tends to be lower than (or similar to) the freshwater and terrestrial N:P ratios (Guildford and Hecky 2000; Sterner and Elser 2002). Marine phytoplankton that have low N:P ratios (lots of phosphorus relative to nitrogen) typically absorb nitrogen preferentially, and down-regulate the

absorption of phosphorus and other nutrients so as to absorb these elements in proportion (the Redfield ratio) to the nitrogen that is available in their environment (Valiela 1995; Falkowski 2000; Guildford and Hecky 2000; Sterner and Elser 2002; Mann and Lazier 2006). Variability in N:P ratio among marine phytoplankton reflects variation in nutrient ratios in the environment (which is surprisingly modest in the ocean), variation in storage of "excess nutrients" in vacuoles (which is also modest, given the small size of marine phytoplankton cells), and variation in the physiological requirements of phytoplankton, which differ among species and growth conditions. Rapidly growing cells have a high phosphorus requirement (low N:P ratio) because of the high phosphorus content of ribosomes (the cellular machinery for protein synthesis; Sterner and Elser 2002). Similar patterns are observed in terrestrial plants (Güsewell 2004). The rapid growth characteristic of the bloom and bust cycles of marine phytoplankton is consistent with their generally low N:P ratios.

Marine phytoplankton growth usually responds to additions of either nitrogen or phosphorus in short-term bioassays, although the nitrogen response is usually stronger (Fig. 8.3; Tyrrell 1999; Elser et al. 2007). Also, nitrogen is usually drawn down more rapidly than phosphorus in spring blooms (Valiela 1995; Tyrrell 1999), again suggesting short-term nitrogen limitation of marine phytoplankton production. Over the longer term, however, phosphorus inputs from rivers and dust appear to define the productive potential of much of the ocean. The difference between short- and long-term nutrient limitation may reflect differences in nutrient response between nitrogen-fixing and non-fixing phytoplankton. Nitrogen-fixing phytoplankton typically grow more slowly than non-fixing taxa because of the high energy cost of nitrogen fixation. Phosphorus inputs, however, stimulate nitrogen fixers more strongly than non-fixers, allowing their production to increase until the growth of nitrogen fixers is again phosphorus-limited. The net result is proximate (short-term) limitation of marine phytoplankton production by nitrogen and ultimate (long-term) limitation by phosphorus (Tyrrell 1999).

Trace elements that constrain nitrogen fixation also contribute to nitrogen limitation in the ocean. Iron is a cofactor for nitrogenase, the nitrogen-fixing enzyme, and is also required by non-nitrogen-fixing phytoplankton. In the subequatorial gyres, the Subarctic Pacific, and the Southern Ocean surrounding Antarctica, surface nitrogen and phosphorus concentrations are relatively high, and about half of the available nitrogen and phosphorus are mixed to depth without being absorbed by phytoplankton (Falkowski et al. 1998). In these regions, production fails to respond to addition of these nutrients, leading to a syndrome known as "high-nutrient, low-chlorophyll" (HNLC; Valiela 1995; Falkowski et al. 1998; Mann and Lazier 2006). Large-scale iron-addition experiments in these regions have caused phytoplankton blooms large enough to be seen from satellites, indicating that iron, which is required for nitrogenase activity, limits the capacity of phytoplankton to use nitrogen and phosphorus. During glacial periods, there may have been tenfold greater input of iron- and phosphorus-bearing dust to the ocean, thus stimulating ocean productivity and in turn lowering atmospheric CO_2 concentrations (Martin 1990; Falkowski et al. 1998). The key role of iron in regulating production in some sectors of the open ocean has led to the suggestion that large-scale iron fertilization might stimulate ocean production enough to scavenge large amounts of CO_2 from the atmosphere and sequester it in the deep ocean as dead organic matter. The iron-addition experiments, however, show that this stimulation of production is relatively short-lived, presumably because other elements quickly become limiting to production, as soon as the iron demands of phytoplankton are met. Silica, a key constituent in the frustules (glass shells) of diatoms, is another nutrient that has been implicated in some HNLC zones (Dugdale et al. 1995; Mann and Lazier 2006). Grazing contributes to low phytoplankton biomass and productivity in other HNLC areas, suggesting that there may sometimes simply not be enough phytoplankton biomass to use the nutrients that are available (Valiela 1995).

Lake Ecosystems

Many of the nutrient effects on phytoplankton nutrient absorption and production in lakes are similar to those in the ocean. With respect to many of its properties, the ocean is just a large salty lake. Both have a surface mixed layer separated by a pycnocline from a denser, more nutrient-rich deep layer. These layers are stratified by surface heating and mixed by winds. Except near the shore, most primary producers in lakes and the ocean are single-celled phytoplankton whose growth is strongly constrained by nutrient diffusion to the cell surface. These single-celled organisms are generally extremely small, which maximizes their surface-to-volume ratio and minimizes the limitation by nutrient diffusion. Phytoplankon production is strongly affected by both nutrient availability and grazing, with nutrient availability explaining much of the geographic patterns of variation in lakes and the ocean (Kalff 2002; Mann and Lazier 2006).

Both phosphorus and nitrogen limit the primary production of most unpolluted lakes in the short term. Short-term bioassays generally show strong responses of lake phytoplankton production to additions of either nitrogen or phosphorus and a synergistic response to the two elements in combination, just as in the ocean

(Fig. 8.3; Guildford and Hecky 2000; Kalff 2002, Elser et al. 2007; Sterner 2008). Benthic phytoplankton of lakes also respond to both nitrogen and phosphorus but respond more strongly to phosphorus. In the relatively small number of whole-lake experiments that have been conducted (all in oligotrophic lakes), however, production tends to respond more strongly to phosphorus than to nitrogen. Why might nitrogen be less limiting in lakes than the ocean? Perhaps lakes have greater access to the micronutrients that limit marine nitrogen fixation. Certainly iron is more abundant in most lake water than in the pelagic ocean and more readily replenished by runoff, dust, or annual mixing. Nitrate concentrations are typically an order of magnitude higher in lake than in ocean water (Valiela 1995), and lake phytoplankton may have higher N:P ratios than marine phytoplankton, again suggesting generally greater availability of nitrogen in lakes than the ocean. As in the ocean, most of the variation in phytoplankton N:P ratios in lakes reflects variation in phosphorus concentration.

The relative importance of nitrogen vs. phosphorus limitation of phytoplankton growth in the ocean and lakes is actively debated (Sterner and Elser 2002; Elser et al. 2007; Schindler et al. 2008; Sterner 2008; Howarth et al. 2011). Part of the challenge is that phytoplankton are so small that they cannot easily be separated from bacteria and detritus, making it difficult to measure phytoplankton chemistry and nutrient response separately from that of decomposers and detritus. In addition, short-term responses, which tend to show phytoplankton growth responses to multiple nutrients in many aquatic environments, often differ from longer-term responses that generally show greater phosphorus limitation in lakes. These differences can be analyzed in terms of proximate vs. ultimate factors that control primary production (Vitousek et al. 2010). In lakes, nitrogen addition often stimulates phytoplankton growth, just as does addition of phosphorus and even carbon in oligotrophic lakes (Schindler 1974), but it can do little to increase the supply of phosphorus, which is controlled by phosphorus inputs from outside the surface water of the lake. In contrast, adding phosphorus can stimulate phytoplankton growth

directly and also favor the growth of cyanobacteria that fix nitrogen and increase its supply to all of the organisms in a lake. In that sense, both nitrogen and phosphorus represent proximate limiting nutrients, but only a change in phosphorus supply can ultimately transform most lakes from oligotrophic to eutrophic (Schindler 1971).

Rivers and Streams

Phytoplankton growth in streams and rivers can be limited by nitrogen or phosphorus or both, depending on the terrestrial matrix (Fig. 8.3; Elser et al. 2007). Many streams, particularly headwater streams, are not strongly nutrient-limited, in part because turbulence reduces diffusion limitation, although responses are often seen in heterotrophic components. The relative importance of nitrogen and phosphorus limitation depends on climate, hydrologic flow paths, watershed parent material, landscape age, and land use (Green and Finlay 2010). Phosphorus limitation of stream production, for example, is more common in the southeastern U.S., where nitrogen deposition from atmospheric pollution is high, and the parent material is relatively old and depleted of phosphorus inputs to watersheds (Horne and Goldman 1994). Nitrogen limitation occurs more often in lands that are less weathered and receive less nitrogen deposition.

Streams generally have much higher nitrate than ammonium concentrations, even when they occur in ammonium-dominated watersheds for at least three reasons (Peterson et al. 2001): (1) Nitrate is more mobile in soils than ammonium and therefore is preferentially transported in groundwater to streams. (2) Riparian zones and streams often have high nitrification rates. (3) Stream organisms preferentially absorb ammonium over nitrate. Thus nitrate is more mobile than ammonium in streams, as on land, but for somewhat different reasons. Because of high rates of nitrogen absorption and cycling by the stream bed, most nitrogen that enters streams from terrestrial ecosystems is absorbed within minutes to hours and is processed multiple times before it reaches the ocean (Peterson et al. 2001).

River-basin patterns of land use strongly influence nitrogen absorption in rivers and streams. Agricultural and urban streams have higher nitrate concentrations, and their algae absorb larger quantities of nitrate than in less polluted waters (Mulholland et al. 2008). However, their stream biota are less efficient in removing nitrate from the water (i.e., remove a smaller proportion) and therefore export more nitrate downstream than in nutrient-poor streams. At the river-basin scale, small streams account for the largest quantity of the nitrate absorption in unpolluted river systems because nitrate absorption by large rivers is limited by nitrate delivery from upstream. With intermediate nitrogen loading, small streams decline in their efficiency of nitrogen absorption, allowing export to larger rivers that absorb most of the nitrate. With high nitrogen loading, stream export exceeds the capacity of all stream reaches to absorb nitrate, and nitrate is exported to estuaries and the ocean (Mulholland et al. 2008). On average, 20–25% of nitrogen deposition on land is transported to the ocean or inland basins.

Terrestrial Ecosystems

Nutrient cycling in terrestrial ecosystems involves highly localized exchanges between plants, microbes, and their physical environment. In contrast to carbon, which is exchanged with a well-mixed atmospheric pool, nutrients in terrestrial ecosystems are absorbed by plants and returned to the soil largely within the extent of the root system of an individual plant. More than 90% of the nitrogen and phosphorus absorbed by plants of most terrestrial ecosystems comes from the recycling of nutrients that were returned from vegetation to soils in previous years (Table 8.2). The controls over nutrient absorption and use must therefore be examined at a more local scale than for carbon. Individual ecosystems, and indeed individual plants, have strong local effects on nutrient supply (Hobbie 1992; Van Breemen and Finzi 1998). Deep-rooted oaks and dogwoods that absorb calcium from depth and produce a cation-rich litter, for example, alter surface soil chemistry, leading to a very different ground flora than beneath an adjacent shallow-rooted pine that

Table 8.2 Major sources of nutrients that are absorbed by terrestrial plants

Nutrient	Source of plant nutrient (% of total)		
	Deposition/ fixation	Weathering	Recycling
Temperate forest (Hubbard Brook)			
Nitrogen	7	0	93
Phosphorus	1	< 10?	> 89
Potassium	2	10	88
Calcium	4	31	65
Tundra (Barrow)			
Nitrogen	4	0	96
Phosphorus	4	< 1	96

Data from Chapin (1991b)

absorbs less cations and produces more acidic litter (Thomas 1969; Andersson 1991).

Nutrient Movement to the Root

Roots gain access to nutrients by three mechanisms: diffusion, mass flow, and root interception. Roots absorb only those dissolved nutrients that come in contact with live root cells. Because roots constitute only a small proportion (<1%) of the belowground volume, dissolved nutrients must first move from the **bulk soil** (i.e., the soil that is not in direct contact with roots) to the root surface before plants can absorb them.

Diffusion

Diffusion is the process that delivers most nutrients to plant roots. Diffusion is the movement of molecules or ions along a concentration gradient. **Nutrient absorption** and mineralization provide the driving forces for diffusion to the root surface by reducing nutrient concentration at the root surface (absorption) and increasing the concentration elsewhere in the soil (mineralization). Mineralization and other inputs to the pool of soluble nutrients are the main controls over the quantity of nutrients available to diffuse to the root surface (see Chap. 9).

Cation exchange capacity (CEC) of soils also influences the pool of nutrients available to diffuse to the root and the volume of soil that the root exploits. Soils with a high CEC store more available cations per unit soil volume, that is, they have a high **buffering capacity**, but retard the rate of nutrient movement to the root surface through exchange reactions. These reactions remove cations from the soil solution at times of high solution concentration and return the cations at times of low concentration in the soil solution (see Chap. 3). The root can therefore tap more nutrients than are actually dissolved in the soil solution at any point in time, particularly in soils with a high base saturation, that is, where the exchange complex has abundant cations. Anion exchange capacity is generally much lower than cation exchange capacity, so most anions, like nitrate, diffuse more rapidly through soils than do cations. Chemically reactive anions like phosphate, however, tend to precipitate, reducing their solution concentration and therefore their rate of diffusion to the root surface.

Rates of diffusion differ strikingly among ions, due to differences in **charge density** (i.e., the charge per unit hydrated volume of the ion). Charge density, in turn, depends on the number of charges per ion and the hydrated radius of the ion. Divalent cations like calcium and magnesium are bound more tightly to the exchange complex and diffuse more slowly than do monovalent cations like ammonium and potassium. Ions of a given charge also differ slightly in diffusion rates because of differences in radius and number of water molecules that are loosely bound to the ion.

Soil particle size and moisture determine the length of the diffusion path from the bulk soil to the root surface. Ions diffuse through water films that coat the surface of soil particles. The higher the water content and the smaller the particle size, the more direct is the diffusion path from the bulk soil to the root surface. Diffusion is therefore faster in moist than in dry soils and in clay-rich than in coarse-textured sandy soils.

Each absorbing root creates a **diffusion shell**, that is, a cylinder of soil that is depleted in the nutrients absorbed by the root. This diffusion shell constitutes the zone of soil directly influenced by root absorption. The root accesses a relatively large volume of soil for those ions that diffuse rapidly. Nitrate, for example, which diffuses rapidly,

is typically depleted in a shell approximately 6–10 mm in radius around each absorbing root, whereas ammonium is depleted over a radius of < 1–2 mm, and phosphate is depleted over a radius of < 1 mm. It therefore takes a higher root density to exploit fully the soil for phosphate or ammonium than for nitrate. The root densities in many ecosystems are high enough to exploit most of the soil volume for nitrate but only a small proportion of the soil volume for ammonium or phosphate. The major way in which a plant can enhance absorption of ions that diffuse slowly is to increase root length and therefore the proportion of the soil that it exploits.

Mass Flow

Mass flow of nutrients to the root surface augments the supply of ions provided by diffusion. Mass flow is the movement of dissolved nutrients to the root surface in flowing soil water. Transpirational water loss by plants is the major mechanism that causes mass flow of soil solution to the root surface. Mass flow can be an important mechanism supplying those nutrients that are abundant in the soil solution or that the plant needs in small quantities. Calcium, for example, is present in such a high concentration in many soils that the plant requirements for calcium are completely met by mass flow of calcium from the bulk soil to the root surface (Table 8.3). Corn, for example, receives fourfold more calcium by mass flow to the root than the root actually acquires. Plants that receive too much calcium by mass flow actively secrete calcium from roots into the soil solution, creating a diffusion gradient *away* from the root surface toward the bulk soil. Other nutrients are required in such small quantities by plants (micronutrients) that mass flow meets the entire requirement (Table 8.3). Mass flow is, however, insufficient to supply those nutrients, such as nitrogen, phosphorus, and potassium that are required by plants in large quantities but present at low concentrations in the soil solution. These macronutrients (i.e., nutrients required in large quantities) are supplied primarily by diffusion. Even in agricultural soils, where soil solution concentrations are much higher, mass flow

Table 8.3 Mechanisms by which nutrients move to the root surface

Nutrient	Quantity absorbed by the plant (g m^{-2})	Mechanism of nutrient supply (% of total absorbed)		
		Root interception	Mass flow	Diffusion
Sedge tundra (Natural ecosystem)				
Nitrogen	2.2	–	0.5	99.5
Phosphorus	0.14	–	0.7	99.3
Potassium	1.0	–	6	94
Calcium[a]	2.1	–	250	0
Magnesium	4.7	–	83	17
Corn crop (Agricultural ecosystem)				
Nitrogen	19	1	79	20
Phosphorus	4	2	4	94
Potassium	20	2	18	80
Calcium[a]	4	150	413	0
Magnesium[a]	4.5	33	244	0
Sulfur	2.2	5	95	0
Iron	0.2	–	53	–
Manganese[a]	0.03	–	133	0
Zinc	0.03	–	33	–
Boron[a]	0.02	–	350	0
Copper[a]	0.01	–	400	0
Molybdenum[a]	0.001	–	200	0

[a]Mass flow of these elements is sufficient to meet the total plant requirement, so no additional nutrients must be supplied by diffusion. The amount supplied by mass flow was calculated from the concentration of the nutrients in the bulk soil solution multiplied by the rate of transpiration. The amount supplied by diffusion is calculated by difference; other forms of transport to the root (e.g., mycorrhizae) may also be important but are not included in these estimates
Data from Barber (1984), Chapin et al. (1980), and Lambers et al. (2008)

supplies less than 10% of those nutrients that typically limit plant production. Diffusion, rather than mass flow, is therefore the major mechanism that supplies potentially limiting nutrients (nitrogen, phosphorus, and potassium) to plants. Diffusion becomes even more important in supplying nutrients as soil fertility declines (Table 8.3).

Saturated flow of water through soils supplies additional nutrients and replenishes diffusion shells. Saturated flow is the movement of water through soil in response to gravity (see Chap. 4). After a rain, water drains vertically through the soil by saturated flow whenever the water content exceeds the soil water-holding capacity. Because nutrient availability and mineralization rates are generally highest in the uppermost soils, this vertical flow of water redistributes nutrients and replenishes diffusion shells surrounding roots. Both root growth and vertical soil water movement occur preferentially in soil cracks, quickly eliminating diffusion shells around these roots. Saturated flow

is also important in ecosystems where there is regular horizontal flow of ground water across an impermeable soil layer. Deep-rooted species in tundra underlain by permafrost, for example, have tenfold greater nutrient absorption and productivity in areas of rapid subsurface flow than in areas without lateral groundwater flow (Chapin et al. 1988). The high productivity of trees and shrubs in riparian ecosystems results in part because their roots often extend to the water table and to groundwater beneath the stream (the hyporheic zone), where roots tap the saturated flow of nutrients through the rooting zone.

Root Interception

Root interception is *not* an important mechanism of supplying nutrients to roots. As roots elongate into new soil, they intercept available nutrients in this unoccupied soil. The quantity of available nitrogen, phosphorus, and potassium per unit soil volume is, however, always less than the quantity of nutrients required to construct the root, so root interception can never be an important mechanism of nutrient supply to the shoot. Root growth is critical, *not* because it intercepts nutrients, but because it explores new soil volume and creates new root surface to which nutrients can move by diffusion and mass flow.

Nutrient Absorption

Nutrient absorption. Who is in charge? Three factors control nutrient absorption by vegetation: nutrient supply rate from the soil, root length, and root activity per unit root. Just as with photosynthesis, several factors influence nutrient absorption at the ecosystem scale. Our main conclusions in this section are: (1) Nutrient supply rate is *the* major factor accounting for differences among ecosystems in nutrient absorption at steady state. In other words, nutrient supply by the soil rather than plant traits determines biome differences in nutrient absorption by vegetation. (2) Plant traits such as root length, root depth, and root activity influence total nutrient absorption by vegetation mainly in situations where supply rate exceeds plant nutrient requirements (e.g., some recently logged sites or heavily fertilized agricultural fields) or where plant traits provide access to soil pools that would otherwise be inaccessible (e.g., deep soil pools). Given enough time, plant species sort themselves into sites where their capacity to absorb nutrients matches the soil supply. (3) Root length is the major factor governing *which* plants in an ecosystem are most successful in competing for a limited supply of nutrients.

Nutrient Supply

Across a broad range of nutrient availability, nutrient absorption by vegetation is driven primarily by nutrient supply. The most compelling evidence that nutrient supply drives absorption by vegetation is that most ecosystems, even those with relatively fertile soils, respond to nutrient addition with increased nutrient absorption and NPP (Fig. 8.1), just as observed in aquatic ecosystems.

Development of Root Length

When vegetation is recovering after disturbance or at high soil fertility, root length strongly influences nutrient absorption and NPP. Under these circumstances, the production of new root length allows the plant to explore soil volumes where diffusion shells among adjacent roots do not overlap and nutrients are not fully exploited by the existing root system. This is particularly likely to occur after disturbance. Even with a fully developed "root canopy," increased root growth by an individual plant may be advantageous because it increases the proportion of the total nutrient supply captured by that plant (Craine 2009).

Simulation models show quantitatively the role of different plant and soil parameters in determining nutrient absorption by vegetation (Nye and Tinker 1977). These models show that nutrient absorption is more sensitive to nutrient supply and to the volume of soil exploited by roots than to the kinetics of nutrient absorption, particularly for immobile ions like phosphate. At low nutrient supply rates, for example, variation in factors affecting diffusion (diffusion coefficient and buffering capacity) and root length (elongation rate) have a much greater effect on nutrient absorption than do kinetics (maximum and minimum capacity for absorption or affinity of roots for nutrients) or factors influencing mass flow (transpiration rate; Fig. 8.4). Absorption kinetics is more important in determining which root gets the nutrients, not the total absorption by vegetation.

Root length is a better predictor of nutrient absorption than is root biomass. Root length correlates closely with nutrient acquisition in short-term studies of nutrient absorption by plants from soils. Roots with a high **specific root length** (SRL, i.e., root length per unit mass) maximize their root length per unit root mass and therefore the volume of soil that can be explored by a given

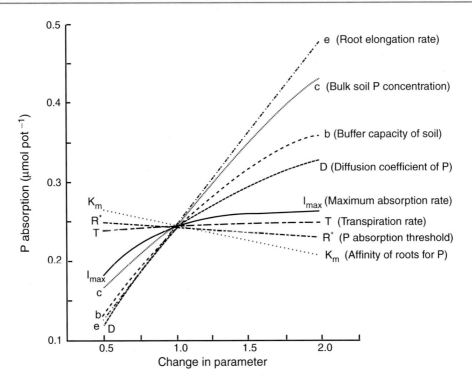

Fig. 8.4 Effect of changing parameter values (from 0.5 to 2.0 times the standard value) in a model that simulates phosphate absorption by roots of soybean. The factors that have greatest influence on phosphate absorption are plant parameters that determine the quantity of roots (*e*) and soil parameters influencing phosphate supply from the soil (*c*, *b*, and *D*). Redrawn from Clarkson (1985)

investment in root biomass. We know much less about the morphology and physiology of roots than of leaves. The limited available data suggest, however, that herbaceous plants (especially grasses) often have a greater SRL than woody plants and that there is a wide range in SRL among roots in any ecosystem. Much of the variation in SRL reflects the multiple functions of belowground organs. Roots can have a high SRL either because they have a small diameter or because they have a low tissue density (mass per unit volume). Some belowground stems and coarse roots have large diameters to store carbohydrates and nutrients or to transport water and nutrients and play a minor role in nutrient absorption. There may also be a tradeoff between SRL and the longevity of fine roots, with high-density roots being less prone to desiccation and herbivory than low-density roots. Both the leaves and roots of slowly growing species often have high tissue density, low rates of resource acquisition (carbon

and nutrients, respectively) but greater longevity than do leaves and roots of more rapidly growing species (Craine 2009; Freschet et al. 2010).

Roots grow preferentially in areas of high resource availability. Root growth in the soil is not random. Roots that encounter microsites of high nutrient availability branch profusely (Hodge et al. 1999), allowing plants to exploit preferentially zones of high nutrient availability. This explains why root length is greatest in surface soils (see Fig. 7.5), where nutrient inputs and mineralization are greatest, even though roots tend to be **geotropic** (i.e., grow vertically downward). This exploitation of nutrient hot spots ensures that plants maximize the nutrient return for a given investment in roots. This pattern of root growth also reduces the fine-scale heterogeneity in soil nutrient concentration. At a finer scale, **root hairs**, the elongate epidermal cells of the root that extend out into the soil, increase in length (e.g., from 0.1 to 0.8 mm) in response to a reduction in the supply of nitrate or phosphate (Bates and Lynch 1996). Both of these responses increase the length and surface area of roots available for nutrient absorption. Exploitation of hot spots does not always occur (Robinson 1994), however, and may be more pronounced in rapidly growing than in slowly growing species (Huante et al. 1998). Plants extend their length of nutrient-absorbing organs through growth of roots or root hairs or association with mycorrhizal fungi. Each of these modes of exploring new soil is more pronounced under conditions of low nutrient supply, although we focus here on root elongation because this process is best documented.

Mycorrhizae

Mycorrhizae increase the volume of soil exploited by plants. Mycorrhizae are symbiotic relationships between plant roots and fungal hyphae, in which the plant acquires nutrients from the fungus in return for carbohydrates that are the major carbon source for the fungus. About 80% of angiosperm plants, all gymnosperms, and many ferns are mycorrhizal (Wilcox 1991). These mycorrhizal relationships are important across a broad range of environmental and nutritional conditions, including fertilized crops (Allen 1991; Smith and Read 1997). With respect to nutrient absorption, mycorrhizal hyphae basically serve as an extension of the root system into the bulk soil, often providing 1–15 m of hyphal length per cm of root, that is, an increase in absorbing length of 2–3 orders of magnitude. Because the nutrient transport through hyphae occurs more rapidly than by diffusion along a tortuous path through soil-water films, mycorrhizae reduce the diffusion limitation of absorption by plants. The small diameter of mycorrhizal hyphae (< 0.01 mm) compared to roots (generally 0.1–1 mm) enables plants to exploit more soil with a given biomass investment in mycorrhizal hyphae than for the same biomass invested in roots. Plants typically invest 4–20% of GPP in supporting mycorrhizal hyphae (Lambers et al. 1996). Most of this carbon supports mycorrhizal respiration rather than fungal biomass, so a given carbon investment in mycorrhizal biomass

can represent a large carbon cost to the plant. Mycorrhizae are most important in supplementing those nutrients that diffuse slowly through soils, particularly phosphate and potentially ammonium in those ecosystems with low rates of nitrification. Although laboratory experiments show that plants consistently exclude mycorrhizae from roots under high-nutrient conditions, the extensive distribution of mycorrhizae across a wide range of soil fertilities, including most crop ecosystems, suggests that mycorrhizae continue to provide a net benefit to plants even in relatively fertile soils.

There are several types of mycorrhizae, the most common being **arbuscular mycorrhizae** (AM; also termed vesicular arbuscular mycorrhizae, VAM) and **ectomycorrhizae**. AM fungi grow through the cell walls of the **root cortex**, that is, the layers of root cells involved in nutrient absorption, much like a root pathogenic fungus. In contrast to root pathogens, AM produce **arbuscules**, which are highly branched treelike structures produced by the fungus and surrounded by the plasma membrane of the root cortical cells. Arbuscules are the structures that exchange nutrients and carbohydrates between the fungus and the plant. AM are most common in herbaceous communities, such as grasslands, in phosphorus-limited tropical forests, and in early successional temperate forests. Many AM associations are relatively nonspecific and can occur even with "ectomycorrhizal plant species" shortly after disturbance. AM are generally eliminated from these species after ectomycorrhizae colonize the roots.

In a given ecosystem type, AM associations are best developed under conditions of phosphorus limitation, where they short-circuit the diffusion limitation of absorption (Allen 1991; Read 1991). The AM symbiosis is a dynamic interaction between plant and fungus, in which both roots and hyphae turn over rapidly. Under conditions where plant growth is carbon-limited, as in young seedlings or in shaded or highly fertile conditions, mycorrhizae may act as parasites and reduce plant growth (Koide 1991; Lekberg and Koide 2005). Under these conditions, the plant reduces the number of infection points in new roots. As older roots die, this reduces the proportion of colonized roots, thus decreasing the carbon drain from the plant. AM associations might be viewed as a balanced parasitism between root and fungus that is carefully regulated by both partners.

Ectomycorrhizae are relatively stable associations between roots and fungi that occur primarily in temperate and high-latitude woody plants. The exchange organ is a **mantle** or sheath of fungal hyphae that surrounds the root plus additional hyphae that grow through the cell walls of the cortex (the **Hartig net**). Roots respond to ectomycorrhizal colonization by reducing root elongation and increasing branching, forming short, highly branched rootlets. Fungal tissue accounts for about 40% of the volume of these root tips. As with AM, ectomycorrhizae involve an exchange of nutrients and carbohydrates between the fungus and the plant. In contrast to AM, ectomycorrhizae generally prolong root longevity. Ectomycorrhizae also differ from AM in that they have proteases and other enzymes that attack organic nitrogen compounds. The fungus then absorbs the resulting amino acids and transfers them to the plant (Read 1991). Ectomycorrhizae therefore enhance both nitrogen and phosphorus absorption by plants.

Other mycorrhizal associations differ functionally from AM and ectomycorrhizae. Fine-rooted heath plants in the families Ericaceae and Epacridaceae, for example, form mycorrhizae in which the fungal tissue accounts for 80% of the root volume. These mycorrhizae, like ectomycorrhizae, hydrolyze organic nitrogen and transfer the resulting amino acids to their host plants. Many non-photosynthetic orchids depend on their mycorrhizae for carbon as well as nutrients. Their mycorrhizal fungi generally form links between the orchid and some photosynthetic plant species, especially conifers. In this case, the non-photosynthetic plant is clearly parasitic on the fungus.

As with the orchid-fungal association, ectomycorrhizae and AM often attach to several host plants, often of different species. Carbon and nutrients can be transferred among plants through this fungal network, although relatively few studies have shown a *net* transfer of carbon among plants (Simard et al. 1997), altering competitive interactions and promoting establishment of shade-tolerant tree seedlings in the understory (Booth and Hoeksema 2010). The quantitative and functional significance of these transfers in forest ecosystems is poorly known.

Nitrogen Fixation

Nitrogen-fixing plants access large quantities of nitrogen in high-light, nitrogen-limiting environments. Plants that form symbiotic relationships with nitrogen-fixing bacteria trade carbohydrates for nitrogen, just as with many mycorrhizal associations (see Chap. 9). Through this association, plants are able to tap the abundant pool of atmospheric di-nitrogen, which is otherwise unavailable to organisms. Nitrogen fixation is energetically expensive and therefore most frequent in habitats with abundant light and low nitrogen availability. These include many dry environments such as savannas or areas with minimal soil development. We discuss nitrogen fixation in greater detail in the next chapter.

Root Absorption Properties

Active transport is the major mechanism by which plants absorb potentially limiting nutrients from the soil solution at the root surface. Plant roots acquire nutrients from the soil solution primarily by **active transport**, an energy-dependent transport of ions across cell membranes against a concentration gradient. Due to the high concentrations of ions and metabolites inside plant cells, there is a constant leakage out of the root along a concentration gradient. Phosphate, for example, leaks from roots at about a third of the rate at which it is absorbed from the soil. This passive leakage of ions, sugars, and other metabolites may account for much of the exudation from fine roots. Ions that enter the root move passively by mass flow and diffusion through the cell walls of the **cortex** toward to the interior of the root (Fig. 8.5). As nutrients move through the cortical cell walls toward the center of the root, adjacent cortical cells absorb these nutrients by

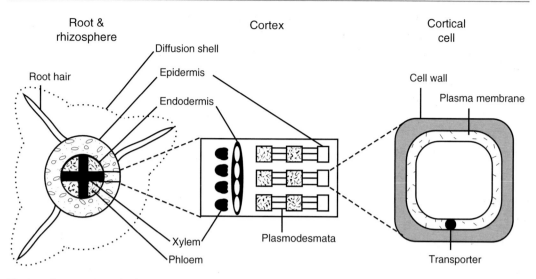

Fig. 8.5 Cross section of a root at three scales. The rhizosphere (or diffusion shell) is the zone of soil influenced by the root. The cortex has an outer layer of cells (the **epidermis**), some of which are elongated to form root hairs. The cortex is separated from the transport tissues (xylem and phloem) by a layer of wax-impregnated cells (the **endodermis**). Each cortical cell absorbs ions that diffuse through the pore spaces in the cell wall to the cell membrane. Membrane-bound proteins (**transporters**) transport ions across the cell membrane by active transport. Ions move from the outermost cortical cells toward the endodermis either through the cell walls or through the cytoplasmic connections between adjacent cortical cells (**plasmodesmata**)

active transport. Nutrients can move through the cell walls only as far as the **endodermis**, a **suberin** (wax)-coated layer of cells between the cortex and the xylem. Once nutrients are absorbed by cortical cells, they move through a chain of interconnected cells to the endodermis, where they are secreted into the dead xylem cells that transport water and nutrients to the shoot in the transpiration stream. As much as 30–50% of the carbon budget of the root supports nutrient absorption, indicating the large energetic cost of nutrient absorption (Lambers et al. 2008). Elements required in small quantities are often absorbed simply by mass flow or diffusion into the root cortical cells (Table 8.3).

Some plant species tap pools of nutrients that are unavailable to other plants. Although all plants require the same suite of nutrients in similar proportions, nitrogen is available in several forms (nitrate, ammonium, amino acids, etc.) that differ in availability among ecosystems. Species differ in their relative preference for these nitrogen forms (Table 8.4) and often show a high capacity to absorb those forms that are most abundant in the ecosystems to which they are adapted. Many species that occupy highly organic soils of tundra and boreal forest ecosystems, for example, preferentially absorb amino acids (Näsholm et al. 1998; Kielland et al. 2006), although even agricultural species utilize amino acid nitrogen (Näsholm et al. 2000). An important community consequence of species differences in nitrogen preference is that nitrogen represents several distinct resources for which species can compete. Species in the same community often have quite different isotopic signatures of tissue nitrogen because they acquire nitrogen from different sources – either different chemical fractions (nitrate, ammonium, organic nitrogen), different pathways (different mycorrhizal symbionts), or different soil depths (Fig. 8.6; McKane et al. 2002; Kahmen et al. 2008).

The three major forms of nitrogen differ in their carbon cost of incorporation into biomass. The carbon cost of incorporating amino acids is minimal, whereas ammonium must be attached to a carbon skeleton (the process of **assimilation**) before it is useful to the nitrogen economy of the plant. Finally, nitrate must be reduced to ammonium

Table 8.4 Preference ratios for plant absorption of different forms of nitrogen, when all forms are equally available

Species	$NH_4^+:NO_3^-$ preference[a]	Glycine:NH_4^+ preference[a]	References
Arctic vascular plants	1.1	2.1 ± 0.6 (12)	Chapin et al. 1993, Kielland 1994
Arctic nonvascular plants	–	5.0 ± 1.5 (2)	Kielland 1997
Boreal trees	19.3 ± 5.8 (4)	1.3	Chapin et al. 1986a, Kronzucker et al. 1997, Näsholm et al. 1998
Alpine sedges	3.9 ± 1.3 (12)	1.5 ± 0.4 (11)	Raab et al. 1999
Temperate heath	–	1.0	Read and Bajwa 1985
Salt marsh	1.3	–	Morris 1980
Mediterranean shrub	1.2	–	Stock and Lewis 1984
Barley	2.5 (2)	0.5	Chapin et al. 1993, Bloom and Chapin 1981
Tomato	0.6	–	Smart and Bloom 1988

[a]A preference ratio > 1 indicates that the first form of nitrogen is absorbed preferentially over the second. Numbers in parenthesis are the number of species or varieties studied. These studies show that many plants preferentially absorb glycine (a highly mobile amino acid) over ammonium and preferentially absorb ammonium over nitrate, when all forms are equally available

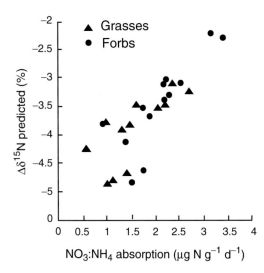

Fig. 8.6 The relationship between the ratio of predicted absorption of nitrate to ammonium and the predicted $\partial^{15}N$ concentration in seven forb species and eight grass species. Nitrate in these grassland sites had a high $\partial^{15}N$ concentration, so species that absorbed more nitrate had a higher tissue $\partial^{15}N$ concentration (less negative $\partial^{15}N$). Redrawn from Kahmen et al. (2008)

before it can be assimilated. Nitrate reduction is energetically expensive. Most plants transport some of the nitrate to leaves, where they use excess reducing power from the light reaction to reduce nitrate. In this case, the high energy cost of nitrate reduction does not detract from energy

available for other plant processes (Smirnoff et al. 1984). High availability of light and nitrate usually increases the proportion of nitrate reduced in leaves. Species also differ in their capacity to reduce nitrate in leaves, with species adapted to high-nitrate environments usually having a higher capacity to reduce nitrate in their leaves. Tropical and subtropical perennials and many annual plants typical of disturbed habitats, for example, reduce a substantial proportion of their nitrate in leaves (Lambers et al. 2008), whereas temperate gymnosperms and heath plants (family Ericaceae) reduce most nitrate in the roots (Smirnoff et al. 1984). Nitrogen availability is usually so limited in temperate and high-latitude terrestrial environments that the relative availability of nitrogen forms in the soil is more important than cost of assimilation in determining the forms of nitrogen absorbed and used by plants. Plants usually absorb whatever they can get.

Plant species also differ in the pools of phosphorus they can tap. Roots of some plant species produce phosphatase enzymes that release inorganic phosphate for absorption by plant roots (Richardson et al. 2007). The dominant sedge in arctic tussock tundra, for example, meets about 75% of its phosphorus requirement by absorbing the products of its root phosphatase enzymes (Kroehler and Linkins 1991). Other plants, particularly those in dry environments, secrete chelates

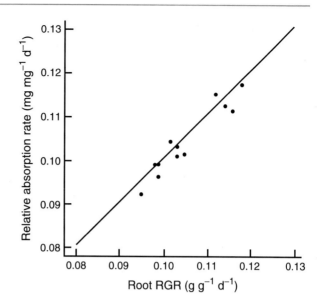

Fig. 8.7 The rate of nitrogen absorption in tobacco as a function of the relative growth rate of roots (RGR). Redrawn from Raper et al. (1978)

such as citrate or malate that diffuse from the root into the bulk soil. These chelates bind iron from insoluble iron–phosphate complexes, thereby solubilizing phosphate. Soluble phosphate then diffuses to the root, where it is absorbed (Lambers et al. 2008). Some plants, particularly Australian and South African heath plants in the Proteaceae, produce dense clusters of roots (**proteoid roots**) that are particularly effective in secreting chelates and solubilizing iron phosphate. There are many classes of chelates (**siderophores**) produced by plant roots, although the benefit of these secretions to the plant is poorly known. Plants therefore differ in the soil phosphorus pools they can exploit, but we have only a rudimentary understanding of the ecosystem consequences of these species differences.

Species differences in rooting depth and density influence the pool of nutrients that can be absorbed by vegetation. Grasslands and forests growing adjacent to one another on the same soil often differ greatly in annual nutrient absorption and productivity because the more deeply rooted forest trees exploit a larger soil volume and therefore a larger pool of water and available nutrients than do shallow-rooted species (see Chap. 11). In summary, there are several mechanisms by which species composition influences the quantity and form of nutrients acquired by vegetation.

Root absorption capacity increases in response to plant demand for nutrients. When the aboveground environment favors rapid growth and associated high **demand** for nutrients, plant roots respond by synthesizing more transport proteins in root cortical cells, thus increasing the capacity of the root to absorb nutrients. Species that have an inherently high relative growth rate or experience conditions that support rapid growth therefore have a high capacity per unit of root to absorb nutrients (Chapin 1980). High light and warm air temperatures, for example, increase root absorption capacity, whereas shade, drought, and phenologically programmed periods of reduced growth lead to a low absorption capacity. Rapidly growing roots, however, have a high capacity to absorb nutrients (Fig. 8.7). The rates of nutrient absorption by vegetation are therefore influenced by both soil factors that determine nutrient supply and plant factors that determine nutrient demand. In field studies, nutrient absorption correlates closely with NPP (Fig. 8.8). It is difficult, however, to separate cause from effect in explaining this correlation.

Changes in root absorption kinetics fine-tune the capacity of plants to acquire specific nutrients. Ion transport proteins are specific for particular ions. In other words, ammonium, nitrate, phosphate, potassium, and sulfate are

Fig. 8.8 Relationship
between nitrogen
absorption of temperate
and boreal coniferous and
deciduous forests and NPP.
Redrawn from Chapin
(1993b)

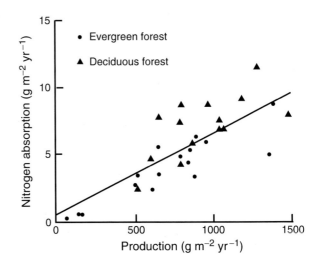

Table 8.5 Effect of environmental stresses on rate of
nutrient absorption by barley

Stress	Ion absorbed	Absorption rate by stressed plant (% of control)
Nitrogen	Ammonium	209
	Nitrate	206
	Phosphate	56
	Sulfate	56
Phosphorus	Phosphate	400
	Nitrate	35
	Sulfate	70
Sulfur	Sulfate	895
	Nitrate	69
	Phosphate	32
Water	Phosphate	32
Light	Nitrate	73

Data are from Lee (1982), Lee and Rudge (1987), and
Chapin (1991b)

each transported by a different membrane-bound
protein that is individually regulated (Clarkson
1985; Lambers et al. 2008). Plants induce the
synthesis of additional transport proteins for
those ions that specifically limit plant growth.
Roots of a phosphorus-limited plant therefore
have a high capacity to absorb phosphate, whereas
roots of a nitrogen-limited plant have a high
capacity to absorb nitrate and ammonium
(Table 8.5). Nitrate reductase, the enzyme that
reduces nitrate to ammonium (the first step before
nitrate-nitrogen can be incorporated into amino

acids for biosynthesis) is also specifically induced
by presence of nitrate.

In summary, there are several adjustments that
plants make to improve resource balance. These
include (1) changes in root:shoot ratio to improve
the balance between acquisition of belowground
and aboveground resources, (2) preferential root
growth in sites of high nutrient availability, and (3)
adjustment of the capacity of roots to absorb
specific nutrients, which brings the plant nutrient
ratios closer to values that are optimal for growth.

**Nutrient absorption alters the chemical
properties of the rhizosphere.** Nutrient absorption by plant roots reduces soluble nutrient concentrations in the soil and is a critical control over
the retention by ecosystems of mobile nutrients
such as nitrate. Forest clearing or crop removal,
for example, makes soils more prone to nitrate
leaching into groundwater and streams (see Fig.
9.14; Bormann and Likens 1979).

A second major consequence of plant nutrient
absorption is a change in rhizosphere pH.
Whenever a root absorbs an excess of cations, it
secretes hydrogen ions (H^+) into the rhizosphere
to maintain electrical neutrality. This H^+ secretion acidifies the rhizosphere. Except for nitrogen, which can be absorbed either as a cation
(NH_4^+) or an anion (NO_3^-), the ions absorbed in
greatest quantities by plants are cations (e.g.,
Ca^{2+}, K^+, Mg^{2+}), with phosphate and sulfate being

the major anions (Table 8.3). When plants absorb most nitrogen as NH_4^+, their cation absorption greatly exceeds anion absorption, and they secrete H^+ into the rhizosphere to maintain charge balance, causing acidification of the rhizosphere. When plants absorb most nitrogen as NO_3^-, their cation–anion absorption is more nearly balanced, and roots have less effect on rhizosphere pH. Ammonium tends to be the dominant form of inorganic nitrogen in acidic soils, whereas nitrate makes up a larger proportion of inorganic nitrogen in basic soils (see Chap. 9). The absorption process therefore tends to make acidic soils more acidic.

Roots also alter the nutrient dynamics of the rhizosphere through large carbon inputs from root death, the sloughing of mucilaginous carbohydrates from **root caps**, and the exudation of organic compounds by roots. These carbon inputs to soil may account for 10–30% of NPP (see Table 6.2). These labile carbon sources stimulate the growth of bacteria, which acquire their nitrogen by mineralizing organic matter in the rhizosphere (see Chap. 7). This nitrogen becomes available to plant roots when bacteria are grazed by protozoa or become energy starved due to a reduction in root exudation (see Fig. 7.9). Plants are sometimes effective competitors with microbes for soil nutrients, for example when plant carbon status is enhanced by added CO_2 (Hu et al. 2001). We know relatively little, however, about factors that govern competition for nutrients between plants and microbes (Schimel and Bennett 2004).

Nutrient Use

Nitrogen and phosphorus co-limit plant growth in most terrestrial ecosystems in the short term, just as in aquatic ecosystems. On average, nitrogen and phosphorus are about equally limiting to plant growth on land in the short term (Fig. 8.3; Elser et al. 2007), although the relative degree of limitation by nitrogen and phosphorus differs within and among ecosystems (Güsewell 2004). Lowland tropical forests on ancient weathered soils, for example, tend to respond most strongly to phosphorus, whereas tundra plants on recently glaciated soils tend to respond more

strongly to nitrogen. This is consistent with the higher N:P ratios in leaves of tropical than of high-latitude plants (Reich and Oleksyn 2004). The high N:P ratio of tropical plants is primarily a consequence of low tissue-phosphorus concentrations (Sterner and Elser 2002; Reich and Oleksyn 2004), just as in lakes and the ocean. In montane tropical forests of Hawai'i, there was a shift from nitrogen as the most limiting element on young soils to phosphorus as the most limiting element on older soils, supporting Walker and Syers' (1976) hypothesis that phosphorus should become less available and ecosystems should become more phosphorus-limited as soils weather (see Chap. 3; Vitousek 2004). Nonetheless, production in most ecosystems responds in the short term to both nitrogen and phosphorus and especially to the two nutrients in combination, suggesting co-limitation (Elser et al. 2007; LeBauer and Treseder 2008; Craine 2009). Whether co-limitation is equally important in the long term, or alternatively, whether only one of these nutrients is capable of transforming ecosystems, as is the case in many lakes, is more difficult to determine because the relatively long life span of many terrestrial plants makes it challenging to carry out experiments for long enough to allow species replacement and hence (potentially) adjustment of nutrient inputs. There is good evidence that abundances of nitrogen-fixing organisms and rates of nitrogen fixation can respond to added phosphorus in some terrestrial ecosystems, suggesting that nitrogen supply could adjust to that of phosphorus. Conversely, human enhancement of nitrogen inputs is pervasive in much of the temperate zone where most nutrient enrichment experiments have been carried out, and so results of many experiments could overstate the long-term importance of phosphorus limitation.

The short-term pattern of co-limitation suggests that (1) plants adjust physiologically to minimize limitation by any single nutrient, just as they adjust allocation to minimize limitation by water, nutrients, or light, and (2) these adjustments are seldom completely effective, so a given ecosystem often responds to one nutrient more strongly than to others, a pattern similar to that seen in relative responses to water, nutrients, and light.

What accounts for the frequent short-term response of production by multiple nutrients in terrestrial ecosystems?

Several plant and ecosystem processes contribute to co-limitation of NPP by multiple nutrients. Just as allocation adjustments reduce strong single-factor limitation by water, nutrients, or light, terrestrial plants adjust their nutritional properties to minimize overwhelming limitation by any single nutrient. As described earlier, plants adjust nutrient absorption to maximize absorption of growth-limiting nutrients and reduce capacity to absorb non-limiting nutrients. Symbiotic associations also reduce nutrient limitation by specific elements. Endomycorrhizal associations reduce phosphorus limitation, and ectomycorrhizal associations reduce both nitrogen and phosphorus limitation. Analogously, symbiotic association with nitrogen-fixing bacteria reduces nitrogen limitation for the host plant and indirectly for other plants in the ecosystem (see Chap. 9). Some plants in strongly phosphorus-limiting environments produce phosphatases or chelates that solubilize phosphorus and reduce the degree of phosphorus limitation. All of these traits alter the rate of nutrient acquisition and are regulated by the relative demand by the plant for nitrogen vs. phosphorus. These processes adjust acquisition rates over the short term to meet the needs of the plants, so they are no longer a simple function of the balance of nutrients supplied by the environment. However, there are limits to this flexibility; while nitrogen fixation can bring nitrogen from outside ecosystem boundaries, there is no biotic process that can bring new phosphorus into a system. Where rocks and minerals within the system are weathering (see Chap. 3), there is an important source of new phosphorus unmatched by nitrogen; most (but not all) rocks contain phosphorus but very little nitrogen. Here, if nitrogen fixation is constrained, nitrogen can be an ultimate limiting resource. However, where the weathering source is depleted in old, high rainfall, often tropical soils, phosphorus is likely to represent the ultimate limitation, although both nitrogen and phosphorus may be limiting in the short term in both situations.

Potentially limiting nutrients absorbed by plants are used primarily to support the production of metabolically active tissues (NPP). Carbon derived from photosynthesis comprises about half of the dry mass of all plant parts and therefore mirrors the distribution of biomass among plant parts (see Chap. 6). Most nutrients, in contrast, are concentrated in metabolically active tissues, although any new tissue requires some nutrient investment. Enzymatic proteins and the nucleic acids involved in protein formation, for example, have high nitrogen concentrations. Energy transformations (e.g., photosynthesis and respiration), nucleic acids, and membrane lipids all require phosphorus. Potassium is also concentrated in metabolically active tissues because of its importance in osmotic regulation. Other cations (e.g., magnesium and manganese) serve as cofactors for enzymes. Only calcium plays a primarily structural role, as a component of cell walls (calcium pectate; Marschner 1995).

As a consequence of their important metabolic roles, nitrogen, phosphorus, and potassium have high concentrations in leaves and to a lesser extent in fine roots, so changes in the supply of these nutrients to plants have powerful multiplier effects on the capacity of vegetation to acquire additional carbon and nutrients. Plants therefore respond to increased accumulation of a growth-limiting nutrient with a linear increase in plant growth rate in laboratory experiments (Ingestad and Ågren 1988) or an increase in NPP in the field (Fig. 8.8). This is similar to the light response curve of entire ecosystems, where ecosystem carbon gain (GPP) increases linearly with light over a broad range of light availability (see Fig. 5.23).

Terrestrial plants accumulate nutrients in storage organs (e.g., stems) and organelles (e.g., vacuoles) at times when nutrient supply exceeds demand. In this way, plants exploit brief pulses of nutrient supply, for example when recently shed autumn leaves are leached by rain (see Chap. 9). This **luxury consumption**, as it is sometimes called, alters element ratios in tissues because nutrient concentration increases more strongly with increasing supply for non-limiting nutrients than for growth-limiting nutrients. Stored nutrients are then drawn upon at times when the demands for growth exceed absorption from the soil (Chapin et al. 1990; Sterner and Elser 2002).

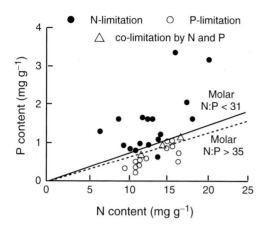

Fig. 8.9 Relationship between nitrogen and phosphorus concentration of leaves in heath plants. Each data point represents a site where nutrient-addition experiments show that plant growth is limited by nitrogen (solid circles), phosphorus (open circles), or both (open triangles). Plants with an N:P mass ratio < 14 (molar ratio of 31) respond primarily to nitrogen, whereas plants with an N:P mass ratio > 16 (molar ratio of 35) respond primarily to phosphorus. Redrawn from Koerselman and Mueleman (1996)

In arctic tundra, for example, cotton sedge can complete a full season's normal growth without any nutrient absorption from soil by drawing on stores acquired in previous years (Jonasson and Chapin 1985). Sometimes, variation in nutrient ratios in plant tissues reflects the relative degree of limitation by different elements (Güsewell 2004) and can be used, for example, to decide the optimal ratio of nutrients in fertilizers that are applied to a crop (Ulrich and Hills 1973). In other cases, nutrient ratios have little relationship to the magnitude of nutrient limitation. Some species, for example, synthesize nitrogen-based defensive compounds like alkaloids. These species have a relatively high N:P ratio but are not necessarily less nitrogen-limited than species with a lower N:P ratio. Finally, soils are chemically heterogeneous at scales ranging from millimeters to continents, leading to differences in nutrient supply rates and therefore plant nutrient ratios. Together these factors modify the nutrient ratios of plants from values that might be considered optimal for growth and cause plant growth and ecosystem NPP to respond to multiple nutrients (Fig. 8.9; Güsewell 2004; Craine et al. 2008).

The sorting of species by habitat contributes to the responsiveness of nutrient absorption and NPP to variations in nutrient supply observed across habitats. Species such as trees that have a large capacity to use nutrients for growth dominate sites with high nutrient supply rates, whereas infertile habitats are dominated by species with extensive root systems but lower capacity to absorb nutrients per unit root length.

Nutrient-use efficiency is greatest where production is nutrient-limited. Differences among plants in tissue-nutrient concentration provide insight into the quantity of biomass that an ecosystem can produce per unit of nutrient. Nutrient use efficiency is the amount of production per unit of nutrient acquired. A useful index of **nutrient use efficiency** (NUE) is the ratio of nutrients to biomass lost in litterfall (i.e., the inverse of nutrient concentration in plant litter; Vitousek 1982). This ratio is highest in unproductive sites (Fig. 8.10), suggesting that plants are more efficient in producing biomass per unit of nutrient acquired and lost if nutrients are in short supply. There are at least two ways in which a plant might maximize biomass gained per unit of nutrient (Berendse and Aerts 1987): through (1) a high **nutrient productivity** (a_n), that is, a high instantaneous rate of carbon absorption per unit nutrient or (2) a long **residence time** (t_r), that is, the average time that the nutrient remains in the plant.

$$\text{NUE} = a_n \times l_t \text{ [g biomass (gN)}^{-1}$$
$$= \text{g biomass (gN)}^{-1} \text{yr}^{-1} \times \text{yr} \quad (8.1)$$

Species characteristic of infertile soils have a long residence time of nutrients but a low nutrient productivity (Table 8.6; Chapin 1980; Lambers and Poorter 1992), suggesting that the high NUE in unproductive sites results primarily from traits that reduce nutrient loss rather than traits promoting a high instantaneous rate of biomass gain per unit of nutrient (Table 8.6). Similarly, shading reduces tissue loss more strongly than it reduces the capacity to gain carbon (Walters and Reich 1999).

There is an innate physiological tradeoff between nutrient residence time and nutrient

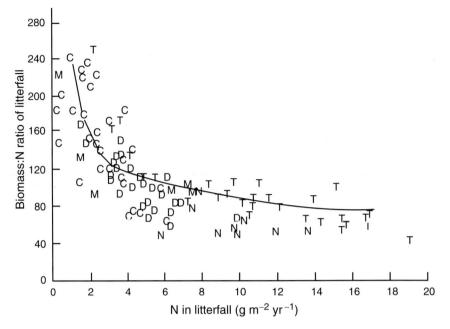

Fig. 8.10 Relationship between the amount of nitrogen in litterfall and nitrogen use efficiency (ratio of dry mass to nitrogen in that litterfall). Each symbol is a different stand, including conifer forests (C), temperate deciduous forests (D), tropical evergreen forests (T), Mediterranean ecosystems (M), and temperate forests dominated by nitrogen fixers (N). Redrawn from Vitousek (1982)

Table 8.6 Nitrogen use efficiency and its physiological components in a heathland evergreen shrub and a grass

Process	Evergreen shrub[a]	Grass[a]
Nitrogen productivity (g biomass (gN)$^{-1}$ yr^{-1})	77	110
Mean residence time (yr)	1.2	0.8
Nitrogen use efficiency (g biomass (gN)$^{-1}$)	90	89

[a]Species are a low-nutrient-adapted evergreen shrub (*Erica tetralix*) and a co-occurring deciduous grass (*Molinia caerulea*) that is adapted to higher soil fertility. Although these two species have similar nitrogen use efficiency, this is achieved by high nitrogen productivity in the high-nutrient-adapted species and by high mean residence time in the low-nutrient-adapted species
Data are from Berendse and Aerts (1987)

productivity. This occurs because the traits that allow plants to retain nutrients reduce their capacity to grow rapidly (Chapin 1980; Lambers and Poorter 1992). Plants with a high nutrient productivity grow rapidly and have high photosynthetic rates, which are associated with low tissue density, a high specific leaf area, and a high tissue-nitrogen concentration (see Chap. 5).

Conversely, a long nutrient residence time is achieved primarily through slow rates of replacement of leaves and roots. In order for leaves to survive a long time, they must have more structural cells to withstand unfavorable conditions and higher concentrations of lignin and other secondary metabolites to deter pathogens and herbivores. Together these traits result in dense leaves with low tissue-nutrient concentrations and therefore low photosynthetic rates per gram of biomass. The high NUE of plants on infertile soils therefore reflects their capacity to retain tissues for a long time rather than a capacity to use nutrients more efficiently in photosynthesis (Craine 2009; Freschet et al. 2010). A high NUE also reduces rates of decomposition and nutrient mineralization because well-defended, low-nutrient tissues decompose slowly when they senesce and induce immobilization of nutrients by microorganisms (Fig. 8.11).

Less is known about the tradeoffs between root longevity and nutrient absorption rate. Nutrient absorption declines as roots age, lose root hairs, and become suberized, so tradeoffs

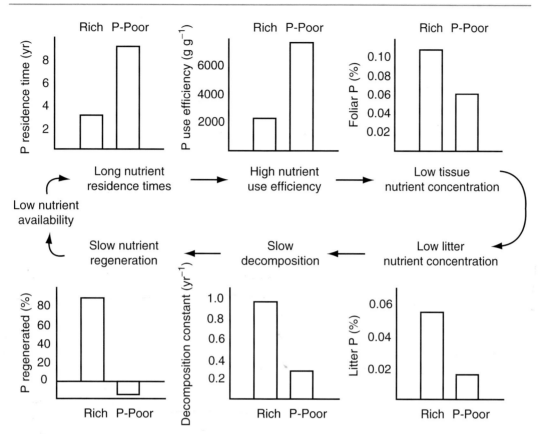

Fig. 8.11 Components of a plant-soil microbial positive feedback, based on data from *Meterosideros* forests on phosphorus (P)-rich and P-poor soils in Hawai'i. Forests on P-poor soils have long P residence times, high P-use efficiency, low leaf-P concentration, low litter-P concentration, slow decomposition, and slow P regeneration, relative to P-cycling patterns on P-rich sites. Data from Vitousek (2004)

between physiological activity and longevity that have been well documented for leaves probably also exist for roots (Craine 2009; Freschet et al. 2010). Slow-growing plants often have low nutrient concentrations in their roots as well as low rates of respiration (Tjoelker et al. 2005), which is consistent with their low capacity for nutrient absorption.

The tradeoff between NUE and rate of resource capture explains the diversity of plant types along resource gradients. Low-nutrient environments are dominated by species that conserve nutrients through low rates of tissue turnover, high NUE, and the physical and chemical properties necessary for tissues to persist for a long time. These stress-tolerant plants outcompete plants that retain less nutrients in infertile environments (Chapin 1980; Craine 2009). A high NUE and associated traits constrain the capacity of plants

to capture carbon and nutrients. In high-nutrient environments, species with high rates of resource capture, rapid growth rates, rapid tissue turnover, and consequently low NUE therefore outcompete plants with high NUE. In other words, neither a rapid growth rate nor a high NUE is universally advantageous because of inherent physiological tradeoffs between these traits. The relative benefit to the plant of efficiency vs. rapid growth depends on environment.

Nutrient Loss from Plants

The nutrient budget of plants, particularly long-lived plants, is determined just as much by nutrient loss as by nutrient absorption. The potential avenues of nutrient loss from plants include tissue senescence and death, leaching of

Table 8.7 Nitrogen and phosphorus resorption efficiency of different growth forms

| Growth form | Resorption efficiency (% of maximum pool)[a] | |
	Nitrogen	Phosphorus
All data	50.3 ± 1.0 (287)	52.2 ± 1.5 (226)
Evergreen trees and shrubs	46.7 ± 1.6 (108)[b]	51.4 ± 2.3 (88)[b]
Deciduous trees and shrubs	54.0 ± 1.5 (115)[c]	50.4 ± 2.0 (98)[b]
Forbs	41.4 ± 3.7 (33)[b]	42.4 ± 7.1 (18)[b]
Graminoids	58.5 ± 2.6 (31)[c]	71.5 ± 3.4 (22)[c]

[a]Data are averages ± SE, with number of species in parenthesis. Different letters within a column (b or c) indicate statistical difference between growth forms (P < 0.05)
Data from Aerts (1995)

dissolved nutrients from plants, consumption of tissues by herbivores and pathogens, exudation of nutrients into soils, and catastrophic loss of nutrients from vegetation by fire, windthrow, and other disturbances. Nutrient loss from plants is an *internal transfer* within ecosystems (the transfer from plants to soil) rather than a *loss from the ecosystem*. After this transfer to soil, nutrients are potentially available for absorption by microbes or plants or may be lost from the ecosystem. Nutrient loss from *plants* to soil therefore has very different consequences than nutrient loss from the *ecosystem* to the atmosphere or to groundwater.

Senescence

Tissue senescence is the major avenue of nutrient loss from plants. Plants reduce loss of nutrients through senescence primarily by reducing tissue turnover, particularly in low-resource environments. The leaves of grasses and evergreen woody plants, for example, show greater leaf longevity in low-nutrient or low-water environments than in high-resource environments (Fig. 8.11; Chapin 1980). Similarly, root longevity of grasses is greatest in low-nutrient sites (Craine 2009). Species differences in tissue turnover strengthen this pattern of high tissue longevity in low-resource environments. The proportion of evergreen woody species increases with decreasing soil fertility, reducing the rate of leaf turnover at the ecosystem level. All else being equal, a reduction in tissue turnover causes a cor-

responding reduction in the loss of the associated tissue nutrients. This reduction in tissue turnover is probably the single most important adaptation for nutrient retention in low-nutrient habitats (Chapin 1980; Lambers and Poorter 1992; Craine 2009).

Nutrient **resorption** is the transfer of soluble nutrients out of a senescing tissue through the phloem. It plays a crucial, but poorly understood, role in nutrient retention by plants. Plants resorb, on average, about half of their nitrogen, phosphorus, and potassium from leaves before leaves are shed at senescence (Table 8.7), so nutrient resorption is quantitatively important to plant nutrient budgets. Resorption efficiency tends to be greatest in plants with a low initial leaf nitrogen and phosphorus concentration (i.e., low-nutrient plants). However, these patterns do not always occur (Aerts and Chapin 2000; Kobe et al. 2005; Craine 2009), and there is a wide range of resorption efficiencies (0–90%) observed among studies. The reasons for this variation are poorly known. Efficient nutrient resorption is promoted by presence of an active sink, for example when new leaf production coincides with senescence of older leaves, as occurs in graminoids (grasses and sedges; Table 8.7) and many evergreens. Drought reduces the efficiency of nutrient resorption (Pugnaire and Chapin 1992; Aerts and Chapin 2000), and wind often dislodges a leaf before resorption is complete in non-graminoid plants. In summary, nutrient resorption efficiency may be such an important trait that most plants have a similar capacity to resorb nutrients. Environmental factors, such as nutrient pulses, drought, and wind, may influence the extent to which this resorption capacity is realized. Resorbed nutrients are transferred to other plant parts (e.g., seeds, storage organs, or leaves at the top of the canopy) to support growth at other times or parts of the plant. Some nutrients, such as calcium and iron, are immobile in the phloem, so plants cannot resorb these nutrients from senescing tissues. Because these nutrients seldom limit plant growth, their lack of resorption has little direct nutritional impact on plants, except where acid rain greatly reduces their availability in soil (Aber et al. 1998; Driscoll et al. 2001).

Plants appear to have no phenologically programmed pattern of senescence and resorption from roots (Craine 2009), which simply stop functioning when physical stresses, root herbivores, or pathogens degrade their physiological capacity below some threshold. However, this lack of apparent senescence and resorption from roots may reflect our ignorance, since very few studies have been done on root senescence (Aerts and Chapin 2000).

Leaching Loss from Plants

Leaching of nutrients from leaves is an important secondary avenue of nutrient loss from plants. Leaching accounts for about 15% of the annual nutrient return from aboveground plant parts to the soil. Rain dissolves nutrients on leaf and stem surfaces and carries these to the soil as **throughfall** (water that drips from the canopy) or **stemflow** (water that flows down stems). Stemflow typically has high concentrations of nutrients due to leaching of the stem surface; however, only a small amount of water moves by this pathway. Throughfall typically accounts for 90% of the nutrients leached from plants. Although plants with high nutrient status lose more nutrients per leaf, the *proportion* of nutrients recycled by leaching is surprisingly similar across a wide range of ecosystems (Table 8.8). Leaching loss is most pronounced for those nutrients that are highly soluble or are not resorbed. As much as 50% of the calcium and 80% of the potassium in an apple leaf, for example, can be leached within 24 h.

Table 8.8 Nutrients leached from the canopy (throughfall) as a percentage of the total aboveground nutrient return from plants to the soil

Nutrient	Throughfall (% of annual return)[a]	
	Evergreen forests	Deciduous forests
Nitrogen	14 ± 3	15 ± 3
Phosphorus	15 ± 3	15 ± 3
Potassium	59 ± 6	48 ± 4
Calcium	27 ± 6	24 ± 5
Magnesium	33 ± 6	38 ± 5

[a]Data are averages ± SE, for 12 deciduous and 12 evergreen forests
Data from Chapin (1991b)

Leaching rate is highest when rain first contacts a leaf, then declines exponentially with time. Ecosystems with very different rainfall regimes may therefore return similar proportions of nutrients to the soil through leaching vs. senescence. Although leaching loss is quantitatively important to plant nutrient budgets, there are no clear adaptations to minimize leaching loss. The thick cuticle of evergreen leaves was once thought to reduce leaching loss and explain the presence of evergreen leaves in wet, nutrient-poor forests. There is no evidence, however, that leaching loss is related to cuticle thickness. Like nutrient resorption, leaching loss from plants is a quantitatively important term in plant nutrient budgets that is not well understood. The acquisition of carbon and nutrients by plants is much better understood by biologists than is the loss of these resources.

Plant canopies can also absorb soluble nutrients from precipitation. Canopy absorption from precipitation is greatest in ecosystems where growth is strongly nutrient-limited.

Herbivory

Herbivores are sometimes a major avenue of nutrient loss from plants. Herbivores consume a relatively small proportion (1–10%) of plant production in many terrestrial ecosystems. In ecosystems such as productive grasslands, however, herbivores regularly eat a large proportion of plant production, and, during herbivore population outbreaks, herbivores may consume most aboveground production (see Chap. 10). Herbivory has a much larger impact on plant nutrient budgets than the biomass losses would suggest because herbivory precedes resorption, so vegetation loses approximately twice as much nitrogen and phosphorus per unit biomass to herbivores as it would through senescence. Animals also generally feed preferentially on tissues that are rich in nitrogen and phosphorus, thus maximizing the nutritional impact of herbivory on plants. There has therefore been strong selection for chemical and morphological defenses that deter herbivores and pathogens. These defenses occur in largest quantities in tissues that are long lived and in environments where nutrient supply is inadequate to readily replace nutrients lost to herbivores (Coley et al. 1985; Gulmon and Mooney 1986; Herms and Mattson 1992). Most nutrients transferred from plants to herbivores are rapidly returned to the soil in feces and urine, where they quickly become available to plants. In this way, herbivory accelerates nutrient cycling (see Chap. 10), especially in ecosystems that are managed for grazing. Nutrients are susceptible to loss from the ecosystem in situations where overgrazing reduces plant biomass to the point that plants cannot absorb the nutrients returned to the soil by herbivores.

Other Avenues of Nutrient Loss from Plants

Other avenues of nutrient loss are poorly known. Although laboratory studies suggest that root exudates containing amino acids may be a significant component of the plant carbon budget (Rovira 1969), the magnitude of nitrogen loss from plants by this avenue is unknown. Other avenues of nutrient loss from plants include plant parasites such as mistletoe and nutrient transfers by mycorrhizae from one plant to another. Although these nutrient transfers may be critical to the nutrient distribution among species in the community, they do not greatly alter nutrient retention or loss by vegetation as a whole.

Disturbances cause occasional large pulses of nutrient loss from vegetation. Fire, wind, disease epidemics, and other catastrophic disturbances cause massive nutrient losses from vegetation when they occur. With the exception of fire and human harvest, the nutrient loss from vegetation represents a nutrient transfer from vegetation to soil rather than a loss from the ecosystem. The pulse of decomposition and mineralization that accompanies this large litter input leads to both rapid nutrient absorption by early successional vegetation and the potential for leaching losses from the ecosystem (see Chap. 9). Nutrient losses from vegetation during wildfire vary with both the nutrient and fire intensity. Nitrogen and sulfur volatilize in fires more than do potassium and phosphorus, for example, whereas calcium and magnesium are largely transferred in ash. Nitrogen losses range from nearly 80% in stand-replacing forest fires to modest in fire-prone savannas and grasslands, where fires generally burn during the dry season after senescence and resorption have occurred and burn more litter than live plant biomass. Most plant nutrients in these ecosystems are stored below ground during times when fires are likely to occur.

Summary

Most of the open ocean is a nutritional desert, remote from the benthic supply of nutrients and distant from terrestrial inputs. The dominant primary producers are single-celled phytoplankton that reduce nutrient limitation through their extremely small size and high surface-to-volume ratio, which speeds nutrient diffusion to the cell surface. The degree of marine nutrient limitation reflects a balance between stratification from surface heating and turbulent mixing by winds and ocean currents. Highly stratified tropical ocean basins have extremely low nutrient availability and productivity, whereas turbulent mixing supports seasonal pulses of productivity in temperate and high-latitude ocean basins. Most pelagic production is co-limited by nitrogen and phosphorus in the short term. Nitrogen limitation in many parts of the ocean is amplified by low availability of iron that limits the activity of nitrogen-fixing cyanobacteria. The nutrient controls over lake productivity are similar to those in the ocean, except that, in the short term, most of the ocean responds most strongly to nitrogen and oligotrophic lakes respond more strongly to phosphorus. In the long term, however, phosphorus may, in many cases, be the ultimate limiting nutrient to both lakes and the ocean. The productivity of rivers and streams can be limited by either nitrogen or phosphorus, depending on the nature of the terrestrial matrix.

Nutrient availability is a major constraint to the productivity of the terrestrial biosphere. Whereas carbon acquisition by plants is determined primarily by plant traits (leaf area and photosynthetic capacity), nutrient absorption is usually governed more strongly by environment (the rate of supply by the soil) than by plant traits. In early succession, however, plant traits can have a significant impact on nutrient absorption by vegetation at the ecosystem level. Diffusion is the major process that delivers nutrients from the bulk soil to the root surface. Mass flow of nutrients in moving soil water is primarily important in replenishing diffusion shells and in supplying those nutrients that are abundant in soils or are required in small amounts by plants.

Plants adjust their capacity to acquire nutrients in several ways. Preferential allocation to roots under conditions of nutrient limitation maximizes the root length available to absorb nutrients. Root growth is concentrated in hot spots of relatively high nutrient availability, maximizing the nutrient return for roots that are produced. Plants further increase their capacity to acquire nutrients through symbiotic associations with mycorrhizal fungi. Plants that grow rapidly, due either to a favorable environment or a high relative growth rate, have a high capacity to absorb nutrients. Plants alter the kinetics of nutrient absorption to absorb those nutrients that most strongly limit growth. In the case of nitrogen, which is the most strongly limiting nutrient in many terrestrial ecosystems, plants typically absorb whatever forms are available in the soil. When all forms are equally available, most plants preferentially absorb ammonium or amino acids rather than nitrate. Nitrate absorption is often important, however, because of its high mobility in soil.

There is an inevitable tradeoff between the maximum rate of nutrient investment in new growth and the efficiency with which nutrients are used to produce biomass. Plants produce biomass most efficiently per unit of nutrient under nutrient-limiting conditions that constrain productivity. Nutrient use efficiency is maximized by prolonging tissue longevity, that is, by reducing the rate at which nutrients are lost. Senescence is the major avenue by which nutrients are lost from plants. Plants minimize the loss of growth-limiting nutrients by resorbing about half of the nitrogen, phosphorus, and potassium from a leaf before it is shed. About 15% of the annual nutrient return from aboveground plant parts to the soil comes as leachates, primarily as throughfall that drips from the canopy. Herbivores can also be important avenues of nutrient loss because they feed preferentially on nutrient-rich tissues and consume these tissues before resorption can occur. For these reasons, plants lose more than twice as much nutrients per unit of biomass to herbivores compared to losses through senescence. Other factors that cause occasional large nutrient losses from vegetation include disturbances (e.g., fire and wind) and diseases that kill tissues or plants.

Review Questions

1. How do oceanographic controls over stratification and mixing influence nutrient absorption and use in marine phytoplankton?
2. Why do phytoplankton use so little of the available nitrogen and phosphorus in HNLC regions of the ocean?
3. Mass flow, diffusion, and root interception are three processes that deliver nutrients to the root surface. How does each process work, and what is their relative importance in supplying nutrients to plants?
4. What is the major mechanism by which plants acquire nutrients that reach the root surface?

5. How do plants compensate for (a) low availability of all nutrients, (b) spatial variability of nutrients in the soil (localized hot spots), (c) imbalance among nutrients required by plants (e.g., nitrogen vs. phosphorus availability)?

6. How does plant growth rate affect nutrient absorption?

7. What are the major mechanisms by which mycorrhizae increase nutrient absorption by plants? Under what circumstances are mycorrhizae most strongly developed?

8. What are the major processes involved in converting nitrogen from nitrate to a form that is biochemically useful to the plant?

9. Why are nutrient and carbon flows in plants so tightly linked? What happens to nutrient absorption when carbon gain is restricted? What happens to carbon gain when nutrient absorption is restricted? What are the mechanisms by which these adjustments occur?

10. What is nutrient use efficiency (NUE)? What are the physiological causes of differences in NUE, and what are the ecosystem consequences?

11. What are the major differences in types of species that occur on fertile vs. infertile soils? What are the advantages and disadvantages of each plant strategy in each soil type?

12. What are the major avenues of nutrient loss from plants? How do all plants minimize this nutrient loss? What additional adaptations minimize nutrient loss from plants that are adapted to infertile soils?

Additional Reading

Aerts, R. and F.S. Chapin, III. 2000. The mineral nutrition of wild plants revisited: A re-evaluation of processes and patterns. *Advances in Ecological Research* 30:1–67.

Chapin, F.S., III. 1980. The mineral nutrition of wild plants. *Annual Review of Ecology and Systematics* 11:233–260.

Craine, J.M. 2009. *Resource Strategies of Wild Plants.* Princeton University Press, Princeton.

Hobbie, S.E. 1992. Effects of plant species on nutrient cycling. *Trends in Ecology & Evolution* 7:336–339.

Kalff, J. 2002. *Limnology.* Prentice-Hall, Upper Saddle River, NJ.

Lambers, H., F.S. Chapin, III, and T.L. Pons. 2008. *Plant Physiological Ecology.* 2nd edition. Springer, New York.

Mann, K.H. and J.R.N. Lazier. 2006. *Dynamics of Marine Ecosystems: Biological-Physical Interactions in the Oceans.* 3rd edition. Blackwell Publishing, Victoria, Australia.

Read, D.J. 1991. Mycorrhizas in ecosystems. *Experientia* 47:376–391.

Sterner, R.W. and J.J. Elser. 2002. *Ecological Stoichiometry: The Biology of Elements from Molecules to the Biosphere.* Princeton University Press, Princeton.

Vitousek, P.M. 1982. Nutrient cycling and nutrient use efficiency. *American Naturalist* 119:553–572.

Vitousek, P.M. 2004. *Nutrient Cycling and Limitation: Hawai'i as a Model System.* Princeton University Press, Princeton.

Nutrient Cycling

9

Nutrient cycling involves nutrient inputs to and outputs from ecosystems and the internal transfers of nutrients within ecosystems. This chapter describes these nutrient dynamics.

Introduction

Human impacts on nutrient cycles have fundamentally altered the regulation of ecosystem processes. Rates of cycling of carbon (see Chaps. 5– 7) and water (see Chap. 4) are ultimately regulated by energy and the availability of a few chemical resources, so changes in availability of these resources fundamentally alter all ecosystem processes. The combustion of fossil fuels has released large quantities of nitrogen and sulfur oxides to the atmosphere and increased their inputs to ecosystems (see Chap. 14). Fertilizer use and the cultivation of nitrogen-fixing crops have further increased the fluxes of nitrogen in agricultural and downstream aquatic ecosystems (Galloway et al. 1995; Vitousek et al. 1997a; Gruber and Galloway 2008). Together these human impacts have doubled the natural background rate of nitrogen inputs to the biosphere and quadrupled the rate of phosphorus inputs (Falkowski et al. 2000). The resulting increases in plant production may be large enough to affect the global carbon cycle. Human disturbances such as forest conversion, harvest, and fire increase the proportion of the nutrient pool that is available and therefore vulnerable to loss. Some of these losses occur by leaching of dissolved elements to groundwater, causing a depletion of soil cations, an increase in soil acidity, and increases in nutrient inputs to aquatic ecosystems. Gaseous losses of nitrogen influence the chemical and radiative properties of the atmosphere, causing air pollution and enhancing the greenhouse effect (see Chap. 2). Changes in the cycling of nutrients therefore dramatically affect the interactions among ecosystems (see Chap. 13) as well as the carbon cycle and climate of Earth.

A Focal Point

Nutrient runoff from freshwater systems to the ocean has created or intensified dead zones in two-thirds of the world's estuaries. Agriculturally derived nutrients delivered to estuaries and coastal zones stimulate production and rain of dead organic matter to depth. This depletes oxygen, leading to extensive death of fish, shrimp, and other invertebrates (Fig. 9.1). How can these effects be reduced by more careful management of nutrient sources in agricultural lands and cities? How can fertilizer applications be matched with crop nutrient demands to reduce fertilizer requirements and reduce offsite impacts of pollution? What is the fate of excess nutrients delivered to the coastal zone? Understanding controls on nutrient fluxes in ecosystems provides insights that can help answer these important management questions.

F.S. Chapin, III et al., *Principles of Terrestrial Ecosystem Ecology*,
DOI 10.1007/978-1-4419-9504-9_9, © Springer Science+Business Media, LLC 2011

Fig. 9.1 Dead zone in the Gulf of Mexico, magnified by nutrient inputs from agricultural runoff from the Mississippi river drainage. Reds and oranges represent high concentrations of phytoplankton and sediments (http://www.nasa.gov/vision/earth/environment/dead_zone.html)

Overview of Nutrient Cycling

Nutrient cycling involves the entry of nutrients to ecosystems, their internal transfers among plants, microbes, consumers, and the environment, and their loss from ecosystems. Some elements, for example, nitrogen, may move either by water or air, while others, for example, phosphorus, lack a significant gaseous phase and generally move only downhill in aqueous solution or as dust particles in the atmosphere. Nutrients become available to ecosystems through lateral transport, the chemical weathering of rocks, the biological fixation of atmospheric nitrogen, and the deposition of nutrients from the atmosphere in rain, wind-blown particles, or gases. Anthropogenic fertilization is an additional nutrient input in managed ecosystems. Internal cycling processes include the interconversion of organic and inorganic forms, chemical reactions that change elements from one ionic form to another, biological absorption by plants and microbes, and exchange of nutrients on surfaces within the soil matrix. Nutrients are lost from ecosystems by leaching, trace gas emission, wind and water erosion, fire, outflow, burial, and the removal of materials in harvest.

Most of the nitrogen and phosphorus required for plant growth in unmanaged ecosystems is supplied by the decomposition of past primary production, including plant litter and soil organic matter (SOM) in terrestrial environments and mineralization of organic matter in the water column or sediments of aquatic ecosystems. Inputs and outputs to or from these ecosystems are a small fraction of the quantity of nutrients that cycle internally, producing relatively **closed systems** with conservative nutrient cycles. Human activities tend to increase inputs and outputs relative to the internal transfers and make the element cycles more open.

We have already described the cycling of nutrients through plants (see Chap. 8). In this chapter, we focus on the nutrient inputs and losses from ecosystems and on the processes within ecosystems that regenerate available nutrients from dead organic matter.

Marine Nutrient Cycling

Large-Scale Nutrient Cycles

Pelagic nutrient cycling in the open ocean is closely coupled to the flow of carbon. The extremely small size of marine primary producers (submicroscopic algal cells and photosynthetic bacteria) dictates that the processes of photosynthesis, nutrient absorption, growth, and reproduction are tightly integrated at the cellular level. We have therefore already described many of the basic features of pelagic nutrient cycling in the context of plant carbon and nutrient absorption (see Chaps. 5 and 8) and growth (see Chap. 6). Key features of pelagic nutrient cycling through phytoplankton include:

- Large-scale patterns of nutrient availability to phytoplankton in the surface ocean depend on the balance of three processes (see Chaps. 6 and 8): (1) Stratification driven by surface heating restricts nutrient delivery from deep water to the surface. (2) Wind-driven mixing disrupts stratification and deepens the mixed layer, increasing nutrient supply but reducing average light availability through the mixed layer. (3) Upwelling supplements nutrient supply and keeps phytoplankton in shallow well-lighted surface waters, supporting high gross primary production (GPP) and NPP.

- Primary production in the open ocean is generally limited in the short term by both nitrogen and phosphorus, with production usually responding most strongly to nitrogen over seasonal-to-annual cycles and to phosphorus or micronutrients over the long term (see Chap. 8).

- Grazing accounts for most of the nutrient return from phytoplankton to the environment (see Chaps. 8 and 10).

- Sedimentation of zooplankton feces and phytoplankton causes a continuous nutrient loss from the pelagic zone that is replenished by nitrogen fixation, upwelling, and mixing (see Chap. 7).

Nitrogen is mineralized (converted from organic nitrogen to ammonium) by several processes in the ocean. Grazers and their predators excrete nitrogen when they breakdown nitrogenous compounds to meet their energetic demands for growth and movement or maintain element **stoichiometry** (nutrient balance), just like protozoans in the rhizosphere (see Chaps. 8 and 10). Grazing is a more prominent pathway of nutrient mineralization in the ocean than on land because of the high proportion of phytoplankton biomass that is grazed rather than dying and decomposing (see Chap. 7). In addition, decomposer bacteria excrete ammonium when their growth is energy-limited. Much of this bacterial nitrogen mineralization occurs on particles to which algae and cyanobacteria are also attached or in micro-patches of high nutrient concentration (Stocker et al. 2008), facilitating efficient recycling of ammonium back to primary producers. This **regenerated production** based on ammonium that is produced within the water column contributes to tight nutrient recycling in the pelagic zone (Dugdale and Goering 1967).

Those dead cells and fecal pellets that sink beneath the pycnocline continue to decompose and mineralize nitrogen. Due to the absence of phytoplankton in these deep dark waters, much of the resulting ammonium is absorbed by nitrifying bacteria that use it as an energy source, releasing nitrate as a waste product (the process of **nitrification**). Thus deep waters tend to have a higher nitrate-to-ammonium ratio than surface waters. In the open ocean, most organic carbon and nitrogen are mineralized in the water column before reaching the sediments (Mann and Lazier 2006). Rates of organic matter inputs and decomposition in the sediments are therefore relatively low, causing sediments to remain relatively well oxygenated. These aerobic conditions favor nitrification (an aerobic process of nitrate release) rather than **denitrification** (anaerobic release of nitrogen trace gases).

In the coastal zone, by contrast, greater productivity and shallower water allow more organic matter to reach the sediments, where it is decomposed or buried. Decomposition of this organic matter in deep water and sediments consumes some or all of the available oxygen, creating an anaerobic environment where sulfate-reducing and denitrifying bacteria use dead organic matter as an energy source and sulfate or nitrate, respectively, as an electron acceptor, producing hydrogen sulfide or nitrogen trace gases (N_2O and N_2) as waste products (see Chap. 3). The gaseous release of N_2O and N_2 by denitrification depletes ocean waters of nitrogen relative to other nutrients such as phosphorus, contributing to the frequent occurrence of nitrogen limitation in coastal waters. Sulfate reduction, however, usually accounts for most of the anaerobic decomposition in coastal sediments (Howarth 1984).

Estuaries

Horizontal flows of water and nutrients govern the nutrient cycling and productivity of estuaries. Estuaries, where rivers enter the ocean, are interfaces between fresh and saline water. Estuaries tend to become stratified by the inflow of low-density fresh water from rivers. This water **entrains** (carries with it) surface ocean water as it flows from the river mouth out into the coastal ocean. Phosphorus-rich bottom water that has been depleted of nitrogen by denitrification flows up bay to replace this surface water. The extent of mixing of phosphorus-rich bottom water with surface water depends primarily on tidal mixing, which is greatest in long or shallow estuaries, and on surface turbulence caused by river discharge, winds, and storms. The Chesapeake Bay, for example, receives about 25% of its phosphorus from the coastal ocean but most of its nitrogen from rivers (Nixon et al. 1996). The balance between stratification and turbulence favors much more mixing in estuaries than in the open ocean, creating an environment that supports very high productivity (Mann and Lazier 2006). Productivity is particularly high at "fronts" between relatively well-mixed estuarine water and deeper, more stratified zones of the coastal ocean. Plumes of estuarine water spread the influence of estuarine mixing well beyond the bay where the river enters the ocean.

Estuaries receive most of their nutrients from the land, an input that has increased substantially in the last century. Outflows of nitrate and phosphate from the Mississippi River doubled in the last half of the twentieth century (Lohrenz et al. 1999), and nitrate movement to the North Atlantic Ocean from major rivers has increased 6–20-fold in the past century (Howarth et al. 1996a). Two-thirds of the estuaries in the U.S. are degraded by nutrient pollution (Howarth et al. 2011). This pollution by rivers reflects increased inputs of fertilizer, atmospheric nitrogen deposition, nitrogen fixation by crops, and food imports (see Chap. 14). The nutrients support extremely high productivity in the estuary and generate large quantities of organic matter that sinks to depth. The resulting stimulation of bacterial activity depletes oxygen in the lower 20 m of the water column, especially in summer. This creates zones of **hypoxia** (low oxygen) and **anoxia** (zero oxygen) thousands of square kilometers in area (Fig. 9.1; Rabalais et al. 2002; Díaz and Rosenberg 2008). Anoxia in these **dead zones** kills benthic organisms and bottom-feeding shrimp and fish and dramatically alters nutrient cycling at the sediment–water interface (Howarth et al. 2011). A combination of increasing land-use change, intensification of agriculture, and warming ocean temperatures has increased the frequency and extent of dead zones in the world's estuaries and coastal waters, threatening many of Earth's most productive fisheries. In addition, dead zones have created a new climate feedback, in which climate warming intensifies stratification that augments the low-oxygen, high-nitrate conditions that favor denitrification and the production of N_2O, a powerful greenhouse gas that contributes to warming climate (Mann and Lazier 2006; Stramma et al. 2008; Codispoti 2010). This exemplifies the unintended global consequences of massive human modification of the global nitrogen cycle (see Chap. 14).

Construction of dams and reservoirs has modified the flow regime of estuaries. Reservoirs accumulate water at times of peak flows and release the water in dry seasons to meet demands for agriculture, hydropower, and other human uses (Carpenter and Biggs 2009). This reduces peak inputs to estuaries that drive mixing and support spring blooms of productivity. This homogenization of flow regime is counterbalanced by levees that prevent floodwaters from spreading over the floodplain and increase peak discharges to estuaries during floods. Surface evaporation from reservoirs and water withdrawals for agriculture reduce discharge and mixing at other times of year. The life history of many fish is linked to the predictable seasonality of estuarine flows and blooms and is often disrupted when dams alter the seasonal flow regime of rivers. Reservoirs also retain substantial amounts of nitrogen and especially phosphorus in sediments (Friedl and Wüest 2002).

Coastal Currents

Upwelling drives the high productivity of coastal currents. There are broad areas of the ocean, especially on the western edges of continents, where surface waters move away from the coast toward the open ocean and are replaced by deep waters that move toward the coast (see Chap. 2). This circulation moves deep nutrient-rich waters to the surface and buoys phytoplankton up to the surface, where light availability is high. Many factors influence the location and strength of coastal upwelling. The strength of offshore winds, for example, is generally strongest during La Niña conditions, and the stability of the surface layer that counterbalances upwelling is generally strongest during summer.

In coastal areas unaffected by upwelling, diurnal tidal fluctuations generate turbulence that mixes deep nutrients upward. The mixing front brings together a low-salinity coastal water mass that is stratified enough to keep phytoplankton in a well-lighted surface zone and a more saline deep-water mass that provides nutrients. The relatively stable location of this front and regular diurnal cycles of tidal mixing provide the conditions that sustain high plankton productivity and support large populations of fish, sea birds, and marine mammals (Mann and Lazier 2006). Upwelling and tidal mixing generate complex temporal and spatial patterns of coastal productivity and trophic dynamics that are often linked to long archeological records of human use.

Lake Nutrient Cycling

As in the ocean, active absorption of nitrogen and phosphorus by phytoplankton often maintains extremely low nutrient concentrations in surface waters of unpolluted lakes. Also, as in the open ocean, nutrient delivery from more nutrient-rich deep waters is minimized by thermal stratification that is occasionally disrupted by mixing events. The isolation of surface waters from nutrient supplies in sediments, however, is less extreme in lakes than in the open ocean for several reasons. (1) The small size of most lakes and ponds fosters tight coupling between primary production (much of which is rooted vascular plants or benthic algae) and resupply of nutrients from sediments. The centers of large lakes have surface waters that are less well coupled to sediments, and the open ocean is extremely disconnected from its sediments. (2) Stratification in lakes reflects only a thermal gradient, whereas the ocean thermocline is reinforced by a salinity gradient, making it more difficult for nutrients to mix to the surface. Storms are therefore more effective in mixing nutrients from depth to the surface in lakes than in the ocean. (3) Finally, due to expected scaling relationships of edges to volume, smaller lakes are more exposed to their surroundings than is the open ocean. Streams and the atmosphere are therefore additional nutrient sources that range from being unimportant to dominant influences in the annual nutrient budgets of lakes.

Nutrient mineralization in lakes has both similarities and differences to that in the ocean. In both lakes and the ocean, grazing and bacterial mineralization on particles of dead organic water recirculates nutrients rapidly within the water column. Dead cells and the feces of zooplankton reach the sediments more readily in lakes than in the ocean because organic matter has only a short distance to travel before reaching the bottom. Although lakes, ponds, and reservoirs cover a very small fraction of Earth's surface (Downing et al. 2006), they are globally important locations for carbon burial (Dean and Gorham 1998). Rates and pathways of nutrient mineralization in sediments

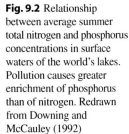

Fig. 9.2 Relationship between average summer total nitrogen and phosphorus concentrations in surface waters of the world's lakes. Pollution causes greater enrichment of phosphorus than of nitrogen. Redrawn from Downing and McCauley (1992)

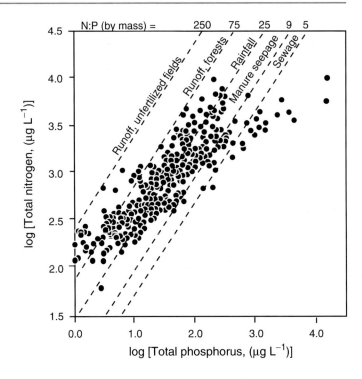

differ strikingly between lakes and the ocean. In unpolluted lakes, phosphorus binds tightly to clay and silt particles in the sediments. In contrast, phosphorus is desorbed from ocean sediments due to competition with sulfate and other anions for exchange sites (Howarth et al. 2011). In addition, nitrogen is depleted by denitrification in anaerobic sediments of estuaries and coastal waters, and phosphorus is resupplied to surface waters, leading to a relatively phosphorus-rich environment.

Even among lakes, there is tremendous diversity in nutrient dynamics that reflect differences in origin and watershed geology, human impact on watersheds, and current biota. Lakes make up about 3% of the global terrestrial land surface. Most lakes and ponds are small and have closer contact with terrestrial ecosystems than the large lakes that have been most intensively studied. Ponds and small lakes <1 km² in area, for example, may account for about 40% of global lake area (Downing et al. 2006). Glacial lakes, which account for about half of the remaining lake area, exhibit a wide range of depths and sizes. Other important lake types include large deep tectonic

lakes such as Lake Baikal and the African rift lakes and small shallow riverine lakes such as oxbows (Kalff 2002).

Deep lakes do not mix seasonally, especially in the tropics where there is little seasonal temperature variation. Deep lakes also have anoxic hypolimnia, where much of the nitrogen reaching the sediments is denitrified and returned to the atmosphere. At the opposite extreme, shallow lakes often have an extensive littoral zone dominated by vascular plants with high productivity, rapid rates of nutrient cycling, and tight coupling between plant production and sediment resupply of nutrients. Nutrient addition from agricultural runoff and sewage has substantially increased the nutrient content of many lakes, changing them from clear blue to a turbid green color (see Fig. 8.2; Carpenter and Biggs 2009). In general, oligotrophic lakes tend to have high N:P ratios, suggesting phosphorus limitation, and N:P ratio in the water decreases in more nutrient-rich lakes (Fig. 9.2).

Water residence time (the time required to replace the water volume of a system) influences many ecosystem properties of aquatic ecosystems (Kalff 2002). The open ocean has a longer water residence time, and estuaries have shorter

Fig. 9.3 Log–log relationship between water residence time and phosphorus input to moist-temperate lakes. Redrawn from Kalff (2002)

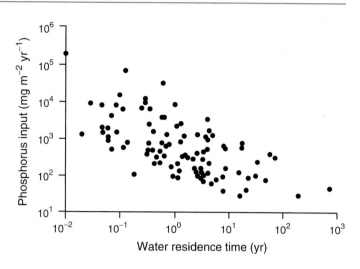

residence times than most lakes. Among lakes, water residence time tends to be long in lakes that are deep (e.g., hundreds to thousands of years in tectonic rift lakes), have small watersheds (e.g., lakes that are high in a drainage basin), or low river-input rates. These lakes are dominated by internal recycling processes, have small inputs of organic carbon and nutrients, support relatively low productivity and rates of nutrient cycling, and are vulnerable to the direct impacts of atmospheric deposition on the lake. Lakes that flush more rapidly (water residence times less than a decade) are particularly vulnerable to pollution from land-use change within the watershed (Fig. 9.3). Pollution that exceeds the capacity of sediments to sequester phosphorus, for example, can cause the sediments to switch from being a sink to a source of phosphorus, at which point it becomes very difficult to control phytoplankton production and maintain water clarity (see Fig. 12.6; Carpenter 2003).

Stream Nutrient Cycling

Carbon and nutrients spiral down streams and rivers and the groundwater beneath them. Streams are not passive channels that carry materials from land to the ocean but process much of the material that enters them (Cole et al. 2007; Mulholland et al. 2008). The strong horizontal flow of water in streams and rivers carries the resulting products downstream, where they are repeatedly reprocessed in successive stream sections (Fisher et al. 1998). This leads to open patterns of nutrient cycling, in which the lateral transfers are much larger than the internal recycling (Giller and Malmqvist 1998). Stream productivity therefore depends on regular subsidies from the surrounding terrestrial matrix and is quite sensitive to changes in these inputs that result from pollution or land-use change (Mulholland et al. 2008). The **spiraling length** of a stream is the average horizontal distance between successive uptake events. It depends on the **turnover length** (the downstream distance moved while an element is in organic form) and the **uptake length** (the average distance that an atom moves from the time it is released until it is absorbed again). A representative spiraling length of a woodland stream is about 200 m. Of this distance, about 10% occurs as microorganisms flow downstream attached to CPOM and FPOM, 1% as consumers move downstream, and the remaining 89% after release of the nutrient by mineralization (Giller and Malmqvist 1998). A unit of nutrient therefore spends most of its time with relatively little movement, but moves rapidly once it is mineralized and soluble in the water. Spiraling is therefore not a gradual process but occurs in pulses. The patterns of **drift** of stream invertebrates are consistent with these generalizations. Invertebrates drift downstream when they are dislodged from substrates or disperse. Drift is

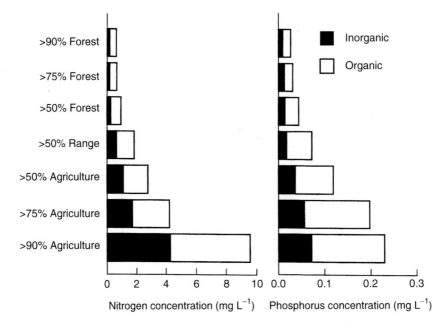

Fig. 9.4 Concentrations of organic and inorganic nitrogen and phosphorus in 928 relatively unpolluted U.S. streams in watersheds with varying degrees of conversion from forest to agriculture. Redrawn from Allan and Castillo (2007)

an important food source for fish but represents only about 0.01% of the invertebrate biomass in a stream at any point in time. In other words, stream invertebrates are so strongly attached to their substrates that carbon and nutrients spiral downstream primarily in the dissolved phase.

Headwater streams less than 10-m wide are particularly important in nutrient processing because they are the immediate recipient of most terrestrial inputs and account for up to 85% of the stream length within most drainage networks (Peterson et al. 2001). Small streams cycle nitrogen efficiently (have shorter uptake lengths) because their shallow depths and high surface–volume ratios enhance nitrogen absorption by algae and bacteria that are attached to rocks and sediments. Large rivers are also important, but for different reasons. Their relatively low velocities, long stream reaches, and high nitrate concentrations allow large quantities of nitrate to be absorbed (Wolheim et al. 2006; Mulholland et al. 2008). Uptake lengths for ammonium range from 10 to 1,000 m and increase exponentially with increases in stream discharge (Peterson et al. 2001).

In unpolluted rivers, a large proportion of the dissolved nitrogen is organic, and nitrate consti-

tutes the bulk of the inorganic nitrogen (Allan and Castillo 2007). Nitrogen fixation by cyanobacteria supplements terrestrial nitrogen inputs in those streams (e.g., desert streams) that are nitrogen-limited and have enough light to support nitrogen fixation (Grimm and Petrone 1997). Land-use change and agricultural intensification increase both the quantity of dissolved nitrogen and phosphorus entering streams and the proportion of it that is inorganic (Fig. 9.4; Seitzinger et al. 2005). Of the nitrogen that enters polluted rivers, 60–75% is denitrified, particularly in the hyporheic zone. In contrast, phosphorus tends to be trapped in sediments, especially in reservoirs, or be transported to the ocean. The N:P ratio of water entering the ocean is typically much lower than that which enters the river (Howarth et al. 1996a).

Nitrogen Inputs to Terrestrial Ecosystems

Biological nitrogen fixation is the main pathway by which new nitrogen enters unpolluted terrestrial ecosystems. Only nitrogen-fixing bacteria

Table 9.1 Organisms and associations involved in di-nitrogen fixation

Type of association[a]	Key characteristics	Representative genera
Heterotrophic N fixers		Bacteria
Associative		
Nodulated (symbiotic)	Legume	*Rhizobium*
	Nonlegume woody plants	*Frankia*
Non-nodulated	Rhizosphere	*Azotobacter, Bacillus*
	Phyllosphere	*Klebsiella*
Free-living	Aerobic	*Azotobacter, Rhizobium*
	Facultative aerobic	*Bacillus*
	Anaerobic	*Clostridium*
Phototrophic N fixers		Cyanobacteria
Associative	Lichens	*Nostoc, Calothrix*
	Liverworts (*Marchantia*)	*Nostoc*
	Mosses	*Holosiphon*
	Gymnosperms (*Cycas*)	*Nostoc*
	Water fern (*Azolla*)	*Nostoc*
Free-living	Cyanobacteria	*Nostoc, Anabaena*
	Purple non-sulfur bacteria	*Rhodospirillium*
	Sulfur bacteria	*Chromarium*

[a]Nitrogen-fixing microbes are heterotrophic bacteria, if they get their organic carbon from the environment. They are phototrophic bluegreen algae, if they produce it themselves through photosynthesis. Some forms of both microbial groups are typically associated with plants, whereas others are free living. Note that the same microbial genus can have both associative and free-living forms
Data from Paul and Clark (1996)

have the capacity to break the triple bonds of N_2 and reduce it to ammonium (NH_4^+), which supports their own growth. Nitrogen fixed by nitrogen-fixing plants becomes available to other plants in the community primarily through the production and decomposition of nitrogen-rich litter.

Biological Nitrogen Fixation

The characteristics of nitrogenase, the enzyme that catalyzes the reduction of N_2 to NH_4^+, dictate much of the biology of nitrogen fixation. The reduction of N_2 catalyzed by nitrogenase has a high energy requirement and therefore occurs only where the bacterium has an abundant carbohydrate supply and adequate phosphorus. The enzyme is denatured in the presence of oxygen, so organisms must protect the enzyme from contact with oxygen. Finally, temperature often constrains the carbon supply and activity of nitrogenase enzymes, so nitrogen fixation is most prominent in tropical environments and constrained at high latitudes (Houlton et al. 2008).

Groups of Nitrogen Fixers

Nitrogen-fixing bacteria in symbiotic association with plants have the highest rates of nitrogen fixation. This occurs because plants can provide the abundant carbohydrates needed to meet the high energy demand of nitrogen fixation. The most common symbiotic nitrogen fixers are *Rhizobium* species associated with legumes (soybeans, peas, etc.) and *Frankia* species (actinomycete bacteria) associated with alder, *Ceanothus*, and other nonlegume woody species (Table 9.1). These plant-associated symbiotic nitrogen-fixing bacteria usually reside in root nodules, where the nitrogenase enzyme is protected from oxygen. Legumes, for example, have leghemoglobin, an oxygen-binding pigment similar to the hemoglobin that transports oxygen in the bloodstream of vertebrate animals. Nitrogen-fixing bacteria in nodules are heterotrophic and depend on carbohydrates from plants to meet the energy requirements of nitrogen fixation. The energetic requirement for nitrogen fixation can be about 25% of GPP under laboratory conditions, two to four times higher than the cost of absorbing inorganic nitrogen from soils (Lambers et al. 2008). The relative costs of nitrogen fixation and nitrogen absorption under field conditions are more difficult to estimate because of the uncertain costs of mycorrhizal association, nitrate reduction, and root exudation. When inorganic nitrogen is naturally abundant or is added to soils, nitrogen-fixing plants generally reduce their capacity for nitrogen fixation and absorb nitrogen from the soil. Phosphorus availability often limits the growth of nitrogen-fixing plants. Moreover, high phosphatase activities in soils associated with nitrogen fixers often supplement supplies of inorganic phosphorus to nitrogen fixers (Houlton et al. 2008).

Free-living heterotrophic nitrogen-fixing bacteria typically have the lowest rates of nitrogen fixation. These bacteria get their organic carbon from the environment and are most active in soils or sediments that have high concentrations of organic matter to provide the carbon substrate that fuels nitrogen reduction (Table 9.1). Other heterotrophic nitrogen fixers occur in the rhizosphere and depend on root exudation and root turnover for their carbon supply. Nitrogen fixers in the anaerobic hindguts of termites provide an important nitrogen source that facilitates the decomposition of wood in the tropics (Yamada et al. 2006). Aerobic heterotrophs have various mechanisms that reduce oxygen concentration in the vicinity of nitrogenase, including high rates of bacterial respiration that depletes oxygen around the bacterial cells or production of slime that reduces oxygen diffusion to the enzyme.

Many free-living nitrogen-fixing **phototrophs** produce their own organic carbon by photosynthesis. These include cyanobacteria (bluegreen bacteria) that occur in aquatic systems and on the surface of many soils. Many phototrophs have specialized non-photosynthetic cells called **heterocysts** that protect nitrogenase from denaturation by the oxygen produced during photosynthesis in adjacent photosynthetic cells.

There are also associative (symbiotic) nitrogen-fixing phototrophs. For example, nitrogen-fixing lichens are composed of green algae or cyanobacteria as the photosynthetic symbiont, cyanobacteria that fix nitrogen, and fungi that provide physical protection. These lichens provide an important nitrogen input in many early successional ecosystems. The small freshwater fern *Azolla* and cyanobacteria such as *Nostoc* form a phototrophic association that is common in rice paddies and tropical aquatic systems.

Legumes and other symbiotic nitrogen fixers have the highest rates of nitrogen fixation, often 5–20 g m^{-2} year^{-1}. Phototrophic symbionts such as *Nostoc* in association with *Azolla* in rice paddies may fix 10 g m^{-2} year^{-1}. When *Nostoc* is a free-living phototroph, it typically fixes about 2.5 g m^{-2} year^{-1}. In contrast, free-living heterotrophs fix only 0.1–0.5 g m^{-2} year^{-1}, a quantity similar to the input from nitrogen deposition in unpolluted environments.

Causes of Variation in Nitrogen Fixation

Biotic and abiotic constraints on nitrogen fixation lead to nitrogen limitation or co-limitation in many ecosystems. The rate of nitrogen fixation varies widely among ecosystems, in part reflecting the types of nitrogen fixers that are present. Even within a single type of nitrogen-fixing system, however, nitrogen fixation rates vary widely. What causes this variation? If nitrogen limits growth in many ecosystems, why does nitrogen fixation not occur almost everywhere? One would expect nitrogen fixers to have a competitive advantage over other plants and microbes that cannot fix their own nitrogen. Why don't nitrogen fixers respond to nitrogen limitation by fixing nitrogen until nitrogen is no longer limiting in the ecosystem? Several factors constrain nitrogen fixation, thereby maintaining nitrogen limitation or co-limitation in many ecosystems (Vitousek and Howarth 1991; Vitousek and Field 1999; Vitousek et al. 2002; Houlton et al. 2008; Hedin et al. 2009).

Energy availability constrains nitrogen fixation rates in closed-canopy ecosystems. The cost of nitrogen fixation (3–6 g carbon g^{-1} N, not including the cost of nodule production) by symbiotic and autotrophic nitrogen fixers is high relative to that of absorbing ammonium or nitrate. Nitrogen fixation is therefore largely restricted to high-light environments where light is less limiting than nitrogen. As canopies close during succession, energy becomes limiting to the establishment of nitrogen-fixing plants. These plants could fix nitrogen if they were in the canopy, but the cost of nitrogen fixation makes it difficult for them to grow through shade to the canopy. Leguminous trees are common in tropical forests and savannas. In savannas, where fires cause large nitrogen losses, leguminous trees are heavily nodulated and fix substantial quantities of nitrogen (Högberg and Alexander 1995). Leguminous trees in tropical forests are less extensively nodulated, but their nitrogen-rich lifestyle is accommodated by the high nitrogen availability of these ecosystems (Vitousek et al. 2002). Here they contribute modestly to annual nitrogen inputs but are important to the long-term nitrogen economy of forests (Pons et al. 2006; Hedin et al. 2009). Nitrogen fixation in aquatic systems is most common in shallow waters or waters with low turbidity where light reaches benthic cyanobacterial mats. When phosphorus availability is adequate, these mats have high fixation rates.

Non-symbiotic heterotrophic nitrogen-fixing bacteria are also limited by the availability of labile organic carbon. When available carbon is scarce, there is no benefit to heterotrophic nitrogen fixation. Decaying wood, which has low nitrogen and high levels of organic carbon, often has substantial rates of heterotrophic nitrogen fixation, including that which occurs in the guts of tropical termites (Yamada et al. 2006). Heterotrophic nitrogen fixation also occurs in anaerobic sediments, but the gaseous loss of nitrogen by **denitrification**, that is, the conversion of nitrate to gaseous forms, usually exceeds the gains from nitrogen fixation.

Nitrogen fixation in many ecosystems is limited by the availability of other nutrients, such as phosphorus. Due to their ready access to nitrogen, the growth of nitrogen-fixing plants is often limited by other nutrients, particularly by phosphorus, which co-limits or secondarily limits plant production in most ecosystems (Elser et al. 2007). Nitrogen fixers often have a nutrient-rich stoichiometry; they use large amounts of phosphorus as well as nitrogen. The growth of nitrogen fixers therefore often becomes phosphorus-limited before that of other plants. Other elements

that can limit nitrogen fixation include molybdenum, iron, and sulfur, which are essential co-factors of nitrogenase (Barron et al. 2009). Molybdenum, for example, often limits nitrogen fixation on highly weathered soils of Australian pastures and lowland tropical forests. Nitrogen fixers may be limited by iron in marine ecosystems, as discussed earlier. Phosphorus, iron, sulfur, or molybdenum may, in these cases, be the ultimate "master element" that limits production, even though nitrogen is the factor to which primary production responds most strongly in short-term experiments.

Consumption of nitrogen-fixing organisms often constrains their capacity to support continuously high nitrogen fixation rates. The high protein content typical of nitrogen fixers enhances their palatability to many herbivores, although nitrogen-based defenses such as alkaloids, which occur in many nitrogen-fixing plants, deter generalist herbivores (see Chap. 10). The resulting intense herbivory on many nitrogen-fixing plants reduces their capacity to compete with other plants, constraining their abundance and nitrogen inputs to the ecosystem (Vitousek and Field 1999; Vitousek et al. 2002). Areas from which grazers are excluded often have more nitrogen-fixing plants and greater nitrogen inputs to the ecosystem and ultimately more productivity and biomass (Ritchie et al. 1998).

Nitrogen Deposition

Nitrogen is deposited in ecosystems in particulate, dissolved, and gaseous forms. All ecosystems receive nitrogen inputs from atmospheric deposition. These inputs are smallest, often 0.1–0.5 g m^{-2} year^{-1}, in ecosystems downwind from pollution-free open-ocean waters (Hedin et al. 1995). Nitrogen inputs to coastal ecosystems derive primarily from organic particulates and nitrate (NO_3^-) in sea-spray evaporites and from ammonia (NH_3) volatilized from seawater. In inland areas, nitrogen derives from the volatilization of NH_3 from soils and vegetation and from dust produced by wind erosion of deserts, unplanted agricultural fields, and other sparsely vegetated ecosystems. Lightning also fixes nitrogen that ultimately contributes to atmospheric deposition.

Human activities are now the major source of nitrogen deposited in many areas of the world (Vitousek et al. 1997a; Gruber and Galloway 2008). The application of urea or ammonia fertilizer leads to volatilization of NH_3, which is then converted to NH_4^+ in the atmosphere and deposited in rainfall. Domestic animal husbandry has also substantially increased emissions of NH_3 to the atmosphere. The emission of nitric oxides (NO and NO_2, together known as NO_x) from fossil fuel combustion, biomass burning, and volatilization from fertilized agricultural systems has dwarfed natural sources at the global scale: 80% of all NO_x flux is anthropogenic (Delmas et al. 1997). Nitrogen derived from these sources can be transported long distances downwind from industrial or agricultural areas before being deposited. "Arctic haze" over the Arctic Ocean and Canadian High Arctic islands, for example, derives primarily from pollutants produced in China and Eastern Europe. Inputs of anthropogenic sources of nitrogen to ecosystems can be quite large, for example 1–2 g m^{-2} year^{-1} in the northeastern U.S. or 5–10 g m^{-2} year^{-1} in northern China, 10–100-fold greater than background levels of nitrogen deposition. The highest rates are similar to the amounts annually absorbed by vegetation and cycled through litterfall (see Chap. 8). Most ecosystems have a substantial capacity to store added nitrogen in soils and vegetation. Once these reservoirs become **nitrogen saturated**, however, nitrogen losses to the atmosphere and groundwater can be substantial. The nitrogen cycle in some polluted ecosystems has changed from being >90% closed (see Table 8.2) to being almost as open as the carbon cycle, in which the amount of nitrogen or carbon annually cycled by vegetation is similar to the amount that is annually gained and lost from the ecosystem. Agricultural systems are often nitrogen-saturated and release substantial quantities of nitrogen to aquifers and aquatic ecosystems; we discuss nutrient cycling in agricultural systems in more detail later. Most forests, in contrast, increase their carbon sequestration in response to nitrogen deposition, indicating that these forests are not

yet nitrogen-saturated (Magnani et al. 2007). The role of nitrogen deposition on carbon sequestration at the global scale, however, appears to be modest, suggesting that anthropogenic nitrogen inputs are unlikely to "solve the climate problem" by enhancing carbon sequestration (Gruber and Galloway 2008).

Climate and ecosystem structure determine the processes by which nitrogen is deposited in ecosystems. Deposition occurs by three processes. (1) Wet deposition delivers nutrients dissolved in precipitation. (2) Dry deposition delivers compounds as dust or aerosols by sedimentation (vertical deposition) or impaction (horizontal deposition or direct absorption of gases such as HNO_3 vapor). (3) Cloud-water deposition delivers nutrients in water droplets onto plant surfaces immersed in fog. Although data are most available for wet deposition because it is most easily measured, wet and dry deposition are often equally important sources of nitrogen inputs (Fig. 9.5). Wet deposition of nitrogen is typically greater in wet than in dry ecosystems. Dry deposition of nitrogen, however, shows no clear correlation with climate, although arid ecosystems receive a larger *proportion* of their nitrogen inputs by dry deposition. Cloud water deposition is greatest on cloud-covered mountaintops and areas of coastal fog. The relative importance of wet, dry, and cloud-water deposition also depends on ecosystem structure. Conifer canopies, for example, tend to collect more dry deposition and cloud-water deposition than do deciduous canopies because of their greater leaf surface area. Their rough canopies also cause moisture-laden air to penetrate more deeply within the forest canopy and therefore to contact more leaf surfaces (see Chap. 4).

The form of nitrogen deposition determines its ecosystem consequences. NO_3^- and NH_4^+ are immediately available for biological absorption by plants and microbes, whereas some organic nitrogen must first be mineralized. Nitrate inputs as nitric acid (and ammonium inputs, if followed by **nitrification**, the conversion of ammonium to nitrate) acidify the soil when nitrate accompanied by base cations leaches from the ecosystem. Organic nitrogen compounds make up about a third of the total nitrogen deposition, but their

Fig. 9.5 Wet, dry, and cloud-water deposition of nitrogen in a variety of ecosystems. These ecosystems are (from high to low elevation): Clingman's Dome NC (CD), Pawnee CO (PW), Whiteface Mountain NY (WF), Coweta NC (CW), Huntington Forest NY (HF), State College PA (SC), Oak Ridge TN (OR), Argonne IL (AR), Thompson WA (TH), and Panola GA (PN). Data from Lovett (1994)

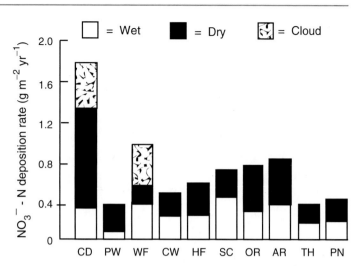

chemical nature varies among ecosystems (Neff et al. 2002). In coastal areas, for example, organic nitrogen is deposited primarily as marine-derived reduced compounds such as amines. In inland areas affected by air pollution, most organic nitrogen enters as oxidized organic nitrogen compounds that result from the reaction of organic compounds with NO_x in the atmosphere.

Weathering of sedimentary rocks may contribute to the nitrogen budgets of some ecosystems. Sedimentary rocks, which make up 75% of the exposed rocks on Earth's surface, sometimes contain substantial nitrogen. In some watersheds underlain by high-nitrogen sedimentary rocks, rock weathering contributes significant nitrogen inputs to ecosystems (Holloway et al. 1998; Thompson et al. 2001). In most ecosystems, however, rock weathering is thought to provide only a small nitrogen input to ecosystems.

Internal Cycling of Nitrogen

Overview of Mineralization

In natural ecosystems, most nitrogen absorbed by plants becomes available through the decomposition of organic matter. In most ecosystems, most (> 99%) soil nitrogen is contained in dead organic matter derived from plants, animals, and microbes. As microbes break down this dead organic matter during decomposition (see Chap. 7), the nitrogen is released as **dissolved**

organic nitrogen (DON) through the action of exoenzymes (Fig. 9.6). Plants and mycorrhizal fungi absorb some DON, using it to support plant growth. Decomposer microbes also absorb DON, using it to support their nitrogen or their carbon requirements for growth. When DON is insufficient to meet the microbial nitrogen requirement, microbes absorb additional inorganic nitrogen, primarily as NH_4^+, from the soil solution (Vitousek and Matson 1988; Fenn et al. 1998). **Immobilization** is the removal of inorganic nitrogen from the available pool by microbial absorption and chemical fixation. Microbial growth is often carbon-limited. Under these circumstances, microbes break down DON, use the carbon skeleton to support their energy requirements for growth and maintenance, and secrete NH_4^+ into the soil. This process is termed **nitrogen mineralization** or **ammonification** because ammonium is the immediate product of this process. In some ecosystems, some or all NH_4^+ is converted to nitrite (NO_2^-) and then to nitrate (NO_3^-), the process of **nitrification**.

Production and Fate of Dissolved Organic Nitrogen

The conversion from insoluble organic nitrogen to dissolved organic nitrogen (DON) makes nitrogen available to plants and microbes (Fig. 9.6). The large pool of particulate organic nitrogen in soils, relative to the sizes of

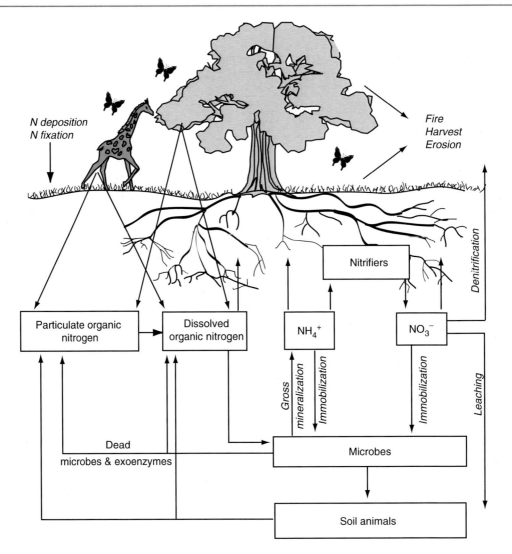

Fig. 9.6 Simplified diagram of the terrestrial nitrogen cycle. Both plants and microbes take up dissolved organic nitrogen (DON), NH_4^+, and NO_3^- and release particulate organic nitrogen (as dead organic matter) and DON. Microbes also release ammonium when they absorb more nitrogen than they require for growth. Nitrifiers are a specialized microbial group that either converts ammonium to nitrite or nitrite to nitrate. Nitrogen is consumed by animals when they eat plants or soil microbes and is returned to the soil as particulate organic nitrogen and DON. Nitrogen is lost from the ecosystem by denitrification, leaching, erosion, harvest, or fire. Nitrogen enters the ecosystem through nitrogen deposition or nitrogen fixation

inorganic pools suggests that this initial step in nitrogen mineralization is the rate-limiting step. All of the organic nitrogen that is eventually mineralized to NH_4^+ or NO_3^- must first be converted to soluble organic forms that can be absorbed by microbes and mineralized (Fig. 9.7). The flux through the DON pool is therefore large, relative to other nitrogen fluxes, even in ecosystems where its concentration is low (Schimel and Bennett 2004). The breakdown of particulate organic nitrogen is carried out in parallel with the breakdown and use of particulate organic carbon and is therefore controlled by the same organisms and factors that control decomposition (see Fig. 7.15). These controls include the quantity and chemical nature of the substrate, the

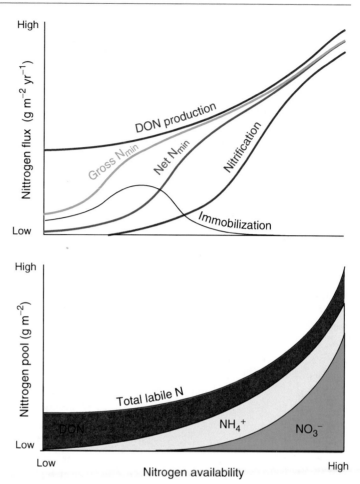

Fig. 9.7 Effect of nitrogen availability on the pools and fluxes of major forms of available nitrogen (dissolved organic nitrogen [DON], ammonium, and nitrate)

environmental factors regulating the activity of soil microbes and animals, and the composition of the microbial community (see Chap. 7; Booth et al. 2005).

Most nitrogen in dead organic matter is contained in complex polymers such as proteins, nucleic acids, and chitin (from fungal cell walls and insect exoskeletons) that are too large to pass through microbial membranes. Microbes must therefore secrete exoenzymes such as proteases, ribonucleases, and chitinases to break down the large polymers into small water-soluble subunits such as amino acids and nucleotides that can be absorbed by microbial cells. Urease is an exoenzyme that breaks down urea from animal urine or fertilizer into CO_2 and NH_3. The microbial enzymes are themselves subject to attack by microbial proteases, so microbes must continu-

ally invest nitrogen in exoenzymes to acquire nitrogen from their environment, a potentially costly process. Exoenzymes often bind to soil minerals and organic matter. This can inactivate the enzyme, if the shape of the active site is altered, or can protect the enzyme against attack from other exoenzymes, lengthening the time that the enzyme remains active in the soil (see Chap. 7). Proteases are produced by mycorrhizal and saprophytic fungi and by bacteria.

Plants, mycorrhizal fungi, or decomposer microbes all absorb DON. This is an important source of both nitrogen and carbon for soil microbes. Plants that absorb DON directly or through their mycorrhizal fungi require no mineralization to acquire this nitrogen. This direct absorption of organic nitrogen by plants occurs in most ecosystems (Read 1991; Kielland 1994; Näsholm et al.

1998; Lipson et al. 1999; Raab et al. 1999) and can meet a significant proportion of the plant nitrogen requirement (see Chap. 8; Lipson et al. 2001), particularly in nitrogen-limited ecosystems. Even crop plants absorb and use DON.

DON is a chemically complex mixture of compounds, only a few percent of which consists of amino acids and other labile forms of nitrogen. Most soils exhibit a similar balance of amino acids (Sowden et al. 1977). The labile DON that is absorbed by microbial cells can be incorporated directly into microbial proteins and nucleic acids. These and other DON compounds can also be metabolized to provide carbon or nitrogen to support microbial growth and maintenance. DON can also be adsorbed onto the soil exchange complex, incorporated into humus, or leached from the ecosystem in groundwater. Amino acids have both positively and negatively charged groups (NH_2^+ and COO^-, respectively). Small neutrally charged amino acids, such as glycine, are most mobile in soils and are therefore most readily absorbed by both plants and microbes (Kielland 1994).

Production and Fate of Ammonium

The net absorption or release of ammonium by microbes depends on their carbon status. When microbial growth is carbon-limited, microbes use the carbon from DON to support growth and respiration and secrete NH_4^+ as a waste product into the soil solution. This process of ammonification is the mechanism by which DON is mineralized to ammonium in soils. Other nitrogen-limited microbes may absorb, or **immobilize**, some of this ammonium and use it for growth. For example, the nitrogen mineralized in nitrogen-rich microsites may diffuse to adjacent nitrogen-limiting microsites, where it is absorbed by plants or other microbes (Schimel and Bennett 2004). Because of this fine-scale heterogeneity in soil nitrogen availability, a given unit of nitrogen can cycle between microbial release and absorption many times before it is absorbed by plants or undergoes some other fate. **Gross mineralization** is the *total* amount of nitrogen released via mineralization (regardless of whether it is subsequently

immobilized or not). Its rate depends primarily on the quantity of microbial food (soil organic matter) and microbial biomass in the soil (Booth et al. 2005). **Net mineralization** is the *net* accumulation of inorganic nitrogen (ammonium plus nitrate) in the soil solution over a given time interval. Net mineralization occurs when microbial growth is limited more strongly by carbon than by nitrogen, whereas net immobilization occurs when microbial growth is nitrogen-limited (Schimel and Bennett 2004). Net mineralization of nitrogen is rapid when either biological processes such as grazing by microbivores or abiotic processes such as freeze–thaw and wet–dry cycles cause a crash of decomposer populations. In either case, surviving microbes have access to large quantities of nutrient-rich tissues.

The form of labile nitrogen that is most available to plants depends primarily on the relative abundance of microsites where microbial growth is nitrogen-limited (immobilization > mineralization) or carbon-limited (nitrogen mineralization > immobilization). In extremely nitrogen-limited soils, such as arctic and alpine tundra and boreal forest, where immobilization predominates, DON produced by exoenzymes of both mycorrhizal and saprophytic microbes is the predominant N form available in the soil and accounts for most nitrogen absorbed by plants. As nitrogen availability increases, so does the proportion of nitrogen-mineralizing microsites, and ammonium diffusing from these microsites becomes available to plants and to microbes (Schimel and Bennett 2004). In addition, nitrifying bacteria, which use ammonium as an energy source, convert increasing proportions of ammonium to nitrate, as ammonium availability increases. In summary, as nitrogen availability increases, microbial growth shifts from nitrogen to carbon limitation; an increasing proportion of the DON absorbed by microbes supports energy demands for growth, with excess nitrogen excreted as ammonium; and nitrifying bacteria use much of the available ammonium as an energy source to support their growth, so nitrification becomes the predominant process.

Net nitrogen mineralization is an excellent measure of the nitrogen supply to plants in ecosystems with high nitrogen availability, where

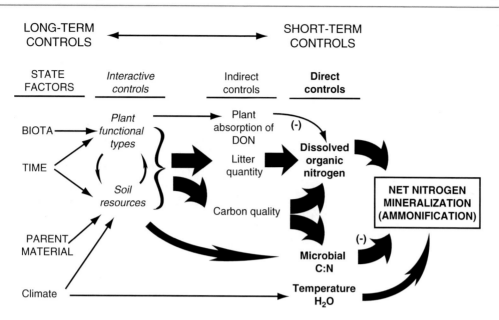

Fig. 9.8 The major factors governing temporal and spatial variation in ammonification (net nitrogen mineralization) in soils. These controls range from the proximate control over nitrogen mineralization (the concentration of dissolved organic nitrogen, physical environment, and microbial C:N ratio) to the state factors and interactive controls that ultimately determine the differences among ecosystems in mineralization rates. The influence of one factor on another is positive unless otherwise indicated (−), and the thickness of the arrows indicates the strength of the direct and indirect effects

microbial growth is primarily carbon-limited, and microbes use DON as a carbon source and excrete the excess nitrogen as ammonium. Under these circumstances, there is little competition for nitrogen between plants and soil microbes. The annual net mineralization in the deciduous forests of eastern North America, for example, approximately equals nitrogen absorption by vegetation (Nadelhoffer et al. 1992). In less fertile ecosystems, such as arctic tundra, plants actively absorb DON, and net nitrogen mineralization rate underestimates the amount of nitrogen that is annually acquired by plants (Nadelhoffer et al. 1992; Schimel and Bennett 2004).

Nitrogen mineralization rate is controlled by the availability of DON and inorganic nitrogen, the activity of soil microbes, and their relative demands for carbon and nitrogen. The quantity and quality of organic matter that enter the soil are the major determinants of the substrate available for both decomposition (see Fig. 7.15) and nitrogen mineralization (Fig. 9.8), so the ecological controls over these inputs govern the rates of both decomposition and nitrogen mineralization (Booth et al. 2005).

Nitrogen mineralization rate responds to two dimensions of substrate quality: (1) carbon quality, which governs the breakdown of dead organic matter to soluble forms (see Chap. 7), and (2) C:N ratio which determines the balance between carbon and nitrogen limitation of microbial growth. The C:N ratio in microbial biomass is about 10:1 (Cleveland and Liptzin 2007). As microbes break down organic matter, they incorporate about 40% of the carbon from their substrates into microbial biomass and return the remaining 60% of the carbon to the atmosphere as CO_2 through respiration. With this 40% growth efficiency, microbes require substrates with a C:N ratio of about 25:1 to meet their nitrogen requirement (Box 9.1). At higher C:N ratios, microbes import nitrogen to meet their growth requirements, and at lower C:N ratios nitrogen exceeds microbial growth requirements and is excreted into the litter and soil. In practice, microbes vary in their C:N ratio (5–10 in bacteria

Box 9.1 Estimation of Critical C:N Ratio for Net Nitrogen Mineralization

The critical C:N ratio that marks the dividing line between net nitrogen mineralization and net nitrogen absorption by microbes can be calculated from the growth efficiency of microbial populations and the C:N ratios of the microbial biomass and their substrate. Assume, for example, that the microbial biomass has a growth efficiency of 40% and a C:N ratio of 10:1. If the microbes break down 100 units of carbon, they will incorporate 40 units of carbon into microbial biomass and respire 60 units of carbon as CO_2. The 40 units of microbial carbon require 4 units of nitrogen to produce a microbial C:N ratio of 10:1 (= 40:4). If the 100 units of original substrate are to supply all of this nitrogen, the initial C:N ratio must have been 25:1 (= 100:4). At higher C:N ratios, microbes must absorb additional inorganic nitrogen from the soil to meet their growth demands. At lower C:N ratios, microbes excrete excess nitrogen into the soil.

and 8–15 in fungi; Paul and Clark 1996), although it is not clear that this translates into any systematic variation in growth efficiency (Thiet et al. 2006). All microbes convert substrates into biomass less efficiently when carbon or nutrient substrates limit their growth, in stressful environments (greater maintenance respiration) or when confronted with more recalcitrant substrates (greater maintenance respiration and more exoenzymes required; Thiet et al. 2006; Manzoni et al. 2008). Nonetheless, 25:1 is often considered the critical C:N ratio above which there is no net nitrogen release from decomposing organic matter. C:N ratio is typically highest in fresh litter, especially woody litter, and declines with time, approaching a C:N ratio of 14 in soil of relatively undisturbed ecosystems and a C:N ratio of 10 in agricultural systems (Stevenson 1994; Cleveland and Liptzin 2007; Fierer et al. 2009a). Thus there is a shift from immobilization (or mineralization, depending on initial C:N ratio) in fresh litter to mineralization as litter is decomposed. Note that, although C:N ratio only indirectly affects decomposition, reflecting its correlation with substrate carbon quality (see Chap. 7), it has a clear mechanistic effect on the net immobilization or mineralization of nitrogen.

There appears to be a universal relationship between litter C:N ratio and nitrogen mineralization or immobilization that depends on substrate quality but is independent of climate (Parton et al. 2007; Manzoni et al. 2008). Climate simply influences the rate at which mineralization or immobilization of nitrogen occurs. Favorable environmental conditions often promote nitrogen immobilization in recent or woody litter with a high C:N ratio but promote mineralization in later stages of decomposition or in long-term studies, where C:N ratio is likely to be lower. Long-term laboratory incubations, for example, show a generally positive effect of temperature on net nitrogen mineralization under favorable moisture conditions. This occurs because temperature stimulates maintenance respiration more strongly than microbial growth, leading to carbon limitation to microbial growth at warm temperatures and excretion of ammonium. In addition, both warm temperatures and microbial production promote predation by soil animals, causing greater microbial turnover and excretion of nitrogen into the soil. Moisture effects are more complex, with nitrogen mineralization generally increasing with soil moisture up to a threshold, above which high moisture restricts oxygen diffusion, microbial activity, and net nitrogen mineralization (Stanford and Epstein 1974). Due to their more favorable soil temperature and moisture and other factors, recently deforested areas typically have higher rates of net nitrogen mineralization than do undisturbed forests (Matson and Vitousek 1981). Across a moisture gradient in the Central Great Plains of the U.S., however, high moisture retarded decomposition and nitrogen mineralization, so the large plant nitrogen pools at the wet end of the gradient reflected greater nitrogen retention by plants and ecosystems rather than a moisture stimulation of nitrogen mineralization (McCulley et al. 2009). Clearly, predictions of environmental effects on nitrogen mineralization require attention to multiple plant and microbial processes, including microbial growth, respiration, and substrate-determined balance between immobilization and mineralization.

The ammonium produced by nitrogen mineralization has several potential fates. In addition to being absorbed by plants or microbes, ammonium readily adsorbs to the negatively charged surfaces of soil minerals and organic matter (see Chap. 3), reducing the concentration of NH_4^+ in the soil solution (often less than 1 ppm). Plant and microbial absorption of NH_4^+ depletes its concentration in the soil solution. This shifts the equilibrium between dissolved and exchangeable pools, causing adsorbed ions to go back into solution from the exchange complex. The cation exchange complex thus serves as a storage reservoir of readily available NH_4^+ and other cations. NH_4^+ can also be fixed in the interlayer portions of certain aluminosilicate clays or complexed with stabilized soil organic matter, which reduces its availability to plants and microbes as long as the organic mineral complex remains intact. Finally, NH_4^+ can be oxidized, mainly by bacteria, to NO_2^- and NO_3^- or converted to ammonia gas (NH_3), and lost to the atmosphere, as described in the next sections.

Production and Fate of Nitrate

Nitrification is the process by which NH_4^+ is oxidized to NO_2^- and subsequently to NO_3^-. Unlike ammonification, which is carried out by a broad suite of decomposers, most nitrification is carried out by a restricted group of **nitrifying bacteria**. There are two general classes of nitrifiers. **Autotrophic nitrifiers** use the energy yield from NH_4^+ oxidation to fix carbon that supports their growth and maintenance, analogous to the use by plants of solar energy to fix carbon via

photosynthesis. **Heterotrophic nitrifiers** gain their energy from breakdown of organic matter.

Autotrophic nitrifiers include two groups, one that converts ammonium to nitrite, for example *Nitrosolobus* and other "Nitroso-" genera, and another that converts nitrite to nitrate, for example *Nitrobacter* and other "Nitro-" genera. These autotrophic nitrifiers are obligate aerobes that synthesize structural and metabolic carbon compounds by reducing CO_2 using energy from NH_4^+ or NO_2^- oxidation to drive CO_2 fixation. In most systems, these two groups occur together, so NO_2^- typically does not accumulate in soils. NO_2^- is most likely to accumulate in dry forest and savanna ecosystems during the dry season, when the activity of *Nitrobacter* is restricted, and in some fertilized ecosystems, where nitrogen inputs are high relative to plant and microbial demands.

Although autotrophic nitrification predominates in many ecosystems, heterotrophic nitrification can be important in ecosystems with low nitrogen availability or acidic soils. Many heterotrophic fungi and bacteria, including actinomycetes, produce NO_2^- or NO_3^- from NH_4^+. Some also use organic nitrogen in the process. Because heterotrophs obtain their energy from organic materials, it is not clear what advantage they gain from the oxidation of NH_4^+ to NO_3^-.

Nitrification has multiple effects on ecosystem processes. The oxidation of NH_4^+ to NO_2^- in the first step of nitrification produces two moles of H^+ for each mole of NH_4^+ consumed and therefore tends to acidify soils. The monooxygenase that catalyzes this step has a broad substrate specificity and also oxidizes many chlorinated hydrocarbons, suggesting a role of nitrifiers in the breakdown of pesticide residues. Finally, nitric oxide (NO) and nitrous oxide (N_2O), which are produced during nitrification (Fig. 9.9), are gases that have important effects on atmospheric chemistry.

The availability of NH_4^+ is the most important direct determinant of nitrification rate (Fig. 9.10; Robertson 1989; Booth et al. 2005). The NH_4^+ concentration must be high enough, at least in some soil microsites, to allow nitrifiers to compete with other soil microbes. This is particularly important for autotrophic nitrifiers, which

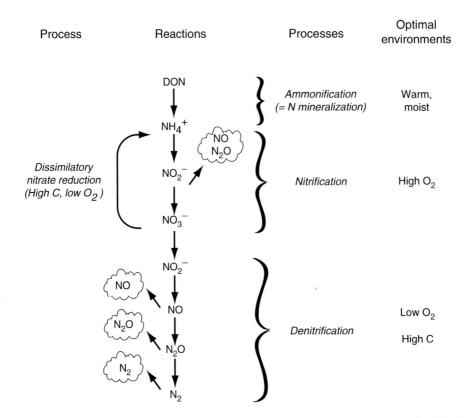

Fig. 9.9 Pathways of autotrophic nitrification and of denitrification and the nitrogen trace gases emitted by these pathways (Firestone and Davidson 1989)

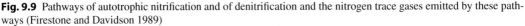
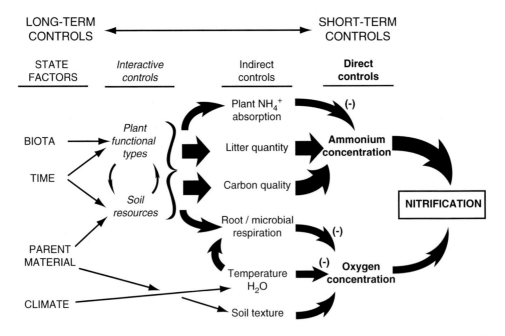

Fig. 9.10 The major factors governing temporal and spatial variation in nitrification in soils (Robertson 1989). These controls range from concentrations of reactants that directly control nitrification to the interactive controls such as climate and disturbance regime that are the ultimate determinants of nitrification rate. The influence of one factor on another is positive unless otherwise indicated (−), and the thickness of the arrows indicates the strength of the direct and indirect effects

rely on NH_4^+ as their sole energy source. NH_4^+ supply, in turn, is regulated by the effects of substrate quality and environment on ammonification rate, as described earlier (Fig. 9.8). Fertilizer inputs and ammonium deposition are additional sources of ammonium to many ecosystems. Conversely, plant roots lower NH_4^+ concentration in the soil solution, thereby competing with nitrifiers for NH_4^+. Productive ecosystems generally have high nitrification rates because high mineralization rates provide abundant ammonium as a substrate for nitrification (Booth et al. 2005). The resulting nitrate is, however, relatively mobile (see Chap. 3) and is often rapidly absorbed by plants or denitrified, so soil nitrate *concentrations* are not necessarily a good indicator of nitrification rate.

Nitrifier populations are often too small in infertile soils to support significant nitrification. When ammonium substrate becomes available (e.g., through additions of nitrogen, or increases in mineralization rates), nitrifier populations and nitrification rates can increase. The response can be rapid in some soils but show a long delay in others (Vitousek et al. 1982). Secondary metabolites, such as tannins, have been hypothesized to inhibit nitrification in some ecosystems, including those in late succession (Rice 1979), but the decline in nitrification in late succession is generally best explained by a decline in ammonium supply rather than through phenolic toxicity to nitrifiers (Pastor et al. 1984; Schimel et al. 1996). Limitation of nitrifier activity by other resources is another possible cause of slow or delayed nitrification. In most cases, however, the availability of ammonium ultimately governs nitrification rate through its effects on both the population density and activity of nitrifying bacteria.

Oxygen is an important additional factor controlling nitrification because most nitrifiers require oxygen as an electron acceptor for the oxidation of NH_4^+. Oxygen availability, in turn, is influenced by many factors, including soil moisture, soil texture, soil structure, and respiration by microbes and roots (Fig. 9.10; see Chap. 3).

Nitrifier activity is sensitive to temperature. It does, however, continue at low rates at low temperatures, so over a long winter season, substantial nitrification can occur, particularly in

nitrogen-rich agricultural soils. Nitrification rates are slow in dry soils primarily because thin water films restrict NH_4^+ diffusion to nitrifiers (Stark and Firestone 1995). Under extremely dry conditions, low water potential further restricts the activity of nitrifiers. The importance of acidity in regulating nitrification rates is uncertain. In laboratory cultures of agricultural soils, maximum nitrification rates occur between pH 6.6 and 8.0 and are negligible below pH 4.5 (Paul and Clark 1996). Many natural ecosystems with acidic soils, however, have substantial nitrification rates, even at pH 4 (Stark and Hart 1997; Booth et al. 2005).

The fraction of mineralized nitrogen that is oxidized to nitrate varies widely among ecosystems. In many unpolluted temperate coniferous and deciduous systems, nitrification is only a small proportion of net mineralization (e.g., 0–4%) because plants and decomposer organisms compete with nitrifiers for ammonium. Nitrogen deposition can increase the fraction of mineralized nitrogen that is nitrified to 23% (McNulty et al. 1990). In tropical forests, in contrast, net nitrification is typically nearly 100% of net mineralization, even in sites with low rates of net mineralization and without inputs of additional nitrogen (Fig. 9.11; Vitousek and Matson 1988). In tropical ecosystems, plant and microbial growth are often limited by nutrients other than nitrogen, and their demand for nitrogen is low, so nitrifiers have ready access to NH_4^+.

The potential fates of nitrate are absorption by plants and microbes, exchange on anion exchange sites, or loss from ecosystems via denitrification or leaching. Nitrate is relatively mobile in soil solutions because it is negatively charged and does not bind to cation exchange sites. It therefore moves readily to plant roots by mass flow or diffusion (see Chap. 8) or can be leached from the soil. Some microbes also absorb nitrate and reduce it to ammonium through **dissimilatory nitrate reduction**, that is, nitrate reduction that does not involve assimilation (immobilization) by microbes (Fig. 9.9). This process is energetically expensive and occurs primarily when microbes are exposed to abundant nitrate and labile carbon under anaerobic conditions, as in tropical wet forests (Silver et al. 2001). Since this combination of conditions also facilitates

Fig. 9.11 The relationship between net nitrogen mineralization and net nitrification (g nitrogen g^{-1} of dry soil for a 10-day incubation) across a range of tropical forest ecosystems (Vitousek and Matson 1984). Nearly all nitrogen that is mineralized in these systems is immediately nitrified. In contrast, nitrification is often less than 25% of net mineralization in temperate ecosystems

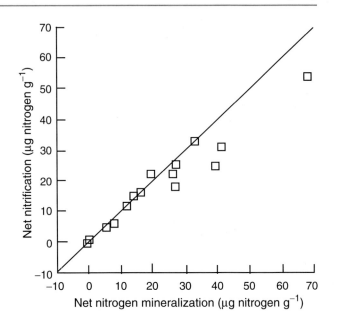

Temporal and Spatial Variability

Fine-scale ecological controls cause large temporal and spatial variability in nitrogen cycling. Nitrogen transformation rates in soils are notoriously variable, with rates often differing by an order of magnitude between adjacent soil samples or sampling dates (Robertson et al. 1997; Schimel and Bennett 2004; Fierer et al. 2009a). This variability reflects the fine temporal and spatial scales over which controlling factors vary. Anaerobic conditions that support denitrification, as described later, dissimilatory nitrate reduction can be an important mechanism of nitrogen retention in wet environments. The low nitrate concentrations observed in many acidic conifer forest soils reflect a combination of low nitrification rates and nitrate absorption by soil microbes and plants (Stark and Hart 1997).

Although NO_3^- is more mobile than most cations, it can be held on exchange sites of soils with a high anion exchange capacity (see Chap. 3). Soils with enough anion exchange capacity can prevent leaching losses of nitrate after disturbance (Matson et al. 1987). In most soils, the strength of the anion adsorption is $PO_4^{3-} > SO_4^{3-} > Cl^- > NO_3^-$, so NO_3^- is desorbed and leached relatively easily.

(see below) in the interiors of soil aggregates, for example, can occur within a millimeter of aerobic soil pores. Fine roots create rhizospheres with high carbon and low soluble nitrogen concentrations adjacent to bulk soil, where carbon-limited soil microbes mineralize organic nitrogen to meet their energy demands. In densely rooted microsites, plants deplete concentrations of NH_4^+ below levels that can sustain nitrification, whereas nitrification can be substantial in adjacent root-free microsites. The impacts of this fine-scale spatial heterogeneity on nitrogen cycling are difficult to study, so we know only qualitatively of their importance (Schimel and Bennett 2004).

Temporal variability in environment and extreme events have a strong influence on nitrogen mineralization. Drying–wetting events and freeze–thaw events, for example, burst many microbial cells and release pulses of nutrients. For this reason, the first rains after a long dry season often causes a pulse of nitrification and nitrate leaching (Davidson et al. 1993). The spring runoff after snowmelt in northern or mountain ecosystems also often carries with it a pulse of nutrient loss to streams because of both freeze–thaw events and the absence of plant absorption of nitrogen during winter. Ninety percent of the annual nitrogen input to Toolik Lake in arctic Alaska, for example, occurs in the first 10 days of snowmelt (Whalen and Cornwell 1985).

The seasonality of nitrogen mineralization often differs from the seasonality of plant nitrogen absorption. In those ecosystems where plants are dormant for part of the year, soil microbes continue to mineralize nitrogen during the dormant season. This temporal asynchrony between microbial activity and plant absorption leads to an accumulation of available nitrogen during the season of plant dormancy that plants use when they become active. In temperate forests, for example, mineralization during winter (even beneath a snowpack) creates a substantial pool of available nitrogen that is not absorbed by plants until the following spring. This asynchrony is particularly important in low-nutrient environments, where microbes may immobilize nitrogen during the season of most active plant growth, effectively competing with plants for nitrogen (Jaeger et al. 1999). In soils that freeze or dry, the death of microbial cells provides additional labile substrates that support net mineralization by the remaining microbes when conditions again become suitable for microbial activity.

Pathways of Nitrogen Loss

Gaseous Losses of Nitrogen

Ammonia volatilization, nitrification, and denitrification are the major avenues of gaseous nitrogen loss from ecosystems. These processes release nitrogen as ammonia gas, nitrous oxide, nitric oxide, and di-nitrogen. Gas fluxes are controlled by the rates of soil processes and by soil and environmental characteristics that regulate diffusion rates through soils. Once in the atmosphere, these gases can be chemically modified and deposited downwind.

Ecological Controls
Ammonia gas (NH_3) can be emitted from soils and senescing leaves. In soils, it is emitted as a consequence of the pH-dependent equilibrium between NH_4^+ and NH_3. At pH values greater than 7, a significant fraction of NH_4^+ is converted to NH_3 gas.

$$NH_4^+ + OH^- \leftrightarrow NH_3 + H_2O \qquad (9.1)$$

Ammonia then diffuses from the soil to the atmosphere. This diffusion is most rapid in coarse dry soils with large air spaces. In dense canopies, some of the NH_3 emitted from soils is absorbed by plant leaves and incorporated into amino acids.

NH_3 flux is low from most ecosystems because NH_4^+ is maintained at low concentrations by plant and microbial absorption and by binding to the soil exchange complex. NH_3 fluxes are substantial, however, in ecosystems where NH_4^+ accumulates due to large nitrogen inputs. In grazed ecosystems, for example, urine patches dominate the aerial flux of NH_3. Agricultural fields that are fertilized with ammonium-based fertilizers or urea often lose 20–30% of the added nitrogen as NH_3, especially if fertilizers are placed on the surface. Nitrogen-rich basic soils are particularly prone to NH_3 volatilization because of the pH effect on the equilibrium between NH_4^+ and NH_3. Leaves also emit NH_3 during senescence, when nitrogen-containing compounds are broken down for transport to storage organs. Fertilization and domestic animal husbandry have substantially increased the flux of NH_3 to the atmosphere (see Chap. 14).

The production of NO and N_2O during nitrification depends primarily on the rate of nitrification. The conversion of NH_4^+ to NO_3^- by nitrification produces some NO and N_2O as by-products (Fig. 9.9), typically at a NO to N_2O ratio of 10–20. The quantities of NO and N_2O released during nitrification are correlated with the total flux through the nitrification pathway, suggesting that nitrification acts like a leaky pipe (Firestone and Davidson 1989), in which a small proportion (perhaps 0.1–10%) of the nitrogen "leaks out" as trace gases during nitrification.

The reduction of nitrate or nitrite to gaseous nitrogen by denitrification occurs under conditions of high nitrate and low oxygen. Many types of bacteria contribute to biological denitrification. They use NO_3^- or NO_2^- as an electron acceptor to oxidize organic carbon for energy when oxygen concentration is low. Most denitrifiers are facultative anaerobes and use oxygen rather than NO_3^-, when oxygen is available. In addition to biological denitrification, **chemodenitrification** converts NO_2^- (nitrite) abiotically to nitric oxide gas (NO) where NO_2^- accumulates

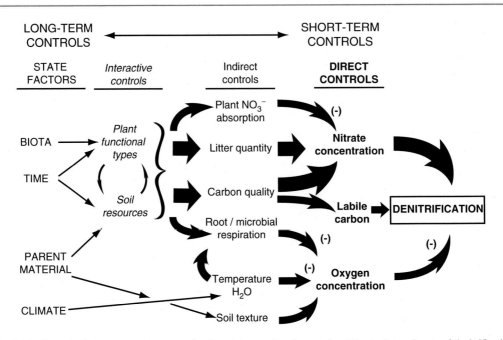

Fig. 9.12 The major factors governing temporal and spatial variation denitrification in soils. These controls range from concentrations of substrates that directly control nitrification to the interactive controls such as climate and disturbance

regime that are the ultimate determinants of denitrification rate. The influence of one factor on another is positive unless otherwise indicated (−), and the thickness of the arrows indicates the strength of the direct and indirect effects

in the soil at low pH. Chemodenitrification is typically much less important than biological denitrification.

The sequence of NO_3^- reduction is: $NO_3^- \rightarrow NO_2^- \rightarrow NO \rightarrow N_2O \rightarrow N_2$, with the last three products, particularly N_2O and N_2, being released as gases to the atmosphere (Fig. 9.9). Most denitrifiers have the enzymatic potential to carry out the entire reductive sequence, but produce variable proportions of N_2O and N_2, depending in part on the relative availability of oxidant (NO_3^-) versus reductant (organic carbon). When NO_3^- is relatively more abundant than labile organic carbon, the reaction goes only partially to completion, and relatively more N_2O than N_2 is produced. Other factors that favor N_2O over N_2 production include low pH, low temperature, and high oxygen. Although NO is often released during denitrification in laboratory incubations, there is seldom a net release in nature because its diffusion to the air is impeded by water-filled pore spaces. Some of the NO that is produced serves as a substrate for further reduction to N_2O or N_2 by denitrifying bacteria.

The three conditions required for significant denitrification are low oxygen, high nitrate concentration, and a supply of organic carbon (Fig. 9.12; Del Grosso et al. 2000). In most non-flooded soils, oxygen and nitrate availabilities exert the strongest control over denitrification. Oxygen supply is reduced by high soil water content, which impedes the diffusion of oxygen through soil pores. Soil moisture, in turn, is controlled by other environmental factors such as slope position, soil texture, and the balance between precipitation and evapotranspiration. Soil oxygen concentration is also sensitive to its rate of consumption by soil microbes and roots. It is consumed most quickly in warm, moist environments.

The second major control over denitrification is an adequate supply of the substrate NO_3^-. Because nitrification is primarily an *aerobic* process, the low-oxygen conditions that favor denitrification often limit NO_3^- supply. Some wetlands, for example, have low denitrification rates despite their saturated soils and large quantities of organic matter due to low availability of nitrate. Wetlands support high denitrification rates only if (1) they receive NO_3^- from outside the system (lateral transfer), (2) they have an aerobic zone above an anaerobic zone (vertical transfer), as in partially drained wetlands, or (3) go through cycles of flooding and drainage (temporal separation) as in many rice paddies. At a finer scale, denitrification can occur within soil aggregates or other anaerobic microsites (e.g., pieces of soil organic matter) in moderately well-drained soils due to fine-scale heterogeneity in soil oxygen concentration and nitrification rate.

Finally, the availability of organic carbon substrates can limit denitrification because the process is carried out primarily by heterotrophic bacteria. Long-term cultivation of agricultural soils, for example, can reduce soil organic matter concentrations enough to limit denitrification. Denitrification, as estimated from major components of the global nitrogen budget (Box 9.2), is quantitatively important, accounting for about a third of the nitrogen loss from the unmanaged terrestrial biosphere (Houlton and Bai 2009).

Box 9.2 Nitrogen Isotopes
Joseph M. Craine

The two isotopic forms of nitrogen (^{14}N and ^{15}N) differ in their number of neutrons but have the same number of protons. As with carbon isotopes (see Box 5.1), the ∂ notation represents the ratio of ^{15}N to ^{14}N relative to an atmospheric standard. Like carbon, the additional atomic mass causes the heavier isotope to react more slowly in some reactions. For the nitrogen cycle, three steps strongly discriminate against molecules that have the heavier isotope (Fig. 9.13). The first is nitrification, which leaves NH_4^+ enriched and NO_3^- depleted in the heavier isotope whenever only a portion of the NH_4^+ pool is nitrified. Second, gaseous nitrogen loss discriminates strongly, whether it is NH_3 volatilization, losses during nitrification, or denitrification, just as the evaporation of water discriminates against the heavier isotopes of hydrogen and oxygen (see Box 4.2). Lastly, the transfer of nitrogen from mycorrhizal fungi to plants leaves the fungi relatively enriched in nitrogen and the plants depleted in ^{15}N.

The changes in the isotopic composition of the different forms of nitrogen in different ecosystems have little functional significance, but the isotopic differences among plants provide key insights into the functioning of plants and the workings of the nitrogen cycle. The differences in ∂^{15}N among plants in the same ecosystem can be used to infer the relative dependence on NH_4^+ vs. NO_3^-. All else being equal, plants that absorb more NH_4^+ are enriched relative to plants that absorb more NO_3^- because of fractionation during nitrification. At the stand level, the relative dependence of different plants on different forms of nitrogen cancel each other out, and stand-level ^{15}N signatures can be used as an index of nitrogen availability. When nitrogen availability is low, nitrogen tends to cycle as organic nitrogen, and plants rely more on mycorrhizal fungi and are relatively depleted in ^{15}N. As nitrogen availability increases, mineralization and inorganic nitrogen pools increase, leading to greater gaseous nitrogen loss and leaving behind enriched forms of nitrogen for plants. Under these conditions, plants also rely less on mycorrhizal fungi for nitrogen. Together, the enrichment of nitrogen pools and the decreasing reliance on mycorrhizal fungi leads to increases in plant ∂^{15}N with increasing nitrogen availability, when ecosystems are compared.

At global scales, non-mycorrhizal plants with high nitrogen concentrations that occupy hot–dry ecosystems have the highest ∂^{15}N, while ectomycorrhizal plants from cold–wet ecosystems have the lowest ∂^{15}N. These patterns are also reflected in soil ∂^{15}N as plant organic matter is returned to the soil and incorporated into soil organic matter. Besides understanding modern patterns of nitrogen availability, the signature of ∂^{15}N remains in plant wood over time and therefore can be used to reconstruct past changes in ecosystem N availability.

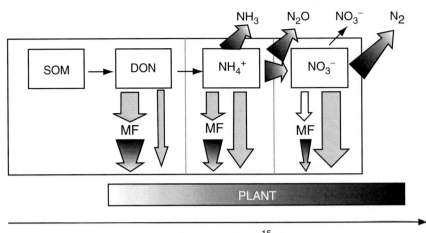

N availability, Plant δ^{15}N

Fig. 9.13 Effect of isotopic fractionation on the ∂^{15}N of ecosystems. As nitrogen availability increases, plants shift from tapping predominantly dissolved organic nitrogen (DON) to ammonium to nitrate, while relying less on mycorrhizal fungi (MF). The width of each arrow indicates the relative contribution of mycorrhizal transfers and direct absorption by roots to total plant absorption. Steps that discriminate against ^{15}N are shown with shaded arrows, leading to a product that is less enriched in ^{15}N (lighter in color) and a substrate that is more enriched in ^{15}N (darker in color). Gaseous nitrogen loss leads to a progressive enrichment of soil available nitrogen from DON to NH_4^+ to NO_3^-. Plants tap progressively more enriched pools as nitrogen availability increases, causing plant ∂^{15}N to be a useful indicator of nitrogen availability when comparing ecosystems

Fires also account for large gaseous losses of nitrogen. The amount and forms of nitrogen volatilized during fire depend on the temperature of the fire. Fires with active flames produce considerable turbulence, are well supplied with oxygen, and release nitrogen primarily as NO_x. Smoldering fires release nitrogen in more reduced forms, such as ammonia (Goode et al. 2000). About a third of the nitrogen is emitted as N_2. Severe stand-replacing fires can cause loss of most of the ecosystem nitrogen, which is gradually replaced during post-fire succession (see Chap. 12). In cooler ground fires, less organic matter is combusted, and less nitrogen is lost. Fire suppression in some areas and biomass burning in others have altered the natural patterns of nitrogen cycling in many ecosystems.

Atmospheric Roles of Nitrogen Gases

The four nitrogen gases have different roles and consequences for the atmosphere. NH_3 that enters the atmosphere reacts with acids and thus neutralizes atmospheric acidity.

$$NH_3 + H_2SO_4 \leftrightarrow (NH_4)_2SO_4 \qquad (9.2)$$

With this reaction, NH_3 is converted back to NH_4^+, which can be deposited downwind on the surface of dry particles or as NH_4^+ dissolved in precipitation. Ammonia volatilization and deposition transfer nitrogen from one ecosystem to another. Ammonia gas itself also can be taken up through the stomates of plant leaves. Indeed, plants typically have an ammonia compensation point, analogous to their CO_2 compensation point for photosynthesis (see Chap. 5).

In the atmosphere, the nitrogen oxides (NO and NO_2, together known as NO_x) are in equilibrium with one another due to their rapid interconversion. NO_x is very reactive, and its concentration regulates several important atmospheric chemical reactions. High NO_x concentrations, for example, direct the oxidation of carbon monoxide, methane, and non-methane hydrocarbons into reactions that produce tropospheric ozone (O_3), an important component of photochemical smog in urban, industrial, and agricultural areas.

$$CO + 2O_2 \leftrightarrow CO_2 + O_3 \qquad (9.3)$$

At low NO_x concentrations, the oxidation of CO consumes O_3.

$$CO + O_3 \leftrightarrow CO_2 + O_2 \qquad (9.4)$$

In addition to its role as a catalyst that alters atmospheric chemistry and generates pollution, NO_x can be transported long distances and alter the functioning of ecosystems downwind. In the form of nitric acid, it is a principal component of acid deposition and adds both available nitrogen and acidity to the soil. In its gaseous NO_2 or HNO_3 forms, it can be absorbed through the stomata of leaves and be used in metabolism (see Chap. 5). It can also be deposited in particulate form, another type of inadvertent fertilization.

In contrast to the highly reactive NO_x, nitrous oxide (N_2O) has an atmospheric lifetime of 150 years and is not chemically reactive in troposphere. The low reactivity of N_2O contributes to a different environmental problem. N_2O is a greenhouse gas that is more than 200 times more efficient per molecule than is CO_2 in absorbing infrared radiation (see Chap. 2). In addition, N_2O in the stratosphere reacts with excited oxygen in presence of ultraviolet radiation to produce NO, which catalyzes the destruction of stratospheric ozone (O_3).

Given that the atmosphere is already 78% N_2, N_2 emissions to the atmosphere via denitrification have no significant atmospheric effects, although these losses may influence ecosystem nitrogen pools. Atmospheric N_2 has a turnover time of thousands of years.

Solution Losses

Nitrogen is lost by leaching as dissolved organic nitrogen from all ecosystems and as nitrate from nitrate-rich ecosystems. Undisturbed and unpolluted ecosystems lose relatively little nitrogen, primarily in the form of dissolved organic nitrogen (Hedin et al. 1995; Perakis and Hedin 2002). Although nitrate is also highly mobile in soils, plants and microbes absorb much

of the nitrate before it leaches below the rooting zone of many ecosystems. Disturbance, however, often augments nitrate leaching from ecosystems by creating environmental conditions that stimulate nitrogen mineralization and by reducing the biomass of vegetation available to absorb nutrients (see Chap. 8). At the Hubbard Brook Forest in the northeastern U.S., for example, experimental removal of all vegetation caused large losses of nitrate, calcium, and potassium to the groundwater and streams (Fig. 9.14; Bormann and Likens 1979). Once vegetation began to regrow, however, the accumulating plant biomass absorbed most of the mineralized nutrients, and stream nutrient concentrations returned to their pre-harvest levels. Nitrate leaching also occurs when additions of fertilizer nitrogen or nitrogen deposition exceed plant and microbial nitrogen demands. Nitrate leaching can therefore be an indicator of **nitrogen saturation**, the changes that occur in ecosystem functioning when anthropogenic nitrogen additions relieve nitrogen limitation to plants and microbes (Aber et al. 1998; Driscoll et al. 2001). In general, the proportional increase in nitrogen losses via leaching and denitrification are larger than the increases in nitrogen pools retained within the ecosystem (Lu et al. 2010). In other words, nitrogen addition makes ecosystems more leaky.

Nitrate loss to groundwater can have important consequences for human health and for the ecological integrity of aquatic ecosystems. Under reducing conditions, nitrate is converted to nitrite, which can reduce the capacity of hemoglobin in animals to transport oxygen, producing anemia, especially in infants. Groundwater in areas of intensive agriculture often has nitrate concentrations that exceed public health standards.

Nitrogen leached from terrestrial ecosystems moves in groundwater to lakes and rivers, and is subsequently lost to the atmosphere through denitrification or transported to the ocean, as discussed earlier.

Solutions that move through the soil must maintain a balanced charge, with negatively charged ions like nitrate balanced by cations or protons. Therefore, every nitrate ion that leaches from soil carries with it a cation such as calcium,

Fig. 9.14 Losses of calcium, potassium, nitrate, and particulate organic matter in stream water before and after deforestation of an experimental watershed at Hubbard Brook in the northeastern U.S. The shaded area shows the time interval during which vegetation was absent due to cutting of trees and herbicide application. Redrawn from Bormann and Likens (1979)

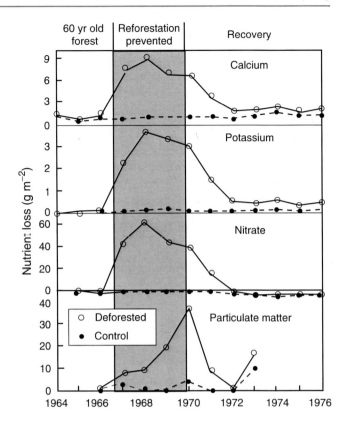

potassium, and ammonium to maintain charge balance. When cation loss by leaching exceeds the rate of cation supply by weathering plus deposition, the net loss of cations can lead to cation deficiency (Driscoll et al. 2001). After these nutrient cations are depleted, nitrate takes with it H^+ or Al^{3+}, which are deleterious to downstream ecosystems. Nitrification also generates acidity:

$$2NH_4^+ + 3O_2 \leftrightarrow 2NO_2^- + H_2O + H^+ \quad (9.5)$$

The hydrogen ion released in this reaction exchanges with other ions on cation exchange sites in the soil, making these cations more vulnerable to leaching loss.

Erosional Losses

Erosion is a natural pathway of nitrogen loss that often increases dramatically after land-use changes. As with leaching, erosional losses of nitrogen include both organic and inorganic forms, although organic forms associated with

soil aggregates and particles are the most important erosional fluxes.

Other Element Cycles

Differences among elements in source (rocks or atmosphere), chemical properties, and plant demand lead to predictable patterns and rates of element cycling. Because most plants have similar stoichiometric ratios of elements (see Chap. 8), there are broad similarities in the patterns of cycling of all essential elements that cycle through ecosystems (Sterner and Elser 2002). This stoichiometry creates a functional linkage, as these elements cycle through vegetation, just as observed in aquatic ecosystems. Productive ecosystems, for example, cycle larger quantities of all essential nutrients through vegetation than do less productive ecosystems. Despite these broad similarities among element cycles, there are important differences in cycling patterns among both elements and ecosystems that depend on the source (rocks or atmosphere), chemical properties, and quantities of different elements required by vegetation. The abiotic processes that provide elements to ecosystems (especially weathering and atmospheric deposition) generally have very different element ratios than those that govern cycling through organisms, and this interplay of biological and geological stoichiometries adds richness and complexity to our analysis of element cycling in terrestrial ecosystems.

Ecosystems differ substantially in the availability of various rock-derived nutrients, depending on parent material and the erosional and weathering history of the site. Limestone, for example, which is derived from marine sediments, often contains substantial phosphorus and is less likely to give rise to phosphorus-limited ecosystems than rocks containing less phosphorus. In contrast, the availability of atmospherically derived nutrients like nitrogen depends strongly on the biological interactions among organisms. The tightness of element cycling within ecosystems also depends on both their solubility in water and the quantities required by vegetation. Chloride, for example, which is highly soluble and required in small quantities by vegetation, has a much more open cycle than do sparingly soluble essential macronutrients like phosphorus.

Beyond these broad generalities, however, the specific properties of elements and their use by organisms generate important differences among elemental cycles. We briefly sketch the major features of the cycling of macronutrients that most often limit the productivity of ecosystems (nitrogen, phosphorus, and potassium) and give examples of macronutrients that less frequently limit productivity (calcium and sulfur), micronutrients that are required in very small quantities (chloride), and elements that are not required and are potentially toxic to organisms (lead).

Phosphorus

Phosphorus is the nutrient whose cycling through vegetation is most tightly coupled to nitrogen. These two nutrients are usually least available in the soil solution relative to annual plant requirement (see Table 8.3) and therefore

most often limit or co-limit plant productivity (Elser et al. 2007). Nitrogen and phosphorus are essential components of the energetic engines of plant production (photosynthesis and respiration). It is therefore not surprising that there are many similarities in their patterns of cycling through vegetation. Mycorrhizal fungi play an important role in the absorption of both nutrients by breaking down nitrogen- and phosphorus-containing particulate organic compounds and transporting the nutrients to plant roots more rapidly than would occur by diffusion. Ectomycorrhizae typical of temperate and high-latitude forests are particularly important in nitrogen acquisition, and arbuscular mycorrhizae typical of grasslands and tropical forests are particularly important in phosphorus acquisition. Plants allocate both nutrients preferentially to metabolically active, resource-acquiring tissues (leaves and fine roots), creating an amplifying (positive) feedback that enhances the capacity of plants to capture additional resources. About half of leaf nitrogen and phosphorus are resorbed from leaves during senescence.

Although these common features link the nitrogen and phosphorus cycles, some processes strengthen this coupling, and others tend to disrupt it (Chapin and Eviner 2004). Within organisms, this coupling is strengthened by ion-specific nutrient absorption adjustments that up-regulate nitrate and ammonium absorption in nitrogen-limited plants and up-regulate phosphate absorption in phosphorus-limited plants (see Table 8.5). Thus plants and the detritus that they produce tend to cycle nitrogen and phosphorus in a ratio that is favorable for plant growth (N:P molar ratio of about 28; Sterner and Elser 2002; McGroddy et al. 2004), although this ratio is quite variable within and among ecosystems (Sterner and Elser 2002; Townsend et al. 2007). At the ecosystem scale over years to decades, nitrogen fixation tends to add nitrogen to nitrogen-limited ecosystems, and denitrification and nitrate leaching tend to remove nitrogen in anaerobic microsites of ecosystems where available nitrogen accumulates in excess of plant and microbial requirements. These fluxes are quantitatively large and strongly influence the nitrogen concentration and its isotopic composition at global scales (Houlton and Bai 2009). These processes strengthen the coupling between nitrogen and phosphorus cycles and generate N:P ratios that are favorable for plants and microbes.

There is also a relatively consistent ratio of nitrogen to organic phosphorus in soils (13.1 ± 0.8) and microbial biomass (6.9 ± 0.4, geometric mean ± SE) across terrestrial ecosystems (Fig. 9.15; Cleveland and Liptzin 2007). As in the ocean and fresh waters (Sterner and Elser 2002), variation in phosphorus concentration accounts for much of the variation in N:P ratios among ecosystems. Microbial N:P ratio, for example, is higher in forests than in grasslands, due to lower microbial P concentrations in forests (Cleveland and Liptzin 2007).

The higher N:P ratio of plants (28:1) than of microbes (7:1) may reflect differences in their biology. Microbes have a higher growth potential than plants, given the need to respond rapidly in a highly variable soil environment. This should require high phosphorus concentrations (low N:P ratio) to support rapid protein synthesis (Sterner and Elser 2002). Plants, in contrast, have a high nitrogen requirement (high N:P ratio) for photosynthesis (half of the nitrogen in leaves). The differences in observed N:P ratios (McGroddy et al. 2004; Cleveland and Liptzin 2007) therefore make sense. We expect plants to be relatively nitrogen-limited and microbes to be relatively phosphorus-limited in the same environment. Each group should adjust nutrient acquisition and release to meet their requirements and should return dead organic matter with an N:P ratio characteristic of their biomass. Through these processes, we expect soil to have an N:P ratio

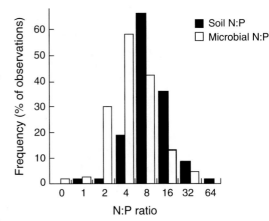

Fig. 9.15 Frequency distribution of N:P ratios in soils and microbial biomass on a $\log_2$ scale. Redrawn from Cleveland and Liptzin (2007)

intermediate between that of plants and microbes, as is observed (Cleveland and Liptzin 2007).

Differences in the chemistry of nitrogen and phosphorus weaken the coupling of their cycles, particularly over long time scales (decades to millennia; Chapin and Eviner 2004). The two elements enter ecosystems through radically different pathways, nitrogen from a constantly available atmosphere pool through nitrogen fixation and phosphorus from the weathering of primary minerals that become depleted by weathering over millions of years (see Fig. 3.5). On young landscapes, for example, weathering of phosphorus-containing apatite by the carbonic acid generated from soil respiration releases phosphorus in available forms at a time when nitrogen often is in short supply (Eq. 9.6).

$$Ca_5(PO_4)_3 + 4H_2CO_3 \leftrightarrow 5Ca^{2+} + 3HPO_4^{2-} + 4HCO_3^- + H_2O \qquad (9.6)$$

This weathering source of phosphorus can be depleted over time, however, especially in wet areas outside the influence of glacial–interglacial cycles, and where geological uplift and erosion are slow. Once it is depleted, there is little or no internal source of phosphorus, and phosphorus inputs are derived primarily from the transport of dust from agricultural or arid areas upwind. Accordingly, the supply of nitrogen vs. phosphorus is decoupled at the ecosystem scale, and ecosystems on ancient soils are more likely to be constrained by phosphorus than by nitrogen (Vitousek 2004).

Microbial processing of dead organic matter can weaken the coupling of phosphorus and nitrogen cycles. Phosphorus turnover is somewhat less tightly linked to decomposition than is nitrogen because the ester linkages that bind phosphorus to carbon (C-O-P) can be cleaved enzymatically without breaking down the carbon skeleton. Nitrogen, in contrast, is more closely coupled to carbon; it is directly bonded to the carbon skeletons of organic matter (C-N) and is generally released by breaking the carbon skeleton into amino acids and other dissolved organic nitrogen-containing compounds. The decomposition process fragments organic matter and exposes the C-O-P bonds to enzymatic attack. Low soil phosphorus availability and high nitrogen availability induces plants and microbes to invest nitrogen in enzymes to acquire phosphorus (Olander and Vitousek 2000). Plant roots and their mycorrhizal associates, particularly arbuscular mycorrhizae, produce phosphatases that cleave ester bonds in organic matter to release phosphate (PO_4^{3-}). Phosphorus therefore cycles quite tightly between organic matter and plant roots in many ecosystems. In tropical forests, for example, mats of mycorrhizal roots in the litter layer produce phosphatases that cleave phosphate from organic matter. Mycorrhizal roots directly absorb much of this phosphate before it interacts with the mineral phase of the soil. Plant and microbial phosphatases are induced by low soil phosphate, as long as there is enough nitrogen to produce these nitrogen-rich enzymes. This contrasts with protease, whose activity correlates more strongly with microbial activity than with concentrations of soil organic nitrogen.

Microbial biomass often accounts for 20–30% of the organic phosphorus in soils (Smith and Paul 1990; Jonasson et al. 1999), much larger than the proportion of microbial carbon (about 2%) or nitrogen (about 4%). Microbial biomass is therefore an important reservoir of potentially available phosphorus, particularly in ecosystems with highly basic or acidic soils that strongly bind phosphorus to mineral surfaces. Microbial phosphorus is potentially more available than inorganic phosphate because it is protected from reactions with the mineral phase of soils, as described later. Although C:N ratios are often considered critical for understanding ecosystem nutrient cycling, C:P ratios of dead organic matter can also be critical in controlling the balance between phosphorus mineralization and immobilization and therefore the supply of phosphorus to plants.

Chemical reactions with soil minerals play a key role in controlling phosphorus availability in soils. Unlike nitrogen, phosphorus undergoes no oxidation–reduction reactions in soils and has no important gas phases. In addition, many of the reactions that control phosphorus availability are geochemical rather than biological in nature. Phosphate (PO_4^{3-}) is the main form of available inorganic phosphorus in soils. Phosphate is initially electrostatically attracted to positively charged sites on minerals through anion exchange. Once there, phosphate can become increasingly tightly bound (and correspondingly unavailable to plants) as it forms one or two covalent bonds with the metals on the mineral surface. Phosphorus can also bind with soluble minerals (especially iron oxides) to form insoluble precipitates. These precipitation reactions help to explain why highly weathered tropical soils (oxisols and ultisols) have extremely low phosphorus availability and why the growth of forests on those soils is often phosphorus-limited (see Chap. 3). The silicate clay minerals that dominate temperate soils fix phosphate to a lesser extent than do the oxides of tropical oxisols.

Phosphate availability is quite sensitive to pH. At low pH, iron, aluminum, and manganese are quite soluble and react with phosphate to form insoluble compounds:

$$Al^{3+} + H_2PO_4^- + 2H_2O \leftrightarrow 2H^+ + Al(OH)_2H_2PO_4$$

soluble   insoluble

(9.7)

Fig. 9.16 Effect of pH on the major forms of phosphorus present in soils. The low solubility of phosphorus compounds at low and high pH result in a relatively narrow window of phosphate availability near pH 6.5. Redrawn from Brady and Weil (2001)

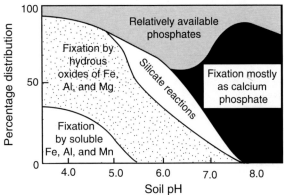

In soils with high concentrations of exchangeable calcium and $CaCO_3$, which typically occur at high pH, calcium phosphate precipitates, reducing phosphate availability in solution:

$$Ca(H_2PO_4)_2 + 2Ca^{++} \leftrightarrow Ca_3(PO_4)_2 + 4H^+$$
$$\text{soluble} \qquad\qquad \text{insoluble} \qquad (9.8)$$

Precipitation of calcium phosphate is one of the main reasons that phosphate fertilizer rapidly becomes unavailable in calcium-rich temperate agricultural ecosystems. Due to the precipitation reactions that occur at high and low pH, phosphorus is most available in a narrow range around pH 6.5 (Fig. 9.16).

Organic compounds in the soil also regulate, both directly and indirectly, phosphorus binding and availability. Charged organic compounds, for example, can compete with phosphate ions for binding sites on the surfaces of oxides or can chelate metals and prevent their reaction with phosphate. Both processes increase phosphate availability in mineral soils. On the other hand, organic compounds form complexes with iron, aluminum, and phosphate that protect these compounds from enzymatic attack. In tropical allophane soils, these complexes constitute a major sink for phosphorus.

Much of the phosphorus that precipitates as iron, aluminum, and calcium compounds is essentially unavailable to plants and is referred to as **occluded phosphorus**. During soil development, primary minerals gradually disappear as a result of weathering and erosional loss. The mass of phosphate in soils tends to shift from mineral,

organic, and non-occluded forms to occluded and organically bound forms, causing a shift from nitrogen to phosphorus limitation in ecosystems over long time scales (see Fig. 3.5; Crews et al. 1995).

The tight binding of phosphate to organic matter or to soil minerals in most soils causes 90% of the phosphorus loss to occur through surface runoff and erosion of particulate phosphorus rather than through leaching of soluble phosphate to groundwater (Tiessen 1995). Two-thirds of the dissolved phosphorus that enters groundwater is organic and therefore less reactive with soil minerals.

Sulfur

Sulfur cycling is tightly coupled to cycling of nitrogen and phosphorus in unpolluted ecosystems, but sulfur pollution uncouples element cycles by enhancing cation loss. Sulfur cycling in unpolluted ecosystems is tightly coupled to the cycling of nitrogen and phosphorus because sulfur is an essential component of proteins and therefore, like nitrogen and phosphorus, is needed to produce metabolically active tissues such as leaves and fine roots. The control over sulfur mineralization from dead organic matter is intermediate between that of nitrogen and phosphorus because sulfur occurs in both carbon-bonded and ester-bonded forms. The ester-bonded forms are sulfur-storage compounds produced by plants under conditions of high sulfur availability. Under sulfur-limiting conditions, plants produce mainly carbon-bonded forms of sulfur, so its mineralization is determined by the carbon demand of microbes, just as with nitrogen (McGill and Cole 1981). Under high-sulfur conditions, however, microbes preferentially mineralize ester-bonded forms of sulfur at a rate that depends on sulfur demand by plants and microbes, just as phosphorus mineralization depends on phosphorus demands of plants and microbes (Chapin and Eviner 2004). Because it is a component of most enzymes, including the nitrogenase of nitrogen fixers, sulfur availability in highly weathered soils of unpolluted areas can limit nitrogen inputs to ecosystems and therefore plant production and nutrient turnover.

Like nitrogen, inorganic sulfur undergoes oxidation–reduction reactions and is therefore sensitive to oxygen availability in the environment. In anaerobic soils, sulfate acts as an electron acceptor that allows microbes to metabolize organic carbon for energy, with hydrogen sulfide being produced as a by-product. In aerobic environments, however, reduced sulfur can be an important energy source for bacteria. The high productivity of deep-sea vents, for example, is based entirely on the oxidation of H_2S from the vents.

Rock weathering, which, together with atmospheric deposition of marine aerosols, is the primary natural source of sulfur in most ecosystems, is increasingly supplemented by atmospheric inputs in the form of acid rain. Combustion of fossil fuels produces gaseous SO_2, which dissolves in cloud droplets to produce H_2SO_4, a strong acid that is a major component of acid rain. As sulfate leaches from soils of ecosystems exposed to acid rain, it carries with it cations such as potassium and magnesium, depleting available pools within the soil and making vegetation demands for these cations increasingly dependent on weathering inputs. In other words, it reduces the tightness of cation recycling in ecosystems. Sulfur compounds in the atmosphere also play critical roles as aerosols, which increase the albedo of the atmosphere and therefore cause climatic cooling (see Chap. 2).

Essential Cations

Rock weathering and atmospheric inputs are the primary inputs of potassium, calcium, and magnesium, the cations required in largest amounts by plants. As with nitrogen, phosphorus, and sulfur, the quantities of these cations cycling in ecosystems from soils to plants and back to soils are much larger than are annual inputs to and losses from ecosystems. Unlike those elements, however, many soils contain a relatively large exchangeably bound pool of cations, whose availability in the soil solution is largely governed by exchange reactions. Their supply depends on the cation exchange capacity of the soil and its base saturation (see Chap. 3), which, in turn, are influenced by parent material and weathering characteristics. Calcium is an important structural component of plant and fungal cell walls. Its release and cycling therefore depends on decomposition in a way somewhat similar to that of nitrogen and phosphorus (Fig. 9.17). Potassium, on the other hand, occurs primarily in cell cytoplasm and is released through the leaching action of water moving through live and dead organic material. Magnesium is intermediate between calcium and potassium in its cycling characteristics. Potassium limits plant production in some ecosystems, but calcium concentration in the soil solution of most ecosystems is so high that it is actively excluded by plant cells during the absorption process (see Chap. 8). Availability of calcium and other cations may be low enough to limit plant production on some old, highly weathered tropical soils.

These cations have no gaseous phase, but atmospheric transfers of these elements (and of essential micronutrients) in dust can be an important pathway of loss by wind erosion from deserts and agricultural areas and an important input to the open ocean and to ecosystems on highly weathered parent materials. Cations can also be lost via leaching. Nitrate, sulfate, and other anions that are leached from ecosystems must be accompanied by cations to maintain electrical neutrality. Intensively fertilized agricultural fields, for example, are prone to cation leaching loss.

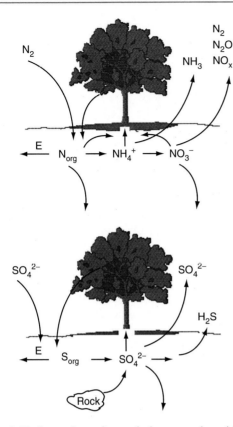

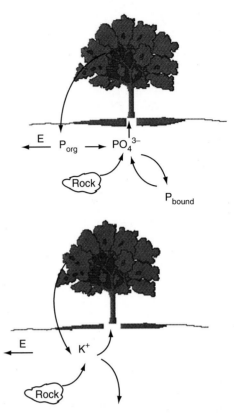

Fig. 9.17 Comparison of natural element cycles with respect to the relative importance of internal recycling, inputs, and outputs. Inputs of nitrogen come primarily from the atmosphere, whereas inputs of phosphorus and potassium come primarily from rocks. Sulfur comes from both the atmosphere and rocks. Over long time scales, atmospheric inputs of all elements can be important. Element losses occur through downward leaching, erosion (E), and, in the case of nitrogen and sulfur, gaseous loss

The declines in forest production observed in Europe and the eastern U.S. in response to acid rain are at least partly a consequence of calcium and magnesium deficiencies induced by cation leaching (Schulze 1989; Aber et al. 1998; Driscoll et al. 2001).

Why does phosphorus rather than rock-derived cations most often limit biological processes in highly weathered sites? The major cations, especially calcium, are absorbed by organisms in much larger quantities than is phosphorus and are more readily leached from soils. In Hawai'i, rock-derived calcium, magnesium, and potassium virtually disappear within 100,000 years but do not limit forest production anywhere on the sequence (Vitousek and Farrington 1997; Vitousek 2004). Atmospheric inputs of cations prevent these elements from becoming limiting in Hawai'i, and likely in many other places. Marine-derived aerosols containing calcium, magnesium, and potassium are deposited on forests in Hawai'i through rain and cloud droplets. Phosphorus concentrations in marine aerosols are low, however, because high phosphorus demands by marine organisms maintain a low concentration in surface waters. The atmospheric inputs of calcium are 10-fold less than weathering inputs in young sites, but are nearly a 100-fold greater than weathering inputs in older sites (Vitousek 2004). In continental interiors, dust from semi-arid and other sparsely vegetated areas is a major source of cations. Even in Hawai'i, dust from Asia, over 6,000 km away, is an important input of phosphorus, especially during glacial times, when vegetation cover was sparse and wind speeds were high (Box 9.3; Chadwick et al. 1999). In situ weathering of parent material is therefore not always the dominant input of minerals to ecosystems.

Box 9.3 Geochemical Tracers to Identify Source of Inputs to Ecosystems

Geochemical tracers have been used to identify dust and determine its rate of input to the Hawaiian Islands. Hawaiian rocks are derived from Earth's mantle, whereas Asian dust comes from the crust. These two sources differ in the ratio of two isotopes of neodynium, in the ratio of europium to other lanthanide elements, and in the ratio of thorium to halfnium. All of these elements are relatively immobile in soils, so changes over time in the isotopic or elemental ratios can be used to calculate time-integrated inputs of Asian dust. Knowing the phosphorus content of the dust, it is then possible to calculate phosphorus inputs by this pathway. Atmospheric inputs of phosphorus are much lower than weathering for the first million years or more of soil development. However, by four million years, rock-derived phosphorus has nearly disappeared, and Asian dust provides most of the phosphorus input to the soil. The biological availability of phosphorus is low in old sites, but it would be much lower were it not for inputs of Asian dust, most of it transported more than 10,000 years ago (Chadwick et al. 1999).

Micronutrients and Nonessential Elements

The cycling of micronutrients and nonessential elements is dominated by the balance between inputs from weathering, precipitation and dust, and outputs in leaching. Vegetation plays relatively little role in the balance between inputs and outputs of elements that are required in small quantities (e.g., chloride) or are not required by organisms (e.g., mercury and lead). Consequently, *external* cycling of elements (ecosystem inputs and outputs) dominates the cycling of nonessential elements, whereas *internal* cycling through vegetation dominates the cycling of essential elements (at least on annual to decadal time scales). The cycling of nonessential elements is therefore not strongly affected by successional changes in vegetation activity, whereas the losses of essential elements decline dramatically during early succession when organic matter and associated nutrients are accumulating in plant and microbial biomass (see Fig. 12.18; Vitousek and Reiners 1975).

Nitrogen and Phosphorus Cycling in Agricultural Systems

Intensive agricultural systems represent an endpoint in terrestrial nutrient cycling and an especially important one for human well-being as well as for their effects on surrounding ecosystems. Harvested crops remove nitrogen, phosphorus, and other nutrients from agricultural soils, and the higher yielding the agricultural system, the greater the removals of essential nutrients. Sustaining agricultural production requires replacing those nutrients, either through biological processes like nitrogen fixation or through the addition of mineral fertilizer or off-site plant or animal wastes to fields. These inputs are a dominant feature of agricultural nutrient cycles (Robertson and Vitousek 2009).

Globally, fertilizer is the major pathway of nutrient addition. These inputs have helped to keep world crop productivity ahead of human population growth. However, environmental costs of nutrient pollution from agriculture have been substantial, including the degradation of downstream water quality and eutrophication of coastal marine ecosystems (Fig. 9.1), the deposition of agriculturally derived nitrogen on downwind terrestrial ecosystems, the development of photochemical smog, and rising global concentrations of the powerful greenhouse gas nitrous oxide.

The fundamental challenge of nutrient management in grain crops in particular is easy to state, but hard to solve. The most economical way to add large quantities of nutrients is a single application of nitrogen-and phosphorus-rich fertilizer once during the cropping cycle, often near planting. At this time, the supply of available nitrogen and phosphorus is much greater than potential plant demand, and much of the fertilizer can be lost to the environment. Alternatively, additions of organically bound nutrients break down more slowly, so supply is less likely to radically exceed demand during the plant-growing season. However, typically they continue to break down during times that annual crops are inactive – after harvest and before planting in subsequent years, so again there are substantial time periods when supply exceeds demand, and high rates of nutrient loss are likely to occur. In contrast, nutrient supply (mainly through decomposition and mineralization) is more closely synchronized with plant nutrient demand in natural systems with perennial plants, and the microbial immobilization of nutrients that often are in short supply further serves to retain essential nutrients. The challenge, then, is to use agricultural practices and biological processes to increase the synchrony of nutrient supply and demand within intensive agricultural systems and to manage the fate of any nutrients that are lost, so they leave in environmentally benign forms (such as N_2) or are recaptured in riparian buffer strips or wetlands.

Crop yields and rates of nutrient input differ markedly among agricultural systems, as do the scientific and policy challenges that must be solved if we are to reduce the environmental footprint of intensive agriculture. The largest differences are associated with different levels of economic development (Vitousek et al. 2009b). In the poorest countries, rates of nitrogen and phosphorus application are less than those removed annually in harvested products – a deficit that contributes to continuing food insecurity in poor countries. These agricultural systems can persist only by drawing down the nutrient capital of soils, thereby decreasing their fertility and over time driving a cycle of degradation. In contrast, many rapidly developing economies have greatly increased both fertilizer applications and agricultural yields in recent decades. The transformation is particularly striking in China, where policy-driven increases in fertilizer use contributed to rising crop yields as China strived for food security. Nutrient additions to many fields far exceed those in the U.S. and Northern Europe, with rates of nitrogen and phosphorus application approaching 700 and 100 kg ha^{-1} year^{-1} (70 and 10 g m^{-2} year^{-1}), respectively. These applications are much greater than the requirements of even the highest-yielding crops, and much of the excess fertilizer is lost to the environment, degrading both air and water quality (Ju et al. 2009). At one time, agricultural production in northwestern Europe followed a similar path. After World War II, national and later European Community policies to boost food security caused many areas to reach nitrogen and phosphorus surpluses within integrated crop/animal production systems as large and damaging as those now observed in China. Since the 1980s, however, increasingly stringent national and European Union regulations and policies have reduced nutrient surpluses. Despite these steps toward nutrient balance, however, agriculturally derived pollution remains substantial in both the air and water of northwestern Europe (Billen et al. 2007; Erisman et al. 2008).

The human costs of inadequate nutrient inputs in the poorest countries are substantial, and research, and policies that address those nutrient deficits can provide substantial human benefits (Sanchez 2010). In contrast, the excessive use of fertilizers in many rapidly developing economies has substantial human and environmental costs and provides equally substantial scientific challenges. In China, research in agricultural biogeochemistry has focused on developing cropping systems in which the supply of nutrients (via fertilizer or other nutrient inputs) is matched as closely as possible in time and space to the demands of growing crops. For example, Ju et al. (2009) demonstrated experimentally that with such practices, additions of nitrogen fertilizer could be cut in half without loss of yield or grain quality, thereby reducing nitrogen losses by >50%. Matson et al. (1998) described a similar solution to excessive fertilizer application to intensive wheat systems in Mexico. In these situations, reducing nutrient inputs, while maintaining or increasing yields, is beneficial agronomically, economically, and environmentally.

Experience in North America and Europe suggests that, even with reduced nutrient inputs, intensive agriculture will cause substantial fluxes of nutrients to downwind and downstream ecosystems. Reducing these losses will require additional efforts. Some practices that can contribute to reducing nutrient losses from agriculture are available now, such as additional technologies for placing or timing nutrient supply to crop needs, modifications to livestock diets, and the preservation or restoration of riparian vegetation strips (Cherry et al. 2008). Bolder efforts to redesign agriculture (e.g., by incorporating perennials into cropping systems) may also be needed. Overall, agricultural systems represent fertile ground for research that is based in and contributes to our fundamental understanding of nutrient cycling and that also contributes to human well-being and environmental quality.

Summary

Nutrients enter ecosystems through inflow from upstream (in aquatic systems), chemical weathering of rocks, the biological fixation of atmospheric nitrogen, and the deposition of nutrients from the atmosphere in rain, windblown particles, or gases. Human activities have greatly increased these inputs, particularly of nitrogen and sulfur, through combustion of fossil fuels, addition of fertilizers, and planting of nitrogen-fixing crops. Unlike carbon, the internal recycling of essential plant nutrients is much larger than the annual inputs and losses from the ecosystem, producing relatively closed nutrient cycles.

Most nutrients that are essential to plant production become available to plants through microbial release of elements from dead organic matter during decomposition. Microbial exoenzymes break down the large polymers in particulate dead organic matter into soluble compounds and ions that can be absorbed by microbes or plant roots. The net mineralization of nutrients depends on the balance between the microbial immobilization of nutrients to support microbial growth and the excretion of nutrients that exceed microbial growth requirements. The first product of nitrogen

mineralization is ammonium. Ammonium can be converted to nitrate by autotrophic nitrifiers that use ammonium as a source of reducing power or by heterotrophic nitrifiers. Both plants and microbes use dissolved organic nitrogen, ammonium, and nitrate in varying proportions as nitrogen sources, when their growth is nitrogen-limited. Soil minerals and organic matter also influence nutrient availability to plants and microbes through exchange reactions (primarily with soil cations, except in some tropical soils that have a substantial anion exchange capacity), the precipitation of phosphorus with soil minerals, and the incorporation of nitrogen into humus.

Nutrients are lost from ecosystems through the leaching of elements out of the ecosystem in solution, emissions of gases, loss of nutrients adsorbed on soil particles in wind or water erosion, and the removal of materials in harvest. Human activities, as with nutrient inputs, often increase nutrient losses from terrestrial ecosystems.

The productivity of most rivers and streams is also co-limited by nitrogen and phosphorus. Nutrients spiral down rivers as they are mineralized from decomposing litter in one stream segment and absorbed by phytoplankton downstream. Nutrients spend 90% of their time in stream organisms attached in place, and 90% of their horizontal distance traveled in the dissolved phase between release from organisms in one place and subsequent absorption by another organism downstream.

Review Questions

1. What are the relative magnitudes of atmospheric inputs and mineralization from dead organic matter in supplying the annual nitrogen absorption by vegetation?

2. If Earth is bathed in di-nitrogen gas, why is the productivity of so many ecosystems limited by availability of nitrogen? What is biological nitrogen fixation? What factors influence the times and places where it occurs?

3. What are the mechanisms by which nitrogen moves from the atmosphere into terrestrial ecosystems?

4. What are the major steps in the mineralization of litter nitrogen to inorganic forms? What microbial processes mediate each step and what are the products of each step? Which of these processes are extracellular and which are intracellular?

5. What ecological factors account for differences among ecosystems in annual net nitrogen mineralization? How does each of these factors influence microbial activity?

6. What determines the balance between nitrogen mineralization and nitrogen immobilization in soils?

7. What factors determine the balance between plant absorption and microbial absorption of dissolved organic and inorganic nitrogen in soils?

8. How do ammonium and nitrate differ in mobility in the soil? Why? How does this influence plant absorption and susceptibility to leaching loss?

9. What is denitrification and what regulates it? What are the gases that can be produced, and what are their roles in the atmosphere?

10. What is the main mechanism by which phosphorus enters ecosystems?

11. What factors control availability of phosphorus for plant absorption? Why is phosphorus availability low in many tropical soils?

12. Why are mycorrhizae so important for plant acquisition of phosphorus?

13. What is the main pathway of phosphorus loss from terrestrial ecosystems?

Additional Reading

Andreae, M.O. and D.S. Schimel. 1989. *Exchange of Trace Gases between Terrestrial Ecosystems and the Atmosphere*. Wiley, New York.

Elser, J.J., M.E.S. Bracken, E. Cleland, D.S. Gruner, W.S. Harpole, et al. 2007. Global analysis of nitrogen and phosphorus limitation of primary producers in freshwater, marine and terrestrial ecosystems. *Ecology Letters* 10:1135–1142.

Fierer, N., A.S. Grandy, J. Six, and E.A. Paul. 2009. Searching for unifying principles in soil ecology. *Soil Biology and Biochemistry* 41:2249–2256.

Gruber, N. and J.N. Galloway. 2008. An Earth-system perspective of the global nitrogen cycle. *Nature* 451:293–296.

Howarth, R.W. (editor) 1996. *Nitrogen Cycling in the North Atlantic Ocean and its Watershed*. Kluwer, Dordrecht.

Mann, K.H. and J.R.N. Lazier. 2006. *Dynamics of Marine Ecosystems: Biological-Physical Interactions in the Oceans*. 3rd edition. Blackwell Publishing, Victoria, Australia.

Paul, E.A. and F.E. Clark. 1996. *Soil Microbiology and Biochemistry*. 2nd Edition Academic Press, San Diego.

Schlesinger, W.H. 1997. *Biogeochemistry. An Analysis of Global Change*. Academic Press.

Sterner, R.W., and J.J. Elser. 2002. *Ecological Stoichiometry: The Biology of Elements from Molecules to the Biosphere*. Princeton University Press, Princeton.

Tiessen, H. 1995. *Phosphorus in the Global Environment: Transfers, Cycles and Management*. John Wiley & Sons, Chichester.

Vitousek, P.M., J.D. Aber, R.W. Howarth, G.E. Likens, P.A. Matson, et al. 1997. Human alteration of the global nitrogen cycle: Sources and consequences. *Ecological Applications* 7:737–750.

Vitousek, P.M., R.L. Naylor, T. Crews, M.B. David, L.E. Drinkwater, et al. 2009. Agriculture: Nutrient imbalances in agricultural development. *Science* 324: 1519–1520.

Trophic Dynamics

10

Trophic dynamics govern the movement of carbon, nutrients, and energy among organisms in an ecosystem. This chapter describes the controls over the trophic dynamics of ecosystems.

Introduction

Although terrestrial animals consume a relatively small proportion of net primary production (NPP), they strongly influence energy flow and nutrient cycling in most ecosystems. In earlier chapters, we emphasized the interactions between plants and soil microbes because these two groups directly account for about 95% of the energy transfers in most terrestrial ecosystems. Plants use solar energy to reduce CO_2 to organic matter, most of which senesces, dies, and directly enters the soil, where it is decomposed by bacteria and fungi. Similarly, most nutrient transfers in ecosystems involve absorption by plants and return to the soil in dead organic matter, from which nutrients are released by microbial breakdown. In most ecosystems, the uncertainties in our estimates of primary production and decomposition exceed the total energy transfers from plants to animals. It is perhaps for this reason that many terrestrial ecosystem ecologists have ignored animals in classical studies of production and biogeochemical cycles. Aquatic ecologists, in contrast, have been unable to ignore animals because herbivory accounts for a much larger proportion of the carbon and nutrient transfer

than in terrestrial ecosystems (Fig. 10.1; Cyr and Pace 1993). Perhaps for this reason, aquatic ecosystem ecologists have generally led the theoretical developments relating to the roles of trophic dynamics in the functioning of ecosystems.

The factors governing energy and nutrient transfer to animals have important societal implications. Many human populations depend heavily on high-protein animal products for food. The rising human population and its diet shift toward greater consumption of meat places increasing pressure on the world's food supply. An ecologically viable strategy for efficiently providing food to a growing human population requires a good understanding of the ecological principles regulating the efficiency of converting plants into biomass of animals – including people.

A Focal Issue

Intense herbivory, due either to overstocking of domestic animals or to removal of predators from less intensively managed systems, reduces the density and diversity of palatable plants. This is one of the most extensive human impacts on the planet, operating through removal of large predatory fish from most of the world's oceans, removal of predators from lands that are intensively managed for human habitation and use, and extensive stocking of grasslands and savannas with domestic livestock (Fig. 10.2). Why do herbivores eat more of some plant species than others? How do interactions between plants, herbivores,

F.S. Chapin, III et al., *Principles of Terrestrial Ecosystem Ecology*,
DOI 10.1007/978-1-4419-9504-9_10, © Springer Science+Business Media, LLC 2011

Fig. 10.1 Comparison of rates of primary productivity and herbivory between aquatic and terrestrial ecosystems. Redrawn from Cyr and Pace (1993)

Fig. 10.2 Intensive herbivory reduces the density of palatable plants, altering ecosystem structure. On Australian rangelands, overstocking of cattle can transform grassland savannas to shrublands (Ludwig and Tongway 1995). Photographs by David Tongway

and their predators influence the structure and functioning of ecosystems? What happens to the energy and nutrients that are consumed by an animal? How does human choice of the proportion of meat and plants consumed influence the land base required to meet the food needs of a growing human population? Answers to these questions provide a framework to address some of the most contentious ecological issues facing society.

Overview of Trophic Dynamics

Energy and nutrient transfers define the trophic structure of ecosystems. The simplest way to visualize the energetic interactions among organisms in an ecosystem is to trace the fate of a packet of energy from the time it enters the ecosystem until it leaves (Lindeman 1942). **Trophic transfers** involve the feeding by one organism on another or on dead organic matter. Plants are called **primary producers** or **autotrophs** because they convert CO_2, water, and solar energy into biomass (see Chaps. 5 and 6). **Heterotrophs** are organisms that derive energy by eating live or dead organic matter. Heterotrophs function as part of two major trophic pathways, one based on live plants (the **plant-based trophic system**) and another based on dead organic matter (the **detritus-based trophic system**). The detritus-based trophic system usually accounts for most of the energy transfer through animals in an ecosystem.

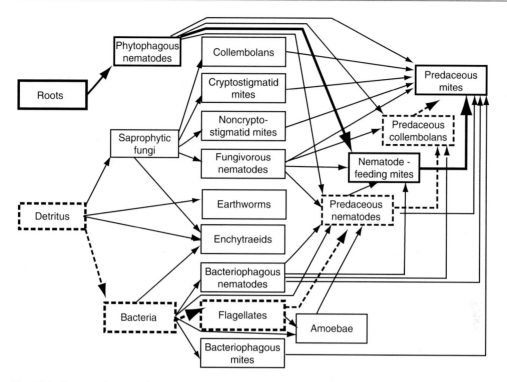

Fig. 10.3 Pattern of energy flow through belowground portions of a grassland food web. Food webs consist of many interconnecting food chains. A plant-based (solid line) and a detritus-based (dashed line) food chain are shown in bold. Modified from Hunt et al. (1987)

Consumers are organisms that eat other live organisms. These include plant-eating **herbivores**, microbe-eating **microbivores**, and animal-eating **carnivores**. A **food chain** is a group of organisms linked together by the process of consumption. Grass, grasshoppers, and birds, for example, form a food chain. Those organisms that obtain their energy with the same number of transfers from plants or detritus belong to the same **trophic level**. Thus in a plant-based trophic system, plants constitute the first trophic level, herbivores the second, primary carnivores the third, secondary carnivores that eat mainly primary carnivores the fourth, etc. (Lindeman 1942; Odum 1959). Similarly, in the detritus-based trophic system, bacteria and fungi directly break down dead soil organic matter and absorb the breakdown products for their own growth and maintenance. These **primary detritivores** are the first trophic level in the detritus-based food chain and are fed on by animals in a series of trophic levels analogous to those in the plant-based trophic system (Fig. 10.3).

Although food chains are an easy way to conceptualize the trophic dynamics of an ecosystem, they are a gross oversimplification for the many organisms that eat more than one kind of food. People, for example, eat food from several trophic levels, including plants (first trophic level), cows (second trophic level), fish (second and often higher trophic levels), and mushrooms (detritivores). Many other mammals and birds also consume both herbivorous and detritus-feeding insects and other animals. The actual energy transfers that occur in all ecosystems are therefore complex **food webs** (Fig. 10.3). We can trace the energy transfers through these food webs only by knowing the contribution of each trophic level to the diet of each animal in the ecosystem. Although food web structures have been partially described for many ecosystems (Pimm 1984), the quantitative patterns of energy flow through food webs are generally poorly known, especially for detritus-based food webs.

Food consists of much more than energy. In fact, animals often select food based as much on protein as on digestible energy content because animals require more nitrogen than do plants (tissue concentrations of 7–14% vs. 0.5–4%; Ayres 1993; Pastor et al. 2006; Barboza et al. 2009). Phosphorus concentrations are also generally higher in animals than plants, so either nitrogen or phosphorus can constrain animal production (Sterner and Elser 2002). Feeding is also strongly influenced by concentrations of plant defensive compounds that are toxic or reduce digestibility. The concentrations of these positive and negative determinants of food quality strongly influence the temporal and spatial patterns of trophic transfer.

The regulation of energy and nutrient flow through food webs is complex and varies considerably among ecosystems. Two theoretical patterns, however, bracket the range of possible controls. (1) The availability of food at the base of the food chain (either plants or detritus) limits the production of upper trophic levels through **bottom-up controls**. In this case, the quantity and quality of food, including the concentrations of nitrogen, phosphorus, and defensive chemicals, determine the amount of food that is eaten and therefore the animal production that can be supported. (2) Alternatively, predators that regulate the abundance of their prey exert **top-down control** on food webs. Most trophic systems exhibit some combination of bottom-up and top-down controls, with the relative importance of these controls varying temporally and spatially (Polis 1999; Allison 2006). In pelagic ecosystems, for example, nutrients, light, and temperature explain much of the geographic and seasonal patterns of production (bottom-up controls), but once a phytoplankton bloom is initiated, zooplankton rapidly grow and reproduce, reducing phytoplankton biomass (top-down controls).

Trophic transfers of energy and nutrients have profound effects on the functioning of ecosystems. They reduce plant biomass, thereby altering all the ecosystem processes that are mediated by plants, including the cycling of water, energy, and nutrients. Consumption of plants and detritus also accelerates the return of nutrients to the environment, although, as we shall see, the effects of herbivory on nutrient cycling depend on initial nutrient availability (Pastor et al. 2006).

Controls Over Energy Flow through Ecosystems

Bottom-Up Controls

Plant production places an upper limit to the energy flow through both plant-based and detritus-based webs. The energy consumed by animals in the plant-based trophic system, on average, cannot exceed the energy that initially enters the ecosystem through primary production. This constitutes a fundamental constraint on the animal production that an ecosystem can support. When all terrestrial ecosystems are compared, herbivore biomass and production tends to increase with increasing primary production (Fig. 10.4). The relationship between primary production and herbivore biomass is particularly strong, when comparisons are made among similar types of ecosystems. In the grasslands of Argentina, for example, the biomass of mammalian herbivores increases with increasing aboveground production along a gradient of water availability in both natural and managed grasslands (Fig. 10.5; Osterheld et al. 1992). In the Serengeti grasslands of Africa, the large herds of ungulates also acquire most of their food in the more productive grasslands (Sinclair 1979; McNaughton 1985). Similarly, productive forests generally have greater insect herbivory than do unproductive forests. When forests are fertilized to increase their production, this usually increases feeding by herbivores (Niemelä et al. 2001).

The world's large fisheries depend on the strong relationship between primary production and animal production, particularly in the coastal zone where the upwelling of nutrient-rich bottom waters supports a high productivity of phytoplankton, zooplankton, and fish (see Chap. 6). At the opposite extreme, productivity is low in the central gyres of tropical oceans that are isolated from nutrient-rich bottom waters and in **oligotrophic** (nutrient-poor) lakes on the Canadian Shield, whose soils were scraped away by Pleistocene glaciers.

Subsidies can supplement secondary production above levels that could be supported by NPP. Most of the energetic base for headwater streams in forests, for example, comes from inputs of

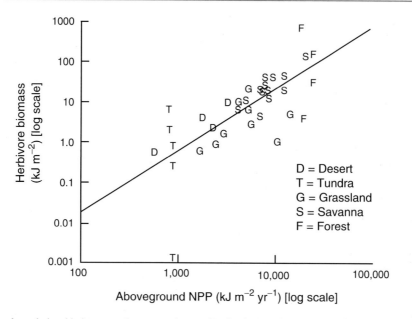

Fig. 10.4 Log–log relationship between aboveground net primary production (NPP) and herbivore biomass. One gram of ash-free biomass is equivalent to 20 kJ of energy. Production and biomass of aboveground herbivores correlates with aboveground NPP across a wide range of ecosystems. Redrawn from McNaughton et al. (1989)

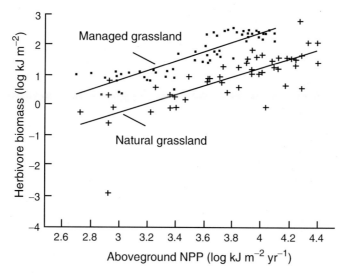

Fig. 10.5 Log–log relationship between mammalian herbivore biomass and aboveground plant production in natural and managed grazing systems of South America. Herbivore biomass increased with increasing NPP. Animal biomass on the managed grassland was 10-fold greater than on the natural grassland at a given level of plant production because managers control predation, parasitism, and disease and provide supplemental drinking water and minerals in managed systems. This difference in herbivore biomass between managed and unmanaged systems indicates that NPP is not the only constraint on animal production. Redrawn from Osterheld et al. (1992)

terrestrial litter. This **allochthonous** input (i.e., an input from outside the stream ecosystem) constitutes a subsidy that, together with **autochthonous production** (i.e., production occurring within the stream), provides the energy that supports aquatic food webs (see Chap. 7). At a finer scale, filter-feeding invertebrates in stream riffles derive most of their energy from algal production in upstream pools (Finlay et al. 2002). Terrestrial food webs near the ocean, rivers, and lakes are often subsidized by inputs of aquatic energy, for example when birds or bears feed on fish, or spiders feed on marine detritus (Polis and Hurd 1996; Milner et al. 2007). High-intensity agricultural production is strongly subsidized by human inputs of nutrients, water, and fossil fuels (Schlesinger 2000).

Biome differences in herbivory reflect differences in NPP, nutrient balance, and plant allocation to structural and chemical defenses. The most dramatic differences in herbivory among ecosystem types are consequences of variation in plant allocation to physical support. Lakes, the ocean, and many rivers and streams are dominated by phytoplankton that allocate most of their energy to cytoplasm rather than to structural support. Most phytoplankton are readily digested by zooplankton, so animals eat a large proportion of primary production and convert it into animal biomass. Even among phytoplankton, chlorophytes (naked green algae) are generally consumed more readily than phytoplankton that produce a protective outer coating, such as diatoms, dinoflagellates, and chrysophytes. At the opposite extreme, forests have a substantial proportion of production allocated to cellulose- and lignin-rich woody tissue that cannot be directly digested by animals. Some animals, however, like ruminants (e.g., cows), caecal digesters (e.g., rabbits), and some insects (e.g., termites) with symbiotic gut microbes are capable of cellulose breakdown. These animals can assimilate some of the energy released by this microbial breakdown of cell walls. Consequently, the fraction on NPP consumed by animals is much lower in forests, where plants allocate much of their biomass to structural material (Barboza et al. 2009; Craine 2009).

Among terrestrial ecosystems, there is a 1,000-fold variation in the quantity of plant biomass consumed by herbivores (McNaughton et al. 1989). Herbivores consume the least biomass per unit land area in unproductive ecosystems such as tundra (Fig. 10.6a). However, the energy consumed by herbivores is quite variable within and among biomes. Consumption by herbivores shows a much stronger relationship with production of edible tissue (e.g., leaves; Fig. 10.6b) than with total aboveground NPP (Fig. 10.6a) because the woody

support structures produced by many plants contribute relatively little to herbivore consumption.

Plant chemical and physical defenses reduce the proportion of energy transferred to herbivores. It has been argued that predation rather than food availability must limit the abundance of herbivores because the world is covered by green biomass that has not been eaten by animals (Hairston et al. 1960). Not all green biomass, however, is digestible enough to serve as food. Ruminants and insects, for example, need plant biomass with at least a 1% nitrogen concentration to gain weight, with even higher requirements for reproducing animals (Craine 2009). In low-nutrient habitats, plants have not only low nitrogen and phosphorus concentrations but also high concentrations of chemical defenses (Bryant and Kuropat 1980; Pastor et al. 2006). In Africa, for example, fertile grasslands support higher diversity and production of herbivores than do the less fertile grasslands. The same pattern is seen in tropical forests, where higher levels of chemical defense and lower levels of insect herbivory occur on infertile than on fertile soils (McKey et al. 1978). Three factors govern the allocation to defense in plants: (1) genetic potential, (2) the environment in which a plant grows, and (3) the seasonal program of allocation.

1. Ecosystem differences in plant defense are determined most strongly by species composition. Terrestrial and aquatic species vary substantially in the type and quantity of defensive compounds produced. Terrestrial plants and marine kelps adapted to low-nutrient environments generally produce long-lived tissues with high concentrations of **carbon-based defense** compounds (i.e., organic compounds that contain no nitrogen, such as tannins, resins, and essential oils; see Chap. 6). These compounds deter feeding by most herbivores (Coley et al. 1985; Hay and Fenical 1988). Tissue loss to herbivores is often similar (1–10%) to the annual allocation to reproduction (i.e., the allocation that most directly determines fitness), suggesting that natural selection for chemical defenses against herbivores must be strong. When genotypes of a species are compared, for example, those individuals that allocate most strongly to defense grow most slowly (Fig. 10.7),

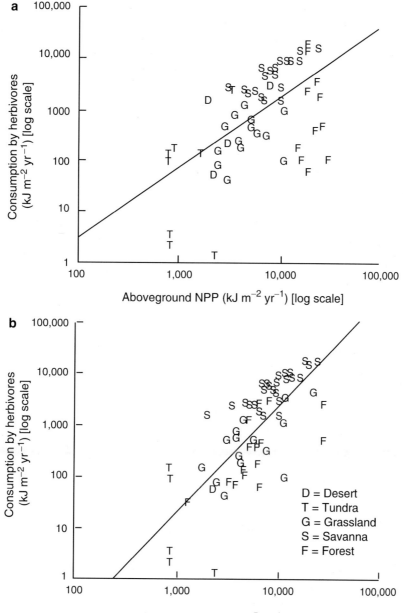

Fig. 10.6 Log–log relationship between (**a**) aboveground NPP or (**b**) foliage production and consumption by herbivores. One gram of ash-free biomass is equivalent to 20 kJ of energy. Consumption by herbivores is more closely related to foliage production than to total aboveground NPP because much of the aboveground NPP is inedible by most herbivores. Redrawn from McNaughton et al. (1989)

suggesting a tradeoff between allocation to growth vs. defense (Coley 1986). Plant species typical of high-nitrogen environments, particularly nitrogen-fixing species, often produce **nitrogen-based defenses** (i.e., organic compounds containing nitrogen, such as alkaloids) that are toxic in relatively small quantities to generalist herbivores. Nitrogen-based defenses are well developed, for example, in terrestrial legumes and freshwater cyanobacteria. Other types of defenses include sulfur-containing defenses, accumulation of selenium or silica,

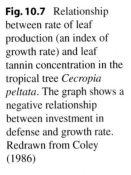

Fig. 10.7 Relationship between rate of leaf production (an index of growth rate) and leaf tannin concentration in the tropical tree *Cecropia peltata*. The graph shows a negative relationship between investment in defense and growth rate. Redrawn from Coley (1986)

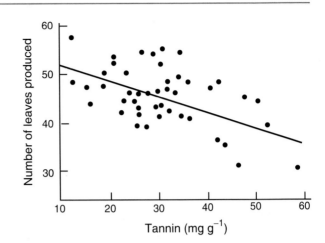

and physical defenses like thorns (Boyd 2004). Reproductive tissues which have high value to the plant and constitute a modest proportion of total production are often protected by nitrogen- or sulfur-based toxic compounds (Zangerl and Berenbaum 2006).

2. Any plant is less palatable when grown in infertile than in fertile soils, due to a lower protein content and a higher level of carbon-based defenses (Ayres 1993). Under conditions of low nutrient availability, growth is constrained more strongly than is photosynthesis, so carbon tends to accumulate (see Chap. 6; Bryant et al. 1983). Under these circumstances, carbon allocated to chemical defense may have only modest negative effects on growth rate.

3. In a given environment, plants vary seasonally in their allocation to defense, with allocation to growth occurring when conditions are favorable and allocation to tissue differentiation and defense when conditions deteriorate (Lorio 1986; Herms and Mattson 1992). Newly expanding leaves, especially those that expand rapidly, are poorly defended and are particularly vulnerable to herbivory (Kursar and Coley 2003).

The first two causes of variation in allocation to plant defense (genetics and environment) lead to high levels of plant defense on infertile soils. Plant defenses are either directly toxic, or reduce the availability of limiting resources to herbivores during ingestion or digestion (Barboza et al. 2009). Tannins, for example, bind with proteins,

reducing N availability to herbivores; alkaloids can act as neurotoxins; and thorns reduce the feeding rate of mammals.

The balance of nitrogen, phosphorus, and digestible energy influence the efficiency with which these resources support animal production (Sterner and Elser 2002). Nonliving materials have a wide range of ratios of carbon to nitrogen to phosphorus. Living protoplasm, however, is much more constrained in these ratios because of the fundamental similarity of biochemical processes in all living cells (Reiners 1986; Sterner and Elser 2002). In general, phosphorus concentration is more variable than nitrogen in both plants and animals. Just as observed in plants (see Chap. 8) and microbes (see Chap. 9), animal production is constrained by the resource (nitrogen, phosphorus, digestible energy) that is most limiting in its food, and animals strengthen the coupling of nitrogen and phosphorus cycles by preferential acquisition of the most limiting element. For example, animals extract nitrogen most efficiently from low-nitrogen food through selective foraging, high rates of nitrogen absorption from the gut, or reduced rates of loss. Similarly, animals extract phosphorus most efficiently from low-phosphorus food. Elements that are less limiting are extracted from food less efficiently and are preferentially released to the environment. These **stoichiometric relationships** (element ratios) are important determinants of element cycling rates in all ecosystems (Sterner and Elser 2002), as discussed earlier and again in this chapter.

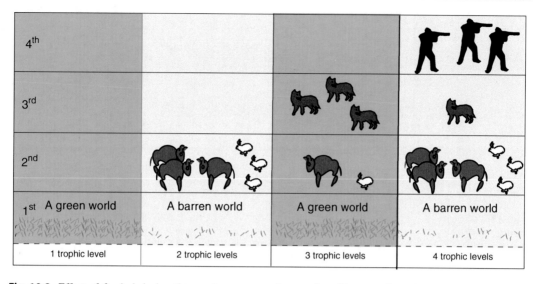

Fig. 10.8 Effect of food chain length on primary producer biomass in situations where trophic cascades operate. Plant biomass is abundant where there are odd numbers of trophic levels (1, 3, 5, etc.) because these have a low biomass of herbivores; plant biomass is reduced where there are even numbers of trophic levels (2, 4, 6, etc.) because these have a large biomass of herbivores

Top-Down Controls

Consumption by predators often alters the abundance of organisms across more than one link in a food web (trophic cascade; Pace et al. 1999). A predator, for example, may reduce the density of its prey, which releases the prey's prey from consumer control (Carpenter et al. 1985; Pace et al. 1999; Beschta and Ripple 2009; Schmitz 2009). Trophic cascades cause an alternation among trophic levels in biomass of organisms (Power 1990). In many streams, for example, if only algae are present, they grow until their biomass becomes nutrient-limited, producing a "green" surface (Fig. 10.8). If there are two trophic levels (plants and herbivores), the herbivores graze the plants to a low biomass level, leaving a barren surface with sparse, fast-growing algae. With three trophic levels, the secondary consumer reduces the biomass and grazing pressure of herbivores, which again allows algae to achieve a high biomass. Algal biomass is generally low when there is an even number (2, 4, etc.) of trophic levels. An odd number of trophic levels in a trophic cascade reduces the biomass of herbivores and releases the algae, producing a "green" world (Fretwell 1977).

Trophic cascades have been demonstrated in a wide range of ecosystems, ranging from the open ocean to tropical rainforests and microbial food webs (Pace et al. 1999; Schmitz et al. 2000; Borer et al. 2005; Beschta and Ripple 2009). Trophic cascades generally result from strong interactions between individual species and are therefore best documented at the level of species rather than ecosystems (Paine 1980; Polis 1999). Because of the species-specific nature of trophic cascades, they are most likely to emerge at the ecosystem scale when a single species dominates a trophic level, for example when *Daphnia* is the dominant herbivore or a minnow-eating fish is the dominant carnivore in a lake (Polis 1999). Similarly, removal of wolves in the western U.S. caused population explosions of elk and other ungulates, which overbrowsed their food supply. Wolf reintroductions reversed this effect through both predation and ungulate avoidance of areas with high predation risk (Frank 2008; Beschta and Ripple 2009).

Eutrophication of fresh waters often leads to strong species dominance, thereby providing conditions where trophic cascades can emerge (Pace et al. 1999). Trophic cascades have important practical implications; introduction of minnow-eating fish, under the right circumstances,

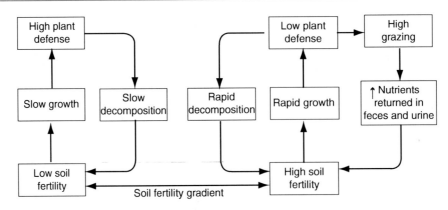

Fig. 10.9 Feedbacks by which grazing and plant defense magnify differences among sites in soil fertility. In infertile soils, herbivory selects for plant defenses, which reduce litter quality, decomposition, and nutrient supply rate. In fertile soils, herbivory speeds the return of available nutrients to the soil. Based on Chapin (1991b)

can release populations of zooplankton grazers, which graze down algal blooms and increase water clarity. Trophic cascades that involve invertebrate herbivores and homeothermic vertebrate predators are particularly strong (Schmitz et al. 2000; Borer et al. 2005). Unfortunately, unanticipated species interactions often become important when trophic dynamics are altered, leading to unexpected responses to species introductions or removal (Kitchell 1992). Manipulation of trophic cascades to address management issues therefore requires a sophisticated understanding and careful testing of the ecology of the species involved and the factors governing their interactions.

Trophic Effects on Nutrient Cycling

Herbivores enhance the productivity of productive ecosystems and reduce the productivity of unproductive ones (Frank 2006; Pastor et al. 2006). Dominance by plants with well-developed defenses in low-nutrient environments tends to reduce the frequency of herbivory in these ecosystems because herbivores select against patches in the landscape where plant palatability is low (see Chap. 13; Frank 2006). Herbivores in these environments (like the plants themselves) efficiently retain and recycle nitrogen and phosphorus and therefore produce feces with very low nutrient concentrations (Barboza

et al. 2009), promoting nutrient immobilization by soil microbes. Herbivores indirectly reduce nutrient cycling in these environments by preferentially eating poorly defended plant species, leading to an increase in the abundance of well-defended plants that produce litter with low nutrient concentrations and high concentrations of plant defenses. The toxicity of many plant species to soil microbes causes reductions in decomposition rates (see Chap. 7) and further reduces soil fertility in low-nutrient environments (Fig. 10.9; Pastor et al. 1988; Northup et al. 1995; Pastor et al. 2006).

Herbivores are more abundant in fertile environments, where plants are more productive and more palatable. Their feeding speeds the turnover of plant biomass and the return of available nutrients to the soil as feces and urine. This short circuits decomposition and nitrogen mineralization and enhances plant production (Ruess and McNaughton 1987; Frank 2006; Pastor et al. 2006). Tissue nitrogen concentrations of about 1.5% appear to separate those infertile ecosystems where herbivory drives a decline in nutrient cycling from those more fertile ecosystems where herbivory enhances nutrient cycling (Pastor et al. 2006).

Plants in fertile environments are often well adapted to herbivory. Fertile grasslands are often more productive when moderately grazed than in the absence of grazers (McNaughton 1979; Milchunas and Lauenroth 1993; Hobbs 1996).

Grazing in many managed ecosystems, however, exceeds that which would occur naturally (Figs. 10.2 and 10.5) because people control animal densities through stocking rates and predator control. High levels of grazing, whether natural or managed, can reduce production and plant cover and increase soil erosion, leading to a decline in soil fertility and the productive potential of an ecosystem (Milchunas and Lauenroth 1993).

Ecological Efficiencies

Trophic Efficiency and Energy Flow

Energy loss with each trophic transfer limits the production of higher trophic levels. Not all of the biomass that is produced at one trophic level is consumed at the next level. Moreover, only some of the consumed biomass is digested and assimilated, and only some of the assimilated energy is converted into animal production (Fig. 10.10). Consequently, a relatively small fraction (generally <1–25%) of the energy available as food at one trophic level is converted into production at the next link in a food chain. This has profound consequences for the trophic structure of ecosystems because each link in the food chain has less energy available to it than did the preceding trophic link. In any plant-based trophic system, plants process the largest quantity of energy, with progressively less energy processed by herbivores, primary carnivores, secondary carnivores, etc. This leads inevitably to an **energy pyramid** (Fig. 10.11; Elton 1927) in which the production at each trophic link ($Prod_n$) depends on the production at the preceding trophic level ($Prod_{n-1}$) and the **trophic efficiency** (E_{troph}) with which the production of the prey ($Prod_{n-1}$) is converted into production of consumers ($Prod_n$).

$$Prod_n = Prod_{n-1} \times E_{troph} = Prod_{n-1} \times \left(\frac{Prod_n}{Prod_{n-1}} \right)$$

$$(10.1)$$

The trophic efficiency of each link in a food chain can be broken down into three ecological efficiencies (Fig. 10.10) related to the efficiencies of consumption ($E_{consump}$), assimilation (E_{assim}), and production (E_{prod}; Lindeman 1942; Odum 1959; Kozlovsky 1968).

$$E_{troph} = E_{consump} \times E_{assim} \times E_{prod} \qquad (10.2)$$

In terrestrial ecosystems, the distribution of biomass among trophic levels can be visualized as a **biomass pyramid** that is similar in structure to the energy pyramid, with greatest biomass in primary producers and progressively less biomass in higher trophic levels (Fig. 10.11). This occurs for at least two reasons. First, as described earlier, the energy pyramid results in less energy available at each successive trophic link. Second, the large allocation to structural tissue and chemical defense in many terrestrial plants minimizes the proportion of plant production that can be converted to secondary production. The decrease in biomass with successive links is most pronounced in forests, where the dominant plants are long lived and produce a large proportion of biomass that is inedible or out of reach of ground-based herbivores. Biomass pyramids are less broad in grasslands where plants have a lower allocation to woody structures, and there is a relatively large biomass of herbivores and higher trophic levels.

In contrast to terrestrial ecosystems, freshwater and marine pelagic ecosystems have *less* biomass of primary producers than of higher trophic levels, leading to an **inverted biomass pyramid** (Fig. 10.11). This difference in trophic structure between terrestrial and pelagic ecosystems reflects the relative turnover rate of biomass among trophic levels. Phytoplankton in aquatic ecosystems have less structure and are more edible than their terrestrial counterparts. They are therefore rapidly grazed, and their biomass does not accumulate. Fish turn over more slowly and accumulate a larger biomass. In summary, terrestrial ecosystems are characterized by large, long-lived

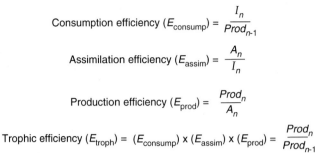

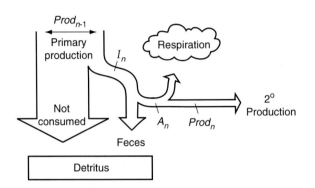

Fig. 10.10 Components of trophic efficiency, which is the product of consumption efficiency, assimilation efficiency, and production efficiency. Production efficiency is the proportion of primary production that is ingested (I_n) by animals. Assimilation efficiency is the proportion of ingested food (I_n) that is assimilated into the blood stream (A_n). Production efficiency is the proportion of assimilated energy (A_n) that is converted to animal production ($Prod_n$). Most primary production is not consumed by animals and passes directly to the soil as detritus. Of the plant material consumed by herbivores, most is transferred to the soils as feces. Of the material assimilated by animals, most supports the energetic demands of growth and maintenance (respiration), and the remainder is converted to new animal biomass (secondary production)

plants, leading to a large plant biomass and relatively small biomass of higher trophic levels. Aquatic ecosystems, in contrast, are characterized by rapidly reproducing phytoplankton that are smaller and more short lived than higher trophic levels (Fig. 10.12).

Regardless of the biomass distribution among trophic levels, there must always be more energy flow through the base of a trophic chain than at higher trophic levels. It is the *energy pyramid* rather than the biomass pyramid that describes the fundamental energetic relationships among trophic levels because energy is lost at each trophic transfer, so there must always be a decline in energy available at each successive trophic level. Trophic efficiencies with respect to nitrogen and phosphorus are discussed later.

Consumption Efficiency

Consumption efficiency is determined primarily by food quality and secondarily by predation. Consumption efficiency ($E_{consump}$) is the proportion of the production at one trophic level ($Prod_{n-1}$) that is ingested by the next trophic level (I_n; Fig. 10.10).

$$E_{consump} = \frac{I_n}{Prod_{n-1}} \qquad (10.3)$$

Unconsumed material eventually enters the detritus-based food chain as dead organic matter. On average, the quantity of food consumed by a given trophic level must be less than the production of the preceding trophic level, or the prey will be driven to extinction. There are, however, often short time periods when the consumption by one

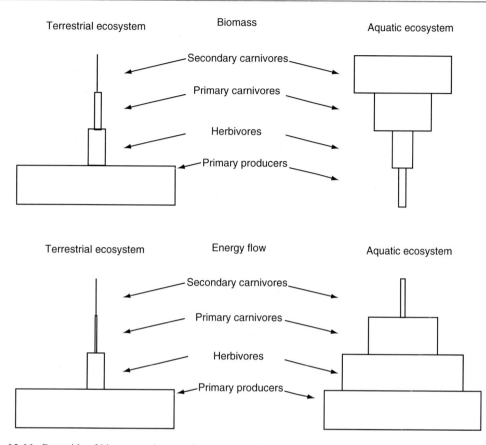

Fig. 10.11 Pyramids of biomass and energy in a terrestrial and an aquatic food chain. The width of each box is proportional to its biomass or energy content. Pyramids of energy are structurally similar in terrestrial and aquatic food chains because energy is lost at each trophic transfer. Biomass pyramids differ between terrestrial and aquatic food chains because most plant biomass (phytoplankton) is eaten in aquatic ecosystems, but not on land

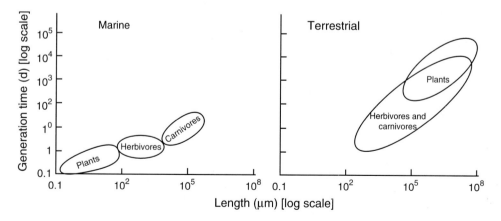

Fig. 10.12 Body size and generation time for organisms in the ocean and on land of dominant plants, herbivores, and carnivores. In the ocean, the dominant plants (pico- and nano-plankton) are generally smaller than the herbivores that feed on them, whereas on land, the dominant plants are often as large or larger than their herbivores. Redrawn from Steele (1991)

Table 10.1 Consumption efficiency of the herbivore trophic level in selected ecosystem types[a]

Ecosystem type	Consumption efficiency (% of aboveground NPP)
Ocean	60–99
Managed rangelands	30–45
African grasslands	28–60
Herbaceous old fields (1–7 year)	5–15
Herbaceous old fields (30 year)	1.1
Mature deciduous forests	1.5–2.5

[a]Data from Wiegert and Owen (1971) and Detling (1988). Terrestrial estimates emphasize consumption by aboveground herbivores and may not accurately reflect the total ecosystem-scale consumption efficiency

trophic level exceeds that in the preceding level. Vertebrate herbivores, for example, consume plants during winter, when there is no plant production. This is, however, offset by other seasons when plants produce more biomass than animals can consume. Situations where consumption efficiency is greater than 100% for prolonged periods lead to dramatic ecosystem changes (Fig. 10.2). If predator control, for example, leads to a large deer population that consumes more plant biomass than is produced, this will reduce plant biomass and alter plant species composition in ways that profoundly affect all ecosystem processes (see Chap. 12; Pastor et al. 1988; Kielland and Bryant 1998; Paine 2000). Similarly, insect outbreaks can substantially reduce the biomass and productivity of their host plants (Allen et al. 2006; Raffa et al. 2008). Sometimes trophic imbalances occur naturally. Some herbivores, such as beavers, typically overexploit their local food supply and move to new areas when their food is depleted. In snowshoe hare or lemming cycles, cyclic variations in herbivore abundance alter the balance between top-down and bottom-up controls.

The proportion of aboveground NPP consumed by herbivores varies at least 100-fold among ecosystems, from less than 1% to greater than 40% (Table 10.1), due primarily to differences in plant allocation to woody structures and chemical defense. Herbivore consumption efficiency is generally lowest in forests (<1–5%), where chemically defended woody biomass accounts for much of the production, and much

of the biomass is out of reach of ground-dwelling herbivores. Herbivore aboveground consumption efficiencies are higher in grasslands (10–60%), where most aboveground material is non-woody, and highest (generally >40%) in pelagic aquatic ecosystems, where most phytoplankton biomass is cell contents rather than cell walls. In these ecosystems, more phytoplankton biomass is often consumed by herbivores than dies and decomposes; this pattern contributes to inverted biomass pyramids (Fig. 10.11). In grasslands, aboveground consumption efficiencies are generally greater for ecosystems dominated by large mammals (25–50%) than those dominated by insects and small mammals (5–15%; Detling 1988). The toxic nature of some plant tissues (due to presence of plant defenses) and inaccessibility of other tissues (e.g., roots to aboveground herbivores) constrain the herbivore consumption efficiency of terrestrial ecosystems. Nematodes, which are important belowground herbivores, consume 5–15% of belowground NPP in grasslands (Detling 1988). The highest aboveground consumption efficiencies in terrestrial ecosystems, ~90%, are on **grazing lawns**, such as those found in some African savannas (McNaughton 1985) and arctic wetlands (Jefferies 1988). These highly productive grasslands are maintained as a lawn of short grass by repeated herbivore grazing. Nutrient inputs in urine and feces from these herbivores promote rapid recycling of nutrients and the high productivity of these grasslands (Fig. 10.9; Ruess et al. 1989).

Consumption efficiencies of carnivores are subject to the same general constraints as herbivores, but carnivores are less constrained by the quality of their food. Consequently, efficiencies are often higher than those of herbivores, ranging from 5% to 100%. Vertebrate predators that feed on vertebrate prey, for example, often have a consumption efficiency greater than 50%, indicating that more of their prey is eaten than enters the soil pool as detritus. Invertebrate carnivores often have a lower consumption efficiency (5–25%) than vertebrate carnivores. Consumption efficiency of a trophic level at the ecosystem scale must integrate vertebrate and invertebrate consumption, including animals that feed below ground, but these efficiencies are not well documented at the ecosystem scale.

More often, consumption efficiency is documented for a single large herbivore in an ecosystem where it is abundant.

The consumption efficiency of a trophic level depends on its biomass and food intake, which are influenced by the quantity and quality of available food (bottom-up controls) and predation controls on consumer biomass (top-down controls). Bottom-up and top-down controls often interact. Rising atmospheric CO_2 concentration, for example, reduces leaf nitrogen concentration and increases the concentration of digestibility-reducing tannins (Ayres 1993). A caterpillar must therefore eat more food over a longer time period to meet its energetic requirements for development, extending the time that it is vulnerable to predators and parasites (Lindroth 1996). Bottom-up controls related to NPP and food quality often explain ecosystem differences in average consumer biomass and consumption, with greater consumer biomass in more productive ecosystems (Figs. 10.4 and 10.5). Predation and weather, however, explain much of the interannual variation in consumer biomass and the quantity of food consumed.

People have substantially altered the trophic dynamics of ecosystems through their effects on consumer biomass. Stocking of lakes with salmonids, for example, increases predation on smaller fish. Removal of fish can have a variety of trophic effects, depending on the trophic level of the target fish. Overfishing of herbivorous fish in coral reefs, for example, allows macroalgae to escape grazing pressure and overgrow the corals, killing them in places. On land, stocking of cattle at densities higher than can be supported by primary production causes overgrazing and a decrease in plant biomass; this has led to the loss of productive capacity in many arid lands (Fig. 10.2; Schlesinger et al. 1990). The consequences of human impacts on trophic systems are highly variable, but they often have profound effects on trophic levels up and down the food chain, as well as on the target species (Pauly and Christensen 1995; Pauly et al. 2005).

The bottom-up controls over consumption efficiency can be described in terms of the factors regulating food intake. Consumption by individual animals depends on the time available for eating, the time spent looking for food, the proportion of food that is eaten, and the rate at which food is consumed and digested. Each of these four determinants of consumption has important ecological, physiological, morphological, and behavioral controls that differ among animal species (Barboza et al. 2009).

Animals do many things other than eating, including predator avoidance, digestion, reproduction, and sleeping. In addition, unfavorable conditions often restrict the time available for foraging, especially for **poikilothermic** animals such as insects, amphibians, and reptiles, whose body temperature depends on the environment. Because of this constraint, desert rodents feed primarily at night; bears hibernate most of the winter; and mosquitoes feed most actively under conditions of low wind, moderate temperatures, and high humidity. **Activity budgets** describe the proportion of the time that an animal spends in various activities. Activity budgets differ among species, seasons, and habitats, but many animals spend a relatively small proportion of their time consuming food. Changes in climate or predator risk that influence activity budgets of an animal can profoundly alter food intake and therefore the energy available for animal production and maintenance. These effects can propagate through food webs. Reintroduction of wolves in Yellowstone National Park in the western U.S., for example, caused elk to concentrate their activity in less productive ecosystems, shifting the landscape patterns of consumption and soil carbon turnover (Frank 2008; Beschta and Ripple 2009).

Animals must find their food before they eat it. Most predators such as wolves spend more time looking for food than ingesting it. Other animals, including most herbivores, search for favorable habitats within a landscape, then spend most of their time ingesting food. Animals generally consume food faster than they can digest it, so some of the time spent in other activities simultaneously contributes to digestion of food.

Once an animal finds its food, it generally consumes only some of it. Many herbivores, for example, select only the youngest leaves of certain plant species and avoid other plant species, older leaves, stems, and roots. Similarly, carnivores may eat only certain parts of an animal and leave behind parts such as skin and large bones.

This selectivity places an upper limit on consumption efficiency. Many animals become more selective as food availability increases. Lions and bears, for example, eat less of their prey when food is abundant. Gypsy moths and snowshoe hares also preferentially feed on certain plant species, given the opportunity, but will feed on almost any plant during population outbreaks, after palatable species have been depleted.

Selectivity also depends on the nutritional demands of an animal. Caribou and reindeer, for example, have a gut flora that is adapted to digest lichens, which are avoided by most other herbivores. These animals eat lichens in winter when low temperatures impose a high energy demand for **homeothermy** (maintenance of a constant body temperature). Lichens have a high content of digestible energy but little protein. In summer, however, when these animals have a high protein requirement for growth and lactation, they increase the proportion of nitrogen-rich vascular plant species in their diet (Klein 1982). Other herbivores may select plant species to minimize the accumulation of plant toxins. Moose or snowshoe hares in the boreal forest, for example, can consume only a certain amount of particular plant species before accumulation of plant toxins has detrimental physiological effects (Bryant and Kuropat 1980; Feng et al. 2009). They therefore tend to avoid plant species with high levels of toxic **secondary metabolites**, that is, compounds that are not essential for normal growth and development. Selectivity by herbivores also depends on the community context. Mammalian generalist herbivores preferentially select plant species when they are uncommon because rare species are consumed too infrequently to reach a threshold of toxicity. Selectivity by these generalist browsers therefore tends to eliminate rare plant species and reduce plant diversity (Feng et al. 2009).

Selectivity differs among animal species. Some grazers, like wildebeest in African savannas, are almost like lawnmowers. They follow the pulse of grass growth that occurs after rains and consume most plants that they encounter. Other animals, like impala, select leaves of relatively high nitrogen and low fiber content, especially in the dry season. Among mammals, there is a continuum from large-bodied **generalist herbivores**, which

are relatively nonselective, to small-bodied **specialist herbivores**, which are highly specific in their food requirements (Barboza et al. 2009). Similar patterns are seen among freshwater zooplankton; large-bodied cladocerans like *Daphnia* are generalist filter feeders, whereas same-sized or smaller copepods are more selective (Thorp and Covich 2001). Specialization is even more pronounced among terrestrial insects. Some tropical insects, for example, eat only one part of a single plant species. The abundance of specialist insects could contribute to the high diversity of tropical forests, by preventing any one plant species from becoming extremely abundant.

Assimilation Efficiency

Assimilation efficiency depends on both the quality of the food and the physiology of the consumer. Assimilation efficiency (E_{assim}) is the proportion of ingested energy (I_n) that is digested and assimilated (A_n) into the bloodstream (Fig. 10.10).

$$E_{assim} = \frac{A_n}{I_n} \qquad (10.4)$$

Unassimilated material returns to the soil as feces, a component of the detrital input to ecosystems.

Assimilation efficiencies are often higher (5–80%) than consumption efficiencies (0.1–50%). Carnivores feeding on vertebrates tend to have higher assimilation efficiencies (about 80%) than do terrestrial herbivores (5–20%) because carnivores eat food that has less structural material and is more digestible than in terrestrial plants. Carnivores that kill large prey can avoid eating indigestible parts such as bones, whereas most terrestrial herbivores consume low-quality cell walls in combination with high-quality cell contents. Among herbivores, species that feed on seeds, which have high concentrations of digestible, energy-rich storage reserves, have a higher assimilation efficiency than those feeding on leaves. Leaf-feeding herbivores, in turn, have higher assimilation efficiencies than those feeding on wood, which has higher concentrations of cellulose and lignin. Many aquatic herbivores have particularly high assimilation efficiency (up to 80%) because of the low allocation to structure in many phytoplankton and other aquatic plants. Even in aquatic ecosystems, however, herbivores that feed on well-defended species have low assimilation efficiencies. Assimilation efficiencies of herbivores feeding on cyanobacteria, for example, can be as low as 20%.

The physiological properties of a consumer strongly influence assimilation efficiency. Ruminants, which carry a vat of cellulose-digesting microbes (the rumen), have a higher assimilation efficiency (about 50%) than do most nonruminant herbivores (Barboza et al. 2009). One reason for the high assimilation efficiency of ruminants is the greater processing time than in nonruminants of similar size, giving more time for microbial breakdown of food. Homeotherms typically have higher assimilation efficiencies than do poikilotherms due to the warmer, more constant gut temperature, which promotes digestion and assimilation. Homeotherms therefore have an advantage over poikilotherms in both consumption and assimilation efficiency.

Production Efficiency

Production efficiency is determined primarily by animal metabolism. Production efficiency (E_{prod}) is the proportion of assimilated energy (A_n) that is converted to animal production ($Prod_n$; Fig. 10.10). Production efficiency includes both growth of individuals and reproduction to produce new individuals.

$$E_{prod} = \frac{Prod_n}{A_n} \qquad (10.5)$$

Assimilated energy that is not incorporated into production is lost to the environment as respiratory heat. Production efficiencies for individual animals vary 50-fold from less than 1% to greater than 50% (Table 10.2) and differ most dramatically between homeotherms (E_{prod} 1–3%) and poikilotherms (E_{prod} 10–50%). Homeotherms expend most of their assimilated energy maintaining a relatively constant body temperature. This high constant body temperature makes their activity less dependent on environmental temperature and increases their capacity to catch prey and avoid

Table 10.2 Production efficiency of selected animals[a]

Animal type	Production efficiency (% of assimilation)
Homeotherms	
Birds	1.3
Small mammals	1.5
Large mammals	3.1
Poikilotherms	
Fish and social insects	9.8
Nonsocial insects	40.7
Herbivores	38.8
Carnivores	55.6
Detritus-based insects	47.0
Noninsect invertebrates	25.0
Herbivores	20.9
Carnivores	27.6
Detritus-based invertebrates	36.2

[a]Data from Humphreys (1979)

predation, but makes homeotherms extremely inefficient in producing new animal biomass. Among homeotherms, production efficiency decreases with decreasing body size because a small size results in a high surface/volume ratio and therefore a high rate of heat loss from the warm animal to the cold environment. In contrast, the production efficiency of poikilotherms is relatively high (about 25%) and tends to decrease with increasing body size. Some large-bodied animals, such as tuna, that belong to groups usually considered poikilotherms are partially homeothermic. Among poikilotherms, production efficiency is lowest in fish and social insects (about 10%), intermediate in noninsect invertebrates (about 25%), and highest in nonsocial insects (about 40%; Table 10.2). Production efficiency often decreases with increasing age because of changes in allocation to maintenance, growth, and reproduction.

Note that belowground NPP, including exudates and transfers to mycorrhizae, is large, poorly quantified, and usually ignored in estimating trophic efficiencies. Our views of trophic efficiencies may change considerably as our understanding of belowground trophic dynamics improves. Fine roots, mycorrhizae, and exudates, for example, turn over quickly and may support high belowground consumption and assimilation efficiencies for herbivores such as nematodes that specialize on these carbon sources (Detling et al. 1980).

Food Chain Length

Production interacts with other factors to determine length of food chains and trophic structure of communities. Both the NPP and the inefficiencies of energy transfer at each trophic link constrain the amount of energy that is available at successive trophic levels and could therefore influence the number of trophic levels that an ecosystem can support. The least productive ecosystems, for example, may have only plants and herbivores, whereas more productive habitats might also support multiple levels of carnivores (Fretwell 1977; Oksanen 1990). Detritus-based food chains also tend to be longer in more productive ecosystems (Moore and de Ruiter 2000). In some aquatic ecosystems, however, the trend can go in the opposite direction. Oligotrophic habitats can support inverted biomass pyramids in which large long-lived fish are more conspicuous than the phytoplankton and invertebrate populations that support them. When ecosystems are compared across broad productivity gradients, there is no simple relationship between NPP and the number of trophic levels (Pimm 1982; Post et al. 2000). Other factors such as environmental variability and the physical structure of the environment often have greater impact on the number of trophic levels than does the energy available at the base of the food chain (Post et al. 2000).

Seasonal and Interannual Patterns

In terrestrial ecosystems, production by one trophic level seldom coincides in time with consumption by the next. The temporal relationship between predator and prey is highly variable, but some common patterns emerge. Plants and their insect predators often use similar temperature and photoperiodic cues to initiate spring growth. However, insects cannot afford to emerge before their food, so there is often a brief window in spring when plants are relatively free of invertebrate herbivory (Fig. 10.13). After insect emergence, there is often a brief window before leaves become too tough or toxic for insects to feed (Feeny 1970; Ayres and MacLean 1987). In contrast to insects, homeotherm herbivores continue to consume food during the cold season, when plants are dormant. In addition, many herbivores migrate seasonally in response to seasonal variation in food quality and environment (Frank 2006; Pastor et al. 2006). These are, however, only three of many highly specific seasonal patterns of interaction between plants and their herbivores. Predation by higher trophic levels often focuses at times when prey are most vulnerable, such as when vertebrates are giving birth to young, when salmon are migrating, or when insects are moving actively in search of food. Again, the specific patterns are quite diverse and depend on the biology of predator and prey. The important point is that production by one trophic level and consumption by the next are seldom equal at any time in the annual cycle.

Predator–prey interactions also vary among years, in part because predators and prey often differ in their responses to interannual variation in weather or long-term trends in climate. Long-term warming and drying trends in the western U.S., for example, have contributed to an extensive outbreak of the mountain pine beetle due to increased

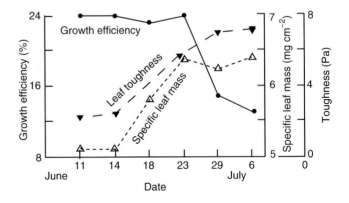

Fig. 10.13 Seasonal pattern of specific leaf mass and leaf toughness of Finnish birch leaves and of growth efficiency of fourth instar larvae of birch moths. The herbivore grows at maximal efficiency until leaves become tough and mature. After this 2-week window of leaf development, the herbivore grows slowly. Data from Ayres and MacLean (1987)

overwinter survival of the insect and a drought-induced decline in tree resistance (Allen et al. 2006; Raffa et al. 2008). Extensive tree mortality has altered virtually all ecosystem processes and shifted these forests from being a regional carbon sink to a source (Kurz et al. 2008). Predator–prey interactions can also drive population cycles of small mammals (Hanski et al. 1991) that cause changes in their food supply and vegetation-related ecosystem processes (see Chap. 12).

Nutrient Transfers

The *pathway* of nutrients through food chains is usually similar to that of energy. Nitrogen, phosphorus, and other nutrients in plants and animals are either organically bound or are dissolved in the cell contents. Nutrients contained in biomass eaten by animals therefore generally follow the same path through food chains as does energy, from plants to herbivores to primary carnivores to secondary carnivores, etc. At each link in the food chain, nutrients are digested and assimilated by animals, just as energy is digested and assimilated, although the efficiencies may differ substantially. As with energy, nutrient losses occur with each trophic transfer in the form of uneaten food, feces, and urine, so the quantity of nutrients transferred must decline with each successive trophic link. The pyramids of nutrient transfers are therefore similar in shape to those of energy flow, although the quantitative dynamics generally differ.

An important exception to this rule is sodium, which is required by animals for transmission of impulses in nerves and muscles. In contrast to animals, most plants do not require sodium and actively exclude it from roots and leaves, so tissue concentrations are lower in plants than would be expected based on soil solution concentrations (see Chap. 8). Sodium is therefore sometimes limiting to herbivores. Many terrestrial herbivores supplement the sodium and other minerals acquired from food by ingesting soil or salts from **salt licks**, which are mineral-rich springs or outcrops. Minerals may therefore show a different pathway of trophic transfer than do other nutrients.

A larger proportion of the nutrients contained in plant production pass through terrestrial herbivores than is the case for energy. Most terrestrial herbivores selectively feed on young tissues with high concentrations of nutrients and digestible energy and low concentrations of cellulose and lignin. Because of selective herbivory on nutrient-rich tissues, a larger proportion of plant-derived nutrients cycle through plant-based trophic systems than is the case for carbon.

Terrestrial herbivores not only select nutrient-rich tissues; they cycle nutrients more rapidly than do plants. Plants resorb about half the nitrogen and phosphorus from leaves during senescence, so plant litter generally has only half the nitrogen and phosphorus concentrations compared to the live tissue eaten by herbivores (see Chap. 8). For this reason, herbivory on leaves is at least twice as important an avenue for nitrogen and phosphorus cycling in terrestrial ecosystems as it is for biomass and energy. The rate of nutrient turnover by animals depends on the relative limitation by nutrients and energy (Sterner and Elser 2002). Terrestrial grazers excrete nutrients that are in excess of their growth requirements in inorganic form or as simple organics such as urea and uric acid that are quickly hydrolyzed in soils (see Chap. 9). In summary, terrestrial herbivores speed nutrient cycling in at least three ways: (1) by removing plants tissues that are more nutrient-rich than would otherwise return to the soil in litterfall, (2) returning nutrients to the soil faster than they would be recycled by plants, and (3) returning nutrients to the soil in forms that can be directly absorbed by plants (Fig. 10.9).

The ratio of elements required by plants and herbivores determines the nature of element limitation in organisms and the patterns of nutrient cycling in ecosystems. Both freshwater and terrestrial plants require nitrogen and phosphorus in a molar ratio of about 30:1 (Fig. 10.14; see Fig. 8.9; Sterner and Elser 2002). The N:P ratio in herbivorous zooplankton and insects is similar (about 26:1) to that in plants. N:P ratio is, however, quite variable among both plants and animals, reflecting both storage of nitrogen or phosphorus that is accumulated in excess of immediate requirement (see Chap. 8) and differences among organisms in their requirements for

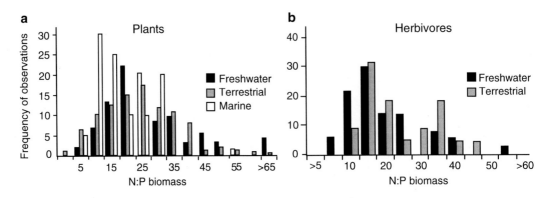

Fig. 10.14 Frequency distribution of N:P molar ratios in freshwater, terrestrial, and marine plants (*left*) and terrestrial and freshwater herbivores (insects and crustaceans, respectively; *right*). Redrawn from Elser et al. (2000)

the two elements. Rapidly growing or actively reproducing zooplankton, for example, have high concentrations of phosphorus-rich ribosomes to support protein synthesis, and therefore a lower N:P ratio (higher phosphorus requirement). Large vertebrates also have a low N:P ratio because their high proportional allocation to bones entails a high phosphorus investment (Sterner and Elser 2002). Fish in an oligotrophic lake, for example, may account for 75% of the water-column phosphorus, and moose antlers may account for 10% of the phosphorus turnover in the boreal forest (Moen et al. 1998; Sterner and Elser 2002).

Given the wide range in N:P ratios of plants and animals, herbivores often confront food resources with a quite different element balance than their own bodies. This imbalance is corrected by efficiently acquiring the most strongly limiting element and returning to the environment a disproportionate share of elements that do not limit their growth. This tends to reinforce the patterns of nutrient limitation in the ecosystem. Differences in N:P ratios among grazers in lakes illustrate the importance of this effect. *Daphnia* is a rapidly growing cladoceran grazer that has a higher phosphorus requirement to support its rapid growth (lower N:P ratio) than more slowly growing copepods. Under conditions of *Daphnia* dominance, grazers accumulate more phosphorus and excrete more nitrogen than when copepods are the dominant grazer; this leads to short-term phosphorus limitation of phytoplankton growth when *Daphnia* dominates and short-term nitrogen limitation when copepods dominate (Sterner and Elser 2002).

The turnover of nutrients in terrestrial vegetation is quite variable (see Chap. 8). Although herbivory accounts for a smaller proportion of the total nutrient return from plants to the environment in terrestrial than in aquatic ecosystems, it could still have important effects on soil and plant N:P ratios. Elk in Yellowstone Park, U.S., for example, retain substantial phosphorus to support bone and antler growth, excreting nitrogen, and raising the N:P ratios of grazed vegetation (Frank 2008). Stoichiometric analyses provide an exciting theoretical framework for linking the nutrient requirements of organisms to element cycling patterns in ecosystems (Sterner and Elser 2002).

Trophic cascades propagate downward to affect carbon and nutrient turnover in soils. Animals affect soil carbon and nutrient turnover through effects on both the quantity and quality of organic material that enters the soil (Fig. 10.9). Reintroduction of wolves to Yellowstone Park, for example, reduced the abundance of elk and shifted their distribution from productive predator-prone lowland habitats to higher elevations, resulting in reduced herbivory and nitrogen mineralization in lowland sites (Frank 2008; Beschta and Ripple 2009). Grazing by herbivores was more important than hillslope position in governing landscape patterns soil carbon turnover (Frank et al. 2011). Similarly, removal of conspicuous spiders in old fields in the northeastern U.S. increased grasshopper herbivory, altered plant species composition, and increased litter quality and nitrogen mineralization rate, indicating the importance of trophic dynamics for ecosystem biogeochemistry (Schmitz 2009).

Detritus-Based Trophic Systems

Detritus-based trophic systems convert a much larger proportion of available energy into production than do plant-based trophic systems. Decomposer organisms (primarily bacteria and fungi) feed on plant, animal, and microbial detritus, just as herbivores feed on live plants. As in the plant-based trophic system, there is a food chain of animals that feed on these decomposer organisms (Fig. 10.15). The principles governing this energy flow are similar to those in the plant-based food chain.

The rate of input and quality of dead organic matter are the major determinants of the quantity of energy that flows through the detritus-based system. The detritus-based food chain exhibits losses of energy to growth and maintenance respiration and to feces, just as in plant-based food chains (Fig. 10.15). Moreover, each trophic transfer entails

the excretion of inorganic N and P, which become available to plants, just as in the plant-based trophic system.

The major structural distinction between plant- and detritus-based systems is that the plant-based system involves a one-way flow of energy, as energy is either transferred up the food chain or is lost from the food chain as respiration, unconsumed production, or feces. In the detritus-based food chain, however, uneaten food, feces, and dead organisms again become substrate for decomposers at the base of the food chain (Fig. 10.15; Heal and MacLean 1975). Energy flow in the detritus-based system therefore has a strong recycling component. Energy is conserved and is available to support detritus-based production until it is respired away or is converted to recalcitrant humic material. Due to the efficient use of carbon that enters the base of the food chain, the detritus-based food

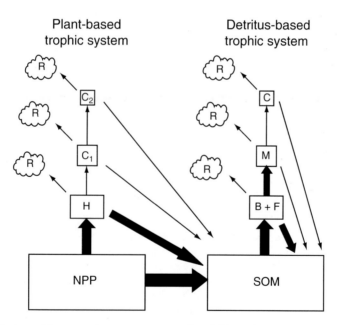

Fig. 10.15 The two basic trophic systems in ecosystems. In the plant-based trophic system, some energy is transferred from live plants to herbivores (H), primary carnivores (C₁), secondary carnivores (C₂), etc. In the detritus-based trophic system, energy is transferred from dead organic matter to bacteria (B) and fungi (F), microbivores (M), carnivores (C), etc. In both trophic systems, energy that is not assimilated at each trophic transfer passes to the detritus pool (as unconsumed organisms or feces). The major difference between these two trophic systems is that energy passes in a one-directional flow through the plant-based trophic system to herbivores and carnivores or to the detrital pool. In the detritus-based trophic system, however, material that is not consumed returns to the base of the food chain and can recycle multiple times through the food chain before it is respired away or converted to recalcitrant humus. Redrawn from Heal and MacLean (1975)

web accounts for most of the energy flow and supports the greatest animal diversity in ecosystems (Heal and MacLean 1975).

The trophic efficiencies of the detritus-based trophic system are generally higher than in the plant-based trophic system. The consumption efficiency of detritus-based food chains is high because all of the potential "food" is consumed several times until it is eventually respired away. Assimilation efficiency is also high in decomposers (bacteria and fungi) because their digestion is extracellular, so, by definition, all the material that is consumed by decomposers is assimilated. Production efficiencies of decomposers (40–60%; see Chap. 9) and animals in detritus-based food chains (35–45%) are also higher than in plant-based trophic systems (Table 10.2). Together these high trophic efficiencies explain why the detritus-based trophic system accounts for most of the secondary production in ecosystems.

Integrated Food Webs

Food webs blur the trophic position of each species in an ecosystem. In the real world, many animals feed on prey from more than one trophic level, often from both the plant-based and detritus-based trophic systems and at multiple trophic levels within each system (Polis 1991). For this reason it is difficult to assign most organisms to a single trophic level. In pelagic ecosystems, for example, zooplankton select food based on size and shape more than on species identity and consume phytoplankton, detrital particles, and small animals. On land, fungivores feed on a mixture of mycorrhizal fungi that derive their energy from plants and saprophytic fungi that decompose dead organic matter. Bacteria also derive energy from root exudates (a component of NPP) and from dead organic matter. Soil animals that eat bacteria and fungi are therefore part of both the plant-based and detritus-based trophic systems. Root-feeding mites and nematodes fall prey to animals that also eat detritus-based animals (Fig. 10.3). All soil food webs therefore process a mixture of plant and detrital energy and nutrients in ways that are difficult to untangle. Aboveground animals also eat

substantial detrital material such as fungi or soil animals. Robins, for example, feed on both earthworms and herbivorous insects. Bears eat plant roots and ants of terrestrial origin (plant-based and largely detritus-based food chains, respectively), and fish from aquatic food webs. Many insects are detrital feeders at the larval stage but as adults drink nectar or blood (plant-based trophic system). About 75% of food webs contain both plant- and detritus-based components (Moore and Hunt 1988), so mixed trophic systems are the rule rather than the exception.

Scavengers such as vultures, hyenas, crabs, and many beetles are technically part of the detritus-based food web, although their consumption, assimilation, and production efficiencies are similar to those of carnivores. Scavengers often kill weakened animals, and many predators feed on prey that have been recently killed by other animals, further blurring the distinction between plant-based and detritus-based food chains.

Parasites, pathogens, and diseases are trophically similar to predators. They derive their energy from host tissues and use the products of these cells for their own growth and reproduction, just like predators. It is difficult in practice, however, to separate the biomass of parasites, pathogens, and diseases from that of their hosts, so the concepts of consumption and assimilation efficiencies are seldom applied to these organisms. Parasites, pathogens, and diseases are therefore often treated as agents of mortality rather than as consumers.

Mutualists also confound the trophic picture. Mycorrhizal fungi can change from being mutualistic to parasitic, depending on environmental conditions and the nutritional status of the host plant (Koide 1991). Under mutualistic conditions, mycorrhizal fungi act as herbivores in transferring carbohydrates from plants to the fungus, whereas nutrient transfer occurs in the opposite direction (detritus-based food chain). The trophic role of these two organisms therefore depends on the constituent of interest. Although the broad outlines of trophic dynamics have a clear conceptual basis, the complexities of nature and our poor understanding of belowground processes often make it difficult to describe these food webs quantitatively.

Summary

Nutrient supply and other factors controlling NPP constrain the energy that is available to higher trophic levels in plant-based trophic systems. These same factors govern the quantity and quality of litter input to the soil and therefore the energy available to the detritus-based trophic system. These factors constitute the bottom-up controls over trophic dynamics. The trophic efficiency with which energy is transferred from one trophic level to the next depends on the efficiencies of consumption, assimilation, and production. Consumption efficiency depends on the interaction of food quantity and quality with predation by higher trophic levels. Consumption efficiency of herbivores is lowest in unproductive habitats dominated by plants that are woody or well-defended. Carnivores generally have higher consumption efficiency than herbivores. Assimilation efficiency is determined primarily by food quality. It is lower in unproductive than in productive habitats and lower for herbivores than for carnivores. In contrast to the other components of trophic efficiency, production efficiency is determined primarily by animal physiology; poikilotherms, for example, have a higher production efficiency than do homeotherms. Most secondary production in terrestrial ecosystems occurs in the detritus-based trophic system. In this system, material that is not consumed or assimilated returns to the base of the food chain and continues to recycle through the food chain until it is respired or converted to recalcitrant humus. Most food webs contain both plant- and detritus-based components. Impacts, including those resulting from human activities, on any link in food webs often propagate to other links in food webs.

Review Questions

1. Describe the pathways of carbon flow in an herbivore-based food chain. How does the efficiency of conversion of food into consumer biomass differ between herbivores and carnivores? What determines the partitioning of assimilated energy between respiration and production?
2. What is the major structural difference between plant-based and detritus-based food chains? Which food chain can support the greatest total production? Why?
3. What are the major structural differences between terrestrial and aquatic food chains? Why do these differences occur?
4. What plant traits determine the amount of herbivory that occurs? What ecological factors influence these plant traits?
5. What are the effects of herbivores on nitrogen cycling?
6. What are the mechanisms by which top predators influence abundance of primary producers in aquatic food chains? How does the number of trophic links affect ecosystem structure?

Additional Reading

Barboza, P.S., K.L. Parker, and I.D. Hume. 2009. *Integrative Wildlife Nutrition*. Springer, Berlin.

Coley, P.D., J.P. Bryant, and F.S. Chapin, III. 1985. Resource availability and plant anti-herbivore defense. *Science* 230:895–899.

Cyr, H. and M.L. Pace. 1993. Magnitude and patterns of herbivory in aquatic and terrestrial ecosystems. *Nature* 343:148–150.

Danell, K., R. Bergstrom, P. Duncan, and J. Pastor. 2006. *Large Herbivore Ecology, Ecosystem Dynamics and Conservation*. Cambridge University Press, Cambridge.

Heal, O.W., and J. MacLean, S.F. 1975. Comparative productivity in ecosystems-secondary productivity. Pages 89–108 *in* W.H. van Dobben, and R.H. Lowe-McConnell, editors. *Unifying Concepts in Ecology*. Junk, The Hague.

Herms, D.A., and W.J. Mattson. 1992. The dilemma of plants: To grow or defend. *Quarterly Review of Biology* 67:283–335.

Lindeman, R.L. 1942. The trophic-dynamic aspects of ecology. *Ecology* 23:399–418.

Oksanen, L. 1990. Predation, herbivory, and plant strategies along gradients of primary productivity. Pages 445–474 *in* J.B. Grace, and D. Tilman, editors. *Perspectives on Plant Competition*. Academic Press, San Diego.

Paine, R.T. 2000. Phycology for the mammalogist: Marine rocky shores and mammal-dominated communities. How different are the structuring processes? *Journal of Mammalogy* 81:637–648.

Pastor, J., R.J. Naiman, B. Dewey, and P. McInnes. 1988. Moose, microbes, and the boreal forest. *BioScience* 38:770–777.

Polis, G.A. 1999. Why are parts of the world green? Multiple factors control productivity and the distribution of biomass. *Oikos* 86:3–15.

Power, M.E. 1992. Top-down and bottom-up forces in food webs: Do plants have primacy? *Ecology* 73:733–746.

Sterner, R.W. and J.J. Elser. 2002. *Ecological Stoichiometry: The Biology of Elements from Molecules to the Biosphere*. Princeton University Press, Princeton.

Species Effects on Ecosystem Processes

11

The nature and diversity of species traits and the interactions among organisms strongly affect ecosystems. This chapter describes the patterns of species effects on ecosystem processes.

Introduction

People have massively altered the species composition of the biosphere. Human activities have modified about 75% of the ice-free surface of Earth (see Fig. 1.8; Ellis and Ramankutty 2008) through changes in land use, disturbance regime, and ecosystem management (Foley et al. 2005; MEA 2005). Human ignitions and fire suppression, for example, have altered fire frequency; many shrublands and grasslands are intensively grazed; and pollution has altered nutrient availability throughout the planet. These changes have altered plant, animal, and microbial species composition and have directly affected ecosystem processes such as primary production and nutrient cycling.

People have also deliberately or unintentionally moved thousands of species around the globe, leading toward a homogenization of the global biota (D'Antonio and Vitousek 1992). Where these species establish sustained, expanding populations in their new habitat, they represent human-caused biological invasions. As this chapter will illustrate, invasions that alter biological properties or processes can change many aspects of ecosystem structure and functioning, underscoring the importance of the organism state factor (see Chap. 1). Biological invasions are not unique in their influence on ecosystems; native species can have equivalent effects, but the rapid changes that often occur after biological invasion can be documented more clearly than can the effects of long-standing components of native communities.

A Focal Issue

Exotic species sometimes change the physical and biotic environment enough to alter the abundance of or even eliminate native species from an ecosystem. People, for example, introduced to New Zealand all of its terrestrial mammals and half of its plant species in the last 200 years (Kelly and Sullivan 2010). Mammalian introductions caused extinction of 25% of New Zealand's original bird fauna, which was rich in ground-nesting species (Tennyson 2010). Similarly, recent expansion of exotic grasses into the Sonoran Desert of the southwestern U.S. outcompetes native species and increases fuel loads. Together these changes threaten to eliminate long-lived fire-sensitive species such as the Saguaro cactus (Fig. 11.1).

Aquatic ecosystems have been even more extensively modified by species introductions. Accidental introductions of species in ballast water and fishing gear or deliberate introduction of fish and other organisms have altered the species composition of most estuaries, rivers, and

F.S. Chapin, III et al., *Principles of Terrestrial Ecosystem Ecology*,
DOI 10.1007/978-1-4419-9504-9_11, © Springer Science+Business Media, LLC 2011

Fig. 11.1 Buffel grass is a European grass that has transformed Sonoran desert of the Southwestern U.S. by outcompeting native species, including seedlings of Saguaro cactus (Olsson et al. in press). Over the longer term, the grass also represents a fire hazard that could eliminate adults of the fire-sensitive Saguaro cactus from its current range. Photograph courtesy of Aaryn Olsson

lakes. Fishless lakes, for example, tend to have a high diversity of birds, plants, amphibians, and invertebrates. All these groups decline in abundance and diversity when fish are introduced (Scheffer et al. 2006).

Although extinction and immigration of species are natural ecological processes, the dramatic increase in the frequency of these events (often greater than 100-fold) in recent decades is rapidly changing the patterns of biodiversity of the planet. It is therefore critical to understand which species changes are most likely to have large ecosystem consequences and to develop strategies to minimize the likelihood of introducing these species to new places.

Overview of Species Effects on Ecosystem Processes

No single species can perform all of the functional roles of organisms within a trophic level of an ecosystem. Up to this point, we have emphasized only the most general properties of organisms. We discussed primary producers, for example, as if they were a homogeneous group of organisms whose traits, such as photosynthetic rate, could be broadly predicted from climate and parent material. Under what circumstances is the diversity of organisms *within* a trophic level important to understanding ecosystem processes?

Biodiversity is the biological diversity present in a system, including genetic diversity within populations, **species diversity** within functionally similar groups of species, and the diversity of ecosystems on a landscape. From an ecosystem perspective, biodiversity can be characterized as the sum of the biological traits of all the species in the ecosystem, weighted by the abundance of each species (Grime 1998). When species are lost, the range of traits represented within the ecosystem declines, which reduces the range of conditions under which ecosystem properties can be sustained. In addition, since each species packages traits in somewhat different combinations, loss or gain of a species changes the ways in which traits interact to influence ecosystem processes.

Fig. 11.2 Expected relationship between ecosystem processes and the number of species, their relative abundance, and the type of species in an ecosystem. (**a**) Some processes (or stocks) may increase (as shown) with increasing species number; others may show an exponential decrease (Vitousek and Hooper 1993). (**b**) Removal of dominant species from an ecosystem has greater impact on ecosystem processes than does removal of rare species. (**c**) Similarly, the removal of keystone species has large ecosystem effects, whereas removal of one species of a functional type allows other species in that functional type to increase in abundance; this compensation would cause only moderate impact on ecosystem processes, until most species from that functional type have been removed. The arrows show the expected change in ecosystem processes in response to species loss. Based on Sala et al. (1996)

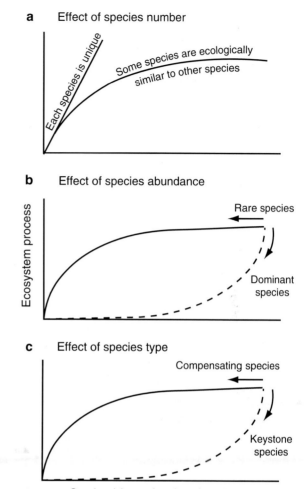

Functional traits are the characteristics of individual organisms that impact their fitness through effects on growth, reproduction, or survival (Díaz and Cabido 2001; Violle et al. 2007).

As a first approximation, the impact of a species depends on its abundance, the geographical range that it occupies, and its per capita impact (Parker et al. 1999; Suding et al. 2008). A change in the abundance of a dominant or widespread species is more likely to affect ecosystems than is a change in abundance of a rare species (Fig. 11.2b; Sala et al. 1996) because dominant species account for most of the carbon and nutrient flow through an ecosystem and have the greatest impact on the environment (Grime 1998). Loss of dominant conifers due to pathogen or insect outbreak, for example, alters microclimate

and plant biomass strongly enough to affect most ecosystem processes (Matson and Waring 1984; Kurz et al. 2008; Raffa et al. 2008). However, rare species can also play important functional roles. In a New Zealand floodplain, for example, nonnative plant species that accounted for only 3% of biomass significantly increased soil carbon, microbial biomass, and abundance of microbial-feeding and predatory nematodes (Peltzer et al. 2009). Rare species become particularly important when extreme events (e.g., insect outbreaks, wildfire, or overgrazing) or environmental changes reduce the biomass of ecologically similar dominant species (Grime 1998; Walker et al. 1999).

If all species were equally abundant and functionally different (i.e., contributed in unique ways to a given process), rates of ecosystem processes might change linearly as the number of species increased (Fig. 11.2a; Vitousek and Hooper 1993; Sala et al. 1996). Nitrogen retention, for example, might increase as species with different rooting depths or preferred forms of nitrogen absorption are added to the ecosystem. In practice, however, the relationship between species number and rate of any given ecosystem process tends to saturate with increasing number of species because some species that are added are ecologically similar to species already present in the community (Fig. 11.2a).

The degree of functional similarity among species is ecologically important (Hooper et al. 2005). A **keystone species** is ecologically distinct from all other species in the ecosystem and has a much greater impact on ecosystem or community processes than would be expected from its bio-mass (Fig. 11.2c; Power et al. 1996). The tsetse fly in Africa, for example, has a large effect on ecosystem processes per unit of tsetse fly biomass because it limits the density of people and their impacts (Sinclair and Norton-Griffiths 1979). Loss of a keystone species has a greater ecological impact than does the loss of a species that is functionally similar to other species because, in the latter case, the remaining species can sustain the relevant ecological functions.

Functional types are groups of species that are "ecologically similar" with respect to either their *effects* on ecosystems (**effect functional types**) or their *response* to environmental change (**response functional types**) (Díaz and Cabido 2001; Elmqvist et al. 2003; Hooper et al. 2005; Suding et al. 2008). Nitrifying bacteria, evergreen shrubs, and termites are examples of functional types that have predictable *effects* on ecosystem processes. Nitrifiers increase the mobility of available nitrogen in soils; evergreen shrubs produce well-defended leaves that have low palat-ability to herbivores and decompose relatively slowly; termites mix the soil vertically and redis-tribute surface litter to depth.

C_4 grasses and fire-adapted species are exam-ples of functional types that may *respond* predict-ably to specific environmental changes. C_4 grasses outperform C_3 grasses at warm temperatures; fire-adapted species survive and resprout rapidly after fire. Ultimately, we want to know how response and effect functional properties relate to one another because this provides a mechanistic basis for understanding how changes in species composition influence ecosystem responses to environmental change. Most evergreen shrubs, for example, not only have predictable effects on the ecosystem but also show predictable responses to the environment, such as growing well at low soil nutrient availability. In contrast, C_4 grass spe-cies exhibit a wide range of growth rates and nutrient responses, making it more difficult to assess the functional consequences of climate-driven changes in their distribution.

The more species of a functional type that are present, the less likely it is that gain or loss of a single species from that functional type will have large ecosystem impacts. Our challenge, as ecol-ogists, is to identify the traits of organisms that have strong effects on ecosystems (Paine 2000) and to predict what environmental changes might alter the abundance of these species.

Effect Functional Types

Species are most likely to have strong ecosys-tem effects when they alter the interactive con-trols (e.g., resource supply or occurrence of disturbance) that directly regulate ecosystem processes (see Chap. 1). These controls influence biogeochemical processes, biophysical processes, trophic interactions, and disturbance regime (Vitousek 1990; Chapin 2003; S.E. Hobbie, per-sonal communication). Species that influence interactive controls indirectly affect all aspects of ecosystem functioning.

Species Effects on Biogeochemistry

Nutrient Supply
Species traits that influence nutrient inputs or losses have important ecosystem effects. The introduction of an active nitrogen fixer into a community that lacks such species augments nitrogen availability and cycling. The introduction of the exotic nitrogen-fixing tree, *Morella faya*

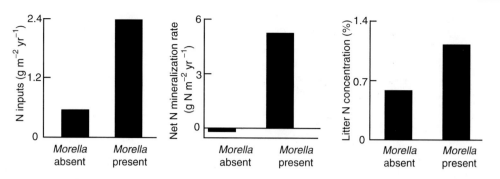

Fig. 11.3 Impact of the nitrogen-fixing tree *Morella faya* on nitrogen inputs, litter nitrogen concentration, and nitrogen mineralization rate in a Hawaiian montane forest. Data are averages ± SE (Vitousek et al. 1987)

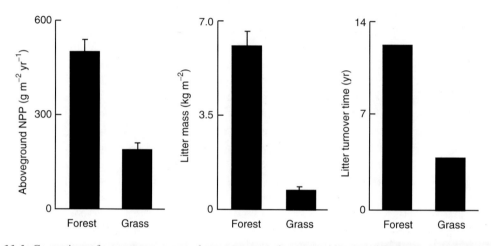

Fig. 11.4 Comparison of ecosystem processes between two exotic communities that differ in rooting depth: annual grassland and *Eucalyptus* forest in California. Data are averages ± SE (Robles and Chapin 1995)

(formerly *Myrica faya*) in Hawai'i, for example, increased nitrogen inputs, litter nitrogen concentration, nitrogen availability, and the composition of both the plant and soil faunal communities (Fig. 11.3; Vitousek et al. 1987; Vitousek 2004). A nitrogen-fixing invader is most likely to be successful in ecosystems that are nitrogen-limited, have no symbiotic nitrogen fixers, and have adequate phosphorus, micronutrients, and light (see Chap. 9; Vitousek and Howarth 1991). Thus, we expect large ecosystem impacts from invasion of nitrogen-fixing species in combinations of the following circumstances: (1) low nitrogen supply (early succession on degraded lands and in other low-nitrogen environments), (2) low competition for light or phosphorus (e.g., early in succession, canopy reduction by grazing of pastures, or phosphorus enrichment of lakes or soils), (3) prefer-

ential grazing on nitrogen-fixing species, or (4) lack of resident nitrogen-fixing species (e.g., islands that are distant from source populations) (Vitousek et al. 2002).

Deep-rooted species can increase the volume of soil tapped by an ecosystem and therefore the supply of water and nutrients available to support production. The perennial bunch grasses that once dominated California grasslands, for example, have been largely replaced by either introduced European annual grasses or planted forests; among those forests are stands of Australian *Eucalyptus*. The deep-rooted *Eucalyptus* trees access a deeper soil profile than do annual grasses, so the forest absorbs more water and nutrients. In dry, nutrient-limited ecosystems, this substantially enhances ecosystem productivity and nutrient cycling (Fig. 11.4) but reduces species diversity.

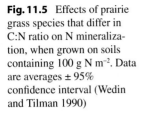

Fig. 11.5 Effects of prairie grass species that differ in C:N ratio on N mineralization, when grown on soils containing 100 g N m⁻². Data are averages ± 95% confidence interval (Wedin and Tilman 1990)

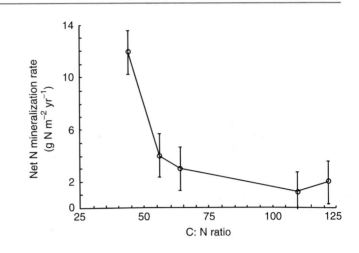

At a more subtle level, species coexistence in arid grasslands depends on species differences in rooting depth and the water sources that they tap (Fargione and Tilman 2005; Nippert and Knapp 2007a, b). Species may also tap resources that might otherwise be unused. The alpine snowbed species *Corydalis conorhiza*, for example, produces "snow roots" that grow upward into the snowpack, where they absorb nitrogen that would otherwise flow downslope at snowmelt and be lost from the system (Onipchenko et al. 2009).

Mycorrhizal fungi also influence the quantity of nutrients that are available to vegetation (see Chap. 8). Absence of appropriate mycorrhizae can restrict the establishment of plantations of exotic forest species.

Animals can influence the resource base of the ecosystem by foraging in one area and depositing nutrients elsewhere in feces and urine (see Chap. 10). Sheep, for example, enrich soils on hilltops where they bed down at night. Migrating salmon perform a similar nutrient-transport role in streams. They feed primarily in the open ocean, then return to small streams where they spawn, die, and decompose. The nutrients carried by the salmon from the ocean can sustain a substantial proportion of the algal and insect productivity of small streams. These nutrient subsidies are transported to adjoining terrestrial habitats by bears and otters that feed on salmon or by predators of insects that emerge from streams (Naiman et al. 2005).

Nutrient Turnover

Species differences in litter quality magnify site differences in soil fertility. Differences among plant species in tissue quality strongly influence litter decomposition rates (see Chap. 7). Litter from low-nutrient-adapted species decomposes slowly because of the negative effects on soil microbes of low concentrations of nitrogen and phosphorus and high concentrations of lignin, tannins, waxes, and other recalcitrant or toxic compounds. This slow decomposition of litter from species characteristic of nutrient-poor sites reinforces the low nutrient availability of these sites (see Fig. 10.9; Hobbie 1992; Wilson and Agnew 1992). Species adapted to high-resource sites, in contrast, produce rapidly decomposing litter due to its higher nitrogen and phosphorus content and lower concentration of recalcitrant compounds, enhancing rates of nutrient turnover in nutrient-rich sites.

Experimental planting of species on a common soil shows that species differences in litter quality can alter soil fertility quite quickly. Early successional prairie grasses, whose litter has a low C:N ratio, for example, enhance net nitrogen mineralization rate of soil within 3 years, compared to the same soil planted with late-successional species whose litter has a high C:N ratio (Fig. 11.5; Wedin and Tilman 1990).

The species composition of lakes strongly influences their biogeochemistry. Zebra mussels, for example, which have spread through

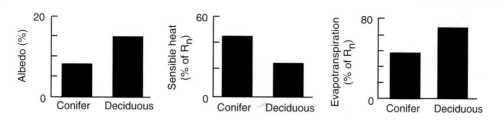

Fig. 11.6 Sensible and latent heat fluxes from deciduous and conifer boreal forests. Data are from Baldocchi et al. (2000)

freshwater systems in the Midwestern U.S., are more effective filter feeders than their native counterparts, filtering from 10% to 100% of the water column per day (Strayer et al. 1999). The resulting increase in turnover of phytoplankton and other edible particles reduces zooplankton abundance and shifts energy flow from the water column to the sediments.

Species Effects on Biophysical Processes

Species effects on microclimate influence ecosystem processes most strongly in extreme environments (Wilson and Agnew 1992; Callaway 1995; Hobbie 1995). Boreal mosses, for example, form thick mats that insulate the soil from warm summer air temperatures (Heijmans et al. 2004). The resulting low soil temperature retards decomposition, contributing to the slow rates of nutrient cycling that characterize these ecosystems (Van Cleve et al. 1991; Turetsky et al. 2010). The sequestration of nitrogen and phosphorus in undecomposed peat reduces growth of vascular plants. In hot environments, the shading of soil by plants is an important factor governing soil microclimate. Establishment of many desert cactuses, for example, often occurs in the shade of "nurse plants."

Species effects on water and energy exchange influence regional climate. The height, rooting depth, and density of the dominant species in an ecosystem govern surface roughness, which strongly influences aerodynamic conductance and therefore the efficiency of water and energy exchange between ecosystems and the atmosphere (see Chap. 4). Rough canopies generate mechanical turbulence, allowing eddies of air from the free atmosphere to penetrate deep within the plant canopy. These eddies efficiently carry water vapor from the ecosystem to the atmosphere. Individuals or species that are taller than surrounding vegetation generate canopy roughness that increases water flux from ecosystems.

Species differences in albedo and water and energy exchange can have effects that are important to the climate system. Conifers that dominate late-successional boreal forests have a low albedo and stomatal conductance and therefore transfer large amounts of sensible heat to the atmosphere. Postfire deciduous forests, in contrast, absorb less energy, due to their high albedo, and transmit more of this energy to the atmosphere as latent rather than sensible heat, resulting in less immediate warming of the atmosphere and more moisture available to support precipitation (Fig. 11.6).

Changes in vegetation caused by overgrazing can also alter regional climate. In the Middle East, for example, overgrazing reduced the cover of plant biomass. Model simulations suggest that the resulting increase in albedo reduced the total energy absorbed, the amount of sensible heat released to the atmosphere, and consequently the amount of convective uplift of the overlying air. Less moisture was therefore drawn inland from the Mediterranean Sea, resulting in less precipitation and reinforcing the vegetation changes (Charney et al. 1977). These vegetation-induced climate feedbacks could have contributed to the desertification of the Fertile Crescent.

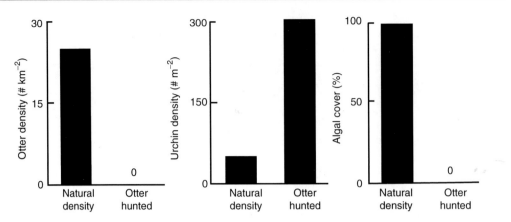

Fig. 11.7 Density of sea otters and sea urchin, and percentage cover of macroalgae in the Aleutian Islands of Alaska. Sites differed in otter density due to differential hunting pressure 300 year previously. Data are from Estes and Palmisano (1974)

Species Effects on Trophic Interactions

Species that alter trophic dynamics can have large ecosystem impacts. When top predators are removed, prey populations sometimes explode and deplete their food resources, leading to a cascade of ecological effects (see Chap. 10). These **top-down controls** are particularly well developed in aquatic systems. The removal of sea otters by Russian fur traders, for example, caused a population explosion of sea urchins that overgrazed kelp (Figs. 11.7 and 11.8; Estes and Palmisano 1974). Recent overfishing in the North Pacific may have triggered similar sea urchin outbreaks, as killer whales moved closer to shore in search of food and switched to sea otters as an alternate prey (Estes et al. 1998). In the absence of dense sea urchin populations, kelp provides the physical structure for diverse subtidal communities and attenuates waves that otherwise cause coastal erosion during storms. Similarly, on land, introduction of arctic foxes to islands reduced seabird populations and the inputs of marine-derived nutrients, causing a shift from grassland to shrubland (Croll et al. 2005).

The addition or removal of a fish species from lakes often has large keystone effects that cascade up or down the food chain (Carpenter et al. 1992; Power et al. 1996). Many nonaquatic ecosystems also exhibit strong responses to changes in predator abundance (Hairston et al. 1960; Strong 1992; Hobbs 1996). Removal of wolves,

for example, releases elk populations that graze down vegetation (Beschta and Ripple 2009), and the removal of elephants or other keystone mammalian herbivores leads to encroachment of woody plants into savannas (Owen-Smith 1988). Disease organisms, such as rinderpest that attacks ungulates in Africa, can also act as a keystone species by greatly modifying competitive interactions and community structure (Bond 1993). Plant species that are introduced without their host-specific insect herbivores or pathogens often become aggressive invaders. The cactus *Opuntia*, for example, became surprisingly abundant when introduced to Australia, in part due to overgrazing, but was reduced to manageable levels by a cactus-specific herbivore *Cactoblastis* that was introduced to control it. Other species that have become aggressive in the absence of their specialist herbivores include goldenrod (*Solidago spp.*) in Europe, wild rose (*Rosa spp.*) in Argentina, and star thistle (*Centaurea spp.*) in California.

Often these top-down controls by predators or pathogens have a much greater effect on biomass and species composition of lower trophic levels than on the total flow of energy or nutrients through the ecosystem (Carpenter et al. 1985) because of greater turnover at the producer level. Intensely grazed grassland systems such as the southern and southeastern Serengeti, for example, have a low plant biomass but rapid cycling of carbon and nutrients due to rapid turnover of

Fig. 11.8 Kelp forest characteristic of otter-occupied subtidal habitat in the Aleutian Islands of Alaska compared to urchin-dominated barrens resulting from elimination of sea otters by Russian fur traders. The three dominant kelps are *Eularia* (*Alaria*), an annual species that extends toward the surface, *Laminaria*, which forms the lower canopy, and *Agarum*, which has holes in the blades. Photographs courtesy of Jim Estes and Mike Kenner

plant biomass and excretion by large mammals. Grazing prevents the accumulation of standing dead litter and hastens the return of nutrients to soil in plant-available forms (McNaughton 1985, 1988). Keystone predators or grazers thus alter the *pathway* of energy and nutrient flow, modifying the balance between plant-based or detritus-based food chains, but we know less about their effects on total energy and nutrient cycling through ecosystems.

Species Effects on Disturbance Regime

Organisms that alter disturbance regime change the relative importance of colonization and species interactions in controlling ecosystem processes. After disturbance, there are substantial changes in most ecological processes, including increased opportunities for colonization by new individuals and often an imbalance between inputs to, and outputs from, ecosystems (see Chap. 12). For this reason, animals or plants that alter disturbance frequency or severity increase the importance of processes, such as colonization, that determine community composition under nonequilibrium conditions. Plants that colonize after disturbance, in turn, affect all aspects of the subsequent functioning of ecosystems.

One of the major mechanisms by which animals affect ecosystem processes is through their action as **ecosystem engineers**, by which they create or modify habitat (Jones et al. 1994; Lawton and Jones 1995; Hobbs 1996). Gophers, pigs, and ants, for example, physically disturb the soil, creating sites for seedling establishment and favoring early successional species (Hobbs and Mooney 1991). African elephants have a similar effect, trampling vegetation and removing portions of trees (Owen-Smith 1988). By analogy, the Pleistocene megafauna may have promoted steppe grassland vegetation by trampling mosses and stimulating nutrient cycling (Zimov et al. 1995).

The shift toward early successional or less woody vegetation generally leads to a lower biomass, a higher ratio of production to biomass, and a litter quality and microenvironment that favor decomposition (see Chap. 12). The associated enhancement of mineralization can either stimulate production (Zimov et al. 1995) or promote ecosystem nitrogen loss (Singer et al. 1984), depending on the magnitude of disturbance.

Beavers in North America are ecosystem engineers that modify the physical environment at a landscape scale (Jones et al. 1994). The associated flooding of organic-rich riparian soils produces anaerobic conditions that promote methanogenesis, so beaver ponds become hot spots of methane emissions (see Chap. 13; Roulet et al. 1997). The recent recovery of beaver populations in North America after intensive trapping during the 19th and early 20th centuries has substantially altered boreal landscapes, leading to a fourfold increase in methane emissions in regions where beaver are abundant (Bridgham et al. 1995).

The major ecosystem engineers in soils are earthworms in the temperate zone and termites in the tropics (Lavelle et al. 1997). Soil mixing by these animals alters soil development and most soil processes by disrupting the formation of distinct soil horizons, reducing soil compaction, and transporting organic matter to depth (see Chap. 7). The associated soil disturbance can greatly reduce soil carbon storage and understory plant diversity (Bohlen et al. 2004).

Plants also alter disturbance regime through effects on flammability. The introduction of grasses into a forest or shrubland, for example, can increase fire frequency and cause the replacement of forest or shrubland by grassland (D'Antonio and Vitousek 1992; Mack et al. 2001; Grigulis et al. 2005). Similarly, boreal conifers are more flammable than deciduous trees because of their large leaf and twig surface area, canopies that extend to the ground surface (acting as ladders for fire to move into the canopy), low moisture content, and high resin content (Johnson 1992). The resins in boreal conifers that promote fire also retard decomposition (Flanagan and Van Cleve 1983) and contribute to fuel accumulation.

In other situations, plants are critical in reducing disturbance by stabilizing soils and reducing wind and soil erosion in early succession. This allows successional development to proceed and retains the soil resources that determine the structure and productivity of late-successional stages. Introduced dune grasses, for example, have altered soil accumulation patterns and dune morphology in the western U.S. (D'Antonio and Vitousek 1992), while introduced acacia to South Africa stabilized sand dunes and aided in the settlement of the area by Europeans. Early successional alpine vegetation stabilizes soils and reduces probability of landslides.

Response Functional Types

Species differences in environmental response broaden the range of environmental conditions under which characteristic ecosystem process rates can be sustained. The species that occupy any given ecosystem typically differ in their geographic ranges and historical responses to past climate variability (Webb and Bartlein 1992). They are therefore likely to also differ in their responses to current seasonal and interannual variation in environment and to directional changes in environment. Species in an ecosystem occur together not because they are adapted to the identical range of environmental conditions but because they can survive, compete, and reproduce in the environments where they co-occur. Therefore different species may improve their performance and be stronger competitors under cool vs. warm conditions, wet vs. dry conditions, fertile vs. unfertile conditions, or in response to changes in frequency of various disturbances or pest outbreaks. The greater the breadth of environmental tolerance represented by the suite of species in an ecosystem, the broader will be the range of conditions under which ecosystem processes such as primary and secondary production and decomposition are sustained at their characteristic rates. In this way, a diversity of environmental responses fosters resilience of ecosystem functioning to environmental variation and change (Elmqvist et al. 2003).

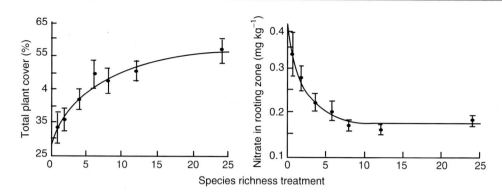

Fig. 11.9 Effect of the number of plant species sown on a plot on total plant cover and nitrate concentration in the rooting zone. Measurements were made 3 years after plots were sown. Data are averages±SE. Redrawn from Tilman et al. (1996)

Response diversity may also enhance the efficiency of resource use and retention in ecosystems. In experimental grassland communities, for example, plots that were planted with many species had greater plant cover and lower concentrations of potentially leachable soil nitrate than did low-diversity plots (Fig. 11.9; Tilman et al. 1996). This could reflect the greater probability of encountering a productive species in more diverse communities (Hooper et al. 2005). Alternatively, the more diverse plots might use more resources if species have **complementary patterns of resource use** (e.g., each species using different types of resources, rooting depths, or seasons of absorption; Tilman 1988; Dimitrakopoulos and Schmid 2004). In the Netherlands, for example, more species-rich heathlands are productive, not because of a single productive species, but because several low-productivity species together account for substantial production (van Ruijven and Berendse 2003). Complementarity tends to develop through natural selection or sorting of species to use resources that are not fully exploited by other species.

Temperate grasslands provide field evidence for complementary patterns of resource use. C_4 grasses are generally active at warmer temperatures than are C_3 grasses. Consequently, C_3 grasses account for most early-season grass production, and C_4 species for more mid-season production. Similarly, in the Sonoran desert, a different suite of annual plants becomes active after winter vs. summer rains. In both cases,

species differences in environmental response enhance annual production. In mixed-cropping agricultural ecosystems, phenological specialization to use different times of year enhances production than do species differences in rooting depth (Steiner 1982).

Diverse ecosystems are not always more productive or more efficient in using resources. Crop or forest monocultures, for example, are often just as productive as mixed cropping systems (Ewel 1986; Vandermeer 1995) or mixed-species forests (Rodin and Bazilevich 1967). The effect of **species richness** on some ecosystem process in experiments often saturates at a much lower number of species (5–10) than characterize most natural communities (Fig. 11.9). Determining the circumstances and mechanisms in which species number influences ecosystem processes is an active area of ecosystem research (Hooper et al. 2005; Naeem et al. 2009).

Response diversity is also important among animals. In Western Polynesia, a large proportion of forest trees produce fleshy fruits that are dispersed by large bats (flying foxes). There is a 60–80% overlap in diet among the bats, so, when populations of several dominant bat species were decimated by a cyclone, other bat species increased in abundance and continued dispersing fruits (Elmqvist et al. 2003). Response diversity among seed dispersers becomes increasingly important as land-use change fragments forest habitats and makes plant establishment more important to species persistence.

Integrating the Effects of Traits on Ecosystems

Functional Matrix of Multiple Traits

Organisms affect ecosystems in multiple ways through the actions of multiple traits. Functional types are a convenient simplification that enables ecologists to consider the effects of a single trait or highly correlated suite of traits on ecosystem processes. For example, we can describe functional types with respect to *either* fire tolerance, growth-related traits, temperature tolerance, rooting depth, or dispersal ability. However, many of these traits vary independently from one another, making it impossible to define a single functional type that captures all of the ways in which species affect ecosystems. For example, species effects on decomposition are mediated by several traits that vary independently of one another, including litter chemistry, labile carbon exudation, and effects on soil moisture. A **functional matrix** of traits extends the functional-types approach to consider all the traits present in an ecosystem (Eviner and Chapin 2003). Each trait (e.g., leaf lignin concentration or growth rate or rooting depth) can be treated as a continuous variable with each species in the ecosystem having a particular value for that trait. Although more complex than a one-dimensional functional-type classification, a functional matrix provides a more accurate description of species effects on ecosystems, particularly for processes that are affected by multiple species traits. In general, functional types are most useful in describing large-scale patterns of species effects, whereas a more inclusive consideration of species traits improves understanding of interactions within a specific ecosystem.

A functional matrix provides useful guidance in ecosystem restoration. Response traits identify the species that tolerate and grow well in a particular environment (Grime 2001). The suite of species that thrive in a particular environment will likely differ in their effects on the environment. By selecting appropriate species, ecologists can shape the trajectory of ecosystem development (Whisenant 1999). For example, cover crops are often selected based on their capacity to add nitrogen (Eviner and Chapin 2001). Similarly, stream restoration may require a riparian species assemblage that resists erosion (response trait) and accumulates nitrate from groundwater (response/effect trait). Once the matrix of traits is known that enable species to thrive in an environment and to have desired effects, it may be possible to identify a set of locally adapted species with the appropriate combination of traits (Eviner and Hawkes 2008). Species interactions and other (often unknown) factors create a local context that governs the relative success of species with a high restoration potential. In addition, inevitable tradeoffs (e.g., between rapid growth and resistance to drought and low soil fertility) limit the combinations of traits that can be assembled.

Linkages Between Response and Effect Traits

The effects of environmental variability and change on ecosystem processes depend on the linkages between the environmental response and the ecosystem effects of species (Suding et al. 2008). The traits that are present in an ecosystem are packaged into distinct species, each of which has a particular set of response and effect traits. If response and effect traits are tightly linked, the ecosystem will respond sensitively to environmental changes that influence these traits. Species with a high capacity for nitrogen absorption, photosynthesis, and growth, for example, respond sensitively to nitrogen supply, produce rapidly decomposing litter, and occupy nitrogen-rich sites, whereas species with low rates of these processes occupy nitrogen-poor sites. In part because of the strong linkages between response and effect traits, ecosystems respond sensitively to variation in nitrogen supply.

In other cases, however, there is little or no correlation between species response and species effect, as in the C_3–C_4 and fruit bat examples given earlier. In these cases, the coexistence of many similar species minimizes ecosystem sensitivity to environmental variation and change because the *effect* functional type (e.g., grasses) includes some species that are productive under warm, dry conditions and others that are productive under cool, wet conditions (Suding et al. 2008). Similarly, the productivity of a grassland that has both palatable and unpalatable grasses will be less sensitive to periods of intense grazing than a grassland that lacks unpalatable grasses (Walker et al. 1999).

Diversity as Insurance

Earth is currently in the midst of the sixth major extinction event in the history of life (Pimm et al. 1995). Although the causes of some of the earlier extinction events are uncertain, they probably resulted from sudden changes in physical environment caused by factors such as asteroid impacts or pulses of volcanism. Current extinction rates are at least 100-fold higher than prehuman extinction rates (Fig. 11.10; Mace et al. 2005). The current extinction event is unique in the history of life because it is biologically driven, specifically by the impact of the human species on land use, species invasions, and environmental change. Although human activities affect many processes at global scales (see Chap. 14; Vitousek 1994), the loss of species diversity is of particular concern because it is irreversible. Once a species is gone, it cannot be recovered. For this reason, it is critical to understand the functional consequences of the current large losses in species diversity (Chapin et al. 2000b).

Diversity provides insurance against functional changes under extreme or novel conditions. Conditions that favor some species will likely reduce the competitive advantage of other functionally similar species, thus stabilizing the total biomass or activity of the entire community (McNaughton 1977; Chapin and Shaver 1985; Tilman et al. 2006). In other words, when some species increase resource capture under conditions that are favorable to them, this leaves fewer resources for other species, which therefore respond by growing less. Annual variation in weather, for example, caused at least a twofold variation production by every major vascular

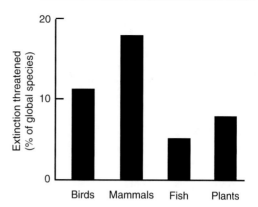

Fig. 11.10 Percentage of major vertebrate and vascular plant species that are currently threatened with extinction. Redrawn from Chapin et al. (2000b)

plant species in arctic tussock tundra. Years that were favorable for some species, however, reduced the productivity of others, so there was no significant variation in productivity at the ecosystem scale among the 5 years of study (Chapin and Shaver 1985). This stabilization of biomass and production by diversity has been observed in many (but not all) studies (Cottingham et al. 2001), including grasslands in response to water and nutrient addition (Lauenroth et al. 1978; Tilman et al. 2006) or grazing (McNaughton 1977), tundra in response to changes in temperature, light, and nutrients (Chapin and Shaver 1985), and lakes in response to acidification (Frost et al. 1995). This stability of processes provided by diversity has societal relevance. Many traditional farmers plant diverse crops, not to maximize productivity in a given year, but to decrease the risk of crop failure in a bad year (Altieri 1990).

Species diversity not only stabilizes ecosystem processes in the face of annual variation in environment but also provides insurance against drastic change in ecosystem structure or processes in response to extreme events (Walker 1992; Chapin et al. 1997). Any change in climate or climatic extremes that is severe enough to cause extinction of one species is unlikely to eliminate all members from a functional type (Walker 1995) because response and effect traits are distributed in various combinations across species (Eviner and Hawkes 2008). The more species there are in a functional-effect type, the less likely it is that any extinction event or series of such events will have serious ecosystem consequences (Holling 1986). In a laboratory experiment that manipulated species diversity of mosses, communities with high species diversity maintained a higher biomass when exposed to drought than did less diverse communities by facilitating the survival of tall dominant mosses (Mulder et al. 2001). Similarly, in field experiments, diversity contributes to sustained community composition and structure of grasslands exposed to manipulated or natural fluctuations in climate and disturbance (Grime et al. 2000; Hobbs et al. 2007; Grime et al. 2008).

Species Interactions and Ecosystem Processes

Species interactions modify the impacts of individual species on ecosystem processes. Most ecosystem processes respond in complex ways to changes in the abundance of species because *interactions* among species generally govern the extent to which species traits are expressed at the ecosystem scale. Species interactions, including mutualism, trophic interactions (predation, parasitism, and herbivory), facilitation, and competition, may affect ecosystem processes directly by modifying pathways of energy and material flow or indirectly by modifying the abundances or traits of species with strong ecosystem effects (Wilson and Agnew 1992; Callaway 1995).

Many species effects on ecosystems are indirect and not easily predicted. Species which themselves have small effects on ecosystem processes can have large indirect effects if they influence the abundance of species with large direct ecosystem effects, as described earlier for trophic interactions. Thus, a seed disperser or pollinator that has little direct effect on ecosystem processes may be essential for the persistence of a canopy species with greater direct ecosystem impact. Stream predatory invertebrates alter the behavior of their prey, making them more vulnerable to fish predation, which leads to an increase in the

weight gain of fish (Soluck and Richardson 1997). In grasslands, a combination of legumes and C_4 grasses augments soil carbon sequestration because legumes promote large nitrogen inputs, and C_4 grasses use this nitrogen efficiently to produce root biomass, which enhances soil carbon storage (Fornara and Tilman 2008). Mixtures of litter from multiple species decompose and mineralize nitrogen at different rates (often more rapid) than would be predicted from each litter type by itself (Gartner and Cardon 2004). The nature of these litter interactions is sensitive to environment (Jonsson and Wardle 2008) and often reflects interactions of nutrients from one litter type with carbon chemistry of other litter types (Dijkstra et al. 2009). Animal–plant–microbe interactions modulate species effects in California grasslands (Eviner and Chapin 2005). Here, experimental plots seeded with goatgrass, which has a low litter quality (high C:N ratio), is associated with a low nitrogen mineralization rate in the absence of disturbance. However, the high root biomass of this species enhances soil cohesion, which reduces the energetic requirement for burrowing by gophers. Gophers are attracted to the goatgrass plots, and the associated disturbance enhances nitrogen mineralization above levels associated with any species in the absence of disturbance. Thus, all types of organism interactions – plant, animal, and microbial – must be considered in understanding the effects of biodiversity on ecosystem functioning. Although each of these examples is unique to a particular ecosystem, the ubiquitous occurrence of species interactions with strong ecosystem effects makes these interactions a general feature of ecosystem functioning (Chapin et al. 2000b). In many cases, changes in these interactions alter the traits that are expressed by species and therefore the effects of species on ecosystem processes. Consequently, simply knowing that a species is present or absent is insufficient to predict its impact on ecosystems. Theoretical frameworks for predicting the types and nature of these interactions are only beginning to emerge (Parker et al. 1999; Polis 1999; Eviner and Hawkes 2008; Cardinale et al. 2009).

Summary

The species diversity of Earth is changing rapidly due to frequent species extinctions (both locally and globally), introductions, and changes in abundance. We are, however, only beginning to understand the ecosystem consequences of these changes. Many species have traits that strongly affect ecosystem processes through their effects on the supply or turnover of limiting resources, microclimate, trophic interactions, and disturbance regime. The impact of these species traits on ecosystem processes depends on the abundance of a species, its functional similarity to other species in the community, and species interactions that influence the expression of important traits at the ecosystem scale.

The effects of species traits on ecosystem processes are generally so strong that changes in the species composition or diversity of ecosystems are likely to alter their functioning, although the exact nature of these changes is often difficult to predict. Functional diversity per se may be ecologically important if it leads to complementary use of resources by different species or increases the probability of including species with particular ecological effects. Because species belonging to the same functional-effect type generally differ in their response to environment, diversity in response within a functional-effect type may stabilize ecosystem processes in the face of temporal variation or directional changes in environment. Introduction of species with different functional effects to an ecosystem, in contrast, may accelerate the rate of ecosystem change.

Review Questions

1. What are functional types? What is the usefulness of the functional-type concept if all species are ecologically distinct?
2. How is the expected ecosystem impact of the loss of a species affected by (a) the number of species in the ecosystem, (b) the abundance or dominance of the species that is eliminated, or (c) the type of species that is eliminated? Explain.
3. If a new species invades or is lost from an ecosystem, which species traits are most likely to cause large changes in productivity and nutrient cycling? Give examples that illustrate the mechanism by which these species effects occur.
4. Which species traits have greatest effects on regional processes such as climate and hydrology?
5. How do species interactions influence the effect of a species on ecosystem processes?
6. How does the diversity of species *within a functional type* affect ecosystem processes? What is the mechanism by which this occurs? Why is it important to distinguish between the effects of changes in species composition within vs. between functional types?
7. What are the mechanisms by which species diversity might affect nutrient absorption or loss in an ecosystem? Suggest an experiment to distinguish between these possible mechanisms. Design an agricultural ecosystem that maintains crop productivity but has tight nutrient cycles.

Additional Reading

Chapin, F.S., III, E.S. Zavaleta, V.T. Eviner, R.L. Naylor, P.M. Vitousek, et al. 2000. Consequences of changing biotic diversity. *Nature* 405: 234–242.

Frost, T.M., S.R. Carpenter, A.R. Ives, and T.K. Kratz. 1995. Species compensation and complementarity in ecosystem function. Pages 224–239 in C.G. Jones, and J.H. Lawton, editors. *Linking Species and Ecosystems*. Chapman and Hall, New York.

Hooper, D.U., F.S. Chapin, III, J.J. Ewel, A. Hector, P. Inchausti, et al. 2005. Effects of biodiversity on ecosystem functioning: A consensus of current knowledge and needs for future research. *Ecological Applications* 75:3–35.

Lawton, J.H., and C.G. Jones. 1995. Linking species and ecosystems: Organisms as ecosystem engineers. Pages 141–150 *in* C.G. Jones, and J.H. Lawton, editors. *Linking Species and Ecosystems*. Chapman and Hall, New York.

Naeem, S., D.E. Bunker, A. Hector, M. Loreau, and C. Perrings, editors. 2009. *Biodiversity, Ecosystem Functioning, and Human Well-being: An Ecological and Economic Perspective*. Oxford University Press, New York.

Parker, I.M., D. Simberloff, W.M. Lonsdale, K. Goodell, M. Wonham, et al. 1999. Impact: Toward a framework for understanding the ecological effects of invaders. *Biological Invasions* 1:3–19.

Power, M.E., D. Tilman, J.A. Estes, B.A. Menge, W.J. Bond, et al. 1996. Challenges in the quest for keystones. *BioScience* 46:609.

Vandermeer, J. 1995. The ecological basis of alternative agriculture. *Annual Review of Ecology and Systematics* 26:201–224.

Vitousek, P.M. 1990. Biological invasions and ecosystem processes: Towards an integration of population biology and ecosystem studies. *Oikos* 57:7–13.

Wilson, J.B., and D.Q. Agnew. 1992. Positive-feedback switches in plant communities. *Advances in Ecological Research* 23:263–336.

Temporal Dynamics

12

Ecosystem processes constantly adjust to temporal variation in environment over all time scales. This chapter describes the major patterns and controls over the temporal dynamics of ecosystems.

Introduction

Ecosystems are always changing in response to past changes as well as responding to current environment (Holling 1973, Wu and Loucks 1995, Turner 2010). In earlier chapters, we emphasized ecosystem responses to the *current* environment. *Past* changes that influence current dynamics include relatively predictable daily and seasonal variations, less predictable or longer-term changes in environment (e.g., passage of weather fronts, el Niño events, and glacial cycles), and disturbances (e.g., treefalls, herbivore outbreaks, logging, and volcanic eruptions). Consequently, the behavior of an ecosystem is always influenced by both the current environment and many previous environmental fluctuations and disturbances. This chapter addresses these temporal dynamics of ecosystems.

A Focal Issue

People have altered ecosystems more rapidly and extensively in the last 50 years than in any comparable time period in human history.

These changes have resulted from an exponentially rising human population, our consumption of resources, and our ever-increasing technological capacity to alter Earth's environment and ecosystems. Perhaps the most urgent need in ecosystem ecology is to improve our understanding of factors governing resilience and change in ecological systems. How do we prepare for changes in the types and severity of disturbances that are occurring? Warming temperatures, for example, are expected to increase sea-surface temperatures and therefore the intensity of hurricanes that impact coastal cities, such as occurred with Hurricane Katrina (see Fig. 2.1). Warmer, drier conditions in dry regions of the world are expected to cause drought and associated wildfires and insect outbreaks, as have occurred in Australia, southern Europe, and the western United States (Fig. 12.1). Flooding is expected to occur more often in wet and low-lying coastal regions. How do ecosystems respond to disturbances that they often encounter? To novel disturbances? What properties of ecosystems enhance their capacity to sustain their structure and functioning in response to changing disturbance regimes? As disturbance regimes move outside their historical patterns due to human-caused climate change, ecosystem ecologists will play a key role in understanding the causes and consequences of altered patterns of disturbance, both for the protection of life and property and to sustain the diversity and other ecological attributes of ecosystems.

F.S. Chapin, III et al., *Principles of Terrestrial Ecosystem Ecology*,
DOI 10.1007/978-1-4419-9504-9_12, © Springer Science+Business Media, LLC 2011

Fig. 12.1 Climate-induced warming has increased the extent of wildfire in many dry areas, often directly threatening life and property in the wildland–urban interface, as in this 2010 fire in Gold Hill, Colorado. Photograph courtesy of Greg Cortopassi @ Cortoimages.com

Ecosystem Resilience and Change

Alternative Stable States

A given environment can often support more than one potential state of an ecosystem. The ecosystems we observe today depend not only on their capacity to thrive under current conditions but also on historical **legacies**, that is, things that happened to them in the past. Legacies such as the past history of land use are important because ecosystems are **complex adaptive systems**. This means the system changes its properties ("adapts") in complex ways in response to changes imposed on it (Levin 1999, Chapin et al. 2009). Large areas of northeastern North America, for example, were deforested for agriculture and since 1850 have reverted to forests (Fig. 12.2). A plow layer is still evident in 150-year-old forests that developed on former agricultural fields. This sharp vertical discontinuity in soil properties and nutrient supply does not occur, however, in forests that developed from previous woodlots (Motzkin et al. 1996, Foster et al. 2010). These alternative histories give rise

to forests with different species composition, drought sensitivity, and services provided to society. A more recent trajectory, which is also sensitive to its historical roots, is toward extensive areas of pavement and other hard surfaces in cities and towns. These hard surfaces also influence species composition, runoff to aquatic ecosystems, and the likely trajectories of future ecosystem change.

The frequent occurrence of **alternative stable states** that can occur in the same current environment is familiar to anyone who has walked through a landscape and observed the bewildering fine-scale variation in ecosystem composition and structure that has no obvious explanation based on spatial variation in the current environment – for example, forest patches dominated by different species due to (often unknown) legacies of past disturbance, colonization by particular species, grazing history, etc. At larger scales, landscape patterns in a watershed may be substantially structured by past fire or land-use history. At continental scales, the historical absence of mammals in New Zealand strongly influenced ecosystem responses to the relatively recent arrival of people and the plants and animals they brought with

Fig. 12.2 Changes in landscape composition and population of New England (Northeastern U.S.) since European colonization. Most land that was cleared for agriculture by 1850 has regrown as forest or more recently been developed as cities and suburbs. Data from Foster et al. (2010)

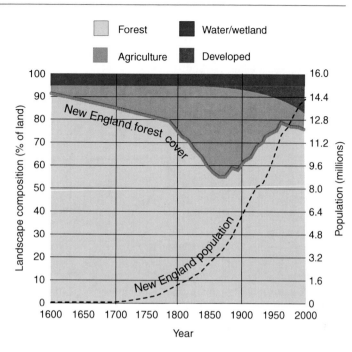

them (Kelly and Sullivan 2010). Extinction of Pleistocene megafauna as a result of climate change and human hunting contributed substantially to the ecosystem changes that occurred 10,000 years ago and are legacies that still structure today's biomes (Flannery 1994, Zimov et al. 1995, Gill et al. 2009). The important role of historical legacies and **path dependence** in explaining current dynamics of ecosystems provides a clear motivation for ecosystem stewardship. Management actions taken today can make a difference in determining the future state of ecosystems (see Chap. 15).

Resilience and Thresholds

Sources of Resilience

Resilience constrains ecosystem responses to perturbations. Although many alternative states of an ecosystem are plausible, ecosystems often maintain relatively stable functional properties for long time periods. Ecosystem **resilience** is the capacity of an ecosystem to sustain its fundamental function, structure, and feedbacks in the face of a spectrum of shocks and perturbations (Holling 1973, Chapin et al. 2009).

Ecosystems are particularly resilient to those fluctuations to which organisms are well-adapted, including day-night or seasonal cycles of light and temperature, El Niño oscillations in weather that recur every 2–10 years, and droughts, fires, or other extreme events that have occurred repeatedly during the evolutionary history of organisms that occupy the ecosystem.

Internal dynamics of ecosystems also generate fluctuations in ecosystem processes. The population density of herbivores, for example, can vary more than 100-fold over a few years, causing large fluctuations in plant biomass, nutrient cycling, and other processes (Fig. 12.3). Fluctuations, outbreaks, or cycles of grasshoppers, gypsy moths, snowshoe hares, and lemmings, for example, are typical of internal dynamics that characterize many ecosystems. These fluctuations and cycles reflect interactions between positive and negative feedbacks among plants, herbivores, predators, and parasites (Hanski et al. 2001). Herbivore populations, for example, often decline after a depletion of their food supply, due to insufficient food or buildup of predators. These feedbacks constrain potential population changes in both predator and prey, conferring resilience to the trophic dynamics of the system (see Chaps. 1 and 10).

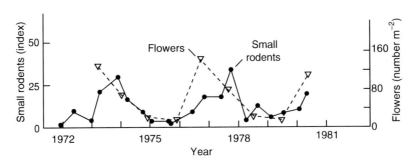

Fig. 12.3 Interannual variation in flowering density of an understory shrub (*Vaccinium myrtillus*) and of small rodents in northern Finland. These herbivores and their food plants show approximately 4-year cycles of abundance. Data from Laine and Henttonen (1983)

Maintenance of slowly changing biogeochemical pools, long-lived organisms, and biodiversity are particularly critical to long-term resilience because these variables structure so many of the interactions in ecosystems (Carpenter and Folke 2006). For example, ecosystems with a high response diversity (see Chap. 11), due to either high plasticity or high genetic or species diversity also exhibit high resilience because of the wide range of environmental or biotic conditions under which particular functions are sustained (Elmqvist et al. 2003). A grassland with both cool-season and warm-season grasses (C_3 and C_4 grasses, respectively), for example, can sustain productivity across a broader range of temperature and moisture conditions than a grassland that contains only one of these grass types. Stabilizing (negative) feedbacks that constrain changes in key slow variables at large temporal and spatial scales confer resilience to the system.

Limits to Resilience

Biological and physical limits to ecosystem resilience make ecosystems vulnerable to large or directional changes. When biotic or environmental changes exceed ecosystem resilience, some trigger for change (e.g., pest outbreak, species invasion, or change in internal dynamics) is increasingly likely to cause path-dependent change to some alternative state. Warmer temperatures that stress trees and increase winter survivorship of mountain pine beetle at high elevations, for example, can cause widespread tree mortality and restrict pine regeneration (Raffa et al. 2008). This increases the likelihood of a shift to a non-forested state. Saturation of the phosphorus-binding capacity of lake sediments can also exceed the resilience of clearwater lakes (Carpenter 2003), as described earlier.

Many of the recent changes in the global environment, including species introductions and extinctions, environmental changes, land-use changes, and introductions of novel chemicals, are likely to exceed the resilience of many ecosystems (Rockström et al. 2009). This can occur if a directional change in environment eventually exceeds the **adaptive range** of the system, that is, the difference between the upper and lower tolerance limits of the system (Fig. 12.4a) or if the environment becomes more variable and exceeds the adaptive range of the system more often (Fig. 12.4b; Smit and Wandel 2006). Alternatively, the adaptive range of the ecosystem may contract (Fig. 12.4c) due to factors such as loss of biotic diversity (e.g., when genetic diversity of crops is reduced), loss of buffering capacity of ecosystems (e.g., due to cation leaching from acid rain), or interactions with other stresses (e.g., exotic pests or high-ozone urban pollution) that constrain the limits to productivity and survivorship of species. These patterns suggest that ecosystem resilience can be enhanced by reducing environmental stresses, fostering biotic response diversity, and minimizing the complexity of interacting stresses that impact ecosystems.

Fig. 12.4 The adaptive range of an ecosystem (difference between upper and lower tolerance limits) relative to temporal variations in an important environmental control (e.g., temperature), when (**a**) the environmental control changes directionally or (**b**) becomes more variable, or (**c**) when the adaptive range declines. Based on Smit and Wandel (2006)

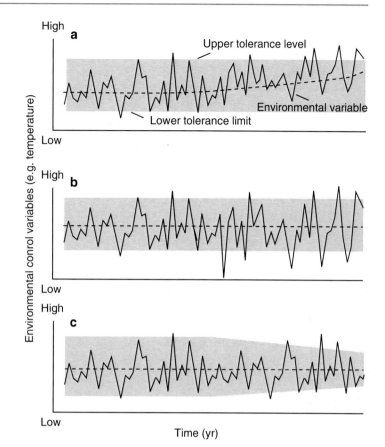

Thresholds and Regime Shifts

When ecosystem resilience is exceeded, regime shifts can occur rapidly and unexpectedly. Most ecosystems exhibit an impressive degree of resilience to natural environmental variability, as a result of their evolutionary and developmental history. They are also often remarkably resilient to the insults imposed by many human-caused changes in environment, structure, and diversity. Consequently, over a broad range of conditions, ecosystems appear to "take care of themselves" through the feedbacks described in earlier chapters. It therefore often comes as quite a surprise when resilience is exceeded, and ecosystems undergo an abrupt (threshold) change to an alternative state. Due to the path-dependent nature of changes in complex adaptive systems, the new state of the system is likely to exhibit different environmental responses and may not readily return to the original system, even when

external stresses are removed and the environment returns to its previous state (Box 12.1). In response to this **regime shift**, a new set of feedbacks and environmental responses emerge, generating resilience of the altered state. In the western U.S., for example, introduction of cheatgrass combined with overgrazing caused widespread replacement of native bunchgrasses by this unpalatable grass. The combination of reduced grazing and increased fire frequency that resulted from cheatgrass invasion maintains this grassland in its new state, which has become quite resilient to a wide variety of management efforts to restore the original grasslands (Brooks et al. 2004). Similarly, once clearwater lakes shift to a turbid state because of phosphorus saturation of sediments, the public becomes concerned and wants to fix the problem. However, it is often extremely difficult to return to the clearwater state, even when phosphorus inputs from

Box 12.1 Resilience and Regime Shifts

The response of an ecosystem to perturbation depends on its resilience and the strength and directionality of perturbations that push it toward alternative states. The behavior of a ball on a surface provides a useful analogy (Fig. 12.5; Holling and Gunderson 2002). The location of the ball represents the state of a system in relationship to some ecological variable (e.g., water availability, as represented by the position along the horizontal axis). Resilience is the tendency for the system to remain in the same state, despite temporal fluctuations in environment. This can be represented by a cup-shaped depression in the surface. If the ecosystem is highly resilient because of adaptations and stabilizing (negative) feedbacks that sustain its properties over a wide range of available moisture conditions, the cup will be broad and deep, and the system will persist in its original state despite substantial moisture perturbations (e.g., floods or droughts; Fig. 12.5a).

If droughts become more frequent or severe, it becomes increasingly likely that some drought, perhaps interacting with another event such as an insect outbreak, may push the system into a different stability domain (a regime shift), where new feedbacks maintain it in the new state (Fig. 12.5b). If ecosystem resilience to drought is eroded, for example by loss of soil organic matter (SOM) or drought-resistant species (shown by the resilience cup becoming less deep), even a modest perturbation may cause a regime shift (Fig. 12.5c).

In practice, stability landscapes are highly dynamic with constant changes in the depth and locations of stability domains (i.e., cups on the landscape that represent alternative stable states of the system). In a directionally changing world, some new stability domains become increasingly likely and current states become increasingly vulnerable (Fig. 12.5d). The challenge for ecosystem ecologists is to enhance the resilience of those stability domains that provide ecosystem integrity and benefits to society and to reduce the resilience

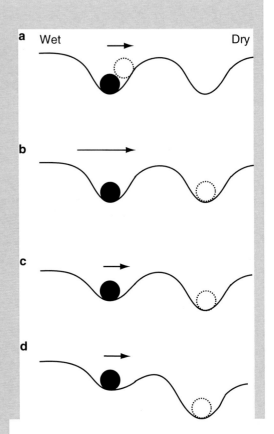

Fig. 12.5 The location of the ball represents the state of an ecosystem in relationship to some ecological variable (e.g., water availability, as represented by the position along the horizontal "water" axis). The depth of each cup defines the resilience of the ecosystem; the breadth of each cup is the range of environmental variation over which the ecosystem tends to remain in the same domain (i.e., is resilient); and the length of the arrow represents the strength of the perturbation (e.g., drought) to which the ecosystem is exposed. The solid ball is the original state of the system and the open ball is the most likely final state. (**a**) Response of a resilient system to a mild drought at steady state; (**b**) response of a resilient system to an extreme drought; (**c**) response of a less resilient system to a mild drought; and (**d**) response of a system to mild drought during a trajectory of declining moisture availability

of undesirable states. In some cases, the historical ecosystem may be feasible to maintain, but increasingly it may become necessary to choose among alternative novel states, if the current system cannot be sustained in the new environment (Hobbs et al. 2009).

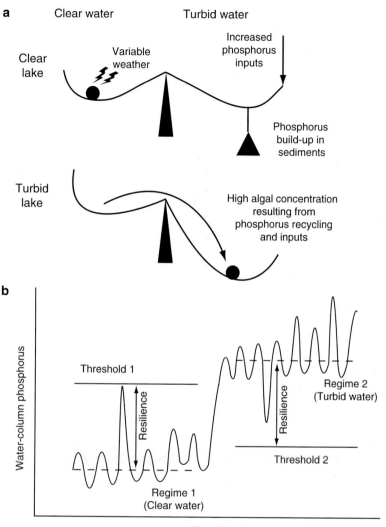

Fig. 12.6 Pan balance model (**a**) and time course (**b**) of changes in lakes in response to increases in phosphorus inputs. In the pan model, the clear lake loses resilience as phosphorus inputs increase. At some point, a stochastic shock such as a wind-driven mixing event shifts the lake from a clearwater to a turbid state. The time course of this change involves fluctuations in water-column phosphorus, initially around concentrations typical of clearwater lakes. When a fluctuation exceeds the resilience of the system, it shifts to a turbid-water regime that has a different average value of water-column phosphorus and a different threshold for return to the clearwater state. Redrawn from Carpenter (2003)

the watershed are greatly reduced (Fig. 12.6; Carpenter 2003).

Disturbances are often the trigger for regime shifts in communities facing gradual changes in conditions (Turner 2010). These shifts are seen in storms that resuspend sediment phosphorus (Carpenter 2003), severe wildfires that alter seedbed characteristics (Johnstone et al. 2010), and windthrow and fires in a warming climate that shift forests to savannas (Frelich and Reich 2009). These regime changes sometimes lead to communities with novel species composition and properties (Williams and Jackson 2007).

Ecologists often understand some of the ecosystem consequences of environmental stresses such as drought, pollution, and warming, but there is currently a very poor ability to predict how much change or stress an ecosystem can withstand before a regime shift occurs or what interacting stresses or events (e.g., insect outbreak) might trigger the shift. Sometimes key parameters like water-column phosphorus become temporally or spatially more variable as the limits of resilience are approached (Scheffer and Carpenter 2003, Carpenter and Brock 2006). A precautionary approach to reducing the risk of regime shifts is to foster resilience and reduce vulnerability by minimizing known stresses (e.g., pollution), maintaining diversity (capacity to deal with a broader range of conditions), and providing conditions where the ecosystem can adjust naturally to persistent environmental changes (Walker et al. 2004).

How can the adaptive range of ecosystems be broadened or shifted to accommodate expected changes in environment? The natural rate of ecosystem response to a changing environment through evolution or migration of genotypes and species may be too slow to keep pace with current rapid rates of change. One approach is to manage migration corridors to maximize opportunities for migration of non-weedy and noninvasive species. A second more controversial approach is **assisted migration**, in which genotypes or species are moved from a region where climate is becoming unfavorable to new places where climate is, or is expected to become, more favorable. Australia, for example, is encouraging the establishment of vineyards in areas where climate is projected to be favorable for grapes 30 years from now, a time when vines reach peak production (NRC 2010). Given the checkered history of efforts to solve management problems by introducing species to new locations, assisted migration raises concerns among many conservation biologists (McLachlan et al. 2007). This approach has received most attention among foresters, who recognize that climate may shift significantly during the lifetimes of individual trees. One approach may be to reforest logged or burned forests with seeds from a wide range of climates and allow whatever climate emerges to select among the tree seedlings that establish (Millar et al. 2007). This contrasts strikingly with current

"best practices" of reseeding with locally adapted genotypes. Assisted migration becomes a publicly attractive alternative in areas where insect outbreaks (e.g., mountain pine beetle) or species shifts (e.g., cheatgrass or junipers) have radically modified the composition of unmanaged ecosystems. Highly flammable invasive grasses have invaded Saguaro National Park, for example, threatening the slow-growing, long-lived species that the park was established to protect (see Fig. 11.1). Should saguaro cactuses be planted beyond their current range in places where grasses have not yet invaded? Ecosystem ecologists can play a constructive role in these debates by exploring the ecosystem consequences of proposed species manipulations (see Chap. 11).

Restoration ecology seeks to trigger regime shifts to alternative, potentially more favorable states. Many terrestrial and aquatic ecosystems that have been degraded by mining, industrial development, stream channelization, or overgrazing are extremely resilient and can remain in a degraded state for a long time due to unfavorable soil or site-moisture conditions. The explicit goal of restoration ecology is to transform these systems to an alternative state that would then generate its own feedbacks to sustain the restored state. For example, nitrogen-fixing trees have been used to speed soil development on mine tailings in the U.K. to generate nutrient cycles similar to those of nearby forested ecosystems (Bradshaw 1983). A valuable new wrinkle in restoration ecology is the goal of transforming degraded ecosystems to a state that is compatible with the projected *future* climate rather than to some historical reference point (Choi 2007, Hobbs and Cramer 2008).

Disturbance

Conceptual Framework

Disturbance is a major cause of long-term fluctuations in the structure and functioning of ecosystems. We define **disturbance** as a relatively discrete event in time that removes plant biomass (Grime 2001). Disturbance has also been described as a relatively discrete event in time and space that alters the structure of populations, communities, and ecosystems and causes changes in resource availability or the physical environment (White and Pickett 1985). Disturbances include herbivore outbreaks, treefalls, fires, hurricanes, floods, glacial advances, and volcanic eruptions. The dividing line between disturbance and normal function is somewhat arbitrary. Herbivory, for example, is often treated as part of the steady-state dynamics of ecosystems, whereas stand-killing insect outbreaks are treated as disturbances. Drought also ranges from minor moisture stress to severe moisture limitation that kills plants and triggers wind erosion. There is a continuum in size, severity, and frequency between normal function and extreme disturbance. Disturbance is not an external event that "happens" to an ecosystem. Like other interactive controls (see Chap. 1), disturbance is an integral part of the functioning of all ecosystems, which responds to and affects most ecosystem processes. Naturally occurring disturbances such as fires and hurricanes are therefore not "bad"; they are normal properties of ecosystems. They are appropriately viewed as disasters when they negatively impact society, often as a result of changes in human interactions with ecosystems.

Human activities have altered the frequency and size of many natural disturbances, such as fires and floods, and have produced new types of disturbance such as large-scale logging, mining, and wars. Many human disturbances have ecological effects that are similar to those of natural disturbances, so the study of either natural or human disturbances provides insights into the regulation of ecosystem processes and human impacts on these processes. Natural and human disturbances interact with environmental gradients to create much of the spatial patterning in landscapes (see Chap. 13; Turner 2010).

After disturbance, ecosystems undergo **succession**, a directional change in ecosystem composition, structure, and functioning. Disturbances that remove live or dead organic matter, for example, are colonized by plants that gradually reduce the availability of light at the soil surface and alter the availability of water and nutrients (Tilman 1985). If there were no further disturbance, succession

would proceed toward a steady state. Because of the path-dependent nature of succession, this steady state might be similar to the pre-disturbance ecosystem or it might move toward some alternative endpoint. Stands of lodgepole pine that burned in the 1988 Yellowstone fires, for example, moved along trajectories of very different stand density, nutrient availability, and productivity, depending on initial seed availability (which depended on seed retention in cones and fire severity) and seedling establishment (Turner et al. 1997, Turner 2010). In practice, however, new disturbances or environmental changes usually occur before succession reaches a steady state. Nonetheless, the concept of directional changes in vegetation after disturbance provides a useful framework for analyzing the role of disturbance in ecosystem processes.

Impact of a Disturbance Event

The impact of a disturbance event depends on three attributes of the disturbance: (1) the type of disturbance, (2) ecosystem sensitivity, and (3) disturbance severity or intensity.

Different **disturbance types** have radically different effects on ecosystems. Fire removes live and dead organic matter and raises environmental temperatures to lethal levels. An unseasonable freeze may also produce lethal temperatures. Floods and landslides remove or add soils and deplete soil oxygen. Hurricanes, storm surges, and logging remove or damage organisms. Species are often adapted to withstand disturbances that occur relatively frequently in their evolutionary history but may be vulnerable to novel disturbances. Benthic communities, for example, may recover slowly from bottom trawling that scrapes surface sediments, although they recover rapidly from severe storms that dislodge individuals. Many upland species are intolerant of flooding, whereas trees from wet environments generally tolerate periodic flooding, but have thin bark and are killed by fire.

Sensitivity to a particular disturbance type depends on system properties at the time of disturbance. Species traits, such as rooting depth

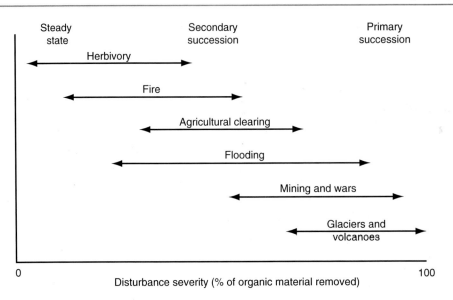

Fig. 12.7 Spectrum of disturbance severity associated with major types of disturbance, ranging from normal steady-state functioning of ecosystems to primary succession

or tolerance to frost, fire, or drought influence sensitivity of individual organisms. In addition, system properties such as density or configuration of plants can influence spread of fire, pathogens, or insect pests and therefore landscape sensitivity to disturbance.

Disturbance severity is magnitude of loss of biomass, soil resources, and species caused by a disturbance. **Intensity** is the energy released per unit area and time. **Primary succession** occurs after severe disturbances that remove or bury most products of ecosystem processes, leaving little or no organic matter or organisms. Disturbances leading to primary succession include volcanic eruptions, glacial retreat, landslides, mining, flooding, coastal dune formation, and lake drainage.

Secondary succession occurs on previously vegetated sites after disturbances such as fires, hurricanes, logging, and agricultural plowing. These disturbances remove or kill substantial live aboveground biomass but leave some soil organic matter and plants or plant propagules in place. Disturbance severity is probably *the* major factor determining the rate and trajectory of vegetation development after disturbance. A severe fire that kills all plants, for example, has a different effect on vegetation recovery than does a fire that burns

only surface litter, allowing surviving vegetation to resprout (Johnstone et al. 2010). There is also a continuum in disturbance severity between large-scale defoliation events and the removal of a single leaf by a caterpillar or between landslides and the burial of surface litter by an earthworm. In other words, there is a continuum in disturbance severity between the day-to-day functioning of ecosystems and events that initiate primary succession (Fig. 12.7).

Recovery and Renewal after Disturbance

Resilience to disturbance and subsequent successional trajectory depend not only on initial disturbance impact but also on disturbance size, pattern, and landscape matrix, which influence post-disturbance recruitment.

The traits and abundance of organisms that survive disturbance are critical to post-disturbance succession. Depending on the type and severity of disturbance and ecosystem sensitivity to the disturbance event, a variable number of individuals and species will survive, grow, and reproduce. Recruitment of new individuals is also important. Some traits, such as heat-induced germination of chaparral post-fire annuals, enable species to respond to specific types of disturbance. Other traits enable species to colonize many types of disturbances. Weedy species, for example, produce abundant small seeds that disperse long distances or remain dormant in the soil from one disturbance to the next. Their germination is often triggered by fluctuations in temperature and nutrients that characterize most disturbed sites (Fenner 1985, Baskin and Baskin 1998), so they are relatively insensitive to disturbance type. Novel disturbances are more likely to lead to slow recovery or trigger a new successional trajectory than are disturbances to which organisms are well adapted.

Disturbance size is highly variable. **Gap-phase succession**, for example, occurs in small gaps created by the death of one or a few plants. Many tropical wet forests or intertidal communities, for example, are mosaics of gaps of different ages. Similarly, gophers create patchy disturbances in grasslands (Yoo et al. 2005). Other ecosystems develop after **stand-replacing disturbances** that can be hundreds of square kilometers in area. Disturbance size influences ecosystems primarily through effects on landscape structure, which influences lateral flow of materials, organisms, and disturbance among patches in the landscape (see Chap. 13). Disturbance size, for example, affects the rate of seed input after fire. Small fires are readily colonized by seeds that blow in from surrounding unburned patches or are carried by mammals and birds. In contrast, regeneration in the middle of large fires, fields, or clearcuts may be limited by seed availability and be colonized primarily by light-seeded species that disperse long distances. Disturbance size also influences the spread of herbivores and pathogens that colonize early successional sites.

Disturbance pattern on the landscape influences the effective size of a disturbance event. Disturbances often leave islands of undisturbed vegetation or create highly irregular shapes with variable distances to propagule sources, causing the effective size of the disturbance to be much smaller than its areal extent would suggest (Turner 2010).

Resilience to disturbance also depends on the properties of the landscape in which the disturbed ecosystem is embedded, particularly its diversity of types and ages of ecosystems that serve as potential propagule sources for post-disturbance colonization. A nature reserve or forest stand that is isolated within an agricultural or urban matrix, for example, has less access to propagules and is less resilient than a similar stand embedded in a matrix of forest stands of varying ages (Fig. 12.8). Similarly, a diverse landscape is more resilient to a broad spectrum of disturbance types than is a uniform landscape, as described in the next section. This suggests that management for harvest efficiency by planting uniform ages of single-species stands reduces landscape resilience (Peterson et al. 1998).

Disturbance Regime

The overall role of disturbance in an ecosystem depends on the frequency and interaction of multiple disturbance types, the nature of individual disturbance events, and the landscape patterns that govern resilience and renewal. Over time, most ecosystems experience a diverse array of disturbance types that occur with differing frequencies and severities. Together these constitute the **disturbance regime** of the ecosystem.

Disturbance frequency varies dramatically among ecosystems and among disturbance types. Herbivory occurs continuously in most ecosystems. At the opposite extreme, volcanic eruptions or floods may never have occurred in some locations. Average fire frequency ranges from once per year in some grasslands to once every several thousand years in some mesic forests. Ecosystems are usually most resilient to disturbances that occur frequently. Ecosystems that experience frequent fire, for example, support fire-adapted species that recover biomass more quickly than in ecosystems in which fire occurs infrequently. Human activities often modify disturbance frequency through initiation or suppression of disturbance. Damming of streams can eliminate spring floods that scour sediments and detritus

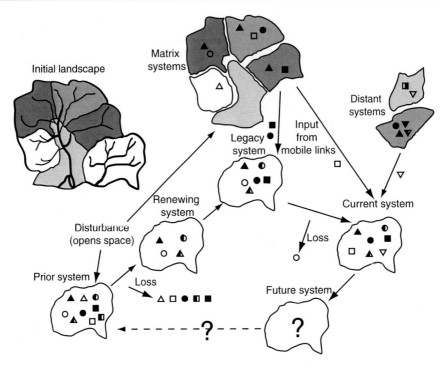

Fig. 12.8 Roles of stand and landscape diversity in ecosystem renewal after disturbance. A disturbance such as a fire, hurricane, volcanic eruption, or war opens space in an ecosystem. In this diagram, each shape represents a different functional group such as algal-grazing herbivores in a coral reef, and the different patterns of shading represent species within a functional group. After disturbance, some species are lost, but an on-site legacy of surviving species serves as the starting point for ecosystem renewal. For example, after boreal fire, about half of the vascular plant species are lost (Bernhardt et al. 2011). The larger the *species diversity* of the pre-disturbance ecosystem, the more species and functional groups are likely to survive the disturbance; the more severe the disturbance the larger the proportion of species lost. (In this figure, all functional groups except "squares" survived the disturbance.) *Landscape diversity* of the matrix surrounding the patch is also important to ecosystem renewal because it provides a reservoir of diversity that can recolonize the disturbed patch. In this figure, the "square" function was renewed by colonization from the matrix surrounding the ecosystem. Through time, some additional species may be gained or lost, and new functional groups (inverted triangles in this diagram) may invade from a distance. Reprinted from Chapin et al. (2009)

from channels, resulting in large changes in stream food webs and capacity to support fish (Power 1992a). Fire suppression in the giant sequoias (*Sequoiadendron gigantea*) of the Sierra Nevada mountains of California made this ecosystem more vulnerable to fire, as a result of the growth of understory trees that formed a ladder for fire to reach from the ground to the canopy. Although the thick-barked sequoias are resistant to ground fires, they are vulnerable to fires that extend into the canopy. In this way, fire suppression increased the risk of catastrophic fires that could eliminate giant sequoias.

The **timing of disturbance** often influences its impact. A strong freeze or fire that occurs during budbreak has greater impact than one that occurs 2 weeks earlier. Similarly, anaerobic conditions associated with flooding of the Mississippi River during the 1993 growing season caused more root and tree mortality than if the flood had occurred when roots were inactive. Hydroelectric dams may eliminate seasonal flooding associated with rain or snowmelt and regulate flow based on electricity demand, often causing a mismatch between disturbance timing and the disturbance regime to which organisms are adapted.

Disturbance is a key interactive control that governs ecosystem processes (see Chap. 1) through its effects on other interactive controls (microenvironment, soil resource supply, and functional types of organisms). Post-fire stands, for example, often have warm, moist soils because of the low albedo of the charred surface and the decrease in leaf area that transpires water and shades the soil. Fire both volatilizes nitrogen, which is lost from the site, and returns inorganic nitrogen and other nutrients to the soil in ash, thus altering soil resource supply. The net effect of fire is usually to enhance nutrient availability, although the magnitude of this effect depends on the nutrient and on fire severity and intensity (Wan et al. 2001, Smithwick et al. 2005). Fire affects the functional types of plants in an ecosystem through its effects on differential survival and competitive balance in the post-fire environment. Because of its sensitivity to, and effect on, other interactive controls, changes in disturbance regime alter the structure and functioning of ecosystems.

Succession

Successional changes occurring over decades to centuries explain much of the local variation among ecosystems. Although climate, soils, and topography explain most of the broad global and regional patterns in ecosystem processes, disturbance regime and post-disturbance succession account for many of the local patterns of spatial variability (see Chap. 13). In this section, we describe common patterns of successional change in major ecosystem processes. These successional changes are most clearly delineated in primary succession, so we begin with a description of primary successional processes and then describe how the patterns differ between primary and secondary succession.

Ecosystem Structure and Composition

Primary Succession
Primary succession occurs after severe disturbances that remove or bury most products of ecosystem processes. Initial species composition on these sites depends on the capacity of plants to deal with the environmental stresses associated with low nitrogen availability and the generally low water-holding capacity of organic-poor soils. Vascular plant species capable of symbiotic nitrogen fixation occur most often (about 75% of sites studied) in early primary succession, although they dominate the vegetation only about 25% of the time (Walker 1993). These species are most common on glacial moraines and mudflows, intermediate on mine tailings, landslides, floodplains, and dunes, and least abundant on volcanoes and rock outcrops. When early successional colonizers fix abundant nitrogen, their net effect is generally to **facilitate** (enhance) the establishment and growth of later successional species (Fig. 12.9; Walker 1993).

Due to their lack of plants and plant propagules, primary successional sites must be colonized by species that disperse to the site. Most initial colonizers have small wind-dispersed seeds. Fresh lava or glacial moraines, for example, are first colonized by wind-dispersed spores of algae, cyanobacteria, and lichens that form soil-stabilizing crusts (Walker and del Moral 2003). These are followed by small-seeded wind-dispersed vascular plants (primarily woody species), whose arrival rates depend largely on distance to seed source (Shiro and del Moral 1995). Late successional species with heavier seeds generally arrive more slowly (Fig. 12.10).

The identity of initial colonizers strongly influences the long-term successional trajectory. After volcanic eruption in Hawai'i, for example, succession usually proceeds slowly from short-statured vegetation dominated by algal crusts, herbaceous plants, and small shrubs to forests dominated by slowly growing tree ferns and trees. An exotic bird-dispersed nitrogen-fixing tree, *Morella faya*, can, however, add enough nitrogen to alter substantially the nitrogen supply, production, species composition, and therefore the successional trajectory of vegetation (Vitousek et al. 1987).

A similar change in successional trajectory occurred after glacial retreat at Glacier Bay, Alaska, but for different reasons. When the glacier first began to retreat in 1800, *Populus* (poplar) and *Picea* (spruce) were the major initial colonizers. Further retreat of the glacier, however, brought early successional habitat within

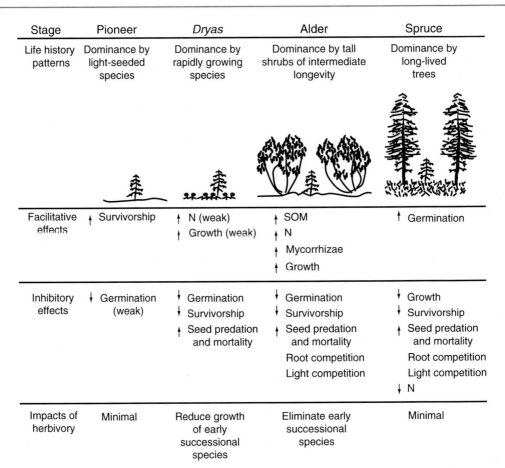

Stage	Pioneer	*Dryas*	Alder	Spruce
Life history patterns	Dominance by light-seeded species	Dominance by rapidly growing species	Dominance by tall shrubs of intermediate longevity	Dominance by long-lived trees
Facilitative effects	↑ Survivorship	↑ N (weak) ↑ Growth (weak)	↑ SOM ↑ N ↑ Mycorrhizae ↑ Growth	↑ Germination
Inhibitory effects	↓ Germination (weak)	↓ Germination ↓ Survivorship ↑ Seed predation and mortality	↓ Germination ↓ Survivorship ↑ Seed predation and mortality Root competition Light competition	↓ Growth ↓ Survivorship ↑ Seed predation and mortality Root competition Light competition ↓ N
Impacts of herbivory	Minimal	Reduce growth of early successional species	Eliminate early successional species	Minimal

Fig. 12.9 Interaction of life-history traits, competition, facilitation, and herbivory in causing successional change after glacial retreat at Glacier Bay, Alaska. Life-history traits determine the pattern of dominance at each successional stage. The rate at which this dominance changes is determined by facilitative or inhibitory effects of the dominant species and by patterns of herbivory. In general, all four of these processes contribute simultaneously to successional change, with the most important processes being life-history traits in the pioneer stage, herbivory in mid-successional stages, facilitation in the alder stage, and competition in late succession. Modified from Chapin et al. (1994)

dispersal distance of nitrogen-fixing alders, which then became an important early successional species (Fastie 1995). Alders increased the nitrogen inputs and long-term productivity of later successional stages (Bormann and Sidle 1990). The late-successional forests on older sites at Glacier Bay therefore followed a different (less productive) successional trajectory than alder-supported forests on younger sites. Human activities strongly affect both the post-disturbance environment and availability of propagules, so future trajectories of succession will likely differ from those that currently predominate.

Secondary Succession

Secondary succession begins on soils that developed beneath vegetation. There is usually a pulse in nutrient availability after disturbance because of the absence of vegetation to absorb nutrients released by mineralization.

Secondary succession also differs from primary succession in having colonizers that are already present on site immediately after disturbance. They may resprout from roots or stems that survived the disturbance or germinate from a soil **seed bank** – seeds produced after previous disturbance events that remain dormant in the

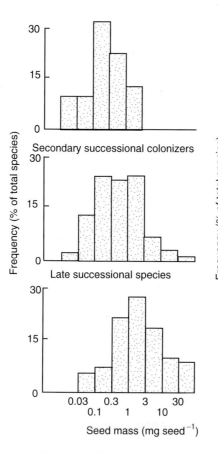

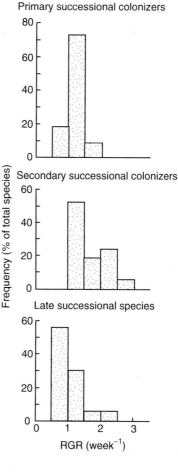

Fig. 12.10 Frequency distribution of log (seed mass) and relative growth rate (RGR) for British species that are primary successional colonizers, secondary successional colonizers, and late-successional species. Data from Grime and Hunt (1975) and Grime et al. (1981). Redrawn from Chapin (1993a)

soil until post-disturbance conditions (light, wide temperature fluctuations, or high soil nitrate) trigger germination (Fenner 1985, Baskin and Baskin 1998). Many forests also have a **seedling bank** (advanced regeneration) of large-seeded species that show negligible growth beneath the dense shade of a forest canopy but grow rapidly when treefall gaps occur. Other colonizers of secondary succession disperse into the disturbed site from adjacent areas. Dispersing species include both small-seeded, wind-dispersed species and large-seeded, animal-dispersed species (Fig. 12.10). Initial colonizers grow rapidly to exploit the resources made available by disturbance.

Gap-phase succession is seldom limited by propagule availability, whereas the successional trajectory of large disturbed sites may depend on the species that disperse to the site (Fastie 1995). Even large disturbances may not be dispersal-limited if the disturbances are so patchy that undisturbed seed sources are well distributed within the disturbed area (Turner 2010).

The changes in species composition that occur after the initial colonization of a site result from a combination of (1) the inherent life-history traits of colonizers, (2) facilitation, (3) competitive interactions, (4) herbivory, and (5) stochastic variation in environment (Connell and Slatyer

Table 12.1 Successional changes in life-history traits after glacial retreat in Glacier Bay, Alaska[a]

Genus	Successional stage	Seed mass (g seed^{-1})	Maximum height (m)	Age at first reproduction (year)	Maximum longevity (year)
Epilobium	Pioneer	72	0.3	1	20
Dryas	Dryas	97	0.1	7	50
Alnus	Alder	494	4	8	100
Picea	Spruce	2,694	40	40	700

[a] Data from Chapin et al. (1994)

1977, Pickett et al. 1987, Walker 1999). **Life-history traits** include seed size and number, potential growth rate, maximum size, and longevity. These traits determine how quickly a species can get to a site, how quickly it grows, how tall it gets, and how long it survives. Most early secondary successional species arrive soon after a disturbance, grow quickly, are relatively short statured, and have a low maximum longevity, compared to late-successional species (Fig. 12.10, Table 12.1; Noble and Slatyer 1980). Even if no species interactions occurred during succession, life-history traits alone would cause a shift in dominance from early to late successional species because of differences in arrival rate, size, and longevity.

Facilitation involves processes in which early successional species make the environment more favorable for the growth of later successional species. Facilitation is particularly important in severe physical environments, such as primary succession, where nitrogen fixation and addition of soil organic matter by early successional species ameliorates the environment and increases the probability that seedlings of other species will establish and grow (Callaway 1995, Brooker and Callaghan 1998). **Competition** is an interaction among two organisms or species that use the same limiting resources (resource competition) or that harm one another in the process of seeking a resource (interference competition). Both competitive and facilitative interactions are widespread in plant communities (Callaway 1995, Bazzaz 1996); their relative importance in causing changes in species composition during succession probably depends on environmental severity (Fig. 12.9; Connell and Slatyer 1977, Callaway 1995). **Herbivores and pathogens**

account for much of the plant mortality during succession. Selective browsing by mammals generally targets early successional species, reducing their competition with later successional species and therefore speeding the rate of successional change (Paine 2000, Walker and del Moral 2003). In intertidal communities, grazing by fish and invertebrates such as limpets exerts a similar effect. Insects exert their greatest impacts during outbreaks that reduce growth or increase mortality of ecologically important plant species. During mid and late succession, for example, when plant demands for water and nutrient are high, periodic drought stress can reduce plant resistance to insects and trigger an outbreak, as in the mountain pine beetle outbreak in western North America (Raffa et al. 2008).

In general, life-history traits determine the *pattern* of species change through succession, and facilitation, competition, and herbivory determine the rate at which this occurs (Chapin et al. 1994). These processes interact with other disturbances to create a diversity of successional pathways in natural ecosystems (Pickett et al. 1987, Walker and del Moral 2003, Turner 2010).

Opportunities for seedling establishment often decline through succession. In many forests, for example, all tree species colonize in early succession, and the successional changes in dominance reflect a gradual transition from small rapidly growing plants to taller, more slowly growing species (Egler 1954, Walker et al. 1986). In other cases, late successional species may establish more gradually. As succession proceeds, the soil becomes covered by leaf litter, creating a less favorable seedbed, and competition increases for light and nutrients among established seedlings.

Fig. 12.11 Runoff from a watershed in a North Carolina forest in the Southeastern U.S. under natural conditions (the calibration period) and after forest harvest. Water yield from the watershed greatly increased in the absence of vegetation and approached pre-harvest levels within 20 years. Redrawn from Hibbert (1967)

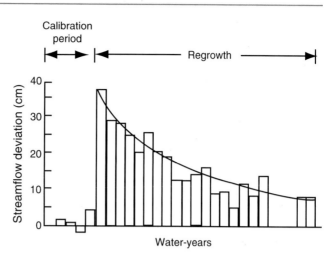

Water and Energy Exchange

Disturbances that eliminate plant biomass increase runoff through a reduction in evapotranspiration. One of the most dramatic consequences of forest cutting or overgrazing is increased runoff to streams and rivers during times of both low flows and flooding (NRC 2008). This has led some resource managers to suggest forest cutting as a way to increase water yields to meet societal demands for water. These increases in discharge are, however, often short lived. As vegetation regrows during succession, runoff declines to pre-harvest levels (or even lower; see Chap. 4), often within 5 years or less (Fig. 12.11; Jones and Post 2004). The rate and pattern of change in runoff after forest harvest depends on patterns of vegetation recovery, relative to the vegetation that was present before harvest (Jones and Post 2004, Brown et al. 2005, NRC 2008). The high nitrogen availability, high photosynthetic rate, and high leaf area early in secondary succession can lead to even higher evapotranspiration and lower runoff than in undisturbed stands (Jones and Post 2004). The short-term gains in discharge after forest harvest are generally smallest in dry ecosystems and dry seasons, that is, the situations where human demands for water are highest (NRC 2008), suggesting that forest harvest is not an effective strategy to increase water yield for human use.

As roots proliferate during succession, more water is absorbed by plants, and less water moves to groundwater and streams. As the canopy increases in height and complexity, a larger proportion of solar energy is trapped, reducing albedo and increasing the energy available to drive evapotranspiration. The high surface roughness of tall complex canopies increases mechanical turbulence and mixing within the canopy. All of these factors contribute to rapid recovery of evapotranspiration during succession.

Successional changes in albedo differ among ecosystems because of the wide range among ecosystems in albedo of bare soil (see Table 4.1). Many recently disturbed sites have a low albedo because of the dark color of moist exposed soils or of charcoal. Albedo increases when vegetation, with its generally higher albedo, begins to cover the soil surface (Fig. 12.12). Albedo probably declines again in late succession due to increased canopy complexity (see Chap. 4). In ecosystems that succeed from deciduous to conifer forest, this species shift causes a further reduction in albedo. The winter energy exchange of northern forests is influenced by snow, which has an albedo three to fivefold higher than vegetation (Betts and Ball 1997). Winter albedo of these forests declines through succession, first as vegetation grows above the snow, then as the canopy becomes denser, and finally when (if) vegetation switches from deciduous to evergreen. All of these changes increase the

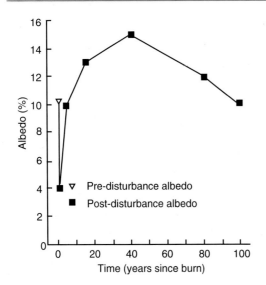

Fig. 12.12 Successional changes in albedo after fire in Alaskan boreal forests. The black post-fire surface causes a decline in albedo. Albedo increases during the herbaceous and deciduous forest phases of succession and declines in late succession due to a switch to conifer vegetation. This successional change occurs more rapidly after moderate fires because of the more rapid replacement of deciduous species by conifers. Data from Chambers and Chapin (2002)

extent to which vegetation masks the snow from incoming solar radiation (Euskirchen et al. 2009).

High surface temperatures that contribute to high emission of longwave radiation dominate energy budgets of early successional sites. Early successional sites often have a high surface temperature for several reasons: (1) The low albedo of recently disturbed sites maximizes radiation absorption and therefore the quantity of energy available at the surface. (2) The low leaf area, small root biomass, and low hydraulic conductance of dry surface soils limits the proportion of energy dissipated by evapotranspiration. (3) The relatively smooth surface of unvegetated or early successional sites minimizes mechanical turbulence that would otherwise transport the heat away from the surface. The resulting high surface temperature promotes both emission of longwave radiation and a high Bowen ratio (ratio of sensible to latent heat flux; see Chap. 4).

The large longwave emission dissipates much of the absorbed radiation after disturbance, so *net*

radiation (the net energy absorbed by the surface) is not as great as we might expect from the low albedo of these sites. For example, net radiation actually declines after fire in the boreal forest despite a reduction in albedo because of the large emission of longwave radiation (Chambers et al. 2005). The soil surface of unvegetated sites is prone to drying between rain events due to the combination of high surface temperatures and the low resupply of water from depth, due to the low hydraulic conductance of dry soils (see Chap. 4). Dry surface soils provide little moisture for surface evaporation and are good thermal insulators, so both evapotranspiration and average ground heat flux are often relatively low on unvegetated surfaces (Oke 1987). Consequently, sensible heat flux accounts for the largest proportion of energy that is dissipated from these sites to the atmosphere. The absolute magnitude of sensible heat flux from early successional sites differs among ecosystems and climate zones and depends on both net radiation (the energy available to be dissipated) and the energy partitioning among sensible, latent, and ground heat fluxes. As succession proceeds, latent heat fluxes become a more prominent component of energy transfer from land to the atmosphere.

Carbon Balance

Primary Succession
In primary succession, productivity and heterotrophic respiration are often greatest in mid-succession. Primary succession begins with little live or dead organic matter, so net primary production (NPP) and heterotrophic respiration are initially close to zero. NPP increases slowly at first because of low plant density, small plant size, and strong nitrogen limitation of growth. NPP and biomass generally increase most dramatically after nitrogen fixers colonize the site. The planting of nitrogen-fixing lupines on English mine wastes (Bradshaw 1983) and the natural establishment of nitrogen-fixing alders after retreat of Alaskan glaciers (Bormann and Sidle 1990), for example, cause sharp increases in plant biomass

and NPP. In primary successional sequences that lack a strong nitrogen fixer, successional increases in biomass and NPP depend on other forms of nitrogen input, including atmospheric deposition, plant and animal detritus, and lateral delivery from flowing groundwater.

Long-term successional trajectories of biomass and NPP differ among ecosystems. A common pattern in forests is that NPP increases from early to mid-succession, then declines after the forest reaches its maximum leaf area index (LAI) (Fig. 12.13; Ryan et al. 1997). Several processes may contribute to these patterns. In some forests, hydraulic conductance declines in late succession, causing water to limit the leaf area that can be supported and therefore gross primary production (GPP) and NPP (see Chap. 6). In other forests, nutrient supply declines in late succession, leading to a corresponding reduction in GPP and NPP (Van Cleve et al. 1991). It is less likely that late-successional declines in NPP reflect increased maintenance respiration to support the increasing biomass, as had been suggested earlier (Odum 1969), because much of forest biomass increase consists of dead cells that do not respire. The mortality of branches and trees often increases in late succession, as trees age. The combination of reduced NPP and increased mortality of plants and plant parts in late succession slows the rate of biomass accumulation, so biomass approaches a relatively constant value (steady state; Fig. 12.14) or declines due to stand thinning. The rate and patterns with which carbon pools and fluxes change through succession depend on both initial conditions and events and climatic fluctuations that occur during succession and are therefore variable within and among ecosystem types (Fig. 12.14; Turner 2010). The long-term endpoints of successional trajectories in biomass and NPP are also uncertain because disturbance usually resets the successional clock before the ecosystem reaches steady state.

Over extremely long time scales, changes in rates of weathering and soil development lead to further changes in biomass and other ecosystem properties (see Chap. 3). Redwoods in California coastal forests, for example, are replaced by a pygmy forest of evergreen trees and shrubs after hundreds of thousands of years due to the formation of a hard pan that prevents drainage and creates anaerobic conditions that retard decomposition and root growth (Westman 1978). The slow-growing plants capable of surviving under these low-nutrient conditions produce litter with high concentrations of phenolics, which further reduce decomposition rate, resulting in an amplifying (positive) feedback that leads to progressively lower biomass, productivity, and nutrient turnover (Northup et al. 1995).

Heterotrophic respiration rate at the start of primary succession is near zero because there is little or no soil organic matter. The low organic content of these soils contributes to their low moisture-holding capacity and CEC (Fig. 12.15; see Chap. 3). The *pattern* of change in heterotrophic respiration through primary succession is similar to the pattern described for NPP. Heterotrophic respiration, however, lags behind the changes in NPP, causing soil organic matter to accumulate (Fig. 12.16). Initially, heterotrophic respiration is low in primary succession because it is limited by the quantity of soil organic matter.

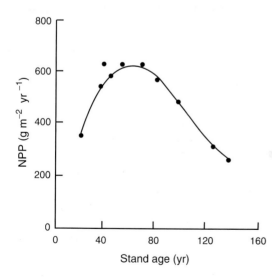

Fig. 12.13 Successional changes in aboveground spruce production in Eastern Russia. NPP declines after the forest reaches maximum LAI at about 60 years of age. Redrawn from Ryan et al. (1997)

Fig. 12.14 Idealized patterns of primary successional changes in plant biomass, *GPP*, *NPP*, plant respiration (R_{plant}), and plant mortality of a forest (*top*). GPP, NPP, and plant respiration often reach a peak in mid-succession and decline in late succession. The actual patterns vary considerably among ecosystems, as illustrated by patterns of aboveground NPP hypothesized for lodgepole pine stands of different initial seedling density in Yellowstone National Park (*bottom*). Redrawn from Turner (2010)

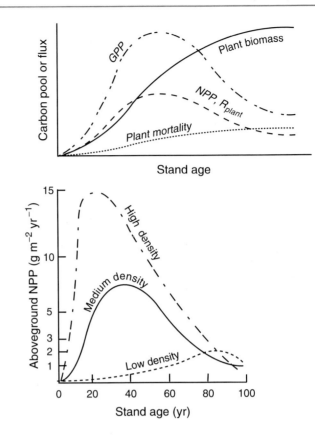

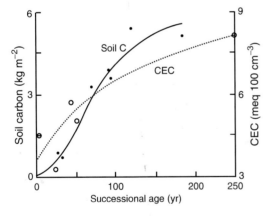

Fig. 12.15 Accumulation during succession of soil organic carbon (Crocker and Major 1955) and associated change in cation exchange capacity (CEC) of mineral soil (Ugolini 1968) after deglaciation at Glacier Bay Alaska. Measurements were made to a depth of 45 cm in mineral soil. The accumulation of soil carbon contributes to the increased CEC, which retains nutrients to support plant growth

Heterotrophic respiration increases substantially in mid-succession in response to increases in the quantity and quality of litter. In forests, the late-successional decline in NPP reduces litter inputs to soils, causing heterotrophic respiration to decline. In those ecosystems where nutrient availability declines in late succession, this reduces litter quality and quantity, further reducing decomposition rate and heterotrophic respiration (Van Cleve et al. 1993).

NEP is the balance between carbon inputs in NPP and carbon losses from heterotrophic respiration. NEP usually increases from early and mid-succession, due to the lag of heterotrophic respiration behind NPP (Figs. 12.16, 12.17). This contributes to the carbon accumulation of mid-latitude north temperate forests that established in abandoned agricultural lands one to two centuries earlier (Goulden et al. 1996, Valentini et al. 2000). NEP typically declines in late succession but generally remains positive, even after many centuries (see Fig. 7.20; Luyssaert et al. 2007, Xiao et al. 2008).

Net ecosystem carbon balance (NECB) reflects not only photosynthesis and heterotrophic respiration (i.e., NEP) but also other carbon transfers, including losses by combustion and leaching and

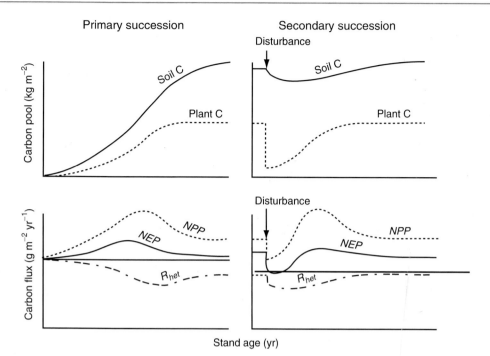

Fig. 12.16 Idealized patterns of change in carbon pools (plants and soils) and fluxes (*NPP*, R_{het}, and *NEP*) in primary and secondary succession. In early primary succession, plant and soil carbon accumulates slowly because NPP is greater than heterotrophic respiration, that is, there is a positive NEP. In early secondary succession, soil carbon declines after disturbance because carbon losses from heterotrophic respiration exceed carbon gain from NPP, leading to a negative NEP. In late succession, plant and soil carbon approach steady state (in this idealized diagram), and NEP approaches zero. In both primary and secondary succession, NPP and NEP are maximal in mid-succession. Net carbon accumulation in the ecosystem (NECB) would differ from the patterns shown, if leaching losses and other carbon fluxes are substantial

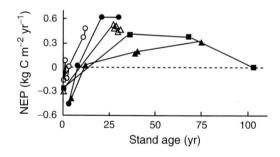

Fig. 12.17 Successional changes in NEP of European forests measured by CO_2 exchange. *Black circles, Picea sitkensis; white triangles, Pinus pinaster; gray squares, Pinus sylvestris; black triangles, Pinus sylvestris; white circles, Quercus cerris.* Redrawn from Magnani et al. (2007)

lateral transfers among ecosystems (e.g., animal movements, forest harvest, and food or waste transfers). Like NEP, NECB typically remains positive in mid to late succession (Magnani et al. 2007, Luyssaert et al. 2008), indicating that eco-

system carbon balance seldom reaches steady state before disturbance recurs. Alternatively, late-successional ecosystems may be responding to recent environmental changes (rising temperature, atmospheric CO_2, and nitrogen deposition) that support continued carbon accumulation (Magnani et al. 2007, Luyssaert et al. 2008) or loss (Oechel et al. 2000). In the boreal forest, climate warming governs NECB more strongly through effects on fire regime than through its effects on NEP (Bond-Lamberty et al. 2007).

Secondary Succession

The initial carbon pools and fluxes are much larger in secondary than in primary succession. Carbon dynamics are dramatically different between secondary and primary succession because secondary succession begins with an initial stock of soil organic matter. Immediately after disturbance, NPP is low in secondary succession because of low plant biomass, just as in primary succession (Fig. 12.16). NPP recovers more quickly in secondary than in primary succession, however, due to the generally rapid colonization and high growth rate of herbs, grasses, and resprouting perennial species. High availability of light, water, and nutrients supports the high growth potential of early successional vegetation in many secondary successional sequences. The herbaceous species that dominate most early secondary successional sites return most of their biomass to the soil each year. Perennial plants, particularly woody species, increase in abundance, biomass, and NPP more rapidly because they retain a larger proportion of their biomass. Changes in biomass and NPP in mid- and late secondary succession are similar to patterns described for primary succession (Figs. 12.13, 12.16) because they are controlled by the same factors and processes – largely the soil resources available to support production and the growth potential of the species typical of the ecosystem.

In contrast to primary succession, heterotrophic respiration in mesic ecosystems is often quite high early in secondary succession (Fig. 12.16) because many disturbances transfer large amounts of labile carbon to soils and create a warm, moist environment that is favorable for decomposition. The size of the initial input to the soil carbon pool depends on the type and severity of the disturbance. After a treefall, hurricane, or insect outbreak, there are large inputs of new labile carbon from leaf and root death. Fire consumes some of the surface SOM but also adds new carbon to the soil through death of roots and unburned aboveground plant material. The large quantity and high quality of litter of early secondary successional plants also promotes heterotrophic respiration. In mid-succession, the regrowing vegetation uses an increasing proportion of the available water and nutrients and reduces soil temperature by shading the soil surface. These changes in environment cause a decline in decomposition. Heterotrophic respiration declines in late succession because the decline in NPP reduces litter input; litter quality often declines; and the environment becomes less favorable than in early succession.

How do these contrasting patterns of NPP and heterotrophic respiration affect NEP? In early

secondary succession, NEP is negative because heterotrophic respiration causes large carbon losses, and there is little NPP (Figs. 12.16, 12.17). In early succession before the peak in NPP, ecosystems begin accumulating carbon again, as soon as NPP outpaces heterotrophic respiration. In late succession, ecosystems typically accumulate carbon at a slow rate that depends on the environmental limitations to NPP and heterotrophic respiration (Magnani et al. 2007, Luyssaert et al. 2008). Other avenues of carbon loss from ecosystems such as leaching of dissolved organic carbon may influence NECB in ways that are not readily predicted from successional dynamics.

Although the successional patterns of NPP, heterotrophic respiration, and carbon stocks in plants and soils that we have described are often observed, the details, timing, and long-term trajectory of these patterns differ substantially within and among ecosystems, depending on factors such as initial ecosystem carbon stocks, resource availability, disturbance severity, and successional pathway (Turner 2010).

Nutrient Cycling

Primary Succession

Nutrient dynamics during succession are both a cause and a consequence of the dynamic interplay between NPP and decomposition. The most dramatic change in nutrient cycling during early primary succession is the accumulation of nitrogen in vegetation and soils. Most parent materials have extremely low nitrogen contents in the absence of biotic influences, so the initial nitrogen pools in the ecosystem are small and depend on atmospheric inputs. At this initial stage of primary succession, nitrogen is the element that most strongly limits plant growth and therefore the rates of accumulation of plant biomass and SOM (Crocker and Major 1955, Vitousek 2004). The rate of nitrogen input, which is often associated with the establishment of nitrogen-fixing plants (both free-living cyanobacteria and symbiotic nitrogen fixers), therefore governs the initial dynamics of nutrient cycling in primary succession. As leaves and roots of nitrogen-fixing plants senesce and are eaten by herbivores, the nitrogen is transferred from plants to the soil, where it is mineralized and absorbed by both nitrogen-fixing and non-fixing plants. Litter from non-nitrogen-fixing plants becomes an increasingly important source for nitrogen mineralization as primary succession proceeds. This causes the ecosystem to shift from an open nitrogen cycle, with substantial input from nitrogen fixation (see Chap. 9), to a more closed nitrogen cycle in which plant growth depends on the mineralization of soil organic nitrogen. During mid-succession, plants and soil microbes are so efficient at accumulating nutrients that losses of nitrogen and other essential elements from ecosystems are often negligible (Fig. 12.18; Vitousek and Reiners 1975). In late-successional ecosystems that approach steady state (NECB approximately zero), nitrogen inputs to the ecosystem may be largely balanced by nitrogen losses from

leaching (especially as dissolved organic nitrogen) and denitrification, causing ecosystem nitrogen pools to approach a relatively stable size. In those ecosystems where NECB remains positive, nitrogen will also likely continue to accumulate.

The accumulation of other essential elements during primary succession depends on accumulation in biotic pools and the formation of secondary minerals. Early in primary succession, biological storage pools in vegetation and soils are small, and so they can retain only a small fraction of the elements mobilized by weathering. Abiotic processes, especially the formation of secondary clay minerals (see Chap. 3), are more important in retaining many elements, both through the incorporation of some elements (e.g., magnesium) into clay lattices and through cation-exchange processes. The formation of secondary clay minerals and the elements they retain vary depending on climate (with a larger fraction of elements retained by the clays that form in dry sites) and parent material (with the formation of highly reactive allophone in volcanic areas). Organic matter is more important as a source of, and sink for, elements later in primary succession and throughout secondary succession.

Later in soil development, additional changes in nutrient cycling occur as the supply of weatherable minerals is depleted or becomes bound in unavailable forms. Availability of phosphorus and cations, for example, typically declines in old, highly weathered sites as they leach or become bound in unavailable forms (see Chaps. 3 and 9). Under these circumstances, phosphorus or other elements may limit plant production (Chadwick et al. 1999), and cycling rates of these limiting elements regulate cycling rates of nitrogen and other minerals.

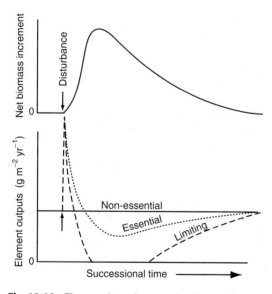

Fig. 12.18 Changes through succession in net biomass increment in vegetation and in the losses of limiting, essential, and nonessential elements. In early succession, when biomass accumulates rapidly, elements that are required for this production (especially growth-limiting elements) accumulate in new plant and microbial biomass, so they are not lost from the ecosystem by leaching. In late succession, when the element requirements for new plant and microbial biomass are balanced by element release from the breakdown of dead organic matter, nutrient inputs to the ecosystem are approximately balanced by nutrient outputs, regardless of whether nutrients are required by vegetation or not. Modified from Vitousek and Reiners (1975)

Secondary Succession

Secondary succession after natural disturbances differs from primary succession because it generally begins with higher nitrogen availability. Natural disturbances that initiate secondary succession produce a pulse of nutrient availability because disturbance-induced changes in environment and litter inputs increase mineralization of dead organic matter and reduce plant biomass and nutrient absorption. Fires, which may volatilize large amounts of nitrogen, also return nutrients in ash, as described earlier, leading to increased nutrient availability after fire (Wan et al. 2001). Plant growth is therefore generally less strongly nutrient-limited early in secondary succession compared to primary succession, nitrogen is usually adequate to support high rates of photosynthesis and growth (Scatena et al. 1996, Smithwick et al. 2005). The pulse of nutrient availability and the reduction in plant biomass and capacity for plant absorption after disturbance also increase the vulnerability of ecosystems to nutrient loss. High rates of nitrogen mineralization and nitrification stimulate the production of nitrate that can be denitrified or leached below the rooting zone. The occurrence or extent of nitrogen loss depends on the balance between nitrogen mineralization and absorption by plants and microbes. Rains that occur immediately after a fire (Minshall et al. 1997, Betts and Jones 2009) or hurricane (Schaefer et al. 2000) often leach nitrate into groundwater and streams. These nutrient losses to streams decline as nutrients are immobilized by microbes and absorbed by regrowing vegetation (Turner 2010).

The vulnerability of ecosystems to nutrient losses after disturbance has been illustrated in many forest-harvest experiments, such as those at Hubbard Brook in the U.S. After stream discharge and chemistry had been monitored for several years, the forest was cut on an entire watershed and regenerating vegetation was killed with herbicides (Bormann and Likens 1979). The combination of high decomposition and mineralization rates and absence of plant absorption after disturbance caused large losses of essential plant nutrients in stream water (see Fig. 9.14). When vegetation was allowed to regrow, the increased plant absorption caused nutrient losses in stream water to decline to pre-harvest levels. These studies show clearly that the dynamics of nutrient loss after disturbance are highly variable, with the extent of nutrient loss often depending on nutrient availability at the time of disturbance and the capacity of regenerating vegetation to absorb nutrients.

Human disturbances create a wide range of initial nutrient availabilities. Some disturbances, such as mining, can produce an initial environ-

ment that is even less favorable than most natural primary successional habitats for initiation of succession. These habitats may have toxic by-products of mining or mineral material with a low capacity for water and nutrient retention. Some agricultural lands are abandoned to secondary succession after erosion or (in the tropics) formation of plinthite (iron- and aluminum-rich) soil horizons (see Chap. 3), reducing the nutrient-supplying power of soils. Secondary succession in degraded lands may therefore be quite slow. At the opposite extreme, abandonment of rich agricultural lands or the logging of productive forests may create conditions of high nutrient availability, leading to the potential loss of nutrients through leaching and denitrification. These nutrient losses are particularly dramatic in the tropics, where rapid mineralization and biomass burning associated with forest clearing release large amounts of nitrogen as trace gases (NO_x and N_2O) and as nitrate in groundwater (Matson et al. 1987). The impact of agricultural nutrient additions is particularly long-lived for phosphorus because of its effective retention by soils. An understanding of the successional controls over nutrient cycling provides the basis for management strategies that minimize undesirable environmental impacts (see Chap. 15). The return of topsoil or planting of nitrogen-fixing plants on mine wastes, for example, greatly speeds successional development on these sites (Bradshaw 1983). Retention of some organic debris after logging may support microbial immobilization of nutrients that would otherwise be lost.

Trophic Dynamics

The proportion of primary production consumed by herbivores is maximal in early to mid-succession. In early primary and secondary succession, rates of herbivory may be low because of low food density, insufficient cover to hide vertebrate herbivores from their predators, and insufficient canopy to create a humid, non-desiccating environment for invertebrate herbivores. Herbivory is often greatest in early to mid-secondary succession because the rapidly growing herbaceous and shrub species that dominate this stage have high nitrogen concentrations and a relatively low allocation to carbon-based plant defenses (see Chap. 10). This explains why abandoned agricultural fields, recent burn scars, or riparian areas are focal points for browsing mammals, insect herbivores, and their predators. In early successional boreal floodplains, for example, moose consume about 30% of aboveground NPP and account for a similar proportion of the nitrogen inputs to soil (Kielland and Bryant 1998). The abundant insect herbivores on these sites support a high diversity of neotropical migrant birds. Similarly, in temperate and tropical regions, early successional forests support large populations of deer and other browsers. In ecosystems in which nutrient availability declines from early to late succession, plants shift allocation from growth to defense (see Chap. 10). The resulting decline in forage quality reduces levels of consumption by most herbivores and higher trophic levels. Some insect outbreak species are an important exception to this successional pattern. They often attack late-successional trees that are weakened by environmental stress (Raffa et al. 2008).

Vertebrate herbivores can either promote or retard succession, depending on their relative impact on early vs. late-successional species. Vertebrate herbivores both respond to (see Chap. 10) and contribute to successional change. The effects of herbivores on succession differ among ecosystems, depending on the nature and specificity of plant–herbivore interactions. However, several common patterns emerge.

In forested regions, birds, rodents, and other vertebrates often enhance the dispersal of early successional species such as blackberries, junipers, and grasses into abandoned agricultural fields and other disturbed sites. Birds and squirrels also disperse the large seeds of late-successional species such as oak and hickory into early successional sites. These animal-mediated dispersal events are particularly important in secondary succession, where the rapid development of herbaceous vegetation makes it difficult for small-seeded woody species to compete and establish successfully.

The relatively low levels of carbon-based plant defenses in species that typically characterize early

secondary forest succession make these plants a nutritious target for generalist insect and vertebrate herbivores. Preferential feeding on these species reduces their height growth and reproductive output. Browsed plants respond to aboveground herbivory by reducing root allocation, making them less competitive for water and nutrients (Ruess et al. 1998). Many late-successional species produce chemical defenses that deter generalist herbivores. Selective herbivory contributes to the **competitive release** of late successional species, enabling them to overtop and shade their early successional competitors. In this way, selective browsing by mammals often speeds successional change in forests (Pastor et al. 1988, Kielland and Bryant 1998, Paine 2000). In tropical rainforests, mammalian herbivores maintain the diversity of understory seedlings that become the next generation of canopy dominants because they feed preferentially on the "weedy" tree seedlings that are most common in the understory (Dirzo and Miranda 1991).

In contrast to forests, many grasslands and savannas are maintained by mammalian herbivores that prevent succession to forests. Elephants, for example, browse and uproot trees in African savannas. These savannas succeed to closed forests in areas where elephant populations have been reduced by overhunting. In North American prairies, browsers and fire restrict the invasion of trees. When these sources of disturbance are reduced, trees often invade and convert the grassland to forest. Similarly, at the end of the Pleistocene, the decline in large mammals that occurred on many continents, in part from human hunting, contributed to the vegetation changes that occurred at that time (Flannery 1994, Zimov et al. 1995, Gill et al. 2009).

Herbivores have multiple effects on nutrient cycling in early succession. In the short term, they enhance nutrient availability by returning available nutrients to the soil in feces and urine, which short-circuits the decomposition process (Kielland and Bryant 1998). Herbivory can also alter the temperature and moisture regime for decomposition at the soil surface by reducing leaf and root biomass. The quality of litter that a given plant produces is also enhanced by herbivory (Irons et al. 1991). Over the long term, however, herbivory accelerates plant succession by removing early successional species, which tends to reduce nutrient cycling rates and nutrient losses (see Fig. 10.9; Pastor et al. 1988, Kielland and Bryant 1998).

Temporal Scaling of Ecological Processes

Temporal extrapolation requires an understanding of the typical time scales of important ecological processes. Ecologists generally measure ecological processes for shorter time periods than the time scales over which we would like to make predictions. No studies, for example, provide detailed information about the functioning of ecosystems over time scales of decades to centuries – the time scale over which ecosystems are likely to respond to global environmental change. **Temporal scaling** is the extrapolation of measurements made at one time interval to longer (or occasionally shorter) time intervals. Simply multiplying an instantaneous flux rate by 24 h to get a daily rate or by 365 days to get an annual rate seldom gives a reasonable approximation because this ignores the temporal variation in driving variables and the time lags and thresholds in ecosystem responses to these drivers. Rates of photosynthesis, for example, differ between night and day and between summer and winter.

One approach to temporal scaling is to select measurements that are consistent with the time scale and question of interest. A second approach is to extrapolate results based on models that simulate processes accounting for important sources of variation over the time scale of interest. The key to temporal scaling is therefore to focus clearly on the processes that are important over the time scales of interest. Entire books have been written on temporal scaling based on isotopic measurements (Ehleringer et al. 1993), long-term measurements (Sala et al. 2000), and modeling (Ehleringer and Field 1993, Waring and Running 2007). Here we provide a brief overview of these approaches.

Isotopic tracers are an important tool for estimating long-term rates of net carbon exchange of plants and ecosystems because they integrate the net effect of carbon and some nutrient inputs and losses throughout the time period of carbon exchange (see Box 5.1). [13]C content of plants in dry environments, for example, provides an integrated measure of water use efficiency (WUE) during the time interval during which the plant material was produced. [13]C content of soils in ecosystems that have changed in dominant vegetation from C_3 to C_4 plants provides an integrated measure of soil carbon turnover since the time that the vegetation change occurred. These measurements are appropriate for estimating long-term rates because they incorporate effects of processes that occur slowly or intermittently that might not be captured in short-term gas-exchange measurements. Seasonally integrated water use efficiency measured with stable isotopes, for example, is affected by dry and wet periods that influence seasonal water and carbon exchange, whereas instantaneous measurements of gas exchange are unlikely to be representative of the entire annual cycle. Similarly, NPP integrates over longer time periods than does photosynthesis or respiration, and successional changes in soil carbon stocks integrate over longer time periods than do measurements of NPP and decomposition.

Process-based models are important tools for temporal scaling because they make projections of the state of the ecosystem over longer time intervals (or at different times or places) than can be measured directly. The challenge in developing models for temporal extrapolation is the selection of the driving variables that account for the most important sources of temporal variation over the time scale of interest. The diurnal pattern of net photosynthesis can often be adequately simulated based on the relationship of net photosynthesis to light and temperature. Annual estimates of photosynthetic flux (GPP), however, also require information on seasonal variation in leaf biomass and photosynthetic capacity. In annual simulations, the diurnal variation in photosynthesis is less important to model explicitly because it is quite predictable, based on the empirical relationship between daily photosynthesis and average daily temperature and light. **Slow variables**, such as successional changes in LAI or nitrogen availability, are often treated as constants in

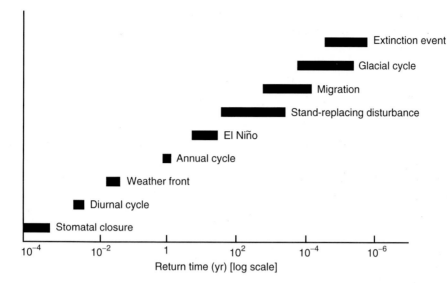

Fig. 12.19 Variation in return time for variables that strongly affect ecosystem processes. For any particular process, such as NPP, fast variables like stomatal closure can be ignored, slow variables like El Niño or stand-replacing disturbance strongly affect the process, and extremely slow variables, such as glacial cycles can be treated as constants

short-term ecological studies, but can become key control variables over longer time scales (Peterson et al. 1998, Carpenter and Turner 2000). We must therefore think carefully about which critical driving variables are likely to change over the time scale of intended predictions and look for evidence of the relationship of ecological processes to these slow variables. Models of carbon flux based on the relationship of GPP and respiration to daily or monthly climate, for example, can be validated by comparing model output to patterns of carbon flux observed over longer time scales (e.g., interannual variation in carbon flux). Ecosystem controls vary over a wide range of temporal scales (Fig. 12.19).

Spatial variation in driving variables sometimes gives hints as to which slow variables are important to include in long-term extrapolations. The spatial relationship between the distribution of biomes or plant functional types and climate, for example, has been used to predict how vegetation might respond to future climatic warming (Prentice et al. 1992, VEMAP-Members 1995, Euskirchen et al. 2009). Spatial relationships with driving variables often reflect quasi-equilibrium relationships. Tropical dry forests, for example, occur where the average climate is warm and has a distinct dry season. Temporal extrapolations should also consider extreme events and time lags that may not be evident from an examination of spatial pattern. Ice-storms, a spring freeze, intense droughts, 100-year floods, and other events with long-lasting effects strongly influence the structure and functioning of ecosystems long after they occur.

Summary

Ecosystem processes are constantly adjusting to past changes that have occurred over all time scales, ranging from sun flecks that last milliseconds to soil development that occurs over millions of years. Ecosystem processes that occur slowly, such as soil organic matter development, deviate most strongly from steady state and are most strongly affected by legacies of past events. Ecosystem processes are highly resilient to predictable changes in environment such as those that occur diurnally and seasonally and in response to disturbances to which organisms are well adapted.

Stand-replacing disturbances greatly reduce evapotranspiration and increase runoff. Evapotranspiration increases through succession more rapidly than might be expected from biomass recovery because early successional vegetation has high transpiration rates. Sensible heat flux tends to show the reverse successional pattern with high sensible heat flux (or longwave radiation) immediately after disturbance and lower sensible heat flux as rapidly growing mid-successional vegetation establishes and transfers energy to the atmosphere as water vapor.

Because disturbance is a natural component of all ecosystems, the successional changes in ecosystem processes after disturbance are important to understanding regional patterns of ecosystem dynamics. Successional changes in ecosystems are particularly sensitive to the severity, frequency, and type of disturbance. Through primary succession, carbon accumulates in vegetation and soils and leads to positive NEP because changes in decomposition lag behind changes in NPP. NPP in forests is often greatest in mid-succession. Secondary succession begins with a large negative NEP due to low NPP and rapid decomposition, but carbon cycling in mid- and late succession is similar to the patterns in primary succession.

Nutrient cycling changes through early primary succession as nitrogen fixers establish and add nitrogen to the ecosystem. Other elements cycle in proportion to the cycling of nitrogen. In secondary succession, however, nitrogen is generally most available in early succession. At this time, nitrogen and other elements are vulnerable to loss until the potential of plants and microbes to absorb nutrients exceeds the rate of net mineralization. This tightens the nitrogen cycle. Recycling within the ecosystem is strongest in mid-succession, when rates of nutrient mineralization constrain the rates of absorption by vegetation.

The role of herbivores in succession differs among ecosystem types and successional stages. Mammals often accelerate the early successional changes in forests by eliminating or reducing the competitive ability of palatable early successional species. In grasslands, however, herbivores prevent the establishment of woody species that might otherwise transform grasslands into shrublands and forests. Some insects have their greatest impact in late succession, particularly in forests, where they can be important agents of mortality.

Review Questions

1. Provide examples of ways in which the carbon and nitrogen cycling of an ecosystem might be influenced by the legacy of events that occurred 1 week ago, 5 years ago, 100 years ago, 2,000 years ago.

2. What properties of disturbance regimes determine the ecological consequences of disturbance? How do these properties differ between treefalls in a tropical wet forest and fire in a dry conifer forest?

3. What are the major processes causing successional change in plant species? How does the relative importance of these processes differ between primary and secondary succession?

4. How do NPP, decomposition, and the carbon pools in plants and soils change through primary succession? At what successional stage does carbon accumulate most rapidly? Why? How do these patterns differ between primary and secondary succession? Why do these differences occur?

5. How does nitrogen cycling differ between primary and secondary succession? At what stages is this difference most pronounced?

6. How do trophic dynamics change through succession? Why?

7. How do water and energy exchange change through succession? What explains these patterns?

8. What are the major issues to consider in extrapolating information from one temporal scale to another? Describe ways in which this temporal extrapolation might be done.

Additional Reading

Bormann, F.H., and G.E. Likens. 1979. *Pattern and Process in a Forested Ecosystem.* Springer-Verlag, New York.

Chapin, F.S., III, L.R. Walker, C.L. Fastie, and L.C. Sharman. 1994. Mechanisms of primary succession following deglaciation at Glacier Bay, Alaska. *Ecological Monographs* 64:149–175.

Connell, J.H., and R.O. Slatyer. 1977. Mechanisms of succession in natural communities and their role in community stability and organization. *American Naturalist* 111:1119–1114.

Crocker, R.L., and J. Major. 1955. Soil development in relation to vegetation and surface age at Glacier Bay, Alaska. *Journal of Ecology* 43:427–448.

Fastie, C.L. 1995. Causes and ecosystem consequences of multiple pathways of primary succession at Glacier Bay, Alaska. *Ecology* 76:1899–1916.

Peters, D.P.C., A.E. Lugo, F.S. Chapin, III, S.T.A. Pickett, M. Duniway, et al. 2011. Cross-system comparisons elucidate disturbance complexities and generalities. *Ecosphere* 2(7):art81. doi:10-1890/ES11-00115.1.

Raffa, K.F., B.H. Aukema, B.J. Bentz, A.L. Carroll, J.A. Hicke, et al. 2008. Cross-scale drivers of natural disturbances prone to anthropogenic amplification: The dynamics of bark beetle eruptions. *BioScience* 58:501–517.

Turner, M.G. 2010. Disturbance and landscape dynamics in a changing world. *Ecology* 91:2833–2849.

Vitousek, P.M., and W.A. Reiners. 1975. Ecosystem succession and nutrient retention: A hypothesis. *BioScience* 25:376–381.

Vitousek, P.M. 2004. *Nutrient Cycling and Limitation: Hawai'i as a Model System.* Princeton University Press, Princeton.

Zimov, S.A., V.I. Chuprynin, A.P. Oreshko, F.S. Chapin, III, J.F. Reynolds, et al. 1995. Steppe-tundra transition: An herbivore-driven biome shift at the end of the Pleistocene. *American Naturalist* 146:765–794.

Landscape Heterogeneity and Ecosystem Dynamics

13

Landscape heterogeneity determines the regional consequences of processes occurring in individual ecosystems. In this chapter, we describe the major causes and consequences of landscape heterogeneity.

aid in understanding and quantifying landscape interactions. We then discuss sources of spatial heterogeneity within and among ecosystems and the consequences of that heterogeneity for interactions among ecosystems on a landscape.

Introduction

Spatial heterogeneity within and among ecosystems affects the functioning of individual ecosystems and entire regions. In previous chapters, we emphasized the controls over ecosystem processes in relatively homogenous units or **patches** of an ecosystem. The spatial pattern of ecosystems in a region, however, also influences ecosystem processes. Riparian ecosystems between upland agricultural systems and streams, for example, may filter nitrate and other pollutants that would otherwise enter streams. At a finer scale, nutrient cycling and organic matter accumulation in arid ecosystems occurs more rapidly beneath than between shrubs. The fragmentation of ecosystems into smaller units separated by other patch types influences the abundance and diversity of animals. All of the processes and mechanisms that operate in ecosystems (see Chaps. 4–11) have important spatial dimensions. In this chapter, we first discuss the concepts and characteristics of landscapes that

A Focal Issue

Human land-use change has fragmented landscapes throughout the world, often shifting the balance so that managed patches become the widespread matrix in which small fragments of less managed lands persist (Fig. 13.1). The increase in ratio of edge to area of these fragments alters physical environment throughout the patch, and the loss of connectivity among patches reduces their capacity to support many species. As global demand for food increases, how do we manage landscapes to meet these needs and to sustain the functioning of natural patches in the landscape? What are sustainable proportions of lands of differing management intensity? What happens if that proportion is exceeded? What configuration of natural and managed patches best meets the needs of nature and society? Careful attention to landscape configuration and dynamics can reduce the regional impacts of human actions in an increasingly human-dominated planet.

F.S. Chapin, III et al., *Principles of Terrestrial Ecosystem Ecology*,
DOI 10.1007/978-1-4419-9504-9_13, © Springer Science+Business Media, LLC 2011

Fig. 13.1 Shifting agriculture in the uplands of Yunan Province of China. The pressure of rising population has reduced the time that lands remain forested and increased the proportion of agricultural lands on the landscape. Photograph by Desmanthus4food (http://upload.wikimedia.org/wikipedia/commons/b/ba/Swidden_agriculture_in_Yunnan_Province_uplands.JPG)

Concepts of Landscape Heterogeneity

Spatial patterns control ecological processes at all scales. Landscapes are mosaics of patches that differ in ecologically important properties. **Landscape ecology** addresses the causes and consequences of spatial heterogeneity (Urban et al. 1987, Forman 1995, Turner et al. 2001, Cadenasso et al. 2007). This field focuses on both the interactions among patches on the landscape and the behavior and functioning of the landscape as a whole. Landscape processes can be studied at any scale, ranging from the mosaic of gopher mounds in a square meter of grassland to biomes that are patchily distributed across the globe (Fig. 13.2). Landscape processes are often studied at scales of watersheds or regions.

Some landscape patches are **biogeochemical hot spots** with high process rates, causing them to be more important than their areal extent would suggest. Beaver ponds, for example, are biogeochemical hot spots for methane emissions in boreal landscapes (Roulet et al. 1997); recently cleared pastures in the central Amazon Basin are hot spots for nitrous oxide emissions (Matson et al. 1987); and cities are hotspots for carbon emissions. Hot spots are defined with respect to a particular process and occur at all spatial scales, from the rhizosphere surrounding a root to urine patches in a grazed pasture, to wetlands in a watershed, to tropical forests on the globe. The environmental controls over biogeochemical hotspots often differ radically from controls in the surrounding **matrix**, that is, the predominant patch type in the landscape. Only by studying processes in hot spots can we understand these processes and extrapolate their consequences to larger scales. Landscape ecology therefore plays an essential role in understanding the Earth System because of the importance of estimating fluxes (and their controls) of energy and materials at regional and global scales.

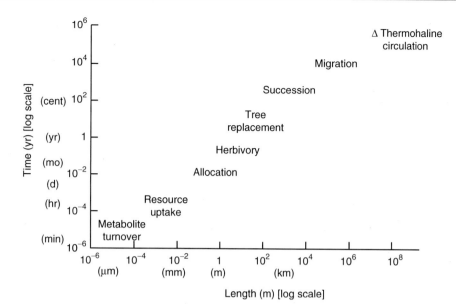

Fig. 13.2 Temporal and spatial scales at which selected ecosystem processes occur. The study of any ecosystem process requires understanding at least one level below (to provide mechanistic understanding) and one level above (to provide context with respect to patterns of temporal and spatial variability)

The size, shape, and spatial distribution of patches in the landscape govern interactions among patches. Patch size influences habitat heterogeneity. Large forest fragments in an agricultural landscape, for example, contain greater habitat heterogeneity and support more species and bird pairs than do small patches (Freemark and Merriam 1986, Wiens 1996). Patch size also influences the spread of propagules and disturbance from one patch to another (see Chap. 12). **Patch shape** influences the effective size of patches by determining the average distance from each point in the patch to an edge. Patch size and shape together determine the ratio of edge to area of the patch. The edge-to-area ratio of lakes, for example, is critical in determining the relative importance of pelagic and lake-margin production in supplying energy to aquatic food webs.

The **configuration**, that is, the spatial arrangement of patches in a landscape, influences landscape properties because it determines which patches interact and the spatial extent of their interactions. Riparian areas are important because they are an interface between terrestrial and aquatic ecosystems. Their linear configuration and location make them much more important than their small areal extent would suggest.

Configuration, together with patch size and shape, influence the **connectivity** among patches. The population dynamics of many organisms depend on movement between patches, which is strongly influenced by their connectivity (Turner et al. 2001). Birds and small animals in an agricultural landscape, for example, use fencerows to travel among patches of suitable habitat. In a patchy environment, local populations may go extinct, and the dynamics of **metapopulations**, that is, populations that consist of partially isolated subpopulations, depend on relative rates of local extinctions in patches and colonization from adjacent patches (Hanski 1999). Species conservation plans often encourage the use of corridors to facilitate movement among suitable habitat patches (Fahrig and Merriam 1985, Chetkiewicz et al. 2006, Saura and Pascual-Hortal 2007), although the effectiveness of corridors is debated (Rosenberg et al. 1997, Turner et al. 2001). Connectivity may be particularly critical at times of climatic change. Isolated nature reserves, for example, may contain species that cannot adapt or migrate in response to rapid environmental change. The effectiveness of corridors among patches depends on the size and mobility of organisms and the nature of disturbances that move among patches (Wu and Loucks 1995). A fencerow, for example, may be a corridor for voles, a barrier for cattle, but invisible to birds. A high connectivity among patches is not always beneficial. The high connectivity of extensive cornfields of the Midwestern U.S., for example, might allow pests to decimate large regions in response to climate change.

Ecological **boundaries** are critical to the interactions among neighboring landscape elements (Gosz 1991). Animals like deer, for example, are edge specialists that forage in one patch type and seek protection from predation in another. The size of the patch and its edge-to-area ratio determine the total habitat available to edge specialists. Edges often experience a different physical environment than do the interiors of patches. Forest boundaries adjacent to clearcuts, for example, experience more wind and solar radiation and are drier than are patch interiors (Chen et al. 1995). In tropical rainforests, the trees within 400 m of an edge experience more frequent blowdowns than do trees farther from an edge (Laurance and Bierregaard 1997). These differences in physical environment affect rates of disturbance and nutrient cycling, which translate into variations in recruitment, productivity, and competitive balance among species. The depths to which these edge effects penetrate differ among processes and ecosystems. Wind effects, for example, may penetrate more deeply from an edge than would availability of mycorrhizal propagules.

The abruptness of boundaries (that is, edge contrast) influences their role in the landscape (McCoy et al. 1986). Relatively broad gradients often occur at the boundaries between biomes, where there is a gradual shift in some controlling variable such as precipitation or temperature. Sharper boundaries tend to occur where steep gradients in physical variables control the distribution of organisms and ecosystem processes (e.g., between a stream and its riparian zone) or where an ecologically important functional type (e.g., trees) reaches its climatic limit. Physically determined boundaries can be stable under climate change whereas climatically determined boundaries can fluctuate or move directionally (Peters et al. 2009). Climatically determined boundaries, such as tree line or the savanna-forest border, for example, are useful places to study the effects of climatic change because species may be sensitive to small changes in climate.

Causes of Spatial Heterogeneity

Landscape heterogeneity stems from environmental variation, population and community processes, and disturbance (Turner 2005). Spatial variation in state factors (e.g., topography and parent material) and interactive controls (e.g., disturbance and dominant plant species) determine the natural matrix of spatial variability in ecosystems (Holling 1992). Human activities are an increasing cause of changes in the spatial heterogeneity of ecosystems.

Detection and Analysis of Spatial Heterogeneity

Remote sensing provides a set of tools to determine the structure and some aspects of the functioning of heterogeneous landscapes. Much of the spatial heterogeneity of interest occurs at spatial scales that cannot be observed from a single point on the ground. Remote sensing provides a suite of techniques – from low-technology aerial observation and photography to repeated satellite imagery – that allow us to visualize ecosystems across a large area all at once, to see them synoptically. Recent developments in remote sensing have transformed our ability to analyze ecosystem heterogeneity, and ongoing developments will continue to do so. For example, the integration of aircraft-based LIDAR (light detection and ranging, which is used to measure topography, canopy height, and vegetation structure) with high-spectral, high-spatial resolution spectrometry (which can measure aspects of canopy chemistry and physiological stress) allows highly resolved measurements of spatial variation in plant structure and chemistry, more or less simultaneously across thousands of hectares (Asner et al. 2007). When applied to natural terrestrial ecosystems, this approach can be used to detect, map, and analyze directly spatial heterogeneity in ecosystem structure and aspects of ecosystem functioning (Vitousek et al. 2009a). This approach can also be applied to understanding the distribution, dynamics, and consequences associated with biogeochemical hot spots, such as those associated with termite mounds in African savannas (Levick et al. 2010) and nitrogen-fixing biological invaders in Hawaiian rainforests (Hall and Asner 2007).

State Factors and Interactive Controls

Differences in abiotic characteristics and associated biotic processes account for the basic matrix of landscape variability. Temperature, precipitation, parent materials, and topography vary independently across Earth's surface. Some of these state factors, such as rock type, exhibit sharp boundaries and can therefore be classified into distinct patches. Others, including climate variables, vary more continuously and generate gradients in ecosystem structure and functioning, although amplifying feedbacks among processes controlled by these underlying gradients often create sharp boundaries in ecosystem structure and functioning. Analysis of these landscape classes and gradients shows that different factors control spatial pattern at different spatial scales. Regional-scale patterns of vegetation, net primary production (NPP), soil organic matter, litter quality, and nutrient availability in grasslands, for example, correlate with regional gradients in precipitation and temperature (Fig. 13.3; Burke et al. 1989). In contrast, topography, soil texture, and land-use history explain most variability at the scale of a few kilometers (Burke et al. 1999). Broad elevational and aspect-related patterns of ecosystem processes in tropical forests on the Hawaiian Islands are also governed largely by climate with local variation reflecting the type and age of parent material (Vitousek et al. 1992, Raich et al. 1997, Vitousek 2004). The resulting differences in soils give rise to consistent differences in nitrogen cycling (Pastor et al. 1984), phosphorus cycling (Lajtha and Klein 1988), and

nitrous oxide emissions (Matson and Vitousek 1987). These comparative studies provide a basis for extrapolating ecosystem processes to regional scales based on the underlying spatial matrix of abiotic factors.

Community Processes and Legacies

Historical legacies, stochastic dispersal events, and other community processes can modify the underlying relationship between environment and the distribution of a species. Ecosystem processes depend not only on the current environment but also on past events that influence the species present at a site (see Chap. 12). In Yellowstone National Park, for example, landscape variation in fire severity and cone **serotiny** (extent to which seeds are retained in cones) caused post-fire seedling recruitment of lodgepole pine to range from 0 to $>500,000$ stems ha^{-1}, which, in turn, strongly influenced post-fire productivity and nutrient cycling (see Fig. 12.14; Turner et al. 1999, Turner 2010). In arid and semi-arid ecosystems, soil processes are strongly influenced by the presence or absence of individual plants, resulting in "resource islands" beneath plant canopies (Schlesinger et al. 1990, Burke and Lauenroth 1995). The distribution of species on a landscape results from a combination of habitat requirements of a species, historical legacies (see Chap. 12), and stochastic events. Once these patterns are established, they can persist for a long time, if the species effects are strong. Fine-scale distribution of hemlock and sugar maple that developed in Michigan several thousand years ago, for example, has been maintained because each tree species produces soil conditions that favor its own persistence (Davis et al. 1998).

Disturbance

Natural disturbances are ubiquitous in ecosystems and create spatial patterning at many scales. The **patch dynamics** of a landscape reflect cycles of disturbance and post-disturbance

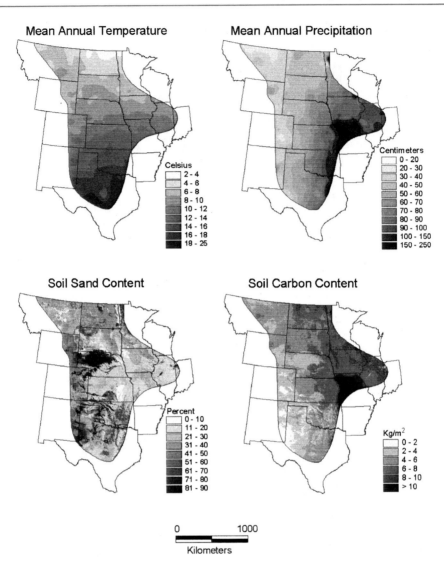

Fig. 13.3 Regional patterns of air temperature, precipitation, soil sand (a measure of the coarseness of soil texture), and soil carbon content across the Great Plains of the U.S. (Burke et al. 1989). Soil carbon content was modeled based on regional databases of the environmental variables using the CENTURY model. Soil carbon content varies regionally in ways that are predictable from climate and soil texture. Figure kindly provided by Indy Burke

succession (see Chap. 12; Pickett and White 1985, Turner 2010). Under relatively stable conditions, this generates a **shifting steady-state mosaic**, in which the vegetation at any point in the landscape is always changing but, averaged over a large enough area, the proportion of the landscape in each successional stage remains relatively constant (Bormann and Likens 1979). Although every point in the landscape may be at a different successional stage, the landscape as a whole may be close to steady state (Turner et al. 1993). Shifting steady-state mosaics develop (1) in environmentally uniform areas, where disturbance is the main source of landscape variability, (2) when disturbances are small relative to the size of the landscape, and (3) when the rate of recovery is similar to the return time of the disturbance (Fig. 13.4). When disturbances are small and recovery is rapid, most of the landscape will be in mid- to late-successional stages. In the

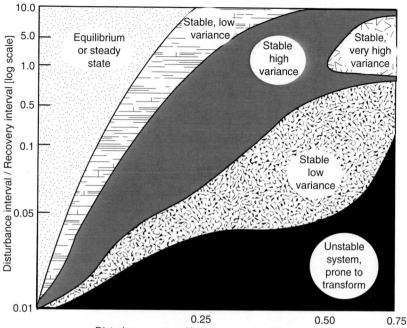

Fig. 13.4 Effect of disturbance size (relative to the size of the landscape) and disturbance frequency (relative to the time required for the ecosystem to recover) on the stability of landscape processes. Landscapes are close to steady state when disturbances are small (relative to the size of the study area) and when they are infrequent (relative to the time required for ecosystem recovery). As disturbances become more frequent or larger, the landscape becomes more heterogeneous, and individual patches are increasingly likely to shift to different successional trajectories. Redrawn from Turner et al. (1993)

primary tropical rainforests of Costa Rica, for example, the regular occurrence of treefalls results in maximum tree age of only 80–140 years (Hartshorn 1980). Gap-phase disturbance contributes to the maintenance of the productivity and nutrient dynamics of the forest. Light, and sometimes nutrient availability, increases in tree-fall gaps, providing resources that allow species with higher resource requirements to grow quickly and maintain themselves in the forest mosaic (Chazdon and Fetcher 1984, Brokaw 1985). Disturbances by animals in grasslands and shrublands can also generate a shifting steady-state mosaic. Gophers, for example, disturb patches of California serpentine grasslands, causing patches to turn over every 3–5 years (Hobbs and Mooney 1991).

Large-scale infrequent disturbances alter the structure and processes of some ecosystems over large areas. These disturbances create non-steady-state mosaics in which large expanses of the landscape are in the same successional stage. After Puerto Rico's Hurricane Hugo in 1989, for example, most of the trees in the hurricane path were broken off or blown over or lost a large proportion of their leaves, resulting in a massive transfer of carbon and nutrients from vegetation to the soils. The large pulse of high-quality litter increased decomposition rates substantially over large areas (Scatena et al. 1996).

Fire can also create large patches of a single successional stage on the landscape (Johnson 1992). In 1988, wildfires burned about a third of Yellowstone National Park. Fires of this magnitude and intensity recur every few centuries (Schoennagel et al. 2004). Long-term human fire suppression has increased the proportion of late-successional communities in many forests characterized by ground fires (e.g., ponderosa pine or sequoia in the western U.S.). This results in a more homogeneous and spatially continuous, fuel-rich environment in which fires can burn large areas. Forests that are characterized by stand-replacing crown fires (e.g., lodgepole pine in Yellowstone) quickly regenerate enough fuel to burn again. Fire suppression therefore has little effect on the fire regime of these forests.

Even large disturbed areas are often internally quite patchy, creating a **functional mosaic**, that is, a landscape with functionally important differences among patches. Fires, for example, usually produce islands of unburned vegetation and patches of varying burn severity that often differ dramatically in the density of regenerating trees, productivity, and rates of nitrogen cycling (Turner 2010). Unburned islands act as seed sources for post-fire succession and protective cover for wildlife, greatly reducing the effective size of the disturbance (Turner et al. 1997). In many cases, patches become less distinct as succession proceeds, so spatial heterogeneity may decline with time in non-steady-state mosaics (see Fig. 12.14b; Turner 2010).

Human-induced disturbances alter the natural patterns and magnitude of landscape heterogeneity. The signature of human influence is readily detectable in landscape patterns (Cardille and Lambois 2010). Isolated land-use changes may augment landscape heterogeneity by creating small patches within a matrix of largely natural vegetation. However, human activities have transformed as much as 75% of the ice-free terrestrial surface (see Fig. 1.8; Turner et al. 1990, Ellis and Ramankutty 2008). We have cleared or selectively harvested forests; converted grasslands and savannas to pastures or agricultural systems; drained wetlands; flooded uplands; and irrigated drylands. As land-use change becomes more extensive, the human-dominated patches become the matrix in which isolated fragments of natural ecosystems are embedded, reducing landscape heterogeneity and causing a qualitative change in landscape structure and functioning. These contrasting impacts of human actions on landscape heterogeneity are illustrated by the practice of shifting agriculture.

Shifting agriculture is a source of landscape heterogeneity at low population densities but reduces landscape heterogeneity as human population increases. Shifting agriculture, also known as **slash-and-burn agriculture** or **swidden agriculture**, involves the clearing of forest for crops followed by a fallow period during which forests regenerate, after which the cycle repeats (Fig. 13.1). Shifting agriculture is practiced extensively in the tropics and in the past played an important role in clearing the forests of Europe and eastern North America. Small areas of forest are typically cleared of most trees and burned to release organically bound nutrients. Crops are planted in species mixtures, with multiple plantings and harvests (Vandermeer 1990). As soil fertility drops, and insect and plant pests encroach, often within 3–5 years, the agricultural plots are abandoned, and the forest regenerates. The regenerating forests provide fuel and may be managed to provide fruits and other useful products for 20–40 years until the cycle repeats. Shifting agriculture generates landscape heterogeneity at many scales, ranging from different aged patches within a forest to different crop species within a field. With moderate human population densities that allowed long enough fallow periods and judicious selection of land for cultivation, shifting agriculture persisted for thousands of years without any progressive change in biogeochemical cycles (Ramakrishnan 1992, Palm et al. 2005).

As population density increases, land becomes scarcer, and the fallow periods are shortened or eliminated, leading to a more homogeneous agricultural landscape. Under these conditions, nutrient and organic matter losses during the agricultural phase cannot be recouped, and the system degrades, requiring larger areas to provide enough food. As the landscape becomes dominated by active cropland or early successional weedy species, the seed sources of mid-successional species are eliminated, preventing forest regrowth and further reducing the potential for landscape heterogeneity. In northeast India, for example, this shifting agriculture appears unsustainable when the rotation cycle declines below 10 years (Table 13.1; Ramakrishnan 1992, Palm et al. 2005).

Table 13.1 Comparison of ecosystem processes among several agricultural systems in Northeast India[a]

Ecosystem	Nitrogen input (g N m^{-2} year^{-1})[b]	NPP (g m^{-2} year^{-1})[c]	Litterfall (g m^{-2} year^{-1})[c]	Soil erosion (g m^{-2} year^{-1})[d]
Natural forest	15.5	2,360	118	800
Shifting agriculture				
5-year cycle	5.7	550	48	6,900
10-year cycle	9.2	670	71	3,000
30-year cycle	10.9	1,480	98	1,500
Mixed crop	–	100	–	100
Intensive agriculture				
Coffee	12.4	50	–	1,200
Tea	28.4	100	–	2,600
Ginger	21.3	190	–	20,000

[a]Note that the natural forest, which was an undisturbed sacred grove, had a higher productivity and litterfall than any of the managed ecosystems, even though the annual nitrogen inputs in litterfall were less than many of the intensively managed crops. When the rotation cycle of shifting agriculture became shorter than 10 years, there was a substantial drop in nitrogen cycling and litterfall and an increase in erosion. The continuous cropping systems (mixed and intensive) were less productive than the shifting agriculture, even for the crop phase. Data from Ramakrishnan (1992)
[b]Nitrogen inputs are from natural litterfall in the natural forest and the fallow phase of shifting agriculture but from fertilizer in the intensive agriculture
[c]Values are grams dry matter for the fallow phase for shifting agriculture
[d]Values are for the total rotational cycle

Interactions Among Sources of Heterogeneity

Landscape heterogeneity and disturbance history interact to influence further disturbance. Disturbance is more than a simple overlay on the spatial patterns governed by environment because even slight variations in topography or edaphic factors can influence the frequency, type, or severity of natural disturbances and the probability that land will be cleared by people. Slope and aspect of a hillside affect solar irradiance, soil moisture, soil temperature, and evapotranspiration rate. These factors, in turn, contribute to variation in biomass accumulation, species composition, and fuel characteristics. Different parts of the landscape therefore differ in susceptibility to fire. The resulting mosaic of patch types with different flammabilities can prevent a small, locally contained fire from moving across large areas. Slope and aspect can also directly influence the exposure of ecosystems to fire spread because fire generally moves uphill and tends to halt at ridgetops. Elevation and topographic position also influence the susceptibility of forest trees to windthrow (Foster 1988). Alternatively, a wildfire can spread across a large, highly connected landscape, even if started by a small ignition source. An untended campfire resulted in the largest fire in the history of Colorado, the Hayman fire in 2002 (Graham 2003).

Patchiness created by disturbance and other legacies influences the probability and spread of disturbance, thereby maintaining the mosaic structure of landscapes. The spread of fire, for example, creates patches of early successional vegetation in fire-prone ecosystems that are less flammable than late-successional vegetation (Rupp et al. 2000). In this way, past disturbances create a legacy that governs the probability and patch size of future disturbances. The effectiveness of these disturbance-generated early successional firebreaks depends on climate. At times of extreme fire weather, almost any vegetation will burn (Turner 2010).

Past history of insect or pathogen outbreaks also generates a spatial pattern that determines the pattern of future outbreaks. In mountain hemlock ecosystems of the Northwestern U.S., low light and nutrient availability in old-growth stands make trees vulnerable to a root pathogen. The resulting tree death increases light, nitrogen

Fig. 13.5 Transect of tree height and net nitrogen mineralization rate across a disturbance caused by root pathogens in a hemlock stand in the Northwestern U.S. Nitrogen mineralization increases dramatically beneath trees recently killed by the pathogen (position *2*). As trees recover (positions *3, 4,* and *5*), net nitrogen mineralization declines toward rates typical of undisturbed forest (position *1*). Data from Matson and Boone (1984)

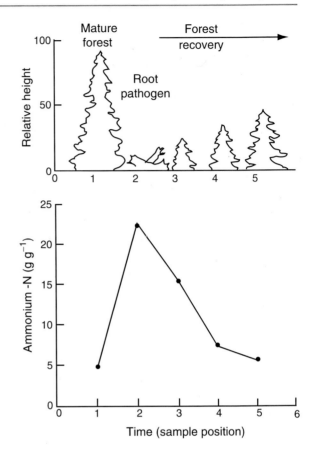

mineralization, and nutrient availability, making the regenerating forest resistant to further attack (Matson and Boone 1984; Fig. 13.5). The infections tend to move through stands in a wave-like pattern, attacking susceptible patches and creating resistant patches in their wake (Sprugel 1976), just as described for fire. Similarly, hurricanes that blow down large patches of trees generate early successional patches of short-statured trees that are less vulnerable to windthrow in the next hurricane. Even the fine-grained steady-state mosaics that characterize gap-phase succession are self-sustaining because young trees that grow in a gap created by treefall are less likely to die than are older trees. In summary, those disturbances that reduce the probability of future disturbance generate a stabilizing (negative) feedback that tends to stabilize the disturbance regime of a landscape, resulting in a shifting steady-state mosaic with a characteristic patch size and return interval. Any long-term trend in

climate or soil resources or extreme events that alters disturbance regime will probably alters the characteristic distribution of patch sizes and shapes on the landscape.

Interactions among processes that are controlled at different scales lead to nonlinear and sometimes catastrophic consequences (Peters et al. 2004). A wildfire starts, for example, when lightning strikes a tree whose structure and chemistry support combustion under ambient weather conditions. Fire spread, however, depends on the arrangement of plant canopies within a stand. Fire may then spread from one stand to another if there is high connectivity among stands, again leading to a nonlinear increase in combustion, once this threshold is passed. Finally, hot extensive fires create their own winds, which can greatly accelerate fire spread. Desertification, shrub encroachment into grasslands, and disease epidemics follow similar nonlinear changes in rates of spread, as new controls come into play at

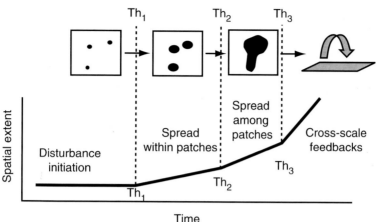

Fig. 13.6 Cross-scale interactions that influence the non-linear spread of disturbance across a landscape. Disturbances (e.g., fire, drought, disease epidemics) are often controlled by different processes at different scales (e.g., at scales of individuals, patches, landscapes, and regions). This produces thresholds (Th_1, Th_2,...) and changes in temporal responses as new processes exert control at each scale. Amplifying feedbacks cause upward shifts in the curve. Stabilizing feedbacks cause downward shifts in the curve. Redrawn from Peters et al. (2004)

progressively larger scales (Fig. 13.6). The changes that occur at progressively larger scales can be either amplifying or stabilizing. Predicting the consequences of these **cross-scale linkages** requires an understanding of the predominant controls at each scale of interest.

Human activities that create novel small-scale disturbances have the potential to produce unanticipated effects at larger scales. The cultivation of extensive areas of drought-sensitive crops on marginal lands in the U.S. in the 1920s created a landscape matrix of cultivated, drought-susceptible patches interspersed with small patches of native grassland (Peters et al. 2004). Hot, dry weather combined with strong winds in the 1930s resulted in reduced plant cover, high plant mortality, and localized wind erosion on the cultivated patches. At the landscape scale, these small dust storms became aggregated among patches to generate massive dust storms ("black blizzards") that disturbed intervening uncultivated patches and spread to affect much of the country (see Fig. 3.1; Peters et al. 2004, Schubert et al. 2004). The impact was extensive enough to reduce rainfall, intensifying the drought. Blowing soil from the Great Plains was documented as far as the east coast, over 1,500 km away.

Disturbances that increase the probability of other disturbances complicate predictions of landscape pattern. Insect outbreaks that kill trees in a fine-scale mosaic, for example, are often thought to increase the overall flammability of the forest, although the evidence for this pattern is sparse, and insects sometimes reduce fire risk by reducing fuel density (Turner 2010, Simard et al. 2011). The public concern about large fires after insect outbreaks then creates public pressure for salvage logging of insect-killed stands. This logging creates patches of clearcuts that are intermediate in size between those created by insects and those that might have been produced by a catastrophic fire. It is difficult to predict in advance what patterns of patch structure will develop. Rule-based models that define conditions under which particular scenarios are likely to occur provide a framework for predictions in the face of multiple potential outcomes (Starfield 1991).

Human activities create amplifying and stabilizing feedbacks to disturbances that alter the patch structure and functioning of landscapes. In principle, the effect of human-induced disturbances, such as land clearing, on landscape structure is no different than that of any other disturbance. However, the novel nature and the increasingly extensive occurrence of human disturbances are rapidly altering the structure of many landscapes. The construction of a road

Lightning   Anthropogenic   Major
fires   fires   roads

Fig. 13.7 Maps of fires caused by naturally and human-caused fires in Alaska. The human-caused fires mirror the road and river transportation corridors, indicating the importance of human access in altering regional fire regime. Redrawn from Gabriel and Tande (1983)

through the tropical wet forests of Rondonia, Brazil, for example, created a simple linear disturbance of negligible size. The sudden increase in human access, however, led to rapid clearing of forest patches that were much larger than natural treefall gaps or the hand-cleared patches created by shifting agriculture. Similarly, road access is the major factor determining the distribution of fire ignitions in the boreal forest of interior Alaska (Fig. 13.7). In general, road access is one of the best predictors of the spread of human-induced disturbances in relatively natural landscapes (Dale et al. 2000).

Socioeconomic factors, such as farmer income, interact with site characteristics to influence human impacts on landscape pattern. Heterogeneous landscapes are often converted to fine-scale mosaics of agricultural and natural vegetation, whereas large areas suitable for mechanized agriculture are more likely to be deforested in large blocks. In northern Argentina, for example, patches of dry deciduous forests on the eastern slopes of the Andes were converted to small patches of cropland, or modified by grazing into thorn-scrub grazing lands and secondary forests (Fig. 13.8; Cabido and Zak 1999). On the adjacent plains, however, larger parcels were initially deforested for grazing and more recently converted to mechanized agriculture. Large holdings on the plains are owned by companies that make land-use decisions based on the global economy. Small family producers in the mountains maintain a more traditional lifestyle that involves smaller, less frequent changes in land use (Zak et al. 2008).

Disturbance is increasingly used as a management tool to generate stand and landscape structures that more closely mimic those of natural ecosystems. Forest harvest varies from 0% to 100% tree removal, and the sizes and shapes of clearcuts can be altered from the standard checkerboard pattern to mimic more natural disturbances (Franklin et al. 1997). Forest harvest regimes can also be designed to retain some of the functional attributes of late-successional forests, such as the filtering function of riparian vegetation, the presence of large woody debris, and the retention of a few large trees as seed source and nesting habitat. Protection of these features can significantly reduce the ecological impact of forest harvest. Prescribed fire is increasingly used as a management tool, particularly in areas where a century of "Smoky the Bear" policy of complete fire suppression led to unnaturally large fuel accumulations in some ecosystems. Prescribed fires are typically lit under weather conditions where fire intensity and severity are low, so the fire can be readily controlled. In populated regions, vegetation may be physically removed as a substitute for fire because prescribed fires are considered unsafe. Natural fire, prescribed fire, and physical removal of vegetation probably differ in their impacts on ecosystem processes due to differences in the quantity of organic matter and nutrients removed; these differences affect subsequent regrowth.

Ecologists are only beginning to understand the long-term consequences of different disturbance regimes for the structure and functioning of ecosystems and landscapes. As this understanding

Fig. 13.8 Satellite-based map of the Cordoba region of Northern Argentina in 1999, showing semi-natural vegetation (*black*), lands that have been modified by grazing (*gray*) and croplands (*white*). The plains to the east are more suitable for mechanized agriculture and are large land holdings with large areas converted to croplands. Lands to the west are more mountainous and less suitable for mechanized agriculture. They are owned by small farmers, each of whom maintains a heterogeneous mosaic of land use. The proportion of area converted to cropland is greater in large land holdings suitable for intensive agriculture (Cabido and Zak 1999). Figure kindly provided by Marcelo Cabido and Marcelo Zak

improves, more informed decisions can be made in using disturbance as a tool in ecosystem management (see Chap. 15). Management of disturbance regime can recreate landscape structures in which natural disturbance regimes can again come into play or can mimic the ecological effects of disturbance under conditions in which the natural disturbance pattern has unacceptable societal consequences.

Patch Interactions on the Landscape

Interactions among patches on the landscape influence the functioning of individual patches and the landscape as a whole. Landscape patches interact through lateral movement of water, energy, nutrients, or organisms across boundaries from one patch to another. This occurs through topographically controlled interactions, transfers through the atmosphere, biotic transfers, and the spread of disturbance. These transfers are critically important to the long-term sustainability of ecosystems because they represent losses from donor ecosystems and subsidies to recipient ecosystems. Large changes in these transfers constitute changes in inputs and outputs of resources and therefore alter the functioning of ecosystems.

Topographic and Land–Water Interactions

Topographically controlled redistribution of materials is the predominant physical pathway by which materials move between ecosystems (Fig. 13.9). Gravity is a potent force for landscape interactions. It causes water to move downhill, carrying dissolved and particulate materials. Gravity is also the driving force for landslides, soil creep, and other forms of soil movement (see Chap. 3). These topographically controlled processes transfer materials from gopher mounds to the surrounding grass matrix, from uplands to lowlands, from terrestrial to aquatic systems, and from freshwater ecosystems

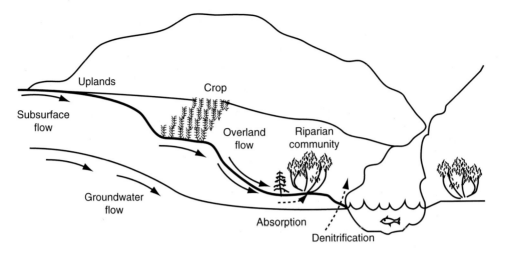

Fig. 13.9 Topographically controlled interactions among ecosystems in a landscape via erosion and solution transfers in subsurface flow or groundwater. Riparian forest trees absorb nutrients primarily from well-aerated soils, whereas denitrification requires anoxic conditions, which generally occur below the water table. Nitrogen absorption and denitrification are the most important mechanisms by which riparian zones filter nitrogen from groundwater between upland ecosystems and streams

Fig. 13.10 Relationship between the total nitrogen input to the major watersheds of the world and nitrate loading in rivers. Redrawn from Howarth et al. (1996a)

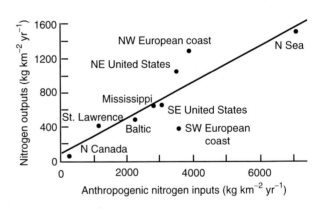

to estuaries and the ocean (Naiman et al. 2005, Yoo et al. 2005).

The nature of donor ecosystems and their management govern the transfer of dissolved materials. Regions with intensive agriculture and those receiving substantial nitrogen deposition transfer substantial quantities of nitrate and phosphorus to rivers, lakes, and groundwater (Carpenter and Biggs 2009). Nitrate loading in rivers, for example, correlates closely with the total nitrogen input to the major watersheds of the world (Fig. 13.10; Howarth et al. 1996a). At more local scales, the patterns of land use and urbanization influence the input of nutrients to lakes and streams. These increased fluxes of dissolved nitrogen have multiple environmental consequences, including health hazards, acidification, eutrophication, and reduced biodiversity of downstream freshwater and marine ecosystems (Howarth et al. 1996a, Nixon et al. 1996).

Erosion moves particulate material containing nutrients and organic matter from one

ecosystem and deposits it in another. Erosion ranges in scale from silt suspended in flowing water to movement of whole mountainsides in landslides. The quantity of material moved depends on many physical factors, including slope position, slope gradient, the types of rocks and unconsolidated material underlying soils, and the types of erosional agents (e.g., amount and intensity of rainfall events; see Chap. 3). The biological characteristics of ecosystems are also critical. Vegetation type, root strength, disturbance, management, and human development can be as important as the vertical gradient or parent material. Forest harvest on steep slopes in the Northwestern U.S., for example, has increased the frequency of landslides. Similarly, upland agriculture often increases sedimentation and the associated transfer of nutrients and contaminants (Comeleo et al. 1996, Syvitski et al. 2005). Proper management of up-slope systems through use of cover crops, reduced tillage, and other management practices can reduce erosional transfers of materials.

Landscape pattern influences the transfer of materials among ecosystems. In managed and unmanaged landscapes, ecosystems interact with one another along topographic sequences, with nutrients leached from uplands providing a nutrient subsidy to mid-slope or lowland ecosystems (Shaver et al. 1991). The configuration of these ecosystems in the landscape determines the pattern of nutrient redistribution and their outputs to groundwater and streams. Riparian vegetation zones, including wetlands and floodplain forests, act as filters and sediment traps for the water and materials moving from uplands to streams (Fig. 13.9). The dominance of riparian zones by disturbance-adapted plants that tolerate soil deposition and have rapid growth rates contributes to their efficiency as landscape filters. Riparian zones play a particularly crucial role in agricultural watersheds, where they remove fertilizer-derived nitrogen and phosphorus and eroding sediments. The fine-textured, organic-rich soils and moist conditions characteristic of most riparian areas also promote denitrification of incoming nitrate. Plant uptake and denitrification together account for the decline in nitrate con-centration as groundwater flows from agricultural fields through riparian forests to streams. Phosphorus is retained in riparian areas primarily by plant and microbial absorption of nutrients and by physical adsorption to soils because phosphorus has no pathway of gaseous loss.

The high productivity and nutrient status of riparian vegetation and the presence of water cause riparian areas to be intensively used by animals, including livestock in managed ecosystems. People also use riparian areas intensively for water, gravel, transportation corridors, and recreation. Long-term elevated inputs from heavily fertilized agricultural areas or from wetlands used for tertiary sewage treatment (i.e., to remove the products of microbial decomposition) can saturate the capacity of riparian areas to filter nutrients from groundwater. Overexploitation of riparian areas can increase sediment and nutrient loading to streams and reduces shading, making freshwater ecosystems more vulnerable to changes in land use within the watershed (Correll 1997, Lowrance et al. 1997, Naiman and Décamps 1997).

In some cases, landscape pattern has no apparent effect on ecosystem processes. During severe fire weather, for example, all stands burn, and landscape patterns of differential flammability are relatively unimportant (Turner et al. 1994). Landscape pattern is most likely to be important when there is a distinct directionality of patch interaction (e.g., nutrient flow from land to water) and when disturbances are of low-to-moderate intensity (Turner 2010).

The properties of recipient ecosystems influence their sensitivity to landscape interactions. The vulnerability of ecosystems to inputs from other patches in the landscape depends largely on their capacity to sequester or transfer the inputs. Riparian areas, for example, may have a higher capacity to retain a pulse of nutrients or transfer them to the atmosphere by denitrification than do upland late-successional forests. Streams characterized by frequent floods are less likely to accumulate sediment inputs than are slow-moving streams and rivers because floods flush sediments from river channels of steep stream reaches. Lakes on calcareous substrates or those that receive abundant groundwater input due to a location low in a watershed are better buffered against inputs of acidity and nutrients than are oligotrophic lakes on granitic substrates or lakes high in a watershed that receive less groundwater input (Webster et al. 1996).

Estuaries, the coastal ecosystems located where rivers mix with seawater, are a striking example of the way in which ecosystem properties influence their sensitivity to inputs from the landscape. They are among the most productive ecosystems on Earth (Howarth et al. 1996b, Nixon et al. 1996). Their high productivity stems in part from the inputs they receive from land and from the physical structure of the ecosystem, which is stabilized by the presence of sea grasses and other rooted plants. This tends to dampen wave and tidal energy, reducing resuspension and increasing sedimentation. Salinity and other geochemical changes that occur as the waters mix lead to flocculation and settling of suspended particles. Nutrient absorption by the rooted vegetation and phytoplankton, burial by sedimentation, and denitrification in anoxic sediments function as sinks for nutrients flowing from upstream watersheds, just as in riparian zones. The stability of the landscape on the Mississippi River Delta, for example, depends on regular delivery of sediments from upstream to replace soils removed by tidal erosion. Channels, levees, and other engineering solutions to flood control and water management may reduce the short-term probability of flooding but also eliminate the sediment supply that builds and maintains these barrier islands that protect the coast from larger storms. The drainage of wetlands to support urban development in New Orleans caused widespread subsidence of the land surface and reduced the capacity of the wetlands to store water from storm surges. Together this shift from dependence on natural landscape interactions to engineered alternatives contributed to the catastrophic impact of Hurricane Katrina in New Orleans in 2005 (Kates et al. 2006). Many estuaries, including the Gulf of Mexico near the entrance of the Mississippi River, are becoming saturated by nutrient enrichment within their watersheds,

resulting in harmful phytoplankton blooms, loss of sea grass, and increasing frequency of anoxia or hypoxia that create dead zones that kill fish and benthic invertebrates such as shrimp (see Fig. 9.1; Rabalais et al. 2002).

Atmospheric Transfers

Atmospheric transport of gases and particles links ecosystems over large distances and coarse spatial scales. Gases emitted from managed or natural ecosystems are processed in the atmosphere and can be transported for distances ranging from kilometers to the globe. Once deposited, they can alter the functioning of the recipient ecosystems (Fig. 13.11), just as with topographically controlled transfers.

In areas downwind of agriculture, NH_3 and NO_x can represent a significant fraction of nitrogen deposition. Dutch heathlands, for example, receive at least 10-fold more nitrogen deposition than would occur naturally. The magnitude of these inputs is similar to the quantity of nitrogen that annually cycles through vegetation, greatly increasing the openness of the nitrogen cycle. Areas downwind of industry and fossil fuel combustion receive nitrogen largely as NO_x. Sulfur gases, including SO_2, are also produced by fossil fuel combustion, although improved regulations have reduced these emissions and deposition relative to NO_x.

The large nitrogen inputs to ecosystems have important consequences for NPP, nutrient cycling, trace gas fluxes, and carbon storage. Chronic nitrogen deposition initially reduces nitrogen limitation by increasing nitrogen cycling rates, foliar nitrogen concentrations, and NPP. Above some threshold, however, the ecosystem becomes saturated with nitrogen (Fig. 13.12; Aber et al. 1998). As excess nitrate and sulfate leach from the soil, they carry with them cations to maintain charge balance, inducing calcium and magnesium deficiency in vegetation (Driscoll et al. 2001). In southern Sweden, for example, over half of the plant-available cations have been lost from the upper 70 cm of soil in the past half-century, due at least in part to chronic exposure

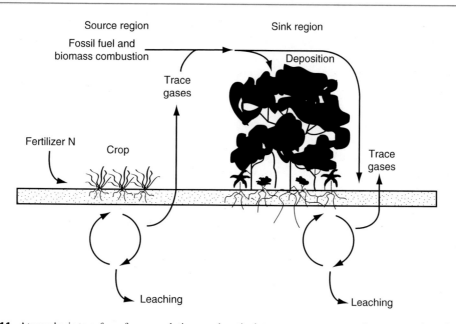

Fig. 13.11 Atmospheric transfers of gases, solutions, and particulates among ecosystems. Inputs come from fossil fuel and biomass combustion and from trace gases originating from natural and managed ecosystems

Fig. 13.12 Changes hypothesized to occur as forests undergo long-term nitrogen deposition and nitrogen saturation. Redrawn from Galloway et al. (2003)

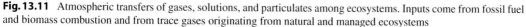

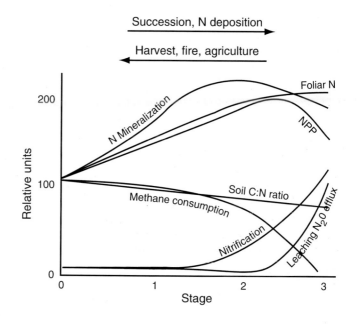

to acid precipitation (Hallbacken 1992). The exchange complex becomes more dominated by manganese, aluminum, and hydrogen ions, increasing soil acidity and the likelihood of aluminum toxicity. Together this suite of soil changes often enhances frost susceptibility, impairs root development, and promotes herbivory, leading to forest decline in many areas of Europe and the Northeastern U.S. (Schulze 1989, Aber et al. 1998). The major surprise, however, has been how resilient some forests have been to acid rain, often retaining most of the nitrogen inputs within the ecosystem for decades and sustaining their productivity and carbon storage (Magnani et al. 2007, de Vries et al. 2009, Janssens et al. 2010).

In other forests, half of the nitrogen inputs are lost in streamflow (Lovett et al. 2000). The vulnerability of ecosystems to acid rain depends in part on the magnitude of inputs (related to distance from pollution sources and amount of precipitation received) and initial soil acidity, which in turn depends on parent material and species composition. For example, in the Northeastern U.S., forest productivity has declined in sites with granitic bedrock, particularly at high elevations where soil pools of base cations are smallest (Likens et al. 1996, Fahey et al. 2005). This is associated with calcium loss from soils and reduced growth and increased mortality of sugar maple, a calcium-sensitive species and important canopy dominant. Calcium addition to a watershed caused increases in mycorrhizal colonization, tissue calcium, and growth of sugar maple seedlings, particularly at high elevations (Juice et al. 2006). Acid rain also increases nitrogen inputs to streams and reduces the acid-neutralizing capacity of lakes (Aber et al. 1998, Carpenter et al. 1998, Driscoll et al. 2001). The increases in lake acidity are most pronounced in watersheds whose bedrock is poor in cations. In these lakes, acidity reduces the size, survival, and density of fish, in part through reductions in their food supply (Driscoll et al. 2001).

Nearly all research on the transport, deposition, and ecosystem consequences of anthropogenic nitrogen has been conducted in the Temperate Zone. Further increases in nitrogen deposition will, however, likely occur primarily in the tropics and subtropics (Galloway et al. 1995), where plant and microbial growth are often limited by elements other than nitrogen. These ecosystems might therefore show more immediate nitrogen loss in trace gases or leaching in response to nitrogen deposition (Matson et al. 1998). On the other hand, soil properties such as high clay content or cation exchange capacity may allow tropical soils to sequester substantial quantities of nitrogen before they become leaky.

Biomass burning transfers nutrients directly from terrestrial pools to the atmosphere and then to downwind ecosystems. Biomass combustion releases a suite of gases that reflect the elemental concentrations in vegetation and fire intensity. About half of dry biomass consists of carbon, so the predominant gases released are carbon compounds in various stages of oxidation, including carbon dioxide, methane, carbon monoxide, and smaller quantities of non-methane hydrocarbons. The atmospheric role of these gases varies. CO_2 and CH_4 are greenhouse gases, whereas carbon monoxide and non-methane hydrocarbons react in the troposphere to produce ozone and other atmospheric pollutants that can affect downwind ecosystems (see Chap. 2). Nitrogen is also released in various oxidation states, including nitrogen oxides (NO and NO_2, together known as NO_x) and ammonia (NH_3). The proportional release of these forms also depends on the intensity of the burn, with NO_x typically accounting for most of the emissions. Sulfur-containing gases, organic soot and other aerosol particles, elemental carbon, and many trace species of carbon, nitrogen, and sulfur also have important regional and global effects. Satellite and aircraft data show that these gases and aerosols in biomass burning plumes can be transported long distances.

Windblown particles of natural and anthropogenic origins link ecosystems on a landscape. The role of the atmosphere as a transport pathway among ecosystems varies among elements. For some base cations (Ca^{2+}, Mg^{2+}, Na^+, and K^+) and for phosphorus, dust transport is the major atmospheric link among ecosystems. At the local-to-regional scale, dust from roads or rivers can alter soil pH and other soil properties that account for regional zonation of vegetation and land–atmosphere exchange (Walker et al. 1998). At the global scale, Saharan dust is transported across the Atlantic Ocean and deposited on the Amazon by tropical easterlies. Although the annual input of dust is small, it contributes substantially to soil development over the long term (Okin et al. 2004). Similarly, dust from the Gobi desert is deposited in wet forests of the Hawaiian Islands at the rate of 1.25 $g\,m^{-2}$ year^{-1}. In old soils, that is, those >2 million years old, dust input can be the largest source of phosphorus (Chadwick et al. 1999, Vitousek 2004). In the

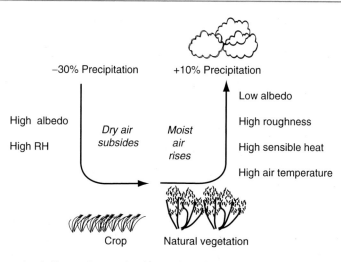

Fig. 13.13 Effects on regional climate of conversion from heathland to barley croplands in southwestern Australia. The heathland absorbs more radiation (low albedo) and transmits a larger proportion of this energy to the atmosphere as sensible heat than does adjacent croplands. This causes air to rise over the heathland and draws in moist air (high relative humidity [RH]) laterally from the irrigated cropland; this causes subsidence of air over the cropland, just as with the air circulation of sea breezes (see Chap. 2). Rising moist air increases precipitation by 10% over heathland, whereas subsiding dry air reduces precipitation by 30% over the cropland. Data from Chambers (1998; see Fig. 4.1)

Western U.S., the building of railroads between 1860 and 1900 combined with intense cattle and sheep grazing in the American Southwest increased dust loads in the atmosphere to result in reduced duration of snow cover in the San Juan Mountains (Painter et al. 2007, Neff et al. 2008). Similarly, large Asian dust storms can travel as far as the Western U.S., causing faster and earlier snow melt with less water available for irrigation, cities, and ski resorts in the Rocky Mountains.

Land–atmosphere exchange of water and energy in one location influences downwind climate. The ocean and large lakes moderate the climate of adjacent land areas by reducing temperature extremes and increasing precipitation (see Chap. 2). Human alteration of the land surface is now occurring so extensively that it also has significant effects on downwind ecosystems. Conversion of Australian heathlands to agriculture has, for example, increased precipitation over heathlands and reduced it by 30% over agricultural areas (Fig. 13.13; see Fig. 4.1). Deforestation in the Amazon reduces regional evapotranspiration. This reduction in water recycling reduces the moisture available for precipitation elsewhere in the basin, making the climate less favorable for tropical rainforests (Foley et al.

2003b). At a global scale, the clearing of land for agriculture has reduced regional albedo and evapotranspiration, leading to greater sensible heat flux (Chase et al. 2000, Foley et al. 2003b, Field et al. 2007). At all spatial scales the atmospheric transfer of heat and water vapor from one ecosystem to another strongly affects ecosystem processes in downwind ecosystems. The climatic impacts on downwind ecosystems of reservoirs, irrigation of arid lands, and land-use change are seldom included in assessments of the potential impacts of these management projects.

Movement of Plants and Animals on the Landscape

The movement and dispersal of plants and animals link ecosystems on a landscape. Large animals typically consume forage from high-quality patches and deposit feces and urine where they rest or sleep. Sheep in New Zealand, for example, often spend nights on ridges, moving nutrients upward and counteracting the downward nutrient transport by gravity. Marine birds transfer so much phosphorus from marine foods to the land that the **guano** deposited in their traditional nesting areas has served as a major source of phosphorus for fertilizer. **Anadromous fish**, that is, marine fish that enter freshwater to breed, also transport marine-derived nutrients to terrestrial ecosystems. These fish carry the nutrients up rivers and streams, where they become an important food item for terrestrial predators, which transport the marine-derived nutrients to riparian and upland terrestrial ecosystems (Willson et al. 1998, Helfield and Naiman 2001). The enhanced tree growth supported by these marine-derived nutrients, in turn, provides more shade, less streambank erosion, larger coarse woody debris, and therefore potentially better salmon habitat than in streams that lack salmon (Helfield and Naiman 2001). Similarly, insects that feed on seaweed and other marine detritus are an important food source for spiders on islands, merging marine and terrestrial food webs (Polis and Hurd 1996).

Animals also transfer plants, especially as seeds, on fur and in feces. Many plants have evolved life history strategies to take advantage of this efficient form of dispersal. This dispersal mechanism has contributed to the spread of invasive plants. Feral pigs, a non-native herbivore in Hawaiian rainforests, for example, transfer seeds of invasive plants such as the passion vine, which alters patterns of nutrient cycling. Similarly, the alien bird white eye spreads the alien nitrogen fixer *Morella faya* (Woodward et al. 1990), which alters the nitrogen status of native ecosystems (Vitousek et al. 1987). Thus, invasions of both plants and animals from one ecosystem to another can contribute to a variety of ecosystem changes.

Animals that move among patches can have effects that differ among patch types. Edge specialists such as deer, for example, may concentrate their browsing in one habitat type but seek protection from predators and deposit nutrients in another (Seagle 2003). Predator–prey dynamics may also structure nutrient heterogeneity based on where predation occurs and carcasses are deposited, for example when moose feed on nitrogen-rich aquatic vegetation and are killed by wolves in upland sites (Bump et al. 2009). At a larger scale, migratory birds move seasonally among different ecosystem types. Lesser snow geese, for example, overwinter in the Southern

U.S. and breed in the Canadian Arctic. Populations of this species have increased by more than an order of magnitude as a result of increased use of agricultural crops (rice, corn, and wheat) on the wintering grounds and reduced hunting pressure. This species now exceeds the carrying capacity of its summer breeding grounds and has converted productive arctic salt marshes into unvegetated barrens (Jefferies and Bryant 1995).

People are an increasing cause of lateral transfers of materials among ecosystems, through addition of fertilizers, pesticides, etc., introduction of propagules of invasive species, removal of crops and forest products, and diversion of water. The resulting nutrient inputs to aquatic systems occur in locations where riparian zones and other ecological filters are often degraded or absent. The nutrient transfers in food from rural to urban areas are substantial. Food imports to Europe are a significant component of the European carbon and especially nitrogen balance (Ciais et al. 2008), and food imports are the major source of nitrogen to coastal watersheds (Driscoll et al. 2003). Water diversion by people has substantially altered rates and patterns of land-use change in arid areas at the expense of rivers and wetlands (see Chap. 4). As water becomes increasingly scarce in the coming decades, pressures for water diversion are likely to increase.

Disturbance Spread

Patch size and arrangement determine the spread of disturbance across a landscape. Disturbance is a critical interactive control over ecosystem processes that is strongly influenced by horizontal spread from one patch to another. Fire and many pests and pathogens move most readily across continuous stretches of disturbance-prone vegetation. Fuelbreaks of nonflammable vegetation, for example, reduce fire risk at the urban–wildland interface. Fires create their own fuelbreaks because post-fire vegetation is generally less flammable than that which precedes a fire. Theoretical models suggest that, when less than half of the landscape is disturbance-prone, severity is more important than frequency in determining the impacts of disturbance. When large proportions of the landscape are susceptible to disturbance, however, the frequency of disturbance becomes increasingly important (Gardner et al. 1987, Turner et al. 1989). The size of patches also influences the spread of disturbance. Landscapes dominated by large patches tend to have a low frequency of large fires, which in turn generate large patches. Landscapes with small patches have greater edge-to-area ratio, so fires tend to spread more frequently into less flammable vegetation (Rupp et al. 2000). In this way, landscapes tend to sustain their characteristic disturbance regime, until modified by other factors (e.g., climate or land-use change).

Patchy agricultural landscapes are less prone to spread of pests and pathogens than are large continuous monocultures. Intensive agriculture has reduced landscape patchiness in several respects. The average size of individual fields and the proportion of the total area devoted to agriculture have generally increased, as has the use of genetically uniform varieties. This can lead to rapid spread of pests across the landscape.

Human Land-Use Change and Landscape Heterogeneity

Human modification of landscapes has fundamentally altered the role of ecosystems in regional and global processes. Much of the land-use change has occurred within the last two to three centuries, a relatively short time in the context of evolution or landscape development. Since 1700, for example, the land area devoted to crop production has increased 466% and now accounts for about 10–20% of the ice-free terrestrial surface (Ellis and Ramankutty 2008). Many areas of the world are therefore dominated by a patchwork of agricultural fields, pastures, and remnant unmanaged ecosystems. Similar patchworks of cut and regenerating forest interspersed with small areas of old-growth forest are common on every continent. Human-dominated landscapes supply large amounts of food, fiber, and other ecosystem services to society. Two general patterns of land-use change emerge: (1) **extensification**, that is,

the increase in *area* affected by human activities, and (2) **intensification**, that is, the increase in *inputs* applied to a given area of land or water.

Extensification

Land-use changes include both land-use conversions and modifications (Meyer and Turner 1992). **Land-use conversion** involves a human-induced change in ecosystem type to one dominated by different physical environment or plant functional type, for example, the change from forest to pasture or from stream to reservoir. **Land-use modification** is the human alteration of an ecosystem in ways that significantly affect ecosystem processes, community structure, and population dynamics without radically changing the physical environment or dominant plant functional type. Examples include alteration of natural forest to managed forest, savanna management as grazing lands, and alteration of traditional low-input agriculture to high-intensity agriculture. In aquatic ecosystems, this includes the alteration of flood frequency by dams and levees or the stocking of lakes for sport fishing. Both types of land-use change alter the functioning of ecosystems, the interaction of patches on the landscape, and the functioning of landscapes as a whole.

Deforestation is an important conversion in terms of spatial extent and ecosystem and global consequences. Forests cover about 25% of the terrestrial surface, 2–3 times the total agricultural land area (see Fig. 1.8, Table 6.6). Globally, forest area has decreased about 15% (i.e., by 9 million km^2) since preagricultural times. Much of the European and the Indian subcontinents, for example, were prehistorically blanketed by forests, but over the last five to ten centuries have supported extensive areas of agriculture. Similarly, North America was once contiguously wooded from the Atlantic seaboard to the Mississippi River, but large areas of this forest were cleared by European settlers at rates similar to those that now characterize tropical forests (Dale et al. 2000).

Today, conversion of forests to pasture or agriculture is one of the dominant land-use changes in the humid tropics. The magnitude of this land-use change is uncertain, but that uncertainty is diminishing as both the technology and the applicability of remote sensing improve (Chambers et al. 2007). As recently as the 1990s, only the largest clearings could be monitored accurately via remote sensing – however, newly developed tools and analyses permit the detection of areas of selective logging as well as those of clearing and conversion. For example, Asner et al. (2010) used a multi-scale approach with aircraft-based LIDAR and satellite systems to evaluate carbon stocks and losses in a 4.3 million hectare region of the Peruvian Amazon, with a spatial resolution of 0.1 ha. Their analysis demonstrated that an amount of carbon equivalent to just over 1% of standing biomass in this relatively remote area was lost to land-use change in a decade and that inclusion of selective logging in the analyses increased the calculated amount of carbon lost by nearly 50%. On a finer spatial scale, repeated LIDAR measurements of forest structure can be used to estimate rates of treefall gap formation and filling directly, on a landscape scale (Kellner et al. 2009). In time, these direct measurements of forest turnover will be applied to larger and more heterogeneous areas (Kellner and Asner 2009). These tools will become increasingly important as efforts to retain or sequester carbon in biomass become more widespread.

The trajectory of landscape change caused by deforestation depends on both the nature of the original forests and the land use that follows. Primary forests are likely to persist longer in remote regions such as much of Amazonia and Central Africa than elsewhere. The permanent or long-term conversion of forests to managed ecosystems involves burning or removal of most of the biomass and often leads to large losses of carbon, nitrogen, and other nutrient elements from the system. Logging, in contrast, removes only the commercially valuable trees and may cause less carbon and nutrient loss from soils. The nutrient losses that accompany deforestation can alter adjacent ecosystems, particularly aquatic ecosystems, and influence the atmosphere and climate through changes in trace gas fluxes and water and energy exchange (see Chap. 4). They

may also affect forest regrowth after disturbance (Davidson et al. 2004).

Reforestation of abandoned agricultural land through natural succession or active tree planting is also changing landscapes, particularly in the Eastern U.S., Europe, China, and Russia. In the Eastern U.S., for example, much of the land that was originally cleared reverted to forest dominated by native species (see Fig. 12.2). In Chile, however, plantations of rapidly growing exotic trees such as *Pinus radiata* are replacing primary forests (Armesto et al. 2001). These plantations have low diversity and a quite different litter chemistry and pattern of nutrient cycling than do the primary forests that they replace. In addition, some tropical areas are now experiencing a "forest transition," in which the past net forest decline has been reversed, and overall forest cover is increasing, similar to patterns that occurred across much of the temperate zone in the twentieth century (Rudel et al. 2005).

The characteristics of the regenerating forests also depend on the previous types of land use (Foster et al. 1996, 2010). Long-term and intensive agricultural practices can compact the soil, alter soil structure and drainage capability, deplete the soil organic matter, reduce soil water-holding capacity, reduce nutrient availability, deplete the seed bank of native species, and introduce new weedy species. The forests that regenerate on such land may therefore differ substantially from the original forest, or from those that regrow on less-intensively managed lands (Motzkin et al. 1996). Grazing intensity and accompanying land management practices also influence potential revegetation, with more intensively grazed systems often taking longer to regain forest biomass. Natural reforestation under these conditions may proceed slowly or not at all.

Use of grasslands, savannas, shrublands, and cleared forests for cattle grazing is the most extensive modification of natural ecosystems occurring today, and, like deforestation, it can now be monitored and analyzed by remote sensing. Globally, thousands of square kilometers of savanna are burned annually to maintain productivity for cattle grazing. Although both fire and grazing are natural components of mesic grasslands, changes in the frequency or severity of burning and grazing alters ecosystem processes (Knapp et al. 1998). Burning releases nutrients and stimulates the production of new leaves that have a higher protein content and are more palatable to grazers. Conversely, grazers reduce fire probability by reducing the accumulation of grass biomass and leaf litter. Fire and browsers both prevent establishment of most trees, which might otherwise convert savannas to woodlands or forests. When fire frequency increases substantially, however, the loss of carbon and nitrogen from the system can reduce soil fertility and water retention and (and therefore productivity). Fire can also affect regional trace-gas budgets and deposition in downwind ecosystems and the transfer of nutrients and sediments to aquatic ecosystems.

Expansion of marine fishing has altered marine food webs globally, with cascading effects on most ecosystem processes. The area of the world's ocean that is actively fished has increased substantially, in part because technological advances allow fish and benthic invertebrates to be harvested more efficiently and stored for longer times before returning to markets. Most of Earth's continental shelves, the most productive marine ecosystems, are now actively fished, as are the productive high-latitude open ocean basins. Removal of fish has cascading effects on pelagic ecosystems because fish predation has large top-down effects on the biomass, and species composition of zooplankton, which in turn impact primary productivity by phytoplankton and the recycling of nutrients within the water column (see Chap. 10; Pauly et al. 1998). Harvesting of benthic invertebrates, such as clams, crabs, and oysters, also has large ecosystem effects because of direct habitat disturbance and the effects of these organisms on detrital food webs and benthic decomposition. The globalization of marine fisheries has a broader impact than we might expect because many large fish are highly mobile and migrate for thousands of kilometers (Berkes et al. 2006). Large changes in these fish populations therefore have ecological effects that diffuse widely throughout the ocean and even into freshwater ecosystems in the case of anadromous fish.

Intensification

Intensification of agriculture often reduces landscape heterogeneity and increases the transfer of nutrients and other pollutants to adjacent ecosystems. Agricultural intensification generally involves the use of high-yield crop varieties combined with tillage, irrigation, industrially produced fertilizers, and often pesticides and herbicides. Intensification has allowed food production to keep pace with the rapid human population growth (see Fig. 8.1; Evans 1980, Naylor 2009). Although this practice has reduced the areal extent of land required for agriculture, it has nearly eliminated some ecosystem types that would naturally occupy areas of high soil fertility. Intensive agriculture is most developed on relatively flat areas such as floodplains and prairies that are suitable for irrigation and use of large farm machinery. The high cost of this equipment requires that large areas be cultivated, largely eliminating natural patterns of landscape heterogeneity (Fig. 13.8).

Agricultural intensification generates biogeochemical hot spots that alter ecosystem processes in ways that impact the local, regional, and global environment (Matson et al. 1997). The large regular inputs of nutrients required to sustain intensive agriculture (see Fig. 8.1) increase the emissions of nitrogen trace gases that play a significant role in the global nitrogen cycle and link these ecosystems with downwind ecosystems (see Chaps. 9 and 14).

Nutrient loading on land increases non-point sources of pollution for neighboring aquatic ecosystems (Strayer et al. 2003, Carpenter and Biggs 2009). Phosphorus additions on land have particular large impacts in lakes for at least two reasons. First, primary production of most lakes is limited most fundamentally by phosphorus and therefore responds sensitively to even small phosphorus additions. Second, much of the phosphorus added to agricultural fields is chemically fixed, so more phosphorus is often added to fields than is absorbed by crops. On the North China Plain, for example, three times more phosphorus is added than is removed in crops. Large additions represent a massive reservoir of phosphorus that will continue to enter aquatic ecosystems long after farmers stop adding fertilizer. Phosphorus inputs from human sewage and live-stock manure have similarly long-lasting effects. In some cases, farmers can "mine" past fertilizer addition to meet crop needs. Many farms in the upper Midwest corn belt, for example, now apply less phosphorus than is removed in crops, and depend on the legacy of previous phosphorus applications (Vitousek et al. 2009b).

Land-use change has caused greater eco-logical impact during the twentieth century than any other global change. Understanding and projecting future changes in land use are therefore critical to predicting and managing future changes in the Earth System. Land-change scientists have developed effective interdisciplin-ary collaborations among climatologists, geogra-phers, ecologists, agronomists, and social scientists to evaluate the rates, causes, and conse-quences of land-use and land-cover change (Lambin et al. 2003). These collaborations permit the development of plausible scenarios for future land-use/land-cover change. Optimistic scenarios that assume that the growing human population will be fed rather than die from famines, wars, or disease epidemics project continued large changes in land use, particularly in developing countries (MEA 2005). What actually occurs in the future is, of course, uncertain, but these and other sce-narios suggest that land-use change will continue to be the major cause of global environmental change in the coming decades. Ecologists work-ing together with policy makers, planners, and managers have the opportunity to develop approaches that will minimize the impact of future landscape changes (see Chap. 15). This vision must recognize the large effects of land-use change on landscape processes, their conse-quences on local-to-global scale, and the relationship with human activities and behaviors.

Extrapolation to Larger Scales

Extrapolation of ecosystem processes to large spatial scales requires an understanding of the role of spatial heterogeneity in ecosystem pro-cesses. Efforts to estimate the cumulative effect of ecosystem changes at landscape to regional and global scales have contributed to increased recognition of the importance of landscape pro-cesses in ecosystem dynamics. Estimates of annual carbon sequestration, for example, require that rates measured (or modeled) in a few loca-tions be extrapolated over large areas or (increas-ingly) that methods be developed that can measure carbon pools over large areas in ways that account for spatial heterogeneity. These esti-mates are economically and politically important in international negotiations to reduce human impacts on the climate system.

Many approaches to spatial extrapolation have been used, each with its advantages and disad-vantages (Miller et al. 2004, Turner and Chapin 2005). A useful starting point is to multiply the rates typical of the most widespread land-cover type by the area of concern to give a "back-of-the-envelope" estimate of regional pools and fluxes. This might suggest whether a process of potential interest (e.g., deforestation effects on regional precipitation) warrants more careful consideration. A "paint-by-numbers" approach identifies potentially important patch types and estimates the flux or pool for the entire area by multiplying the average value for each patch type (e.g., the yield of major types of crops or the car-bon stocks of different forest types) times the areal extent of that patch type. This provides a more realistic approximation that can guide process-based research. This approach requires the selection of representative values of processes and accurate estimates of the area of each patch type. This extrapolation approach can be com-bined with empirical regression relationships (rather than a single representative value) to esti-mate process rates for each patch type. Carbon pools in a given forest type, for example, might be estimated as a function of temperature or NDVI rather than assuming that a single value could represent the carbon stocks of all sites. Improvements in satellite remote sensing tech-nologies and the development of multi-scale sampling that assimilates satellite sensors, air-craft sampling, and ground-based analyses have allowed the development of sampling strategies that incorporate spatial variation in both the state factors and patch dynamics (Asner et al. 2010).

Process-based models can also be used to estimate fluxes or pools over large areas or under novel conditions. These estimates are based on maps of input variables for an area (e.g., maps of climate, elevation, soils, and satellite-based indices of leaf area or more sophisticated measures of ecosystem structure and functioning) and a model that relates input variables to the ecosystem properties simulated by the model (Box 13.1; Potter et al. 1993, VEMAP-Members 1995, Running et al. 2004). Regional evapotranspiration, for example, can be estimated from satellite data on vegetation structure and maps of temperature and precipitation that are used as inputs to an ecosystem model (Running et al. 1989). Estimates from ecosystem models are sensitive to the quality, quantity, and uncertainty of the input data and to the validity and degree of generality of the relationships assumed by the models. The generality of relationships used in ecosystem models can then be tested through comparisons of model output with field data and through intercomparisons of models that differ in their structure but use the same input data (Cramer et al. 2001).

Any extrapolation exercise requires consideration of biogeochemical hot spots with high process rates. Regional extrapolation of methane flux at high latitudes, for example, should consider beaver ponds (Roulet et al. 1997) and thermokarst lakes (Walter et al. 2006) because they have very high fluxes relative to their area, just as analyses of savannas need to consider the distribution of termite mounds. Similarly, estimates of NEP require differentiation between young and old forests because forest age is an important determinant of NEP (see Chap. 12).

Processes that are strongly influenced by interactions among patches on a landscape cannot be extrapolated to large scales without explicitly considering these interactions. The effects of climate change on wildfire risk to communities, for example, is strongly influenced by fire spread, which depends on the configuration of ecosystems on the landscape. Spatially explicit models that incorporate the spread of disturbance among patches on a landscape are critical for projections of long-term changes in vegetation and disturbance regime (Gardner et al. 1987, Rupp et al. 2000, Perry and Enright 2006).

Box 13.1 Spatial Scaling Through Ecological Modeling

The complexity of ecological controls over all the processes that influence ecosystem carbon balance makes long-term projections of terrestrial carbon storage a daunting task. Making these projections is, however, critical to assessing the relative role of different terrestrial ecosystems in the global carbon balance. Experiments that test multiple environmental effects on terrestrial carbon storage are difficult to design. Modeling of complex combinations of environmental–biotic interactions extends what can be learned from a limited amount of empirical information. Ecosystem models have, for example, been used to identify key controls over net ecosystem carbon balance (NECB) to assess the role of the biosphere in regulating atmospheric CO_2 concentration under different scenarios of fossil fuel emissions and climate change (IPCC 2007).

Many of key processes regulating NECB involve changes occurring over decades to centuries. The temporal resolution of the models must therefore be coarse, with **time steps** (the shortest unit of time simulated by the model) of a day, month, or year. Use of relatively long time steps reduces the level of detail that can be considered. The short-term pulses of decomposition associated with drying and wetting cycles or grazing by soil fauna, for example, are subsumed in the shape of the annual temperature and moisture response curves of decomposition and in the decomposition coefficients. Only the more general controls such as temperature, moisture, and chemistry can still be observed with an annual time step.

The basic structure of a model of NECB must include the pools of carbon in soils and vegetation. It must also include carbon fluxes

(continued)

Box 13.1 (continued)

from the atmosphere to plants (GPP or NPP), from plants to the atmosphere (plant respiration, harvest, and combustion), from plants to soil (litterfall), and from soil to the atmosphere (heterotrophic respiration and disturbance). Models differ in the detail with which these and other pools and fluxes are represented. Plants, for example, might be considered a single pool, or be separated into different plant parts (leaves, stems, and roots), functional types of plants (e.g., trees and grasses in a savanna), or chemical fractions such as cell wall and cell contents. Under some circumstances, certain fluxes (e.g., fire and leaching) are ignored. There is no single "best" model of NECB. Each model has a unique set of objectives, and the model structure must be designed to meet these objectives, and results must be interpreted in light of the objectives and the assumptions that are built into the model. We briefly describe how three models incorporate information about controls over NEP, emphasizing how the differences in model structure make each model appropriate to particular questions or ecosystems. NEP models ignore carbon fluxes associated with disturbance and leaching.

Perhaps the biggest challenge in model development is deciding which processes to include. One approach is to use a hierarchical series of models to address different questions at different scales (Reynolds et al. 1993). Models of leaf-level photosynthesis and of microclimate within a canopy have been developed and extensively tested for agricultural crops, based on the basic principles of leaf biochemistry and the physics of radiation transfer within canopies. One output of these models is a regression relationship between environment at the top of the canopy and net photosynthesis by the canopy. This environment–photosynthesis regression relationship can then be incorporated into models operating at larger temporal and spatial scales to simulate NPP, without explicitly including all the details of biochemistry and radiation transfer. This hierarchical

approach to modeling provides an opportunity to **validate** the model output (i.e., to compare the model predictions with data obtained from field observations or experimental manipulations) at several scales of temporal and spatial resolution, providing confidence that the model captures the important underlying processes at each level of resolution.

The Terrestrial Ecosystem Model (TEM; Fig. 13.14) was designed to simulate ecosystem carbon budgets for all locations on Earth at 0.5° longitude × 0.5° latitude resolution (60,000 grid cells) for time periods of a century or more (McGuire et al. 2001). TEM has a relatively simple structure and a monthly time step, so it can run efficiently in large numbers of grid cells for long periods of time. Soil, for example, consists of a single carbon pool. The model assumes simple universal relationships between environment and ecosystem processes based on general principles that have been established in ecosystem studies. The model assumes, for example, that decomposition rate of the soil carbon pool depends on the size of this pool and is influenced by the temperature, moisture, and C:N ratio of the soil. The model incorporates feedbacks that constrain the possible model outcomes. The nitrogen released by decomposition, for example, determines the nitrogen available for NPP, which in turn governs carbon inputs to the soil and therefore the pool of soil carbon available for decomposition. The model is validated by comparison of model output with global patterns of carbon pools and fluxes in natural ecosystems (McGuire et al. 2001), making the model useful in simulating regional and global patterns of soil carbon storage under historical or potential future climatic conditions.

CENTURY (Fig. 13.14) was originally developed to simulate changes in soil carbon storage in grasslands in response to variation in climate, soils, and tillage (Parton et al. 1987, Parton et al. 1993). It has since been adapted to most global ecosystem types. In CENTURY,

(continued)

Box 13.1 (continued)

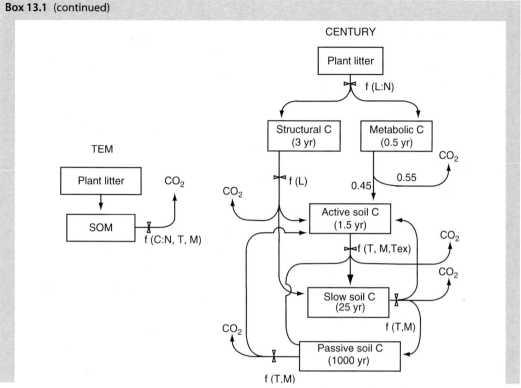

Fig. 13.14 Diagrammatic representation of the decomposition portion of two terrestrial ecosystem models: TEM (McGuire et al. 1995b) and CENTURY (Parton et al. 1987). Inputs from the vegetation component of these models are shown as plant litter. *Arrows* indicate the fluxes of carbon from litter to other pools and eventually to CO_2. The *bow-ties* indicate controls over these fluxes (or the partitioning of the flux between two pools) as functions (f) of C:N ratio (C:N), lignin (L), lignin: N ratio (L:N), temperature (T), and moisture (M). In CENTURY, we show representative residence times of different carbon pools in grassland soils

the soil is subdivided into three compartments (active, slow, and passive soil carbon pools) that are defined empirically by turnover rates observed in soils. The active pool represents microbial biomass and labile carbon in the soil with a turnover time of days to years. The slow pool consists of more recalcitrant materials with a turnover time of years to decades. The passive pool is humified carbon that is stabilized on mineral surfaces with a turnover time of hundreds to thousands of years. The detailed representation of soil pools in CENTURY enables it to estimate changes in decomposition under situations where a change in disturbance regime or climate alters the decomposition of some soil pools more than others. A change in climate, for example, primarily affects the active and slow

pools, with the passive pool remaining protected by clay minerals, whereas tillage enhances the decomposition of all soil pools.

How do we know whether the patterns of NEP estimated by global-scale models are realistic? A comparison of model results with field data for the few locations where NEP has been measured provides one reality check. At these sites, measurements of NEP over several years spanning a range of weather conditions provides a measure of how that ecosystem responds to variation in climate. This allows a test of the model's ability to capture the effects of ecosystem structure and climate on NEP.

The seasonal and interannual patterns of atmospheric CO_2 provide a second reality check for global models of NEP. Atmospheric transport

(continued)

Box 13.1 (continued)

models describe the patterns of redistribution of water, energy, and CO_2 through Earth's atmosphere. These transport models can be run in inverse mode to estimate the spatial and temporal patterns of CO_2 uptake and release from the land and ocean that are required to produce the observed patterns of CO_2 concentration in the atmosphere (Fung et al. 1987, Tans et al. 1990). The global patterns of CO_2 sources and sinks estimated from the atmospheric transport models can then be compared with the patterns estimated from ecosystem models. Any large discrepancy between these two modeling approaches provides hints about processes or locations where either the ecosystem or the atmospheric transport models have not adequately captured the important controls over carbon exchange and transport.

Summary

Spatial heterogeneity within and among ecosystems is critical to the functioning of individual ecosystems and entire regions. Landscapes are mosaics of patches that differ in ecologically important properties. Some patches, for example, are biogeochemical hot spots that are much more important than their area would suggest. The size, shape, connectivity, and configuration of patches on a landscape influence their interactions. Large patches, for example, may have a smaller proportion of edge habitat. The shape and connectivity of patches influences their effective size and heterogeneity in ways that differ among organisms and processes. The distribution of patches on a landscape is important because it determines the nature of transfers of materials and disturbance among adjacent patches. The boundaries between patches have unique properties that are important to edge specialists. Boundaries also have physical and biotic properties that differ from the centers of patches, so differences among patches in edge-to-area ratios, due to patch size and shape, influence the average rates of processes in a patch.

State factors, such as topography and parent material, govern the underlying matrix of spatial variability in landscapes. This physically determined pattern of variability is modified by biotic processes and legacies in situations where species strongly affect their environment. These landscape patterns and processes in turn influence disturbance regime, which further modifies the landscape pattern. Humans are exerting increasing impact on landscape patterns and change. Land-use decisions that convert one land-surface type to another (e.g., deforestation, reforestation, shifting agriculture) or that modify its functioning (e.g., cattle grazing on rangelands) influence both the sites where those activities occur and the functioning of neighboring ecosystems and the landscape as a whole. Human impacts on ecosystems are becoming both more *extensive* (i.e., impacting more area) and more *intensive* (i.e., having greater impact per unit area).

Ecosystems do not exist as isolated units on the landscape. They interact through the movement of water, air, materials, organisms, and disturbance from one patch to another. Topographically controlled movement of water and materials to downslope patches depends on the arrangement of patches on the landscape and the properties of those patches. Riparian areas, for example, are critical filters that reduce the transfer of nutrients and sediments from upland ecosystems to streams, lakes, estuaries, and the ocean. Aerial transport of nutrients, water, and heat strongly influences the nutrient inputs and climate of downwind ecosystems. These aerial transfers among ecosystems are now so large and pervasive as to have strong effects on the functioning of the entire biosphere. Animals transport nutrients and plants at a more local scale and influence patterns of colonization and ecosystem change. The spread of disturbance among patches influences both the temporal dynamics and the average properties of patches on a landscape. The connectivity of ecosystems on the landscape is rarely incorporated into management and planning activities. The increasing human impacts on landscape interactions must be considered in any long-term planning for the sustainability of managed and natural ecosystems.

Review Questions

1. What is a landscape? What properties of patches determine their interactions in a landscape?
2. How do fragmentation and connectivity influence the functioning of a landscape?
3. Give examples of spatial heterogeneity in ecosystem structure at scales of 1 m, 10 m, 1 km, 100 km, and 1,000 km. How does spatial heterogeneity at each of these scales affect the way in which these ecosystems function? In other words, if heterogeneity at each scale disappeared, what would be the differences in the way in which these ecosystems function.
4. What are the major natural and anthropogenic sources of spatial heterogeneity in a landscape? How do these sources of heterogeneity influence the way in which these landscapes function? How do interactions among these sources of heterogeneity affect landscape dynamics?
5. What is the difference between a shifting steady-state mosaic and a non-steady-state mosaic? Give examples of each.
6. What is the difference between intensification and extensification? What has been the role of each in ecosystem and global processes?
7. Which ecosystem processes are most strongly affected by landscape pattern? Why?
8. What properties of boundaries influence the types of interactions that occur between patches within a landscape?
9. Describe how patches within a landscape interact through (1) the flow of water, (2) transfers of materials through the atmosphere, (3) movement of animals, and (4) the movement of disturbance. What properties of landscapes and patches influence the relative importance of these mechanisms of patch interaction?
10. What issues must be considered in extrapolating processes measured at one scale to larger areas? How does the occurrence of hot spots influence approaches to spatial scaling?

Additional Reading

Forman, R.T.T. 1995. *Land Mosaics: The Ecology of Landscapes and Regions.* Cambridge University Press, Cambridge.

Foster, D.R., B. Donahue, D. Kittredge, K.F. Lambert, M. Hunter, et al. 2010. Wildlands and Woodlands: A Vision for the New England Landscape. Harvard University, Petersham, MA. (http://www.wildland sandwoodlands.org/).

Lovett, G.M., C.G. Jones, M.G. Turner, and K.C. Weathers, editors. 2005. *Ecosystem Function in Heterogeneous Landscapes.* Springer, New York.

Matson, P.A., W.J. Parton, A.G. Power, and M.J. Swift. 1997. Agricultural intensification and ecosystem properties. *Science* 227:504–509.

Meyer, W.B., and B.L. Turner, III. 1992. Human population growth and global land-use/cover change. *Annual Review of Ecology and Systematics* 23:39–61.

Naiman, R.J., H. Décamps, and M.E. McClain. 2005. *Riparia: Ecology, Conservation, and Management of Streamside Communities.* Elsevier, Amsterdam.

O'Neill, R.V., D.L. DeAngelis, J.B. Waide, and T.F.H. Allen. 1986. *A Hierarchical Concept of Ecosystems.* Princeton University Press, Princeton.

Pickett, S.T.A., and M.L. Cadenasso. 1995. Landscape ecology: Spatial heterogeneity in ecological systems. *Science* 269:331–334.

Turner, M.G. 2010. Disturbance and landscape dynamics in a changing world. *Ecology* 91:2833–2849.

Turner, M.G., R.H. Gardner, and R.V. O'Neill. 2001. *Landscape Ecology in Theory and Practice: Pattern and Process.* Springer-Verlag, New York.

Urban, D.L., R.V. O'Neill, and H.H. Shugart. 1987. Landscape ecology. *BioScience* 37:119–27.

Vitousek, P.M. 2004. *Nutrient Cycling and Limitation: Hawai'i as a Model System.* Princeton University Press, Princeton.

Waring, R.H., and S.W. Running. 2007. *Forest Ecosystems: Analysis at Multiple Scales.* 3rd Edition. Academic Press, New York.

Changes in the Earth System

14

The magnitude of biotic and human impacts on ecosystem processes becomes clear when summed at the global scale. This chapter describes changes in the biogeochemical cycles of the Earth System that have occurred during the Anthropocene.

Introduction

Human activities have altered biogeochemical cycles at global scales in ways that change the functioning of Earth as an ecosystem. Human activities have dramatically altered element cycles since the beginning of the industrial revolution. Burning of fossil fuels in particular has increased emissions of CO_2, nitric oxides, and several sulfur gases. Mining and agriculture have also altered the availability and mobility of carbon, nitrogen, phosphorus, and sulfur. Changes in these biogeochemical cycles have altered Earth's climate, speeding up the global hydrologic cycle, which in turn feeds back to other biogeochemical cycles. Together these changes alter ecosystems at all scales, ranging from individual organisms to the entire biosphere. In this chapter, we summarize at the global scale the pools and fluxes in key biogeochemical cycles and the factors responsible for change.

A Focal Issue

The aggregate effects of human activities have altered biogeochemical cycles at global scales. Fossil fuel emissions have increased atmospheric CO_2 concentration by 35% and increased ocean acidity by a similar proportion. Ocean acidity is the "other CO_2 problem" that is invisible to most people but it has potentially profound effects by dissolving the calcium carbonate structures of marine organisms as diverse as reef-forming corals, marine invertebrates such as mussels and crabs, and microscopic foraminifera that are an important base to marine food chains. Many marine coral reefs are already threatened by rising temperatures, warming-induced coral bleaching, and nutrient and sediment runoff from land. How do these multiple human impacts interact to affect reef development and the diverse ecosystems that they support (Fig. 14.1)? What are the potential consequences of altering the food base of the world oceans? How might changes in ocean productivity feedback to affect the CO_2 concentration and climate of Earth? Are these effects large or small compared to the changes in the capacity of terrestrial biosphere to influence climate? Understanding of global biogeochemical cycles places these important questions in an integrated context that can inform society of the interactive consequences of human actions.

F.S. Chapin, III et al., *Principles of Terrestrial Ecosystem Ecology*,
DOI 10.1007/978-1-4419-9504-9_14, © Springer Science+Business Media, LLC 2011

Fig. 14.1 Coral reefs are the rainforests of the ocean in the sense of being hot spots of biodiversity. Reefs also provide important food resources and storm protection to local residents and are an important cultural, recreational, and aesthetic resources that benefit society broadly. Photograph of corals near Key Largo, Florida (istockphoto)

Human Drivers of Change

The rising human population and its consumption of resources account for many recent changes in the Earth System. The last 10,000 years (**Holocene** interglacial period), since the end of the last ice advance, constitute a remarkably stable and benign period in Earth's climate history (Fig. 14.2). This stability contributed to the initiation of agriculture that provided people with a more stable food supply, the formation of sedentary communities to tend and use this food, and the founding of diverse civilizations around the globe. Human population increased about 100-fold (from 5 million to 700 million) from the end of the last ice age to the beginning of the industrial revolution in 1750. It increased another 10-fold over the next 250 years to seven billion people in 2010. Throughout human history people have affected their environment, just as all organisms do (see Chap. 11). This included human hunting that

contributed to the extinction of the Pleistocene megafauna (Flannery 1994, Zimov et al. 1995, Gill et al. 2009) and the spread of agriculture and grazing by domestic livestock that have altered land cover on about half of the terrestrial surface (see Chap. 13; Ellis and Ramankutty 2008). However, prior to the industrial revolution, these changes were small enough in scale that they had had only modest effects on the global environment. Since 1750, however, human population and its consumption of resources have had a dramatic impact on the Earth System. These changes have been particularly pronounced since 1950 (the **Great Acceleration**), with projections of even more rapid changes in the first half of the twenty-first century, if human use of resources is not substantially reduced (Steffen et al. 2004, Young and Steffen 2009).

Even modest temperature variations during the Holocene, such as the Medieval Warm Period and the Little Ice Age (Fig. 14.2), had

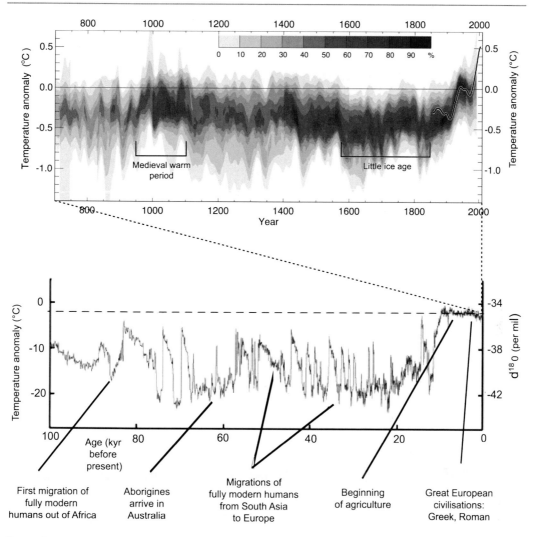

Fig. 14.2 Temperature trends of the last 1,300 and last 100,000 years. Temperatures for the last 1,300 years have been estimated from 10 proxy records collected throughout the world; the percentage of records that show a given temperature (± 1 standard error) is shown by the degree of shading. The solid line shows the temperatures estimated from direct temperature records (IPCC 2007). Temperatures for the last 100,000 years are estimated from ^{18}O concentrations in ice cores; also shown are selected events in human history. Redrawn from IPCC (2007) and Young and Steffen (2009)

major impacts on food production and human migration. The sharp increase in global temperature since 1800 is unprecedented in the last 1,000 years. This raises the distinct possibility that human impacts on the Earth System could push it beyond a threshold to a new state that might be less favorable to human well-being (see Chap. 15; MEA 2005; Rockström et al. 2009).

The Global Water Cycle

Water Pools and Fluxes

Only a tiny fraction of Earth's water (0.01%) is in soils, where it is accessible to plants and available to support the activities of terrestrial organisms. Most of Earth's water is in the ocean

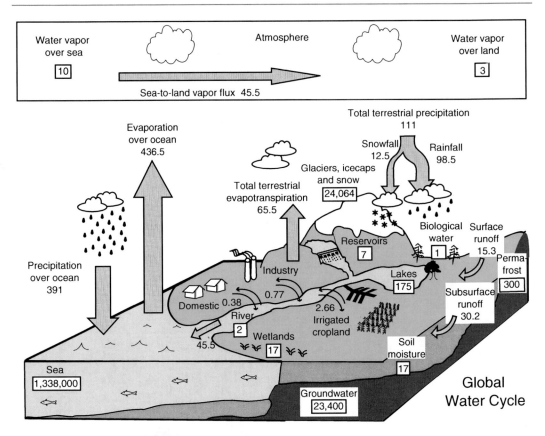

Fig. 14.3 The global water cycle, showing approximate magnitudes of the major pools (1,000 km³; *boxes*) and fluxes (1,000 km³ year⁻¹; *arrows*). Most water is in the ocean, ice, and groundwater, where it is not directly accessible to terrestrial organisms. The major water fluxes are precipitation, evapotranspiration, and runoff. Modified from Carpenter and Biggs (2009)

(96.5%), ice caps and glaciers (2.4%), and groundwater (1%; Fig. 14.3; Oki and Kanae 2006, Carpenter and Biggs 2009). About 90% of the water that evaporates from the ocean returns there as precipitation. Another 10% of ocean evaporation (45,000 km³ year⁻¹) moves over the land, where it falls as precipitation and returns to the ocean as river runoff. The evaporation from land (65,000 km³ year⁻¹) is about 15% of global evaporation, although land occupies about 30% of Earth's surface; this indicates that average evapotranspiration rates are about half as great on land as over the ocean. There are large regional variations in evaporation rate over both land and ocean related to climate and, in the case of land, in water availability and transpiration rates of vegetation. Of the terrestrial precipitation (110,000 km³ year⁻¹), about 40% comes from the ocean (45,000 km³ year⁻¹), and 60% (65,000 km³ year⁻¹) is evaporated from land and recycled. Evaporation and precipitation are highly variable, both regionally and seasonally.

The quantity of water in the atmosphere is only 2.6% of that which annually cycles through the atmosphere in evaporation and transpiration, giving an average **turnover time** (i.e., time required to replenish this pool) of about 10 days. Precipitation is therefore tightly linked to evapotranspiration from upwind ecosystems over time scales of hours to weeks. Soil moisture has an average turnover time of about 2 months, with substantial regional variability, so plant water use is quite sensitive to seasonal variations in precipitation. Groundwater has an average turnover time of about 200 years (Fig. 14.3). This makes it a more dependable water source than surface moisture but also implies that replenishment of groundwater takes a long time, if it is overexploited for irrigation or other uses. In some cases, **fossil groundwater** accumulated in the past, when climate may have been different. In these cases, replenishment of groundwater may not occur in the current climate or may take much longer than its calculated turnover time implies.

Anthropogenic Changes in the Water Cycle

Human-induced climate warming has accelerated the global hydrologic cycle through increases in both evapotranspiration and precipitation. As Earth's air warms, it holds more moisture, driving greater evaporation and increasing the potential for precipitation. Precipitation over land, for example, increased north of 30°N during the twentieth century (IPCC 2007). With continued warming, wet areas are projected to become wetter, with more frequent large floods, and dry areas may become drier.

Land-use changes also alter the hydrologic cycle by changing (1) the quantity of energy absorbed, (2) the pathway of energy loss, and (3) the moisture content and temperature of the atmosphere. Conversion from tropical rainforest to pasture, for example, leads to less energy absorption because of increased albedo and a larger proportion of energy dissipated to the atmosphere as sensible rather than latent heat (Foley et al. 2003b). The warmer drier atmosphere allows less precipitation, favoring the persistence of pastures rather than succession to rainforests (see Fig. 2.14). When land-use changes are extensive, they can have continental-scale effects on temperature and precipitation, often at locations remote from the region of land-cover change, as a result of large-scale adjustments in atmospheric circulation (Chase et al. 2000). Land-cover changes in Southeast Asia, for example, have particularly large effects on global-scale climate through atmospheric teleconnections.

Terrestrial ecosystems are generally more sensitive to soil moisture than to precipitation. Soil moisture will probably decline in areas with reduced precipitation and in regions where evaporation increases more than precipitation. Models generally project increased soil moisture at high latitudes and oceanic islands and reduced summer soil moisture in the interiors of continents due to higher temperatures and insufficient increases (or reductions) of rainfall. Many continental areas that are currently important for agriculture, such as the Ukraine and the mid-western U.S., may be particularly prone to future drought, and grain-producing areas may migrate poleward to areas that are currently too cold to support intensive agriculture. These changes in location of soil moisture suitable for agriculture will have major regional and national economic and societal impacts.

Consequences of Changes in the Water Cycle

Society depends most directly on some of the smallest and most vulnerable pools in the global hydrologic cycle. Nonirrigated agriculture, for example, relies on soil water derived from precipitation, a relatively small pool that responds rapidly to changes in the balance between precipitation and evapotranspiration. In some areas, soil moisture derived from precipitation is supplemented by irrigation, which withdraws water from lakes, rivers, and groundwater. Irrigated croplands have increased fivefold during the twentieth century and support 40% of global crop production (Fig. 14.4; Gleick 1998, Carpenter and Biggs 2009). During the past century, there was an eightfold increase in the water used to support human activities, which paralleled a fourfold increase in human population and a 50% increase in per capita water consumption. People now use 25% of the continental runoff (see Chap. 4). Most of this water is used for hydroelectric power and irrigation. Selective expansion of irrigated agriculture in very poor regions of the world represents an important opportunity to alleviate hunger and poverty (Carpenter and Biggs 2009).

The scarcity of water is only part of the hydrologic challenges facing society. Forty percent of the world's population had no access to adequate

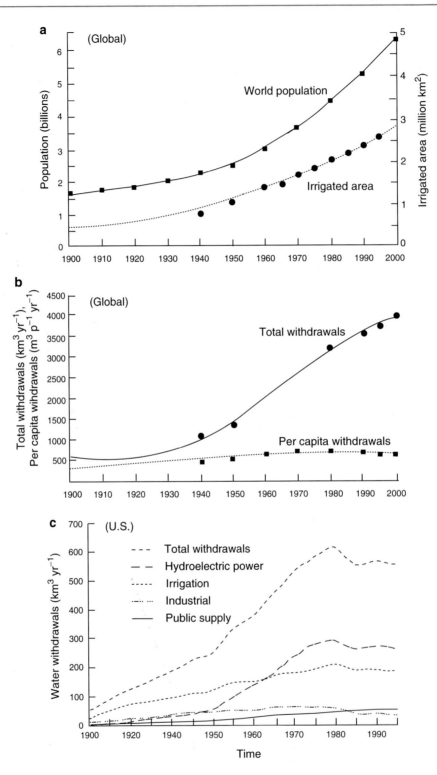

Fig. 14.4 Trends in (**a**) world population and global land area under irrigation, (**b**) water withdrawals to support human activities expressed as a global total and on a per capita (p^{-1}) basis, and (**c**) water withdrawal in the U.S. separated by economic sector. Redrawn from Gleick (1998)

sanitation in 2004, and 17% had no clean drinking water (Vörösmarty et al. 2005). The shortage of clean water is particularly severe in the developing nations of the world, where future population growth and water requirements are likely to be greatest (Postel and Richter 2003).

The projected increases in human demands for fresh water will have strong impacts on aquatic ecosystems through diversion of fresh water for irrigation and modification of flow regimes by dams and reservoirs. These impacts can be minimized by increasing the efficiency with which society uses water and nutrients.

The Global Carbon Cycle

Carbon Pools and Fluxes

Photosynthetic uptake of carbon from the atmosphere and ocean provides the fuel for most biological processes. This reduced carbon comprises about half of the mass of Earth's organic matter. Biological systems, in turn, respire CO_2 when they use organic carbon as an energy source to support maintenance and growth. The controls over the carbon cycle depend on time scale, ranging from seconds, where cycling is controlled by photosynthetic rate and surface–air exchange, to millions of years, where cycling is controlled by movements of Earth's crust (see Chaps. 5–7).

Carbon is distributed among four major pools: the atmosphere, ocean, land (soils and vegetation), and sediments and rocks (Fig. 14.5; Reeburgh 1997, Sarmiento and Gruber 2006, IPCC 2007). Atmospheric carbon, which consists primarily of CO_2, is the smallest but most dynamic of these pools. It turns over about every 5 years, primarily through its removal by photosynthesis and return by respiration. The metabolism of organisms therefore constitutes the engine that drives the global carbon cycle on time scales of seconds to centuries.

Carbon is present in the ocean as dissolved organic carbon (DOC), dissolved inorganic carbon (DIC), and particulate organic carbon (POC), which consists of both live organisms and dead

material. Most (98%) of this carbon is in inorganic form, primarily as bicarbonate (90%), with most of the rest as carbonate. Free CO_2, the form that is directly used by most marine primary producers, accounts for less than 1% of this inorganic pool. These three forms of DIC are in a pH-dependent equilibrium (see Chap. 5). The marine biota account for only 3 Pg (3×10^{15} g) of carbon, although they cycle almost as much carbon annually as does terrestrial vegetation. The carbon in marine biota turns over about every 3 weeks.

The ocean's surface waters that interact with the atmosphere contain about 920 Pg of carbon, similar to the quantity in the atmosphere (Fig. 14.5). The capacity of the ocean to take up carbon is constrained by three categories of processes that operate at different time scales (Schlesinger 1997). In the short term, surface exchange rate depends on wind speed, surface temperature, and the CO_2 concentration of surface waters. On daily to monthly time scales, the CO_2 concentration in surface water depends on photosynthesis and pH-dependent buffering reactions. Finally, the surface waters are a relatively small pool (only 75–200 m deep) of water that exchanges relatively slowly with deeper ocean layers because the warm, low-salinity surface water is less dense than deeper layers (see Fig. 2.10). Carbon that enters surface waters is transported slowly to depth by two major mechanisms. First, organic detritus and its $CaCO_3$ skeletal content, which form in the euphotic zone, sink to deeper waters, a process termed the **biological pump** (see Chap. 7). Second, bottom-water formation in the polar seas transports dissolved carbon to depth, a process termed the **solubility pump** (see Chap. 2). Once carbon reaches intermediate and deep waters, it is stored for hundreds to thousands of years before returning to the surface through upwelling. Most (97%) of the ocean carbon is in the intermediate and deep waters (Fig. 14.5).

The terrestrial biosphere contains the largest biological reservoir of carbon. There is nearly as much carbon in terrestrial vegetation as in the atmosphere, with 2–3 times more organic carbon in soils than in the atmosphere (Fig. 14.5; Jobbágy

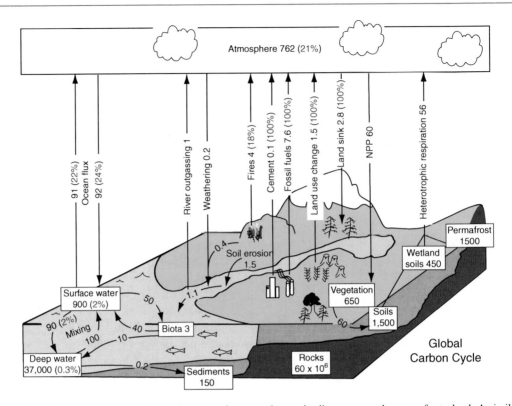

Fig. 14.5 The global carbon cycle, showing approximate magnitudes of the major pools (*boxes*) and fluxes (*arrows*) in units of Pg year^{-1} for the 1990s. A petagram (Pg) is 10^{15} g. *Red numbers in parenthesis* are the anthropogenic contributions to these pools and fluxes relative to preindustrial times (1750). Data are from Sabine et al. (2004), Sarmiento and Gruber (2006), and IPCC (2007), and anthropogenic fluxes for 2000–2006 from Canadell et al. (2007). The carbon pools that contribute to carbon cycling over decades to centuries are the atmosphere, land (vegetation and soils), and surface ocean water. On land, the carbon gain by vegetation due to "fertilization" by elevated CO_2 and nitrogen deposition (i.e., the land sink) is slightly greater than the carbon loss due to land-use change, leading to net carbon transfer to land. A similar quantity of land carbon is eroded into rivers, with half being outgassed to the atmosphere and half transported to the ocean. The carbon input from the atmosphere to the ocean is also slightly greater than the carbon returned to the atmosphere. These terrestrial and ocean sinks are less than half of the carbon emitted to the atmosphere from burning of fossil fuels, leading to CO_2 accumulation in the atmosphere. The terrestrial biosphere accounts for 50–60% of global NPP. Most (80%) of the marine NPP is released to the environment by heterotrophic respiration, with the remaining 20% going to the deep ocean by the biological pump. Ocean upwelling returns most of this carbon to the surface ocean waters

and Jackson 2000, Sabine et al. 2004, IPCC 2007). Permafrost (permanently frozen ground) also contains a large carbon pool that, until recently, turned over very slowly (Zimov et al. 2006, Schuur et al. 2008, Tarnocai et al. 2009). Terrestrial NPP is slightly greater than that in the ocean, but, due to the much larger plant biomass on land, terrestrial plant carbon has a turnover time of about 11 years, compared to 3 weeks in the ocean. NPP is about half of GPP (i.e., photosynthetic carbon gain) on land (60 Pg year^{-1} out of 120 Pg year^{-1}) and in the ocean (45 Pg year^{-1} out of 103 Pg year^{-1}; Prentice et al. 2001, IPCC 2007). Soil carbon turns over on average every 25 years. These average turnover times mask large differences in turnover time among components of the terrestrial carbon cycle. Photosynthetically fixed carbon in chloroplasts turns over on time scales of seconds

Table 14.1 Average (2000–2006) annual emissions and fate of anthropogenic carbon. Adapted from Canadell et al. (2007) and Le Quéré et al. (2009)

Sources and sinks of anthropogenic carbon	Annual net flux (Pg C year^{-1})
Anthropogenic carbon sources	9.1
Fossil fuel and cement production (8.7 in 2008)	7.6
Land-use change	1.5
Carbon sinks (1990–2000)	9.1
Storage in the atmosphere	4.1
Oceanic uptake	2.2
Terrestrial uptake	2.8

through photorespiration (see Chap. 5). Leaves and roots are replaced over weeks to years, and wood is replaced over decades to centuries. Components of soil organic matter also have quite different turnover times, with labile forms turning over in minutes and humus having turnover times of decades to thousands of years (see Chap. 7).

Carbon in rocks and sediments accounts for well over 99% of Earth's carbon (10^7 Pg; Reeburgh 1997, Schlesinger 1997). This carbon pool cycles extremely slowly, with turnover times of millions of years. Factors governing the turnover of these pools are geologic processes associated with the rock cycle, including the movement of continental plates, volcanism, uplift, and weathering (see Chap. 3).

Human activities are now a significant component of the global carbon cycle. Human carbon emissions from combustion of fossil fuels increased 40% from 1990 to 2008 (to 8.7 Pg year^{-1}; Canadell et al. 2007, IPCC 2007, Le Quéré et al. 2009). Land-use conversion releases an additional 1.5 Pg year^{-1} of carbon by biomass burning and enhanced decomposition (Table 14.1; Canadell et al. 2007). Together these anthropogenic fluxes are about 15% of the carbon cycled by terrestrial or by marine production, making human carbon emissions the third largest biologically controlled flux of carbon to the atmosphere. Moreover, unlike primary production, human carbon fluxes represent net additions to the atmosphere.

Changes in Atmospheric CO_2

Critical processes in the carbon cycle respond to environment at multiple time scales. The critical controls over carbon cycling are photosynthesis and respiration on time scales of seconds to years; NPP, SOM turnover, and disturbance on time scales of years to centuries; and uplift, weathering, and ocean sedimentation over thousands to millions of years. Atmospheric CO_2 concentration has varied at least 10-fold through Earth's history, from the preindustrial concentration of 280 ppmv to greater than 3,000 ppmv. Geochemical processes determine variation in atmospheric CO_2 on geological time scales. These include the weathering of silicate rocks (which consumes CO_2 and releases bicarbonate), burial of organic carbon in sediments, and volcanism (which releases CO_2) (Berner 1997, Sundquist and Visser 2004). Biological processes influence geochemical cycling in many ways, for example by increasing weathering rates (see Chap. 3). Although critical on long time scales, the rates of these geochemical processes are so slow compared to anthropogenic changes that they do not influence current trajectories of change in atmospheric CO_2.

Over the last 650,000 years, changes in solar input associated with variations in Earth's orbit (see Chap. 2) caused cyclic variation in atmospheric CO_2 concentrations associated with glacial–interglacial cycles (Fig. 14.6; Petit et al. 1999, Sigman and Boyle 2000, IPCC 2007). CO_2 concentration declined during glacial periods and increased during interglacials. These changes in CO_2 concentration are much larger than can be explained simply by changes in light intensity and temperature in response to altered solar input. The large biospheric changes must result from amplification by biogeochemical feedbacks in the Earth System. Several feedbacks could contribute to these atmospheric changes (Sigman and Boyle 2000, IPCC 2007). (1) Increased transport of dust off the less-vegetated continents during glacial periods may have increased iron, phosphorus, and silica transport and enhanced NPP in high-latitude ocean basins, leading to increased CO_2 uptake and transport to depth via the biological pump

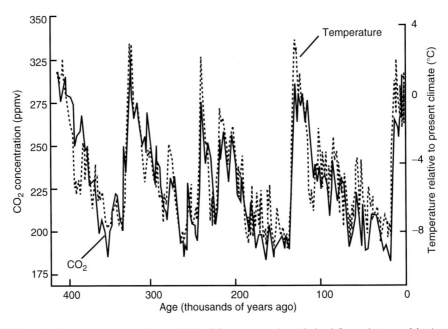

Fig. 14.6 Variations in temperature and atmospheric CO_2 concentrations derived from air trapped in Antarctic ice cores. Redrawn from Folland et al. (2001)

(see Fig. 7.26). (2) Extensive winter sea ice around Antarctica may have reduced out-gassing of CO_2 in locations of upwelling of CO_2-rich deep waters (Stephens and Keeling 2000). (3) Additional carbon may have been stored on land during glacial periods – both on continental shelves exposed by the drop in sea level and in permafrost at high latitudes (Zimov et al. 2006). However, terrestrial systems also lost carbon during glacial periods, due to the replacement of forests by grasslands, deserts, tundra, and ice sheets. Although the net effect of all these changes is uncertain (IPCC 2007), there have been large redistributions of carbon between land, atmosphere, and ocean over the course of glacial cycles (Bird et al. 1994, Crowley 1995). An improved understanding of controls over carbon redistribution among global pools could indicate how the Earth System will respond to current trends of increasing temperature and atmospheric CO_2.

Like air temperature (Fig. 14.2), atmospheric CO_2 concentration has been relatively stable over the last 12,000 years, ranging from about 260 to 280 ppmv in preindustrial times (Fig. 14.7). During the past century, however, CO_2 concentration has risen 10-fold more rapidly than at any time in the previous 20,000 years (Petit et al. 1999). Its concentration of 390 ppmv in 2011 is the highest in at least 650,000 years and probably the last 20 million years (Pearson and Palmer 2000, Canadell et al. 2007). Despite recent efforts to reduce emissions, atmospheric CO_2 continues to rise at an ever-increasing rate (Canadell et al. 2007, IPCC 2007, Solomon et al. 2009). This occurs primarily because of increasing emissions, especially in rapidly developing nations like China and India, and continued high emission rates by developed nations, such as the U.S., Japan, and Europe. These recent changes in the global cycles of carbon and other elements caused by human activities are large enough to indicate that Earth has entered a new geologic epoch, the **Anthropocene** (see Fig. 2.15; Crutzen 2002).

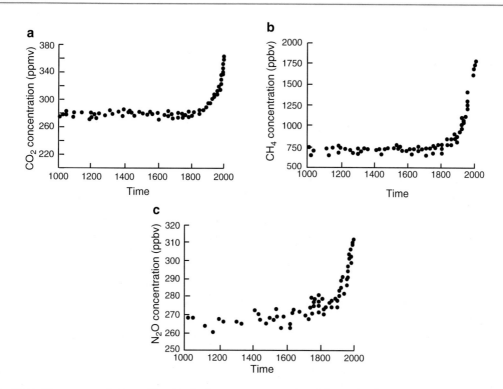

Fig. 14.7 Changes over the last millennium in the atmospheric concentrations of three radiatively active gases that are influenced by human activities. Data shown are a composite of time series from air trapped in Antarctic ice cores and from direct atmospheric measurements. Redrawn from Prentice et al. (2001)

Marine Sinks for CO_2

The ocean removes CO_2 from the atmosphere through dissolution in seawater and photosynthesis by marine organisms. The dissolution of CO_2 in the ocean, which accounts for most movement of CO_2 to the ocean, produces acidity (H^+), as it equilibrates with bicarbonate and carbonate ions (Eq. 14.1; Doney et al. 2009). The rising dissolved CO_2 concentration and resulting 30% increase in ocean acidity have at least two important consequences (Feely et al. 2004, Orr et al. 2005). It tends to dissolve the carbonate (e.g., $CaCO_3$) shells of marine invertebrates (e.g., lobsters, oysters, and corals; Fig. 14.1; Eq. 14.2) and

diatoms, altering the functioning of marine ecosystems. In addition, it reduces the rate at which CO_2 dissolves in the ocean – i.e., makes the ocean a weaker sink for CO_2. This contributes to an increasing proportion of fossil-fuel carbon that remains as a greenhouse gas in the atmosphere (see Fig. 7.28, Box 14.1; Canadell et al. 2007).

$$CO_2 + H_2O \leftrightarrow H^+ + HCO_3^- \leftrightarrow 2H^+ + CO_3^{2-}$$
(14.1)

$$CO_2 + CaCO_3 + H_2O \leftrightarrow 2HCO_3^- + Ca^{2+}$$
(14.2)

Box 14.1 Partitioning of Carbon Uptake Between the Land and Ocean

Only about half of the anthropogenic CO_2 that enters the atmosphere remains there. The land or ocean takes up the remainder (Table 14.1). Changes in the oxygen content of the atmosphere provide a measure of relative importance of land and ocean uptake. Net terrestrial uptake of CO_2 is accompanied by a net release of oxygen, with a 1:1 ratio of moles of CO_2 absorbed to moles of O_2 released. When CO_2 dissolves in ocean water, however, this causes no net release of oxygen. This difference in exchange processes can be used to partition the total CO_2 uptake between terrestrial and ocean components (Keeling et al. 1996b).

The relative abundance of the two stable isotopes of carbon (^{13}C and ^{12}C) in the atmosphere provides a second measure of the relative activity of the terrestrial and oceanic components of the global carbon cycle (Ciais et al. 1995). Fractionation during photosynthesis by C_3 plants discriminates against ^{13}C, causing biospheric carbon to be depleted in ^{13}C by about 18‰ relative to the atmosphere. Exchanges with the ocean, however, involve relatively small fractionation effects. Changes in the $^{13}C/^{12}C$ ratio of atmospheric CO_2 therefore indicate the relative magnitude of terrestrial and oceanic CO_2 uptake.

Measurement of the global pattern and temporal changes in oxygen concentration and the $^{13}C/^{12}C$ ratio of atmospheric CO_2 suggest that the land and ocean contribute about equally to the removal of anthropogenic CO_2 from the atmosphere (IPCC 2007). There are, however, many assumptions and complications in using either of these approaches to estimate the relative magnitudes of terrestrial and oceanic carbon uptake. The advantage of atmospheric measurements is that they give an integrated estimate of all uptake processes on Earth because of the relatively rapid rate at which the atmosphere mixes.

Terrestrial Sinks for CO_2

Land-use change, CO_2 fertilization, nitrogen deposition, and various climate effects contribute to the terrestrial sink for CO_2 (Schimel 1995, Reich et al. 2006, Luo 2007). The conversion of forests to agricultural lands dominated land-use change in the middle and high latitudes until the mid-twentieth century. Today, forest regrowth in abandoned agricultural lands has enhanced carbon storage, particularly in Europe and North America. The widespread suppression of wildfire also enhances the mid-latitude carbon sink because it reduces fire emissions and allows woody plants to encroach into grasslands (Houghton 2004). These are probably the most important reasons why north-temperate terrestrial ecosystems are a net carbon sink (IPCC 2007). Meanwhile, increasing rates of deforestation in the tropics reduce the low-latitude sink for CO_2 (Field et al. 2007).

CO_2 enhancement of photosynthesis also contributes to carbon storage (Norby et al. 2005, Long et al. 2006), although not as much as the short-term CO_2 response of photosynthesis might suggest. Over the longer term, CO_2 uptake becomes nutrient-limited, as nutrients become sequestered in live and dead organic matter (see Chap. 6; Shaver et al. 1992, Norby et al. 2010). The effect of CO_2 fertilization on carbon storage appears to be smaller than that due to reforestation in the temperate zone, but in the tropics, CO_2 fertilization appears sufficient to offset the loss to deforestation of carbon-fixation capacity (Field et al. 2007).

Nitrogen additions through fertilizer applications or atmospheric deposition of air pollutants like NO_x from fossil-fuel burning have stimulated photosynthesis and reduced respiration, leading to greater carbon sequestration in some places (see Chap. 7; Magnani et al. 2007, de Vries et al. 2009, Janssens et al. 2010).

Finally, climate changes (including changes in temperature, moisture, and radiation) affect carbon storage through their effects on carbon inputs (photosynthesis) and outputs (respiration). These effects vary regionally and are difficult to generalize because direct climatic effects (e.g., stimulation of respiratory carbon loss by warmer temperatures) are often offset by indirect effects (e.g., stimulation of NPP by the nutrients released during decomposition; Shaver et al. 2000). Vegetation generally has a much higher C:N ratio (160:1) than does soil organic matter (14:1), so the transfer of a given quantity of nitrogen from the soil to plants enhances carbon storage (Vukicevic et al. 2001). In addition, plant respiration acclimatizes to temperature, so ecosystem respiration increases less in response to warming than might be expected from short-term measurements (Luo et al. 2001).

The relative importance of the various mechanisms of enhanced carbon storage in the terrestrial biosphere is uncertain (Schimel et al. 2001), but together they are probably sufficient to account for the observed movement of a fraction of anthropogenic CO_2 from the atmosphere to land. Just as described for the ocean, the strength of the terrestrial carbon sink appears to be weakening (see Fig. 7.28; Le Quéré et al. 2007), suggesting that the various sink mechanisms (forest regrowth, CO_2 fertilization, nitrogen addition, and climate effects) are beginning to saturate and may remove less CO_2 from the atmosphere in the future. The most effective mechanism of stabilizing atmospheric CO_2 concentration is therefore to reduce anthropogenic emissions.

CO_2 Effects on Climate

Much of the increased concentration of fossil-fuel CO_2 will remain in the atmosphere for hundreds to thousands of years. If all anthropogenic emissions ceased today, about half would be absorbed by lands and the ocean within 30 years, about 30% of it would remain in the atmosphere for several centuries, and the remaining 20% for thousands of years (IPCC 2007, Archer et al. 2009, Solomon et al. 2009). There

are at least four reasons why CO_2 disappears slowly from the atmosphere: (1) the efficiency of the land and ocean sinks is weakening, as described previously; (2) the deep ocean, which is the major long-term sink for CO_2, equilibrates very slowly with the surface ocean and the atmosphere; (3) stabilizing feedbacks minimize changes in ecosystem carbon pools – for example, the increase in decomposition that occurs in response to increased photosynthesis and litter inputs (see Chap. 7); and (4) weathering of silicate rocks, which is the largest long-term sink for CO_2 on land, occurs very slowly. Because CO_2 is the largest anthropogenic contributor to climate warming (see Fig. 2.18), past CO_2 emissions already commit us to a warmer planet, and decisions about future emissions will strongly influence the magnitude of continued climate warming. In addition, much of the heat absorbed as a result of increased concentrations of greenhouse gases has gone into the ocean and will return to the atmosphere, even if natural cycles or some (as yet unknown) technological solution instantly removed all fossil-fuel carbon from the atmosphere (Solomon et al. 2009). This long-term commitment to future warming enhances concerns that Earth has approached, or perhaps exceeded, a threshold of "dangerous climate change" that warrants rapid and vigorous efforts to reduce carbon emissions to the atmosphere (Stern 2007, Rockström et al. 2009).

The Global Methane Budget

Human activities are responsible for increasing methane concentrations in the atmosphere. Although the methane (CH_4) concentration of the atmosphere (1.8 ppmv) is much less than that of CO_2 (390 ppmv), CH_4 is about 23 times more efficient per molecule as a greenhouse gas than is CO_2. Like CO_2, the CH_4 concentration of the atmosphere has increased exponentially since the beginning of the industrial revolution (Fig. 14.7). The CH_4 increase accounts for 20% of the increased greenhouse warming potential of the atmosphere (see Fig. 2.18; Bousquet et al. 2006, IPCC 2007). Documenting the major global

Table 14.2 Global sources and sinks of methane. Data from Wang et al. (2004), Chen and Prinn (2006), and IPCC (2007)

Methane sources and sinks	Annual flux (Tg CH_4 year^{-1})
Natural sources	168
Wetlands	145
Termites and ruminants	23
Anthropogenic sources	428
Coal combustion	48
Oil and gas combustion	36
Landfills and waste	70
Fermentation by cattle	119
Rice agriculture	112
Biomass burning	43
Total sources	596
Sinks	581
Reaction with OH	511
Removal in stratosphere	40
Removal by soils	30
Atmospheric increase	1–22[a]

[a]Annual atmospheric increase declined from about 22 Tg CH_4 year^{-1} in the 1990s to about 1 Tg CH_4 year^{-1} in 2000–2004

sources and sinks of atmospheric CH_4 is therefore important to understanding the recent increases in global temperature and the potential for future climate warming.

Methane is produced only under anaerobic conditions (see Chap. 7). Wetlands account for 85% of the naturally produced CH_4, with the remainder coming primarily from freshwater sediments, fermentation in the guts of animals (e.g., termites and ruminants), and various geological sources (Table 14.2). Anthropogenic methane sources are 2.5 times larger than the natural sources, showing why CH_4 accumulates in the atmosphere despite its high reactivity and rapid turnover (9 years). Fossil-fuel extraction and refining; waste management (landfills, animal wastes, and domestic sewage treatment); and agricultural sources (rice paddies, biomass burning, and fermentation in guts of domestic ruminants like cattle) are each important CH_4 sources. The concentration and rate of accumulation in the atmosphere are known quite precisely, but the relative contributions of different sources and sinks are still topics of active debate. Important new sources are still being identified, including

high-latitude thaw lakes and reservoirs with organic-rich substrates (St. Louis et al. 2000, Friedl and Wüest 2002, Walter et al. 2007).

CH_4 reacts readily with OH radicals in the atmosphere in the presence of sunlight. This photochemical process is the major sink for atmospheric CH_4, accounting for 85% of the CH_4 consumption (Table 14.2). Additional CH_4 mixes into the stratosphere, where it reacts with ozone (see Chap. 2) or is removed by methanotrophs in soils (see Chap. 7). The annual atmospheric accumulation of CH_4 is about 10% of the annual anthropogenic flux, as compared to 50% for CO_2.

The Global Nitrogen Cycle

Nitrogen Pools and Fluxes

The productivity of many ecosystems on both land and sea is limited in part by the supply of available nitrogen. Almost all of the nitrogen that is relevant to biogeochemistry is in a single pool (the atmosphere) with comparatively small quantities in the ocean, rocks, and sediments (Fig. 14.8). Organic nitrogen pools are miniscule relative to the atmospheric pool and occur primarily in soils and terrestrial vegetation. Although nitrogen makes up 78% of the atmosphere, it is nearly all N_2 and is unavailable to most organisms. N_2 is transformed to biologically available forms via nitrogen fixation by bacteria in soils and aquatic systems, or living in association with plants. The global quantity of nitrogen fixed annually by natural ecosystems is quite uncertain, with estimates near 100 Tg year^{-1} for terrestrial ecosystems and between 40 and 200 Tg year^{-1} for marine ecosystems. Lightning probably adds an additional 3–10 Tg year^{-1} of nitrogen to the available pool. Prior to human alteration, the amount of nitrogen entering the biosphere via nitrogen fixation was approximately balanced by return to the unavailable pools via denitrification and burial in sediments. During glacial periods, the input of iron and other micronutrients may have caused nitrogen fixation to exceed denitrification, reducing the degree of nitrogen limitation in the ocean. In interglacial periods such as the present, denitrification may exceed nitrogen fixation. There is

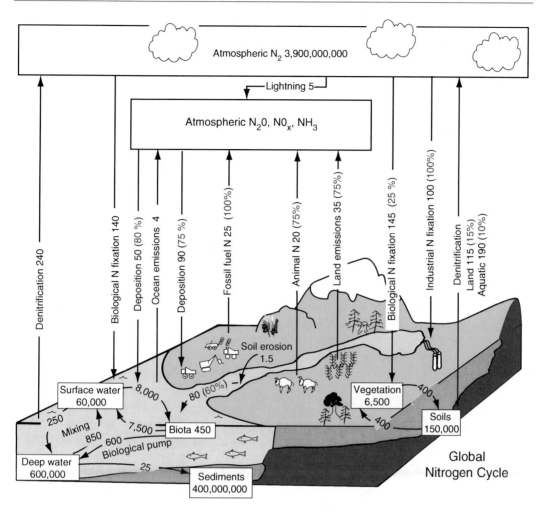

Fig. 14.8 The global nitrogen cycle, showing approximate magnitudes of the major pools (*boxes*) and fluxes (*arrows*) in units of Tg year^{-1}. A teragram (Tg) is 10^{12} g. *Numbers in parenthesis* are the anthropogenic contributions to these pools and fluxes. Data are from Reeburgh (1997), Chapin et al. (2002), Galloway et al. (2004), and Gruber and Galloway (2008). To ensure consistency among global cycles of different elements, pools and fluxes of biota were calculated from the global carbon budget (Fig. 14.5) assuming mass-based C:N ratios (Sterner and Elser 2002) of marine biota (6.6), terrestrial vegetation (100), and terrestrial litter (150) and are close to published estimates for global nitrogen budgets. The

atmosphere contains the vast majority of Earth's nitrogen. The amount of nitrogen that annually cycles through terrestrial non-crop vegetation is fourfold greater than inputs by nitrogen fixation. In the ocean, the annual cycling of nitrogen through the biota is 60-fold greater than inputs by nitrogen fixation. Denitrification is the major output of nitrogen to the atmosphere. Human activities increase nitrogen inputs through fertilizer production, planting of nitrogen-fixing crops, and combustion of fossil fuels. Human activities also increase emissions of nitrogen trace gases (NO_x, N_2O, NH_3) through fossil fuel emissions, land emissions (agriculture, fire, land-use change), and animal husbandry

considerable debate about the current degree of balance or imbalance between marine nitrogen fixation and denitrification (Falkowski et al. 1998). In contrast to carbon, nitrogen is cycled quite tightly within terrestrial ecosystems, with the annual throughput often being at least four-fold greater than inputs and losses.

Anthropogenic Changes in the Nitrogen Cycle

In the past century, human activities have approximately doubled the quantity of nitrogen fixed from the atmosphere into terrestrial systems. The Haber process, which uses energy

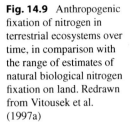

Fig. 14.9 Anthropogenic fixation of nitrogen in terrestrial ecosystems over time, in comparison with the range of estimates of natural biological nitrogen fixation on land. Redrawn from Vitousek et al. (1997a)

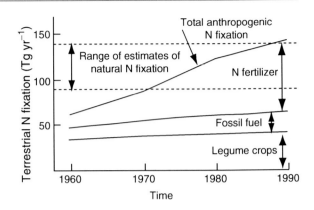

from fossil fuels to convert N_2 to NH_3 to produce fertilizers, fixes more nitrogen than any other anthropogenic process. Industrial fixation of nitrogen by the Haber process began increasing substantially in the 1940s, reaching 30 Tg year^{-1} by 1970 and 100 Tg year^{-1} by 2000 (Fig. 14.9). It is projected to be 165 Tg year^{-1} by 2050 (Galloway et al. 2004). Initially, most nitrogen fertilizer was applied in developed nations, but from 2000 to 2009 about 80% of the global increase in use of nitrogen fertilizer occurred in China and India. Much of the projected increase in fertilizer use is expected to occur in developing nations.

Cultivation of nitrogen-fixing crops such as soybeans, alfalfa, and peas adds fixed nitrogen over and above that which is added via biological fixation in natural ecosystems. Agricultural crops account for about 25% of terrestrial nitrogen fixation (Fig. 14.8). Some nitrogen fixation is also carried out by free-living and associative nitrogen fixers like *Azolla* that commonly occur in rice paddies.

Human activities account for most of the nitrogen trace gases transferred from Earth to the atmosphere. In addition to the large pool of relatively unreactive N_2, the atmosphere contains several nitrogen trace gases, including NO_x (NO and NO_2), N_2O, and NH_3. Although the pools and fluxes of these nitrogen trace gases are much smaller than those of N_2 (Fig. 14.8), they play a very active role in atmospheric chemistry and have been more strongly affected by human activities (see Chap. 9).

Nitrous oxide (N_2O), which is increasing at the rate of 0.2–0.3% year^{-1} (Fig. 14.7), is an inert gas that is 200-fold more efficient than CO_2 as a greenhouse gas and contributes about 6% of the greenhouse warming (IPCC 2007). Nitrification and denitrification in the ocean and in tropical soils are the major natural sources of N_2O (Schlesinger 1997, Galloway et al. 2004). Human activities have nearly doubled N_2O flux from Earth to the atmosphere, primarily through agricultural fertilization. Other anthropogenic N_2O sources include cattle and feedlots, biomass burning, and various industrial sources. N_2O is broken down in the stratosphere, where it catalyzes the destruction of stratospheric ozone.

Human activities have tripled the flux of ammonia (NH_3) from land to the atmosphere (Galloway et al. 2004). Domestic animals are now the single largest global source of ammonia; agricultural fertilization, biomass burning, and human sewage are other important sources. Cultivated soils, which account for only 10% of the ice-free land area (see Table 6.6), account for about half of the ammonia flux from soils to the atmosphere. In summary, activities associated with agriculture (animal husbandry, fertilizer addition, and biomass burning) are the major cause for increased ammonia transport to the atmosphere and account for 60% of the global flux. Ammonia is a reactant in many atmospheric reactions that form aerosols and generate air pollution. Ammonia is also the main acid-neutralizing agent in the atmosphere, raising the pH of rainfall, cloud water, and aerosols. Most of the ammonia emitted to the atmosphere returns to Earth in precipitation.

Human activities have increased NO_x flux to the atmosphere six- to sevenfold, primarily through the combustion of fossil fuels. Nitrification is the largest natural terrestrial source of NO (see Chap. 9). Fertilizer addition has increased the magnitude of this source, with additional NO coming from biomass burning. Preindustrial NO_x fluxes were greater in tropical than temperate ecosystems, due to frequent burning of tropical savannas, soil emissions, and production by lightning (Holland et al. 1999). Most NO_x deposition now occurs in the temperate zone, where deposition rates have increased four-fold since preindustrial times.

Nitrogen deposition affects many ecosystem processes. The widespread nitrogen limitation or co-limitation of plant production in nontropical ecosystems results in retention of a large proportion of anthropogenic nitrogen that is deposited in ecosystems, particularly in young, actively growing forests that are accumulating nutrients in vegetation (see Fig. 12.18). Nitrogen deposition often stimulates carbon storage, by stimulating production in nitrogen-limited sites and reducing heterotrophic respiration in nitrogen-rich sites (Magnani et al. 2007, de Vries et al. 2009, Janssens et al. 2010). Nonetheless, the overall role of nitrogen deposition in explaining the terrestrial land sink for carbon is quite uncertain.

Nitrogen accumulation in production and organic matter storage cannot increase indefinitely. After long-term chronic nitrogen inputs, nitrogen supply may exceed plant and microbial demands, resulting in **nitrogen saturation** (Aber et al. 1998, Driscoll et al. 2001). When ecosystems become nitrogen saturated, nitrogen losses to stream water, groundwater, and the atmosphere increase and should eventually approach nitrogen inputs. Nitrogen saturation is often associated with declines in forest productivity and increased tree mortality in coniferous forests in Europe (Schulze 1989) and the U.S. (Aber et al. 1995, Fahey et al. 2005).

Temperate forests vary regionally in the rate at which they approach nitrogen saturation, depending on rates of nitrogen inputs and the capacity of soils to buffer these inputs (Aber et al. 1995, Fahey et al. 2005, Juice et al. 2006). In tropical forests, where nitrogen availability is typically high relative to plant and microbial demands, anthropogenic nitrogen deposition may lead to immediate nitrogen losses (Hall and Matson 1998), which could have potentially negative effects on plant and soil processes (Matson et al. 1999). In general, the capacity of a forest ecosystem to retain nitrogen is linked to its productive potential and its degree of nitrogen limitation (Aber et al. 1995, Magill et al. 1997, Magnani et al. 2007).

The addition of limiting nutrients can alter species dominance and reduce the diversity of ecosystems. Nitrogen addition to grasslands or heathlands, for example, increases the dominance of nitrogen-demanding grasses, which then suppress other plant species (Berendes et al. 1993). These species changes can convert nutrient-poor, diverse heathlands to species-poor forests and grasslands (Aerts and Berendse 1988, Tilman and Wedin 1991).

Human activities increase the nitrogen transfer from terrestrial to aquatic ecosystems. The massive nitrogen additions to terrestrial ecosystems, in the form of deposition, fertilization, food imports, and growth of nitrogen-fixing crops, have led to a dramatic increase in nitrogen concentrations in surface and ground waters over the past century (see Chap. 9).

The Global Phosphorus Cycle

Phosphorus Pools and Fluxes

Unlike carbon and nitrogen, phosphorus has only a tiny gaseous component and no biotic pathway that brings new phosphorus into ecosystems. Ecosystems, until recently, therefore derived most available phosphorus from organic forms, and phosphorus cycled quite tightly within terrestrial ecosystems. Like nitrogen, phosphorus is an essential nutrient that is often in short supply. Marine and freshwater sediments and terrestrial soils account for most phosphorus on Earth's surface (Fig. 14.10). Most of this store is not directly accessible to the biota but occurs primarily in insoluble forms such as calcium or iron

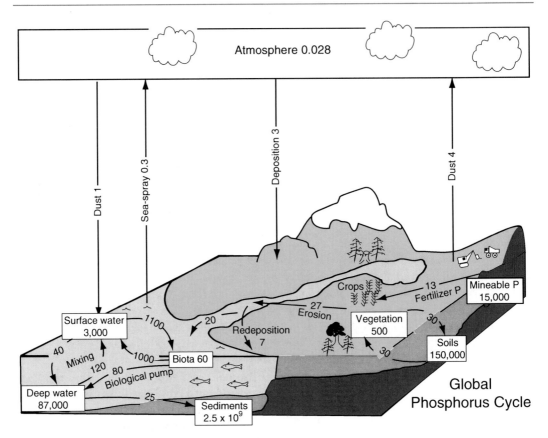

Fig. 14.10 The global phosphorus cycle, showing approximate magnitudes of the major pools (*boxes*) and fluxes (*arrows*) in units of Tg year^{-1}. A teragram (Tg) is 10^{12} g. Data are from Smil (2000) and Ruttenberg (2004). To ensure consistency among global cycles of different elements, pools and fluxes of biota were calculated from the global carbon and nitrogen budgets (Figs. 14.5, 14.8) assuming mass-based N:P ratios (Sterner and Elser 2002) of marine biota (7.2) and terrestrial vegetation and litter (12.6), and are close to published estimates for global phosphorus budgets for marine biota but smaller than published estimates for terrestrial biota (Ruttenberg 2004). Most phosphorus that participates in biogeochemical cycles over decades to centuries is present in soils, sediments, and the ocean. Phosphorus cycles tightly between vegetation and soils on land and between marine biota and surface waters in the ocean. The major human impact on the global phosphorus cycle has been application of fertilizers (equivalent to about a third of that which cycles naturally through vegetation) and erosional loss from crop and grazing lands

phosphate. Most organic phosphorus is in plant or microbial biomass. Recycling of that organic matter when it dies is the major source of phosphorus that is directly available to organisms.

The physical transfers of phosphorus around the global system are constrained by the lack of a major atmospheric gaseous component. Leaching losses in natural ecosystems are also low due to the low solubility of phosphorus. Instead, phosphorus moves around the globe primarily through wind erosion and runoff of particulates in rivers and streams to the ocean. The major flux in the global phosphorus cycle (excluding human activities) is via hydrologic transport from land to the ocean. In the ocean, some of those phosphorus-containing particulates are recycled by marine biota and the rest is buried in sediments. Because there is no atmospheric link from the ocean to land, the flow is one-way on short time scales (Smil 2000). On geological time scales (tens to hundreds of millions of years), phosphorus-containing sedimentary rocks are exposed and weathered, resupplying phosphorus to the biosphere (Ruttenberg 2004).

Fig. 14.11 Changes in the global use of inorganic phosphorus fertilizers during the twentieth century. Redrawn from Smil (2000)

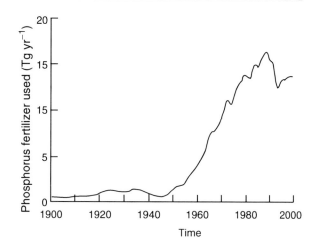

Anthropogenic Changes in the Phosphorus Cycle

Human activities have enhanced the mobility of phosphorus and altered its natural cycling by mining of phosphorus-rich deposits, which accelerates the rate at which phosphorus weathers from rocks, and also by accelerating erosion and wind- and waterborne transport. Inorganic phosphorus fertilizers have been produced since the mid-1800s, but the amount produced and applied has increased dramatically since the mid-twentieth century (Fig. 14.11), coincident with the intensification of agriculture that accompanied the "Green Revolution" (Smil 2000). Between 1850 and 2000, agricultural systems received about 550 Tg of new phosphorus. The annual application of phosphorus to agricultural ecosystems (10–15 Tg year^{-1}) is about a third of that which cycles naturally through all terrestrial ecosystems (Fig. 14.10).

Human land-use change has also increased phosphorus losses from ecosystems. Water and wind erosion cause a 15 Tg year^{-1} phosphorus loss from the world's croplands, an amount similar to the annual fertilizer inputs. Overgrazing has also increased erosional losses, mobilizing about 12 Tg year^{-1} of phosphorus from grazing lands (Smil 2000). About 25% of this is redeposited in floodplains or deposited in reservoirs. The production of human and animal wastes have led to point and nonpoint sources of phosphorus. The total phosphorus transfer from land to the ocean has increased 50–300% due to human activities (Ruttenberg 2004).

Together, these changes have increased the transport of phosphorus around the world (Howarth et al. 1995). Because phosphorus commonly limits production in lakes, the inadvertent phosphorus fertilization of freshwater ecosystems can lead to eutrophication and associated negative consequences for aquatic organisms and society (see Chap. 9). Phosphorus transport by windblown dust can also affect downwind ecosystems such as the Southern Ocean.

The Global Sulfur Cycle

The global cycle of sulfur shares characteristics with the global cycles of nitrogen and phosphorus. The sulfur cycle, like the nitrogen cycle, has a significant atmospheric component. The gaseous forms in the atmosphere have low concentrations but play important roles. Like phosphorus, sulfur is primarily rock derived. Seawater, sediments, and rocks are the largest reservoirs of sulfur (Fig. 14.12). The atmosphere contains little sulfur. Prior to human activities of the past several centuries, sulfur became available to the biosphere primarily through the weathering of sedimentary pyrite. Once weathered, sulfur moves through the global system by hydrologic transport or emission to the atmosphere as a reduced sulfur gas or sulfur-containing particles. About 100 Tg year^{-1} of sulfur, moving mostly as

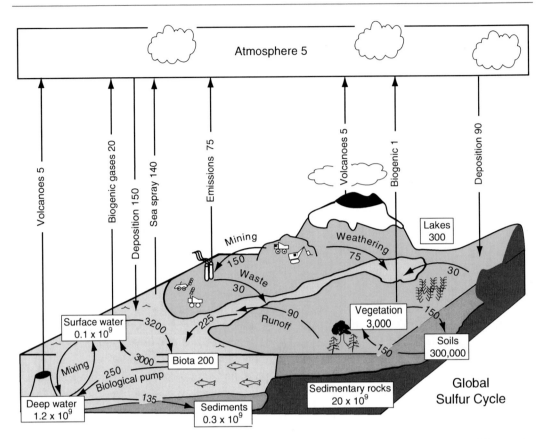

Fig. 14.12 The global sulfur cycle, showing approximate magnitudes of the major pools (*boxes*) and fluxes (*arrows*) in units of Tg year^{-1}. A teragram (Tg) is 10^{12} g. Data are from Galloway (1996), Reeburgh (1997), Schlesinger (1997), and Brimblecombe (2004). To ensure consistency among global cycles of different elements, pools and fluxes of biota were calculated from the global carbon and nitrogen budgets (Figs. 14.5, 14.8) assuming mass-based N:S ratio of 7.4 (Bolin et al. 1983). Most sulfur is in rocks, sediments, and ocean waters. The major fluxes in the sulfur are through the biota and various trace gas fluxes. Human activities have doubled the global fluxes of sulfur through mining and increased gas emissions

dissolved sulfate, was transported through rivers to the coastal margins or open ocean in the preindustrial world (Galloway 1996).

Sulfur can be reduced to sulfide or to other trace sulfur gases in anaerobic environments such as wetlands and coastal sediments. The emission of sulfate from seawater (sea spray) and sulfur trace gases from the ocean (160 Tg year^{-1}) is about 100-fold greater than that from continents (Fig. 14.12). Marine biogenic emissions include dimethylsulfide (DMS), one of the primary sources of atmospheric sulfate; emissions of SO_2 from volcanic eruptions are the other major natural source.

Sulfur emitted to the atmosphere typically has a short residence time. It is oxidized to sulfate by reaction with OH radicals. Sulfate rains out downwind within a few days, generally as sulfuric acid. Sulfuric acid quickly condenses to form sulfate in cloud droplets, which readily evaporate to form sulfate aerosols. These aerosols have both direct and indirect effects on Earth's energy budget. Their direct effect is to backscatter (reflect) incoming shortwave radiation, thus reducing solar inputs and tending to reduce global temperature (see Fig. 2.18). Their indirect effects are more complicated and difficult to predict. As particulates, they act as cloud condensation nuclei by providing a surface on which water can condense, thereby influencing cloud formation, cloud lifetimes, cloud droplet size, and therefore cloud albedo. The uncertainty of the direction and magnitude of the multiple effects of sulfate aerosols on climate is a key reason for concern about the anthropogenic changes in the global sulfur cycle.

Human activities now transfer about 135 Tg year^{-1} of sulfur to the atmosphere and ocean, increasing the natural cycling rate by about 50% (Fig. 14.12). Half of this sulfur arises from fossil-fuel combustion and ore refining, and the rest comes from mobilization of sulfur in dust from farming, animal husbandry, erosion of exposed sediments, and other sources. Much of the anthropogenic sulfur moves through the atmosphere and is deposited on land, where it can accumulate in soils or biota, or is discharged to the ocean in solution.

Reconstruction of global temperature records from ice cores shows that sulfur dioxide from volcanic emissions is a major cause of interannual climate variation over long time scales. Consequently, the dramatic increase in sulfur aerosols due to anthropogenic emissions will undoubtedly play an important role in future climate changes. The cooling effects of sulfur emissions and their associated direct and indirect effects could range from 0 to 1.5 W m^2, partially offsetting the warming due to greenhouse gases (IPCC 2007).

Summary

Ecological processes and human activities play major roles in most biogeochemical cycles. The magnitude of biotic and human impacts on ecosystem processes is substantial when summed at the global scale.

Most water is in the ocean, ice, and groundwater, where it is not directly accessible to terrestrial organisms. The major water fluxes are precipitation, evapotranspiration, and runoff. Human activities have speeded up the global hydrologic cycle by increasing global temperature, which enhances evapotranspiration and therefore precipitation, and by diverting much of the accessible fresh water for human use. Availability of adequate fresh water will be an increasingly scarce resource for society, if current human population trends continue.

Biotic processes (photosynthesis and respiration) constitute the engine that drives the global carbon cycle. The four major carbon pools that contribute to carbon cycling over decades to centuries are the atmosphere, land, ocean, and surface sediments. On land, the carbon gain by vegetation is slightly greater than the carbon loss in respiration, leading to net carbon storage on land. The net carbon input to the ocean is also slightly greater than the net carbon return to the atmosphere. Marine primary production is about the same as that on land. Most (80%) of this marine NPP is released to the environment by respiration, with the remaining 20% going to the deep ocean by the biological pump. Ocean upwelling returns most of this carbon to the surface ocean waters; only small quantities are deposited in sediments. Human activities cause a net carbon flux to the atmosphere through combustion of fossil fuels, cement production, and land-use change. This flux is equivalent to 14% of terrestrial heterotrophic respiration.

The atmosphere contains the vast majority of Earth's nitrogen. The amount of nitrogen that annually cycles through terrestrial vegetation is ninefold greater than inputs by nitrogen fixation. In the ocean, the annual cycling of nitrogen through the biota is 70-fold greater than inputs by nitrogen fixation. Denitrification is the major output of nitrogen to the atmosphere. Human activities have doubled the quantity of nitrogen fixed by the terrestrial biosphere through fertilizer production, planting of nitrogen-fixing crops, and combustion of fossil fuels.

Most phosphorus that participates in biogeochemical cycles over decades to centuries is present in soils, sediments, and the ocean. Phosphorus cycles tightly between vegetation and soils on land and between marine biota and surface waters in the ocean. The major human impact on the global phosphorus cycle has been application of fertilizers (equivalent to about 40% of that which naturally cycles through vegetation) and erosional loss from crop and grazing lands (equivalent to about half of that which annually cycles through vegetation). Most sulfur is in rocks, sediments, and ocean waters. The major fluxes in the sulfur are through the biota and various trace gas fluxes. Human activities have substantially increased global fluxes of sulfur through mining and increased gas emissions.

Review Questions

1. How do the major global cycles (carbon, nitrogen, phosphorus, sulfur, and water) differ from one another in terms of (1) the major pools and (2) the major fluxes? In which cycles are soil pools and fluxes largest? In which cycles are atmospheric pools and fluxes largest?

2. How have human activities changed the global water cycle? If the world has so much water, and this water is replenished so frequently by precipitation, why are people concerned about changes in the global water cycle? In what regions of the world will changes in the quantity and quality of water have greatest societal impact? Why?

3. How do the controls over the global carbon cycle differ between time scales of months, decades, and millennia? How has atmospheric CO_2 varied on each of these time scales, and what has caused this variation?

4. How have human activities altered the global carbon cycle? What are the mechanisms that explain why some of the CO_2 generated by human activities becomes sequestered on land and in the ocean?

5. What are the major causes and the climatic consequences of increased atmospheric concentrations of CO_2, CH_4, and N_2O? What changes in human activities would be required to reduce the rate of increase of these gases, and what would be the societal consequences of these policy changes?

6. What are the major natural sources and sinks of atmospheric methane? How might these be changed by recent changes in climate and atmospheric composition?

7. What are the major natural sources and sinks of atmospheric N_2O? How might these be changed by recent changes in climate and land use?

8. How have human activities changed the global nitrogen cycle? How have these changes affected the nitrogen cycle in unmanaged ecosystems?

9. How do changes in the nitrogen cycle affect the global carbon cycle? How does soil fertility affect the mechanism by which nitrogen affects the carbon cycle?

10. How have human activities changed the global phosphorus and sulfur cycles? How do changes in these cycles affect the global cycles of other elements?

Additional Reading

Bousquet, P., P. Cias, J.B. Miller, E.J. Dlugokenck, D.A. Houglustaine et al. 2006. Contribution of anthropogenic and natural sources of atmospheric methane variability. *Nature* 443:439–443.

Brimblecombe, P. 2004. The global sulfur cycle. Pages 645–682 *in* W.H. Schlesinger, editor. *Biogeochemistry.* Elsevier, Amsterdam.

Canadell, J.G., C. Le Quéré, M.R. Raupach, C.B. Field, E.T. Buitehuls, et al. 2007. Contributions to accelerating atmospheric CO_2 growth from economic activity, carbon intensity, and efficiency of natural sinks. *Proceedings of the National Academy of Sciences, USA* 104:10288–10293.

Galloway, J.N., F.J. Dentener, D.G. Capone, E.W. Boyer, R.W. Howarth et al. 2004. Nitrogen cycles: Past, present, and future. *Biogeochemistry* 70:153–226.

IPCC. 2007. *Climate Change 2007: The Physical Science Basis, Contribution of Working Group I to the Fourth Assessment Report of the Intergovernmental Panel on Climate Change.* Cambridge University Press, Cambridge.

Matson, P.A., W.H. McDowell, A.R. Townsend, and P.M. Vitousek. 1999. The globalization of N deposition: Ecosystem consequences in tropical environments. *Biogeochemistry* 46:67–83.

Oki, T. and S. Kanae. 2006. Global hydrological cycles and world water resources. *Science* 313: 1068–1072.

Ruttenberg, K.C. 2004. The global phosphorus cycle. Pages 585–643 *in* W. H. Schlesinger, editor. *Biogeochemistry.* Elsevier, Amsterdam.

Schlesinger, W.H. 1997. *Biogeochemistry: An Analysis of Global Change.* Academic Press, San Diego.

Smil, V. 2000. Phosphorus in the environment: Natural flows and human interferences. *Annual Review of Energy in the Environment* 25:53–88.

Managing and Sustaining Ecosystems

15

Human activities influence all of Earth's ecosystems. This chapter summarizes the principles by which important ecological properties can be sustained to meet the needs of ecosystems and society.

Introduction

Growth of the human population and our use of resources have altered ecosystems more rapidly and extensively in the last 50 years than in any comparable period of human history (Fig. 15.1; MEA 2005). Accelerating human impacts are causing global changes in most major ecosystem controls: climate (global climate change), soil and water resources (nitrogen deposition, erosion, diversions), disturbance regime (land-use change, fire control), and functional types of organisms (species introductions and extinctions). All ecosystems are therefore experiencing directional changes in ecosystem controls, creating novel conditions and, in many cases, amplifying (positive) feedbacks that accelerate changes to new types of ecosystems. These changes in interactive controls inevitably alter the properties of ecosystems, often to the detriment of society.

A Focal Issue

Given that human activities have and will continue to shape ecosystems of the planet, how can these be managed to sustain ecosystem properties and the services they provide to society (Fig. 15.2)? In this chapter, we describe some general principles that contribute to sound ecosystem management. Maintaining Earth's ecosystems, even the "wild" ones, in the face of anthropogenic changes requires new management approaches that recognize the increasing human domination of the biosphere (Palmer et al. 2004). We review management approaches that draw on ecosystem ecology and other sciences to manage and sustain ecosystems and the benefits we derive from them.

Sustaining Social–Ecological Systems

People and nature are interconnected components of coupled social–ecological systems. People inhabit 80% of the ice-free land surface of the planet and therefore are integral components of most ecological systems (Ellis and Ramankutty 2008). Many of the negative human impacts on ecosystems are unintended, as people seek to meet multiple desires and needs within a social context. Failure to recognize key linkages between ecosystems and society creates vulnerabilities that could be avoided by proper **ecosystem management**, i.e., resource management that promotes long-term sustainability of ecosystems and the delivery of essential ecosystem goods and services to society. The loss of flood control associated with wetland drainage and reduced sediment delivery to

F.S. Chapin, III et al., *Principles of Terrestrial Ecosystem Ecology*,
DOI 10.1007/978-1-4419-9504-9_15, © Springer Science+Business Media, LLC 2011

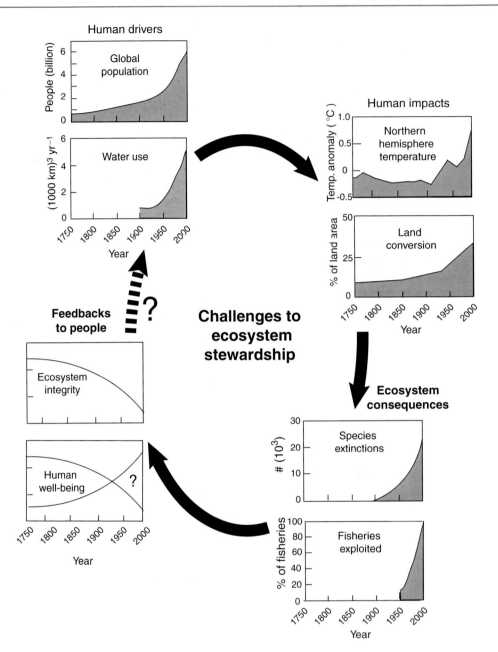

Fig. 15.1 Challenges to ecosystem management and stewardship. Changes in human population and resource consumption alter climate and land cover, which have important ecosystem consequences such as species extinctions and overexploitation of fisheries. These changes reduce ecosystem integrity and have regionally variable effects on well-being, which feed back to further changes in human drivers. Redrawn from Chapin et al. (2009) with panels from Steffen et al. (2004)

barrier islands during urban development of New Orleans, for example, was overlooked until Hurricane Katrina caused major flooding and loss of life and property in 2005 (Box 15.1; Kates et al. 2006). Ecosystem managers must therefore be aware not only of environmental and biological factors that influence ecosystems but also of the social and political forces that influence decisions that cause unintended effects on ecosystems (Fig. 15.3).

Fig. 15.2 Human actions are modifying ecosystems at scales that influence the Earth System. Society now faces the challenge of managing its relationship to the biosphere to sustain and enhance the benefits provided by ecosystems to support human well-being. Photograph from istockphoto

---

Box 15.1 Social-Ecological Interactions and the Flooding of New Orleans

The southeastern coastal plain of the U.S. is low and flat, with much of the land created out of sediments delivered to the coast by major rivers. Louisiana's offshore barrier islands and extensive wetlands that protect cities from storms and floods are products of this fluvial-delta system (NRC 2006). The construction of levees and reservoirs has reduced sediment delivery that maintains these natural protective features. Land subsidence resulting from the extraction of oil and gas, drainage of low-lying areas, and other development activities also contributes to vulnerability to storms and flooding (NRC 2006). New Orleans, for example, has subsided an average of 5 mm year^{-1}, so much of New Orleans is below sea level and persists only because of a system of levees and pumps. Storm surges caused by Hurricane Katrina broke through the system of protective levees surrounding New Orleans in 2005, causing extensive loss of life and property (Kates et al. 2006).

Sustainability

Sustainability requires recognition of tradeoffs resulting from choices that influence social–ecological systems today and in the future. Most decisions that negatively affect ecosystems are not malicious but reflect choices to pursue certain socioeconomic benefits. Mining and over-grazing, for example, generally occur through efforts to meet people's desires for minerals and food, respectively. The ecological consequences of these actions are sometimes less obvious or of

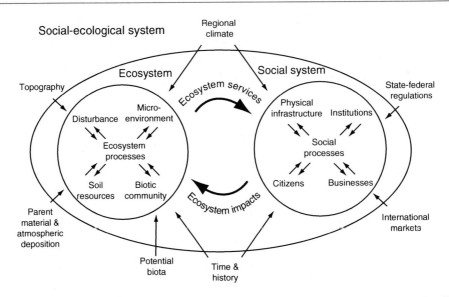

Fig. 15.3 Controls over the functioning of social–ecological systems. Human impacts on ecosystems are mediated by a variety of social processes, and society benefits from the services provided by ecosystems. Modified from Whiteman et al. (2004)

less immediate concern to the decision maker than are short-term social or economic benefits. Ecologists can play an important role in these decisions by documenting potential **tradeoffs** that influence ecological and social risks and opportunities (Matson 2009). Sustainability provides an important framework for clarifying the consequences of choices facing society. These choices are particularly stark for developing nations, where people depend very directly on ecosystems for survival but also seek to escape conditions of persistent poverty and poor quality of life. In 1987, the Brundtland Commission (WCED 1987) proposed to the United Nations a sustainability framework that addressed this twin challenge of meeting needs for ecological conservation and human development. **Sustainability**, as defined in that report, is the use of the environment and resources to meet the needs of the present without compromising the ability of future generations to meet their needs. Sustainability does not require that ecosystems remain unchanged, which would be impossible in a rapidly changing world. Moreover, we cannot know with certainty what future generations will want. Sustainability simply requires that the productive base available to future generations be sustained

in ways that provide them with the opportunities to make their own choices.

The productive base on which society depends has both ecological and socioeconomic dimensions. Sustainability requires that the total **capital**, or productive base (assets), of the system be sustained. This capital has natural, built (manufactured), human, and social components (Arrow et al. 2004). **Natural capital** consists of both nonrenewable resources (e.g., oil reserves) and renewable ecosystem resources (e.g., plants, animals, and water) that support the production of goods and services on which society depends (Daily 1997). **Built capital** consists of the physical means of production beyond that which occurs in nature (e.g., tools, clothing, shelter, dams, and factories). **Human capital** is the capacity of people to accomplish their goals; it can be increased through various forms of learning. Together, these forms of capital constitute the **inclusive wealth** of the system, i.e., the productive base (assets) available to society (Dasgupta 2001, Chapin et al. 2009). Although not included in the formal definition of inclusive wealth, **social capital** is another key societal asset. It is the capacity of groups of people to act collectively to solve problems (Coleman 1990). Components of each of these forms of capital change over time. Natural capital, for example, can increase through improved management of ecosystems, including restoration or renewal of degraded ecosystems or establishment of networks of marine protected areas; built capital through investment in bridges or schools; human capital through education and training; and social capital through development of new partnerships to solve problems. Increases in this productive base constitute **genuine investment**. **Investment** is the increase in the quantity of an asset times its value. Sustainability requires that genuine investment be positive, i.e., that the productive base (inclusive wealth) *not* decline over time (Arrow et al. 2004). This provides an objective criterion for assessing whether management is sustainable.

To some extent, different forms of capital can **substitute** for one another, for example natural wetlands can serve water purification functions that might otherwise require the construction of expensive water-treatment facilities. Well-informed leadership may be able to implement cost-effective solutions to a given problem (a substitution of human capital for economic capital). There are, however, limits to the extent to which different forms of capital can be substituted. Water and food, for example, are **essential** for survival, and no other forms of capital can completely substitute for them. They therefore have extremely high value to society when they become scarce. Similarly, other forms of capital cannot readily compensate for declines in the capacity of agricultural soils to retain enough water for crop production, the presence of species that pollinate critical crops, a sense of cultural identity, or the trust that society has in its leadership. Losses of many forms of human, social, and natural capital are especially problematic because of the impossibility or extremely high costs of providing appropriate substitutes (Folke et al. 1994, Daily 1997). We therefore focus particular attention on ways to sustain these critical components of capital, without which future generations cannot meet their needs (Arrow et al. 2004).

Well-informed managers often have guidelines for sustainably managing the components of inclusive wealth. For example, harvesting rates of renewable natural resources should not exceed regeneration rates; waste emissions should not exceed the assimilative capacity of the environment; nonrenewable resources should not be exploited at a rate that exceeds the creation of renewable substitutes; education and training should provide opportunities for disadvantaged segments of society to improve their quality of life (Barbier 1987, Costanza and Daly 1992, Folke et al. 1994). These guidelines provide a framework for many of the practical decisions faced by ecosystem managers.

The concept of maintaining positive genuine investment as a basis for sustainability is important because it recognizes that the capital assets of social–ecological systems inevitably change over time and that people differ through time and across space in the value that they place on different forms of capital. If the productive base of a system is sustained, future generations can make their own choices about how best to meet their needs. This defines criteria for deciding whether certain practices are sustainable in a changing world. There are substantial challenges in measuring changes in various forms of capital, in terms of both their quantity and value to society. Nonetheless, the best current estimates suggest that manufactured and human capital have increased in the last 50 years in most countries but that natural capital has declined as a result of depletion of renewable and nonrenewable resources and through pollution and loss of the functional benefits of biodiversity (Arrow et al. 2004; MEA 2005). In some of the poorer developing nations, the loss of natural capital has been larger than increases in manufactured and human capital, indicating a clearly unsustainable pathway of development (MEA 2005).

Ecological Dimensions of Sustainability

Ecosystem services provide a pragmatic framework for managing ecological sustainability. Although natural capital is a fundamental

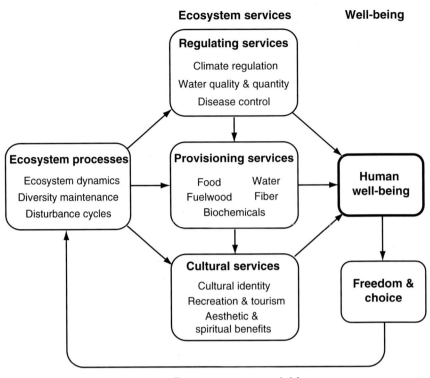

Ecosystem services   **Well-being**

Fig. 15.4 Linkages among ecosystem processes, ecosystem services, and the well-being of society, a framework developed by the Millennium Ecosystem Assessment (MEA 2005). Ecosystem processes are the foundation for ecosystem services that are directly used by society and which strongly influence human well-being. Human actions influence the life-support system of the planet through effects on environment (e.g., climate) and on ecosystems. Redrawn from Chapin (2009)

measure of the capacity of ecosystems to meet society's needs over the long term, it provides little guidance to ecosystem managers seeking to address specific social–ecological issues such as multiple-use forestry or ocean management. Different ecosystem configurations may reflect similar levels of natural capital but provide different patterns of **ecosystem services**, the benefits that society derives from ecosystems. Ecosystems provide well-recognized **provisioning services (goods)**, including water, timber, forage, fuels, medicines, and precursors to industrial products that are harvested from ecosystems. Ecosystems also provide **regulatory services** such as recycling of water and chemicals, mitigation of floods, pollination of crops, and cleansing of the atmosphere, as well as **cultural services** that meet recreational, aesthetic, and spiritual needs (Fig. 15.4; Table 15.1; Daily 1997, MEA

2005). All of these services depend on ecosystem processes, sometimes known as supporting services. These processes include biogeochemical cycles, diversity maintenance, and disturbance cycles. Unless these underlying ecosystem properties are maintained, other services that are more directly recognized and valued by society cannot be sustained.

The overuse or misuse of resources can alter the functioning of ecosystems and the services they provide. Land-use change, for example, can degrade the capacity of watersheds to purify water, leading to large water-treatment costs to cities (Grove 2009). Degradation and loss of wetlands can expose communities to increased damage from floods and storm surges (Kates et al. 2006). Decimation of populations of insect pollinators has reduced yields of many crops (Ricketts et al. 2004). Introductions and invasions of non-native

Table 15.1 General categories of ecosystem services and examples of the societal benefits that are most directly affected. Modified from Chapin (2009)

Ecosystem services	Direct benefits to society
Ecosystem processes (supporting services)	
Maintenance of soil resources	Nutrition, shelter
Water cycling	Health, waste management
Carbon and nutrient cycling	Nutrition, shelter
Maintenance of disturbance regime	Safety, nutrition, health
Maintenance of biological diversity	Nutrition, health, cultural integrity
Provisioning services	
Fresh water	Health, waste management
Food and fiber	Nutrition, shelter
Fuelwood	Warmth, health
Biochemicals	Health
Genetic resources	Nutrition, health, cultural integrity
Regulating services	
Climate regulation	Safety, nutrition, health
Erosion, water quantity/quality, pollution	Health, waste management
Disturbance propagation	Safety
Control of pests, invasions, and diseases	Health
Pollination	Nutrition
Cultural services	
Cultural identity and cultural heritage	Cultural integrity, values
Spiritual, inspirational, and aesthetic benefits	Values
Recreation and ecotourism	Health, values

species such as killer bees, fire ants, and zebra mussels, through the actions of humans, cause enormous damage to living resources and threaten human health (Patz et al. 2005, Díaz et al. 2006). Human activities also indirectly affect ecosystem goods and services through changes in the atmosphere, hydrologic systems, and climate (see Chap. 14).

Management decisions often involve choices that reflect tradeoffs among ecosystem services. Forest harvest, for example, yields forest products at the expense of recreational opportunities provided by the uncut forest. Policies that enhance recreational values to snow machine users may diminish their value to cross-country skiers. Deforestation of tropical forests may provide local users with both forest products and land to support agriculture but degrade soils in ways that diminish the livelihood opportunities for future generations. An important step in ecosystem management is to assess potential impacts of decisions on *multiple* ecosystem services. This is challenging, given the huge number of services

provided by ecosystems and uncertainties in their responses to a particular action (Table 15.1; Box 15.2). It is often pragmatic to focus particularly on a few **critical ecosystem services**, those services that are most vulnerable to change, have fewest options for technological or ecological substitution, and are most valued by society (A. Kinzig, personal communication).

Scenarios of likely outcomes enable managers and other **stakeholders** (people who are affected by outcomes) to compare the effects on ecosystem services of alternative policy options (Peterson et al. 2003, Carpenter et al. 2006). Zoning decisions about development options on lakeshore property, for example, influence not only the type of development that is likely to occur, but also pollutant levels, fish stocks, and the recreational opportunities of current and future users.

Once critical ecosystem services are identified and their likely responses to particular actions are estimated, people are still faced with difficult choices between alternative uses of the environment. Should a wetland be preserved for its cultural

Box 15.2 Assessing Tradeoffs Among Ecosystem Services: Hydropower Versus Conservation in New Zealand

Most policy choices that influence ecosystems involve tradeoffs among different balances of ecosystem services. The political controversy that developed over policies related to the balance between hydropower development and conservation in Fiordland, New Zealand illustrate the role that ecosystem management can play in assessing tradeoffs and negotiating favorable social–ecological outcomes (Mark et al. 2001). The New Zealand Government agreed in 1963 to build a hydropower facility to provide the electricity to a multi-national aluminum smelter that the government considered important to diversifying the national economy, providing local employment, and reversing a population drift within New Zealand. This involved diverting New Zealand's second-largest river (Waiau River) through a tunnel via a hydroelectric station into a pristine marine sound in Fiordland. In order to maximize power generation, the government planned a second phase that would raise by up to 24 m the levels of one of the two major lakes that are the main gateway to Fiordland National Park, New Zealand's largest national park and a World Heritage Site.

Public concern over the ecological and aesthetic implications of raising the lake levels to meet the power demand of the smelter eventually led to ecological studies that documented the consequences of greatly exceeding the maximum historical lake levels (tree mortality) or minimum lake level (lake shore slumping). These findings of substantial and highly detrimental ecological consequences of lake-level manipulation led to a petition signed by about 10% of the nation's population demanding that the hydroelectric contract with the smelter be renegotiated to avoid lake raising and to minimize environmental impacts. The debate over this issue led to a change in central government and establishment of a group of Lake Guardians to recommend ecological and engineering guidelines to minimize environmental impacts and meet industry's power needs within the normal range of lake-level variation. In the context of these findings, new legislation stipulated that this hydroelectric project must manage the water level sustainably. Under this legislation, the government then assembled about 20 stakeholder groups to oversee renegotiation of the resource management of water-related ecosystem services. After 5 years of negotiation and collection of additional information, consensus was reached about water management to maximize ecological integrity and provide acceptable levels of electricity to the smelter. These included maintaining lake levels within their natural historic limits, guaranteed minimum flow of the Waiau River to restore habitat for fish and other biota, restoration of wetlands that had been modified by previous river management, compensation to local indigenous (Maori) peoples for loss of traditional food resources, and maximizing power production within these constraints. The final negotiated agreement sustained most of the ecosystem services that had been discussed and was not contested by any of the 20 stakeholder groups.

This case study illustrates several general issues about ecosystem tradeoffs: (1) Assessing both the ecological and socioeconomic consequences of important policy changes is essential. Decisions that ignore either the ecological or socioeconomic consequences are likely to be unsustainable. (2) Big issues are not easy to resolve and often require enough discussion to develop trust and understanding among user groups. (3) Enduring solutions benefit from long-term environmental monitoring, as well as input and negotiation among multiple users committed to achieving a compromise that is mutually acceptable.

and aesthetic assets, used for sewage treatment, or drained and converted to agriculture? Which services should freshwater systems be managed for? Individuals and societies are constantly making decisions about how to use ecosystem goods and services. These decisions, however, often emphasize short-term economic benefits and assume that ecosystem services that might be lost are "free" and therefore have zero cost if they are degraded (Daily et al. 2000).

Valuation of ecosystem services is one way to organize information to help inform such decisions (Daily et al. 2000). Valuation of ecosystem services requires sound ecological information and a clear understanding of alternatives and impacts. Ecological understanding is critical, for example, to characterize the services provided by ecosystems and the processes by which they are generated. This information is often site specific, so local and traditional ecological knowledge is needed. Ecological and economic information must then be integrated to make sound decisions.

In some cases, the economic worth of ecosystem services can be estimated directly from market values of lands or products or from costs that are avoided by retaining the service (e.g., avoided cost of water treatment by retaining wetlands). In other cases, surveys or other indirect approaches are required that assess the values that people place on alternative outcomes (Goulder and Kennedy 1997). Once estimated, the economic values of ecosystem services (or costs of their degradation) can be considered explicitly in decisions that influence sustainability. The protection of highly valued and well-understood services (such as clean water) through the protection of ecosystems is increasingly viewed as a wise alternative to expensive construction and engineering projects (Box 15.3). With increasing knowledge, the benefits of protecting the less-known ecosystem services will become more widely recognized.

We address human dimensions of sustainability later in the context of managing social–ecological systems.

Box 15.3 Water Purification for New York City

New York City has a long tradition of clean water. This water, which originates in the Catskill Mountains, was once bottled and sold because of its high purity. In recent years, the Catskills natural ecological purification system has been overwhelmed by sewage and agricultural runoff, causing water quality to drop below accepted health standards. The cost of a filtration plant to purify this water was estimated at $6–$8 billion in capital costs, plus annual operating costs of $300 million, a high price to pay for what once could be obtained for free (NRC 2000, Pires 2004).

This high cost prompted investigation of the cost of restoring the integrity of the watershed's natural purification services. The cost of this environmental solution was approximately $1 billion to purchase and halt development on critical lands within the watershed, to compensate landowners for restrictions on private development, and to subsidize the improvement of septic systems. The huge cost savings provided by ecosystem services was selected by the city as the preferred alternative. This choice provided additional valuable services including flood control and sequestration of carbon in plants and soils.

Conceptual Framework for Ecosystem Management

Ecosystem management seeks to sustain or enhance the functional properties of ecosystems that support biodiversity and the ecosystem services on which society depends. Given the continual changes (and often directional trends) in the interactive controls that regulate ecosystem processes, it is more practical to manage ecosystems for sustainability of *general properties* such as productive potential and resilience to change than to attempt to prevent all fluctuations and changes. Soil resources, biodiversity, and disturbance regimes are interactive controls that are often affected by human activities and have particularly strong effects on ecosystems and the services they provide (see Chap. 1).

Sustaining Soil Resources

Soils and sediments are key slow variables that regulate ecosystem processes by providing resources required by organisms. The controls over the formation, degradation, and resource-supplying potential of soils and sediments are therefore central to sound ecosystem management and to sustaining the natural capital on which society depends (see Chap. 3). The quantity of soil in an ecosystem depends largely on the balance between inputs from weathering or deposition and losses from erosion. In addition, organisms, especially plants, add organic matter to soils through death of tissues and individuals, which is offset by losses through decomposition. In general, the presence of a plant canopy and litter layer minimizes erosion by reducing the impact of raindrops on the surface soil and the resulting decline in water infiltration. Human activities that reduce vegetation cover can increase erosion rates by several orders of magnitude, causing soils that may have accumulated over thousands of years to be lost in years to decades. This constitutes an essentially permanent loss of the productive capacity of ecosystems. Similarly, human modification of river channels can alter sediment inputs to floodplains and deltas. In the southern U.S., for example, loss of sediment inputs and subsequent soil subsidence led to the disappearance of barrier islands that had previously protected New Orleans from hurricanes (Box 15.1).

Fine particles such as clay and organic matter are particularly important in water and nutrient retention (Chap. 3). They are typically concentrated near the soil surface, where they are vulnerable to loss by erosion. Human activities that foster wind and water erosion, such as deforestation, overgrazing, plowing, or fallowing of agricultural fields, therefore erode the water- and nutrient-retaining capacity of soils much faster than the total loss of soil volume might suggest. Preventing even modestly augmented erosion rates is therefore critical to sustaining the productive capacity of terrestrial ecosystems.

Accelerated soil erosion is one of the most serious causes of global declines in ecosystem services. The erosional loss of fine soil particles is a direct cause of **desertification**, soil degradation that occurs in drylands (Stafford Smith et al. 2009). Desertification can be triggered by drought, reduced vegetation cover, overgrazing, or their interactions (Reynolds and Stafford Smith 2002, Foley et al. 2003a). When drought reduces vegetation cover, for example, goats and other livestock graze more intensively on the remaining vegetation. Extreme poverty and lack of a secure food supply often constrain options for reducing grazing pressure at times of drought because short-term food needs take precedence over practices that might prevent erosion. Wetter regions can also experience severe erosional loss of soil, especially where vegetation loss exposes soils to overland flow. The Yellow River in China, for example, transports 1.6 billion tons of sediment annually from agricultural areas in the loess plateau at its headwaters. Similar erosional losses occurred when grasslands were plowed for agriculture in the U.S. during droughts of the 1930s, creating the dustbowl. Management that maintains vegetation cover, particularly in steep terrain and adjacent to streams, can reduce erosion potential substantially, thereby maintaining the productive potential of terrestrial ecosystems.

Soil erosion from land represents a sediment input to lakes and estuaries. At a global scale, the increased sediment input to the ocean from accelerated erosion is partially offset by the increased sediment capture by lakes and reservoirs. Therefore lakes, including reservoirs, and estuaries are the aquatic ecosystems most strongly affected by terrestrial erosion. Especially in agricultural areas, these sediment and nutrient inputs to aquatic ecosystems can be just as problematic as the loss of productive potential on land (see Chaps. 9 and 13).

Sustaining Biodiversity

Biodiversity strongly influences the range of environmental and biotic conditions under which ecosystem processes can be sustained. Diverse ecosystems contain species that sustain a wide range of ecosystem processes (**effect diversity**) through their use and cycling of soil resources. Diverse systems also contain organisms likely to sustain ecosystem services under a wide range of environmental and biotic conditions (**response diversity**; see Chap. 11; Elmqvist et al. 2003, Suding et al. 2008). This delivery of ecosystem services depends on the kinds of species present (functional composition), genetic diversity within species, species diversity within stands, and landscape diversity across regions (Table 15.2).

Biodiversity in ecosystems that have not been strongly modified by human activities tends to "take care of itself." Species diversity represents those species that have reached a particular location, can grow and reproduce in that environment, and survive in the face of competition and predation from other species present. If a species disappears from a particular patch, it might recolonize from adjoining patches. Human activities often, however, radically alter the physical and biotic environment through changes in land use and landscape structure or through introduction or elimination of species that govern competitive and trophic interactions among species (Foley et al. 2005). Introduction of rats on islands that historically had no mammals, for example, eliminates flightless birds and many species of native plants (Towns et al. 2006). Introduction of exotic nitrogen-fixing species into low-nitrogen

environments favors competitive domination by fast-growing weedy species. Predator removal can cause an explosion of herbivore densities that reduce plant diversity. Long-term trends in climate, nutrient deposition, and erosion are now altering the physical environment of the entire planet, altering competitive interactions among species, and often eliminating species that cannot compete effectively under these new conditions. These species losses are occurring much more rapidly than migration or evolution can restore diversity to its former levels. These human effects on biodiversity cumulatively explain why the world is now in the sixth major extinction event in the history of life on Earth (Chapin et al. 2000b). Moreover, loss of species diversity is perhaps the least reversible of the many human-caused global changes. Soil or land cover or the composition of the atmosphere may take thousands of years to return toward its predisturbance state, but extinction is literally forever.

Ecosystem management strongly influences the maintenance or loss of biodiversity. On intensively managed forests or agricultural lands, managers usually deliberately minimize diversity in order to produce uniform stands that can be efficiently managed and harvested. There is a tradeoff, however, between harvest efficiency and the vulnerability of these low-diversity stands to environmental and biotic variability and change (see Chap. 11). These low-diversity stands often require suppression of natural pathogens and disturbances to maintain their productivity. Unintentional human impacts can also alter diversity. Addition of resources such as water or nutrients reduces the number of potentially limiting resources for which plants can compete and therefore the diversity of species that can coexist (Harpole and Tilman 2007).

In less intensively managed ecosystems, biodiversity can be fostered by minimizing the magnitude and extent of novel changes in ecosystems. This reduces the likelihood of loss of species that are well adapted to historical environmental and biotic conditions. For example, minimizing land conversion to agriculture or of fire in tropical forests maintains habitat for native species. Proportional cover of native habitat is a strong predictor of biodiversity in a region. Similarly, preventing the introduction or spread of exotic species

Table 15.2 Examples of biodiversity effects on ecosystem services. We separate the diversity effects into those due to functional composition, numbers of spe-cies, genetic diversity within species, and landscape structure and diversity. Modified from Díaz et al. (2006)

Ecosystem service	Diversity component and mechanism
1. Production by societally important plants	*Functional composition*: (1) fast-growing species produce more biomass; (2) species differ in timing and spatial pattern of resource use (complementarity allows more resources to be used)
	Species number: large species pool is more likely to contain productive species
2. Stability of crop production	*Genetic diversity*: buffers production against losses to pests and environmental variability
	Species number: cultivation of multiple species in the same plot maintains high production over a broader range of conditions
	Functional composition: species differ in their response to environment and disturbance, stabilizing production
3. Maintenance of soil resources	*Functional composition*: (1) fast-growing species enhance soil fertility; (2) dense root systems prevent soil erosion
4. Regulation of water quantity and quality	*Landscape diversity*: intact riparian corridors reduce erosion
	Functional composition: fast-growing plants have high transpiration rates, reducing stream flow
5. Pollination for food production and species survival	*Functional composition*: loss of specialized pollinators reduces fruit set and diversity of plants that reproduce successfully
	Species number: loss of pollinator species reduces the diversity of plants that successfully reproduce (genetic impoverishment)
	Landscape diversity: large, well-connected landscape units enable pollinators to facilitate gene flow among habitat patches
6. Resistance to invasive species with negative ecological/cultural effects	*Functional composition*: some competitive species resist the invasion of exotic species
	Landscape structure: roads can serve as corridors for spread of invasive species; natural habitat patches can resist spread
	Species number: species-rich communities are likely to have less unused resources and more competitive species to resist invaders
7. Pest and disease control	*Genetic diversity or species number*: reduces density of suitable hosts for specialized pests and diseases
	Landscape diversity: provides habitat for natural enemies of pests
8. Biophysical climate regulation	*Functional composition*: determines water and energy exchange, thus influencing local air temperature and circulation patterns
	Landscape structure: influences convective movement of air masses and therefore local temperature and precipitation
9. Climate regulation by carbon sequestration	*Landscape structure*: fragmented landscapes have greater edge-to-area ratio; edges have greater carbon loss
	Functional composition: small, short-lived plants store less carbon
	Species number: high species number reduces pest outbreaks that cause carbon loss
10. Protection against natural hazards (e.g., floods, hurricanes, fires)	*Landscape structure*: influences disturbance spread or protection against natural hazards
	Functional composition: (1) extensive root systems prevent erosion and uprooting; (2) deciduous species are less flammable than evergreens

reduces the likelihood of large-scale biodiversity and ecosystem change (Vitousek 1990). Species that have novel ecosystem effects (e.g., nitrogen fixers or highly flammable species) or that have escaped the diseases and predators that control their populations in sites of origin are particularly likely to have strong impacts on biodiversity. Finally, maintaining natural patterns of disturbance and landscape connectivity sustain populations of all successional stages within a landscape and provide pathways for movement and post-disturbance colonization, as described in the next section.

Sustaining Variability and Resilience

Disturbance shapes the long-term fluctuations in the structure and functioning of ecosystems and therefore their resilience and vulnerability to change. Disturbance is not something that "happens" to ecosystems but is an integral part of their functioning and a key source of temporal and spatial variation in landscapes (see Chap. 12). Species are typically adapted to the disturbance regime that shaped their evolutionary histories. Management that alters this disturbance regime, for example by preventing floods, wildfire, or pest outbreaks, can therefore create conditions to which species are poorly adapted. For example, past efforts to prevent these natural disturbances (e.g., "Smokey-the-Bear" efforts to prevent all wildfires) creates homogeneous patches of late-successional habitat that no longer support early successional species. In addition, late-successional ecosystems are often prone to disease and pest outbreaks (Matson and Boone 1984) that can spread extensively in homogeneous late-successional stands (Raffa et al. 2008). Management that allows small naturally occurring disturbances to occur creates spatial heterogeneity that reduces disturbance spread and therefore the likelihood of large catastrophic disturbances (Holling and Meffe 1996). Allowing small-scale disturbances to occur is often politically challenging, however, because small disturbances sometimes reduce or destroy the economic value of resources that people want to harvest (e.g., forest harvest), create risks in inhabited landscapes (e.g., the wildland–urban interface), or reduce the aesthetic value of familiar patches within a landscape. These tradeoffs are best addressed through long-term social–ecological planning, as discussed later.

The landscape diversity generated by small-scale disturbance creates a mosaic of ecosystems with contrasting structure and species composition. Each stand type is likely to differ in its response to various predictable and unforeseen shocks and disturbances, including historically important disturbances and novel conditions caused by changes in climate, pollution, or novel disturbance regimes (e.g., altered frequency and severity of wildfire or flooding). Thus, just as with genetic or species diversity within stands, landscape diversity fosters *resilience* to both historical and novel disturbances (Table 15.2; see Fig. 12.8; see Chaps. 12 and 13).

Management requires a landscape perspective that considers interactions among ecosystems. A lake cannot be managed sustainably, for example, without considering the nutrient inputs from the surrounding landscape, and forest production can be managed most sustainably as a landscape mosaic by taking account of disturbances such as hurricanes, fire, and logging. The resilience and sustainability of lakes depends on a range of process controls that function at different scales to mitigate the effects of disturbance (Carpenter and Biggs 2009). These process controls include the filtration effects of riparian vegetation and wetlands, the role of game fish in trophic dynamics, and the absorption of nutrients by macrophytes. When these components are intact, landscapes containing lakes can withstand perturbations such as droughts, floods, forest fires, and some land-use change (Turner 2010). Management of landscapes at coarse spatial scales requires different information than management of individual lakes, fields, or forest stands. At coarse spatial scales, monitoring of food webs in lakes is not feasible, so land-use records, remote sensing of lake clarity, knowledge of local residents, and surveys of fishing activity and success provide useful input to models. An important implication of a landscape focus is that it requires the recognition of ecosystem response to multiple driving forces.

Applying Ecosystem Principles to Management

Ecosystem management is the application of ecological science to resource management to promote long-term sustainability of ecosystems and the delivery of essential ecosystem goods and services to society. The concept was adopted by the U.S. Forest Service in 1992 and has since been developing in theory and application, using a set of common principles (Table 15.3). In this section, we illustrate the application of these principles to selected resource management issues.

Table 15.3 Attributes of ecosystem management, based on Christensen et al. (1996)

Sustainability	Intergenerational sustainability is the primary objective
Goals	Measurable goals are defined that assess sustainability of outcomes
Ecological understanding	Ecological research at all levels of organization informs management
Ecological complexity	Ecological diversity and connectedness reduces risks of unforeseen change
Dynamic change	Evolution and change are inherent in ecological sustainability
Context and scale	Key ecological processes occur at many scales, linking ecosystems to their matrix
Humans as ecosystem components	People actively participate in determining sustainable management goals
Adaptability	Management approaches will change in response to changes in scientific knowledge and human values

Forest Management

The challenge for sustainable forestry is to define the attributes of forested ecosystems that are ecologically and societally important and to maximize these ecosystem services in the face of change. Forest managers face management challenges that are, in part, logical consequences of the long-lived nature of forest trees (Szaro et al. 1999, Swanson and Chapin 2009):

Managing forests for multiple ecosystem services involves strong tradeoffs among costs and benefits to different users, with choices having implications for multiple human generations. Forests provide many ecosystem services, including fuel wood, timber products, water supply, recreation, species conservation, and aesthetic and spiritual values. To support these services, nutrient supply rates must be sufficient to support rapid growth, yet not so high that they lead to large nutrient losses or changes in species composition. The rate at which stands are harvested must be balanced with their rate of regeneration after logging. Species diversity typical of natural mosaics of forest stands should be maintained. The sizes and arrangement of logged patches should provide a semi-natural landscape mosaic with dependable seed sources and patterns of forest edges that allow natural use and movement of animal populations (Franklin et al. 1997). Since it is difficult to anticipate the long-term consequences of different management approaches, there are benefits to using multiple approaches in different areas to meet different user needs and to increase the likelihood that some of these approaches will have favorable long-term outcomes (Bormann and Kiester 2004).

Managing forests under conditions of rapid change is challenging because a forest stand is likely to encounter novel environmental and socioeconomic conditions during the life of the individual trees in the stand. Forest ecosystems across the globe face threats from both intentional and inadvertent human impacts, including air pollution, invasive species, and, perhaps above all, global climate change. Because most forest trees will reach maturity under quite different conditions than they begin life, it may be appropriate to reseed forests with a range of genotypes from different climate zones (Millar et al. 2007). This differs from best practices of the past in which locally adapted genotypes of trees were preferred for reforestation.

Forest conversion to new land uses is a state change that is difficult and time consuming to reverse, given the long regeneration time of forest trees and ecosystems. Historic and ongoing land use has converted about 40% of preindustrial forest cover to agriculture, built environments, and plantations of a single or narrowly constrained set of species, often exotics (Shvidenko et al. 2005, Foster et al. 2010). Under other conditions, large-scale agricultural abandonment or increased economic value of forests, as for carbon sequestration or aesthetic benefits, can foster **reforestation** or **afforestation** (the regeneration of forests on recently harvested sites or planting of new forests on previously non-forested sites, respectively).

Fisheries Management

Formulation of management options for fisheries requires an understanding of ecosystem

resilience. Management options include marine reserves, quota systems, new approaches for setting fishing limits based on population sizes of fish stocks, and economic incentives for long-term population maintenance. Unrestricted fish harvest can reduce sustainability by replacing the natural stabilizing (negative) feedbacks to population changes with amplifying (positive) feedback responses that drive harvested populations to low levels (Berkes et al. 2006, Walters and Ahrens 2009). Supply-and-demand economics and government subsidies, for example, often maintain or *increase* fishing intensity when fish populations decline (Ludwig et al. 1993, Pauly and Christensen 1995). This contrasts with the *decreasing* predation pressure that would accompany a decline in prey population in an unmanaged ecosystem (Francis 1990, Walters and Ahrens 2009).

Management of the North Pacific salmon fishery has instituted a stabilizing (negative) feedback on fishing pressure through tight regulation of fishing activity. Commercial and subsistence fishing are allowed only after enough fish have moved into spawning streams to ensure adequate recruitment. This negative feedback to fishing pressure may contribute to the record-high salmon catches from this fishery after 40 year of management (Ludwig et al. 1993, Walters and Ahrens 2009). Sustaining the fishery also requires protection of spawning streams from changes in other interactive controls. These changes include dams that prevent winter floods (disturbance regime), warming of streams by removal of riparian vegetation of logged sites (microenvironment), species introductions (functional types), and inputs of silt and nutrients in agricultural and urban runoff and sewage (nutrient resources).

A common approach to sustainable management is to harvest only the production in excess of that which would occur when the fish stock is limited by density-dependent mortality, termed **surplus production** (Rosenberg et al. 1993, Hilborn et al. 1995). The existence and magnitude of surplus production depends on whether the remaining fish increase their growth rate or reproductive success when some fish are harvested. This in turn depends on the stability of interactive controls (e.g., physical environment, nutrients, and

predation pressure) and the extent to which these interactive controls respond to changes in fisheries stocks. The major challenge in fisheries management is to estimate surplus production in the face of fluctuating interactive controls and uncertainty in the relationship between these controls and the fish population size. Fisheries biologists actively debate whether any ecosystem is sustainable when subjected to continuous human harvest (Ludwig et al. 1993, Rosenberg et al. 1993, Walters and Ahrens 2009).

Ecosystem Renewal

Ecosystem renewal often benefits from the introduction of amplifying (positive) feedbacks that push the ecosystem to a new, more desirable state. Many ecosystems become degraded through a combination of human impacts, including soil loss, air and water pollution, habitat fragmentation, water diversion, fire suppression, and introduction of exotic species. In degraded agricultural systems and grazing lands, the challenge is to restore them to a productive enough state to provide goods and services to people. In other cases, the goal is to restore the natural composition, structure, processes, and dynamics of the original ecosystem (Christensen et al. 1996). Advances in restoration practices involve identifying the impediments to recovery of ecosystem structure and function and overcoming these impediments with artificial interventions that often use or mimic natural processes and interactive controls (Meffe et al. 2002).

Interventions can be applied to any component of ecosystems, but hydrology, and soil and plant community characteristics are commonly the focus of effort (Box 15.4; Dobson et al. 1997, Meffe et al. 2002). Low soil fertility and organic content are common problems in heavily managed agricultural and pasture systems and in forests or grasslands reestablishing on mine wastes. Fertilizers and nitrogen-fixing trees can restore soil nutrients and organic inputs (Bradshaw 1983). Once soil characteristics are appropriate, plant species can be reintroduced by seeding, planting, or natural immigration (Dobson et al. 1997). The scientific basis for restoration ecology

Box 15.4 Everglades Restoration Study

Major human impacts on the natural hydrology of the Everglades ecosystem in the southeastern U.S. began in the early twentieth century. In response to hurricanes, flooding, and the resulting loss of human life and property, the U.S. Army Corps of Engineers built levees, canals, pumping stations, and water control structures that separated the remaining Everglades from growing urban and agricultural areas (Davis and Ogden 1994). The water flow to the remaining "natural" Everglades declined sharply and occurred as pulses of nutrient-rich agricultural and urban runoff regulated by water-control structures. These hydrologic changes caused pronounced fluctuations in water levels and increased the frequency of major drying events (DeAngelis et al. 1998). The survival of many species, including birds, alligators, and crocodiles, depends on reasonable regularity in the rise and fall of water level throughout the year. Since the 1940s the nesting populations of wading birds declined by 90% (Davis and Ogden 1994). Land-use change such as agricultural drainage destroyed many high-elevation, short-hydroperiod wetlands. Eutrophication altered competitive interactions and increased the impacts of invasive species (DeAngelis et al. 1998).

The goals of South Florida ecosystem restoration program include the maintenance of ecological processes such as disturbance regimes, hydrologic processes, and nutrient cycles and maintenance of viable populations of all native species. The U.S. Army Corps of Engineers was charged with both improving protection of Everglades National Park and providing enough water to meet the demands of a large urban and agricultural economy. Planned construction projects include the creation of storm-water treatment areas to remove phosphorus from the water and to allow increased water diversion into the Everglades (DeAngelis et al. 1998). Additional land is being purchased to provide areas of water storage and a buffer zone between natural areas and the expanding urban zone.

An ecosystem model was developed to evaluate alternative rehabilitation and management options. This spatially explicit landscape model was linked with individual-based modeling of ten higher trophic-level indicator species to provide quantitative predictions relevant to the goals of the Everglades Restoration (DeAngelis et al. 1998). These indicator species, including the Florida panther, white ibis, and American crocodile, differ in their use of the landscape and resources and span a range of habitat needs and trophic interactions (Davis and Ogden 1994). The simultaneous success of all of these species in a restored Everglades would imply health of the overall ecosystem (DeAngelis et al. 1998). The program has adopted a hierarchical modeling approach in which models of higher trophic-level indicator species use information from models at intermediate trophic levels (fish, aquatic macroinvertebrates such as crayfish, and several reptile and amphibian functional types) and lower trophic levels (periphyton, aggregated mesofauna, and macrophytes). These species-specific models are then layered on a landscape Geographic Information System (**GIS**) model that includes hydrologic and abiotic factors such as surface elevations, vegetation types, soil types, road locations, and water levels (DeAngelis et al. 1998). South Florida provides an example of the incorporation of scientific knowledge of ecosystem processes into long-term state and national ecosystem management efforts.

is actively developing and exploring new challenges (Young et al. 2005). In a rapidly changing world, for example, it may be more practical to target renewal efforts toward ecosystem types that are compatible with emerging climate conditions rather than attempting to restore ecosystems to a historical state that is increasingly out of equilibrium with its environment (Harris et al. 2006, Choi 2007, Hobbs and Cramer 2008). In this context, **renewal ecology** rather than **restoration ecology** may be the most appropriate framework.

Management for Endangered Species

Management for endangered species requires a landscape perspective. The focus of endangered-species protection has generally been the establishment of protected areas containing populations of the target species and vegetation associated with those species. Establishment of parks is, however, insufficient protection for species when people continue to influence important state factors and interactive controls, such as climate, fire regime, water flows, or species introductions (Hobbs et al. 2010). If climate changes, for example, animals may be trapped inside a park that no longer has a suitable climate or vegetation. Selection of parks that have a range of elevations provides an opportunity for organisms to migrate vertically to higher elevations in response to climate warming. Habitat fragmentation and land-use change also alter the natural linkages among ecosystems inside and outside of parks. Nearly all parks therefore require management to compensate for human impact. The boundaries of Yellowstone National Park, for example, block migration of elk to traditional wintering areas, so winter food supplements must be provided. These winter food supplements in combination with the extirpation of natural predators release the elk population from their natural population controls. Managers must therefore allow hunting or relocation of elk as an alternative mechanism of population regulation. Using intensive management to replace interactive controls, rather than working to sustain the interactive controls, is an expensive, complex, and difficult task, especially when the management has multiple, often conflicting goals (Beschta and Ripple 2009).

Socioeconomic Contexts of Ecosystem Management

Effective ecosystem management requires an integrated social–ecological framework to understand and manage for sustainability in a changing world. Ecological sustainability cannot be divorced from economic and cultural sustainability. A policy that promotes ecological sustainability at the expense of its human residents cannot be effectively implemented or sustained. Conversely, programs of economic development that sacrifice long-term ecological or cultural sustainability cannot be sustained over the long term. An emerging challenge is to address regional sustainability in ways that simultaneously consider the ecological, economic, and cultural costs and benefits of particular policies (Berkes et al. 2003, Clark and Dickson 2003, Turner et al. 2003a, Chapin et al. 2009). Design and implementation of policies that foster social–ecological sustainability require close collaboration among many groups, including ecologists, economists, sociologists, anthropologists, policy makers, resource managers, landowners, and industrial and recreational users (Armitage et al. 2007). This comprehensive social–ecological approach (**ecosystem stewardship**) uses both the ecological principles outlined in this book and the principles and understanding developed in many fields of social science. Its objectives, scale, and roles for science and management differ significantly from more traditional management approaches (Table 15.4).

Meeting Human Needs and Wants

Success of ecosystem management depends on the capacity of ecosystem services to meet human needs. Human well-being, or quality of life, reflects a hierarchy of human needs (Maslow 1943): Basic physiological needs such as food and water are the most fundamental, followed by perceptions of safety and security, then sense of belonging through social connections with family and community, then the need for self-esteem and the respect of others, and finally, self-fulfillment through creative actions and efforts to correct social and environmental injustices. Opportunities for social–ecological sustainability increase as more of Maslow's components of well-being are met. People who lack the resources to meet their basic needs for survival will use local ecosystems to meet these needs, regardless of longer-term consequences. As the hierarchy of human needs is increasingly fulfilled, the opportunities for sustainability are thought to improve. However, many societies that have traditionally depended directly on local harvest actively seek to sustain

Table 15.4 Differences between steady-state resource management and ecosystem stewardship. Modified from Chapin et al. (2010)

Characteristic	Steady-state resource management	Ecosystem stewardship
Reference point	Historic condition	Trajectory of change
Central goal	Ecological integrity	Sustain social–ecological systems and delivery of ecosystem services
Predominant approach	Manage resource stocks and condition	Manage stabilizing and amplifying feedbacks
Role of uncertainty	Reduce uncertainty before taking action	Embrace uncertainty: maximize flexibility to adapt to an uncertain future
Role of research	Researchers transfer findings to managers who take action	Researchers and managers collaborate through adaptive management to create continuous learning loops
Role of resource manager	Decision maker who sets course for sustainable management	Facilitator who engages stakeholder groups to respond to and shape social–ecological change and nurture resilience
Response to disturbance	Minimize disturbance probability and impacts	Disturbance cycles used to provide windows of opportunity
Resources of primary concern	Species composition and ecosystem structure	Biodiversity, well-being, and adaptive capacity

their lands, even when Maslow's hierarchy of needs is only modestly met (Dietz et al. 2003, Agrawal et al. 2008, Berkes et al. 2009).

People often consume more resources than are essential to meet their basic needs. Below a per capita income of about $12,000, the wealth of nations correlates closely with the average happiness of their citizens (Diener and Seligman 2004). Similar correlations are observed within countries. Happiness does not significantly increase, however, once an individual's basic material needs are satisfied (Easterlin 2001). As people acquire greater wealth above this level, they aspire to achieve even greater wealth, which, in turn, seems to reduce their happiness and overall satisfaction. These findings suggest two basic approaches to achieving a more sustainable match between the flow of ecosystem services and the material needs of society: (1) assure that the basic material needs of poor people are met and (2) reduce the upward spiral of consumption by people whose material needs are already met.

Managing Flows of Ecosystem Services

Economic costs and benefits strongly influence the sustainability of ecosystem management, as discussed earlier in the context of valuation of ecosystem services. To recap briefly, ecosystem services whose value is uncertain or unknown tend to be undervalued in decisions relative to commodities like wood or fish that can be bought or sold. In addition, greater value is often given to resources that provide immediate benefit than to those resources that are saved for future generations. In traditional economic terms, the value of goods and services received in the future is **discounted** (reduced) by a percentage that reflects the **opportunity cost** (alternative investment) of conserving ecosystem services for the future. Some economists argue that the discount rate of sustaining ecosystem services for use by future generations should be zero, if their use today reduces the capacity of future generations to meet their needs (Heal 2000). Harvest of an old-growth forest, for example, might prevent future generations from enjoying the biodiversity benefits of these forests for several centuries. Ecosystem ecologists can play an important role in decisions involving tradeoffs between present and future generations by documenting the sensitivity of ecosystem services to alternative management actions and their subsequent rates of renewal.

Natural resources that are privately owned and sold in the market place are often challenging to manage sustainably because long-term benefits are likely to be strongly discounted, resulting in greater consideration of short-term costs and benefits. This is particularly true in areas undergoing land development, where rising property values and taxes increase the economic incentives to sell land for development. Privately held timber or ranch lands, for example, are often sold to real estate developers. These land developments not only modify the ecosystem services provided by these lands but also constrain options for maintaining natural disturbance regimes on nearby public lands. Innovative arrangements such as the sale of conservation or agricultural easements, however, allow individuals to continue current land uses at rural tax rates (Ginn 2005, Sayre 2005, Foster et al. 2010). Sometimes, however, decisions continue to follow historical patterns because the **transaction costs** of the time and effort required to learn, negotiate, and enforce new ways of doing things outweigh the benefits to the individual of novel sustainable solutions (Kofinas 2009). Ecosystem managers can sometimes reduce these transaction costs by facilitating the negotiation of conservation and agricultural easements or implementation of other novel sustainable solutions. More generally, policies that align economic incentives with sustainability goals greatly improve the opportunities for sustainable resource use, as in the conservation easements described above. Alternatively, maladaptive subsidies that encourage unsustainable behavior, such as subsidies to fishermen and loggers to maintain harvesting effort when stocks decline below economically profitable levels reduce the likelihood of sustainable resource use.

Publicly owned natural resources are often managed by government agencies whose responsibility is to manage certain flows of ecosystem services. Agencies sometimes prioritize specific ecosystem services. In the U.S., for example, many state Departments of Fish and Game or Forestry prioritize the maximum or optimum sustained yield (MSY or OSY, respectively) and efficient production of the natural resources for which they are responsible. In principle, this should allow sustainable use of these resources over the long term. Despite its sustainability goal, management for MSY or OSY tends to overexploit targeted resources because of overly optimistic assumptions about the capacity to sustain productivity, avoid disturbances, regulate harvesters' behavior, and anticipate extreme economic or environmental events (Holling and Meffe 1996) and ignore the many other benefits that those lands might provide under different management. Ecosystem management that emphasizes multiple use through the delivery of a broader range of ecosystem services is challenging to implement because of tradeoffs among alternative uses. Local timber-based communities, urban residents, and national conservation groups, for example, generally differ in the value placed on different combinations of ecosystem services. Ecosystem ecologists can contribute to well-informed multiple-use resource management by identifying the controls and trends in supporting, provisioning, regulating, and cultural services resulting from different management practices (Meffe et al. 2002).

Cultural services provided by ecosystems often motivate sustainable use. Many traditional societies maintain a spiritual or cultural respect for the species and processes that characterize the lands and waters from which they derive their livelihoods (Berkes 2008, Berkes et al. 2009). Some communities maintain sacred groves that meet spiritual needs but also serve as reservoirs of biodiversity that provide seeds and pollinators for surrounding lands (Ramakrishnan 1992, Brown 2003, Tengö et al. 2007). Many ranchers, farmers, fishermen, and urban residents also value the ecosystem services provided by the lands and waters that they use. Provided the right circumstances, local residents can be articulate spokespersons and stewards for sustainable management of these lands (Armitage et al. 2007). On the other hand, local people can also be outspoken advocates of unsustainable harvest policies. The challenge for ecosystem management is to find a balance of uses that supports local livelihoods at a level that is sustainable over the long term. In many cases, people have a **sense of place** for the lands and waters where they grew up or live. This can be as (or more) powerful than economic incentives in motivating sustainable use of land. Ecosystem ecologists can support this sense of place through engagement of local residents in citizen science and education to learn about the places where they live or of visitors who may value distant places for aesthetic or other reasons. Ecosystem ecologists also have much they can learn about the places they manage from local people based on their observations and cultural knowledge (Berkes 2008).

Sustainable ecosystem management is not restricted to private or publicly owned lands. Resource-dependent societies often sustainably manage natural resources that they hold in common (common-pool resources) even in the absence of private property or government regulation. A variety of informal rules for managing **common-pool resources** have evolved in different societies (Dietz et al. 2003, Ostrom 2009). Management of common-pool resources is most likely to be sustainable when

- The resource used in common occurs within clearly defined boundaries and is managed by resource users (e.g., water in a watershed or fish in a coral reef)
- The benefits that users receive are proportional to labor and costs that users spend in sustaining and harvesting the resource
- Users participate in forming and modifying the rules so that no outsider can make arbitrary rules that determine the distribution of the resource among users
- Users (or their representatives) monitor resource use to make sure that no individual harvests more than their share
- Users who violate harvest rules are punished in proportion to the seriousness of the offense
- Users have easy ways to resolve conflicts
- Users have the right to organize if they are dissatisfied with the way the resource is managed

None of these conditions is essential or guarantees that resource use will be sustainable, but each condition increases the likelihood of sustainable resource use. Examples of apparently sustainable management of common-pool resources for decades to centuries include lobster harvest in Maine, subsistence fisheries in many parts of the world, harvest of hay in Switzerland and of bamboo in Japan (Ostrom 1990). There are also many examples of *unsustainable* management of common-pool resources. The circumstances that influence the success or failure of these informal

management systems usually depend on local conditions and history (Ostrom 2007). A corrupt leader, for example, can undermine a system that might otherwise work well. Similarly, privatization or government efforts to regulate locally managed common-pool resources can disrupt sustainable patterns of local control and use. **Open access**, in which common-pool resources can be harvested by anyone without restrictions, as in many open-ocean fisheries, creates conditions that are least likely to allow sustainable management – the tragedy of the commons (Hardin 1968, Berkes et al. 2006).

Addressing Political Realities

Many of greatest challenges faced by resource managers reflect the social and political environment in which they work. Differences among users in goals and values, power relationships, regulatory and financial constraints, personalities, and other social and political factors often dominate the day-to-day challenges faced by resource managers. Social processes are therefore an integral component of ecosystem management. At times of rapid social or environmental change, frameworks for managing natural resources may become dysfunctional, requiring communication with a broader set of users and managers and openness to new ways of doing things. Management organizations that have become bureaucratic may be resistant to change or slow to adjust. Power hierarchies within these organizations may either facilitate or inhibit efforts to manage ecosystems for multiple ecosystems services during times of change.

Political awareness is crucial to ecosystem ecologists who wish to inform policies for sustainability. Only a small fraction of scientific research actually influences policy (Clark and Holliday 2006, Kristjanson et al. 2009). To be effective, science must, first and foremost, be **credible** in the sense that it is "good science" that is grounded in understanding and facts rather than in arguments of how the world should work. Second, it must be viewed as **legitimate** (unbiased and respectful) by multiple user groups rather than being seen as the agenda of a single advocacy group. This often requires extensive engagement and dialogue with multiple user groups who may have different concerns and views on preferred policy outcomes. Finally, science is most effective when it is **salient**, i.e., presented to the right people at the right time. Scientists often publish their findings in scientific journals without making the extra effort to present it to those managers who are most likely to use the information in a form that is useful for decision making. Also, reports that are published after the window for policy change has closed will not have much policy impact. Some rules of thumb for linking scientific knowledge with action include (Clark and Holliday 2006, Kristjanson et al. 2009):

- Joint definition by scientists and users of the research problems to be addressed.
- Research dialogue and management that leads to "use-inspired science," i.e., science that improves basic understanding while seeking to provide information that solves problems (Stokes 1997).
- Engagement of boundary organizations (e.g., nongovernmental organizations) that help bridge communication gaps between researchers and policy makers.
- Systems framework that recognizes scientific research as just one piece of a broad set of social–ecological considerations.
- Research designed to facilitate learning rather than knowledge production; this often entails greater risk-taking than most scientific research.
- Emphasis on capacity building with flexibility, often involving networks that develop new strategies and develop local capacity for action.
- Manage asymmetries of power to level the playing field among multiple producers and users of knowledge; this may require reaching out to potentially disenfranchised users.

Innovation and Adaptive Management

Adaptive management, involving experimentation in the design and implementation of policies, is central to effective management of ecosystems. It involves "learning by doing."

An **adaptive policy** is one that is designed from the outset to test hypotheses about the ways in which ecosystem behavior is altered by human actions. In this way, if the policy fails, learning occurs, so better policies can be applied in the future. Perhaps as a result of frequent management failures and gaps in scientific knowledge, the concept of adaptive management has become central to the implementation of ecosystem management. One advantage of adaptive management stems from the high degree of uncertainty in real-life complex systems (Levin 1999). Instead of delaying timely action due to the lack of certainty, adaptive management provides the opportunity to learn from management experience. The lack of action in the face of uncertainty is a management decision, and it can have ecosystem and societal consequences that are at least as detrimental as actions based on reasonable hypotheses about how ecosystems function. Hypotheses that underlie adaptive management might consider the probabilities of both desired outcomes and ecological disasters (Starfield and Bleloch 1991). A preferred policy, for example, may be one that has a moderate probability of desirable outcomes and a low probability of causing an ecological disaster.

Adaptive management can be applied to both big questions and small ones. **Single-loop learning**, for example, adjusts actions needed to meet a previously agreed-upon management goal, such as changes in harvest levels needed to sustain populations of a particular fish or tree species (Fig. 15.5). **Double-loop learning**, however, requires that managers evaluate the approach they have used previously before taking further action, for example assessing the costs and benefits of managing forests for multiple ecosystem services rather than for a single product (e.g., trees; Armitage et al. 2007, Kofinas 2009).

Double-loop learning requires "out-of-the-box" thinking and innovation. Approaches to stimulating innovation vary with social context (Westley et al. 2006). When conditions are static and management is relatively rigid, innovation can be stimulated by facilitating communication among different levels in the hierarchy (e.g., among practitioners and mid- and upper-level managers) about the nature of problems and

Fig. 15.5 Single- and double-loop learning. Single-loop learning involves changing actions to meet identified management goals (e.g., modifies harvest rate to conform to specified catch limits), often through trial and error. Double-loop learning includes a reflection process of evaluating underlying assumptions and models that are the basis of defining problems (e.g., revising the indicators and simulation models used to calculate the relationship between fertilizer inputs and crop production based on recent policy outcomes). Reprinted from Kofinas (2009)

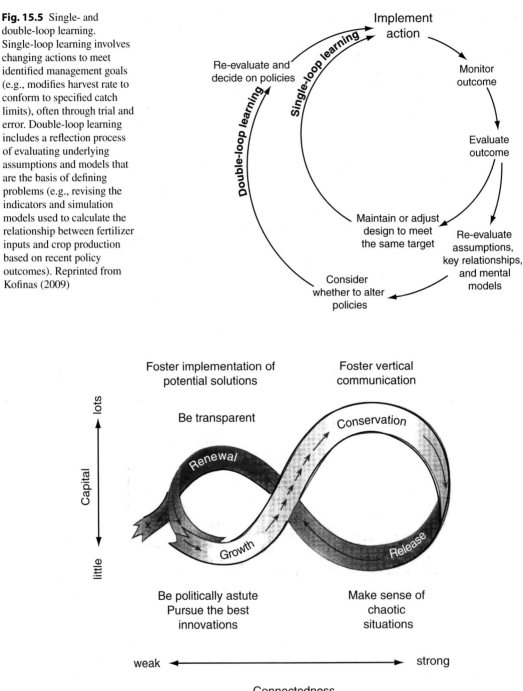

Fig. 15.6 Strategies for effective innovation at different phases of social–ecological disturbance or renewal. Social–ecological systems often go through cycles of growth and development (e.g., ecosystem succession, development of management expertise in an agency), then a conservation phase where conditions are relatively stable, then perhaps a radical change (release) due to a change in ecological or political environment, then a reorganization in a similar or modified condition (Holling and Gunderson 2002). The strategies that are most effective in stimulating innovation and novelty differ among these phases of the cycle of disturbance and renewal. (Westley et al. 2006)

potential solutions (Fig. 15.6). During times of management crisis, innovation that identifies patterns, explains causes, and suggests potential solution may be particularly helpful. As potential solutions begin to emerge, efforts to assess and promote (with transparency) the most appropriate solutions may lead to greatest progress. Finally, after an agreed-upon solution emerges, politically astute efforts to implement it are particularly helpful (Westley et al. 2006). In any case, adaptive management must work hand in hand with careful monitoring of outcomes. Without that monitoring, there is little chance to learn how interventions have or have not worked and therefore little opportunity to improve them.

Sustainable Development: Social–Ecological Transformation

Integrated conservation and development projects (ICDPs) seek to address conservation and human livelihood concerns in the developing world. ICDPs focus equally on biological conservation and human development, typically through externally funded, locally based projects (Wells and Brandon 1993, Kremen et al. 1994, Berkes et al. 2009). In the past, conservation and development projects typically were considered separately, by different organizations, sometimes with conflicting goals and consequences (Sutherland 2000). However, the two directives are more likely to be successful if considered together. The main goal of ICDPs is to link these previously opposing goals. In response to the failure of conservation and development projects to succeed separately, ICDPs emerged in the 1980s and established formal partnerships between conservation organizations and development agencies in an effort to create environmentally sound, economically sustainable alternatives to destructive land-use change (Kremen et al. 1994, Barrett and Arcese 1995, Alpert 1996).

An important objective of ICDPs is to determine the types, intensities, and distribution of resource use that are compatible with the conservation of biodiversity and the maintenance of ecological processes (Alpert 1996). Most ICDPs

therefore have the following characteristics. (1) They link conservation of natural habitats with the improvement of living conditions in the local communities. (2) They are site-based and tailored to specific problems such as impending loss of exceptional habitat. (3) They attract international expertise, local support, and external sources of income, and (4) they adapt to conditions in the developing world such as heavy dependence on natural resources, high population growth, and high opportunity costs of protected areas (Alpert 1996). ICDPs often seek to team a nongovernmental organization, a foreign donor agency, a national agency in charge of forestry, wildlife, or parks with local traditional and official leaders. Projects should couple biological information and scientific knowledge of ecosystem processes with the interests of managers and local communities in their design and implementation.

One major challenge of successful ICDPs is to develop an appropriate research mechanism to collect the scientific data needed to guide the dual objective of conservation and development (Kremen et al. 1994). It is critical to monitor biodiversity and ecosystem processes across space and time and at multiple levels of ecological organization (species, communities, ecosystems, and landscape) and their responses to management (Noss 1990, Kremen et al. 1994). Ecological and socioeconomic indicators can identify the causes and consequences of habitat loss, monitor changes in resource use and harvesting impacts, or evaluate the success of various management programs (Kremen et al. 1994, Barrett and Arcese 1995, Kremen et al. 1998). A successful monitoring program is essential to test the hypothesis that economic development linked to conservation promotes conservation.

In the 1990s, more than 100 ICDP projects were initiated, including over 50 in at least 20 countries in sub-Saharan Africa (Alpert 1996). A review of African projects concluded that ICDPs do not provide a definitive solution to habitat loss, but they can offer medium-term solutions to local conflicts between biological conservation and natural resource use in economically poor, remote areas of exceptional ecological importance (Alpert 1996). Only 16% of World Bank projects in developing nations have been successful in stimulating both ecological sustainability and human development (Tallis et al. 2008). Limited tourist revenue potential, lack of local management capacity, political unrest, large human populations, customary rights to land or resources enclosed by reserves, or the absence of an official protected area can pose significant impediments to the success of a project (Alpert 1996). The ICDPs most successful in promoting conservation contain significant community participation, which fosters improved community attitudes toward conservation (Brown 2003, Liu et al. 2008, Berkes et al. 2009). As with other kinds of ecosystem management, getting the science right is an essential, but insufficient, step. Over the long run, as we learn from successes and failures, the approaches employed in ICDPs will evolve to address remaining impediments and challenges.

Summary

Human activities influence all ecosystems on Earth. Ecosystems are directly impacted by activities such as resource harvests, land conversion, and management and are indirectly influenced by human-caused changes in atmospheric chemistry, hydrology, and climate. Because human activities strongly influence most of Earth's ecosystems, it follows that we should also take responsibility for their care and protection. Part of that responsibility must be to slow the rate and extent of global changes in climate, biogeochemical cycles, and land use. In addition, active management of all ecosystems is required to maintain populations, species, and ecosystem functions in the face of anthropogenic change and to sustain the provision of goods and services that humans receive from them.

State factors and interactive controls exert such strong control over ecosystem processes that changes in these controlling factors inevitably alter ecosystems and reduce the extent to which their current properties can be sustained. Management practices can, however, strongly influence the degree of sustainability. If the goal of management is to enhance sustainability of managed and unmanaged ecosystems, this requires that state factors and interactive controls be conserved as much as possible and that stabilizing (negative) feedbacks, which contribute to maintaining these controls, be strengthened within and among ecosystems. Directional changes in many of these ecosystem controls heighten the challenge of sustainably managing natural resources and threaten the sustainability of natural ecosystems everywhere.

The ecosystem approach to management applies ecological understanding to resource management to promote long-term sustainability of ecosystems and the delivery of essential ecosystem goods and services to society. This requires a landscape or regional perspective to account for interactions among ecosystems and explicitly includes humans as components of this regional system. Ecosystem management acknowledges the importance of stochastic events and our inability to predict future conditions with certainty. Adaptive management takes actions based on hypotheses of how management will affect the ecosystem. Based on the results of these experiments, management policies are modified to improve sustainability.

Integrated conservation and development projects (ICDPs) apply adaptive management to conservation in the developing world. ICDPs focus equally on biological conservation and human development. The main goal of ICDPs is to link these often previously opposing goals, based on the assumption that local human populations will place immediate socioeconomic security before conservation concerns. A fundamental principle underlying ecosystem management in general, and ICDPs in particular, is that people are an integral component of regional systems and that planning for a sustainable future requires solutions that are ecologically, economically, and culturally sustainable.

Review Questions

1. What are the major direct and indirect effects of human activities on ecosystems? Give examples of the magnitude of human impacts on ecosystems.

2. How does the resilience of an ecosystem influence its sustainability in the face of human-induced environmental change? What ecological properties of ecosystems influence their sustainability?

3. Describe a management approach that would maximize ecosystem sustainability. What factors or events are most likely to cause this management approach to fail?

4. What are ecosystem goods and services? How can an understanding of ecosystem services be used in management decisions?

5. What is ecosystem management? What is the role of humans in ecosystems in the context of ecosystem management?

6. What are the advantages and disadvantages of adaptive management as an approach to managing ecosystems?

7. What have integrated conservation and development projects (ICDPs) taught us about the advisability of including humans as components of ecosystems?

Additional Reading

Alpert, P. 1996. Integrated conservation and development projects: Examples from Africa. *BioScience* 46:845–855.

Chapin, F.S., III, G.P. Kofinas, and C. Folke, editors. 2009. *Principles of Ecosystem Stewardship: Resilience-Based Natural Resource Management in a Changing World.* Springer, New York.

Christensen, N.L., A.M. Bartuska, J.H. Brown, S. Carpenter, C. D'Antonio, et al. 1996. The report of the Ecological Society of America committee on the scientific basis for ecosystem management. *Ecological Applications* 6:665–691.

Clark, W.C. and N.M. Dickson. 2003. Sustainability science: The emerging research program. *Proceedings of the National Academy of Sciences, USA* 100: 8059–8061.

Daily, G.C. 1997. *Nature's Services: Societal Dependence on Natural Ecosystems.* Island Press, Washington, D. C.

Foley, J.A., R. DeFries, G.P. Asner, C. Barford, G. Bonan, et al. 2005. Global consequences of land use. *Science* 309:570–574.

Holling, C.S. and G.K. Meffe. 1996. Command and control and the pathology of natural resource management. *Conservation Biology* 10:328–337.

Lubchenco, J., A.M. Olson, L.B. Brubaker, S.R. Carpenter, M.M. Holland, et al. 1991. The sustainable biosphere initiative: An ecological research agenda. *Ecology* 72:371–412.

Matson, P.A. 2009. The sustainability transition. *Issues in Science and Technology* 25:39–42.

MEA (Millennium Ecosystem Assessment). 2005. *Ecosystems and Human Well-being: Synthesis.* Island Press, Washington.

Steffen, W.L., A. Sanderson, P.D. Tyson, J. Jäger, P.A. Matson, et al. 2004. *Global Change and the Earth System: A Planet under Pressure.* Springer-Verlag, New York.

Turner, B.L., II, R.E. Kasperson, P.A. Matson, J.J. McCarthy, R.W. Corell, et al. 2003. A framework for vulnerability analysis in sustainability science. *Proceedings of the National Academy of Sciences, USA* 100:8074–8079.

Abbreviations

A	Soil horizon	$^{13}C_{std}$	^{13}C content of a standard
A_{area}	Photosynthetic rate (per unit leaf area)	C_{V1}	^{13}C content of vegetation from the initial vegetation type
A_{mass}	Photosynthetic rate (per unit leaf mass)		
A_n	Quantity of energy or material assimilated by a trophic level	C_{V2}	^{13}C content of vegetation from the second vegetation type
a_n	Nutrient productivity	Ca^{2+}	Calcium ion
A_{net}	Net photosynthesis (per unit leaf mass or per unit ground area)	cal	Calorie
		CAM	Crassulacean acid metabolism
ABA	Abscisic acid	CEC	Cation exchange capacity
ADP	Adenosine diphosphate	CFC	Chlorofluorocarbon
Al^{3+}	Aluminum	CH_4	Methane
AM	Arbuscular mycorrhizae	CH_2O	Organic matter
ANPP	Aboveground NPP	Cl^-	Chloride ion
APAR	Absorbed photosynthetically active radiation	cm	Centimeter (10^{-2} m)
		cmo	Centimole
Ar	Argon	C:N	Carbon:nitrogen ratio
ATP	Adenosine triphosphate	CO	Carbon monoxide
B	Soil horizon	CO_2	Carbon dioxide
BT	Gene from the bacterium *Bacillus thuringiensis* that has been introduced into the corn genome and is toxic to the European corn borer	CO_3^-	Carbonate ion
		COOH	Carboxyl group
		C:P	Carbon:phosphorus ratio
		cP	Centipoise (unit of viscosity)
C	Celsius degrees; carbon; soil horizon	CPOM	Coarse particulate organic matter
C_3	Photosynthetic pathway whose initial carboxylation products are three-carbon acids	D	Deuterium
		d	Day
		DDT	An insecticide
C_4	Photosynthetic pathway whose initial carboxylation products are four-carbon acids	DIC	Dissolved inorganic carbon
		DIN	Dissolved inorganic nitrogen
		DMS	Dimethylsulfoxide
C_{S1}	Percentage of soil carbon derived from the initial vegetation type	DNA	Deoxyribonucleic acid
		DOC	Dissolved organic carbon
C_{S2}	^{13}C content of second soil type	DON	Dissolved organic nitrogen

F.S. Chapin, III et al., *Principles of Terrestrial Ecosystem Ecology*,
DOI 10.1007/978-1-4419-9504-9, © Springer Science+Business Media, LLC 2011

E	Soil horizon	H_2SO_4	Sulfuric acid
E	Evapotranspiration rate of an ecosystem	I_0	Irradiance above the canopy or at the water surface
e	Exponential base	I_n	Quantity of energy or material ingested by trophic level n
e$^-$	Electron		
E_{assim}	Assimilation efficiency	I_z	Irradiance at depth z (beneath the canopy or water surface)
$E_{consump}$	Consumption efficiency		
E_{prod}	Production efficiency	ICDP	Integrated Conservation and Development Project
E_{troph}	Trophic efficiency		
ENSO	El Niño/Southern Oscillation	ITCZ	Intertropical convergence zone
F_{CH4}	Flux of methane into ecosystem	J	Joule
F_{CO}	Flux of carbon monoxide into ecosystem	J_p	Rate of water flow through plants
		J_s	Rate of water flow through the soil
F_{DIC}	Flux of dissolved inorganic carbon into ecosystem	K	Degrees Kelvin
		K	Equilibrium constant
$F_{disturb}$	Flux of carbon from an ecosystem to the atmosphere due to disturbance	k	Extinction coefficient; decomposition constant
F_{DOC}	Flux of dissolved organic carbon into ecosystem	K$^+$	Potassium ion
		K_{in}	Incoming shortwave radiation
F_n	Component of the gravitational force that is normal to the slope	K_{out}	Outgoing shortwave radiation
		kcal	Kilocalorie
F_p	Component of the gravitational force that is parallel to the slope	kg	Kilogram (10^3 g)
		kJ	Kilojoule (10^3 J)
F_{POC}	Flux of particulate organic carbon into ecosystem	km	Kilometer (10^3 m)
		L	Liter; lignin
F_t	Total gravitational force	L	Latent heat of vaporization; leaf area index
F_{VOC}	Flux of volatile organic carbon into ecosystem		
		l	Path length of a column of soil or xylem; length of organism
f( )	Function of (parameters in parenthesis)		
Fe^{2+}	Ferrous iron	L_{in}	Incoming longwave radiation
Fe^{3+}	Ferric iron	L_{out}	Outgoing longwave radiation
FPAR	Fraction of PAR absorbed by vegetation	L_p	Hydraulic conductivity of plant xylem
FPOM	Fine particulate organic matter	L_s	Hydraulic conductivity of soil
G	Ground heat flux	L_t	Litter mass at time t
g	Gram	L_0	Litter mass at time zero
GIS	Geographic Information System	LAI	Leaf area index
GPP	Gross primary production	Lidar	Light Detection and Ranging
H	Hydrogen; herbivore	L:N	Lignin:nitrogen ratio
H	Sensible heat flux	LUE	Light use efficiency
h	Hour; height	M	Microbivore; moisture
H$^+$	Hydrogen ion	m	Meter
H$_2$	Hydrogen gas	m	Mass
HCO$_3^-$	Bicarbonate ion	M_a	Angular momentum
H_2CO_3	Carbonic acid	Ma	Million years
HNLC	High-nutrient, low-chlorophyll	mg	Milligram (10^{-3} g)
HNO$_3$	Nitric acid	Mg^{2+}	Magnesium ion
H_2O	Water	MJ	Megajoule (10^6 J)
H_2S	Hydrogen sulfide	mL	Milliliter (10^{-3} L)

mm	Millimeter (10^{-3} m)	OH^-	Hydroxide ion
Mn^{4+}	Manganese ion	P	Phosphorus
MODIS	Moderate Resolution Imaging Spectroradiometer	P	Precipitation
Mol	Mole	P_{org}	Organic phosphorus
MPa	Megapascal	p	Person
MRT	Mean residence time	Pa	Pascal
mV	Millivolt	PAR	Photosynthetically active radiation
N	Atomic nitrogen; north	PBL	Planetary boundary layer
N_2	Di-nitrogen gas	PCB	Polychlorinated biphenyl (an industrial class of compounds containing chlorine)
N_{avail}	Available nitrogen		
N_{org}	Organic nitrogen		
Na^+	Sodium ion	PDO	Pacific decadal oscillation
NADP	Nicotinamide adenine dinucleotide phosphate (in oxidized form)	PEP	Phosphoenolpyruvate
		Pg	Petagram (10^{15} g)
NADPH	Nicotinamide adenine dinucleotide phosphate (in reduced form)	pH	Negative log of H^+ activity
		PNA	Pacific North America pattern
NAO	North Atlantic Oscillation	PO_4^{3-}	Phosphate ion
NDVI	Normalized difference vegetation index	POC	Particulate organic carbon
		PON	Particulate organic nitrogen
NECB	Net ecosystem carbon balance	ppbv	Parts per billion by volume
NEE	Net ecosystem exchange	ppmv	Parts per million by volume
NEP	Net ecosystem production	ppt	Parts per thousand
NH_3	Ammonia gas	$Prod_n$	Production by trophic level n
NH_4^+	Ammonium ion	$Prod_{n-1}$	Production by the preceding trophic level
NIR	Near infrared radiation		
nl	Natural log	PW	Petawatt
nm	Nanometer (10^{-9} m)	Q_{10}	Proportional increase in the rate of a process with a 10°C increase in temperature
nmol	nanomole (10^{-9} mol)		
N_2O	Nitrous oxide		
NO	Nitric oxide	R	Bedrock
NO_2	Nitrogen dioxide	R	Runoff; respiration; universal gas constant
NO_2^-	Nitrite ion		
NO_3^-	Nitrate ion	r	Radius
NO_x	Nitric oxides in general (includes NO and NO_2)	$R_{ecosyst}$	Ecosystem respiration
		R_{growth}	Growth respiration
N:P	Nitrogen-to-phosphorus ratio	R_{het}	Heterotrophic respiration
NPP	Net primary production	R_{ion}	Respiration associated with ion uptake
NUE	Nutrient (or nitrogen) use efficiency		
O	Atomic oxygen; soil horizon	R_{maint}	Maintenance respiration
O_2	Molecular oxygen	R_{net}	Net radiation
O_3	Ozone	R_{plant}	Plant respiration
O_a	Highly decomposed organic horizon	R_{sam}	Isotope ratio of a sample
O_e	Moderately decomposed organic horizon	R_{std}	Isotope ratio of a standard
		Re	Reynolds number
O_i	Slightly decomposed organic horizon	RH	Relative humidity
OH	Hydroxyl radical	Rubisco	Ribulose-bis-phosphate carboxylase
		RuBP	Ribulose-bis-phosphate

S	Sulfur; south	VIS	Visible radiation
S	Heat storage by a surface; water storage by an ecosystem	VPD	Vapor pressure deficit
		W	Watt
s	Second	WUE	Water use efficiency
S_{org}	Organic sulfur	yr	Year
SE	Standard error	z	Depth beneath the canopy or water surface
SeaWiFS	Sea-viewing Wide Field-of-view Sensor		
		α	Albedo
Si	Silicon	β	Bowen ratio
SLA	Specific leaf area	δ	del; difference in isotope concentration relative to a standard
SO_2	Sulfur dioxide		
SO_4^{2-}	Sulfate ion	Δ	Change in a quantity
SOM	Soil organic matter	ε	Emissivity
SRL	Specific root length	μg	Microgram (10^{-6} g)
T	Temperature	μm	Micrometer (10^{-6} m)
t	Time	μL	Microliter (10^{-6} L)
t_r	Residence time	μmol	Micromole (10^{-6} moles)
Tg	Teragram (10^{12} g)	ρ	Density
U.K.	United Kingdom	σ	Stefan-Boltzman constant
U.S.	United States	Ψ_m	Matric potential
UV	Ultraviolet	Ψ_o	Osmotic potential
v	Velocity	Ψ_p	Pressure potential
V_k	Kinematic viscosity	Ψ_t	Total water potential
VAM	Vesicular arbuscular mycorrhizae		

Glossary

A horizon Uppermost mineral horizon of soils.

Abiotic Not directly caused or induced by organisms.

Abiotic condensation Non-enzymatic reaction of quinones with other organic materials in soil.

Abscisic acid Plant hormone that is transported from roots to leaves and causes a reduction in stomatal conductance.

Absorbence Fraction of the global solar irradiance incident on a surface that is absorbed.

Acclimation Morphological or physiological adjustment by an individual plant to compensate for the change in performance caused by a change in one environmental factor (e.g., temperature).

Acid rain Rain that has low pH, due to high concentrations of sulfuric and nitric acid released from combustion of fossil fuels.

Active transport Energy-requiring transport of ions or molecules across a membrane against an electrochemical gradient.

Activity budget Proportion of time that an animal spends in various activities.

Actual vegetation Vegetation that actually occurs on a site.

Adaptation Genetic adjustment by a population to maximize performance in a particular environment.

Adaptive management Management involving experimentation in the design and implementation of policies so that subsequent management can be modified based on learning from these experiments; "learning by doing."

Adaptive range Difference between the upper and lower tolerance limits of the system.

Advection Net horizontal transfer of gases or water.

Aerobic Occurring in the presence of oxygen.

Aerodynamic conductance Conductance of water vapor through a canopy from the vegetation or soil surface to the bulk atmosphere. Sometimes termed the boundary layer conductance of a canopy.

Aerosol Small (0.005 to 5 μm) solid or liquid particles suspended in air.

Afforestation Planting of new forests on previously non-forested sites.

Aggregate Clumps of soil particles bound together by polysaccharides, fungal hyphae, or minerals.

Albedo Fraction of the incident shortwave radiation reflected from a surface.

Alfisol Soil order that develops beneath temperate and subtropical forests, characterized by less leaching than spodosol.

Allocation Proportional distribution of photosynthetic products or newly acquired nutrients among different organs or functions in a plant.

Allochthonous input Input of energy and nutrients from outside the ecosystem; synonymous with subsidy.

Alternative stable states Alternative system states, each of which is plausible in a particular environment.

Ammonification See nitrogen mineralization.

Amorphous minerals Minerals with no regular arrangements of atoms.

Amplifying feedback Interaction in which two components of a system both have a positive effect on one another, or both have a negative effect on one another; this accentuates the changes in the system; synonymous with positive feedback.

Anadromous Life cycle in which reproduction occurs in lakes, streams or rivers, and the adult phase occurs primarily in the ocean.

Anaerobic Occurring in the absence of oxygen.

Andisol Soil order characterized by young soils on volcanic substrates.

Angular momentum Intensity of rotational motion.

Anion Negatively charged ion.

Anion exchange capacity Capacity of a soil to hold exchangeable anions on positively charged sites at the surface of soil minerals and organic matter.

Anioshydric plants Plants from dry sites that show little response of stomatal conductance to soil drying and therefore continue to photosynthesize and to absorb and lose water as the soil dries.

Anoxic Without oxygen.

Anthropocene Geologic epoch characterized by human impacts, initiated with the Industrial Revolution.

Anthropogenic Resulting from or caused by people.

Arbuscular mycorrhizae Mycorrhizae that exchange carbohydrates between plant roots and fungal hyphae via arbuscules; also termed vesicular arbuscular mycorrhizae or endomycorrhizae.

Arbuscules Exchange organs between plant and mycorrhizal fungus that occur within plant cells.

Aridisol Soil order that develops in arid climates.

Aspect Compass direction that a slope faces.

Assimilation Incorporation of an inorganic resource (e.g., CO_2 or NH_4^+ into organic compounds; transfer of digested food from the intestine to the bloodstream of an animal.

Assimilation efficiency Proportion of ingested energy that is assimilated into the bloodstream of an animal.

Assisted migration Movement of genotypes or species from a region where climate is becoming unfavorable to new places where climate is, or is expected to become, more favorable.

Autochthonous production Production occurring within the ecosystem.

Autotroph Organism that produces organic matter from CO_2 and environmental energy rather than by consuming organic matter produced by other organisms. Most autotrophs produce organic matter by photosynthesis; synonymous with primary producer.

B horizon Soil horizon with maximum accumulation of iron and aluminum oxides and clays.

Backscatter Reflection from small particles.

Base cations Non-hydrogen, non-aluminum cations.

Base flow Background stream flow from groundwater input in the absence of recent storm events.

Base saturation Percentage of the total exchangeable cation pool that is accounted for by base cations.

Bedrock Unweathered rock at the base of a soil profile.

Beneficial nutrients Elements that enhance growth under specific conditions or for specific groups of plants.

Benthic Associated with aquatic sediments.

Biofilm Microbial community embedded in a matrix of polysaccharides secreted by bacteria.

Biogenic Biologically produced.

Biogeochemical hot spot Zone of high rates of biogeochemical processes in a soil or landscape.

Biogeochemistry Biological interactions with chemical processes in ecosystems.

Biological pump Flux of carbon and nutrients in feces and dead organisms from the euphotic zone to deeper waters and the sediments of the ocean.

Biomass Quantity of living material (e.g., plant biomass).

Biomass burning Combustion of plants and soil organic matter following forest clearing.

Biomass pyramid Quantity of biomass in different trophic levels of an ecosystem.

Biome General class of ecosystems (e.g., tropical rain forest, arctic tundra).

Biosphere Biotic component of Earth, including all ecosystems and living organisms.

Biotic Caused or induced by organisms.

Bloom Rapid increase in phytoplankton biomass.

Blue water Liquid water in rivers, lakes, reservoirs, and groundwater aquifers that is potentially available to society.

Bottom-up controls Regulation of consumer populations by quantity and quality of food.

Bottom water Deep ocean water below about 1000 m depth.

Boundary layer Thin layer around a leaf or root in which the conditions differ from those in the bulk atmosphere or soil, respectively.

Boundary layer conductance Conductance of water vapor across the boundary layer of still air near an individual leaf; sometimes also applied to the aerodynamic conductance of a canopy.

Bowen ratio Ratio of sensible to latent heat flux.

Brine rejection Exclusion of salt during formation of ice crystals in sea ice.

Buffering capacity Capacity of the soil to release cations to replace ions lost from the soil by uptake or leaching.

Bulk air Air above the canopy that is not strongly influenced by the canopy.

Bulk density Mass of soil per unit volume.

Bulk soil Soil outside the rhizosphere.

Bundle sheath cells Cells surrounding the vascular bundle of a leaf; site of C_3 photosynthesis in C_4 plants.

C horizon Soil horizon that is relatively unaffected by the soil-forming processes.

C_3 photosynthesis Photosynthetic pathway in which CO_2 is initially fixed by Rubisco, producing three-carbon acids.

C_4 photosynthesis Photosynthetic pathway in which CO_2 is initially fixed by PEP carboxylase during the day, producing four-carbon organic acids.

Calcic horizon Hard calcium (or magnesium) carbonate-rich horizon formed in deserts; formerly termed caliche.

Caliche See calcic horizon.

Canopy interception Fraction of precipitation that does not reach the ground.

Capital Productive assets of a system. Natural capital consists of both non-renewable resources (e.g., oil reserves) and renewable ecosystem resources (e.g., plants, animals, and water) that support the production of goods and services on which society depends. Built capital consists of the physical means of production beyond that which occurs in nature (e.g., tools, clothing, shelter, dams, and factories). Human capital is the capacity of people to accomplish their goals; it can be increased through various forms of learning. Social capital is the capacity of groups of people to act collectively to solve problems.

Carbon-based defense Organic compounds that contain no nitrogen and defend plants against pathogens and herbivores.

Carbon-fixation reactions Those reactions in photosynthesis that use the products of the light-harvesting reactions to reduce CO_2 to sugars; also termed light-independent reactions or dark reactions.

Carboxylase Enzyme that catalyzes the reaction of a substrate with CO_2.

Carboxylation Attachment of CO_2 to an acceptor molecule.

Carnivore Organism that eats live animals.

Catalyst Molecule that speeds the conversion of substrates to products.

Catchment See drainage basin.

Catena Sequence of soils or ecosystems between hillcrests and valley bottoms, whose characteristics reflect slope position, drainage, and other topographic processes.

Cation Positively charged ion.

Cation exchange capacity Capacity of a soil to hold exchangeable cations on negatively charged sites at the surface of soil minerals and organic matter.

Cavitation Breakage of water columns under tension in the xylem.

Cellobiase Enzyme that breaks down cellobiose to form glucose.

Cellobiose Organic compound composed of two glucose units formed by cellulose breakdown.

Charge density Charge per unit hydrated volume of the ion.

Chelation Reversible chemical combination, usually with high affinity, with a metal ion (e.g., iron, copper).

Chemical alteration Chemical changes in dead organic matter during decomposition.

Chemodenitrification Abiotic conversion of nitrite to nitric oxide (NO).

Chlorofluorocarbon Organic chemicals containing chlorine and/or fluorine; gases that destroy stratospheric ozone.

Chlorophyll Green pigment involved in light capture by photosynthesis.

Chloroplast Organelles that carry out photosynthesis.

Chronosequence Sites that are similar to one another with respect to all state factors except time since disturbance.

Circadian rhythms Innate physiological cycles in organisms that have a period of about 24 hours.

Clay Soil particles less than 0.002 mm diameter.

Climate modes Relatively stable patterns of global atmospheric circulation.

Climate system Interactive system made up of the atmosphere, hydrosphere, biosphere, cryosphere, and land surface.

Climatic climax Hypothetical endpoint of succession that is determined only by climate.

Closed-basin lake Lakes in dry climates that have such high evaporation rates that outflow seldom occurs.

Closed system System in which the internal transfers of substances are much greater than inputs and outputs.

Cloud condensation nuclei Aerosols around which water vapor condenses to form cloud droplets.

C:N ratio Ratio of carbon mass to nitrogen mass.

CO_2 compensation point CO_2 concentration at which net photosynthesis equals zero.

Coarse particulate organic matter Organic matter in aquatic ecosystems, including leaves and wood, that is larger than 1 mm diameter.

Collector Benthic macroinvertebrate that feeds on fine organic particles; includes filtering collectors that consume suspended particles and gathering collectors that consume deposited particles.

Common-pool resources Resources that are held in common, are depletable, and from which it is costly to exclude people's use (e.g., the atmosphere, fresh water, marine fish).

Community Group of co-existing organisms in an ecosystem.

Compensation depth Depth at which GPP equals phytoplankton respiration integrated through the water column.

Compensation point Temperature, CO_2 concentration or light level at which net carbon exchange by a leaf is zero (i.e., photosynthesis equals respiration).

Competition Interactions among organisms that use the same limiting resources (resource competition) or that harm one another in the process of seeking a resource (interference competition).

Competitive release Sudden increase in growth, when resource availability increases in response to death or reduced growth of neighboring individuals.

Complementary resource use Use of resources that differ in type, depth, or timing by co-occurring species.

Complex adaptive system System whose components interact in ways that cause the system to adjust (i.e., adapt) in response to changes in conditions.

Conductance Flux per unit driving force (e.g., concentration gradient); inverse of resistance.

Configuration Spatial arrangement of patches in a landscape.

Connectivity Degree of connectedness among patches in a landscape.

Consortium Group of genetically unrelated bacteria, each of which produce only some of the enzymes required to break down complex macromolecules.

Consumer Organism that meets its energetic and nutritional needs by eating other living organisms.

Consumption efficiency Proportion of the production at one trophic level that is ingested by the next tropic level.

Convection Heat transfer by turbulent movement of a fluid (e.g., air or water).

Coriolis effect Tendency, due to Earth's rotation, of objects to be deflected to the right in the northern hemisphere and to the left in the southern hemisphere.

Cortex Layers of root cells outside the endodermis involved in nutrient uptake.

Coupling Effectiveness of atmospheric mixing between the canopy and the atmosphere. Also the linkages among biogeochemical cycles.

Crassulacean acid metabolism Photosynthetic pathway in which stomates open and carbon is fixed at night into four-carbon acids. During the day stomates close, C_4 acids are decarboxylated, and CO_2 is fixed by C_3 photosynthesis.

Credible science Science that is grounded in understanding and facts rather than in arguments of how the world should work.

Critical ecosystem services Those services that are most vulnerable to change, have fewest options for technological or ecological substitution, and are most valued by society.

Cross-scale linkages Processes that connect the dynamics of a system to events occurring at other times and places.

Crystalline minerals Minerals with highly regular arrangements of atoms.

Cultural services Non-material benefits that are important to society's well-being (e.g., recreational, aesthetic, and spiritual benefits).

Cytoplasm Contents of a cell that are contained within its plasma membrane, but outside the vacuole and the nucleus.

Dead zone Coastal zone of anoxic conditions that kill benthic organisms and bottom-feeding shrimp and fish and dramatically alter nutrient cycling at the sediment-water interface; often triggered by eutrophication.

Deciduous Shedding leaves in response to specific environmental cues.

Decomposer Organism that breaks down dead organic matter and consumes the resulting energy and nutrients for its own production.

Decomposition Breakdown of dead organic matter through fragmentation, chemical alteration, and leaching.

Decomposition rate constant Constant (k) that describes the exponential breakdown of a tissue.

Decoupling coefficient Measure of the extent to which the canopy is decoupled from the bulk atmosphere.

Deep water Ocean water greater than 1,000 m depth.

Deforestation Conversion of forest to a non-forest ecosystem type.

Demand Requirement; the term is used in the context of the control of the rate of a process (e.g., nutrient uptake) by the amount needed.

Denitrification Conversion of nitrate to gaseous forms (N_2, NO, and N_2O).

Deposition Atmospheric input of materials to an ecosystem.

Depositional zone Portion of a drainage basin, where deposition rate exceeds erosion rate.

Desertification Soil degradation that occurs in drylands.

Detritivore Organism that derives energy from breakdown of dead organic matter.

Detritus Dead plant and animal material, including leaves, stems, roots, dead animals, and animal feces.

Detritus-based trophic system Organisms that consume detritus or energy derived from detritus.

Diffuse radiation Shortwave radiation that is scattered by particles and gases in the atmosphere.

Diffusion Net movement of molecules or ions along a concentration gradient due to their random kinetic activity.

Diffusion shell Zone of nutrient depletion around individual roots caused by active nutrient uptake at the root surface.

Direct radiation Radiation that comes directly from the sun without scattering or reradiation by the atmosphere or objects in the environment.

Discrimination Preferential reaction (or diffusive flux) of the lighter isotope of an element or compound containing that element.

Dissimilatory nitrate reduction Microbial reduction of nitrate to ammonium.

Dissolved inorganic carbon CO_2, bicarbonate, and carbonate dissolved in water.

Dissolved organic carbon Water-soluble organic carbon.

Dissolved organic nitrogen Water-soluble organic nitrogen compounds.

Disturbance Relatively discrete event in time that removes plant biomass.

Disturbance intensity Energy released per unit area and time.

Disturbance regime Range of severity, frequency, type, size, timing, and intensity of disturbances characteristic of an ecosystem.

Disturbance severity Magnitude of loss of biomass, soil resources, and species caused by a disturbance.

Divalent Ions with two charges.

Doldrums Region near the equator with light winds and high humidity.

Double-loop learning Learning that requires that managers evaluate the approach they have used previously before taking further action, for example assessing the costs and benefits of managing forests for multiple ecosystem services rather than for a single product (e.g., trees).

Down-regulation Decrease in capacity to carry out a reaction; for example down-regulation of CO_2 uptake in response to elevated CO_2.

Downwelling Downward movement of surface ocean water, due to high density associated with high salinity and low temperature.

Drainage basin A river or stream and all the terrestrial surfaces that drain into it; synonymous with catchment or watershed.

Drift Invertebrates that move downstream in flowing water.

E horizon Heavily leached horizon beneath the A horizon that is formed in humid climates.

Eccentricity Degree of ellipticity of Earth's orbit around the sun.

Ecosystem Ecological system consisting of all the organisms in an area and the physical environment with which they interact.

Ecosystem ecology Study of the interactions between organisms and their environment as an integrated system.

Ecosystem engineer Organism that alters resource availability by modifying the physical properties of soils and litter.

Ecosystem goods See provisioning services.

Ecosystem management Application of ecological science to resource management to promote long-term sustainability of ecosystems and the delivery of essential ecosystem goods and services to society.

Ecosystem model Framework that describes the major pools and fluxes in an ecosystem and the factors that regulate these fluxes.

Ecosystem processes Inputs or losses of materials and energy to and from the ecosystem and the transfers of these substances among components of the system.

Ecosystem respiration Sum of plant respiration and heterotrophic respiration.

Ecosystem services Benefits that people derive from ecosystems, including provisioning, regulating, and cultural services.

Ectomycorrhizae Mycorrhizal association in some woody plants in which a large part of the fungal tissue is found outside the root.

Eddy covariance Method of estimating flux of energy and materials (e.g., water vapor and CO_2) between the ecosystem and the atmosphere by measuring their transfer in eddies of air.

El Niño Warming of surface water throughout the central and eastern tropical Pacific Ocean.

Electron-transport chain Series of membrane-bound enzymes that produce ATP and NADPH as a result of passing electrons down an electropotential gradient.

Emissivity Coefficient that describes the maximum rate at which a body emits radiation, relative to a perfect (black body) radiator, which has a value of 1.0.

Endocellulase Enzyme that breaks down the internal bonds to disrupt the crystalline structure of cellulose.

Endodermis Layer of suberin-coated cells between the cortex and xylem of roots; water penetrates this layer only by moving through the cytoplasm of these cells.

Energy pyramid Quantity of energy transferred between successive trophic levels.

Entisol Soil order characterized by minimal soil development.

Environmental stress Environmental factor that reduces plant performance; physical force that promotes mass wasting of soils.

Enzyme Organic molecule produced by an organism that catalyzes a chemical reaction.

Epidermis Layer of cells on the surface of a leaf or root.

Epilimnion Surface water layer that is heated by absorbed radiation and mixed by wind.

Epiphytic Attached to plant surfaces.

Equilibrium Condition of a system that remains unchanged over time because of a balance among opposing forces.

Equinox Date when the sun is directly overhead at the equator, and the entire earth surface receives approximately twelve hours of daylight.

Erosional zone Portion of a drainage basin, where erosion dominates over deposition.

Estuary Coastal ecosystem where a river mixes with seawater.

Euphotic zone Uppermost layer of water in aquatic ecosystems where there is enough light to support phytoplankton growth, i.e., where algal photosynthesis exceeds algal respiration.

Eutrophic Nutrient-rich.

Eutrophication Nutrient-induced increase in phytoplankton productivity.

Evapotranspiration Water loss from an ecosystem by transpiration and surface evaporation; equivalent to latent heat flux, but expressed in water units.

Evergreen Retention of green leaves throughout the year.

Exocellulase Enzyme that cleaves off disaccharide units from the ends of cellulose chains, forming cellobiose.

Exoenzyme Enzyme that is secreted by an organism into the environment.

Extensification Expansion of the aerial extent of land-cover change due to human activities.

Extinction coefficient Constant that describes the exponential decrease in irradiance through a canopy or medium (e.g., water).

Exudation Secretion of soluble organic compounds by organisms into the environment.

Facilitation Processes by which some species make the environment more favorable for the growth of other species.

Fast variables Variables that change rapidly and are often the focus of resource managers.

Feedback Response in which the product of one of the final steps in a chain of events affects one of the first steps in this chain; fluctuations in rate or concentration are minimized with stabilizing feedbacks or magnified with amplifying feedbacks.

Fermentation Anaerobic process that breaks down labile organic matter to produce organic acids and CO_2.

Ferrell cell Atmospheric circulation cell between 30° and 60° N or S latitude that is driven indirectly by dynamical processes.

Field capacity Water held by a soil after gravitational water has drained.

Filter feeder Aquatic animal that feeds on suspended particles.

Fine particulate organic matter Particulate organic matter in aquatic ecosystems that is smaller than 1 mm diameter.

Fire intensity Rate of heat production.

Fixation Covalent binding of an ion to a mineral surface.

Flow path Subsurface pathway of water movement.

Flux Flow of energy or materials from one pool to another.

Food chain Group of organisms that are linked together by the linear transfer of energy and nutrients from one organism to another.

Food web Group of organisms that are linked together by the transfer of energy and nutrients that originates from the same source.

Fossil groundwater Groundwater that accumulated during a wetter climate and is no longer being replenished at a significant rate.

Fractionation Preferential incorporation of a light isotope (e.g., ^{12}C vs. ^{13}C).

Fragmentation Breaking up of intact litter into small pieces.

Fulvic acids Humic compounds that are relatively water-soluble due to their extensive side chains and many charged groups.

Functional matrix Matrix of all the functional traits represented by the species in an ecosystem.

Functional mosaic Landscape with functionally important differences among patches.

Functional type Group of species that are similar with respect to their impacts on community or ecosystem processes (effects functional type) and/or their response to a given environmental change, such as elevated CO_2 (response functional types).

Gap-phase succession Succession that occurs in small patches within a stand due to death of individual plants or plant parts.

Gelisol Soil order characterized by presence of permafrost.

Generalist herbivore Herbivore that is relatively non-selective in its choice of plant species.

Geotropism Growth direction of plant organs with respect to gravity.

Gley soil Blue-gray soil due to conversion of ferric to ferrous iron; formed under anaerobic conditions.

Glomalin Glycoprotein produced by many mycorrhizal fungi that cements microaggregates together to form macroaggregates.

Graminoid Grass-like plant (grasses, sedges, and rushes).

Grazer Herbivore that consumes herbaceous plants (terrestrial ecosystems) or periphyton (aquatic ecosystems).

Grazing lawn Productive grassland or wetland ecosystem in which plants are heavily grazed but supported by large nutrient inputs from grazers.

Great acceleration Rapid increase in human impacts on Earth's life support system since 1950.

Greenhouse effect Warming of the atmosphere due to atmospheric absorption of longwave radiation.

Greenhouse gas Atmospheric gas that absorbs longwave radiation.

Green water Water that evaporates from the soil surface or is transpired by plants.

Gross primary production Net carbon input to ecosystems, i.e., net photosynthesis expressed at the ecosystem scale ($g\ C\ m^{-2}\ yr^{-1}$).

Ground heat flux Heat transferred from the surface into the soil.

Groundwater Water in soil and rocks beneath the rooting zone.

Growth Production of new biomass.

Guano Large accumulations seabird feces.

Gyres Large circulation systems in surface ocean waters.

Hadley cell Atmospheric circulation cell between the equator and 30°N or S latitude, driven by expansion and uplift of equatorial air and subsidence of cool dense subtropical air.

Halocline Relatively sharp vertical gradient in salinity in a lake or ocean.

Hard pan Soil horizon with low hydraulic conductivity.

Hartig net Hyphae that penetrate cell walls of root cortical cells in ectomycorrhizae.

Heat capacity Amount of energy required to raise the temperature of unit volume of a body by 1°C.

Herbivore Organism that eats live plants.

Herbivory Consumption of plants by animals.

Heterocyst Specialized non-photosynthetic cells of phototrophs that protect nitrogenase from denaturation by oxygen.

Heterotroph Organism that consumes organic matter produced by other organisms rather than producing organic matter from CO_2 and environmental energy. Hetertrophs include decomposers, consumers, and parasites.

Heterotrophic respiration Respiration by non-autotrophic organisms (i.e., microbes and animals).

Histosol Soil order characterized by highly organic soils due to poor drainage and low oxygen.

Homeothermy Maintenance of a constant body temperature.

Horizon Layer in a soil profile. The horizons, from top to bottom, are the O horizon, which consists of organic matter above mineral soil; the A horizon, a dark layer with substantial organic matter; the E horizon, which is heavily leached; a B horizon, where iron and aluminum oxides and clays accumulate; and a C horizon, which is relatively unaffected by soil-forming processes.

Horse latitudes Latitudes 30°N and S, characterized by weak winds and high temperatures.

Human well-being Quality of life; basic material needs for a good life, freedom and choice, good social relations, and personal security.

Humic acid Relatively insoluble humic compounds with extensive networks of aromatic rings and few side chains.

Humification Non-enzymatic process by which recalcitrant breakdown products of decomposition are complexed to form humus.

Humus Complex mixture of soil organic compounds with highly irregular structure.

Hydraulic conductivity Capacity of a given volume of a substance (such as soil) to conduct water; this defines the relationship between discharge and the hydraulic gradient causing it.

Hydraulic lift Upward movement of water through roots from deep moist soils to dry surface soils along a gradient in water potential.

Hydrothermal vent Vent that emits reduced gases such as H_2S in zones of sea-floor spreading.

Hyphae Filamentous structures that make up the vegetative body of fungi.

Hypolimnion Deep water layer that is unaffected by surface turbulence.

Hyporheic zone Zone of flowing groundwater within the streambed.

Hypoxic Weakly oxygenated.

Ice-albedo feedback Atmospheric warming caused by warming-induced decrease in albedo due to earlier melting of sea ice.

Igneous rocks Rocks formed when magma from Earth's core cools near the surface.

Immobilization Removal of inorganic nutrients from the available pool by microbial uptake and chemical fixation.

Inceptisol Soil order characterized by weak soil development.

Infiltration Movement of water into the soil.

Integrated Conservation and Development Project Project in a developing nation that focuses simultaneously on biological conservation and human development.

Intensification Intensive application of water, energy, and fertilizers to agricultural ecosystems to enhance their productivity.

Interactive controls Factors that control and respond to ecosystem characteristics, including resource supply, microenvironment, functional types of organisms, and disturbance regime.

Interception Contact of nutrients with roots due to the growth of roots to the nutrients; fraction of precipitation that does not reach the ground (canopy interception).

Intermediate water Middle layer of ocean water between about 200 and 1000 m depth.

Intertropical convergence zone Region of low pressure and rising air where surface air from the northern and southern hemispheres converges.

Inversion Increase in atmospheric temperature with height.

Inverted biomass pyramid Biomass pyramid in which there is a smaller biomass of primary producers than of upper trophic levels; typical of pelagic ecosystems of lakes, streams, and oceans.

Investment Increase in the quantity of an asset times its value. Genuine investment constitutes an increase in the total capital of the system.

Ionic binding Electrostatic attraction between oppositely charged ions or surfaces.

Irradiance Radiant energy flux density received at a surface, i.e., the quantity of radiant energy received at a surface per unit time.

Isohydric plants Plants from moist sites that close their stomata at relatively high soil moisture before they experience large changes in plant water potential.

Jet stream Strong winds over a broad height range in the upper troposphere.

Katabatic winds Downslope winds that occur at night when air cools, becomes more dense, and flows downhill.

Kelvin waves Large-scale ocean waves that travel back and forth across the ocean.

Keystone species Species that has a much greater impact on ecosystem processes than would be expected from its biomass; functional type represented by a single species.

La Niña Sea surface temperatures in the equatorial Pacific Ocean associated with strong upwelling of cold water off South America and warm currents in the western Pacific.

Labile Easily decomposed.

Land breeze Night breeze from the land to the ocean caused by the higher surface temperature over the ocean at night.

Landscape Mosaic of patches that differ in ecologically important properties.

Land-use conversion Human-induced change of an ecosystem to one that is dominated by a different physical environment or different plant functional types.

Land-use modification Human alteration of an ecosystem in ways that significantly affect ecosystem processes, community structure and population dynamics without changing the physical environment or the dominant plant functional type of the ecosystem.

Lapse rate Rate at which air temperature decreases with height above Earth's surface; averages about 6.5°C km^{-1}.

Latent heat flux Energy transferred between a surface and the atmosphere by the evaporation of water or the condensation of water vapor; equivalent to evapotranspiration, but expressed in energy units.

Latent heat of vaporization Energy required to change a gram of a substance from a liquid to a vapor without change in temperature.

Laterite See plinthite layer.

Law of the minimum Plant growth is limited by a single resource at any one time; another resource becomes limiting only when the supply of the first resource is increased above the point of limitation.

Leaching Downward movement of materials in solution. This can occur from the canopy to the soil, from soil organic matter to the soil solution, from one soil horizon to another, or from the ecosystem to ground water or aquatic ecosystems.

Leaf area index (LAI) Projected (i.e., one side of a flat leaf) leaf area per unit ground area.

Legacy Effect of past events on the current functioning of an ecosystem.

Legitimate science Science that is unbiased and respectful of multiple user groups with different concerns and views on preferred policy outcomes.

Life history traits Traits (e.g., seed size and number, potential growth rate, maximum size, and longevity) of an organism that determine how quickly a species can get to a site, how quickly it grows, how tall it gets, and how long it survives.

Light compensation point Irradiance at which net photosynthesis equals zero.

Light-harvesting reactions Reactions of photosynthesis that transform light energy into chemical energy; also termed light-dependent reactions.

Light saturation Range of light availabilities over which the rate of photosynthesis is insensitive to irradiance.

Light use efficiency Ratio of GPP to absorbed photosynthetically active radiation at the leaf or ecosystem scale.

Limitation Reduced rate of a process (e.g., NPP, growth or photosynthesis) due to inadequate supply of a resource (e.g., nutrient or light) or low temperature. Proximate limitation reflects the immediate response to addition of the resource. Ultimate limitation reflects long-term transformation of the system when the resource is added.

Lithosphere Hard outermost shell of Earth

Litter Dead plant material that is sufficiently intact to be recognizable.

Litterbag Mesh bag used to measure decomposition rate of detritus.

Litterfall Shedding of aboveground plant parts and death of plants.

Littoral zone Shore of a lake or ocean.

Loam Soil with substantial proportions of at least two size classes of soil particles.

Loess Soil derived primarily from wind-blown silt particles.

Longwave radiation Radiation with wavelengths of about 4–30 μm.

Lotic Characterized by flowing water.

Luxury consumption Accumulation of nutrients in excess of immediate needs for growth (storage).

Macrofauna Soil animals > 2 mm in width.

Macronutrients Nutrients that are required in large quantities by organisms.

Macrophyte Large aquatic plant (not phytoplankton).

Macropores Large pores between soil aggregates that allow rapid movement of water, roots, and soil animals.

Magma Molten rock in Earth's crust.

Mantle Fungal hyphae that surround the root in ectomycorrhizae; also termed sheath.

Mass flow Bulk transport of solutes due to the movement of soil solution.

Mass wasting Downslope movement of soil or rock material under the influence of gravity without the direct aid of other media such as water, air, or ice.

Matric potential Component of water potential caused by adsorption of water to surfaces; it is considered a component of pressure potential in some treatments.

Matrix Predominant patch type in a landscape.

Mean residence time Mass divided by the flux into or out of the pool over a given time period; synonymous with turnover time.

Mesofauna Soil animals 0.1–2 mm in width.

Mesopause Boundary between the mesosphere and thermosphere.

Mesophyll cells Photosynthetic cells in a leaf.

Mesosphere Atmospheric layer between the stratosphere and the thermosphere, which is characterized by a decrease in temperature with height.

Metalimnion Water layer of intermediate depth (between the epilimnion and the hypolimnion).

Metamorphic rocks Sedimentary or igneous rocks that are modified by exposure to heat or pressure.

Metapopulations Populations of a species that consist of partially isolated subpopulations.

Methanogen Methane-producing bacteria.

Methanotroph Methane-consuming bacteria.

Microbial loop Microbial food web (including both plant- and detritus-based organic material) that recycles carbon and nutrients within the euphotic zone.

Microbial transformation Transformation of plant-derived substrates into microbial-derived substrates as a result of microbial turnover.

Microbivore Organism that eats microbes.

Microenvironment Local environmental conditions (e.g., temperature, pH) that influence the rates of ecosystem processes. It influences

organism activity but is not consumed nor depleted by organisms.

Microfauna Soil animals < 0.1 mm in width.

Micronutrients Nutrients that are required in small quantities by organisms.

Micropores Small pores between soil particles, often within soil aggregates.

Milankovitch cycles Cycles of solar input to Earth caused by regular variations in Earth's orbit (eccentricity, tilt, and precession).

Mineralization Conversion of carbon and nutrients from organic to inorganic forms due to the breakdown of litter and soil organic matter. Gross mineralization is the total amount of nutrients released via mineralization (regardless of whether they are subsequently immobilized or not). Net mineralization is the *net* accumulation of inorganic nutrients in the soil solution over a given time interval.

Mollisol Soil order characterized by an organic-rich, fertile A horizon that grades into a B horizon.

Monovalent Ions with a single charge.

Monsoon Tropical or subtropical system of air flow characterized by a seasonal shift between prevailing onshore and offshore winds.

Mutualism Symbiotic relationship between two species that benefits both partners.

Mycorrhizae Symbiotic relationship between plant roots and fungal hyphae, in which the plant acquires nutrients from the fungus in return for carbohydrates that constitute the major carbon source for the fungus.

Mycorhizosphere Zone of soil that is directly influenced by mycorrhizal hyphae.

Nanoplankton Plankton 2 to 20 μm in diameter.

Negative feedback See stabilizing feedback.

Net ecosystem carbon balance Net annual carbon accumulation by the ecosystem.

Net ecosystem exchange Net CO_2 exchange between the land or ocean and the atmosphere.

Net ecosystem production Balance between gross primary production and ecosystem respiration (or between net primary production and heterotrophic respiration).

Net photosynthesis Net rate of carbon gain

measured at the level of individual cells or leaves. It is the balance between simultaneous CO_2 fixation and respiration of photosynthetic cells in the light (including both photorespiration and mitochondrial respiration).

Net primary production Quantity of new plant material produced annually (GPP minus plant respiration); includes new biomass, hydrocarbon emissions, root exudates, and transfers to mycorrhizae.

Net radiation Balance between the inputs and outputs of shortwave and longwave radiation.

New production Phytoplankton production supported by nutrients mixed upward from below the euphotic zone.

Niche Ecological role of an organism in an ecosystem.

Nitrification Conversion of ammonium to nitrate in the soil. Autotrophic nitrifiers use the energy yield from NH_4^+ oxidation to fix carbon used in growth and maintenance, analogous to the way plants use solar energy to fix carbon via photosynthesis. Heterotrophic nitrifiers gain their energy from breakdown of organic matter.

Nitrogenase Enzyme that converts to di-nitrogen to ammonium.

Nitrogen-based defense Plant defensive compound containing nitrogen.

Nitrogen fixation Conversion of di-nitrogen gas to ammonium.

Nitrogen mineralization Conversion of dissolved organic nitrogen to ammonium; synonymous with ammonification.

Nitrogen saturation Ecosystem condition in which nitrogen inputs exceed plant and microbial nitrogen requirements so that the system loses nitrogen to the atmosphere and to groundwater and streams.

Non-steady-state mosaic Landscape that is not in equilibrium with the current environment because large-scale disturbances cause large proportions of the landscape to be in one or a few successional stages.

Normalized difference vegetation index (NDVI) Index of vegetation greenness.

North Atlantic Drift Poleward extension of the Gulf Stream.

Nutrients Material resources in addition to carbon, oxygen, and water that are required for life.

Nutrient cycling Mineralization and uptake of nutrients within an ecosystem patch.

Nutrient limitation Limitation of plant growth due to insufficient supply of a nutrient. See proximate and ultimate nutrient limitation.

Nutrient productivity Instantaneous rate of carbon gain per unit nutrient.

Nutrient spiraling Mineralization and uptake of nutrients that occurs as dead organic matter, dissolved nutrients, and organisms move along a section of a stream or river.

Nutrient uptake Nutrient absorption by plant roots.

Nutrient use efficiency Growth per unit of plant nutrient; ratio of nutrients to biomass lost in litterfall; also calculated as nutrient productivity times residence time.

O horizon Organic horizon above mineral soil.

Occluded phosphorus Unavailable phosphate that is most tightly bound to oxides of iron and aluminum.

Oligotrophic Nutrient-poor.

Omnivore Organism that eats food from several trophic levels.

Open access Situation in which potential users are not excluded from using a resource.

Orographic effects Effects due to presence of mountains.

Osmotic potential Component of water potential due to the presence of substances dissolved in water.

Overland flow Movement of water over the soil surface.

Oxidation Loss of electrons by an electron donor in oxidation-reduction reactions.

Oxisol Soil order found in the wet tropics characterized by highly weathered, leached soils.

Oxygenase Enzyme that catalyzes a reaction with oxygen.

Ozone Molecular form of oxygen (O_3) that is a reactive component of pollution in the troposphere and an absorber of UV radiation in the stratosphere.

Ozone hole Zone of destruction of stratospheric ozone at high southern and high northern latitudes. This hole allows increased penetration of UV radiation to Earth's surface.

Parent material Rocks or other substrates that generate soils through weathering.

Patch Relatively homogeneous stand of an ecosystem in a landscape.

Path dependence Effects of historical legacies on the future trajectory of a system.

Pelagic Open water.

PEP carboxylase Initial carboxylating enzyme in C_4 photosynthesis.

Periphyton Assemblages of algae, bacteria, and invertebrates that attach to stable surfaces such as rocks and vascular plants.

Permafrost Permanently frozen ground, i.e., soil that remains frozen for at least two years.

Permanent wilting point Water held by a soil that cannot be extracted by plant uptake.

pH Negative log of the hydrogen ion (H^+) activity (effective concentration) in solution and is a measure of the active acidity of the system.

Phagocytosis Consumption of material by a cell by enclosing it in a membrane-bound structure that enters the cell.

Phenology Time course of periodic events in organisms that are correlated with climate (e.g., budbreak).

Phloem Long-distance transport system in plants for flow of carbohydrates and other solutes.

Phosphatase Enzyme that hydrolyzes phosphate from a phosphate-containing organic compound.

Phosphorus fixation Binding of phosphorus in soils by strong chemical bonds.

Photodestruction Breakdown of photosynthetic pigments under high light.

Photo-oxidation Oxidation of compounds by light energy; photosynthetic enzymes can be photo-oxidized under conditions of high light.

Photoperiod Daylength.

Photoprotection Protection of photosynthetic pigments from destruction by high light.

Photorespiration Production of CO_2 due to the oxygenation reaction catalyzed by Rubisco; best viewed as a process that recovers much of the products of the oxygenase activity of Rubisco.

Photosynthesis Biochemical process that uses light energy to convert CO_2 to sugars. Net photosynthesis is the net carbon input to ecosystems; synonymous at the ecosystem scale with gross primary production.

Photosynthetic capacity Photosynthetic rate per unit leaf mass measured under favorable conditions of light, moisture, and temperature.

Photosynthetically active radiation Visible light that supports photosynthesis; radiation with wavelengths between 400 and 700 nm.

Phototroph Nitrogen-fixing microorganism that produces its own organic carbon through photosynthesis.

Phreatophyte Deep-rooted plant that taps groundwater.

Phyllosphere decomposition Decomposition that occurs on leaves prior to senescence.

Phytoplankton Microscopic primary producers suspended in the surface water of aquatic ecosystems.

Picoplankton Plankton < 2 μm in diameter.

Pixel Individual cell of a satellite image that provides a generalized spectral response for that area.

Planetary boundary layer Lower portion of the troposphere that is directly affected by the fluxes and friction of Earth's surface.

Planetary waves Large (>1500 km length) waves in the atmosphere that are influenced by the Coriolis effect, land-ocean heating contrasts, and the locations of large mountain ranges.

Plankton Microscopic organisms suspended in the surface water of aquatic ecosystems.

Plant-based trophic system Plants, herbivores, and organisms that consume herbivores and their predators.

Plant defense Chemical or physical property of plants that deters herbivores.

Plasmodesmata Cytoplasmic connections between adjacent cortical cells.

Plate tectonics Theory describing the large-scale motions of continental and ocean plates across Earth's surface.

Plinthite layer Iron-rich layer in tropical soils that have hardened irreversibly on exposure to repeated saturation and drying cycles; formerly termed laterite.

Podzol See spodosol.

Poikilothermic Organism whose body temperature depends on the environment.

Polar cell Atmospheric circulation cell between 60° and the pole driven by subsidence of cold converging air at the poles.

Polar front Boundary between the polar and subtropical air masses characterized by rising air and frequent storms.

Polyphenol Soluble organic compound with multiple phenolic groups.

Pool Quantity of energy or material in an ecosystem compartment such as plants or soil.

Positive feedback See amplifying feedback.

Potential biota Organisms that are present in a region and could potentially occupy the site.

Potential vegetation Vegetation that would occur in the absence of human disturbance.

Precession A "wobbling" in Earth's axis of rotation with respect to the stars, determining the date during the year when solstices and equinoxes occur.

Precipitation Water input to an ecosystem as rain and snow.

Pressure potential Component of water potential generated by gravitational forces and by physiological processes of organisms.

Prevailing wind Most frequent wind direction.

Primary detritivore Organisms (bacteria and fungi) that eat dead organic matter.

Primary minerals Minerals present in the rock or unconsolidated parent material before chemical changes have taken place.

Primary producers Organisms that convert CO_2, water, and solar energy into biomass (i.e., plants); synonymous with autotroph.

Primary production Conversion of CO_2, water, and solar energy into biomass. Gross primary production (GPP) is the net carbon input to ecosystems, i.e., net photosynthesis expressed at the ecosystem scale (g C m^{-2} yr^{-1}). Net primary production is the net carbon accumulation by vegetation (GPP minus plant respiration).

Primary succession Succession following severe disturbances that remove or bury most products of ecosystem processes, leaving little or no organic matter or organisms.

Production efficiency Proportion of assimilated energy that is converted to animal production, including both growth and reproduction.

Profile Vertical cross-section of soil.

Protease Protein-hydrolyzing enzyme.

Proteoid roots Dense clusters of fine roots produced by certain families such as the Proteaceae.

Protozoan Single-celled animal.

Provisioning services Products of ecosystems that are directly harvested by people (e.g., food, fiber, and water); synonymous with ecosystem goods or renewable resources.

Proximate limiting nutrient Nutrient that immediately enhances plant growth after it is added (short-term nutrient limitation).

Pseudosand Stable aggregates of clay particles cemented together by iron oxides in clay-rich oxisols and ultisols.

Pycnocline Relatively sharp vertical gradient in water density in a lake or ocean.

Quality Chemical nature of live or dead organic matter that determines the ease with which it is broken down by herbivores or decomposers, respectively.

Quantum yield Moles of CO_2 fixed per mole of light quanta absorbed; the initial slope of the light-response curve.

Quinone Highly reactive class of compounds produced from polyphenols.

Radiatively active gases Gases that absorb infrared radiation (water vapor, CO_2, CH_4, N_2O and industrial products like chlorofluorocarbons [CFCs]).

Rain shadow Zone of low precipitation downwind of a mountain range.

Reach Stream segment.

Recalcitrant Resistant to microbial breakdown.

Redfield ratio Ratio of nitrogen to phosphorus atoms ($\approx$16) giving optimal growth of algae.

Redox potential Electrical potential of a system due to the tendency of substances in it to lose or gain electrons.

Reduction The gain of electrons by an electron acceptor in oxidation-reduction reactions.

Reflected radiation Shortwave radiation that is reflected from clouds and objects in the landscape.

Regenerated production Phytoplankton production supported by nutrients regenerated within the euphotic zone.

Regime shift Abrupt large-scale change to a new state characterized by very different structure and feedbacks.

Regulating services Effects of ecosystems on processes that extend beyond their boundaries (e.g., regulation of climate, water quantity and quality, disease, wildfire spread, and pollination).

Relative growth rate Growth per unit plant biomass.

Relative humidity Ratio of the actual amount of water held in the atmosphere compared to maximum that could be held at that temperature.

Renewable resources See provisioning services.

Renewal ecology Enhancement of the natural capital of a system to provide ecosystem services in the context of the current or desired future state of the system.

Residence time Average time that an element or tissue remains in a system, calculated as the pool size divided by the input; synonymous with turnover time.

Resilience Capacity of a social-ecological system to maintain similar structure, functioning, and feedbacks despite shocks and perturbations.

Resorption Withdrawal of nutrients from tissues during their senescence.

Resorption efficiency Proportion of the maximum tissue nutrient pool that is resorbed prior to tissue senescence or death.

Resources Substances that are taken up from the environment and used by organisms to support their growth and maintenance (e.g., light, CO_2, water, nutrients).

Respiration Biochemical process that converts carbohydrates into CO_2 and water, releasing energy that can be used for growth and maintenance. Respiration can be associated with trophic groups (plant respiration, animal respiration, microbial respiration) or combinations of groups (heterotrophic respiration: animal plus microbial respiration; ecosystem respiration: heterotrophic plus plant respiration). Alternatively, respiration can be defined by the way in which the resultant energy is used (maintenance respiration, growth respiration, respiration to support ion uptake).

Rhizosphere Zone of soil that is directly influenced by roots.

Riparian Located along a streambank.

River continuum concept Idealized transition in ecosystem structure and functioning that integrates stream size, energy sources, food webs, and nutrient processing into a longitudinal model of river metabolism from headwaters to the ocean.

Rock cycle Formation, transformation, and weathering of rocks.

Root cap Cells at the tips of roots that produce mucilaginous carbohydrates that lubricate the movement of roots through soil.

Root cortex Layers of root cells involved in nutrient absorption.

Root exudation Diffusion and secretion of organic compounds from roots into the soil.

Root hair Elongate epidermal cell of the root that extends out into the soil.

Root:shoot ratio Ratio of root biomass to shoot biomass.

Roughness element Obstacle to air flow (e.g., a tree) that creates mechanical turbulence.

Rubisco Ribulose bisphosphate carboxylase; photosynthetic enzyme that catalyzes the initial carboxylation in C_3 photosynthesis.

Runoff Water loss from an ecosystem in streams and rivers.

Salient science Science that is presented to the right people at the right time.

Saline Salty.

Salinization Salt accumulation due to evaporation of surface water.

Salt flat Depression in an arid area that accumulates salt because it receives runoff but has no outlet; see also closed-basin lake.

Salt lick Mineral-rich springs or outcrops that are used by animals as a source of minerals.

Salt pan Surface salt accumulation in desert depressions.

Sand Soil particles 0.05 to 2 mm diameter.

Saprotrophic Eating dead organic matter (as with non-mycorrhizal fungi).

Sapwood Total quantity of functional conducting tissue of the xylem.

Saturated flow Drainage of water under the influence of gravity.

Savanna Grassland with scattered trees or shrubs.

Sea breeze Daytime onshore breeze that occurs on coastlines due to greater heating of the land than the water.

Secondary metabolites Compounds produced by plants that are not essential for normal growth and development.

Secondary minerals Crystalline and amorphous products that are formed through the reaction of materials released during weathering.

Secondary succession Succession that occurs on previously vegetated sites after a disturbance in which there are residual effects of organisms and organic matter from organisms present before the disturbance.

Sedimentary rocks Rocks formed from sediments.

Seed bank Seeds produced after previous disturbances that remain dormant in the soil until post-disturbance conditions (light, wide temperature fluctuations, and/or high soil nitrate) trigger germination.

Seedling bank Seedlings beneath a canopy that show negligible growth in the dense shade of a forest canopy but grow rapidly in tree-fall gaps.

Selective preservation Increase in concentration of recalcitrant material as a result of decomposition of labile substrates.

Senescence Programmed breakdown of plant tissues.

Sense of place Self-identification with a particular location or region.

Sensible heat Heat energy that can be sensed (e.g., by a thermometer) and involves no change in state.

Sensible heat flux Energy that is conducted from a warm surface to the air immediately above it and then moved upward to the bulk atmosphere by convection.

Serotiny Extent to which seeds are retained in cones.

Seston Particles suspended in the water column, including algae, bacteria, detritus, and mineral particles.

Shade leaf Leaf that is acclimated to shade or is produced by a plant adapted to shade.

Shear strength of soil Shear stress that a soil can sustain without slope failure.

Shear stress of soil Force parallel to the slope that drives mass wasting events such as landslides.

Shifting agriculture Clearing of forest for crops followed by a fallow period during which forests regrow, after which the cycle repeats; synonymous with slash-and-burn or swidden agriculture.

Shifting steady-state mosaic Landscape in which the vegetation at any point in the landscape is always changing but, averaged over a large enough area, the proportion of the landscape in each successional stage remains relatively constant.

Shortwave radiation Radiation with wavelengths of about 0.2–4.0 μm, including ultraviolet, visible, and near infrared radiation.

Shredder Invertebrate that breaks leaves and other detritus into pieces and digests the microbial jam on the surface of these particles.

Siderophore Organic chelate produced by plant roots.

Silt Soil particles 0.002–0.05 mm diameter.

Single-loop learning Learning that adjusts actions to meet previously agreed-upon management goals, such as changes in harvest levels needed to sustain populations of a particular fish or tree species.

Sink strength Demand of a plant organ or process for carbohydrates.

Slash-and-burn agriculture See shifting agriculture.

Slow variables Variables that change slowly and are key control variables over longer time scales.

Snow-albedo feedback Atmospheric warming caused by warming-induced decrease in albedo due to earlier snowmelt.

Social-ecological stewardship Strategy for shaping the trajectory of social-ecological change to enhance ecosystem resilience and human well-being.

Soil The weathered portion of Earth's crust between the litter layer and bedrock; see also horizons.

Soil creep Downhill movement of soil; dubious character covered with dirt.

Soil order Major soil groupings in the U.S. soil taxonomic classification.

Soil organic matter Dead organic matter in the soil that has decomposed to the point that its original identity is uncertain.

Soil phase Soils belonging to the same soil type that differ in landscape position, stoniness, or other soil properties.

Soil resources Water and nutrients available in the soil.

Soil series Soils belonging to the same order that differ in profile characteristics, such as number and types of horizons, thickness, and horizon properties.

Soil structure Binding together of soil particles to form aggregates.

Soil texture Proportional distribution of soil particle sizes.

Soil types Soils belonging to the same soil series but having different textures of the A horizon.

Solstice Date of maximum or minimum daylength.

Solubility pump Downward flux of carbon from surface to deep waters due to the downwelling of CO_2-rich North Atlantic or Antarctic waters.

Sorption Binding of an ion to a mineral surface, ranging from electrostatic attraction to covalent binding.

Source Part of a system (e.g., plant, landscape, or climate system) that shows a net export of a compound.

Southern Oscillation Atmospheric pressure changes over the southeastern Pacific and Indian Ocean.

Specialist herbivore Herbivore that specializes on consumption of one or a few plant species or tissues.

Species diversity Number, evenness, and composition of species in an ecosystem; the total range of biological attributes of all species present in an ecosystem.

Species richness Number of species in an ecosystem.

Specific heat Energy required to warm a gram of a substance by 1°C.

Specific leaf area Ratio of leaf area to leaf mass.

Specific root length Root length per unit root mass.

Spiraling length Average horizontal distance that a nutrient moves between successive uptake events.

Spodosol Soil order characterized by highly leached soils that develop in cold climates; also termed podzol in European terminology.

Stabilizing feedback Interaction in which two components of a system have opposite effects on one another; this reduces the rate of change in the system; synonymous with negative feedback.

Stakeholders People who are affected by the outcomes of a policy or action.

Stand-replacing disturbance Large disturbance that affects an entire stand of vegetation.

State factors Independent variables that control the characteristics of soils and ecosystems (climate, parent material, topography, potential biota, time, and human activities).

Steady state State of a system in which increments are approximately equal to losses, when averaged over a long time (e.g., the turnover time of the system); there are no *net* changes in the major pools in a system at steady state.

Stemflow Water that flows down stems to the ground.

Stoichiometric relationship Element ratio.

Stomata Pores in the leaf surface through which water and CO_2 are exchanged between the leaf and the atmosphere.

Stomatal conductance Flux of water vapor or CO_2 per unit driving force between the interior of a leaf and the atmosphere.

Stratification Separation of lake or ocean water into 2–3 layers of differing density due to differences in temperature and/or salinity.

Stratopause Boundary between the stratosphere and the mesosphere.

Stratosphere Atmospheric layer above the troposphere, which is heated from the top and characterized by an increase in temperature with height.

Stroma Gel matrix within the chloroplast in which the carbon-fixation reactions occur.

Subduction Downward movement of a plate margin beneath another plate.

Suberin Hydrophobic waxy substance that occurs in the cell walls of the endodermis and exodermis of plant roots.

Sublimation Vaporization of a solid such as snow.

Subsidy Energy or nutrient transfers from one ecosystem to another; synonymous with allochthonous input.

Substitutibility Capacity of one form of capital (e.g., a wetland) to provide the function that might be provided by another (e.g., water treatment plant).

Succession Directional change in ecosystem structure and functioning after disturbance.

Sunfleck Short period of high irradiance that interrupts a general background of low diffuse radiation.

Sun leaf Leaf that is acclimated to high light or is produced by a plant adapted to high light.

Surface conductance Potential of the leaf and soil surfaces in the ecosystem to lose water. Similar to stomatal conductance but applied at the canopy scale.

Surface roughness Vertical irregularities in the height of the canopy surface.

Surface water Surface layer of the ocean heated by the sun and mixed by winds, typically 75–200 m deep.

Surplus production Production in excess of that which would occur when the fish stock is limited by density-dependent mortality.

Sustainability Use of the environment and resources to meet the needs of the present without compromising the ability of future generations to meet their needs.

Swidden agriculture See shifting agriculture.

Systems ecology Study of the ecosystem as a group of components linked by fluxes of materials or energy.

Taiga Boreal forest.

Teleconnections Dynamic interactions that interconnect distant regions of the atmosphere.

Temporal scaling Extrapolation of measurements made at one time interval to longer (or occasionally shorter) time intervals.

Thermocline Relatively sharp vertical temperature gradient in a lake or ocean.

Thermohaline circulation Global circulation of deep and intermediate ocean waters driven by downwelling of cold saline surface water off of Greenland and Antarctica.

Thermosphere Outermost layer of the atmosphere, which is characterized by an increase in temperature with height.

Threshold Critical level of one or more ecosystem controls that, when crossed, cause abrupt ecosystem changes.

Throughfall Water that drops from the canopy to the ground.

Thylakoids Membrane-bound vesicles in chloroplasts in which the light-harvesting reactions of photosynthesis occur.

Tilt Angle of Earth's axis of rotation and the plane of its orbit around the sun.

Time step Shortest time interval simulated by a model.

Top-down controls Regulation of population dynamics by predation.

Toposequence Series of ecosystems that are similar except with respect to their topographic position.

Tradeoffs Alternative choices, for example among management regimes that offer different bundles of ecosystem services.

Tradewinds Easterly winds between 30°N and 30°S latitudes.

Transfer zone Portion of a drainage basin, where erosion and deposition are in dynamic balance over long time scales.

Transformation Conversion of the organic compounds contained in litter to recalcitrant organic compounds in soil humus. Also, fundamental change in the state of a system that results in different control variables and feedbacks defining the state of the system.

Transpiration Water movement through stomata from plants to the atmosphere.

Transporter Membrane-bound proteins that transport ions across cell membranes.

Trophic cascade top-down effect of predators on the biomass of organisms at lower trophic levels; results in alternation of high and low biomass of organisms in successive trophic levels.

Trophic efficiency Proportion of production of prey that is converted to production of consumers at the next trophic level.

Trophic interactions Feeding relationships among organisms.

Trophic level organisms that obtain their energy with the same number of steps removed from plants or detritus.

Trophic transfer Flux of energy or materials due to consumption of one organism by another.

Tropopause Boundary between the troposphere and the stratosphere.

Troposphere Lowest atmospheric layer, which is heated from the bottom, continually mixed by weather systems, and characterized by a decrease in temperature with height.

Tundra Ecosystem type that is too cold to support growth of trees.

Turbulence Irregular velocities of air or water movement that can transport heat and materials much more rapidly than by diffusion. Mechanical turbulence is caused by the uneven slowing of air by a rough surface. Convective turbulence is caused by the increased buoyancy of surface air caused by heat transfer from the surface.

Turnover Replacement of a pool; ratio of the flux to the pool size; lake mixing that occurs when surface waters become more dense than deep waters.

Turnover length Downstream distance traveled by a particle of carbon or nutrient between entering the stream and being respired to CO_2.

Turnover time Average time that an element spends in a system (pool/input); synonymous with mean residence time.

Ultimate limiting nutrient Nutrient whose sustained addition stimulates production and transforms a community or ecosystem.

Ultisol Soil order characterized by substantial leaching a warm, humid environment.

Unsaturated flow Water movement through soils with a water content less than field capacity.

Uplift Upward movement of Earth's surface.

Uptake Absorption of water or mineral by an organism or tissue.

Uptake length Average distance that an atom moves from the time it is released by mineralization until it is absorbed again.

Upwelling Upward movement of deep and intermediate ocean water, usually driven by offshore winds near coasts.

Validation Comparison of model predictions with data.

Vapor density Mass of water per volume of air; absolute humidity.

Vapor pressure Partial pressure exerted by water molecules in the air.

Vapor pressure deficit Difference in actual vapor pressure and the vapor pressure in air of the same temperature and pressure that is saturated with water vapor; loosely used to describe the difference in vapor pressure in air immediately adjacent to an evaporating surface and the bulk atmosphere, although strictly speaking the air masses are at different temperatures.

Vertisol Soil order characterized by swelling and shrinking clays.

Vesicular arbuscular mycorrhizae See arbuscular mycorrhizae.

Voids Spaces between soil particles.

Water holding capacity Difference in soil water content between field capacity and permanent wilting point.

Water potential Potential energy of water relative to pure water at the soil surface.

Water residence time Time required to replace the water volume of a system.

Water-saturated All soil pores filled with water.

Watershed See drainage basin. In England the term refers to the ridge that separates two drainages.

Water use efficiency Ratio of GPP to water loss; also sometimes calculated as the ratio of NPP to cumulative transpiration (growth water use efficiency).

Water vapor feedback Greenhouse effect provided by water vapor, when the atmosphere warms and increases its water vapor content.

Weathering Processes by which parent rocks and minerals are altered to more stable forms. Physical weathering breaks rocks into smaller fragments with greater surface area. Chemical weathering results from chemical reactions between rock minerals and the atmosphere or water.

Westerlies Surface winds that blow from the west.

Xanthophyll cycle Transfer of absorbed energy to xanthophyll and eventually to heat at times when electron acceptors are not available to transfer electrons to carbon-fixation reactions.

Xeric Characterized by plants that are tolerant of dry conditions.

Xylem Water-conducting tissue of plants.

Zooplankton Microscopic animals suspended in the surface water of aquatic ecosystems.

References

Aber, J., W. McDowell, K. Nadelhoffer, A. Magill, G. Bernstson, et al. 1998. Nitrogen saturation in temperate forest ecosystems. *BioScience* 48:921–934.

Aber, J.D., A. Magill, S.G. McNulty, R.D. Boone, K.J. Nadelhoffer, et al. 1995. Forest biogeochemistry and primary production altered by nitrogen saturation. *Water Air and Soil Pollution* 85:1665–1670.

Adair, E.C., W.J. Parton, S.J. Del Grosso, W.L. Silver, M.E. Harmon, et al. 2008. A simple three-pool model accurately describes patterns of long-term, global litter decomposition in the Long-term Intersite Decomposition Experiment Team (LIDET) data set. *Global Change Biology* 14:2636–2660.

Aerts, R. and F. Berendse. 1988. The effect of increased nutrient availability on vegetation dynamics in wet heathlands. *Vegetatio* 76:63–69.

Aerts, R. 1995. Nutrient resorption from senescing leaves of perennials: Are there general patterns? *Journal of Ecology* 84:597–608.

Aerts, R. 1997. Climate, leaf litter chemistry and leaf litter decomposition in terrestrial ecosystems: A triangular relationship. *Oikos* 79:439–449.

Aerts, R. and F.S. Chapin, III. 2000. The mineral nutrition of wild plants revisited: A re-evaluation of processes and patterns. *Advances in Ecological Research* 30:1–67.

Agrawal, A., A. Chhatre, and R. Hardin. 2008. Changing governance of the world's forests. *Science* 320:1460–1462.

Ahrens, C.D. 1998. *Essentials of Meteorology: An Invitation to the Atmosphere.* 2nd edition. Wadsworth, Belmont, California.

Ainsworth, E.A. and S.P. Long. 2005. What have we learned from 15 years of free-air CO_2 enrichment (FACE)? A meta-analytic review of the responses of photosynthesis, canopy properties and plant production to rising CO_2. *New Phytologist* 165: 351–371.

Algesten, G., S. Sobek, A.K. Bergström, A. Ågren, L.J. Tranvik, et al. 2003. The role of lakes in organic carbon cycling in the boreal zone. *Global Change Biology* 10:141–147.

Allan, J.D. and M.M. Castillo. 2007. *Stream Ecology: Structure and Function of Running Waters.* 2nd edition. Springer, Dordrecht.

Allen, J.L., S. Wesser, C.J. Markon, and K.C. Winterberger. 2006. Stand and landscape level effects of a major outbreak of spruce beetles on forest vegetation in the Copper River Basin, Alaska. *Forest Ecology and Management* 227:257–266.

Allen, M.F. 1991. *The Ecology of Mycorrhizae.* Cambridge University Press, Cambridge.

Allison, S.D. 2006. Brown ground: A soil carbon analog for the Green World Hypothesis? *American Naturalist* 167:619–627.

Alpert, P. 1996. Integrated conservation and development projects: Examples from Africa. *BioScience* 46: 845–855.

Altieri, M.A. 1990. Why study traditional agriculture? Pages 551–564 *in* C.R. Carrol, J.H. Vandermeer, and P.M. Rosset, editors. *Agroecology.* McGraw Hill, New York.

Amthor, J.S. 2000. The McCree-deWit-Penning de Vries-Thornley respiration paradigms: 30 years later. *Annals of Botany* 86:1–20.

Amundson, R. and H. Jenny. 1997. On a state factor model of ecosystems. *BioScience* 47:536–543.

Amundson, R., D.D. Richter, G.S. Humphreys, E.G. Jobbágy, and J. Gaillardet. 2007. Coupling between biota and earth materials in the critical zone. *Elements* 3:327–332.

Anderson, R.V., D.C. Coleman, and C.V. Cole. 1981. Effects of saprotrophic grazing on net mineralisation. Pages 201–216 *in* F.E. Clark and T. Rosswall, editors. *Terrestrial Nitrogen Cycles: Processes, Ecosystem Strategies and Management Impacts.* Ecological Bulletins, Stockholm.

Andersson, T. 1991. Influence of stemflow and throughfall from common oak (*Quercus robur*) on soil chemistry and vegetation patterns. *Canadian Journal of Forest Research* 21:917–924.

Archer, D., M. Eby, V. Brovkin, A. Ridgwell, L. Cao, et al. 2009. Atmospheric lifetime of fossil fuel carbon dioxide. *Annual Review of Earth and Planetary Science* 37:117–134.

Armesto, J.J., R. Rozzi, and J. Caspersen. 2001. Temperate forests of North and South America. Pages 223–249 *in* F.S. Chapin, III, O.E. Sala, and E. Huber-Sannwald, editors. *Global Biodiversity in a Changing Environment: Scenarios for the 21st Century.* Springer-Verlag, New York.

Armitage, D., F. Berkes, and N. Doubleday, editors. 2007. *Adaptive Co-Management: Collaboration, Learning, and Multi-Level Governance.* University of British Columbia Press, Vancouver.

Arrow, K., L. Goulder, P. Dasgupta, G. Daily, P. Ehrlich, et al. 2004. Are we consuming too much? *Journal of Economic Perspectives* 18:147–172.

Asner, G.P., D.E. Knapp, M.O. Jones, T. Kennedy-Bowdoin, R.E. Martin, et al. 2007. Carnegie Airborne Observatory: In-flight fusion of hyperspectral imaging and waveform light detection and ranging (wLiDAR) for three-dimensional studies of ecosystems. *Journal of Applied Remote Sensing* 1:DOI: 10.1117/1111.2794018.

Asner, G.P., G.V.N. Powell, J. Mascaro, D.E. Knapp, J.K. Clark, et al. 2010. High resolution forest carbon stocks and emissions in the Amazon. *Proceedings of the National Academy of Sciences, USA* 107:16738–16742.

Aston, A.R. 1979. Rainfall interception by eight small trees. *Journal of Hydrology* 42:383–396.

Ataroff, M. and M.E. Naranjo. 2009. Interception of water by pastures of *Pennisetum clandestinum* Hochst ex Chiov. and *Melinus minutiflora* Beauv. *Agricultural and Forest Meteorology* 149:1616–1620.

Ayres, E., H. Steltzer, B.L. Simmons, R.T. Simpson, J.M. Steinweg, et al. 2009. Home-field advantage accelerates leaf litter decomposition in forests. *Soil Biology and Biochemistry* 41:606–610.

Ayres, M.P. and S.F. MacLean, Jr. 1987. Development of birch leaves and the growth energetics of *Epirrita autumnata* (Geometridae). *Ecology* 68:558–568.

Ayres, M.P. 1993. Plant defense, herbivory, and climate change. Pages 75–94 *in* P.M. Kareiva, J.G. Kingsolver, and R.B. Huey, editors. *Biotic Interactions and Global Change.* Sinauer Assoc., Sunderland, MA.

Bailey, R.G. 1998. *Ecoregions: The Ecosystem Geography of the Oceans and Continents.* Springer-Verlag, New York.

Baines, S.B. and M.L. Pace. 1994. Sinking fluxes along a trophic gradient: Patterns and their implications for the fate of primary production. *Canadian Journal of Fisheries and Aquatic Sciences* 51:25–36.

Baker, J.M., T.E. Ochsner, R.T. Veterea, and T.J. Griffis. 2007. Tillage and soil carbon sequestration. What do we really know? *Agriculture, Ecosystems & Environment* 118:1–5.

Baldocchi, D., F.M. Kelliher, T.A. Black, and P.G. Jarvis. 2000. Climate and vegetation controls on boreal zone energy exchange. *Global Change Biology* 6 (Suppl. 1): 69–83.

Baldocchi, D., E. Falge, L. Gu, R. Olson, D. Hollinger, et al. 2001. FLUXNET: A new tool to study the temporal and spatial variability of ecosystem-scale carbon dioxide water vapor, and energy flux densities. *Bulletin of the American Meteorological Society* 82: 2415–2434.

Baldocchi, D.D. and J.S. Amthor. 2001. Canopy photosynthesis: History, measurements, and models. *in* J. Roy, B. Saugier, and H.A. Mooney, editors. *Terrestrial Global Productivity.* Academic Press, San Diego.

Baldocchi, D.D. 2003. Assessing the eddy covariance technique for evaluating carbon dioxide exchange rates of ecosystems: past, present and future. *Global Change Biology* 9:479–492.

Baldocchi, D.D., L. Xu, and N. Kiang. 2004. How plant functional-type, weather, seasonal drought, and soil physical properties alter water and energy fluxes of an oak-grass savanna and an annual grassland. *Agricultural and Forest Meteorology* 123:13–39.

Barber, S.A. 1984. *Soil Nutrient Bioavailability.* John Wiley & Sons, New York.

Barbier, E.B. 1987. The concept of sustainable economic development. *Environmental Conservation* 14: 101–110.

Barboza, P.S., K.L. Parker, and I.D. Hume. 2009. *Integrative Wildlife Nutrition.* Springer, Berlin.

Bardgett, R.D. 2005. *The Biology of Soil: A Community and Ecosystem Approach.* Oxford University Press, Oxford.

Bardgett, R.D., W.D. Bowman, R. Kaufmann, and S.K. Schmidt. 2005a. A temporal approach to linking aboveground and belowground ecology. *Trends in Ecology & Evolution* 20:634–641.

Bardgett, R.D., M.B. Usher, and D.W. Hopkins, editors. 2005b. *Biological Diversity and Function in Soils.* Cambridge University Press, Cambridge.

Barrett, C.B. and P. Arcese. 1995. Are integrated conservation-development projects sustainable? On the conservation of large mammals in sub-Saharan Africa. *World Development* 23:1073–1084.

Barron, A.R., N. Wurzburger, J.P. Bellenger, S.J. Wright, A.M.L. Kraepiel, et al. 2009. Molybdenum limitation of asymbiotic nitrogen fixation in tropical forest soils. *Nature Geoscience* 2:42–45.

Barry, R.G. and R.J. Chorley. 2003. *Atmosphere, Weather and Climate.* 8th edition. Routledge, London.

Baskin, C.C. and J.M. Baskin. 1998. *Seeds: Ecology, Biogeography, and Evolution of Dormancy and Germination.* Academic Press, San Diego.

Bates, T.R. and J.P. Lynch. 1996. Stimulation of root hair elongation in *Arabidopsis thaliana* by low phosphorus availability. *Plant Cell and Environment* 19:529–538.

Bayley, P.B. 1989. Aquatic environments in the Amazon Basin, with an analysis of carbon sources, fish production, and yield. Pages 399–408 *in* D.P. Dodge, editor. *Proceedings of the International Large River Symposium, Canadian Special Publications of Fisheries and Aquatic Sciences.*

Bazzaz, F.A. 1996. *Plants in Changing Environments. Linking Physiological, Population, and Community Ecology.* Cambridge University Press, Cambridge.

Beare, M.H., R.W. Parmelee, P.F. Hendrix, W. Cheng, D.C. Coleman, et al. 1992. Microbial and faunal interactions and effects on litter nitrogen and decomposition in agroecosystems. *Ecological Monographs* 62:569–591.

Berendse, F. and R. Aerts. 1987. Nitrogen-use efficiency: A biologically meaningful definition? *Functional Ecology* 1:293–296.

Berendse, F., R. Aerts, and R. Bobbink. 1993. Atmospheric nitrogen deposition and its impact on terrestrial ecosystems. Pages 104–121 *in* C.C. Vos and P. Opdam, editors. *Landscape Ecology of a Stressed Environment.* Chapman and Hall, London.

Berg, B. and H. Staaf. 1980. Decomposition rate and chemical changes of Scots pine needle litter. II. Influence of chemical composition. Pages 373–390 *in* T. Persson, editor. *Structure and Function of Northern Coniferous Forests: An Ecosystem Study.* Ecological Bulletins, Stockholm.

Berg, B., M.-B. Johansson, V. Meentemeyer, and W. Kratz. 1998. Decomposition of tree root litter in a climatic transect of coniferous forests in Northern Europe: A synthesis. *Scandinavian Journal of Forest Research* 13:202–212.

Berg, B. and V. Meentemeyer. 2002. Litter quality in a north European transect vs. carbon storage potential. *Plant and Soil* 242:83–92.

Bergh, J. and S. Linder. 1999. Effects of soil warming during spring on photosynthetic recovery in boreal Norway spruce stands. *Global Change Biology* 5: 245–253.

Berkes, F., J. Colding, and C. Folke, editors. 2003. *Navigating Social-Ecological Systems: Building Resilience for Complexity and Change.* Cambridge University Press, Cambridge.

Berkes, F., T.P. Hughes, and R.S. Steneck. 2006. Globalization, roving bandits, and marine resources. *Science* 311:1557–1558.

Berkes, F. 2008. *Sacred Ecology: Traditional Ecological Knowledge and Resource Management.* 2nd edition. Taylor & Francis, Philadelphia.

Berkes, F., G.P. Kofinas, and F.S. Chapin, III. 2009. Conservation, community, and livelihoods: Sustaining, renewing, and adapting cultural connections to the land. Pages 129–147 *in* F.S. Chapin, III, G.P. Kofinas, and C. Folke, editors. *Principles of Ecosystem Stewardship: Resilience-Based Natural Resource Management in a Changing World.* Springer, New York.

Berner, E.K., R.A. Berner, and K.L. Moulton. 2004. Plants and mineral weathering: Past and present. Pages 169–188 *in* J.I. Drever, editor. *Surface and Ground Water, Weathering, and Soils. Treatise on GeoChemistry 5.* Elsevier, San Diego.

Berner, R.A. 1997. The rise of plants and their effect on weathering and atmosphere CO_2. *Science* 276: 544–546.

Bernhardt, E.L., T.N. Hollingsworth, and F.S. Chapin, III. 2011. Fire severity mediates climate-driven shifts in understory composition of black spruce stands in interior Alaska. *Journal of Vegetation Science* 22:32–44.

Berry, J. and O. Björkman. 1980. Photosynthetic response and adaptation to temperature in higher plants. *Annual Review of Plant Physiology* 31:491–543.

Berry, W.L. 1970. Characteristics of salts secreted by *Tamarix aphylla. American Journal of Botany* 57: 1226–1230.

Beschta, R.L. and W.J. Ripple. 2009. Large predators and trophic cascades in terrestrial ecosystems in the western United States. *Biological Conservation* 142: 2401–2414.

Bettis, E.A., III, D.R. Muhs, H.M. Roberts, and A.G. Wintle. 2003. Last Glacial loess in the conterminous USA. *Quaternary Science Reviews* 22:1907–1946.

Betts, A.K. and J.H. Ball. 1997. Albedo over the boreal forest. *Journal of Geophysical Research-Atmospheres* 102:28901–28909.

Betts, E.F. and J.B. Jones. 2009. Impact of wildfire on stream nutrient chemistry and ecosystem metabolism in boreal forest catchments of interior Alaska. *Arctic, Antarctic, and Alpine Research* 41:407–417.

Bhupinderpal-Singh, A. Nordgren, M.O. Lövfvenius, M.N. Högberg, P.-E. Mellander, et al. 2003. Tree root and soil heterotrophic respiration as revealed by girdling of boreal Scots pine forest: Extending observations beyond the first year. *Plant, Cell & Environment* 26:1287–1296.

Biggs, B.J.F. 1996. Patterns in benthic algae of streams. Pages 31–56 *in* R.J. Stevenson, M.L. Bothwell, and R.L. Lowe, editors. *Algal Ecology.* Academic Press, San Diego.

Bilby, R.E. and G.E. Likens. 1980. Importance of organic debris dams in the structure and function of stream ecosystems. *Ecology* 61:1107–1113.

Billen, G., J. Garnier, J. Némery, M. Sebilo, A. Sferratore, et al. 2007. Human activity and material fluxes in a regional river basin: The Seine River watershed. *Science of the Total Environment* 375:80–97.

Billings, W.D. and H.A. Mooney. 1968. The ecology of arctic and alpine plants. *Biological Review* 43:481–529.

Bird, M.I., J. Lloyd, and G.D. Farquhar. 1994. Terrestrial carbon storage at the LGM. *Nature* 371:585.

Birkeland, P.W. 1999. *Soils and Geomorphology.* 3rd edition. Oxford University Press, New York.

Bloom, A.J., F.S. Chapin, III, and H.A. Mooney. 1985. Resource limitation in plants: An economic analogy. *Annual Review of Ecology and Systematics* 16: 363–392.

Bloom, A.J. and F.S. Chapin, III. 1981. Differences in steady-state net ammonium and nitrate influx by cold and warm-adapted barley varieties. *Plant Physiology* 68:1064–11067.

Bohlen, P.J., S. Scheu, C.M. Hale, M.A. McLean, S. Migge, et al. 2004. Non-native invasive earthworms as agents of change in northern temperate forests. *Frontiers in Ecology and the Environment* 2:427–435.

Bolin, B., P. Crutzen, P. Vitousek, R. Woodmansee, E. Goldberg, et al. 1983. Interactions of biogeochemical cycles. Pages 1–39 *in* B. Bolin and R. Cook, editors. *The Major Biogeochemical Cycles and Their Interactions*. Wiley, New York.

Bonan, G.B. 1993. Physiological controls of the carbon balance of boreal forest ecosystems. *Canadian Journal of Forest Research* 23:1453–1471.

Bonan, G.B. 2008. *Ecological Climatology: Principles and Applications*. 2nd edition. Cambridge University Press, Cambridge.

Bond, W.J. 1993. Keystone species. Pages 237–253 *in* E.-D. Schulze and H.A. Mooney, editors. *Ecosystem Function and Biodiversity*. Springer-Verlag, Berlin.

Bond-Lamberty, B.P., S.D. Peckham, D.E. Ahl, and S.T. Gower. 2007. Fire as the dominant driver of central Canadian boreal forest carbon balance. *Nature* 450:89.

Booth, M.G. and J.D. Hoeksema. 2010. Mycorrhizal networks counteract competitive effects of canopy trees on seedling survival. *Ecology* 91:2294–2302.

Booth, M.S., J.M. Stark, and E.B. Rastetter. 2005. Controls on nitrogen cycling in terrestrial ecosystems: A synthetic analysis of literature data. *Ecological Monographs* 75:139–157.

Borchert, R. 1994. Soil and stem water storage determine phenology and distribution of tropical dry forest trees. *Ecology* 75:1437–1449.

Borer, E.T., E.W. Seabloom, J.B. Shurin, K.E. Anderson, C.A. Blanchette, et al. 2005. What determines the strength of a trophic cascade? *Ecology* 86: 528–537.

Bormann, B.T. and R.C. Sidle. 1990. Changes in productivity and distribution of nutrients in a chronosequence at Glacier Bay National Park, Alaska. *Journal of Ecology* 78:561–578.

Bormann, B.T. and A.R. Kiester. 2004. Options forestry: Acting on uncertainty. *Journal of Forestry* 102: 22–27.

Bormann, F.H. and G.E. Likens. 1979. *Pattern and Process in a Forested Ecosystem*. Springer-Verlag, New York.

Bousquet, P., P. Cias, J.B. Miller, E.J. Dlugokenck, D.A. Houglustaine, et al. 2006. Contribution of anthropogenic and natural sources of atmospheric methane variability. *Nature* 443:439–443.

Boyd, R.S. 2004. Ecology of metal hyperaccumulation. *New Phytologist* 162:563–567.

Bradford, M.A., C.A. Davies, S.D. Frey, T.R. Maddox, J.M. Melillo, et al. 2008. Thermal adaptation of soil microbial respiration to elevated temperature. *Ecology Letters* 11:1316–1327.

Bradley, B.A. and J.F. Mustard. 2005. Identifying land cover variability distinct from land cover change: Cheatgrass in the Great Basin. *Remote Sensing of Environment* 94:204–213.

Bradshaw, A.D. 1983. The reconstruction of ecosystems. *Journal of Ecology* 20:1–17.

Brady, N.C. and R.R. Weil. 2001. *The Nature and Properties of Soils*. 13th Edition edition. Prentice Hall, Upper Saddle River, New Jersey.

Brady, N.C. and R.R. Weil. 2008. *The Nature and Properties of Soils*. 14th edition. Pearson Education, Inc., Upper Saddle River, NJ.

Bridgham, S.D., C.A. Johnston, J. Pastor, and K. Updegraff. 1995. Potential feedbacks of northern wetlands on climate change. *BioScience* 45:262–274.

Brimblecombe, P. 2004. The global sulfur cycle. Pages 645–682 *in* W.H. Schlesinger, editor. *Biogeochemistry*. Elsevier, Amsterdam.

Brokaw, N.V.L. 1985. Gap-phase regeneration in a tropical forest. *Ecology* 66:682–687.

Brooker, R.W. and T.V. Callaghan. 1998. The balance between positive and negative plant interactions and its relationship to environmental gradients: A model. *Oikos* 81:196–207.

Brooks, M.L., C.M. D'Antonio, D.M. Richardson, J.B. Grace, J.E. Keeley, et al. 2004. Effects of invasive alien plants on fire regimes. *BioScience* 54:677–688.

Brown, A.E., L. Zhang, T.A. McMahon, A.W. Western, and R.A. Vertessy. 2005. A review of paired catchment studies for determining changes in water yield resulting from alterations in vegetation. *Journal of Hydrology* 310:28–61.

Brown, K. 2003. Integrating conservation and development: A case of institutional misfit. *Frontiers of Ecology and the Environment* 1:479–487.

Bryant, J.P. and P.J. Kuropat. 1980. Selection of winter forage by subarctic browsing vertebrates: The role of plant chemistry. *Annual Review of Ecology and Systematics* 11:261–285.

Bryant, J.P., F.S. Chapin, III, and D.R. Klein. 1983. Carbon/nutrient balance of boreal plants in relation to vertebrate herbivory. *Oikos* 40:357–368.

Bump, J.K., K.B. Tischler, A.J. Schrank, R.O. Peterson, and J.A. Vucetich. 2009. Large herbivores and aquatic-terrestrial links in southern boreal forests. *Journal of Animal Ecology* 78:338–345.

Burgess, S.S.O., M.A. Adams, N.C. Turner, and C.K. Ong. 1998. The redistribution of soil water by tree root systems. *Oecologia* 115:306–311.

Burke, I.C., C.M. Yonker, W.J. Parton, C.V. Cole, K. Flach, et al. 1989. Texture, climate, and cultivation effects on soil organic matter content in U.S. grassland soils. *Soil Science Society of America Journal* 53:800–805.

Burke, I.C., D.S. Schimel, C.M. Yonker, W.J. Parton, L.A. Joyce, et al. 1990. Regional modeling of grassland biogeochemistry using GIS. *Landscape Ecology* 4:45–54.

Burke, I.C. and W.K. Lauenroth. 1995. Biodiversity at landscape to regional scales. Pages 304–311 *in* V.H. Heywood, editor. *Global Biodiversity Assessment*. Cambridge University Press, Cambridge.

Burke, I.C., W.K. Lauenroth, R. Riggle, P. Brannen, B. Madigan, et al. 1999. Spatial variability of soil properties in the shortgrass steppe: The relative importance of topography, grazing, microsite, and plant species in controlling spatial patterns. *Ecosystems* 2:422–438.

Cabido, M.R. and M.R. Zak. 1999. *Vegetación del Norte de Córdoba*. Imprenta Nico, Córdoba, Argentina.

Cadenasso, M.L., S.T.A. Pickett, and K. Schwarz. 2007. Spatial heterogeneity in urban ecosystems: Reconceptualizing land cover and a framework for classification. *Frontiers in Ecology and the Environment* 5:80–88.

Caldwell, M.M. and J.H. Richards. 1989. Hydraulic lift: Water efflux from upper roots improves effectiveness of water uptake from deep roots. *Oecologia* 79:1–5.

Callaway, R.M. 1995. Positive interactions among plants. *Botanical Review* 61:306–349.

Canadell, J., R.B. Jackson, J.R. Ehleringer, H.A. Mooney, O.E. Sala, et al. 1996. Maximum rooting depth of vegetation types at the global scale. *Oecologia* 108: 585–595.

Canadell, J.G., C. Le Quéré, M.R. Raupach, C.B. Field, E.T. Buitehuls, et al. 2007. Contributions to accelerating atmospheric CO_2 growth from economic activity, carbon intensity, and efficiency of natural sinks. *Proceedings of the National Academy of Sciences, USA* 104:10288–10293.

Cardille, J.A. and M. Lambois. 2010. From the redwood forest to the Gulf Stream waters: human signature nearly ubiquitous in representative US landscapes. *Frontiers in Ecology and the Environment* 8:130–134.

Cardinale, B.J., E. Duffy, D. Srivastava, M. Loreau, M. Thomas, et al. 2009. Towards a food-web perspective on biodiversity and ecosystem functioning Pages 105–120 *in* S. Naeem, D.E. Bunker, M. Loreau, A. Hector, and C. Perring, editors. *Biodiversity, Ecosystem Functioning, and Human Well-being: An Ecological and Economic Perspective.* Oxford University Press, New York.

Carmack, E. and D.C. Chapman. 2003. Wind-driven shelf/basin exchange on an Arctic shelf: The joint roles of ice cover extent and shelf-break bathymetry. *Geophysical Research Letters* 30:1778, doi:1710.1029/2003GL017526.

Carpenter, S.R., J.F. Kitchell, and J.R. Hodgson. 1985. Cascading trophic interactions and lake productivity. *BioScience* 35:634–639.

Carpenter, S.R., S.G. Fisher, N.B. Grimm, and J.F. Kitchell. 1992. Global change and freshwater ecosystems. *Annual Review of Ecology and Systematics* 23:119–139.

Carpenter, S.R., N.F. Caraco, D.L. Correll, R.W. Howarth, A.N. Sharpley, et al. 1998. Nonpoint pollution of surface waters with phosphorus and nitrogen. *Ecological Applications* 9:559–568.

Carpenter, S.R. and M.G. Turner. 2000. Hares and tortoises: Interactions of fast and slow variables in ecosystems. *Ecosystems* 3:495–497.

Carpenter, S.R., J.J. Hodgson, J.F. Kitchell, M.L. Pace, D. Bade, et al. 2001. Trophic cascades, nutrients, and lake productivity: Whole-lake experiments. *Ecological Monographs* 71:163–186.

Carpenter, S.R. 2003. *Regime Shifts in Lake Ecosystems: Pattern and Variation.* International Ecology Institute, Lodendorf/Luhe, Germany.

Carpenter, S.R., E.M. Bennett, and G.D. Peterson. 2006. Scenarios for ecosystem services: An overview. *Ecology and Society* 11:http://www.ecologyandsociety.org/vol11/iss11/art29/.

Carpenter, S.R. and W.A. Brock. 2006. Rising variance: A leading indicator of ecological transition. *Ecology Letters* 9:308–315.

Carpenter, S.R. and C. Folke. 2006. Ecology for transformation. *Trends in Ecology & Evolution* 21:309–315.

Carpenter, S.R. and R. Biggs. 2009. Freshwaters: Managing across scales in space and time. Pages 197–220 *in* F.S. Chapin, III, G.P. Kofinas, and C. Folke, editors. *Principles of Ecosystem Stewardship: Resilience-Based Natural Resource Management in a Changing World.* Springer, New York.

Carson, R. 1962. *Silent Spring.* Crest, New York.

Cerling, T.E. 1999. Paleorecords of C_4 plants and ecosystems. Pages 445–469 *in* R.F. Sage and R.K. Monson, editors. C_4 *Plant Biology.* Academic Press, San Diego.

Chabot, B.F. and D.J. Hicks. 1982. The ecology of leaf life spans. *Annual Review of Ecology and Systematics* 13:229–259.

Chadwick, O.A., L.A. Derry, P.M. Vitousek, B.J. Huebert, and L.O. Hedin. 1999. Changing sources of nutrients during 4 million years of soil and ecosystem development. *Nature* 397:491–497.

Chambers, J.Q., G.P. Asner, D.C. Morton, L.O. Anderson, S.S. Saatchi, et al. 2007. Regional ecosystem structure and function: Ecological insights from remote sensing of tropical forests. *Trends in Ecology & Evolution* 22:414–423.

Chambers, S. 1998. *Short- and Long-Term Effects of Clearing Native Vegetation for Agricultural Purposes.* PhD. Flinders University of South Australia, Adelaide, Australia.

Chambers, S. and F.S. Chapin, III. 2002. Fire effects on surface-atmosphere energy exchange in Alaskan black spruce ecosystems: Implications for feedbacks to regional climate. *Journal of Geophysical Research* 108:8145, doi:8110.1029/2001JD000530.

Chambers, S., J. Beringer, J. Randerson, and F.S. Chapin, III. 2005. Fire effects on net radiation and energy partitioning: Contrasting responses of tundra and boreal forest ecosystems. *Journal of Geophysical Research – Atmospheres* 110:D09106.

Chapin, F.S., III. 1980. The mineral nutrition of wild plants. *Annual Review of Ecology and Systematics* 11:233–260.

Chapin, F.S., III and G.R. Shaver. 1985. Individualistic growth response of tundra plant species to environmental manipulations in the field. *Ecology* 66:564–576.

Chapin, F.S., III, K. Van Cleve, and P.R. Tryon. 1986a. Relationship of ion absorption to growth rate in taiga trees. *Oecologia* 69:238–242.

Chapin, F.S., III, P.M. Vitousek, and K. Van Cleve. 1986b. The nature of nutrient limitation in plant communities. *American Naturalist* 127:48–58.

Chapin, F.S., III, N. Fetcher, K. Kielland, K.R. Everett, and A.E. Linkins. 1988. Productivity and nutrient cycling of Alaskan tundra: Enhancement by flowing soil water. *Ecology* 69:693–702.

Chapin, F.S., III. 1989. The cost of tundra plant structures: Evaluation of concepts and currencies. *American Naturalist* 133:1–19.

Chapin, F.S., III, E.-D. Schulze, and H.A. Mooney. 1990. The ecology and economics of storage in plants. *Annual Review of Ecology and Systematics* 21: 423–448.

Chapin, F.S., III. 1991a. Integrated responses of plants to stress. *BioScience* 41:29–36.

Chapin, F.S., III. 1991b. Effects of multiple environmental stresses on nutrient availability and use. Pages 67–88 *in* H.A. Mooney, W.E. Winner, and E.J. Pell, editors. *Response of Plants to Multiple Stresses.* Academic Press, San Diego.

Chapin, F.S., III and L. Moilanen. 1991. Nutritional controls over nitrogen and phosphorus resorption from Alaskan birch leaves. *Ecology* 72:709–715.

Chapin, F.S., III. 1993a. Physiological controls over plant establishment in primary succession. Pages 161–178 *in* J. Miles and D.W.H. Walton, editors. *Primary Succession.* Blackwell Scientific Publishers, Ltd., Oxford.

Chapin, F.S., III. 1993b. Functional role of growth forms in ecosystem and global processes. Pages 287–312 *in* J.R. Ehleringer and C.B. Field, editors. *Scaling Physiological Processes: Leaf to Globe.* Academic Press, San Diego.

Chapin, F.S., III, L. Moilanen, and K. Kielland. 1993. Preferential use of organic nitrogen for growth by a non-mycorrhizal arctic sedge. *Nature* 361:150–153.

Chapin, F.S., III, L.R. Walker, C.L. Fastie, and L.C. Sharman. 1994. Mechanisms of primary succession following deglaciation at Glacier Bay, Alaska. *Ecological Monographs* 64:149–175.

Chapin, F.S., III, G.R. Shaver, A.E. Giblin, K.G. Nadelhoffer, and J.A. Laundre. 1995. Response of arctic tundra to experimental and observed changes in climate. *Ecology* 76:694–711.

Chapin, F.S., III, M.S. Torn, and M. Tateno. 1996. Principles of ecosystem sustainability. *American Naturalist* 148:1016–1037.

Chapin, F.S., III, B.H. Walker, R.J. Hobbs, D.U. Hooper, J.H. Lawton, et al. 1997. Biotic control over the functioning of ecosystems. *Science* 277:500–504.

Chapin, F.S., III, W. Eugster, J.P. McFadden, A.H. Lynch, and D.A. Walker. 2000a. Summer differences among arctic ecosystems in regional climate forcing. *Journal of Climate* 13:2002–2010.

Chapin, F.S., III, E.S. Zavaleta, V.T. Eviner, R.L. Naylor, P.M. Vitousek, et al. 2000b. Consequences of changing biotic diversity. *Nature* 405:234–242.

Chapin, F.S., III, P.A. Matson, and H.A. Mooney. 2002. *Principles of Terrestrial Ecosystem Ecology.* Springer-Verlag, New York.

Chapin, F.S., III. 2003. Effects of plant traits on ecosystem and regional processes: A conceptual framework for predicting the consequences of global change. *Annals of Botany* 91:455–463.

Chapin, F.S., III and V.T. Eviner. 2004. Biogeochemistry of terrestrial net primary production. Pages 215–247 *in* W.H. Schlesinger, editor. *Treatise on Geochemistry.* Elsevier, Amsterdam.

Chapin, F.S., III, M. Sturm, M.C. Serreze, J.P. McFadden, J.R. Key, et al. 2005. Role of land-surface changes in arctic summer warming. *Science* 310:657–660.

Chapin, F.S., III, G.M. Woodwell, J.T. Randerson, G.M. Lovett, E.B. Rastetter, et al. 2006a. Reconciling carbon-cycle concepts, terminology, and methods. *Ecosystems* 9:1041–1050.

Chapin, F.S., III, J. Yarie, K. Van Cleve, and L.A. Viereck. 2006b. The conceptual basis of LTER studies in the Alaskan boreal forest. Pages 3–11 *in* F.S. Chapin, III, M.W. Oswood, K. Van Cleve, L.A. Viereck, and D.L. Verbyla, editors. *Alaska's Changing Boreal Forest.* Oxford University Press, New York.

Chapin, F.S., III, J.T. Randerson, A.D. McGuire, J.A. Foley, and C.B. Field. 2008. Changing feedbacks in the earth-climate system. *Frontiers in Ecology and the Environment* 6:313–320.

Chapin, F.S., III. 2009. Managing ecosystems sustainably: The key role of resilience. Pages 29–53 *in* F.S. Chapin, III, G.P. Kofinas, and C. Folke, editors. *Principles of Ecosystem Stewardship: Resilience-Based Natural Resource Management in a Changing World.* Springer, New York.

Chapin, F.S., III, G.P. Kofinas, and C. Folke, editors. 2009. *Principles of Ecosystem Stewardship: Resilience-Based Natural Resource Management in a Changing World.* Springer, New York.

Chapin, F.S., III, S.R. Carpenter, G.P. Kofinas, C. Folke, N. Abel, et al. 2010. Ecosystem stewardship: Sustainability strategies for a rapidly changing planet. *Trends in Ecology & Evolution* 25:241–249.

Charney, J.G., W.J. Quirk, S.-H. Chow, and J. Kornfield. 1977. A comparative study of effects of albedo change on drought in semiarid regions. *Journal of Atmospheric Sciences* 34:1366–1385.

Chase, T.N., R.A. Pielke, Sr., T.G.F. Kittel, R.R. Nemani, and S.W. Running. 2000. Simulated impacts of historical land cover changes on global climate in northern winter. *Climate Dynamics* 16:93–105.

Chazdon, R.L. and N. Fetcher. 1984. Photosynthetic light environments in a lowland rain forest in Costa Rica. *Journal of Ecology* 72:553–564.

Chazdon, R.L. and R.W. Pearcy. 1991. The importance of sunflecks for forest understory plants. *BioScience* 41:760–766.

Chen, J., J.F. Franklin, and T.A. Spies. 1995. Growing-season microclimatic gradients from clearcut edges into old-growth Douglas-fir forests. *Ecological Applications* 5:74–86.

Chen, Y.-H. and R.G. Prinn. 2006. Estimation of atmospheric methane emission between 1996–2001 using a 3-D global chemical transport model. *Journal of Geophysical Research – Atmospheres* 110:D10307, doi:10310.11029/12005JD006058.

Cheng, W., Q. Zhang, D.C. Coleman, C.R. Carroll, and C.A. Hoffman. 1996. Is available carbon limiting microbial respiration in the rhizosphere? *Soil Biology and Biochemistry* 28:1283–1288.

Cheng, W., D.W. Johnson, and S. Fu. 2003. Rhizosphere effects on decomposition. *Soil Science Society of America Journal* 67:1418–1427.

Cherry, K.A., M. Shepherd, P.J.A. Withers, and S.J. Mooney. 2008. Assessing the effectiveness of actions to mitigate nutrient loss from agriculture: A review of methods. *Science of the Total Environment* 406:1–23.

Chetkiewicz, C.-L.B., C.C. St. Clair, and M.S. Boyce. 2006. Corridors for conservation: integrating pattern and process. *Annual Review of Ecology, Evolution and Systematics* 37:317–342.

Chivers, M.R., M.R. Turetsky, J.M. Waddington, J.W. Harden, and A.D. McGuire. 2009. Effects of experimental water table and temperature manipulations on ecosystem CO_2 fluxes in an Alaskan boreal peatland. *Ecosystems* 12:1329–1342.

Choi, Y.D. 2007. Restoration ecology to the future: A call for a new paradigm. *Restoration Ecology* 15:351–353.

Christensen, N.L., A.M. Bartuska, J.H. Brown, S. Carpenter, C. D'Antonio, et al. 1996. The report of the Ecological Society of America committee on the scientific basis for ecosystem management. *Ecological Applications* 6:665–691.

Church, M. 2002. Geomorphic thresholds in riverine landscapes. *Freshwater Biology* 47:541–557.

Ciais, P., P.P. Tans, M. Trolier, J.W.C. White, and R.J. Francey. 1995. A large northern hemisphere terrestrial CO_2 sink indicated by the $^{13}C/^{12}C$ ratio of atmospheric CO_2. *Nature* 269:1098–1102.

Ciais, P., I. Janssens, A. Shvidenko, C. Wirth, Y. Malhi, et al. 2005a. The potential for rising CO_2 to account for the observed uptake of carbon by tropical, temperate and boreal forest biomes. Pages 109–150 *in* H. Griffith and P. Jarvis, editors. *The Carbon Balance of Forest Biomes.* Taylor and Francis, Milton Park, UK.

Ciais, P., M. Reichstein, N. Viovy, A. Granier, J. Ogée, et al. 2005b. Europe-wide reduction in primary productivity caused by heat and drought in 2003. *Nature* 437:529–533.

Ciais, P., A.V. Borges, G. Abril, G. Meybeck, G. Folberth, et al. 2008. The lateral carbon pump and the European carbon balance. Pages 341–360 *in* A.J. Dolman, R. Valentini, and A. Freibauer, editors. *The Continental-Scale Greenhouse Gas Balance of Europe.* Springer-Verlag, New York.

Clarholm, M. 1985. Interactions of bacteria, protozoa and plants leading to mineralization of soil nitrogen. *Soil Biology and Biochemistry* 17:181–187.

Clark, D.A., S. Brown, D.W. Kicklighter, J.Q. Chambers, J.R. Thomlinson, et al. 2001. Measuring net primary production in forests: Concepts and field methods. *Ecological Applications* 11:356–370.

Clark, W.C. and N.M. Dickson. 2003. Sustainability science: The emerging research program. *Proceedings of the National Academy of Sciences, USA* 100:8059–8061.

Clark, W.C. and L. Holliday. 2006. *Linking Knowledge with Action for Sustainable Development: The Role of Program Management.* National Academies Press, Washington.

Clarkson, D.T. 1985. Factors affecting mineral nutrient acquisition by plants. *Annual Review of Plant Physiology* 36:77–115.

Clay, K. 1990. Fungal endophytes of grasses. *Annual Review of Ecology and Systematics* 21:275–297.

Clein, J.S. and J.P. Schimel. 1994. Reduction in microbial activity in birch litter due to drying and rewetting events. *Soil Biology and Biochemistry* 26: 403–406.

Clein, J.S., B.L. Kwiatkowski, A.D. McGuire, J.E. Hobbie, E.B. Rastetter, et al. 2000. Modeling carbon responses of tundra ecosystems to historical and projected climate: A comparison of a plot- and a global-scale ecosystem model to identify process-based uncertainties. *Global Change Biology* 6(Suppl. 1): 127–140.

Clements, F.E. 1916. *Plant Succession: An Analysis of the Development of Vegetation.* Carnegie Institution of Washington Publication 242, Washington, D. C.

Cleveland, C.C. and D. Liptzin. 2007. C:N:P stoichiometry in soil: Is there a "Redfield ratio" for the microbial biomass? *Biogeochemistry* 85:235–252.

Codispoti, L.A. 2010. Interesting times for marine N_2O. *Science* 327:1339–1340.

Cody, M.L. and H.A. Mooney. 1978. Convergence versus nonconvergence in mediterranean-climate ecosystems. *Annual Review of Ecology and Systematics* 9:265–321.

Cohen, J.E. 1994. Marine and continental food webs: Three paradoxes. *Philosophical Transactions of the Royal Society of London, Series B* 343:57–69.

COHMAP. 1988. Climatic changes of the last 18,000 years: Observations and model simulations. *Science* 241:1043–1052.

Cole, J.J., N.F. Caracao, G.W. Kling, and T.K. Kratz. 1994. Carbon dioxide supersaturation in the surface waters of lakes. *Science* 265:1568–1570.

Cole, J.J., Y.T. Prairie, N.F. Caraco, W.H. McDowell, L.J. Tranvik, et al. 2007. Plumbing the global carbon cycle: Integrating inland waters into the terrestrial carbon budget. *Ecosystems* 10:171–184.

Coleman, D.C. 1994. The microbial loop concept as used in terrestrial soil ecology studies. *Microbial Ecology* 28:245–250.

Coleman, J. 1990. *Foundations of Social Theory.* Harvard University Press, Cambridge, MA.

Coley, P.D., J.P. Bryant, and F.S. Chapin, III. 1985. Resource availability and plant anti-herbivore defense. *Science* 230:895–899.

Coley, P.D. 1986. Costs and benefits of defense by tannins in a neotropical tree. *Oecologia* 70:238–241.

Comeleo, R.L., J.F. Paul, P.V. August, J. Copeland, C. Baker, et al. 1996. Relationships between watershed stressors and sediment contamination in Chesapeake Bay estuaries. *Landscape Ecology* 11:307–319.

Connell, J.H. and R.O. Slatyer. 1977. Mechanisms of succession in natural communities and their role in community stability and organization. *American Naturalist* 111:1119–1144.

Cornelissen, J.H.C. 1996. An experimental comparison of leaf decomposition rates in a wide range of temperate plant species and types. *Journal of Ecology* 84: 573–582.

Correll, D.L. 1997. Buffer zones and water quality protection: General principles.*in* N.E. Haycock, T.P. Burt, K.W.T. Goulding, and G. Pinay, editors. *Buffer Zones: Their Processes and Potential in Water Protection.* Quest Environmental, Harpenden.

Costa, M.H. and J.A. Foley. 1999. Trends in the hydrological cycle of the Amazon basin. *Journal of Geophysical Research* 104:14189–14198.

Costanza, R. and H. Daly. 1992. Natural capital and sustainable development. *Conservation Biology* 6:37–46.

Cottingham, K.L., B.L. Brown, and J.T. Lennon. 2001. Biodiversity may regulate the temporal variability of ecological processes. *Ecology Letters* 4:72–85.

Coûteaux, M.-M., P. Bottner, and B. Berg. 1995. Litter decomposition, climate and litter quality. *Trends in Ecology & Evolution* 10:63–66.

Cowles, H.C. 1899. The ecological relations of the vegetation on the sand dunes of Lake Michigan. *Botanical Gazette* 27:95–117.

Craine, J.M., D.A. Wedin, and F.S. Chapin, III. 1999. Predominance of ecophysiological over environmental controls over CO_2 flux in a Minnesota grassland. *Plant and Soil* 207:77–86.

Craine, J.M. and P.B. Reich. 2005. Leaf-level light compensation points are lower in shade-tolerant woody seedlings: Evidence from a synthesis of 115 species. *New Phytologist* 166:710–713.

Craine, J.M., C. Morrow, and W.D. Stock. 2008. Nutrient concentration ratios and co-limitation of aboveground production by nitrogen and phosphorus in Kruger National Park, South Africa. *New Phytologist* 179: 829–836.

Craine, J.M. 2009. *Resource Strategies of Wild Plants.* Princeton University Press, Princeton.

Cramer, W., A. Bondeau, F.I. Woodward, I.C. Prentice, R.A. Betts, et al. 2001. Global response of terrestrial ecosystem structure and function to CO_2 and climate change: Results from six dynamic global vegetation models. *Global Change Biology* 7:357–373.

Crews, T.E., K. Kitayama, J.H. Fownes, R.H. Riley, D.A. Herbert, et al. 1995. Changes in soil phosphorus fractions and ecosystem dynamics across a long chronosequence in Hawaii. *Ecology* 76:1407–1424.

Crocker, R.L. and J. Major. 1955. Soil development in relation to vegetation and surface age at Glacier Bay, Alaska. *Journal of Ecology* 43:427–448.

Croll, D.A., J.L. Maron, J.A. Estes, E.M. Danner, and G.V. Byrd. 2005. Introduced predators transform subarctic islands from grassland to tundra. *Science* 307:1959–1961.

Crowley, T.J. 1995. Ice-age terrestrial carbon changes revisited. *Global Biogeochemical Cycles* 9: 377–389.

Crutzen, P.J. 2002. Geology of mankind. *Nature* 415:23.

Cunningham, S.A., B. Summerhayes, and M. Westoby. 1999. Evolutionary divergences in leaf structure and chemistry, comparing rainfall and soil nutrient gradients. *Ecological Monographs* 69:569–588.

Currie, W.S., M.E. Harmon, I.C. Burke, S.C. Hart, W.J. Parton, et al. 2010. Cross-biome transplants of plant litter over a decade reveal both extension and limitation of the climate-litter quality paradigm. *Global Change Biology* 16:1744–1761.

Curtis, P.S. and X. Wang. 1998. A meta-analysis of elevated CO_2 effects on woody plant mass, form, and physiology. *Oecologia* 113:299–313.

Cyr, H. and M.L. Pace. 1993. Magnitude and patterns of herbivory in aquatic and terrestrial ecosystems. *Nature* 343:148–150.

D'Antonio, C.M. and P.M. Vitousek. 1992. Biological invasions by exotic grasses, the grass-fire cycle, and global change. *Annual Review of Ecology and Systematics* 23:63–87.

Daily, G.C. 1997. *Nature's Services: Societal Dependence on Natural Ecosystems.* Island Press, Washington.

Daily, G.C., T. Soderqvist, S. Aniyar, K. Arrow, P. Dasgupta, et al. 2000. Ecology: The value of nature and the nature of value. *Science* 289:395–396.

Dale, V.H., S. Brown, R. Haeuber, N.T. Hobbs, N. Huntly, et al. 2000. Ecological principles and guidelines for managing the use of land. *Ecological Applications* 10:639–670.

Dasgupta, P. 2001. *Human Well-Being and the Natural Environment.* Oxford University Press, Oxford.

Davidson, E.A., P.A. Matson, P.M. Vitousek, R. Riley, K. Dunkin, et al. 1993. Process regulation of soil emissions of NO and N_2O in a seasonally dry tropical forest. *Ecology* 74:130–139.

Davidson, E.A., C.J.R. de Carvalho, I.C.G. Vieira, R.D. Figueiredo, P. Moutinho, et al. 2004. Nitrogen and phosphorus limitation of biomass growth in a tropical secondary forest. *Ecological Applications* 14: 150–163.

Davidson, E.A. and I.A. Janssens. 2006. Temperature sensitivity of soil carbon decomposition and feedbacks to climate change. *Nature* 440:165–173.

Davies, W.J. and J. Zhang. 1991. Root signals and the regulation of growth and development of plants in drying soil. *Annual Review of Plant Physiology and Molecular Biology* 42:55–76.

Davis, M.B., R.R. Calcote, S. Sugita, and H. Takahara. 1998. Patchy invasion and the origin of a hemlock-hardwood forest mosaic. *Ecology* 79:2641–2659.

Davis, S.M. and J.C. Ogden, editors. 1994. *Everglades: The Ecosystem and Its Restoration.* St Lucie, Delray Beach, Florida.

Dawson, T.E. 1993. Water sources of plants as determined from xylem-water isotopic composition: Perspectives on plant competition, distribution, and water relations. Pages 465–496 *in* J.R. Ehleringer, A.E. Hall, and G.D. Farquhar, editors. *Stable Isotopes and Plant Carbon-Water Relations.* Academic Press, San Diego.

Dawson, T.E. and T.W. Siegwolf. 2007. *Stable Isotopes as Indicators of Ecological Change.* Academic Press-Elsevier, San Diego.

De Deyn, G.B., J.H.C. Cornelissen, and R.D. Bardgett. 2008. Plant functional traits and soil carbon sequestration in contrasting biomes. *Ecology Letters* 11:516–531.

de Vries, W., S. Solberg, M. Dobbertin, H. Sterba, D. Laubhann, et al. 2009. The impact of nitrogen deposition on carbon sequestration by European forests and heathlands. *Forest Ecology and Management* 258:1814–1823.

Dean, W.E. and E. Gorham. 1998. Magnitude and significance of carbon burial in lakes, reservoirs, and peatlands. *Geology* 26:535–538.

DeAngelis, D.L. and W.M. Post. 1991. Positive feedback and ecosystem organization. Pages 155–178 *in* M. Higashi and T.P. Burns, editors. *Theoretical Studies of Ecosystems: The Network Perspective.* Cambridge University Press, Cambridge.

DeAngelis, D.L., L.J. Gross, M.A. Huston, W.F. Wolff, D.M. Fleming, et al. 1998. Landscape modeling for Everglades ecosystem restoration. *Ecosystems* 1:64–75.

Del Grosso, S.J., W.J. Parton, A.R. Mosier, D.S. Ojima, A.E. Kulmala, et al. 2000. General model for N_2O and N_2 gas emissions from soils due to denitrification. *Global Biogeochemical Cycles* 14:1045–1060.

Delmas, R., C. Jambert, and Serga. 1997. Global inventory of NO_x sources. *Nutrient Cycling in Agroecosystems* 48:51–60.

Demming-Adams, B. and W.W. Adams. 1996. The role of xanthophyll cycle carotenoids in the protection of photosynthesis. *Trends in Plant Sciences* 1:21–26.

Detling, J.K., D.T. Winn, C. Procter-Gregg, and E.L. Painter. 1980. Effects of simulated grazing by belowground herbivores on growth, CO_2 exchange, and carbon allocation patterns of *Bouteloua gracilis. Journal of Applied Ecology* 17:771–778.

Detling, J.K. 1988. Grasslands and savannas: Regulation of energy flow and nutrient cycling by herbivores. Pages 131–148 *in* L.R. Pomeroy and J.J. Alberts, editors. *Concepts of Ecosystem Ecology.* Springer-Verlag, New York.

Díaz, R.J. and R. Rosenberg. 2008. Spreading dead zones and consequences for marine ecosystems. *Science* 321:926–929.

Díaz, S. and M. Cabido. 2001. Vive la différence: Plant functional diversity matters to ecosystem processes. *Trends in Ecology & Evolution* 16:646–655.

Díaz, S., J. Fargione, F.S. Chapin, III, and D. Tilman. 2006. Biodiversity loss threatens human well-being. *Plant Library of Science (PLoS)* 4:1300–1305.

Diener, E. and M.E.P. Seligman. 2004. Beyond money: Toward an economy of well-being. *Psychological Science in the Public Interest* 5:1–31.

Dietrich, W.E. and J.T. Perron. 2006. The search for a topographic signature of life. *Nature* 439:411–418.

Dietz, T., E. Ostrom, and P.C. Stern. 2003. The struggle to govern the commons. *Science* 302:1907–1912.

Dijkstra, F.A., J.B. West, S.E. Hobbie, and P.B. Reich. 2009. Antagonistic effects of species on C respiration and net N mineralization in soils from mixed coniferous plantations. *Forest Ecology and Management* 257:1112–1118.

Dimitrakopoulos, P.G. and B. Schmid. 2004. Biodiversity effects increase linearly with biotope space. *Ecology Letters* 7:574–583.

Dingman, S.L. 2001. *Physical Hydrology.* 2nd edition. Prentice Hall, Upper Saddle River, NJ.

Dirzo, R. and A. Miranda. 1991. Altered patterns of herbivory and diversity in the forest understory: A case study of the possible consequences of contemporary defaunation. Pages 273–287 *in* P.W. Price, T.M. Lewinsohn, G.W. Fernandes, and W.W. Benson, editors. *Plant-Animal Interactions: Evolutionary Ecology in Tropical and Temperate Regions.* Wiley, New York.

Dobson, A.P., A.D. Bradshaw, and A.J.M. Baker. 1997. Hopes for the future: Restoration ecology and conservation biology. *Science* 277:515–522.

Dokuchaev, V.V. 1879. Abridged historical account and critical examination of the principal soil classifications existing. *Transactions of the Petersburg Society of Naturalists* 1:64–67.

Doney, S.C., V.J. Fabry, R.A. Feely, and J.A. Kleypas. 2009. Ocean acidification: The other CO_2 problem. *Annual Review of Marine Science* 1:169–192.

Downing, J.A. and E. McCauley. 1992. The nitrogen: phosphorus relationship in lakes. *Limnology and Oceanography* 37:936–945.

Downing, J.A., Y.T. Prairie, J.J. Cole, C.M. Duarte, L.J. Tranvik, et al. 2006. The global abundance and size distribution of lakes, ponds, and impounds. *Limnology and Oceanography* 51:2388–2397.

Drake, B.G., G. Peresta, E. Beugeling, and R. Matamala. 1996. Long-term elevated CO_2 exposure in a Chesapeake Bay wetland: Ecosystem gas exchange, primary production, and tissue nitrogen. Pages 197–214 *in* G.W. Koch and H.A. Mooney, editors. *Carbon Dioxide and Terrestrial Ecosystems.* Academic Press, San Diego.

Driscoll, C.T., G.B. Lawrence, A.J. Bulger, T.J. Butler, C.S. Cronan, et al. 2001. Acidic deposition in the northeastern United States: Sources and inputs, ecosystem effects and management strategies. *BioScience* 51:180–198.

Driscoll, C.T., D. Whitall, J. Aber, E. Boyer, M. Castro, et al. 2003. Nitrogen pollution in the northeastern United States: Sources, effects, and management opitons. *BioScience* 53:357–374.

Dugdale, R.C. and J.J. Goering. 1967. Uptake of new and regenerated forms of nitrogen in primary productivity. *Limnology and Oceanography* 12:196–206.

Dugdale, R.C. 1976. Nutrient cycles. Pages 141–172 *in* D.H. Cushing and J.J. Walsh, editors. *The Ecology of the Seas.* W. B. Saunders, Philadelphia.

Dugdale, R.C., F.P. Wilkerson, and H.J. Minas. 1995. The role of a silicate pump in driving new production. *Deep Sea Research (Part I, Oceanographic Research Papers)* 42:697–719.

Dunne, T., L.A.K. Mertes, R.H. Meade, J.E. Richey, and B.R. Forsberg. 1998. Exchanges of sediment between the flood plain and channel of the Amazon River in Brazil. *Geological Society of America Bulletin* 110:450–467.

Easterlin, R.A. 2001. Income and happiness: Towards a unified theory. *The Economic Journal* 111:465–484.

Edwards, E.J. and S.A. Smith. 2010. Phylogenetic analyses reveal the shady history of C$_4$ grasses. *Proceedings of the National Academy of Sciences, USA* 107: 2532–2537.

Egler, F.E. 1954. Vegetation science concepts. I. Initial floristic composition, a factor in old-field vegetation development. *Vegetatio* 4:414–417.

Ehleringer, J.R. and H.A. Mooney. 1978. Leaf hairs: Effects on physiological activity and adaptive value to a desert shrub. *Oecologia* 37:183–200.

Ehleringer, J.R. and C.B. Osmond. 1989. Stable isotopes. Pages 281–300 *in* R.W. Pearcy, J. Ehleringer, H.A. Mooney, and P.W. Rundel, editors. *Plant Physiological Ecology: Field Methods and Instrumentation*. Chapman and Hall, London.

Ehleringer, J.R. 1993. Carbon and water relations in desert plants: An isotopic perspective. Pages 155–172 *in* J.R. Ehleringer, A.E. Hall, and G.D. Farquhar, editors. *Stable Isotopes and Plant Carbon-Water Relations*. Academic Press, San Diego.

Ehleringer, J.R. and C.B. Field, editors. 1993. *Scaling Physiological Processes: Leaf to Globe*. Academic Press, San Diego.

Ehleringer, J.R., A.E. Hall, and G.D. Farquhar, editors. 1993. *Stable Isotopes and Plant Carbon-Water Relations*. Academic Press, San Diego.

Ehleringer, J.R., N. Buchmann, and L.B. Flanagan. 2000. Carbon isotope ratios in belowground carbon cycle processes. *Ecological Applications* 10:412–422.

Ellenberg, H. 1978. *Vegetation von Mittleuropa*. Eugen Ulmer, Stuttgart.

Ellenberg, H. 1979. Man's influence on tropical mountain ecosystems in South-America: 2nd Tansley lecture. *Journal of Ecology* 67:401–416.

Ellis, E.C. and N. Ramankutty. 2008. Putting people on the map: Anthropogenic biomes of the world. *Frontiers in Ecology and the Environment* 6:439–447.

Elmqvist, T., C. Folke, M. Nyström, G. Peterson, J. Bengtsson, et al. 2003. Response diversity, ecosystem change, and resilience. *Frontiers in Ecology and the Environment* 1:488–494.

Elser, J.J., W.F. Fagan, R.F. Denno, D.R. Dobberfuhl, A. Folarin, et al. 2000. Nutritional constraints in terrestrial and freshwater food webs. *Nature* 408:578–580.

Elser, J.J., M.E.S. Bracken, E. Cleland, D.S. Gruner, W.S. Harpole, et al. 2007. Global analysis of nitrogen and phosphorus limitation of primary producers in freshwater, marine and terrestrial ecosystems. *Ecology Letters* 10:1135–1142.

Elton, C.S. 1927. *Animal Ecology*. Macmillan, New York.

Enquist, B.J., A.J. Kerkhoff, S.C. Stark, N.G. Swenson, M.C. McCarthy, et al. 2007. A general integrative model for scaling plant growth, carbon flux, and functional trait spectra. *Nature* 449:218–222.

Enríquez, S., C.M. Duarte, and K. Sand-Jensen. 1993. Patterns in decomposition rates among photosynthetic organisms: The importance of detritus C:N:P content. *Oecologia* 94:457–471.

Erisman, J.W., A. Bleeker, A. Hensen, and A. Vermeulen. 2008. Agricultural air quality in Europe and the future perspective. *Atmospheric Environment* 42: 3209–3217.

Estes, J.A. and J.F. Palmisano. 1974. Sea otters: Their role in structuring nearshore communities. *Science* 185: 1058–1060.

Estes, J.A., M.T. Tinker, T.M. Williams, and D.F. Doak. 1998. Killer whale predation on sea otters linking oceanic and nearshore ecosystems. *Science* 282: 473–476.

Eugster, W., W.R. Rouse, R.A. Pielke, J.P. McFadden, D.D. Baldocchi, et al. 2000. Land-atmosphere energy exchange in arctic tundra and boreal forest: Available data and feedbacks to climate. *Global Change Biology* 6 (Suppl. 1):84–115.

Euskirchen, E.S., A.D. McGuire, D.W. Kicklighter, Q. Zhuang, J.S. Clein, et al. 2006. Importance of recent shifts in soil thermal dynamics on growing season length, productivity, and carbon sequestration in terrestrial high-latitude ecosystems. *Global Change Biology* 12:731–750.

Euskirchen, E.S., A.D. McGuire, and F.S. Chapin, III. 2007. Energy feedbacks to the climate system due to reduced high latitude snow cover during 20th century warming. *Global Change Biology* 13: 2425–2438.

Euskirchen, E.S., A.D. McGuire, F.S. Chapin, III, S. Yi, and C.C. Thompson. 2009. Changes in vegetation in northern Alaska under scenarios of climate change, 2003–2100: Implications for climate feedbacks. *Ecological Applications* 19:1022–1043.

Evans, J.R. 1989. Photosynthesis and nitrogen relationships in leaves of C$_3$ plants. *Oecologia* 78:9–19.

Evans, L.T. 1980. The natural history of crop yield. *American Scientist* 68:388–397.

Eviner, V.T. and F.S. Chapin, III. 2001. Plant species provide vital ecosystem functions for sustainable agriculture, rangeland management and restoration. *California Agriculture* 55:54–59.

Eviner, V.T. and F.S. Chapin, III. 2003. Functional matrix: A conceptual framework for predicting multiple plant effects on ecosystem processes. *Annual Review of Ecology and Systematics* 34:455–485.

Eviner, V.T. and F.S. Chapin, III. 2005. Selective gopher disturbance influences plant species effects on nitrogen cycling. *Oikos* 109:154–166.

Eviner, V.T. and C.V. Hawkes. 2008. Embracing variability in the application of plant-soil interactions to the restoration of communities and ecosystems. *Restoration Ecology* 16:713–729.

Ewel, J.J. 1986. Designing agricultural ecosystems for the humid tropics. *Annual Review of Ecology and Systematics* 17:245–271.

Ewing, H.A., K.C. Weathers, P.H. Templer, T.E. Dawson, M.K. Firestone, et al. 2009. Fog water and ecosystem function: Heterogeneity in a California redwood forest. *Ecosystems* 12:417–433.

Fahey, T., C. Bledsoe, R. Day, R. Ruess, and A. Smucker. 1998. *Fine Root Production and Demography*. CRC Press, Boca Raton, FL.

Fahey, T.J., T.G. Siccama, C.T. Driscoll, G.E. Likens, J.L. Campbell, et al. 2005. The biogeochemistry of carbon at Hubbard Brook. *Biogeochemistry* 75:109–176.

Fahrig, L. and G. Merriam. 1985. Habitat patch connectivity and population survival. *Ecology* 66:1762–1768.

Falkenmark, M. and J. Rockström. 2004. *Balancing Water for Humans and Nature: The New Approach in Ecohydrology*. Earthscan, London.

Falkowski, P.G., R.T. Barber, and V. Smetacek. 1998. Biogeochemical controls and feedbacks on ocean primary production. *Science* 281:200–206.

Falkowski, P.G. 2000. Rationalizing elemental ratios in unicellular algae. *Journal of Phycology* 36:3–6.

Falkowski, P.G., R.J. Scholes, E. Boyle, J. Canadell, D. Canfield, et al. 2000. The global carbon cycle: A test of our knowlege of Earth as a system. *Science* 290:291–296.

Fan, S., M. Gloor, J. Mahlman, S. Pacala, J. Sarmiento, et al. 1998. A large terrestrial carbon sink in North America implied by atmospheric and oceanic carbon dioxide data and models. *Science* 282:442–446.

Fargione, J. and D. Tilman. 2005. Niche differences in phenology and rooting depth promote coexistence with a dominant C_4 bunchgrass *Oecologia* 143:598–606.

Farquhar, G.D. and T.D. Sharkey. 1982. Stomatal conductance and photosynthesis. *Annual Review of Plant Physiology* 33:317–345.

Fastie, C.L. 1995. Causes and ecosystem consequences of multiple pathways of primary succession at Glacier Bay, Alaska. *Ecology* 76:1899–1916.

Fasullo, J.T. and K.E. Trenberth. 2008. The annual cycle of the energy budget. Part II: Meridional structures and poleward transport. *Journal of Climate* 21:2313–2325.

Federov, A.V. and S.G. Philander. 2000. Is El Niño changing? *Science* 288:1997–2002.

Feely, R.A., C.L. Sabine, K. Lee, W. Berelson, J. Kleyas, et al. 2004. Impact of anthropogenic CO_2 on the $CaCO_3$ system in the oceans. *Science* 305:362–366.

Feeny, P.P. 1970. Seasonal changes in oak leaf tannins and nutrients as cause of spring feeding by winter moth caterpillars. *Ecology* 51:565–581.

Fenchel, T. 1994. Microbial ecology on land and sea. *Philosophical Transactions of the Royal Society of London, Series B* 343:51–56.

Feng, Z., R. Liu, D.L. DeAngelis, J.P. Bryant, K. Kielland, et al. 2009. Plant toxicity, adaptive herbivory, and plant community dynamics. *Ecosystems* 12:534–547.

Fenn, M.E., M.A. Poth, J.D. Aber, J.S. Baron, B.T. Bormann, et al. 1998. Nitrogen excess in North American ecosystems: Predisposing factors, ecosystem responses and management strategies. *Ecological Applications* 8:706–733.

Fenner, M. 1985. *Seed Ecology*. Chapman and Hall, London.

Field, C. 1983. Allocating leaf nitrogen for the maximization of carbon gain: Leaf age as a control on the allocation program. *Oecologia* 56:341–347.

Field, C. and H.A. Mooney. 1986. The photosynthesis-nitrogen relationship in wild plants. Pages 25–55 *in* T.J. Givnish, editor. *On the Economy of Plant Form and Function*. Cambridge University Press, Cambridge.

Field, C., F.S. Chapin, III, P.A. Matson, and H.A. Mooney. 1992. Responses of terrestrial ecosystems to the changing atmosphere: A resource-based approach. *Annual Review of Ecology and Systematics* 23:201–235.

Field, C.B. 1991. Ecological scaling of carbon gain to stress and resource availability. Pages 35–65 *in* H.A. Mooney, W.E. Winner, and E.J. Pell, editors. *Integrated Responses of Plants to Stress*. Academic Press, San Diego.

Field, C.B., D.B. Lobell, H.A. Peters, and N.R. Chiariello. 2007. Feedbacks of terrestrial ecosystems to climate change. *Annual Review of Environment and Resources* 32:1–29.

Fierer, N., J.M. Craine, K. McLauchghan, and J.P. Schimel. 2005. Litter quality and the temperature sensitivity of decomposition. *Ecology* 86:320–326.

Fierer, N., M. Breitbart, J. Nulton, P. Salamon, C. Lozupone, et al. 2007. Metagenomic and small-subunit rRNA analyses reveal the genetic diversity of bacteria, Archaea, fungi, and viruses in soil. *Applied & Environmental Microbiology* 73:7059–7066.

Fierer, N., A.S. Grandy, J. Six, and E.A. Paul. 2009a. Searching for unifying principles in soil ecology. *Soil Biology and Biochemistry* 41:2249–2256.

Fierer, N., M.S. Strickland, D. Liptzin, M.A. Bradford, and C.C. Cleveland. 2009b. Global patterns in belowground communities. *Ecology Letters* 12: 1238–1249.

Findlay, S.E.G., J.L. Tank, S. Dye, H.M. Valett, P.J. Mulholland, et al. 2002. A cross-system comparison of bacterial and fungal biomass in detritus pools of headwater streams. *Microbial Ecology* 43:55–66.

Finlay, J.C., S. Khandwala, and M.E. Power. 2002. Spatial scales of carbon flow in a river food web. *Ecology* 83:1845–1859.

Finlay, J.C. 2011. Stream size and human influences on ecosystem production in river networks. Ecosphere 2(7):artXX. doi:10.1890/ES11-00071.1.

Finlay, R.D. 2008. Ecological aspects of mycorrhizal symbiosis: With special emphasis on the functional diversity of interactions involving the extraradical mycelium. *Journal of Experimental Botany* 59:1115–1126.

Firestone, M.K. and E.A. Davidson. 1989. Microbiological basis of NO and N_2O production and consumption in soil. Pages 7–21 *in* M.O. Andreae and D.S. Schimel, editors. *Exchange of Trace Gases Between Terrestrial Ecosystems and the Atmosphere*. John Wiley and Sons, Ltd., New York.

Fisher, R.F. and D. Binkley. 2000. *Ecology and Management of Forest Soils*. 3rd edition. John Wiley & Sons, Inc., New York.

Fisher, S.G., L.J. Gray, N.B. Grimm, and D.E. Busch. 1982. Temporal succession in a desert stream ecosystem following flash flooding. *Ecological Monographs* 52:92–110.

Fisher, S.G., N.B. Grimm, E. Martí, R.M. Holmes, and J.B. Jones, Jr. 1998. Material spiraling in stream corridors: A telescoping ecosystem model. *Ecosystems* 1:19–34.

Flanagan, P.W. and A.K. Veum. 1974. Relationships between respiration, weight loss, temperature, and moisture in organic residues on tundra. Pages 249–277 *in* A.J. Holding, O.W. Heal, S.F. Maclean, Jr., and P.W. Flanagan, editors. *Soil Organisms and Decomposition in Tundra.* Tundra Biome Steering Committee, Stockholm.

Flanagan, P.W. and K. Van Cleve. 1983. Nutrient cycling in relation to decomposition and organic matter quality in taiga ecosystems. *Canadian Journal of Forest Research* 13:795–817.

Flannery, T.F. 1994. *The Future Eaters.* Reed Books, Victoria.

Flynn, K.J. 2003. Do we need complex mechanistic phytoacclimation models for phytoplankton? *Limnology and Oceanography* 48:2243–2249.

Fog, K. 1988. The effect of added nitrogen on the rate of decomposition of organic matter. *Biological Review* 63:433–462.

Foley, J.A., J.E. Kutzbach, M.T. Coe, and S. Levis. 1994. Feedbacks between climate and boreal forests during the Holocene epoch. *Nature* 371:52–54.

Foley, J.A., I.C. Prentice, N. Ramankutty, S. Levis, D. Pollard, et al. 1996. An integrated biosphere model of land surface processes, terrestrial carbon balance, and vegetation dynamics. *Global Biogeochemical Cycles* 10:603–628.

Foley, J.A., M.T. Coe, M. Scheffer, and G. Wang. 2003a. Regime shifts in the Sahara and Sahel: Interactions between ecological and climatic systems in Northern Africa. *Ecosystems* 6:524–539.

Foley, J.A., M.H. Costa, C. Delire, N. Ramankutty, and P. Snyder. 2003b. Green surprise? How terrestrial ecosystems could affect earth's climate. *Frontiers of Ecology and the Environment* 1:38–44.

Foley, J.A., R. DeFries, G.P. Asner, C. Barford, G. Bonan, et al. 2005. Global consequences of land use. *Science* 309:570–574.

Folke, C., M. Hammer, R. Costanza, and A. Jansson. 1994. Investing in natural capital: Why, what, and how? Pages 1–20 *in* A. Jansson, M. Hammer, C. Folke, and R. Costanza, editors. *Investing in Natural Capital.* Island Press, Washington.

Folland, C.K., T.R. Karl, J.R. Christy, R.A. Clarke, G.V. Gruza, et al. 2001. Observed climate variability and change. Pages 99–181 *in* J.T. Houghton, Y. Ding, D.J. Griggs, M. Noguer, P.J. van der Linden, et al., editors. *Climate Change 2001: The Scientific Basis.* Cambridge University Press, Cambridge.

Forman, R.T.T. 1995. *Land Mosaics: The Ecology of Landscapes and Regions.* Cambridge University Press, Cambridge.

Fornara, D.A. and D. Tilman. 2008. Plant functional composition influences rates of soil carbon and nitrogen accumulation. *Journal of Ecology* 96:314–322.

Foster, D.R. 1988. Disturbance history, community organization and vegetation dynamics of the old-growth Pisgah Forest, southwestern New Hampshire. *Journal of Ecology* 76:105–134.

Foster, D.R., D.A. Orwig, and J.S. McLachlan. 1996. Ecological and conservation insights from reconstruc-

tive studies of temperate old-growth forests. *Trends in Ecology & Evolution* 11:419–424.

Foster, D.R., B. Donahue, D. Kittredge, K.F. Lambert, M. Hunter, et al. 2010. *Wildlands and Woodlands: A Vision for the New England Landscape.* Harvard University, Petersham, MA.

Francis, R.C. 1990. Fisheries science and modeling: A look to the future. *Natural Resource Modeling* 4:1–10.

Frank, D.A. 2006. Large herbivores in heterogeneous grassland ecosystems. Pages 326–347 *in* K. Danell, R. Bergström, P. Duncan, and J. Pastor, editors. *Large Mammalian Herbivores, Ecosystem Dynamics, and Conservation.* Cambridge University Press, Cambridge.

Frank, D.A. 2008. Ungulate and topographic control of nitrogen:phosphorus stoichiometry in a temperate grassland: Soils, plants and mineralization rates. *Oikos* 117:591–601.

Frank, D.A., T. Depriest, K. McLauchlan, and A. Risch. 2011. Topographic and ungulate regulation of soil C turnover in a temperate trassland ecosystem. *Global Change Biology* 17:495–504.

Franklin, J.F., D.R. Berg, D.A. Thornburgh, and J.C. Tappeiner. 1997. Alternative silvicultural approaches to timber harvesting: Variable retention harvest systems. Pages 111–140 *in* K.A. Kohm and J.F. Franklin, editors. *Creating a Forestry for the 21ˢᵗ Century: The Science of Ecosystem Management.* Island Press, Washington.

Freemark, K.E. and H.G. Merriam. 1986. Importance of area and habitat heterogeneity to bird assemblages in temperate forest fragments. *Biological Conservation* 31:95–105.

Frelich, L.E. and P.B. Reich. 2009. Will environmental change reinforce the impact of global warming on the prairie-forest border of central North America? *Frontiers in Ecology and the Environment.*

Freschet, G.T., J.H.C. Cornelissen, R.S.P. van Longtestijn, and R. Aerts. 2010. Evidence of the 'plant economics spectrum' in a subarctic flora. *Journal of Ecology* 98:362–373.

Fretwell, S.D. 1977. The regulation of plant communities by food chains exploiting them. *Perspectives in Biology and Medicine* 20:169–185.

Friedl, G. and A. Wüest. 2002. Disrupting biogeochemical cycles: Consequences of damming. *Aquatic Sciences* 64:55–65.

Frost, T.M., S.R. Carpenter, A.R. Ives, and T.K. Kratz. 1995. Species compensation and complementarity in ecosystem function. Pages 224–239 *in* C.G. Jones and J.H. Lawton, editors. *Linking Species and Ecosystems.* Chapman and Hall, New York.

Fung, I.Y., C.J. Tucker, and K.C. Prentice. 1987. Application of advanced very high resolution radiometer vegetation index to study atmosphere-biosphere exchange of CO_2. *Journal of Geophysical Research* 92D:2999–3015.

Gabriel, H.W. and G.F. Tande. 1983. *A Regional Approach to Fire History in Alaska.* BLM-Alaska Technical Report 9, U.S.D.I. Bureau of Land Management.

Galloway, J.N., W.H. Schlesinger, H. Levy, II, A. Michaels, and J.L. Schnoor. 1995. Nitrogen fixation: Anthropogenic enhancement-environmental response. *Global Biogeochemical Cycles* 9:235–252.

Galloway, J.N. 1996. Anthropogenic mobilization of sulfur and nitrogen: Immediate and delayed consequences. *Annual Review of Energy in the Environment* 21:261–292.

Galloway, J.N., J.D. Aber, J.W. Erisman, S.P. Seitzinger, R.W. Howarth, et al. 2003. The nitrogen cascade. *BioScience* 53:341–356.

Galloway, J.N., F.J. Dentener, D.G. Capone, E.W. Boyer, R.W. Howarth, et al. 2004. Nitrogen cycles: Past, present, and future. *Biogeochemistry* 70:153–226.

Gardner, R.H., B.T. Milne, M.G. Turner, and R.V. O'Neil. 1987. Neutral models for the analysis of broad-scale landscape pattern. *Landscape Ecology* 1:19–28.

Gardner, W.R. 1983. Soil properties and efficient water use: An overview. Pages 45–64 *in* H.M. Taylor, W.R. Jordan, and T.R. Sinclair, editors. *Limitations to Efficient Water Use in Crop Production*. American Society of Agronomy, Madison.

Garnier, E. 1991. Resource capture, biomass allocation and growth in herbaceous plants. *Trends in Ecology & Evolution* 6:126–131.

Gartner, T.B. and Z.G. Cardon. 2004. Decomposition dynamics in mixed-species leaf litter. *Oikos* 104:230–246.

Gessner, M.O., C.M. Swan, C.K. Dang, B.G. McKie, R.D. Bardgett, et al. 2010. Diversity meets decomposition. *Trends in Ecology & Evolution* 25:325–331.

Gholz, H.L., D.A. Wedin, S.M. Smitherman, M.E. Harmon, and W.J. Parton. 2000. Long-term dynamics of pine and hardwood litter in contrasting environments: Toward a global model of decomposition. *Global Change Biology* 6:751–765.

Gilichinsky, D., T. Vishnivetskaya, M. Petrova, E. Spirina, V. Mamykin, et al. 2008. Bacteria in permafrost. Pages 83–102 *in* R. Margesin, F. Schinner, J.-C. Marx, and C. Gerday, editors. *Psychrophiles: From Biodiversity to Biotechnology*. Springer-Verlag, Berlin.

Gill, J.L., J.W. Williams, S.T. Jackson, K.B. Lininger, and G.S. Robinson. 2009. Pleistocene megafaunal collapse, novel plant communities, and enhanced fire regimes in North America. *Science* 326: 1100–1103.

Giller, P.S. and B. Malmqvist. 1998. *The Biology of Streams and Rivers*. Oxford University Press, Oxford.

Gilliam, F.S., T.R. Seastedt, and A.K. Knapp. 1987. Canopy rainfall interception and throughfall in burned and unburned tallgrass prairie. *Southwestern Naturalist* 32:267–271.

Ginn, W.J. 2005. *Investing in Nature: Case Studies of Land Conservation in Collaboration with Business*. Island Press, Washington.

Gleason, H.A. 1926. The individualistic concept of the plant association. *Bulletin of the Torrey Botanical Club* 53:7–26.

Gleick, P.H. 1998. *The World's Water 1998–1999. The Biennial Report on Freshwater Resources*. Island Press, Washington.

Goetz, S.J., A.G. Bunn, G.A. Fiske, and R.A. Houghton. 2005. Satellite-observed photosynthetic trends across boreal North America associated with climate and fire disturbance. *Proceedings of the National Academy of Sciences, USA* 102:13521–13525.

Gollan, T., N.C. Turner, and E.D. Schulze. 1985. The responses of stomata and leaf gas exchange to vapor pressure deficits and soil water content. III. In the sclerophyllous woody species *Nerium oleander*. *Oecologia* 65:356–362.

Golley, F. 1961. Energy values of ecological materials. *Ecology* 42:581–584.

Golley, F.B. 1993. *A History of the Ecosystem Concept in Ecology: More than the Sum of the Parts*. Yale University Press, New Haven.

Goode, J.G., R.J. Yokelson, D.E. Ward, R.A. Susott, R.E. Babbitt, et al. 2000. Measurements of excess O_2, CO_2, CO, CH_4, C_2H_4, C_2H_2, HCN, NO, NH_3, HCOOH, CH_3COOH, HCHO, and CH_3OH in 1997 Alaskan biomass burning plumes by airborne Fourier transform infrared spectroscopy (AFTIR). *Journal of Geophysical Research* 105:22147–22166.

Gorham, E. 1991. Biogeochemistry: Its origins and development. *Biogeochemistry* 13:199–239.

Gosz, J.R. 1991. Fundamental ecological characteristics of landscape boundaries. Pages 8–30 *in* M.M. Holland, P.G. Risser, and R.J. Naiman, editors. *Ecotones: The Role of Landscape Boundaries in the Management and Restoration of Changing Environments*. Chapman and Hall, New York.

Goulden, M.L., J.W. Munger, S.-M. Fan, B.C. Daube, and S.C. Wofsy. 1996. CO_2 exchange by a deciduous forest: Response to interannual climate variability. *Science* 271:1576–1578.

Goulder, L.H. and D. Kennedy. 1997. Valuing ecosystem services: Philosophical bases and empirical methods. Pages 23–48 *in* G.C. Daily, editor. *Nature's Services: Societal Dependence on Natural Ecosystems*. Island Press, Washington, D. C.

Gower, S.T., C.J. Kucharik, and J.M. Norman. 1999. Direct and indirect estimation of leaf area index, f(APAR), and net primary production of terrestrial ecosystems. *Remote Sensing of the Environment* 70:29–51.

Gower, S.T. 2002. Productivity of terrestrial ecosystems. Pages 516–521 *in* H.A. Mooney and J. Canadell, editors. *Encyclopedia of Global Change*. Blackwell Scientific, Oxford.

Graedel, T.E. and P.J. Crutzen. 1995. *Atmosphere, Climate, and Change*. Scientific American Library, New York.

Graetz, R.D. 1991. The nature and significance of the feedback of change in terrestrial vegetation on global atmospheric and climatic change. *Climatic Change* 18:147–173.

Graham, R.T. 2003. *Hayman Fire Case Study. General Technical Report RMRS-GTR-114*. U.S. Forest Service. Rocky Mountain Research Station, Ogden, UT.

Green, M.B. and J.C. Finlay. 2010. Patterns of hydrologic control over stream water total nitrogen to phosphorus ratios. *Biogeochemistry* 99:15–30.

Grigulis, K., S. Lavorel, I.D. Davies, A. Dossantos, F. Lloret, et al. 2005. Landscape-scale positive feedbacks between fire and expansion of the large tussock grass, *Ampelodesmos mauritanica*, in Catalan shrublands. *Global Change Biology* 11:1042–1053.

Grim, R.E. 1968. *Clay Mineralogy*. McGraw-Hill, New York.

Grime, J.P. and R. Hunt. 1975. Relative growth rate: Its range and adaptive significance in a local flora. *Journal of Ecology* 63:393–422.

Grime, J.P., G. Mason, A.V. Curtis, J. Rodman, S.R. Band, et al. 1981. A comparative study of germination characteristics in a local flora. *Journal of Ecology* 69:1017–1059.

Grime, J.P. 1998. Benefits of plant diversity to ecosystems: Immediate, filter and founder effects. *Journal of Ecology* 86:902–910.

Grime, J.P., V.K. Brown, K. Thompson, G.J. Masters, S.H. Hillier, et al. 2000. The response of two contrasting limestone grasslands to simulated climate change. *Science* 289:762–765.

Grime, J.P. 2001. *Plant Strategies, Vegetation Processes, and Ecosystem Properties*. John Wiley & Sons, Chichester, UK.

Grime, J.P., J.D. Fridley, A.P. Askew, K. Thompson, J.G. Hodgson, et al. 2008. Long-term resistance to simulated climate change in an infertile grassland. *Proceedings of the National Academy of Sciences, USA* 105:10029–10032.

Grimm, N.B. and K.C. Petrone. 1997. Nitrogen fixation in a desert stream ecosystem. *Biogeochemistry* 37:33–61.

Groendahl, L., T. Fribort, and H. Soegaard. 2007. Temperature and snow-melt controls on interannual variability in carbon exchange in the high Arctic. *Theoretical and Applied Climatology* 88:111–125.

Gross, M.R., R.M. Coleman, and R.M. McDowell. 1988. Aquatic productivity and the evolution of diadromous fish migration. *Science* 239:1291–1293.

Grove, J.M. 2009. Cities: Managing densely settled social-ecological systems. Pages 281–294 *in* F.S. Chapin, III, G.P. Kofinas, and C. Folke, editors. *Principles of Ecosystem Stewardship: Resilience-Based Natural Resource Management in a Changing World*. Springer, New York.

Gruber, N. and J.N. Galloway. 2008. An Earth-system perspective of the global nitrogen cycle. *Nature* 451: 293–296.

Guenther, A., C. Hewitt, D. Erickson, R. Fall, C. Geron, et al. 1995. A global model of natural volatile organic compound emissions. *Journal of Geophysical Research* 100D:8873–8892.

Guildford, S.J. and R.E. Hecky. 2000. Total nitrogen, total phosphorus, and nutrient limitation in lakes and oceans: Is there a common relationship? *Limnology and Oceanography* 45:1213–1223.

Gulis, V. and K. Suberkropp. 2003. Effect of inorganic nutrients on relative contributions of fungi and bacteria to carbon flow from submerged decomposing leaf litter. *Microbial Ecology* 45:11–19.

Gulledge, J., A. Doyle, and J. Schimel. 1997. Different NH_4^+-inhibition patterns of soil CH_4 consumption:

A result of distinct CH_4 oxidizer populations across sites? *Soil Biology and Biochemistry* 29:13–21.

Gulmon, S.L. and H.A. Mooney. 1986. Costs of defense on plant productivity. Pages 681–698 *in* T.J. Givnish, editor. *On the Economy of Plant Form and Function*. Cambridge University Press, Cambridge, U.K.

Gurney, K.R., R.M. Law, A.S. Denning, P.J. Rayner, D. Baker, et al. 2002. Towards robust regional estimates of CO_2 sources and sinks using atmospheric transport models. *Nature* 415:626–630.

Güsewell, S. 2004. N:P ratios in terrestrial plants: Variation and functional significance. *New Phytologist* 164: 243–266.

Gutierrez, J.R. and W.G. Whitford. 1987. Chihuahuan desert annuals: Importance of water and nitrogen. *Ecology* 68:2032–2045.

Haberl, H., K.H. Erb, F. Krausmann, V. Gaube, A. Bondeau, et al. 2007. Quantifying and mapping the human appropriation of net primary production in Earth's terrestrial ecosystems. *Proceedings of the National Academy of Sciences, USA* 104:12942–12945.

Hagen, J.B. 1992. *An Entangled Bank: The Origins of Ecosystem Ecology*. Rutgers University Press, New Brunswick, New Jersey.

Hairston, N.G., F.E. Smith, and L.B. Slobodkin. 1960. Community structure, population control and competition. *American Naturalist* 94:421–425.

Hall, S.J. and P.A. Matson. 1998. Nitrogen oxide emissions after nitrogen additions in tropical forests. *Nature* 400:152–155.

Hall, S.J. and G.P. Asner. 2007. Biological invasions alter regional N-oxide emissions in Hawaiian rain forests. *Global Change Biology* 13:2143–2160.

Hallbacken, L. 1992. *The Nature and Importance of Long-Term Soil Acidification in Swedish Forest Ecosystems*. Swedish University of Agricultural Sciences, Department of Ecology and Environmental Research, Uppsala.

Hanski, I., L. Hansson, and H. Henttonen. 1991. Specialist predators, generalist predators, and the microtine rodent cycle. *Journal of Animal Ecology* 60:353–367.

Hanski, I. 1999. *Metapopulation Ecology*. Oxford University Press, Oxford.

Hanski, I., H. Henttonen, E. Korpimäki, L. Oksanen, and P. Turchin. 2001. Small-rodent dynamics and predation. *Ecology* 82:1505–1520.

Harden, J.W., S.E. Trumbore, B.J. Stocks, A. Hirsch, S.T. Gower, et al. 2000. The role of fire in the boreal carbon budget. *Global Change Biology* 6 (Suppl. 1): 174–184.

Hardin, G. 1968. The tragedy of the commons. *Science* 162:1243–1248.

Harmon, M.E., W.L. Silver, B. Fasth, H. Chen, I.C. Burke, et al. 2009. Long-term patterns of mass loss during the decomposition of leaf and fine root litter: An intersite comparison. *Global Change Biology* 15:1320–1338.

Harpole, W.S. and D. Tilman. 2007. Grassland species loss resulting from reduced niche dimension. *Nature* 446:791–793.

Harris, J.A., R.J. Hobbs, E. Higgs, and J. Aronson. 2006. Ecological restoration and global climate change. *Restoration Ecology* 14:170–176.

Hartshorn, G.S. 1980. Neotropical forest dynamics. *Biotroprica* 12:23–30.

Hay, M.E. and W. Fenical. 1988. Marine plant-herbivore interactions: The ecology of chemical defense. *Annual Review of Ecology and Systematics* 19:111–145.

Haynes, R.J. 1986. The decomposition process: Mineralization, immobilization, humus formation, and degradation. Pages 52–126 in R.J. Haynes, editor. *Mineral Nitrogen in the Plant-Soil System.* Academic Press, Orlando.

Heal, G. 2000. *Nature and the Marketplace: Capturing the Value of Ecosystem Services.* Island Press, Washington.

Heal, O.W. and J. MacLean, S. F. 1975. Comparative productivity in ecosystems: Secondary productivity. Pages 89–108 in W.H. van Dobben and R.H. Lowe-McConnell, editors. *Unifying Concepts in Ecology.* Junk, The Hague.

Hedin, L.O., J.J. Armesto, and A.H. Johnson. 1995. Patterns of nutrient loss from unpolluted, old-growth temperate forests: Evaluation of biogeochemical theory. *Ecology* 76:493–509.

Hedin, L.O., E.N.J. Brookshire, D.N.L. Menge, and A.R. Barron. 2009. The nitrogen paradox in tropical forest ecosystems. *Annual Review of Ecology and Systematics* 40:613–635.

Heijmans, M.M.P.D., W.J. Arp, and F.S. Chapin, III. 2004. Controls on moss evaporation in a boreal black spruce forest. *Global Biogeochemical Cycles* 18:GB2004, doi:2010.1029/2003GB002128.

Heinsch, F.A., M. Zhao, S.W. Running, J.S. Kimball, R.R. Nemani, et al. 2006. Evaluation of remote sensing based terrestrial productivity from MODIS using regional tower eddy flux network observations. *IEEE Transactions on Geoscience and Remote Sensing* 44:1908–1925.

Helfield, J.M. and R.J. Naiman. 2001. Effects of salmon-derived nitrogen on riparian forest growth and implications for stream productivity. *Ecology* 82:2403–2409.

Herms, D.A. and W.J. Mattson. 1992. The dilemma of plants: To grow or defend. *Quarterly Review of Biology* 67:283–335.

Heywood, V.H. and R.T. Watson, editors. 1995. *Global Biodiversity Assessment.* Cambridge University Press, Cambridge.

Hibbert, A.R. 1967. Forest treatment effects on water yield.*in* W.E. Sopper and H.W. Lull, editors. *International Symposium on Forest Hydrology.* Pergamon Press, New York.

Hicks, W.T. and M.E. Harmon. 2002. Diffusion and seasonal dynamics of O_2 in woody debris from the Pacific Northwest, USA. *Plant and Soil* 243:67–79.

Hilborn, R., C.J. Walters, and D. Ludwig. 1995. Sustainable exploitation of renewable resources. *Annual Review of Ecology and Systematics* 26:45–67.

Hirose, T. and M.J.A. Werger. 1987. Maximizing daily canopy photosynthesis with respect to the leaf nitrogen allocation pattern in the canopy. *Oecologia* 72:520–526.

Hobbie, S.E. 1992. Effects of plant species on nutrient cycling. *Trends in Ecology & Evolution* 7:336–339.

Hobbie, S.E. 1995. Direct and indirect effects of plant species on biogeochemical processes in arctic ecosystems. Pages 213–224 in F.S. Chapin, III and C. Körner, editors. *Arctic and Alpine Biodiversity: Patterns, Causes and Ecosystem Consequences.* Springer-Verlag, Berlin.

Hobbie, S.E. and P.M. Vitousek. 2000. Nutrient regulation of decomposition in Hawaiian montane forests: Do the same nutrients limit production and decomposition? *Ecology* 81:1867–1877.

Hobbie, S.E. 2008. Nitrogen effects on litter decomposition: A five-year experiment in eight temperate grassland and forest sites. *Ecology* 89:2633–2644.

Hobbs, N.T. 1996. Modification of ecosystems by ungulates. *Journal of Wildlife Management* 60:695–713.

Hobbs, R.J. and H.A. Mooney. 1991. Effects of rainfall variability and gopher disturbance on serpentine annual grassland dynamics. *Ecology* 72:59–68.

Hobbs, R.J., S. Yates, and H.A. Mooney. 2007. Long-term data reveal complex dynamics in grassland in relation to climate and long-term disturbance. *Ecological Monographs* 77:545–568.

Hobbs, R.J. and V.A. Cramer. 2008. Restoration ecology: Interventionist approaches for restoring and maintaining ecosystem function in the face of rapid environmental change. *Annual Review of Environment and Resources* 33:39–61.

Hobbs, R.J., E. Higgs, and J.A. Harris. 2009. Novel ecosystems: Implications for conservation and restoration. *Trends in Ecology & Evolution* 24:599–605.

Hobbs, R.J., D.N. Cole, L. Yung, E.S. Zavaleta, G.H. Aplet, et al. 2010. Guiding concepts for park and wilderness stewardship in an era of global environmental change. *Frontiers in Ecology and the Environment* 8:483–490.

Hodge, A., D. Robinson, B. Griffiths, and A. Fitter. 1999. Why plants bother: Root proliferation results in increased nitrogen capture from an organic patch when two grasses compete. *Plant, Cell and Environment* 22:811–820.

Högberg, P. and I.J. Alexander. 1995. Roles of root symbioses in African woodland and forest: Evidence from ^{15}N abundance and foliar nutrient concentrations. *Journal of Ecology* 83:217–224.

Högberg, P., A. Nordgren, N. Buchmann, A.F.S. Taylor, A. Ekblad, et al. 2001. Large-scale forest girdling experiment demonstrates that current photosynthesis drives soil respiration. *Nature* 411:789–792.

Holdridge, L.R. 1947. Determination of world plant formations from simple climatic data. *Science* 105:367–368.

Holland, E.A., F.J. Dentener, B.H. Braswell, and J.M. Sulzman. 1999. Contemporary and pre-industrial global reactive nitrogen budgets. *Biogeochemistry* 46:1–37.

Holling, C.S. 1973. Resilience and stability of ecological systems. *Annual Review of Ecology and Systematics* 4:1–23.

Holling, C.S. 1986. Resilience of ecosystems: Local surprise and global change. Pages 292–317 *in* W.C. Clark and R.E. Munn, editors. *Sustainable Development and the Biosphere*. Cambridge University Press, Cambridge.

Holling, C.S. 1992. The role of forest insects in structuring the boreal landscape. Pages 170–191 *in* H.H. Shugart, R. Leemans, and G.B. Bonan, editors. *A Systems Analysis of the Global Boreal Forest*. Cambridge University Press, Cambridge.

Holling, C.S. and G.K. Meffe. 1996. Command and control and the pathology of natural resource management. *Conservation Biology* 10:328–337.

Holling, C.S. and L.H. Gunderson. 2002. Resilience and adaptive cycles. Pages 25–62 *in* L.H. Gunderson and C.S. Holling, editors. *Panarchy: Understanding Transformations in Human and Natural Systems*. Island Press, Washington.

Hollinger, D.Y., S.V. Ollinger, A.D. Richardson, T.P. Meyers, D.B. Dail, et al. 2010. Albedo estimates for land surface models and support for a new paradigm based on foliage nitrogen concentration. *Global Change Biology* 16:696–710.

Holloway, J.M., R.A. Dahlgren, B. Hansen, and W.H. Casey. 1998. Contribution of bedrock nitrogen to high nitrate concentrations in stream water. *Nature* 395:785–788.

Hooper, D.U. and P.M. Vitousek. 1998. Effects of plant composition and diversity on nutrient cycling. *Ecological Monographs* 68:121–149.

Hooper, D.U., F.S. Chapin, III, J.J. Ewel, A. Hector, P. Inchausti, et al. 2005. Effects of biodiversity on ecosystem functioning: A consensus of current knowledge and needs for future research. *Ecological Applications* 75:3–35.

Horne, A.J. and C.R. Goldman. 1994. *Limnology*. McGraw-Hill, New York.

Houghton, R.A. 2004. The contemporary carbon cycle. Pages 473–513 *in* W.H. Schlesinger, editor. *Biogeochemistry*. Elsevier, Amsterdam.

Houlton, B.Z., Y.-P. Wang, P.M. Vitousek, and C.B. Field. 2008. A unifying framework for dinitrogen fixation in the terrestrial biosphere. *Nature* 454:327–330.

Houlton, B.Z. and E. Bai. 2009. Imprint of denitrifying bacteria on the global terrestrial biosphere. *Proceedings of the National Academy of Sciences, USA* 106:21713–21716.

Howarth, R.W. 1984. The ecological significance of sulfur in the energy dynamics of salt marsh and marine sediments. *Biogeochemistry* 1:5–27.

Howarth, R.W., H. Jensen, R. Marino, and H. Postma. 1995. Transport and processing of phosphorus in nearshore and oceanic waters. Pages 323–345 *in* H. Tiessen, editor. *Phosphorus in the Global Environment: Transfers, Cycles, and Management*. John Wiley & Sons, Chichester.

Howarth, R.W., G. Billen, D. Swaney, A. Townsend, N. Jaworski, et al. 1996a. Regional nitrogen budgets and N and P fluxes for the drainages to the North Atlantic Ocean: Natural and human influences. *Biogeochemistry* 35:75–139.

Howarth, R.W., R. Schneider, and D. Swaney. 1996b. Metabolism and organic carbon fluxes in the tidal, freshwater Hudson River. *Estuaries* 19:848–865.

Howarth, R.W. and R. Marino. 2006. Nitrogen as the limiting nutrient for eutrophication in coastal marine ecosystems: Evolving views over three decades. *Limnology and Oceanography* 51:364–376.

Howarth, R.W., F. Chan, D.J. Conley, J. Garnier, S.C. Doney, et al. 2011. Coupled biogeochemical cycles: Eutrophication and hypoxia in temperate estuaries and coastal marine ecosystems. *Frontiers in Ecology and the Environment* 9:18–26.

Hu, S., F.S. Chapin, III, M.K. Firestone, C.B. Field, and N.R. Chiariello. 2001. Nitrogen limitation of microbial decomposition in a grassland under elevated CO_2. *Nature* 409:188–191.

Huante, P., E. Rincón, and F.S. Chapin, III. 1998. Effect of changing light availability on nutrient foraging in tropical deciduous tree-seedlings. *Oikos* 82:449–458.

Humphreys, W.F. 1979. Production and respiration in animal populations. *Journal of Animal Ecology* 48:427–454.

Hunt, H.W., D.C. Coleman, E.R. Ingham, E.T. Elliott, J.C. Moore, et al. 1987. The detrital food web in a shortgrass prairie. *Biology and Fertility of Soils* 3:57–68.

Hutley, L.B., D. Doley, D.J. Yates, and A. Boonsaner. 1997. Water-balance of an Australian subtropical rainforest at altitude: The ecological and physiological significance of intercepted cloud and fog. *Australian Journal of Botany* 45:311–329.

Huxman, T.E., M.D. Smith, P.A. Fay, A.K. Knapp, M.R. Shaw, et al. 2004. Convergence across biomes to a common rain-use efficiency. *Nature* 429:651–654.

Ingestad, T. and G.I. Ågren. 1988. Nutrient uptake and allocation at steady-state nutrition. *Physiologia Plantarum* 72:450–459.

Insam, H. 1990. Are the soil microbial biomass and basal respiration governed by the climatic regime? *Soil Biology and Biochemistry* 22:525–532.

IPCC. 2007. *Climate Change 2007: The Physical Science Basis, Contribution of Working Group I to the Fourth Assessment Report of the Intergovernmental Panel on Climate Change*. Cambridge University Press, Cambridge.

Irons, J.G., III, J.P. Bryant, and M.W. Oswood. 1991. Effects of moose browsing on decomposition rates of birch leaf litter in a subarctic stream. *Canadian Journal of Fisheries and Aquatic Science* 48:442–444.

Jackson, R.B., J. Canadell, J.R. Ehleringer, H.A. Mooney, O.E. Sala, et al. 1996. A global analysis of root distributions for terrestrial biomes. *Oecologia* 108:389–411.

Jackson, R.B., J.S. Sperry, and T.E. Dawson. 2000. Root water uptake and transport: Using physiological processes in global predictions. *Trends in Plant Science* 5:482–488.

Jackson, R.B., E.G. Jobbágy, R. Avissar, S.B. Roy, D.J. Barrett, et al. 2005. Trading water for carbon with biological carbon sequestration. *Science* 310:1944–1947.

Jaeger, C.H., III, S.E. Lindow, W. Miller, E. Clark, and M.K. Firestone. 1999. Mapping sugar and amino acid availability in soil around roots with bacterial sensors of sucrose and tryptophan. *Applied and Environmental Microbiology* 65:2685–2690.

Janssens, I.A., W. Dieleman, S. Luyssaert, J.-A. Subke, M. Reichstein, et al. 2010. Reduction of forest soil respiration in response to nitrogen deposition. *Nature Geoscience* 3:315–322.

Jarvis, P.G. 1976. The interpretation of the variations in leaf water potential and stomatal conductance found in canopies in the field. *Philosophical Transactions of the Royal Society of London, Series B* 273:593–610.

Jarvis, P.G. and J.W. Leverenz. 1983. Productivity of temperate, deciduous and evergreen forests. Pages 233–280 *in* O.L. Lange, P.S. Nobel, C.B. Osmond, and H. Ziegler, editors. *Encyclodedia of Plant Physiology, New Series*. Springer-Verlag, Berlin.

Jarvis, P.G. and K.G. McNaughton. 1986. Stomatal control of transpiration: Scaling up from leaf to region. *Advances in Ecological Research* 15:1–49.

Jefferies, R.L. 1988. Vegetation mosaics, plant-animal interactions, and resources for plant growth. Pages 341–369 *in* L. Gottlieb and S.K. Kain, editors. *Plant Evolutionary Biology*. Chapman and Hall, London.

Jefferies, R.L. and J.P. Bryant. 1995. The plant-vertebrate herbivore interface in arctic ecosystems. Pages 271–281 *in* F.S. Chapin, III and C. Körner, editors. *Arctic and Alpine Biodiversity: Patterns, Causes, and Ecosystem Consequences*. Springer-Verlag, Berlin.

Jenny, H. 1941. *Factors of Soil Formation*. McGraw-Hill, New York.

Jenny, H. 1980. *The Soil Resources: Origin and Behavior*. Springer-Verlag, New York.

Jenny, H., R.J. Arkley, and A.M. Schultz. 1969. The pigmy forest-podsol ecosystem and its dune associates of the Mendocino Coast. *Madroño*. 20:60–74.

Jobbágy, E.G. and R.B. Jackson. 2000. The vertical distribution of soil organic carbon and its relation to climate and vegetation. *Ecological Applications* 10:423–436.

Johnson, E.A. 1992. *Fire and Vegetation Dynamics. Studies from the North American Boreal Forest*. Cambridge University Press, Cambridge.

Johnson, M.D., J. Völker, H.V. Moeller, E. Laws, K.J. Breslauer, et al. 2009. Universal constant for heat production in protists. *Proceedings of the National Academy of Sciences, USA* 106:6696–6699.

Johnstone, J.F., T.N. Hollingsworth, F.S. Chapin, III, and M.C. Mack. 2010. Changes in fire regime break the legacy lock on successional trajectories in the Alaskan boreal forest. *Global Change Biology* 16:1281–1295.

Jonasson, S. and F.S. Chapin, III. 1985. Significance of sequential leaf development for nutrient balance of the cotton sedge, *Eriophorum vaginatum* L. *Oecologia* 67:511–518.

Jonasson, S., A. Michelsen, and I.K. Schmidt. 1999. Coupling of nutrient cycling and carbon dynamics in the Arctic: Integration of soil microbial and plant processes. *Applied Soil Ecology* 11:135–146.

Jones, C.G., J.H. Lawton, and M. Shachak. 1994. Organisms as ecosystem engineers. *Oikos* 69: 373–386.

Jones, H.G. 1992. *Plants and Microclimate: A Quantitative Approach to Environmental Plant Physiology*. 2nd edition. Cambridge University Press, Cambridge.

Jones, J.A. 2000. Hydrologic processes and peak discharge response to forest removal, regrowth, and roads in ten small experimental basins, western Cascades, Oregon. *Water Resources Research* 36:2621–2642.

Jones, J.A. and D.A. Post. 2004. Seasonal and successional streamflow response to forest cutting and regrowth in the northwest and eastern United States. *Water Resources Research* 40:W05203, doi: 05210.01029/02003WR002952.

Jones, J.B., J.D. Schade, S.G. Fisher, and N.B. Grimm. 1997. Organic matter dynamics in Sycamore Creek, a desert stream in Arizona, USA. *Journal of the North American Benthological Society* 16:78–82.

Jonsson, M. and D.A. Wardle. 2008. Context dependency of litter-mixing effects on decomposition and nutrient release across a long-term chronosequence. *Oikos* 117:1674–1682.

Ju, X.-T., G.-X. Xing, X.-P. Chen, S.-L. Zhang, L.-J. Zhang, et al. 2009. Reducing environmental risk by improving N management in intensive Chinese agricultural systems. *Proceedings of the National Academy of Sciences, USA* 106:3041–3046

Juice, S.M., T.J. Fahey, T.G. Siccama, C.T. Driscoll, E.G. Denny, et al. 2006. Response of sugar maple to calcium addition to northern hardwood forest at Hubbard Brook, NH. *Ecology* 87:1267–1280.

Kahmen, A., W. Wanek, and N. Buchmann. 2008. Foliar $\delta^{15}N$ values characterize soil N cycling and reflect nitrate or ammonium preference of plants along a temperate grassland gradient. *Oecologia* 156:861–870.

Kalff, J. 2002. *Limnology*. Prentice-Hall, Upper Saddle River, NJ.

Kaplan, L.A. and T.L. Bott. 1989. Diel fluctuations in bacterial activity on streambed substrata during vernal alagal blooms: Effects of temperature, water chemistry, and habitat. *Limnology and Oceanography* 34:718–733.

Karlsson, O.M., J.S. Richardson, and P.A. Kiffney. 2005. Modeling organic matter dynamics in headwater streams of south-western British Columbia, Canada. *Ecological Modelling* 183:463–476.

Kates, R.W., C.E. Colten, S. Laska, and S.P. Leatherman. 2006. Reconstruction of New Orleans after Hurricane Katrina: A research perspective. *Proceedings of the National Academy of Sciences, USA* 103: 14653–14660.

Keeley, J.E. 1990. Photosynthetic pathways in freshwater aquatic plants. *Trends in Ecology & Evolution* 5: 330–333.

Keeling, C.D., J.F.S. Chin, and T.P. Whorf. 1996a. Increased activity of northern vegetation inferred from atmospheric CO_2 measurements. *Nature* 382: 146–149.

Keeling, R.F., S.C. Piper, and M. Heimann. 1996b. Global and hemispheric CO_2 sinks deduced from changes in atmospheric O_2 concentration. *Nature* 381:218–221.

Kelliher, F.M., R. Leuning, M.R. Raupach, and E.-D. Schulze. 1995. Maximum conductances for evaporation from global vegetation types. *Agricultural and Forest Meteorology* 73:1–16.

Kelliher, F.M. and R. Jackson. 2001. Evaporation and the water balance. Pages 206–217 *in* A. Sturman and R. Spronken-Smith, editors. *The Physical Environment: A New Zealand Perspective*. Oxford University Press, Melbourne, Australia.

Kellner, J.R. and G.P. Asner. 2009. Convergent structural responses of tropical forests to diverse disturbance regimes. *Ecology Letters* 12:887–897.

Kellner, J.R., D.B. Clark, and S.P. Hubbell. 2009. Pervasive canopy dynamics produce short-term stability in a tropical rain forest landscape. *Ecology Letters* 12:155–164.

Kelly, D. and J.J. Sullivan. 2010. Life histories, dispersal, invasions, and global change: Progress and prospects in New Zealand ecology, 1989–2029. *New Zealand Journal of Ecology* 34:207–217.

Kemmitt, S., C.V. Lanyon, I.S. Waite, Q. Wen, A.G. O'Donnell, et al. 2008. Mineralization of native soil organic matter is not regulated by the size, activity or composition of the soil microbial biomass – a new perspective. *Soil Biology and Biochemistry* 40:61–73.

Kerkhoff, A.J., B.J. Enquist, J.J. Elser, and W.F. Fagan. 2005. Plant allometry, stoichiometry and the temperature-dependence of primary productivity. *Global Ecology and Biogeography* 14:585–598.

Kielland, K. 1994. Amino acid absorption by arctic plants: Implications for plant nutrition and nitrogen cycling. *Ecology* 75:2373–2383.

Kielland, K. 1997. Role of free amino acids in the nitrogen economy of arctic cryptogams. *Ecoscience* 4:75–79.

Kielland, K. and J. Bryant. 1998. Moose herbivory in taiga: Effects on biogeochemistry and vegetation dynamics in primary succession. *Oikos* 82:377–383.

Kielland, K., J.W. McFarland, and K. Olson. 2006. Amino acid uptake in deciduous and coniferous taiga ecosystems. *Plant and Soil* 288:297–307.

Killingbeck, K.T. and W.G. Whitford. 1996. High foliar nitrogen in desert shrubs: An important ecosystem trait or defective desert doctrine? *Ecology* 77:1728–1737.

Kitchell, J.F., editor. 1992. *Food Web Management: A Case Study of Lake Mendota*. Springer-Verlag, New York.

Kleiden, A. and H.A. Mooney. 2000. A global distribution of biodiversity inferred from climatic constraints: Results from a process-based modelling study. *Global Change Biology* 6:507–523.

Klein, D.R. 1982. Fire, lichens, and caribou. *Journal of Range Management* 35:390–395.

Kling, G.W., G.W. Kipphut, and M.C. Miller. 1991. Arctic lakes and streams as gas conduits to the atmosphere: Implications for tundra carbon budgets. *Science* 251:298–301.

Knapp, A.K., J.M. Briggs, D.C. Hartnett, and S.L. Collins, editors. 1998. *Grassland Dynamics: Konza Prairie, and Long-Term Ecological Research in Tallgrass Prairie*. Oxford University Press, New York.

Knapp, A.K. and M.D. Smith. 2001. Variation among biomes in temporal dynamics of aboveground primary production. *Science* 291:481–484.

Knorr, M., S.D. Frey, and P.S. Curtis. 2005. Nitrogen additions and litter decomposition: A meta-analysis. *Ecology* 86:3252–3257.

Kobe, R.K., C.A. Lepczyk, and M. Iyer. 2005. Resorption efficiency decreases with increasing green leaf nutrients in a global data set. *Ecology* 86:2780–2792.

Koerselman, W. and A.F.M. Mueleman. 1996. The vegetation N:P ratio: A new tool to detect the nature of nutrient limitation. *Journal of Applied Ecology* 33:1441–1450.

Kofinas, G.P. 2009. Adaptive co-management in social-ecological governance. Pages 77–101 *in* F.S. Chapin, III, G.P. Kofinas, and C. Folke, editors. *Principles of Ecosystem Stewardship: Resilience-Based Natural Resource Management in a Changing World*. Springer, New York.

Koide, R.T. 1991. Nutrient supply, nutrient demand and plant response to mycorrhizal infection. *New Phytologist* 117:365–386.

Körner, C., J.A. Scheel, and H. Bauer. 1979. Maximum leaf diffusive conductance in vascular plants. *Photosynthetica* 13:45–82.

Körner, C. and W. Larcher. 1988. Plant life in cold climates. *Symposium of the Society of Experimental Biology* 42:25–57.

Körner, C. 1994. Leaf diffusive conductances in the major vegetation types of the globe. Pages 463–490 *in* E.-D. Schulze and M.M. Caldwell, editors. *Ecophysiology of Photosynthesis*. Springer-Verlag, Berlin.

Körner, C. 1999. *Alpine Plant Life*. Springer-Verlag, Berlin.

Kortelainen, P., M. Rantakari, J.T. Huttunen, T. Mattsson, J. Alm, et al. 2006. Sediment respiration and lake trophic state are important predictors of large CO_2 evasion from small boreal lakes. *Global Change Biology* 12:1554–1567.

Kozlovsky, D.G. 1968. A critical evaluation of the trophic level concept. I. Ecological efficiencies. *Ecology* 49:48–60.

Kozlowski, T.T., P.J. Kramer, and S.G. Pallardy. 1991. *The Physiological Ecology of Woody Plants*. Academic Press, San Diego.

Kramer, P.J. and J.S. Boyer. 1995. *Water Relations of Plants and Soils*. Academic Press, San Diego.

Kremen, C.K., A.M. Merenlender, and D.D. Murphy. 1994. Ecological monitoring: A vital need for integrated conservation and development programs in the tropics. *Conservation Biology* 8:388–397.

Kremen, C.K., I. Raymond, and K. Lance. 1998. An interdisciplinary tool for monitoring conservation impacts in Madagascar. *Conservation Biology* 12:549–563.

Kristjanson, P., R.S. Reid, N.M. Dickson, W.C. Clark, D. Romney, et al. 2009. Linking international agricultural research knowledge with action for sustainable development. *Proceedings of the National Academy of Sciences, USA* 106:5047–5052.

Kroehler, C.J. and A.E. Linkins. 1991. The absorption of inorganic phosphate from ^{32}P-labeled inositol hexaphosphate by *Eriophorum vaginatum*. *Oecologia* 85:424–428.

Kronzucker, H.J., M.Y. Siddiqi, and A.M. Glass. 1997. Conifer root discrimination against soil nitrate and the ecology of forest succession. *Nature* 385:59–61.

Kucharik, C.J., J.A. Foley, C. Delire, V.A. Fisher, M.T. Coe, et al. 2000. Testing the performance of a dynamic global ecosystem model: Water balance, carbon balance and vegetation structure. *Global Biogeochemical Cycles* 14:795–825.

Kummerow, J., B.A. Ellis, S. Kummerow, and F.S. Chapin, III. 1983. Spring growth of shoots and roots in shrubs of an Alaskan muskeg. *American Journal of Botany* 70:1509–1515.

Kursar, T.A. and P.D. Coley. 2003. Convergence in defense syndromes of young leaves in tropical rainforests. *Biochemical Systematics and Ecology* 31: 929–949.

Kurz, W.A., C.C. Dymond, G. Stinson, G.J. Rampley, E.T. Neilson, et al. 2008. Mountain pine beetle and forest carbon feedback to climate change. *Nature* 452:987–990.

Kuzyakov, Y., J.K. Friedel, and K. Stahr. 2000. Review of mechanisms and quantification of priming effects. *Soil Biology and Biochemistry* 32:1485–1498.

Lafleur, P.M. and E.R. Humphreys. 2007. Spring warming and carbon dioxide exchange over low Arctic tundra in central Canada. *Global Change Biology* 14:740–756.

Lafont, S., L. Kergoat, G. Dedieu, A. Chevillard, U. Karstens, et al. 2002. Spatial and temporal variability of land CO_2 fluxes estimated with remote sensing and analysis data over western Eurasia. *Tellus B* 4:820–833.

Laiho, R., H. Vasander, T. Penttila, and J. Laine. 2003. Dynamics of plant-mediated organic matter and nutrient cycling following water-level draw-down in boreal peatlands. *Global Biogeochemical Cycles* 17:17:1053. doi:1010.1029/2002GB002015.

Laine, K. and H. Henttonen. 1983. The role of plant production in microtine cycles in northern Fennoscandia. *Oikos* 40:407–418.

Lajtha, K. and M. Klein. 1988. The effect of varying phosphorus availability on nutrient use by *Larrea tridentata*, a desert evergreen shrub. *Oecologia* 75:348–353.

Lambers, H. and H. Poorter. 1992. Inherent variation in growth rate between higher plants: A search for physiological causes and ecological consequences. *Advances in Ecological Research* 23:187–261.

Lambers, H., O.K. Atkin, and I. Scheurwater. 1996. Respiratory patterns in roots in relation to their functioning. Pages 323–362 *in* Y. Waisel, A. Eshel, and U. Kafkaki, editors. *Plant Roots: The Hidden Half*. Marcel Dekker, New York.

Lambers, H., F.S. Chapin, III, and T.L. Pons. 2008. *Plant Physiological Ecology*. 2nd edition. Springer, New York.

Lambin, E.F., H.J. Geist, and E. Lepers. 2003. Dynamics of land use and land cover in tropical regions. *Annual Review of Environment and Resources* 28:205–241.

Landsberg, J.J. and S.T. Gower. 1997. *Applications of Physiological Ecology to Forest Management*. Academic Press, San Diego.

Larcher, W. 2003. *Physiological Plant Ecology: Ecophysiology and Stress Physiology of Functional Groups*. 4th edition. Springer-Verlag, Berlin.

Lauenroth, W.K., J.L. Dodd, and P.L. Simms. 1978. The effects of water- and nitrogen-induced stresses on plant community structure in a semiarid grassland. *Oecologia* 36:211–222.

Lauenroth, W.K. and O.E. Sala. 1992. Long-term forage production of North American shortgrass steppe. *Ecological Applications* 2:397–403.

Laurance, W.F. and R.O. Bierregaard, editors. 1997. *Tropical Forest Remnants: Ecology, Management and Conservation of Fragmented Forests*. University of Chicago Press, Chicago.

Lavelle, P., D. Bignell, and M. Lepage. 1997. Soil function in a changing world: The role of invertebrate ecosystem engineers. *European Journal of Soil Biology* 33:159–193.

Law, B.E., E. Falge, L. Gu, D.D. Baldocchi, P. Bakwin, et al. 2002. Environmental controls over carbon dioxide and water vapor exchange of terrestrial vegetation. *Agricultural and Forest Meteorology* 113:97–120.

Lawton, J.H. and C.G. Jones. 1995. Linking species and ecosystems: Organisms as ecosystem engineers. Pages 141–150 *in* C.G. Jones and J.H. Lawton, editors. *Linking species and ecosystems*. Chapman and Hall, New York.

Le Quéré, C., C. Rödenbeck, E. T. Buitenhuis, T. J. Conway, R. L. Langenfelds, et al. 2007. Saturation of the Southern Ocean CO_2 sink due to recent climate change. *Science* 316:1735–1738.

Le Quéré, C., M.R. Raupach, J.G. Canadell, G. Marland, L. Bopp, et al. 2009. Trends in the sources and sinks of carbon dioxide. *Nature Geoscience* 2:831–836.

LeBauer, D.S. and K.K. Treseder. 2008. Nitrogen limitation of net primary production in terrestrial ecosystems is globally distributed. *Ecology* 89:371–379.

Lee, R.B. 1982. Selectivity and kinetics of ion uptake by barley plants following nutrient deficiency. *Annals of Botany* 50:429–449.

Lee, R.B. and K.A. Rudge. 1987. Effects of nitrogen deficiency on the absorption of nitrate and ammonium by barley plants. *Annals of Botany* 57:471–486.

Lekberg, Y. and R.T. Koide. 2005. Is plant performance limited by abundance of arbuscular mycorrhizal fungi? A meta-analysis of studies published between 1988 and 2003. *New Phytologist* 168:189–204.

Levick, S.R., G.P. Asner, O.A. Chadwick, L.M. Khomo, K.H. Rogers, et al. 2010. Regional insight into savanna hydrogeomorphology from termite mounds. *Nature Communications* 1:doi:10.1038/ncomms1066.

Levin, S.A. 1999. *Fragile Dominion: Complexity and the Commons*. Perseus Books, Reading, MA.

Lewis, W.M., Jr., S.K. Hamilton, M.A. Rodriguez, J.F. Saunders, and M.A. Lasi. 2001. Foodweb analysis of the Orinoco floodplain based on production estimates and stable isotope data. *Journal of the North American Benthological Society* 20:241–254.

Lieth, H. 1975. Modeling the primary productivity of the world. Pages 237–263 *in* H. Lieth and R.H. Whittaker, editors. *Primary Productivity of the Biosphere*. Springer-Verlag, Berlin.

Likens, G.E., F.H. Bormann, R.S. Pierce, J.S. Eaton, and N.M. Johnson. 1977. *Biogeochemistry of a Forested Ecosystem*. Springer-Verlag, New York.

Likens, G.E., C.T. Driscoll, and D.C. Buso. 1996. Long-term effects of acid rain: Response and recovery of a forest ecosystem. *Science* 272:244–246.

Limm, E.B., K.A. Simonin, A.G. Bothman, and T.E. Dawson. 2009. Foliar water uptake: A common water acquisition strategy for plants of the redwood forest. *Oecologia* 161:449–459.

Lindeman, R.L. 1942. The trophic-dynamic aspects of ecology. *Ecology* 23:399–418.

Lindroth, R.L. 1996. CO_2-mediated changes in tree chemistry and tree-Lepidopteran interactions. Pages 105–120 *in* G.W. Koch and H.A. Mooney, editors. *Carbon Dioxide and Terrestrial Ecosystems*. Academic Press, San Diego.

Lipson, D.A., S.K. Schmidt, and R.K. Monson. 1999. Links between microbial population dynamics and nitrogen availability in an alpine ecosystem. *Ecology* 80:1623–1631.

Lipson, D.A., T.K. Raab, S.K. Schmidt, and R.K. Monson. 2001. An empirical model of amino acid transformations in an alpine soil. *Soil Biology and Biochemistry* 33:189–198.

Liston, G.E. and M. Sturm. 1998. A snow-transport model for complex terrain. *Journal of Glaciology* 44:498–516.

Liu, J., S.X. Li, Z.Y. Ouyang, C. Tam, and X. Chen. 2008. Ecological and socioeconomic effects of China's policies for ecosystem services. *Proceedings of the National Academy of Sciences, USA* 105:9477–9482.

Liu, L. and T.L. Greaver. 2009. A review of nitrogen enrichment effects on three biogenic GHGs: The CO_2 sink may be largely offset by stimulated N_2O and CH_4 emission. *Ecology Letters* 12:1103–1117.

Livingston, G.P. and G.L. Hutchinson. 1995. Enclosure-based measurement of trace gas exchange: Applications and sources of error. Pages 14–51 *in* P.A. Matson and R.C. Harriss, editors. *Biogenic Trace Gases: Measuring Emissions from Soil and Water*. Blackwell Scientific, Oxford.

Lloyd, J. and J.A. Taylor. 1994. On the temperature dependence of soil respiration. *Functional Ecology* 8:315–323.

Lohrenz, S.E., G.L. Fahnenstiel, D.G. Redalje, G.A. Lang, M.J. Dagg, et al. 1999. Nutrients, irradiance, and mixing as factors regulating primary production in coastal waters impacted by the Mississippi River plume. *Continental Shelf Research* 19:1113–1141.

Long, S.P., E.A. Ainsworth, A.D.B. Leakey, J. Nosberger, and D.R. Ort. 2006. Food for thought: Lower-than-expected crop yield stimulation with rising CO_2 concentrations. *Science* 312:1918–1921.

Longhurst, A.R. 1998. *Ecological Geography of the Sea*. Academic Press, San Diego.

Lorio, P.L., Jr. 1986. Growth-differentiation balance: A basis for understanding southern pine beetle-tree interactions. *Forest Ecology and Management* 14:259–273.

Los, S.O., G.J. Collatz, P.J. Sellers, C.M. Malmström, N.H. Pollack, et al. 2000. A global 9-yr biophysical land surface dataset from NOAA AVHRR data. *Journal of Hydrometeorology* 1:183–199.

Lousier, J.D. and S.S. Bamforth. 1990. Soil protozoa. Pages 97–136 *in* D.L. Dindal, editor. *Soil Biology Guide*. John Wiley and Sons, New York.

Lovett, G.M. 1994. Atmospheric deposition of nutrients and pollutants in North America: An ecological perspective. *Ecological Applications* 4:629–650.

Lovett, G.M., K.C. Weathers, and W.V. Sobczak. 2000. Nitrogen saturation and retention in forested watersheds of the Catskill Mountains, New Hampshire. *Ecological Applications* 10:73–84.

Lowrance, R., L.S. Altier, J.D. Newbold, R.R. Schnabel, P.M. Groffman, et al. 1997. Water quality functions of riparian forest buffer systems in the Chesapeake Bay watershed. *Environmental Management* 21:687–712.

Ludwig, D., R. Hilborn, and C. Walters. 1993. Uncertainty, resource exploitation, and conservation: Lessons from history. *Science* 260:17, 36.

Lu, M., Y. Yang, Y. Luo, C. Fang, X. Zhou, et al. 2010. Responses of ecosystem nitrogen cycle to nitrogen addition: A meta-analysis. New Phytologist:doi: 10.1111/j.1469-8137.2010.03563.x.

Ludwig, J.A. and D.J. Tongway. 1995. Spatial organisation of landscapes and its function in semi-arid woodlands, Australia. *Landscape Ecology* 10:51–63.

Luo, Y., S. Wan, D. Hui, and L.L. Wallace. 2001. Acclimatization of soil respiration to warming in a tall grass prairie. *Nature* 413:622–625.

Luo, Y. 2007. Terrestrial carbon-cycle feedback to climate warming. *Annual Review of Ecology, Evolution and Systematics* 38:683–712.

Luyssaert, S., I. Inglima, M. Jung, A.D. Richardson, M. Reichstein, et al. 2007. CO_2 balance of boreal, temperate, and tropical forests derived from a global database. *Global Change Biology* 13:2509–2537.

Luyssaert, S., E.-D. Schulze, A. Börner, A. Knohl, D. Hessenmöller, et al. 2008. Old-growth forests as global carbon sinks. *Nature* 455:213–215.

Lytle, D.A. and N.L. Poff. 2004. Adaptation to natural flow regimes. *Trends in Ecology & Evolution* 19:94–100.

Lyver, P.O.B., H. Moller, and C. Thompson. 1999. Changes in sooty shearwater *Puffinus griseus* chick production and harvest precede ENSO events. *Marine Ecology Progress Series* 188:237–248.

MacArthur, R.H. and E.O. Wilson. 1967. *The Theory of Island Biogeography*. Princeton University Press, Princeton.

Mace, G., H. Masundire, J. Baillie, T. Ricketts, T. Brooks, et al. 2005. Biodiversity. Pages 77–122 *in* R. Hassan,

R.J. Scholes, and N. Ash, editors. *Ecosystems and Human Well-Being: Current State and Trends, Volume 1*. Island Press, Washington.

Mack, M.C., C.M. D'Antonio, and R.E. Ley. 2001. Pathways through which exotic grasses alter N cycling in a seasonally dry Hawaiian woodland. *Ecological Applications* 11:1323–1335.

MacLean, D.A. and R.W. Wein. 1978. Weight loss and nutrient changes in decomposing litter and forest floor material in New Brunswick forest stands. *Canadian Journal of Botany* 56:2730–2749.

Magill, A.H., J.D. Aber, J.J. Hendricks, R.D. Bowden, J.M. Melillo, et al. 1997. Biogeochemical response of forest ecosystems to simulated chronic nitrogen deposition. *Ecological Applications* 7:402–415.

Magnani, F., M. Mencuccini, M. Borghetti, P. Berbigier, F. Berninger, et al. 2007. The human footprint in the carbon cycle of temperate and boreal forests. *Nature* 447:849–851.

Mann, K.H. and J.R.N. Lazier. 2006. *Dynamics of Marine Ecosystems: Biological-Physical Interactions in the Oceans*. 3rd edition. Blackwell Publishing, Victoria, Australia.

Manzoni, S., R.B. Jackson, J.A. Trofymow, and A. Porporato. 2008. The global stoichiometry of litter nitrogen mineralization. *Science* 321:684–686.

Margalef, R. 1968. *Perspectives in Ecological Theory*. University of Chicago Press, Chicago.

Margolis, H., R. Oren, D. Whitehead, and M.R. Kaufmann. 1995. Leaf area dynamics of conifer forests. Pages 181–223 *in* W.K. Smith and T.M. Hinckley, editors. *Ecophysiology of Coniferous Forests*. Academic Press, San Diego.

Mark, A.F., K.S. Turner, and C.J. West. 2001. Integrating nature conservation with hydro-electric development: Conflict resolution with Lakes Manapouri and Te Anau, Fiordland National Park, New Zealand. *Lake and Reservoir Management* 17:1–16.

Mark, A.F. and K.J.M. Dickinson. 2008. Maximizing water yield with indigenous non-forest vegetation: A New Zealand perspective. *Frontiers in Ecology and the Environment* 6:25–34.

Marschner, H. 1995. *Mineral Nutrition in Higher Plants*. 2nd edition. Academic Press, London.

Martin, J.H. 1990. Glacial-interglacial CO_2 exchange: The iron hypothesis. *Paleoceanography* 5:1–13.

Marvier, M., C. McCreedy, J. Regetz, and P. Kareiva. 2007. A meta-analysis of effects of Bt cotton and maize on nontarget invertebrates. *Science* 316: 1475–1477.

Mary, B., S. Recous, D. Darwis, and D. Robin. 1996. Interactions between decomposition of plant residues and nitrogen cycling in soil. *Plant and Soil* 181:71–82.

Maslow, A.H. 1943. A theory of human motivation. *Psychological Review* 50:370–396.

Matson, P.A. and P.M. Vitousek. 1981. Nitrogen mineralization and nitrification potentials following clearcutting in the Hoosier National Forest, Indiana. *Forest Science* 27:781–791.

Matson, P.A. and R.D. Boone. 1984. Natural disturbance and nitrogen mineralization: Wave-form dieback of mountain hemlock in the Oregon Cascades. *Ecology* 65:1511–1516.

Matson, P.A. and R.H. Waring. 1984. Effects of nutrient and light limitation on mountain hemlock: Susceptibility to laminated root rot. *Ecology* 65: 1517–1524.

Matson, P.A. and P.M. Vitousek. 1987. Cross-system comparisons of soil nitrogen transformations and nitrous oxide flux in tropical forest ecosystems. *Global Biogeochemical Cycles* 1:163–170.

Matson, P.A., P.M. Vitousek, J. Ewel, M. Mazzarino, and G. Robertson. 1987. Nitrogen transformations following tropical forest felling and burning on a volcanic soil. *Ecology* 68:491–502.

Matson, P.A. and R.C. Harriss. 1988. Prospects for aircraft-based gas exchange measurements in ecosystem studies. *Ecology* 69:1318–1325.

Matson, P.A., C. Volkmann, K. Coppinger, and W.A. Reiners. 1991. Annual nitrous oxide flux and soil nitrogen characteristics in sagebrush steppe ecosystems. *Biogeochemistry* 14:1–12.

Matson, P.A., W.J. Parton, A.G. Power, and M.J. Swift. 1997. Agricultural intensification and ecosystem properties. *Science* 227:504–509.

Matson, P.A., R.L. Naylor, and I. Ortiz-Monasterio. 1998. The integration of environmental, agronomic, and economic aspects of fertilizer management. *Science* 280:112–115.

Matson, P.A., W.H. McDowell, A.R. Townsend, and P.M. Vitousek. 1999. The globalization of N deposition: Ecosystem consequences in tropical environments. *Biogeochemistry* 46:67–83.

Matson, P.A. 2009. The sustainability transition. *Issues in Science and Technology* 25:39–42.

McAndrews, J.H. 1966. Postglacial history of prairie, savanna, and forest in northwestern Minnesota. *Torrey Botanical Club Memoir* 22:1–72.

McCoy, E.D., S.S. Bell, and K. Walters. 1986. Identifying biotic boundaries along environmental gradients. *Ecology* 67:749–759.

McCulley, R.L., I.C. Burke, and W.K. Lauenroth. 2009. Conservation of nitrogen increases with precipitation across a major grassland gradient in the Central Great Plains of North America. *Oecologia* 159:571–581.

McDowell, N., W.T. Pockman, C.D. Allen, D.D. Breshears, N. Cobb, et al. 2008. Mechanisms of plant survival and mortality during drought: Why do some plants survive while others succumb to drought? *New Phytologist* 178:719–739.

McElroy, M.B. 2002. *The Atmospheric Environment: Effects of Human Activity*. Princeton University Press, Princeton.

McGill, W. and C.V. Cole. 1981. Comparative aspects of cycling of organic C, N, S, and P through soil organic matter. *Geoderma* 26:267–286.

McGroddy, M., T. Deaufresne, and L.O. Hedin. 2004. Scaling of C:N:P stoichiometry in forests worldwide:

Implications of terrestrial Redfield-type ratios. *Ecology* 85:2390–2401.

McGuire, A.D., J.M. Melillo, and L.A. Joyce. 1995a. The role of nitrogen in the response of forest net primary production to elevated atmospheric carbon dioxide. *Annual Review of Ecology and Systematics* 26:473–503.

McGuire, A.D., J.W. Melillo, D.W. Kicklighter, and L.A. Joyce. 1995b. Equilibrium responses of soil carbon to climate change: Empirical and process-based estimates. *Journal of Biogeography* 22:785–796.

McGuire, A.D., S. Sitch, J.S. Clein, R. Dargaville, G. Esser, et al. 2001. Carbon balance of the terrestrial biosphere in the twentieth century: Analyses of CO_2, climate and land-use effects with four process-based models. *Global Biogeochemical Cycles* 15:183–206.

McGuire, A.D., L.G. Anderson, T.R. Christensen, S. Dallimore, L. Guo, et al. 2009. Sensitivity of the carbon cycle in the Arctic to climate change. *Ecological Monographs* 79:523–555.

McGuire, A.D., D.J. Hayes, D.W. Kicklighter, M. Manizza, Q. Zhuang, et al. 2010. An analysis of the carbon balance of the Arctic Basin from 1997 to 2006. *Tellus Series B: Chemical and Physical Meteorology.*

McKane, R.B., E.B. Rastetter, G.R. Shaver, K.J. Nadelhoffer, A.E. Giblin, et al. 1997. Climatic effects on tundra carbon storage inferred from experimental data and a model. *Ecology* 78:1170–1187.

McKane, R.B., L.C. Johnson, G.R. Shaver, K.J. Nadelhoffer, E.B. Rastetter, et al. 2002. Resource-based niches provide a basis for plant species diversity and dominance in arctic tundra. *Nature* 415:68–71.

McKey, D., P.G. Waterman, C.N. Mbi, J.S. Gartlan, and T.T. Struhsaker. 1978. Phenolic content of vegetation in two African rain forests: Ecological implications. *Science* 202:61–63.

McLachlan, J.S., J. Hellmann, and M. Schwartz. 2007. A framework for debate of assisted migration in an era of climate change. *Conservation Biology* 21:297–302.

McNaughton, K.G. 1976. Evaporation and advection I: Evaporation from extensive homogeneous surfaces. *Quarterly Journal of the Royal Meteorological Society* 102:181–191.

McNaughton, K.G. and P.G. Jarvis. 1991. Effects of spatial scale on stomatal control of transpiration. *Agricultural and Forest Meteorology* 54:279–302.

McNaughton, S.J. 1977. Diversity and stability of ecological communities: A comment on the role of empiricism in ecology. *American Naturalist* 111:515–525.

McNaughton, S.J. 1979. Grazing as an optimization process: Grass-ungulate relationships in the Serengeti. *American Naturalist* 113:691–703.

McNaughton, S.J. 1985. Ecology of a grazing ecosystem: The Serengeti. *Ecological Monographs* 53:259–294.

McNaughton, S.J. 1988. Mineral nutrition and spatial concentrations of African ungulates. *Nature* 334:343–345.

McNaughton, S.J., M. Oesterheld, D.A. Frank, and K.J. Williams. 1989. Ecosystem-level patterns of primary productivity and herbivory in terrestrial habitats. *Nature* 341:142–144.

McNulty, S.G., J.D. Aber, T.M. McLellan, and S.M. Katt. 1990. Nitrogen cycling in high elevation forests of the northeastern U.S. in relation to nitrogen deposition. *Ambio* 19:38–40.

McTammany, M.E., J.R. Webster, E.F. Benfield, and M.A. Neatrour. 2003. Longitudinal patterns of metabolism in a southern Appalachian river. *Journal of the North American Benthological Society* 22:359–370.

MEA (Millennium Ecosystem Assessment). 2005. *Ecosystems and Human Well-being: Synthesis.* Island Press, Washington.

Meffe, G.K., L.A. Nielsen, R.L. Knight, and D.A. Schenborn. 2002. *Ecosystem Management: Adaptive, Community-Based Conservation* Island Press, Washington.

Melillo, J.M., J.D. Aber, and J.F. Muratore. 1982. Nitrogen and lignin control of hardwood leaf litter decomposition dynamics. *Ecology* 63:621–626.

Meyer, J.L. and R.T. Edwards. 1990. Ecosystem metabolism and turnover of organic carbon along a blackwater river continuum. *Ecology* 71:668–677.

Meyer, W.B. and B.L. Turner, III. 1992. Human population growth and global land-use/cover change. *Annual Review of Ecology and Systematics* 23:39–61.

Migliavacca, M., M. Reichstein, A.D. Richardson, R. Colombo, M.A. Sutton, et al. 2010. Semiempirical modeling of abiotic and biotic factors controlling ecosystem respiration across eddy covariance sites. *Global Change Biology* 16:187–208.

Milchunas, D.G. and W.K. Lauenroth. 1993. Quantitative effects of grazing on vegetation and soils over a global range of environments. *Ecological Monographs* 63:327–366.

Millar, C.I., N.L. Stephenson, and S.L. Stephens. 2007. Climate change and forests of the future: Managing in the face of uncertainty. *Ecological Applications* 17:2145–2151.

Miller, J.R., M.G. Turner, E.A.H. Smithwick, C.L. Dent, and E.H. Stanley. 2004. Spatial extrapolation: The science of predicting ecological patterns and processes. *BioScience* 54:310–320.

Miller, R.W. and R.L. Donahue. 1990. *Soils. An Introduction to Soils and Plant Growth.* 6th edition. Prentice Hall, Englewood, USA.

Milliman, J.D. and J.P.M. Syvitski. 1992. Geomorphic/tectonic control of sediment discharge to the ocean: The importance of small mountains and rivers. *Journal of Geology* 100:525–544.

Milner, A.M., C. Fastie, F.S. Chapin, III, D.R. Engstrom, and L. Sharman. 2007. Interactions and linkages among ecosystems during landscape evolution. *BioScience* 57:237–247.

Minkkinen, K., R. Korhonen, I. Savolainen, and J. Laine. 2002. Carbon balance and radiative forcing of Finnish peatlands 1900–2100: The impact of forestry drainage. *Global Change Biology* 8:785–799.

Minshall, G.W., C.T. Robinson, and D.E. Lawrence. 1997. Postfire responses of lotic ecosystems in Yellowstone National Park, U.S.A. *Canadian Journal of Fisheries and Aquatic Sciences* 54:2509–2525.

Moen, R., Y. Cohen, and J. Pastor. 1998. Linking moose population and plant growth models with a moose energetics model. *Ecosystems* 1:52–63.

Monserud, R.A. and J.D. Marshall. 1999. Allometric crown relations in three northern Idaho conifer species. *Canadian Journal of Forest Research* 29:521–535.

Monteith, J.L. and M.H. Unsworth. 2008. *Principles of Environmental Physics.* 3rd edition. Elsevier, Amsterdam.

Mooney, H.A. and E.L. Dunn. 1970. Convergent evolution of mediterranean-climate evergreen sclerophyll shrubs. *Evolution* 24:292–303.

Mooney, H.A. 1972. The carbon balance of plants. *Annual Review of Ecology and Systematics* 3:315–346.

Mooney, H.A. 1986. Photosynthesis. Pages 345–373 *in* M.J. Crawley, editor. *Plant Ecology.* Blackwell, Oxford.

Mooney, H.A., J. Canadell, F.S. Chapin, III, J.R. Ehleringer, C. Körner, et al. 1999. Ecosystem physiology responses to global change. Pages 141–189 *in* B. Walker, W. Steffen, J. Canadell, and J. Ingram, editors. *The Terrestrial Biosphere and Global Change: Implications for Natural and Managed Ecosystems.* Cambridge University Press, Cambridge.

Moore, J.C. and H.W. Hunt. 1988. Resource compartmentation and the stability of real ecosystems. *Nature* 333:261–263.

Moore, J.C. and P.C. de Ruiter. 2000. Invertebrates in detrital food webs along gradients of productivity *in* D.C. Coleman and P.F. Hendrix, editors. *Invertebrates as Webmasters in Ecosystems.* CABI Publishing, Oxford.

Moore, J.C., K. McCann, H. Setälä, and P.C. de Ruiter. 2003. Top-down is bottom-up: Does predation in the rhizosphere regulate aboveground dynamics? *Ecology* 84:846–857.

Moore, R.D. and S.M. Wondzell. 2005. Physical hydrology and the effects of forest harvesting in the Pacific Northwest: A review. *Journal of the American Water Resources Association* 41:763–784.

Morris, J.T. 1980. The nitrogen uptake kinetics of *Spartina alterniflora* in culture. *Ecology* 61:1114–1121.

Moss, B. 1998. *Ecology of Fresh Waters: Man and Medium, Past to Future.* 3rd edition. Blackwell Scientific, Oxford.

Motzkin, G., D. Foster, A. Allen, J. Harrod, and R. Boone. 1996. Controlling site to evaluate history: Vegetation patterns of a New England sand plain. *Ecological Monographs* 66:345–365.

Mulder, C.P.H., D.D. Uliassi, and D.F. Doak. 2001. Physical stress and diversity-productivity relationships: The role of positive interactions. *Proceedings of the National Academy of Sciences, USA* 98:6704–6708.

Mulholland, P.J., C.S. Fellows, J.L. Tank, N.B. Grimm, J.R. Webster, et al. 2001. Inter-biome comparison of factors controlling stream metabolism. *Freshwater Biology* 46:1503–1517.

Mulholland, P.J., A.M. Helton, G.C. Poole, R.O. Hall, Jr., S.K. Hamilton, et al. 2008. Stream denitrification across biomes and its response to anthropogenic nitrate loading. *Nature* 452:202–205.

Mullon, C., P. Freon, and P. Cury. 2005. The dynamics of collapse in world fisheries. *Fish and Fisheries* 6:111–120.

Nadelhoffer, K.J., A.E. Giblin, G.R. Shaver, and A.E. Linkins. 1992. Microbial processes and plant nutrient availability in arctic soils. Pages 281–300 *in* F.S. Chapin, III, R.L. Jefferies, J.F. Reynolds, G.R. Shaver, and J. Svoboda, editors. *Arctic Ecosystems in a Changing Climate: An Ecophysiological Perspective.* Academic Press, San Diego.

Nadkarni, N. 1981. Canopy roots: Convergent evolution in rainforest nutrient cycles. *Science* 214:1023–1024.

Naeem, S., D.E. Bunker, A. Hector, M. Loreau, and C. Perrings, editors. 2009. *Biodiversity, Ecosystem Functioning, and Human Well-being: An Ecological and Economic Perspective.* Oxford University Press, Oxford.

Naiman, R.J. and H. Décamps. 1997. The ecology of interfaces: Riparian zones. *Annual Review of Ecology and Systematics* 28:621–658.

Naiman, R.J., H. Décamps, and M.E. McClain. 2005. *Riparia: Ecology, Conservation, and Management of Streamside Communities.* Elsevier, Amsterdam.

Näsholm, T., A. Ekblad, A. Nordin, R. Giesler, M. Högberg, et al. 1998. Boreal forest plants take up organic nitrogen. *Nature* 392:914–916.

Näsholm, T., K. Huss-Danell, and P. Högberg. 2000. Uptake of organic nitrogen in the field by four agriculturally important plant species. *Ecology* 81:1155–1161.

Naylor, R.L. 2009. Managing food production systems for resilience. Pages 259–280 *in* F.S. Chapin, III, G.P. Kofinas, and C. Folke, editors. *Principles of Ecosystem Stewardship: Resilience-Based Natural Resource Management in a Changing World.* Springer, New York.

Neff, J.C., E.A. Holland, F.J. Dentener, W.H. McDowell, and K.M. Russell. 2002. The origin, composition and rates of organic nitrogen deposition: A missing piece of the nitrogen cycle? *Biogeochemistry* 57:99–136.

Neff, J.C. and D.U. Hooper. 2002. Vegetation and climate controls on potential CO_2, DOC and DON production in northern latitude soils. *Global Change Biology* 8:872–884.

Neff, J.C., A.P. Ballantyne, G.L. Famer, N.M. Mahowald, J.L. Conroy, et al. 2008. Recent increase in eolian dust deposition related to human activity in the Western United States. *Nature Geosciences* 1:189–195.

Nepstad, D.C., C.R. deCarvalho, E.A. Davidson, P.H. Jipp, P.A. Lefebvre, et al. 1994. The role of deep roots in the hydrological and carbon cycles of Amazonian forests and pastures. *Nature* 372:666–669.

New, M.G., M. Hulme, and P.D. Jones. 1999. Representing 20th century space-time climate variability, I: Development of a 1961–1990 mean monthly terrestrial climatology. *Journal of Climate* 12:829–856.

New, M.G., D. Lister, M. Hulme, and I. Makin. 2002. A high-resolution data set of surface climate over global land areas. *Climate Research* 21:1–25.

Newman, E.I. 1985. The rhizosphere: Carbon sources and microbial populations. Pages 107–121 *in* A.H. Fitter, D. Atkinson, D.J. Read, and M. Busher, editors. *Ecological Interactions in Soil.* Blackwell, Oxford.

Niemelä, P., F.S. Chapin, III, K. Danell, and J.P. Bryant. 2001. Animal-mediated responses of boreal forest to climatic change. *Climatic Change* 48:427–440.

Nippert, J.B. and A.K. Knapp. 2007a. Linking water uptake with rooting patterns in grassland species. *Oecologia* 153:261–272.

Nippert, J.B. and A.K. Knapp. 2007b. Soil water partitioning contributes to species coexistence in tallgrass prairie. *Oikos* 116:1017–1029.

Nixon, S.W. 1988. Physical energy inputs and the comparative ecology of lake and marine ecosystems. *Limnology and Oceanography* 33:1005–1025.

Nixon, S.W., J.W. Ammerman, L.P. Atkinson, V.M. Berounsky, G. Billen, et al. 1996. The fate of nitrogen and phosphorus at the land-sea margin of the North Atlantic Ocean. *Biogeochemistry* 35:141–180.

Noble, I.R. and R.O. Slatyer. 1980. The use of vital attributes to predict successional changes in plant communities subject to recurrent disturbances. *Vegetatio* 43:5–21.

Norby, R.J., E.H. DeLucia, B. Gielen, C. Calfapietra, C.P. Giardina, et al. 2005. Forest response to elevated CO_2 is conserved across a broad range of productivity. *Proceedings of the National Academy of Sciences, USA* 102:18052–18056.

Norby, R.J., J.M. Warren, C.M. Iversen, B.E. Medlyn, and R.E. McMurtrie. 2010. CO_2 enhancement of forest productivity constrained by limited nitrogen availability. *Proceedings of the National Academy of Sciences USA* 107:19368–19373.

Northup, R.R., Z. Yu, R.A. Dahlgren, and K.A. Vogt. 1995. Polyphenol control of nitrogen release from pine litter. *Nature* 377:227–229.

Norton, J.M., J.L. Smith, and M.K. Firestone. 1990. Carbon flow in the rhizosphere of Ponderosa pine seedlings. *Soil Biology and Biochemistry* 22:449–445.

Norton, J.M. and M.K. Firestone. 1991. Metabolic status of bacteria and fungi in the rhizosphere of ponderosa pine seedlings. *Applied and Environmental Microbiology* 57:1161–1167.

Noss, R.F. 1990. Indicators for monitoring biodiversity: A hierarchical approach. *Conservation Biology* 4: 355–364.

NRC (National Research Council). 2000. *Watershed Management for Potable Water Supply: Assessing New York City's Approach*. National Academies Press, Washington.

NRC (National Research Council). 2006. *Drawing Louisiana's New Map: Addressing Land Loss in Coastal Louisiana*. National Academies Press, Washington.

NRC (National Research Council). 2008. *Hydrologic Effects of a Changing Forest Landscape*. National Academies Press, Washington.

NRC (National Research Council). 2010. *America's Climate Choices: Adapting to the Impacts of Climate Change*. National Academies Press, Washington.

Nulsen, R.A., K.J. Bligh, I.N. Baxter, E.J. Solin, and D.H. Imrie. 1986. The fate of rainfall in a mallee and heath vegetated catchment in southern Western Australia. *Australian Journal of Ecology* 11:361–371.

Nye, P.H. and P.B. Tinker. 1977. *Solute Movement in the Soil-Root System*. University of California Press, Berkeley.

O'Leary, M.H. 1988. Carbon isotopes in photosynthesis. *BioScience* 38:325–336.

O'Neill, R.V., D.L. DeAngelis, J.B. Waide, and T.F.H. Allen. 1986. *A Hierarchical Concept of Ecosystems*. Princeton University Press, Princeton.

Oades, J.M. 1989. An introduction to organic matter in mineral soils. Pages 89–160 *in* J.B. Dixon and S.B. Weed, editors. *Minerals in Soil Environments*. Soil Science Society of America, Madison.

Odum, E.P. 1959. *Fundamentals of Ecology*. W. B. Saunders, Philadelphia.

Odum, E.P. 1969. The strategy of ecosystem development. *Science* 164:262–270.

Oechel, W.C., G.L. Vourlitis, S.J. Hastings, R.C. Zulueta, L. Hinzman, et al. 2000. Acclimation of ecosystem CO_2 exchange in the Alaskan Arctic in response to decadal climate warming. *Nature* 406:978–981.

Oke, T.R. 1987. *Boundary Layer Climates*. 2nd edition. Methuen, London.

Oki, T. and S. Kanae. 2006. Global hydrological cycles and world water resources. *Science* 313:1068–1072.

Okin, G.S., N.M. Mahowald, O.A. Chadwick, and P. Artaxo. 2004. Impact of desert dust on the biogeochemistry of phosphorus in terrestrial ecosystems. *Global Biogeochemical Cycles* 18:GB2005, doi:2010.1029/2003GB002145.

Oksanen, L. 1990. Predation, herbivory, and plant strategies along gradients of primary productivity. Pages 445–474 *in* J.B. Grace and D. Tilman, editors. *Perspectives on Plant Competition*. Academic Press, San Diego.

Olander, L.P. and P.M. Vitousek. 2000. Asymmetry in N and P mineralization: Regulation of extracellular phosphatase and chitinase activity by N and P availability. *Biogeochemistry* 49:175–190.

Olsen, J.S. 1963. Energy storage and the balance of producers and decomposers in ecological systems. *Ecology* 44:322–331.

Olsson, A.D., M.P. McClaran, J. Betancourt, and S.E. Marsh. In press. Sonoran Desert ecosystem transformation by a C_4 grass without the grass-fire cycle. *Diversity and Distributions*.

Onipchenko, V.G., M. Makarov, R.S.P. van Logtestijn, V. Ivanov, A.A. Akhmetzhanova, et al. 2009. New nitrogen uptake strategy: Specialized snow roots. *Ecology Letters* 12:758–764.

Orlove, B.S., J.C.H. Chiang, and M.A. Cane. 2000. Forecasting Andean rainfall and crop yield from the influence of El Niño on Pleiades visibility. *Nature* 403:68–71.

Orr, J.C., V.J. Fabry, O. Aumont, L. Bopp, S.C. Doney, et al. 2005. Anthropogenic ocean acidification over the twenty-first century and its impact on calcifying organisms. *Nature* 437:681–686.

Osborne, C.P. and R.P. Freckleton. 2009. Ecological selection pressures for C_4 photosynthesis. *Proceedings of The Royal Society, Series B* 276:1753–1760.

Osterheld, M., O.E. Sala, and S.J. McNaughton. 1992. Effect of animal husbandry on herbivore-carrying capacity at a regional scale. *Nature* 356:234–236.

Ostfeld, R.S. and F. Keesing. 2000. Biodiversity and disease risk: The case of Lyme disease. *Conservation Biology* 14:722–728.

Ostrom, E. 1990. *Governing the Commons: The Evolution of Institutions for Collective Action.* Cambridge University Press, Cambridge.

Ostrom, E. 2007. A diagnostic approach for going beyond panaceas. *Proceedings of the National Academy of Sciences, USA* 104:15181–15187.

Ostrom, E. 2009. A general framework for analyzing sustainability of social-ecological systems. *Science* 325:419–422.

Ovington, J.D. 1962. Quantitative ecology and the woodland ecosystem concept. *Advances in Ecological Research* 1:103–192.

Owen-Smith, R.N. 1988. *Megaherbivores: The Influence of Very Large Body Size on Ecology.* Cambridge University Press, Cambridge.

Pace, M.L., J.J. Cole, S.R. Carpenter, and J.F. Kitchell. 1999. Trophic cascades revealed in diverse ecosystems. *Trends in Ecology & Evolution* 14:483–488.

Paine, R.T. 1980. Food webs: Linkage, interaction strength and community infrastructure. *Journal of Animal Ecology* 49:667–685.

Paine, R.T. 2000. Phycology for the mammalogist: Marine rocky shores and mammal-dominated communities. How different are the structuring processes? *Journal of Mammalogy* 81:637–648.

Painter, T.H., A.P. Barrett, C.C. Landry, J.C. Neff, M.P. Cassidy, et al. 2007. Impact of disturbed desert soils on duration of mountain snowcover. *Geophysical Research Letters* 34:12, L12502, 12510.11029/12007GL030208.

Palm, C.A., S.A. Vosti, P.A. Sanchez, and P.J. Ericksen. 2005. *Slash-and-Burn Agriculture: The Search for Alternatives.* Columbia University Press, New York.

Palmer, M., E. Bernhardt, E. Chornesky, S. Collins, A. Dobson, et al. 2004. Ecology for a crowded planet. *Science* 304:1251–1252.

Parker, I.M., D. Simberloff, W.M. Lonsdale, K. Goodell, M. Wonham, et al. 1999. Impact: Toward a framework for understanding the ecological effects of invaders. *Biological Invasions* 1:3–19.

Parton, W.J., D.S. Schimel, C.V. Cole, and D.S. Ojima. 1987. Analysis of factors controlling soil organic matter levels in Great Plains grasslands. *Soil Science Society of America Journal* 51:1173–1179.

Parton, W.J., J.M.O. Scurlock, D.S. Ojima, T.G. Gilmanov, R.J. Scholes, et al. 1993. Observations and modeling of biomass and soil organic matter dynamics for the grassland biome worldwide. *Global Biogeochemical Cycles* 7:785–809.

Parton, W.J., W.L. Silver, I.C. Burke, L. Grassens, M.E. Harmon, et al. 2007. Global-scale similarities in nitrogen release patterns during long-term decomposition. *Science* 315:361–364.

Passioura, J.B. 1988. Response to Dr. P. J. Kramer's article, 'Changing concepts regarding plant water relations'. *Plant, Cell and Environment* 11:569–571.

Pastor, J., J.D. Aber, C.A. McClaugherty, and J.M. Melillo. 1984. Aboveground primary production and N and P cycling along a nitrogen mineralization gradient on Blackhawk Island, Wisconsin. *Ecology* 65:256–268.

Pastor, J., R.J. Naiman, B. Dewey, and P. McInnes. 1988. Moose, microbes, and the boreal forest. *BioScience* 38:770–777.

Pastor, J., Y. Cohen, and N.T. Hobbs. 2006. The roles of large herbivores in ecosystem nutrient cycles. Pages 289–325 in K. Danell, editor. *Large Herbivore Ecology, Ecosystem Dynamics and Conservation.* Cambridge University Press, Cambridge.

Paton, T.R., G.S. Humphreys, and P.B. Mitchell. 1995. *Soils: A New Global View.* Yale University Press, New Haven.

Patz, J.A., U.E.C. Confalonieri, F.P. Amerasinghe, K.B. Chua, P. Daszak, et al. 2005. Human health: Ecosystem regulation of infectious diseases. Pages 391–415 in R. Hassan, R.J. Scholes, and N. Ash, editors. *Ecosystems and Well-Being: Current State and Trends, Volume 1.* Island Press, Washington.

Paul, E.A. and F.E. Clark. 1996. *Soil Microbiology and Biochemistry.* 2nd edition. Academic Press, San Diego.

Pauly, D. and V. Christensen. 1995. Primary production required to sustain global fisheries. *Nature* 374:255–257.

Pauly, D., V. Christensen, J. Dalsgaard, R. Froese, and F. Torres, Jr. 1998. Fishing down marine food webs. *Science* 279:860–863.

Pauly, D., J. Alder, A. Bakun, S. Heileman, K.-H. Kock, et al. 2005. Marine fisheries systems. Pages 477–511 in R. Hassan, R.J. Scholes, and N. Ash, editor. *Ecosystems and Human Well-Being: Current State and Trends.* Island Press, Washington.

Payette, S. and L. Filion. 1985. White spruce expansion at the tree line and recent climatic change. *Canadian Journal of Forest Research* 15:241–251.

Pearcy, R.W. 1990. Sunflecks and photosynthesis in plant canopies. *Annual Review of Plant Physiology* 41:421–453.

Pearson, P.N. and M.R. Palmer. 2000. Atmosphere carbon dioxide concentrations over the past 60 million years. *Nature* 406:695–699.

Peltzer, D.A., P.J. Bellingham, H. Kurokawa, L.R. Walker, D.A. Wardle, et al. 2009. Punching above their weight: Low-biomass non-native plant species alter soil properties during primary succession. *Oikos* 118:1001–1014.

Penning de Vries, F.W.T., A.H.M. Brunsting, and H.H. van Laar. 1974. Products, requirements, and efficiency of biosynthesis: A quantitative approach. *Journal of Theoretical Biology* 45:339–377.

Penning de Vries, F.W.T. 1975. The cost of maintenance processes in plant cells. *Annals of Botany* 39:77–92.

Perakis, S.S. and L.O. Hedin. 2002. Nitrogen loss from unpolluted South American forests mainly via dissolved organic compounds. *Nature* 415:416–419.

Perez-Harguindeguy, N., S. Díaz, J.H.C. Cornelissen, F. Vendramini, M. Cabido, et al. 2000. Chemistry and toughness predict leaf litter decomposition rates over a wide spectrum of functional types and taxa in central Argentina. *Plant and Soil* 218:21–30.

Perry, G.L.W. and N.J. Enright. 2006. Spatial modelling of vegetation change in dynamic landscapes: A review of methods and applications. *Progress in Physical Geography* 30:43–72.

Peters, D.P.C., R.A. Pielke, Sr., B.T. Bestelmeyer, C.D. Allen, S. Munson-McGee, et al. 2004. Cross-scale interactions, nonlinearities, and forecasting catastrophic events. *Proceedings of the National Academy of Sciences, USA* 101:15130–15135.

Peters, D.P.C., J.R. Gosz, and S.L. Collins. 2009. Boundary dynamics in landscapes. Pages 458–463 *in* S.A. Levin, S.R. Carpenter, H.C.J. Godfray, A.P. Kinzig, M. Loreau, et al., editors. *Princeton Guide to Ecology*. Princeton University Press, Princeton.

Peters, D.P.C., A.E. Lugo, F.S. Chapin, III, S.T.A. Pickett, M. Duniway, et al. 2011. Cross-system comparisons elucidate disturbance complexities and generalities. *Ecosphere* 2(7):art81. doi:10-1890/ES11-00115.1.

Peterson, B.J., J.E. Hobbie, and T.J. Corliss. 1986. Carbon flow in a tundra stream ecosystem. *Canadian Journal of Fisheries and Aquatic Sciences* 43:1259–1270.

Peterson, B.J., W.M. Wolheim, P.J. Mujlholland, J.R. Webster, J.L. Meyer, et al. 2001. Control of nitrogen export from watersheds by headwater streams. *Science* 292:86–90.

Peterson, G.D., C.R. Allen, and C.S. Holling. 1998. Ecological resilience, biodiversity, and scale. *Ecosystems* 1:6–18.

Peterson, G.D., G.S. Cumming, and S.R. Carpenter. 2003. Scenario planning: A tool for conservation in an uncertain world. *Conservation Biology* 17:358–366.

Petit, J.R., J. Jouzel, D. Raynaud, N.I. Barkov, J.M. Barnola, et al. 1999. Climate and atmospheric history of the past 420,000 years from the Vostok ice core, Antarctica. *Nature* 399:429–436.

Piao, S., P. Ciais, P. Friedlingstein, P. Peylin, M. Reichstein, et al. 2008. Net carbon dioxide losses of northern ecosystems in response to autumn warming. *Nature* 451:49–53.

Piao, S., P. Friedlingstein, P. Ciais, P. Peylin, B. Zhu, et al. 2009. Footprint of temperature changes in the temperate and boreal forest carbon balance. *Geophysical Research Letters* 36:L07404, doi:07410.01029/02009GL037381.

Pickett, S.T.A. and P.S. White. 1985. *The Ecology of Natural Disturbance as Patch Dynamics*. Academic Press, New York.

Pickett, S.T.A., S.L. Collins, and J.J. Armesto. 1987. A hierarchical consideration of causes and mechanisms of succession. *Vegetatio* 69:109–114.

Pickett, S.T.A., J. Kolasa, and C.G. Jones. 1994. *Ecological Understanding. The Nature of Theory and the Theory of Nature*. Academic Press, New York.

Pielke, R.A., Sr. and R. Avisar. 1990. Influence of landscape structure on local and regional climate. *Landscape Ecology* 4:133–156.

Pimm, S.L. 1982. *Food Webs*. Chapman and Hall, New York.

Pimm, S.L. 1984. The complexity and stability of ecosystems. *Nature* 307:321–326.

Pimm, S.L., G.J. Russell, J.L. Gittleman, and T.M. Brooks. 1995. The future of biodiversity. *Science* 269:347–350.

Pires, M. 2004. Watershed protection for a world city: The case of New York. *Land Use Policy* 21:161–175.

Poff, N.L., J.D. Allan, M.B. Bain, J.R. Karr, K.L. Prestegaard, et al. 1997. The natural flow regime: A paradigm for river conservation and restoration. *BioScience* 47:769–784.

Polis, G.A. 1991. Complex trophic interactions in deserts: An empirical critique of food-web theory. *American Naturalist* 138:123–155.

Polis, G.A. and S.D. Hurd. 1996. Linking marine and terrestrial food webs: Allochthonous input from the ocean supports high secondary productivity on small islands and coastal land communities. *American Naturalist* 147:396–423.

Polis, G.A. 1999. Why are parts of the world green? Multiple factors control productivity and the distribution of biomass. *Oikos* 86:3–15.

Pomeroy, J., N. Hedstrom, and J. Parviainen. 1999. The snow mass balance of Wolf Creek, Yukon: Effects of snow sublimation and redistribution. Pages 15–30 *in* J.W. Pomeroy and R.J. Granger, editors. *Wolf Creek Research Basin: Hydrology, Ecology, Environment*. National Water Research Institute, Environment Canada, Saskatoon.

Pons, T.L., K. Perreijn, C. van Kessel, and M.J.A. Werger. 2006. Symbiotic nitrogen fixation in a tropical rainforest: ^{15}N natural abundance measurements supported by experimental isotopic enrichment. *New Phytologist* 173:154–167.

Poorter, H. 1994. Construction costs and payback time of biomass: A whole-plant perspective. Pages 111–127 *in* J. Roy and E. Garnier, editors. *A Whole-Plant Perspective on Carbon-Nitrogen Interactions*. SPB Academic Publishing, The Hague.

Porder, S., A. Paytan, and P.M. Vitousek. 2005. Erosion and landscape development affect plant nutrient status in the Hawaiian Islands. *Oecologia* 142:440–449.

Post, D.M., M.L. Pace, and N.G. Hairston. 2000. Ecosystem size determines food-chain length in lakes. *Nature* 405:1047–1049.

Post, W.M., W.R. Emanuel, P.J. Zinke, and A.G. Stangenberger. 1982. Soil carbon pools and world life zones. *Nature* 298:156–159.

Postel, S.L., G.C. Daily, and P.R. Ehrlich. 1996. Human appropriation of renewable fresh water. *Science* 271:785–788.

Postel, S.L. and B. Richter. 2003. *Rivers for Life: Managing Water for People and Nature*. Island Press, Washington.

Potter, C.S., J.T. Randerson, C.B. Field, P.A. Matson, P.M. Vitousek, et al. 1993. Terrestrial ecosystem production:

A process model based on global satellite and surface data. *Global Biogeochemical Cycles* 7:811–841.

Power, M.E. 1990. Effects of fish in river food webs. *Science* 250:411–415.

Power, M.E. 1992a. Hydrologic and trophic controls of seasonal algal blooms in northern California rivers. *Archivs fur Hydrobiologie* 125:385–410.

Power, M.E. 1992b. Top-down and bottom-up forces in food webs: Do plants have primacy? *Ecology* 73:733–746.

Power, M.E., D. Tilman, J.A. Estes, B.A. Menge, W.J. Bond, et al. 1996. Challenges in the quest for keystones. *BioScience* 46:609–620.

Prentice, I.C., W. Cramer, S.P. Harrison, R. Leemans, R.A. Monserud, et al. 1992. A global biome model based on plant physiology and dominance, soil properties and climate. *Journal of Biogeography* 19:117–134.

Prentice, I.C., G.D. Farquhar, M.J.R. Fasham, M.L. Goulden, M. Heimann, et al. 2001. The carbon cycle and atmospheric carbon dioxide. Pages 183–237 *in* J.T. Houghton, Y. Ding, D.J. Griggs, M. Noguer, P.J. van der Linden, et al., editors. *Climate Change 2001: The Scientific Basis.* Cambridge University Press, Cambridge.

Prescott, C.E. 1995. Does nitrogen availability control rates of litter decomposition in forests? *Plant and Soil* 168–169:83–88.

Prescott, C.E., R. Kabzems, and L.M. Zabek. 1999. Effects of fertilization on decomposition rate of *Populus tremuloides* foliar litter in a boreal forest. *Canadian Journal of Forest Research* 29:393–397.

Press, F. and R. Siever. 1986. *Earth.* 4th edition. W. H. Freeman and Company, New York.

Pugnaire, F.I. and F.S. Chapin, III. 1992. Environmental and physiological factors governing nutrient resorption efficiency in barley. *Oecologia* 90:120–126.

Raab, T.K., D.A. Lipson, and R.K. Monson. 1999. Soil amino acid utilization among species of the Cyperaceae: Plant and soil processes. *Ecology* 80:2408–2419.

Rabalais, N.N., R.E. Turner, and W.J. Wiseman, Jr. 2002. Gulf of Mexico hypoxia, A.K.A. "The dead zone". *Annual Review of Ecology and Systematics* 33: 235–263.

Raffa, K.F., B.H. Aukema, B.J. Bentz, A.L. Carroll, J.A. Hicke, et al. 2008. Cross-scale drivers of natural disturbances prone to anthropogenic amplification: The dynamics of bark beetle eruptions. *BioScience* 58:501–517.

Raich, J.W. and W.H. Schlesinger. 1992. The global carbon dioxide flux in soil respiration and its relationship to climate. *Tellus* 44B:81–99.

Raich, J.W., A.E. Russell, and P.M. Vitousek. 1997. Primary production and ecosystem development along an elevational gradient in Hawaii. *Ecology* 78:707–721.

Ramakrishnan, P.S. 1992. *Shifting Agriculture and Sustainable Development: An Interdisciplinary Study from North-Eastern India.* Parthenon Publishing Group, Park Ridge, NJ.

Randerson, J.T., F.S. Chapin, III, J. Harden, J.C. Neff, and M.E. Harmon. 2002. Net ecosystem production:

A comprehensive measure of net carbon accumulation by ecosystems. *Ecological Applications* 12:937–947.

Raper, C.D., Jr., D.L. Osmond, M. Wann, and W.W. Weeks. 1978. Interdependence of root and shoot activities in determining nitrogen uptake rate of roots. *Botanical Gazette* 139:289–294.

Rastetter, E.B. and G.R. Shaver. 1992. A model of multiple-element limitation for acclimating vegetation. *Ecology* 73:1157–1174.

Read, D.J. and R. Bajwa. 1985. Some nutritional aspects of the biology of ericaceous mycorrhizas. *Proceedings of the Royal Society of Edinburgh* 85B:317–332.

Read, D.J. 1991. Mycorrhizas in ecosystems. *Experientia* 47:376–391.

Redfield, A.C. 1958. The biological control of chemical factors in the environment. *American Scientist* 46:205–221.

Reeburgh, W.S. 1997. Figures summarizing the global cycles of biogeochemically important elements. *Bulletin of the Ecological Society of America* 78:260–267.

Reich, P.B., M.B. Walters, and D.S. Ellsworth. 1997. From tropics to tundra: Global convergence in plant functioning. *Proceedings of the National Academy of Sciences, USA* 94:13730–13734.

Reich, P.B., D.S. Ellsworth, M.B. Walters, J.M. Vose, C. Gresham, et al. 1999. Generality of leaf trait relationships: A test across six biomes. *Ecology* 80: 1955–1969.

Reich, P.B. and J. Oleksyn. 2004. Global patterns of plant leaf N and P in relation to temperature and latitude. *Proceedings of the National Academy of Sciences, USA* 101:11001–11006.

Reich, P.B., S.E. Hobbie, T. Lee, D.W. Ellsworth, J.B. West, et al. 2006. Nitrogen limitation constrains sustainability of ecosystem response to CO_2. *Nature* 440:922–925.

Reichstein, M., D. Papale, R. Valentini, M. Aubinet, C. Bernhofer, et al. 2007. Determinants of terrestrial ecosystem carbon balance inferred from European eddy covariance flux sites. *Geophysical Research Letters* 34:L01402, doi:01410.01029/02006GL027880.

Reiners, W.A. 1986. Complementary models for ecosystems. *American Naturalist* 127:59–73.

Reynolds, J.F. and J.H.M. Thornley. 1982. A shoot:root partitioning model. *Annals of Botany* 49:585–597.

Reynolds, J.F., D.W. Hilbert, and P.R. Kemp. 1993. Scaling ecophysiology from the plant to the ecosystem: A conceptual framework. Pages 127–140 *in* J.R. Ehleringer and C.B. Field, editors. *Scaling Physiological Processes: Leaf to Globe.* Academic Press, San Diego.

Reynolds, J.F. and D.M. Stafford Smith, editors. 2002. *Global Desertification: Do Humans Cause Deserts?* Dahlem University Press, Berlin.

Rice, E.L. 1979. Allelopathy: An update. *Botanical Review* 45:15–109.

Richardson, A.E., T.S. George, I. Jakobsen, and R.J. Simpson. 2007. Plant utilization of inositol phosphates. Pages 242–260 *in* B.L. Turner, A.E. Richardson, and E.J. Mullaney, editors. *Inositol Phosphates: Linking Agriculture and the Environment.* CABI Publishing, Wallingford.

Richter, D.D., Jr. and D. Markewitz. 2001. *Understanding Soil Change: Soil Sustainability over Millennia, Centuries, and Decades.* Cambridge University Press, Cambridge.

Ricketts, T.H., G.C. Daily, P.R. Ehrlich, and C.D. Michener. 2004. Economic value of tropical forest to coffee production. *Proceedings of the National Academy of Sciences, USA* 101:12579–12582.

Ritchie, M.E., D. Tilman, and J.M.H. Knops. 1998. Herbivore effects on plant and nitrogen dynamics in oak savanna. *Ecology* 79:165–177.

Robertson, G.P. 1989. Nitrification and denitrification in humid tropical ecosystems: Potential controls on nitrogen retention. Pages 55–69 *in* J. Proctor, editor. *Mineral Nutrients in Tropical Forest and Savanna Ecosystems.* Blackwell Scientific, Oxford.

Robertson, G.P., K.M. Klingensmith, M.J. Klug, E.A. Paul, J.R. Crum, et al. 1997. Soil resources, microbial activity, and primary production across an agricultural ecosystem. *Ecological Applications* 7:158–170.

Robertson, G.P. and E.A. Paul. 2000. Decomposition and soil organic matter dynamics. Pages 104–116 *in* O.E. Sala, R.B. Jackson, H.A. Mooney, and R.W. Howarth, editors. *Methods in Ecosystem Science.* Springer-Verlag, New York.

Robertson, G.P. and P.M. Vitousek. 2009. Nitrogen in agriculture: Balancing the cost of an essential resource. *Annual Review of Environment and Resources* 34:97–125.

Robinson, D. 1994. The responses of plants to non-uniform supplies of nutrients. *New Phytologist* 127:635–674.

Robles, M. and F.S. Chapin, III. 1995. Comparison of the influence of two exotic species on ecosystem processes in the Berkeley Hills. *Madrono* 42: 349–357.

Rockström, J., L. Gordon, C. Folke, M. Falkenmark, and M. Engvall. 1999. Linkages among water vapor flows, food production, and terrestrial ecosystem services. *Conservation Ecology* 3:[online] http://www.consecol.org/vol3/iss2/art5.

Rockström, J., W. Steffen, K. Noone, Å. Persson, F.S. Chapin, III, et al. 2009. A safe operating space for humanity. *Nature* 461:472–475.

Rodin, L.E. and N.I. Bazilevich. 1967. *Production and Mineral Cycling in Terrestrial Vegetation.* Oliver and Boyd, Edinburgh, Scotland.

Rosenberg, A.A., M.J. Fogarty, M.P. Sissenwine, J.R. Beddington, and J.G. Shepherd. 1993. Achieving sustainable use of renewable resources. *Science* 262:828–829.

Rosenberg, D.K., B.R. Noon, and E.C. Meslow. 1997. Biological corridors: Form, function, and efficacy. *BioScience* 47:677–687.

Roulet, N.T., P.M. Crill, N.T. Comer, A. Dove, and R.A. Boubonniere. 1997. CO_2 and CH_4 flux between a boreal beaver pond and the atmosphere. *Journal of Geophysical Research* 102:29313–29319.

Rovira, A.D. 1969. Plant root exudates. *Botanical Review* 35:35–56.

Rudel, T.K., O.T. Coomes, E. Moran, F. Achard, A. Angelsen, et al. 2005. Forest transitions: Towards a global understanding of land use change. *Global Environmental Change* 15:23–31.

Ruess, R.W. and S.J. McNaughton. 1987. Grazing and the dynamics of nutrient and energy regulated microbial processes in the Serengeti grasslands. *Oikos* 49:101–110.

Ruess, R.W., D.S. Hik, and R.L. Jefferies. 1989. The role of lesser snow geese as nitrogen processors in a sub-arctic salt marsh. *Oecologia* 89:23–29.

Ruess, R.W., K. Van Cleve, J. Yarie, and L.A. Viereck. 1996. Contributions of fine root production and turnover to the carbon and nitrogen cycling in taiga forests of the Alaskan interior. *Canadian Journal of Forest Research* 26:1326–1336.

Ruess, R.W., R.L. Hendrick, and J.P. Bryant. 1998. Regulation of fine root dynamics by mammalian browsers in early successional Alaskan taiga forests. *Ecology* 79:2706–2720.

Ruimy, A., P.G. Jarvis, D.D. Baldocchi, and B. Saugier. 1995. CO_2 fluxes over plant canopies and solar radiation: A review. *Advances in Ecological Research* 26:1–53.

Running, S.W., R.R. Nemani, D.L. Peterson, L.E. Band, D.F. Potts, et al. 1989. Mapping regional forest evapotranspiration and photosynthesis by coupling satellite data with ecosystem simulation. *Ecology* 70:1090–1101.

Running, S.W., P.E. Thornton, R.R. Nemani, and J.M. Glassy. 2000. Global terrestrial gross and net primary productivity from the earth observing system. Pages 44–57 *in* O. Sala, R.B. Jackson, and H.A. Mooney, editors. *Methods in Ecosystem Science.* Springer-Verlag, New York.

Running, S.W., R.R. Nemani, F.A. Heinsch, M. Zhao, M. Reeves, et al. 2004. A continuous satellite-derived measure of global terrestrial primary production. *BioScience* 54:547–560.

Rupp, T.S., A.M. Starfield, and F.S. Chapin, III. 2000. A frame-based spatially explicit model of subarctic vegetation response to climatic change: Comparison with a point model. *Landscape Ecology* 15:383–400.

Ruttenberg, K.C. 2004. The global phosphorus cycle. Pages 585–643 *in* W.H. Schlesinger, editor. *Biogeochemistry.* Elsevier, Amsterdam.

Rutter, A.J., P.C. Robins, A.J. Morton, and K.A. Kershaw. 1971. Predictive model of rainfall interception in forests, 1. Derivation of the model from observations in a plantation of Corsican pine. *Agricultural Meteorology* 9:367–384.

Ryan, M.G., S. Linder, J.M. Vose, and R.M. Hubbard. 1994. Respiration of pine forests. *Ecological Bulletin* 43:50–63.

Ryan, M.G., D. Binkley, and J.H. Fownes. 1997. Age-related decline in forest productivity: Pattern and process. *Advances in Ecological Research* 27:213–262.

Sabine, C.L., M. Heiman, P. Artaxo, D.C.E. Bakker, C.-T.A. Chen, et al. 2004. Current status and past trends of the carbon cycle. Pages 17–44 *in* C.B. Field and M.R. Raupach, editors. *The Global Carbon Cycle:*

Integrating Humans, Climate and the Natural World. Island Press, Washington.

Sage, R.F. 2004. The evolution of C_4 photosynthesis. *New Phytologist* 161:341–370.

Sala, O.E., W.K. Lauenroth, S.J. McNaughton, G. Rusch, and X. Zhang. 1996. Biodiversity and ecosystem functioning in grasslands. Pages 129–149 *in* H.A. Mooney, J.H. Cushman, E. Medina, O.E. Sala, and E.-D. Schulze, editors. *Functional Role of Biodiversity: A Global Perspective.* John Wiley & Sons, Chichester.

Sala, O.E., R.B. Jackson, H.A. Mooney, and R.W. Howarth, editors. 2000. *Methods in Ecosystem Science.* Springer-Verlag, New York.

Saleska, S.R., K. Didan, A.R. Huete, and H.R. da Rocha. 2007. Amazon forests green-up during 2005 drought. *Science* 318:612.

Sampson, R.N., N. Bystriakova, S. Brown, P. Gonzalez, L.C. Irland, et al. 2005. Timber, fuel, and fiber. Pages 585–621 *in* R. Hassan, R. J. Scholes, and N. Ash, editors. Ecosystems and Human Well-Being: Current State and Trends. Cambridge University Press, Cambridge.

Sanchez, P.A. 2010. Tripling crop yields in tropical Africa. *Nature Geoscience* 3:299–300.

Sarmiento, J.L. and N. Gruber. 2006. *Ocean Biogeochemical Dynamics.* Princeton University Press, Princeton.

Saugier, B., J. Roy, and H.A. Mooney. 2001. Estimations of global terrestrial productivity: Converging toward a single number? Pages 543–557 *in* J. Roy, B. Saugier, and H.A. Mooney, editors. *Terrestrial Global Productivity.* Academic Press, San Diego.

Saura, S. and L. Pascual-Hortal. 2007. A new habitat availability index to integrate connectivity in landscape conservation planning: comparison with existing indices and application to a case study. *Landscape and Urban Planning* 83:91–103.

Sayre, N. 2005. *Working Wilderness: The Malpai Borderlands Group Story and the Future of the Western Range.* Rio Nuevo Press, Tucson, AZ.

Scatena, F.N., S. Moya, C. Estrada, and J.D. Chinea. 1996. The first five years in the reorganization of aboveground biomass and nutrient use following Hurricane Hugo in the Bisley Experimental Watersheds, Luquillo Experimental Forest, Puerto Rico. *Biotropica* 28:424–440.

Schaefer, D.A., W.H. McDowell, F.N. Scatena, and C.E. Asbury. 2000. Effects of hurricane disturbance on stream water concentrations and fluxes in eight tropical forest watersheds of the Luquillo Experimental Forest, Puerto Rico. *Journal of Tropical Ecology* 16:189–207.

Scheffer, M. and S.R. Carpenter. 2003. Catastrophic regime shifts in ecosystems: Linking theory to observation. *Trends in Ecology & Evolution* 18:648–656.

Scheffer, M., G.J. van Geest, K. Zimmer, E. Jeppsen, M. Sondergaard, et al. 2006. Small habitat size and isolation can promote species richness: Second-order effects on biodiversity in shallow lakes and ponds. *Oikos* 112:227–231.

Schenk, H.J. and R.B. Jackson. 2002. Rooting depths, lateral root spreads and below-ground/above-ground allometries of plants in water-limited ecosystems. *Journal of Ecology* 90:480–494.

Schimel, D.S. 1995. Terrestrial ecosystems and the carbon cycle. *Global Change Biology* 1:77–91.

Schimel, D.S., J.I. House, K.A. Hibbard, P. Bousquet, P. Ciais, et al. 2001. Recent patterns and mechanisms of carbon exchange by terrestrial ecosystems. *Nature* 414:169–172.

Schimel, J.P., K. Van Cleve, R.G. Cates, T.P. Clausen, and P.B. Reichardt. 1996. Effects of balsam poplar *(Populus balsamifera)* tannins and low molecular weight phenolics on microbial activity in taiga floodplain soil: Implications for changes in N cycling during succession. *Canadian Journal of Botany* 74:84–90.

Schimel, J.P. 2001. Biogeochemical models: Implicit vs. explicit microbiology. Pages 177–183 *in* E.-D. Schulze, S.P. Harrison, M. Heimann, E.A. Holland, J.J. Lloyd, et al., editors. *Global Biogeochemical Cycles in the Climate System.* Academic Press, San Diego.

Schimel, J.P. and J. Bennett. 2004. Nitrogen mineralization: Challenges of a changing paradigm. *Ecology* 85:591–602.

Schimper, A.F.W. 1898. *Pflanzengeographie auf Physiologischer Grundlage.* Fisher, Jena, Germany.

Schindler, D.W. 1971. Carbon, nitrogen, and phosphorus and the eutrophication of freshwater lakes. *Journal of Phycology* 7:321–329.

Schindler, D.W. 1974. Eutrophication and recovery in experimental lakes: Implications for lake management. *Science* 184:897–899.

Schindler, D.W. 1978. Factors regulating phytoplankton production and standing crop in the world's lakes. *Limnology and Oceanography* 23:478–486.

Schindler, D.W. 1985. The coupling of elemental cycles by organisms: Evidence from whole-lake chemical perturbations. Pages 225–250 *in* W. Stumm, editor. *Chemical Processes in Lakes.* Wiley Press, New York.

Schindler, D.W., R.E. Hecky, D.L. Findlay, M.P. Stainton, B.R. Parker, et al. 2008. Eutrophication of lakes cannot be controlled by reducing nitrogen inputs: Results of a 37-year whole-ecosystem experiment. *Proceedings of the National Academy of Sciences, USA* 105: 11254–11258.

Schippers, P., M. Lurling, and M. Scheffer. 2004. Increase in atmospheric CO_2 promotes phytoplankton productivity. *Ecology Letters* 7:446–451.

Schlesinger, W.H. 1977. Carbon balance in terrestrial detritus. *Annual Review of Ecology and Systematics* 8:51–81.

Schlesinger, W.H., J.F. Reynolds, G.L. Cunningham, L.F. Huenneke, W.M. Jarrell, et al. 1990. Biological feedbacks in global desertification. *Science* 247: 1043–1048.

Schlesinger, W.H. 1997. *Biogeochemistry: An Analysis of Global Change.* 2nd edition. Academic Press, San Diego.

Schlesinger, W.H. 2000. Carbon sequestration in soils: Some cautions amidst optimism. *Agriculture, Ecosystems and Environment* 82:121–127.

Schmidt, M.W.I., M.S. Torn, S. Abiven, T. Dittmar, G. Guggenberger, et al. in press. Environment as a dominant control in soil organic matter dynamics. *Nature*.

Schmitz, O.A., P.A. Hambäck, and A.P. Beckerman. 2000. Trophic cascades in terrestial systems: A review of the effects of carnivore removals on plants. *American Naturalist* 155:141–153.

Schmitz, O.J. 2009. Effects of predator functional diversity on grassland ecosystem function. *Ecology* 90:2339–2345.

Schoennagel, T., T.T. Veblen, and W.H. Romme. 2004. The interaction of fire, fuels, and climate across Rocky Mountain forests. *BioScience* 54:661–676.

Schubert, S.D., M.J. Suarez, P.J. Pegion, R.D. Koster, and J.T. Bacmeister. 2004. On the cause of the 1930s dust bowl. *Science* 303:1855–1859.

Schuh, A.E., A.S. Denning, K.D. Corbin, I.T. Baker, M. Uliasz, et al. 2010. A regional high-resolution carbon flux inversion of North America for 2004. *Biogeosciences* 7:1625–1644.

Schulze, E.-D. and F.S. Chapin, III. 1987. Plant specialization to environments of different resource availability. Pages 120–148 *in* E.-D. Schulze and H. Zwolfer, editors. *Potentials and Limitations in Ecosystem Analysis*. Springer-Verlag, Berlin.

Schulze, E.-D., R.H. Robichaux, J. Grace, P.W. Rundel, and J.R. Ehleringer. 1987. Plant water balance. *BioScience* 37:30–37.

Schulze, E.-D. 1989. Air pollution and forest decline in a spruce (*Picea abies*) forest. *Science* 244:776–783.

Schulze, E.-D., F.M. Kelliher, C. Körner, J. Lloyd, and R. Leuning. 1994. Relationship among maximum stomatal conductance, ecosystem surface conductance, carbon assimilation rate, and plant nitrogen nutrition: A global ecology scaling exercise. *Annual Review of Ecology and Systematics* 25:629–660.

Schuur, E.A.G. 2003. Productivity and global climate revisited: The sensitivity of tropical forest growth to precipitation. *Ecology* 84:1165–1170.

Schuur, E.A.G., J. Bockheim, J. Canadell, E.S. Euskirchen, C. Field, et al. 2008. The vulnerability of permafrost carbon to climate change: Implications for the global carbon cycle. *BioScience* 58:701–714.

Schuur, E.A.G., J.G. Vogel, K.G. Crummer, H. Lee, J.O. Sickman, et al. 2009. The effect of permafrost thaw on old carbon release and net carbon exchange from tundra. *Nature* 459:556–559.

Schwoerbel, J. 1987. *Handbook of Limnology*. 5th edition. Halsted Press, New York.

Scurlock, J.M.O. and R.J. Olson. 2002. Terrestrial net primary productivity: A brief history and a new worldwide database. *Environmental Reviews* 10: 91–109.

Seagle, S.W. 2003. Can deer foraging in multiple-use landscapes alter forest nitrogen budgets? *Oikos* 103:230–234.

Seastedt, T.R. 1985. Canopy interception of nitrogen in bulk precipitation by annually burned and unburned tallgrass prairie. *Oecologia* 66:88–92.

Seitzinger, S.P., J.A. Harrison, E. Dumont, A.H.W. Beusen, and A.F. Bouwman. 2005. Sources and delivery of carbon, nitrogen, and phosphorus to the coastal zone: An overview of Global Nutrient Export from Watersheds (NEWS) models and their application. *Global Biogeochemical Cycles* 19:GB4S01, doi: 10.1029/2005GB002606.

Selby, M.J. 1993. *Hillslope Materials and Processes*. 2nd edition. Oxford University Press, Oxford.

Semikhatova, O.A. 2000. Ecological physiology of plant dark respiration: Its past, present and future. *Botanishcheskii Zhurnal* 85:15–32.

Serreze, M.C. 2010. Understanding recent climate change. *Conservation Biology* 24:10–17.

Shaver, G.R., K.J. Nadelhoffer, and A.E. Giblin. 1991. Biogeochemical diversity and element transport in a heterogeneous landscape, the North Slope of Alaska. Pages 105–126 *in* M.G. Turner and R.H. Gardner, editors. *Quantitative Methods in Landscape Ecology*. Springer-Verlag, New York.

Shaver, G.R., W.D. Billings, F.S. Chapin, III, A.E. Giblin, K.J. Nadelhoffer, et al. 1992. Global change and the carbon balance of arctic ecosystems. *BioScience* 61:415–435.

Shaver, G.R., J. Canadell, F.S. Chapin, III, J. Gurevitch, J. Harte, et al. 2000. Global warming and terrestrial ecosystems: A conceptual framework for analysis. *BioScience* 50:871–882.

Shiro, T. and R. del Moral. 1995. Species attributes in early primary succession on volcanoes. *Journal of Vegetation Science* 6:517–522.

Shvidenko, A., D.V. Barber, and R. Persson. 2005. Forest and woodland systems. Pages 585–621 *in* R. Hassan, R.J. Scholes, and N. Ash, editors. *Ecosystems and Human Well-Being: Current State and Trends, Volume 1*. Island Press, Washington.

Sigman, D.M. and E.A. Boyle. 2000. Glacial/interglacial variations in atmospheric carbon dioxide. *Nature* 407:859–869.

Silver, W.L., D.J. Herman, and M.K. Firestone. 2001. Dissimilatory nitrate reduction to ammonium in upland tropical rain forest soils. *Ecology* 82:2410–2416.

Silvola, J., J. Alm, U. Ahlholm, H. Nykanen, and P.J. Martikainen. 1996. CO_2 fluxes from peat in boreal mires under varying temperature and moisture conditions. *Journal of Ecology* 84:219–228.

Simard, M., W.H. Romme, J.M. Griffin, and M.G. Turner. 2011. Do mountain pine beetle outbreaks change the probability of active crown fire in lodgepole pine forests? *Ecological Monographs* 81:3–24.

Simard, S.W., D.A. Perry, M.D. Jones, D.D. Myrold, D.M. Durall, et al. 1997. Net transfer of carbon between ectomycorrhizal tree species in the field. *Nature* 388:579–582.

Sinclair, A.R.E. 1979. The eruption of the ruminants. Pages 82–103 *in* A.R.E. Sinclair and M. Norton-Griffiths, editors. *Serengeti: Dynamics of an Ecosystem*. University of Chicago Press, Chicago.

Sinclair, A.R.E. and M. Norton-Griffiths. 1979. Dynamics of the Serengeti Ecosystem. Pages 1–30 *in* A.R.E. Sinclair and M. Norton-Griffiths, editors. *Serengeti: Dynamics of an Ecosystem*. University of Chicago, Chicago.

Singer, F., W. Swank, and E. Clebsch. 1984. Effects of wild pig rooting in a deciduous forest. *Journal of Wildlife Management* 48:464–473.

Sinsabaugh, R.L., C.L. Lauber, M.N. Weintraub, B. Ahmed, S.D. Allison, et al. 2008. Stoichiometry of soil enzyme activity at global scale. *Ecology Letters* 11:1252–1264.

Smart, D.R. and A.J. Bloom. 1988. Kinetics of ammonium and nitrate uptake among wild and cultivated tomatoes. *Oecologia* 76:336–340.

Smil, V. 2000. Phosphorus in the environment: Natural flows and human interferences. *Annual Review of Energy in the Environment* 25:53–88.

Smirnoff, N., P. Todd, and G.R. Stewart. 1984. The occurrence of nitrate reduction in the leaves of woody plants. *Annals of Botany* 54:363–374.

Smit, B. and J. Wandel. 2006. Adaptation, adaptive capacity and vulnerability. *Global Environmental Change* 16:282–292.

Smith, J.L. and E.A. Paul. 1990. The significance of soil microbial biomass estimations. Pages 357–396 *in* J. Bollag and G. Stotsky, editors. *Soil Biochemistry*. Marcel Dekker, New York.

Smith, S.E. and D.J. Read. 1997. *Mycorrhizal Symbiosis*. Academic Press, London.

Smithwick, E.A.H., M.C. Mack, M.G. Turner, F.S. Chapin, III, J. Zhu, et al. 2005. Spatial heterogeneity and soil nitrogen dynamics in a burned black spruce forest stand: Distinct controls at different scales. *Biogeochemistry* 76:517–537.

Solomon, S., G.-K. Plattner, R. Knutti, and P. Friedlingstein. 2009. Irreversible climate change due to carbon dioxide emissions. *Proceedings of the National Academy of Sciences, USA* 106:1704–1709.

Soluck, D.A. and J.S. Richardson. 1997. The role of stoneflies in enhancing growth of trout: A test of the importance of predator-predator facilitation within a stream community. *Oikos* 80:214–219.

Sowden, F., Y. Chen, and M. Schnitzer. 1977. The nitrogen distribution in soils formed under widely differing climatic conditions. *Geochimica et Cosmochimica Acta* 41:1524–1526.

Sperry, J.S. 1995. Limitations on stem water transport and their consequences. Pages 105–124 *in* B.L. Gartner, editor. *Plant Stems: Physiology and Functional Morphology*. Academic Press, San Diego.

Sperry, J.S., F.C. Meinzer, and K.A. McCulloh. 2008. Safety and efficiency conflicts in hydraulic architecture: Scaling from tissues to trees. *Plant, Cell & Environment* 31:632–635.

Sprugel, D.G. 1976. Dynamic structure of wave-generated *Abies balsamea* forests in the northeastern United States. *Journal of Ecology* 64:880–911.

St. Louis, V.L., C.A. Kelly, E. Duchemin, J.W.M. Rudd, and D.M. Rosenberg. 2000. Reservoir surfaces as sources of greenhouse gases to the atmosphere: A global estimate. *BioScience* 50:766–775.

Stafford Smith, D.M., N. Abel, B. Walker, and F.S. Chapin, III. 2009. Drylands: Coping with uncertainty, thresholds, and changes in state. Pages 171–195 *in* F.S. Chapin, III, G.P. Kofinas, and C. Folke, editors. *Principles of Ecosystem Stewardship: Resilience-Based Natural Resource Management in a Changing World*. Springer, New Work.

Stanford, G. and E. Epstein. 1974. Nitrogen mineralization-water relations in soils. *Soil Science Society of America Journal* 38:103–107.

Starfield, A.M. 1991. Qualitative rule-based modeling. *BioScience* 40:601–604.

Starfield, A.M. and A.L. Bleloch. 1991. *Building Models for Conservation and Wildlife Management*. 2nd edition. Burgess Press, Edina, MN.

Stark, J.M. and M.K. Firestone. 1995. Mechanisms for soil moisture effects on nitrifying bacteria. *Applied Environmental Microbiology* 61:218–221.

Stark, J.M. and S.C. Hart. 1997. High rates of nitrification and nitrate turnover in undisturbed coniferous forests. *Nature* 385:61–64.

Steele, J.H. 1991. Can ecological theory cross the land-sea boundary? *Journal of Theoretical Biology* 153:425–436.

Steffen, W.L., A. Sanderson, P.D. Tyson, J. Jäger, P.A. Matson, et al. 2004. *Global Change and the Earth System: A Planet under Pressure*. Springer-Verlag, New York.

Steiner, K. 1982. *Intercropping in Tropical Smallholder Agriculture with Special Reference to West Africa*. German Agency for Technical Cooperation (GTZ), Eschborn, Germany.

Stephens, B.B. and R.F. Keeling. 2000. The influence of Antarctic sea ice on glacial-interglacial CO_2 variations. *Nature* 404:171–174.

Stern, N. 2007. *The Economics of Climate Change: The Stern Review*. Cambridge University Press, Cambridge.

Sterner, R.W. and J.J. Elser. 2002. *Ecological Stoichiometry: The Biology of Elements from Molecules to the Biosphere*. Princeton University Press, Princeton.

Sterner, R.W. 2008. On the phosphorus limitation paradigm for lakes. *International Review of Hydrobiology* 93:433–445.

Stevenson, F.J. 1994. *Humus Chemistry: Genesis, Composition, Reactions*. 2nd edition. Wiley, New York.

Still, C.J., J.A. Berry, G.J. Collatz, and R.S. DeFries. 2003. Global distribution of C_3 and C_4 vegetation: Carbon cycle implications. *Global Biogeochemical Cycles* 17:doi:10.1029/2001GB001807.

Stock, W.D. and O.A.M. Lewis. 1984. Uptake and assimilation of nitrate and ammonium by an evergreen Fynbos shrub species *Protea repens* L. (Proteaceae). *New Phytologist* 97:261–268.

Stocker, R., J.R. Seymour, A. Samadani, D.E. Hunt, and M.F. Polz. 2008. Rapid chemotactic response enables marine bacteria to exploit microscale nutrient patches. *Proceedings of the National Academy of Sciences, USA* 105:4209–4214.

Stokes, D.E. 1997. *Pasteur's Quadrant: Basic Science and Technological Innovation.* Brookings Institution Press, Washington.

Stramma, L., G.C. Johnson, J. Sprintal, and V. Mohrholz. 2008. Expanding oxygen-minimum zones in the tropical oceans. *Science* 320:655–658.

Strayer, D.L., N.F. Caraco, J.J. Cole, S. Findlay, and M.L. Pace. 1999. Transformation of freshwater ecosystems by bivalves. *BioScience* 49:19–27.

Strayer, D.L., R.E. Beighley, L.C. Thompson, A. Brooks, C. Nilsson, et al. 2003. Effects of land cover on stream ecosystems: Roles of empirical models and scaling issues. *Ecosystems* 6:407–423.

Strong, D.R. 1992. Are trophic cascades all wet? Differentiation and donor-control in speciose ecosystems. *Ecology* 73:747–754.

Sturm, M., J.P. McFadden, G.E. Liston, F.S. Chapin, III, J. Holmgren, et al. 2001. Snow-shrub interactions in arctic tundra: A feedback loop with climatic implications. *Journal of Climate* 14:336–344.

Sturman, A.P. and N.J. Tapper. 1996. *The Weather and Climate of Australia and New Zealand.* Oxford University Press, Oxford.

Sucoff, E. 1972. Water potential in red pine: Soil moisture, evapotranspiration, crown position. *Ecology* 52:681–686.

Suding, K.N., S. Lavorel, F.S. Chapin, III, J.H.C. Cornelissen, S. Díaz, et al. 2008. Scaling environmental change through the community level: A trait-based response- and-effect framework for plants. *Global Change Biology* 14:1125–1140.

Sun, J.M., K.E. Kohfield, and S.P. Harrison. 2000. *Records of aeolian dust deposition on the Chinese Loess Plateau during the Quaternary.* Max-Planck Instutut für Biogeochemie Technical Reports.

Sundquist, E.T. and K. Visser. 2004. The geologic history of the carbon cycle. Pages 425–472 *in* W.H. Schlesinger, editor. *Biogeochemistry.* Elsevier, Amsterdam.

Sutherland, W.J. 2000. *The Conservation Handbook: Research, Management and Policy.* Blackwell Scientific, Oxford.

Sutton, M.A., D. Smpson, P.E. Levy, R.I. Smith, S. Fres, et al. 2008. Uncertainties in the relationship between atmospheric nitrogen deposition and forest carbon sequestration. *Global Change Biology* 14:2057–2063.

Swank, W.T. and J.E. Douglass. 1974. Streamflow greatly reduced by converting deciduous hardwood stands to pine. *Science* 185:857–859.

Swanson, F.J. and F.S. Chapin, III. 2009. Forest Systems: Living with long-term change. Pages 149–170 *in* F.S. Chapin, III, G.P. Kofinas, and C. Folke, editors. *Principles of Ecosystem Stewardship: Resilience-Based Natural Resource Management in a Changing World.* Springer, New York.

Swift, M.J., O.W. Heal, and J.M. Anderson. 1979. *Decomposition in Terrestrial Ecosystems.* Blackwell Scientific Publications, Ltd., Oxford.

Syvitski, J.P.M., C.J. Vörösmarty, A.J. Kettner, and P.A. Green. 2005. Impact of humans on the flux of terrestrial sediment to the global coastal ocean. *Science* 308:376–380.

Szaro, R.C., N.D. Johnson, W.T. Sexton, and A.J. Malk, editors. 1999. *Ecological Stewardship: A Common Reference for Ecosystem Management.* Elsevier Science Ltd, Oxford.

Tallis, H., P. Kareiva, M. Marvier, and A. Chang. 2008. An ecosystem services framework to support both practical conservation and economic development. *Proceedings of the National Academy of Sciences, USA* 105:9457–9464.

Tans, P.P., I.Y. Fung, and T. Takahashi. 1990. Observational constraints on the global CO_2 budget. *Science* 247:1431–1438.

Tansley, A.G. 1935. The use and abuse of vegetational concepts and terms. *Ecology* 16:284–307.

Tarnocai, C., J.G. Canadell, E.A.G. Schuur, P. Kuhry, G. Mazhitova, et al. 2009. Soil organic carbon pools in the northern circumpolar permafrost region. Global Biogeochemical Cycles 23:GB2023, doi:2010.1029/2008GB003327.

Taylor, B.R., D. Parkinson, and W.F.J. Parsons. 1989. Nitrogen and lignin as predictors of litter decay rates: A microcosm test. *Ecology* 70:97–104.

Taylor, D.L., I.C. Herriott, K.E. Stone, J.W. McFarland, M.G. Booth, et al. 2010. Structure and resilience of fungal communities in Alaskan boreal forest soils. *Canadian Journal of Forest Research* 40: 1288–1301.

Tengö, M., K. Johansson, F. Rakotongrasoa, J. Lundberg, J.-A. Andriamaherilala, et al. 2007. Taboos and forest governance: Informal protection of hot spot dry forest in Southern Madagascar. *Ambio* 36:683–691.

Tennyson, A.J.D. 2010. The origin and history of New Zealand's terrestrial vertebrates. *New Zealand Journal of Ecology* 34:6–27.

Terashima, I. and K. Hikosaka. 1995. Comparative ecophysiology of leaf and canopy photosynthesis. *Plant, Cell and Environment* 18:1111–1128.

Teskey, R.O., D.W. Sheriff, and D.Y. Hollinger. 1995. External and internal factors regulating photosynthesis *in* W.K. Smith and T.M. Hinkley, editors. *Resources and Physiology of Conifers: Acquisition, Allocation, and Utilization.* Academic Press, San Diego.

Thiet, R.K., S.D. Frey, and J. Six. 2006. Do growth yield efficiencies differ between soil microbial communities differing in fungal: bacterial ratios? Reality check and methodological issues. *Soil Biology and Biochemistry* 38:837–844.

Thomas, W.A. 1969. Accumulation and cycling of calcium by dogwood trees. *Ecological Monographs* 39:101–120.

Thompson, R.M., C.R. Townsend, D. Craw, R. Frew, and R. Riley. 2001. (Further) links from rocks to plants. *Trends in Ecology & Evolution* 16:543.

Thornton, K.W., B.L. Kimmel, and F.E. Payne. 1990. *Reservoir Limnology: Ecological Perspectives.* John Wiley and Sons, New York.

Thorp, J.H. and A.P. Covich, editors. 2001. *Ecology and Classification of North American Freshwater Invertebrates.* Academic Press, San Diego.

Thurman, H.V. 1991. *Introductory Oceanography.* 6th edition. MacMillan Publishing Company, New York.

Tiessen, H. 1995. Introduction and synthesis. Pages 1–6 *in* H. Tiessen, editor. *Phosphorus in the Global Environment: Transfers, Cycles and Management.* John Wiley & Sons, Chichester.

Tietema, A. and C. Beier. 1995. A correlative evaluation of nitrogen cycling in the forest ecosystems of the EC projects NITREX and EXMAN. *Forest Ecology and Management* 71:143–151.

Tilman, D. 1985. The resource-ratio hypothesis of plant succession. *American Naturalist* 125:827–852.

Tilman, D. 1988. *Plant Strategies and the Dynamics and Function of Plant Communities.* Princeton University Press, Princeton.

Tilman, D. and D. Wedin. 1991. Dynamics of nitrogen competition between successional grasses. *Ecology* 72:1038–1049.

Tilman, D., D. Wedin, and J. Knops. 1996. Productivity and sustainability influenced by biodiversity in grassland ecosystems. *Nature* 379:718–720.

Tilman, D., P.B. Reich, and J.M.H. Knops. 2006. Biodiversity and ecosystem stability in a decade-long grassland experiment. *Nature* 441:629–632.

Tjoelker, M.G., J.M. Craine, D. Wedin, P.B. Reich, and D. Tilman. 2005. Linking leaf and root trait syndromes among 39 grassland and savannah species. *New Phytologist* 167:493–508.

Towns, D.R., I.A.E. Atkinson, and C.H. Daugherty. 2006. Have the harmful effects of introduced rats on islands been exaggerated? *Biological Invasions* 8:863–891.

Townsend, A.R., P.M. Vitousek, and S.E. Trumbore. 1995. Soil organic matter dynamics along gradients in temperature and land use on the island of Hawaii. *Ecology* 76:721–733.

Townsend, A.R., C.C. Cleveland, G.P. Asner, and M.M.C. Bustamante. 2007. Controls of foliar N:P ratios in tropical rain forests. *Ecology* 88:107–118.

Trenberth, K.E., and T.J. Haar. 1996. The 1990–1995 El Niño - Southern Oscillation event: Longest on record. *Geophysical Research Letters* 23:57–60.

Trenberth, K.E. and D.P. Stepaniak. 2003. Coveriability of components of poleward atmospheric energy transports on seasonal and interannual timescales. *Journal of Climate* 16:3691–3705.

Trenberth, K.E., J.T. Fasullo, and J. Kiehl. 2009. Earth's global energy budget. *Bulletin of the American Meteorological Society.*

Treseder, K.K. 2008. Nitrogen additions and microbial biomass: A meta-analysis of ecosystem studies. *Ecology Letters* 11:1111–1120.

Trimble, S.W., F.H. Weirich, and B.L. Hoag. 1987. Reforestation reduces stream flow in the southeastern United States. *Water Resource Research* 23:425–437.

Trumbore, S.E. 1993. Comparison of carbon dynamics in tropical and temperate soils using radiocarbon measurements. *Global Biogeochemical Cycles* 7:275–290.

Trumbore, S.E. and J.W. Harden. 1997. Accumulation and turnover of carbon in organic and mineral soils of the BOREAS northern study area. *Journal of Geophysical Research* 102:28817–28830.

Turetsky, M.R., M.C. Mack, T.N. Hollingsworth, and J.W. Harden. 2010. The role of mosses in ecosystem succession and function in Alaska's boreal forest. *Canadian Journal of Forest Research* 40:1237–1264.

Turner, B.L., II, W.C. Clark, R.W. Kates, J.F. Richards, J.T. Mathews, et al., editors. 1990. *The Earth as Transformed by Human Action: Global and Regional Changes in the Biosphere over the Past 300 Years.* Cambridge University Press, Cambridge.

Turner, B.L., II, R.E. Kasperson, P.A. Matson, J.J. McCarthy, R.W. Corell, et al. 2003a. A framework for vulnerability analysis in sustainability science. *Proceedings of the National Academy of Sciences, USA* 100:8074–8079.

Turner, D.P., S. Urbanski, D. Bremer, S.C. Wofsy, T. Meyers, et al. 2003b. A cross-biome comparison of daily light use efficiency for gross primary production. *Global Change Biology* 9:383–395.

Turner, D.P., W.D. Ritts, W.B. Cohen, T.K. Maeirsperger, S.T. Gower, et al. 2005. Site-level evaluation of satellite-based global terrestrial primary production and net primary production monitoring. *Global Change Biology* 11:666–684.

Turner, M.G., R.V. O'Neill, R.H. Gardner, and B.T. Milne. 1989. Effects of changing spatial scale on the analysis of landscape pattern. *Landscape Ecology* 3:153–162.

Turner, M.G., W.H. Romme, R.H. Gardner, R.V. O'Neill, and T.K. Kratz. 1993. A revised concept of landscape equilibrium: Disturbance and stability on scaled landscapes. *Landscape Ecology* 8:213–227.

Turner, M.G., W.H. Hargrove, R.H. Gardner, and W.H. Romme. 1994. Effects of fire on landscape heterogeneity in Yellowstone National Park, Wyoming. *Journal of Vegetation Science* 5:731–742.

Turner, M.G., W.H. Romme, R.H. Gardner, and W.W. Hargrove. 1997. Effects of fire size and pattern on early succession Yellowstone National Park. *Ecological Monographs* 67:411–433.

Turner, M.G., W.H. Romme, and R.H. Gardner. 1999. Prefire heterogeneity, fire severity and plant reestablishment in subalpine forests of Yellowstone National Park, Wyoming. *International Journal of Wildland Fire* 9:21–36.

Turner, M.G., R.H. Gardner, and R.V. O'Neill. 2001. *Landscape Ecology in Theory and Practice: Pattern and Process.* Springer-Verlag, New York.

Turner, M.G. 2005. Landscape ecology: What is the state of the science? *Annual Review of Ecology, Evolution, and Systematics* 36:319–344.

Turner, M.G. and F.S. Chapin, III. 2005. Causes and consequences of spatial heterogeneity in ecosystem function. Pages 9–30 *in* G.M. Lovett, C.G. Jones, M.G. Turner, and K.C. Weathers, editors. *Ecosystem Function in Heterogeneous Landscapes.* Springer, New York.

Turner, M.G. 2010. Disturbance and landscape dynamics in a changing world. *Ecology* 91:2833–2849.

Tyrrell, T. 1999. The relative influences of nitrogen and phosphorus on oceanic primary production. *Nature* 400:525–531.

Uehara, G. and G. Gillman. 1981. *The Minerology, Chemistry, and Physics of Tropical Soils with Variable Charge Clays*. Westview Press, Boulder.

Ugolini, F.C. 1968. Soil development and alder invasion in a recently deglaciated area of Glacier Bay, Alaska. Pages 115–148 *in* J.M. Trappe, F.F. Franklin, R.F. Tarrant, and G.M. Hansen, editors. *Biology of Alder*. USDA Forest Service, Pacific Northwest Forest and Range Experiment Station, Portland.

Ugolini, F.C. and H. Spaltenstein. 1992. Pedosphere. Pages 123–153 *in* S.S. Butcher, R.J. Charlson, G.H. Orians, and G.V. Wolfe, editors. *Global Biogeochemical Cycles*. Academic Press, London.

Ulrich, A. and J.J. Hills. 1973. Plant analysis as an aid in fertilizing sugar crops: Part I. Sugar beets. Pages 271–288 *in* L.M. Walsh and J.D. Beaton, editors. *Soil Testing and Plant Analysis*. Soil Science Society of America, Madison, WI.

Urban, D.L., R.V. O'Neill, and H.H. Shugart. 1987. Landscape ecology. *BioScience* 37:119–127.

Vadeboncoeur, Y., M.J. Vander Zanden, and D.M. Lodge. 2002. Putting the lake back together: Reintegrating benthic pathways into lake food web models. *BioScience* 52:44–54.

Vadeboncoeur, Y., G.D. Peterson, M.J. Vander Zanden, and J. Kalff. 2008. Benthic algal production across lake size gradients: Interactions among morphometry, nutrients, and light. *Ecology* 89:2542–2552.

Valentini, R., G. Matteucci, A.J. Dolman, E.-D. Schulze, C. Rebmann, et al. 2000. Respiration as the main determinant of carbon balance in European forests. *Nature* 404:861–864.

Valiela, I. 1995. *Marine Ecological Processes*. 2nd edition. Springer-Verlag, New York.

Van Breemen, N. and A.C. Finzi. 1998. Plant-soil interactions: Ecological aspects and evolutionary implications. *Biogeochemistry* 42:1–19.

Van Cleve, K., W.C. Oechel, and J.L. Hom. 1990. Response of black spruce (*Picea mariana*) ecosystems to soil temperature modification in interior Alaska. *Canadian Journal of Forest Research* 20:1530–1535.

Van Cleve, K., F.S. Chapin, III, C.T. Dyrness, and L.A. Viereck. 1991. Element cycling in taiga forest: State-factor control. *BioScience* 41:78–88.

Van Cleve, K., C.T. Dyrness, G.M. Marion, and R. Erickson. 1993. Control of soil development on the Tanana River floodplain, interior Alaska. *Canadian Journal of Forest Research* 23:941–955.

van Ruijven, J. and F. Berendse. 2003. Positive effects of plant species diversity on productivity in the absence of legumes. *Ecology Letters* 6:170–175.

Vance, E.D. and F.S. Chapin, III. 2001. Substrate-environment interactions: Multiple limitations to microbial activity in taiga forest floors. *Soil Biology and Biochemistry* 33:173–188.

Vander Zanden, M.J., T.E. Essington, and Y. Vadeboncoeur. 2005. Is pelagic top-down control in lakes augmented by benthic energy pathways? *Canadian Journal of Fisheries and Aquatic Sciences* 62:1422–1431.

Vander Zanden, M.J., S. Chandra, S.-K. Park, Y. Vadeboncoeur, and C.R. Goldman. 2006. Efficiencies of benthic and pelagic trophic pathways in a subalpine lake. *Canadian Journal of Fisheries and Aquatic Sciences* 63:2608–2620.

Vandermeer, J.H. 1990. Intercropping. Pages 481–516 *in* C.R. Carrol, J.H. Vandermeer, and P.M. Rosset, editors. *Agroecology*. McGraw Hill, New York.

Vandermeer, J.H. 1995. The ecological basis of alternative agriculture. *Annual Review of Ecology and Systematics* 26:201–224.

Vannote, R.I., G.W. Minshall, K.W. Cummings, J.R. Sedell, and C.E. Cushing. 1980. The river continuum concept. *Canadian Journal of Fisheries and Aquatic Sciences* 37:120–137.

VEMAP-Members. 1995. Vegetation/ecosystem modeling and analysis project: Comparing biogeography and biogeochemistry models in a continental-scale study of terrestrial ecosystem responses to climate change and CO_2 doubling. *Global Biogeochemical Cycles* 9:407–437.

Verbyla, D.L. 1995. *Satellite Remote Sensing of Natural Resources*. CRC Press, Boca Raton, Florida.

Verhoef, H.A. and L. Brussaard. 1990. Decomposition and nitrogen mineralization in natural and agro-ecosystems: The contribution of soil animals. *Biogeochemistry* 11:175–211.

Villar, R., J.R. Robleto, Y. De Jong, and H. Poorter. 2006. Differences in construction costs and chemical composition between deciduous and evergreen woody species are small as compared to differences among families. *Plant, Cell and Environment* 29:1629–1643.

Violle, C., M.-L. Navas, D. Vile, E. Kazakou, C. Fortunel, et al. 2007. Let the concept of trait be functional! *Oikos* 116:882–892.

Vitousek, P.M. and W.A. Reiners. 1975. Ecosystem succession and nutrient retention: A hypothesis. *BioScience* 25:376–381.

Vitousek, P.M. 1982. Nutrient cycling and nutrient use efficiency. *American Naturalist* 119:553–572.

Vitousek, P.M., J.R. Gosz, C.C. Grier, J.M. Melillo, and W.A. Reiners. 1982. A comparative analysis of potential nitrification and nitrate mobility in forest ecosystems. *Ecological Monographs* 52:155–177.

Vitousek, P.M. 1984. Litterfall, nutrient cycling, and nutrient limitation in tropical forests. *Ecology* 65:285–298.

Vitousek, P.M. and P.A. Matson. 1984. Mechanisms of nitrogen retention in forest ecosystems: A field experiment. *Science* 225:51–52.

Vitousek, P.M., L.R. Walker, L.D. Whiteaker, D. Mueller-Dombois, and P.A. Matson. 1987. Biological invasion by *Myrica faya* alters ecosystem development in Hawai'i. *Science* 238:802–804.

Vitousek, P.M. and P.A. Matson. 1988. Nitrogen transformations in a range of tropical forest soils. *Soil Biology and Biochemistry* 20:361–367.

Vitousek, P.M. 1990. Biological invasions and ecosystem processes: Towards an integration of population biology and ecosystem studies. *Oikos* 57:7–13.

Vitousek, P.M. and R.W. Howarth. 1991. Nitrogen limitation on land and in the sea: How can it occur? *Biogeochemistry* 13:87–115.

Vitousek, P.M., G. Aplet, D. Turner, and J.J. Lockwood. 1992. The Mauna Loa environmental matrix: Foliar and soil nutrients. *Oecologia* 89:372–382.

Vitousek, P.M. and D.U. Hooper. 1993. Biological diversity and terrestrial ecosystem biogeochemistry. Pages 3–14 *in* E.-D. Schulze and H.A. Mooney, editors. *Biodiversity and Ecosystem Function.* Springer-Verlag, Berlin.

Vitousek, P.M. 1994. Beyond global warming: Ecology and global change. *Ecology* 75:1861–1876.

Vitousek, P.M., J.D. Aber, R.W. Howarth, G.E. Likens, P.A. Matson, et al. 1997a. Human alteration of the global nitrogen cycle: Sources and consequences. *Ecological Applications* 7:737–750.

Vitousek, P.M. and H. Farrington. 1997. Nitrogen limitation and soil development: Experimental test of a biogeochemical theory. *Biogeochemistry* 37:63–75.

Vitousek, P.M., H.A. Mooney, J. Lubchenco, and J.M. Melillo. 1997b. Human domination of Earth's ecosystems. *Science* 277:494–499.

Vitousek, P.M. and C.B. Field. 1999. Ecosystem constraints to symbiotic nitrogen fixers: A simple model and its implications. *Biogeochemistry* 46:179–202.

Vitousek, P.M., K. Cassman, C.C. Cleveland, T. Crews, C.B. Field, et al. 2002. Towards an ecological understanding of biological nitrogen fixation. *Biogeochemistry* 57/58:1–45.

Vitousek, P.M. 2004. *Nutrient Cycling and Limitation: Hawai'i as a Model System.* Princeton University Press, Princeton.

Vitousek, P.M., G.P. Asner, O.A. Chadwick, and S. Hotchkiss. 2009a. Landscape-level variation in forest structure and biogeochemistry across a substrate age gradient in Hawai'i. *Ecology* 90:3074–3086.

Vitousek, P.M., R.L. Naylor, T. Crews, M.B. David, L.E. Drinkwater, et al. 2009b. Agriculture: Nutrient imbalances in agricultural development. *Science* 324:1519–1520.

Vitousek, P.M., S. Porder, B.Z. Houlton, and O.A. Chadwick. 2010. Terrestrial phosphorus limitation: Mechanisms, implications, and nitrogen-phosphorus interactions. *Ecological Applications* 20:5–15.

Vogt, K.A., C.C. Grier, and D.J. Vogt. 1986. Production, turnover, and nutrient dynamics in above- and below-ground detritus of world forests. *Advances in Ecological Research* 15:303–377.

Vörösmarty, C.J., C. Leveque, C. Revenga, R. Bos, C. Caudill, et al. 2005. Fresh water. Pages 165–207 *in* R. Hassan, R.J. Scholes, and N. Ash, editors. *Ecosystems and Human Well-Being: Current State and Trends, Volume 1.* Island Press, Washington.

Vrba, E.S. and S.J. Gould. 1986. The hierarchical expansion of sorting and selection: Sorting and selection cannot be equated. *Paleobiology* 12:217–228.

Vukicevic, T., B.H. Braswell, and D. Schimel. 2001. A diagnostic study of temperature controls on global terrestrial carbon exchange. *Tellus* 53B:150–170.

Wagener, S.M., M.W. Oswood, and J.P. Schimel. 1998. Rivers and soils: Parallels in carbon and nutrient processing. *BioScience* 48:104–108.

Waldrop, M.P. and D.R. Zak. 2006. Response of oxidative enzyme activities to nitrogen deposition affects soil concentrations of dissolved organic carbon. *Ecosystems* 9:921–933.

Walker, B.H. 1992. Biodiversity and ecological redundancy. *Conservation Biology* 6:18–23.

Walker, B.H. 1995. Conserving biological diversity through ecosystem resilience. *Conservation Biology* 9:747–752.

Walker, B.H., A. Kinzig, and J. Langridge. 1999. Plant attribute diversity, resilience, and ecosystem function: The nature and significance of dominant and minor species. *Ecosystems* 2:95–113.

Walker, B.H., C.S. Holling, S.R. Carpenter, and A. Kinzig. 2004. Resilience, adaptability, and transformability in social-ecological systems. *Ecology and Society* 9:http://www.ecologyandsociety.org/vol9/iss2/art5.

Walker, D.A., J.G. Bockheim, F.S. Chapin, III, W. Eugster, J.Y. King, et al. 1998. A major arctic soil pH boundary: Implications for energy and trace-gas fluxes. *Nature* 394:469–472.

Walker, L.R., J.C. Zasada, and F.S. Chapin, III. 1986. The role of life history processes in primary succession on an Alaskan floodplain. *Ecology* 67:1243–1253.

Walker, L.R. 1993. Nitrogen fixers and species replacements in primary succession. Pages 249–272 *in* J. Miles and D.W.H. Walton, editors. *Primary succession on land.* Blackwell, Oxford.

Walker, L.R. 1999. Patterns and processes in primary succession. Pages 585–610 *in* L.R. Walker, editor. *Ecosystems of Disturbed Ground.* Elsevier, Amsterdam.

Walker, L.R. and R. del Moral. 2003. *Primary Succession and Ecosystem Rehabilitation.* Cambridge University Press, Cambridge.

Walker, T.W. and J.K. Syers. 1976. The fate of phosphorus during pedogenesis. *Geoderma* 15:1–19.

Wall, D.H., G. Adams, and A.N. Parsons. 2001. Soil biodiversity. Pages 47–82 *in* F.S. Chapin, III, O.E. Sala, and E. Huber-Sannwald, editors. *Global Biodiversity in a Changing Environment: Scenarios for the 21st century.* Springer-Verlag, New York.

Wall, D.H., M.A. Bradford, M.G. St. John, J.A. Trofymow, V. Behan-Pelletier, et al. 2008. Global decomposition experiment shows soil animal impacts on decomposition are climate-dependent. *Global Change Biology* 14:2661–2677.

Wallwork, J.A. 1976. *The Distribution and Diversity of Soil Fauna.* Academic Press, New York.

Walter, K.M., L.C. Smith, and F.S. Chapin, III. 2007. Methane bubbling from northern lakes: Present and future contributions to the global methane budget. *Philosophical Transactions of the Royal Society of London* A365:1657–1676.

Walter, M.K., S.A. Zimov, J.P. Chanton, D. Verbyla, and F.S. Chapin, III. 2006. Methane bubbling from Siberian thaw lakes as a positive feedback to climate warming. *Nature* 443:71–75.

Walters, C.J. and S.J.D. Martell. 2004. *Fisheries Ecology and Management*. Princeton University Press, Princeton.

Walters, C.J. and R. Ahrens. 2009. Oceans and estuaries: Managing the commons. Pages 221–240 *in* F.S. Chapin, III, G.P. Kofinas, and C. Folke, editors. *Principles of Ecosystem Stewardship: Resilience-Based Natural Resource Manangement in a Changing World*. Springer, New York.

Walters, M.B. and P.B. Reich. 1999. Low-light carbon balance and shade tolerance in the seedlings of woody plants: Do winter deciduous and broad-leaved evergreen species differ? *New Phytologist* 143:143–154.

Wan, S., D. Hui, and Y. Luo. 2001. Fire effects on nitrogen pools and dynamics in terrestrial ecosystems: A meta-analysis. *Ecological Applications* 11:1349–1365.

Wang, J.S., J.A. Logan, M.B. McElroy, B.N. Duncan, I.A. Megretskaia, et al. 2004. A 3-D model analysis of the slowdown and interannual variability in the methane growth rate from 1988 to 1997. *Global Biogeochemical Cycles* 18:GB3011, doi:3010.1029/3003GB002180.

Waring, R.H., J.J. Landsberg, and M. Williams. 1998. Net primary production of forests: A constant fraction of gross primary production? *Tree Physiology* 18:129–134.

Waring, R.H. and S.W. Running. 2007. *Forest Ecosystems: Analysis at Multiple Scales*. 3rd edition. Academic Press, San Diego.

WCED. 1987. *Our Common Future*. World Commission on Environment and Development, Oxford University Press, Oxford.

Weaver, C.P. and R. Avissar. 2001. Atmospheric disturbances caused by human modification of the landscape. *Bulletin of the American Meteorological Society* 82:269–281.

Webb, T. and P.J. Bartlein. 1992. Global changes during the last three million years: Climatic controls and biotic responses. *Annual Review of Ecology and Systematics* 23:141–173.

Webster, J.R. and E.F. Benfield. 1986. Vascular plant breakdown in freshwater ecosystems. *Annual Review of Ecology and Systematics* 17:567–594.

Webster, J.R., S.W. Golliday, E.F. Benfield, D.J. D'Angelo, and G.T. Peters. 1990. Effects of forest disturbance on particulate organic matter budgets of small streams. *Journal of the North American Benthological Society* 9:120–140.

Webster, J.R., J.B. Wallace, and E.F. Benfield. 1995. Organic processes in streams of the eastern United States.*in* C.E. Cushing, G.W. Minshall, and K.W. Cummins, editors. *Ecosystems of the World 22: River and Stream Ecosystems*. Elsevier, Amsterdam.

Webster, J.R. and J.L. Meyer. 1997. Organic matter budgets for streams: A synthesis. *Journal of the North American Benthological Society* 16:141–161.

Webster, J.R. 2007. Spiraling down the river continuum: Stream ecology and the U-shaped curve. *Journal of the North American Benthological Society* 26:375–389.

Webster, K.E., T.K. Kratz, C.J. Bowser, and J.J. Magnuson. 1996. The influence of landscape position on lake chemical responses to drought in northern Wisconsin. *Limnology and Oceanography* 41:977–984.

Webster, P.J. and T.N. Palmer. 1997. The past and future of El Niño. *Nature* 390:562–564.

Wedin, D.A. and D. Tilman. 1990. Species effects on nitrogen cycling: A test with perennial grasses. *Oecologia* 84:433–441.

Wells, M.P. and K.E. Brandon. 1993. The principles and practice of buffer zones and local participation in biodiversity conservation. *Ambio* 22:157–162.

West, T.O. and W.M. Post. 2002. Soil organic carbon sequestration rates by tillage and crop rotation: A global data analysis. *Soil Science Society of America Journal* 66:1930–1946.

Westley, F., B. Zimmerman, and M.Q. Patton. 2006. *Getting to Maybe: How the World is Changed*. Random House, Toronto.

Westman, W.E. 1978. Patterns of nutrient flow in the pygmy forest region of northern California. *Vegetatio* 36:1–15.

Whalen, S.C. and J.C. Cornwell. 1985. Nitrogen, phosphorus, and organic carbon cycling in an arctic lake. *Canadian Journal of Fisheries and Aquatic Sciences* 42:797–808.

Whisenant, S. 1999. *Repairing Damaged Ecosystems*. Cambridge University Press, New York.

White, M.A., P.E. Thornton, S.W. Running, and R.R. Nemani. 2000. Parameterization and sensitivity analysis of the BIOME-BGC terrestrial ecosystem model: Net primary production controls. *Earth Interactions* 4:1–85.

White, P.S. and S.T.A. Pickett. 1985. Natural disturbance and patch dynamics: An introduction. Pages 3–13 *in* S.T.A. Pickett and P.S. White, editors. *The Ecology of Natural Disturbance and Patch Dynamics*. Academic Press, New York.

Whiteman, G., B.C. Forbes, J. Niemelä, and F.S. Chapin, III. 2004. Bringing feedback and resilience of high-latitude ecosystems into the corporate boardroom. *Ambio* 33:371–376.

Whittaker, R.H. and W.A. Niering. 1965. Vegetation of the Santa Catalina Mountains, Arizona. (II) A gradient analysis of the south slope. *Ecology* 46:429–452.

Whittaker, R.H. 1975. *Communities and Ecosystems*. 2nd edition. Macmillan, New York.

Wiegert, R.G. and D.F. Owen. 1971. Trophic structure, available resources and population density in terrestrial versus aquatic ecosystems. *Journal of Theoretical Biology* 30:69–81.

Wiens, J.A. 1996. Wildlife in patchy environments: Metapopulations, mosaics, and management. Pages 53–84 *in* D.R. McCullough, editor. *Metapopulations and Wildlife Conservation*. Island Press, Washington.

Wilcox, H.E. 1991. Mycorrhizae. Pages 731–765 *in* Y. Waisel, A. Eshel, and U. Kafkaki, editors. *Plant Roots: The Hidden Half*. Marcel Dekker, New York.

Williams, J.W. and S.T. Jackson. 2007. Novel climates, no-analog communities, and ecological surprises. *Frontiers in Ecology and the Environment* 5:475–482.

Williamson, C.E., R.S. Sternberger, D.P. Morris, T.M. Frost, and S.G. Paulsen. 1996. Ultraviolet radiation in North American lakes: Attenuation estimates from DOC measurements and implications for plankton communities. *Limnology and Oceanography* 41:1024–1034.

Willson, M.F., S.M. Gende, and B.H. Marston. 1998. Fishes and the forest. *BioScience* 48:455–462.

Wilson, G.W.T., W. Rice, M.C. Rillig, A. Springer, and D.C. Hartnett. 2009. Soil aggregation and carbon sequestration are tightly correlated with the abundance of arbuscular mycorrhizal fungi: Results from long-term field experiments. *Ecology Letters* 12:452–461.

Wilson, J.B. and D.Q. Agnew. 1992. Positive-feedback switches in plant communities. *Advances in Ecological Research* 23:263–336.

Wilson, K.B., D.D. Baldocchi, M. Aubinet, P. Berbigier, C. Bernhofer, et al. 2002. Energy partitioning between latent and sensible heat flux during the warm season at FLUXNET sites. *Water Resources Research* 38:1294, doi:1210.1029/2001WR000989.

Winner, W.E., H.A. Mooney, K. Williams, and S. von Caemmerer. 1985. Measuring and assessing SO_2 effects on photosynthesis and plant growth. Pages 118–132 *in* W.E. Winner and H.A. Mooney, editors. *Sulfur Dioxide and Vegetation*. Stanford University Press, Stanford, California.

Wolheim, W.M., C.J. Vörösmarty, B.J. Peterson, S.P. Seitzinger, and C.S. Hopkinson. 2006. Relationship between river size and nutrient removal. *Geophysical Research Letters* 33:L06410, doi:06410.01029/02006 GL025845.

Woodward, F.I. 1987. *Climate and Plant Distribution*. Cambridge University Press, Cambridge.

Woodward, S., P.M. Vitousek, K. Benvenuto, and P.A. Matson. 1990. Use of the exotic tree *Myrica faya* by native and exotic birds in Hawaii Volcanoes National Park. *Pacific Science* 44:88–93.

Woodwell, G.M. and R.H. Whittaker. 1968. Primary production in terrestrial communities. *American Zoologist* 8:19–30.

Wright, I.J., P.B. Reich, and M. Westoby. 2001. Strategy shifts in leaf physiology, structure and nutrient content between species of high- and low-rainfall and high- and low-nutrient habitats. *Functional Ecology* 15: 423–434.

Wright, I.J., P.B. Reich, M. Westoby, D.D. Ackerly, Z. Barusch, et al. 2004. The world-wide leaf economics spectrum. *Nature* 428:821–827.

Wright, M.S. and A.P. Covich. 2005. Relative importance of bacteria and fungi in a tropical headwater stream: Leaf decomposition and invertebrate feeding preference. *Microbial Ecology* 49:536–546.

Wu, J. and O.L. Loucks. 1995. From balance of nature to hierarchical patch dynamics: A paradigm shift in ecology. *Quarterly Review of Biology* 70:439–466.

Xiao, J., Q. Zhuang, D.D. Baldocchi, B.E. Law, A.D. Richardson, et al. 2008. Estimation of net ecosystem carbon exchange for the conterminous United States by combining MODIS and AmeriFlux data. *Agricultural and Forest Meteorology* 148:1827–1847.

Xiao, J., Q. Zhuang, B.E. Law, J. Chen, D.D. Baldocchi, et al. 2010. A continuous measure of gross primary production for the conterminous United States derived from MODIS and AmeriFlux data. *Remote Sensing of Environment* 114:576–591.

Yamada, A., T. Inoue, D. Wiwatwitaya, M. Ohkuma, T. Kudo, et al. 2006. Nitrogen fixation by termites in tropical forests, Thailand. *Ecosystems* 9:75–83.

Yoo, K., R. Amundson, A.M. Heimsath, and W.E. Dietrich. 2005. Process-based model linking pocket gopher (*Thomomys bottae*) activity to sediment transport and soil thickness. *Geology* 33:917–920.

Young, O.R. and W. Steffen. 2009. The Earth System: Sustaining planetary life-support systems. Pages 295–315 *in* F.S. Chapin, III, G.P. Kofinas, and C. Folke, editors. *Principles of Ecosystem Stewardship: Resilience-Based Natural Resource Management in a Changing World*. Springer, New York.

Young, T.P., D.A. Petersen, and J.J. Clary. 2005. The ecology of restoration: Historical links, emerging issues and unexplored realms. *Ecology Letters* 8:662–673.

Zak, M.R., M. Cabido, D. Cáceres, and S. Díaz. 2008. What drives accelerated land cover change in central Argentina? Synergistic consequences of climatic, socioeconomic and technological factors. *Environmental Management* 42:181–189.

Zangerl, A.R. and M.R. Berenbaum. 2006. Parsnip webworms and host plants at home and abroad: Trophic complexity in a geographic mosaic. *Ecology* 87:3070–3081.

Zech, W. and I. Kogel-Knabner. 1994. Patterns and regulation of organic matter transformation in soils: Litter decomposition and humification. Pages 303–335 *in* E.-D. Schulze, editor. *Flux Control in Biological Systems: From Enzymes to Populations and Ecosystems*. Academic Press, San Diego.

Zheng, D., S. Prince, and R. Wright. 2003. Terrestrial net primary production estimates for 0.5° grid cells from field observations: A contribution to global biogeochemical modeling. *Global Change Biology* 9:46–64.

Zimmermann, M.H. 1983. *Xylem Structure and the Ascent of Sap*. Springer, New York.

Zimov, S.A., V.I. Chuprynin, A.P. Oreshko, F.S. Chapin, III, J.F. Reynolds, et al. 1995. Steppe-tundra transition: An herbivore-driven biome shift at the end of the Pleistocene. *American Naturalist* 146:765–794.

Zimov, S.A., E.A.G. Schuur, and F.S. Chapin, III. 2006. Permafrost and the global carbon budget. *Science* 312:1612–1613.

Index

MIX
Papier aus verantwortungsvollen Quellen
Paper from responsible sources
FSC® C105338
FSC
www.fsc.org

Printed by Books on Demand, Germany